PRINCIPLES AND APPLICATIONS OF ASYMMETRIC SYNTHESIS

PRINCIPLES AND APPLICATIONS OF ASYMMETRIC SYNTHESIS

Guo-Qiang Lin
Yue-Ming Li
Albert S. C. Chan

WILEY-
INTERSCIENCE

A JOHN WILEY & SONS, INC., PUBLICATION

New York · Chichester · Weinheim · Brisbane · Singapore · Toronto

Library of Congress Cataloging in Publication Data

Lin, Guo-Qiang, 1943–
Principles and applications of asymmetric synthesis / Guo-Qiang Lin, Yue-Ming Li, Albert Chan.
    p. cm.
Includes bibliographical references and index.
ISBN 0-471-40027-0 (cloth : alk. paper)
1. Asymmetric synthesis. I. Li, Yue-Ming, 1966– II. Chan, Albert Sun-Chi, 1950–
III. Title.
QD262.L49 2001
547′.2—dc21   00-043691

Printed in Great Britain.

10 9 8 7 6 5 4 3 2

Dedicated to Professors
Chung-Kwong Poon and Wei-Shan Zhou

■■■■ CONTENTS

Asymmetric synthesis has been one of the important topics of research for chemists in both industrial laboratories and the academic world over the past three decades. The subject matter is not only a major challenge to the minds of practicing scientists but also a highly fertile field for the development of technologies for the production of high-value pharmaceuticals and agrochemicals. The significant difference in physiologic properties for enantiomers is now well known in the scientific community. The recent guidelines laid down for new chiral drugs by the Food and Drug Administration in the United States and by similar regulating agencies in other countries serve to make the issue more obvious. In the past 10 years, many excellent monographs, review articles, and multivolume treatises have been published. Journals specializing in chirality and asymmetric synthesis have also gained popularity. All these attest to the importance of chiral compounds and their enantioselective synthesis.

As practitioners of the art of asymmetric synthesis and as teachers of the subject to postgraduate and advanced undergraduate students, we have long felt the need for a one-volume, quick reference on the principles and applications of the art of asymmetric synthesis. It is this strong desire in our daily professional life, which is shared by many of our colleagues and students, that drives us to write this book. The book is intended to be used by practicing scientists as well as research students as a source of basic knowledge and convenient reference. The literature coverage is up to September 1999.

The first chapter covers the basic principles, common nomenclatures, and analytical methods relevant to the subject. The rest of the book is organized based on the types of reactions discussed. Chapters 2 and 3 deal with carbon–carbon bond formations involving carbonyls, enamines, imines, enolates, and so forth. This has been the most prolific area in the field of asymmetric synthesis in the past decade. Chapter 4 discusses the asymmetric C–O bond formations including epoxidation, dihydroxylation, and aminohydroxylation. These reactions are particularly important for the production of pharmaceutical products and intermediates. Chapter 5 describes asymmetric synthesis using the Diels-Alder reactions and other cyclization reactions. Chapter 6 presents the asymmetric catalytic hydrogenation and stoichiometric reduction of various unsaturated functionalities. Asymmetric hydrogenation is the simplest way of creating new chiral centers, and the technology is still an industrial flagship for chiral synthesis. Because asymmetric synthesis is a highly application-oriented science, examples of industrial applications of the relevant technologies are appropriately illustrated throughout the text. Chapter 7 records the applications of the asymmetric synthetic methods in the total synthesis of natural products. Chapter 8 reviews the use of enzymes and other methods and concepts in asymmetric synthesis. Overall, the book is expected to be useful for beginners as well as experienced practitioners of the art.

We are indebted to many of our colleagues and students for their assistance in various aspects of the preparation of this book. Most notably, assistance has been rendered from Jie-Fei Cheng, Wei-Chu Xu, Lu-Yan Zhang, Rong Li, and Fei Liu from Shanghai Institute of Organic Chemistry (SIOC) and Cheng-Chao Pai, Ming Yan, Ling-Yu Huang, Xiao-Wu Yang, Sze-Yin Leung, Jian-Ying Qi, Hua Chen, and Gang Chen from The Hong Kong Polytechnic University (PolyU). We also thank Sima Sengupta and William Purves of PolyU for proofreading and helping with the editing of the manuscript. Strong support and encouragement from Professor Wei-Shan Zhou of SIOC and Professor Chung-Kwong Poon of PolyU are gratefully acknowledged. Very helpful advice from Prof. Tak Hang Chan of McGill University and useful information on the industrial application of ferrocenyl phosphines from Professor Antonio Togni of Swiss Federal Institute of Technology and Dr. Felix Spindler of Solvias AG are greatly appreciated.

GUO-QIANG LIN
Shanghai Institute of Organic Chemistry

YUE-MING LI
ALBERT S. C. CHAN
The Hong Kong Polytechnic University

2ATMA	2-anthrylmethoxyacetic acid
Ac	acetyl group
AD mix-α	commercially available reagent for asymmetric dihydroxylation
AD mix-β	commercially available reagent for asymmetric dihydroxylation
AQN	anthraquinone
Ar	aryl group
ARO	asymmetric ring opening
BINAL–H	BINOL-modified aluminum hydride compound
BINOL	2,2′-dihydroxyl-1,1′-binaphthyl
BINAP	2,2′-bis(diphenylphosphino)-1,1′-binaphthyl
BLA	Brønsted acid–assisted chiral Lewis acid
Bn	benzyl group
BOC	t-butoxycarbonyl group
Bz	benzoyl group
CAB	chiral acyloxy borane
CAN	cerium ammonium nitrate
CBS	chiral oxazaborolidine compound developed by Corey, Bakshi, and Shibata
CCL	*Candida cyclindracea* lipase
CD	circular dichroism
CE	capillary electrophoresis
CIP	Cahn-Ingold-Prelog
COD	1,5-cyclooctadiene
Cp	cyclopentadienyl group
m-CPBA	m-chloroperbenzoic acid
CPL	circularly polarized light
CSA	camphorsulfonic acid
CSR	chemical shift reagent
DAIB	3-*exo*-(dimethylamino)isoborneol
DBNE	N,N-di-n-butylnorephedrine
DBU	1,8-diazobicyclo[5.4.0]undec-7-ene
DDQ	2,3-dichloro-5,6-dicyano-1,4-benzoquinone

de	diastereomeric excess
DEAD	diethyl azodicarboxylate
DET	diethyl tartrate
DHQ	dihydroquinine
DHQD	dihydroquinidine
DIBAL–H	diisobutylaluminum hydride
DIPT	diisopropyl tartrate
DIBT	diisobutyl tartrate
DMAP	4-N,N-dimethylaminopyridine
DME	1,2-dimethoxyethane
DMF	N,N-dimethylformamide
DMI	dimethylimidazole
DMSO	dimethyl sulfoxide
DMT	dimethyl tartrate
L-DOPA	3-(3,4-dihydroxyphenyl)-L-alanine
DPEN	1,2-diphenylethylenediamine
EDA	ethyl diazoacetate
EDTA	ethylenediaminetetraacetic acid
ee	enantiomeric excess
GC	gas chromatography
HMPA	hexamethylphosphoramide
HOMO	highest occupied molecular orbital
HPLC	high-performance liquid chromatography
Ipc	isocamphenyl
IR	infrared spectroscopy
KHMDS	$KN(SiMe_3)_2$
L*	chiral ligand
LDA	lithium diisopropylamide
LHMDS	$LiN(SiMe_3)_2$
LICA	lithium isopropylcyclohexylamide
LPS	lipopolysaccharide
LTMP	lithium tetramethylpiperidide
MAC	methyl α-(acetamido)cinnamate
MEM	methoxyethoxymethyl group
(R)-MNEA	N,N-di-[(1R)-(α-naphthyl)ethyl]-N-methylamine
MOM	methoxymethyl group
MPA	methoxyphenylacetic acid
Ms	methanesulfonyl, mesyl group
MTPA	α-methoxyltrifluoromethylphenylacetic acid

NAD(P)H	nicotinamide adenine dinucleotide (phosphate)
NHMDS	NaN(SiMe$_3$)$_2$
NLE	nonlinear effect
NME	N-methylephedrine
NMI	1-methylimidazole
NMMP	N-methylmorpholine
NMO	4-methylmorpholine N-oxide
NMR	nuclear magnetic resonance
NOE	nuclear Overhauser effect
ORD	optical rotatory dispersion
Oxone®	commercial name for potassium peroxomonosulfate
PCC	pyridinium chlorochromate
PDC	pyridinium dichromate
PLE	pig liver esterase
4-PPNO	4-phenylpyridine N-oxide
PTAB	phenyltrimethylammonium bromide
PTC	phase transfer catalyst
R*	chiral alkyl group
RAMP	(R)-1-amino-2-(methoxymethyl)pyrrolidine
Red-Al	sodium bis(2-methoxyethoxy)aluminum hydride
Salen	N,N'-disalicylidene-ethylenediaminato
SAMEMP	(S)-1-amino-2-(2-methoxyethoxymethyl)pyrrolidine
SAMP	(S)-1-amino-2-(methoxymethyl)pyrrolidine
S/C	substrate-to-catalyst ratio
SRS	self-regeneration of stereocenters
TAPP	$\alpha\alpha\beta\beta$-tetrakis(aminophenyl)porphyrin
TBAF	tetrabutylammonium fluoride
TBHP	t-butyl hydrogen peroxide
TBDPS	t-butyldiphenylsilyl group
TBS	t-butyldimethylsilyl group
TCDI	1,1-thionocarbonyldiimidazole
Teoc	2-trimethylsilylethyl N-chloro-N-sodiocarbamate
TES	triethylsilyl group
Tf	trifluoromethanesulfonyl group
THF	tetrahydrofuran
TMS	trimethylsilyl group
TMSCN	cyanotrimethylsilane, Me$_3$SiCN
TPAP	tetrapropylammonium perruthenate
Ts	toluenesulfonyl, tosyl group

CHAPTER 1

Introduction

> The universe is dissymmetrical; for if the whole of the bodies which compose the
> solar system were placed before a glass moving with their individual movements,
> the image in the glass could not be superimposed on reality.... Life is dominated
> by dissymmetrical actions. I can foresee that all living species are primordially, in
> their structure, in their external forms, functions of cosmic dissymmetry.
>
> —Louis Pasteur

These visionary words of Pasteur, written 100 years ago, have profoundly influenced the development of stereochemistry. It has increasingly become clear that many fundamental phenomena and laws of nature result from dissymmetry. In modern chemistry, an important term to describe dissymmetry is *chirality** or *handedness*. Like a pair of hands, the two enantiomers of a chiral compound are mirror images of each other that cannot be superimposed. Given the fact that within a chiral surrounding two enantiomeric biologically active agents often behave differently, it is not surprising that the synthesis of chiral compounds (which is often called *asymmetric synthesis*) has become an important subject for research. Such study of the principles of asymmetric synthesis can be based on either intramolecular or intermolecular chirality transfer. Intramolecular transfer has been systematically studied and is well understood today. In contrast, the knowledge base in the area of intermolecular chirality transfer is still at the initial stages of development, although significant achievements have been made.

In recent years, stereochemistry, dealing with the three-dimensional behavior of chiral molecules, has become a significant area of research in modern organic chemistry. The development of stereochemistry can, however, be traced as far back as the nineteenth century. In 1801, the French mineralogist Haüy noticed that quartz crystals exhibited hemihedral phenomena, which implied that certain facets of the crystals were disposed as nonsuperimposable species showing a typical relationship between an object and its mirror image. In 1809, the French physicist Malus, who also studied quartz crystals, observed that they could induce the polarization of light.

In 1812, another French physicist, Biot, found that a quartz plate, cut at the

---

*This word comes from the Greek word *cheir*, which means *hand* in English.

1

right angles to one particular crystal axis, rotated the plane of polarized light to an angle proportional to the thickness of the plate. Right and left forms of quartz crystals rotated the plane of the polarized light in different directions. Biot then extended these observations to pure organic liquids and solutions in 1815. He pointed out that there were some differences between the rotation caused by quartz crystals and that caused by the solutions of organic compounds he studied. For example, he noted that optical rotation caused by quartz was due to the whole crystal, whereas optical rotation caused by a solution of organic compound was due to individual molecules.

In 1822, the British astronomer Sir John Herschel observed that there was a correlation between hemihedralism and optical rotation. He found that all quartz crystals having the odd faces inclined in one direction rotated the plane of polarized light in one direction, while the enantiomorphous crystals rotate the polarized light in the opposite direction.

In 1846, Pasteur observed that all the crystals of dextrorotatory tartaric acid had hemihedral faces with the same orientation and thus assumed that the hemihedral structure of a tartaric acid salt was related to its optical rotatory power. In 1848, Pasteur separated enantiomorphous crystals of sodium ammonium salts of tartaric acid from solution. He observed that large crystals were formed by slowly evaporating the aqueous solution of racemic tartaric acid salt. These crystals exhibited significant hemihedral phenomena similar to those appearing in quartz. Pasteur was able to separate the different crystals using a pair of tweezers with the help of a lens. He then found that a solution of enantiomorphous crystals could rotate the plane of polarized light. One solution rotated the polarized light to the right, while the other one rotated the polarized light to the left.

Pasteur thus made the important deduction that the rotation of polarized light caused by different tartaric acid salt crystals was the property of chiral molecules. The (+)- and (−)-tartaric acids were thought to be related as an object to its mirror image in three dimensions. These tartaric acid salts were dissymmetric and enantiomorphous at the molecular level. It was this dissymmetry that provided the power to rotate the polarized light.

The work of these scientists in the nineteenth century led to an initial understanding of chirality. It became clear that the two enantiomers of a chiral molecule rotate the plane of polarized light to a degree that is equal in magnitude, but opposite in direction. An enantiomer that rotates polarized light in a clockwise direction is called a dextrorotatory molecule and is indicated by a plus sign (+) or italic letter "d". The other enantiomer, which rotates the plane of polarized light in a counterclockwise direction, is called levorotatory and is assigned a minus sign (−) or italic letter "l". Enantiomers of a given molecule have specific rotations with the same magnitude but in opposite directions. This fact was first demonstrated experimentally by Emil Fischer through a series of conversions of the compound 2-isobutyl malonic acid mono amide (**1**, see Scheme 1–1). As shown in Scheme 1–1, compound (+)-**1** can be converted to (−)-**1** through a series of reactions. From their projections, one can see that these two

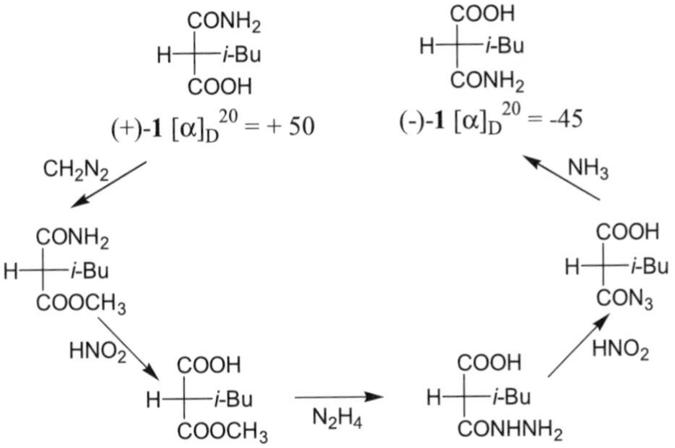

Scheme 1–1. Enantiomers of 2-isobutyl malonic acid mono amide have opposite optical rotations.

compounds are mirror images of each other. Fischer's experimental result easily showed that these two compounds have an opposite specific rotation. The amount of the specific rotation is nearly the same, and the difference may be the result of experimental deviation.

An equal molar mixture of the dextrorotatory and levorotatory enantiomers of a chiral compound is called a racemic mixture or a racemate. Racemates do not show overall optical rotation because the equal and opposite rotations of the two enantiomers cancel each other out. A racemic mixture is designated by adding the prefix ($\pm$) or *rac-* before the name of the molecule.

Within this historical setting, the actual birth of stereochemistry can be dated to independent publications by J. H. van't Hoff and J. A. Le Bel within a few months of each other in 1874. Both scientists suggested a three-dimensional orientation of atoms based on two central assumptions. They assumed that the four bonds attached to a carbon atom were oriented tetrahedrally and that there was a correlation between the spatial arrangement of the four bonds and the properties of molecules. van't Hoff and Le Bell proposed that the tetrahedral model for carbon was the cause of molecular dissymmetry and optical rotation. By arguing that optical activity in a substance was an indication of molecular chirality, they laid the foundation for the study of intramolecular and intermolecular chirality.

1.1 THE SIGNIFICANCE OF CHIRALITY AND STEREOISOMERIC DISCRIMINATION

Chirality is a fundamental property of many three-dimensional objects. An object is chiral if it cannot be superimposed on its mirror image. In such a case, there are two possible forms of the same object, which are called *enantiomers*,

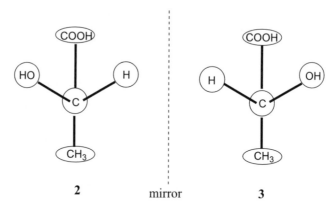

Figure 1–1. Mirror images of lactic acid.

and thus these two forms are said to be enantiomeric with each other. To take a simple example, lactic acid can be obtained in two forms or enantiomers, **2** and **3** in Figure 1–1, which are clearly enantiomeric in that they are related as mirror images that cannot be superimposed on each other.

Enantiomers have identical chemical and physical properties in the absence of an external chiral influence. This means that **2** and **3** have the same melting point, solubility, chromatographic retention time, infrared spectroscopy (IR), and nuclear magnetic resonance (NMR) spectra. However, there is one property in which chiral compounds differ from achiral compounds and in which enantiomers differ from each other. This property is the direction in which they rotate plane-polarized light, and this is called *optical activity* or *optical rotation*. Optical rotation can be interpreted as the outcome of interaction between an enantiomeric compound and polarized light. Thus, enantiomer **3**, which rotates plane-polarized light in a clockwise direction, is described as (+)-lactic acid, while enantiomer **2**, which has an equal and opposite rotation under the same conditions, is described as (−)-lactic acid.

Readers may refer to the latter part of this chapter for the determination of absolute configuration.

Chirality is of prime significance, as most of the biological macromolecules of living systems occur in nature in one enantiomeric form only. A biologically active chiral compound interacts with its receptor site in a chiral manner, and enantiomers may be discriminated by the receptor in very different ways. Thus it is not surprising that the two enantiomers of a drug may interact differently with the receptor, leading to different effects. Indeed, it is very important to keep the idea of chiral discrimination or stereoisomeric discrimination in mind when designing biologically active molecules.

As human enzymes and cell surface receptors are chiral, the two enantiomers of a racemic drug may be absorbed, activated, or degraded in very different ways, both in vivo and in vitro. The two enantiomers may have unequal degrees or different kinds of activity.[1] For example, one may be therapeutically effective, while the other may be ineffective or even toxic.

An interesting example of the above difference is L-DOPA **4**, which is used in the treatment of Parkinson's disease. The active drug is the achiral compound dopamine formed from **4** via in vivo decarboxylation. As dopamine cannot cross the blood–brain barrier to reach the required site of action, the "prodrug" **4** is administered. Enzyme-catalyzed in vivo decarboxylation releases the drug in its active form (dopamine). The enzyme L-DOPA decarboxylase, however, discriminates the stereoisomers of DOPA specifically and only decarboxylates the L-enantiomer of **4**. It is therefore essential to administer DOPA in its pure L-form. Otherwise, the accumulation of D-DOPA, which cannot be metabolized by enzymes in the human body, may be dangerous. Currently L-DOPA is prepared on an industrial scale via asymmetric catalytic hydrogenation.

4

From the above example one can see that stereoisomeric discrimination is very striking in biological systems, and for this reason chirality is recognized as a central concept. If we consider the biological activities of chiral compounds in general, there are four different behaviors: (1) only one enantiomer has the desired biological activity, and the other one does not show significant bioactivity; (2) both enantiomers have identical or nearly identical bioactivity; (3) the enantiomers have quantitatively different activity; and (4) the two enantiomers have different kinds of biological activity. Table 1–1 presents a number of examples of differences in the behavior of enantiomers. The listed enantiomers may have different taste or odor and, more importantly, they may exhibit very different pharmacological properties. For example, D-asparagine has a sweet taste, whereas natural L-asparagine is bitter; (S)-(+)-carvone has an odor of caraway, whereas the (R)-isomer has a spearmint smell; (R)-limonene has an orange odor, and its (S)-isomer has a lemon odor. In the case of disparlure, a sex pheromone for the gypsy moth, one isomer is active in very dilute concentration, whereas the other isomer is inactive even in very high concentration. (S)-propranolol is a β-blocker drug that is 98 times as active as its (R)-counterpart.[2]

Sometimes the inactive isomer may interfere with the active isomer and significantly lower its activity. For example, when the (R)-derivative of the sex pheromone of a Japanese beetle is contaminated with only 2% of its enantiomer, the mixture is three times less active than the optically pure pheromone. The pheromone with as little as 0.5% of the (S)-enantiomer already shows a significant decrease of activity.[3]

A tragedy occurred in Europe during the 1950s involving the drug thalidomide. This is a powerful sedative and antinausea agent that was considered

TABLE 1–1. Examples of the Different Behaviors of Enantiomers

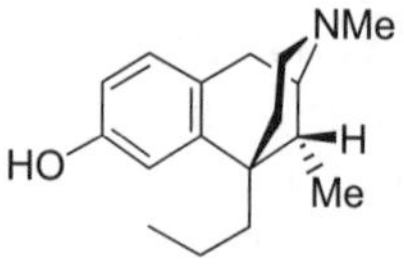

(-)-Benzomorphia
(eases pain, unhabituational)

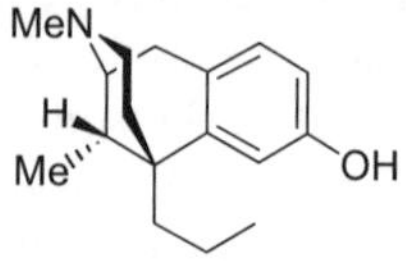

(+)-Benzomorphia
(faintly pain-easing, habituational)

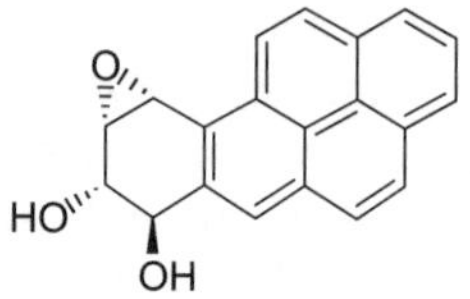

(-)-Benzopyryldiol
(strong carcinogenicity)

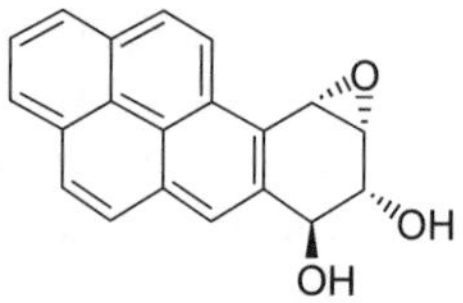

(+)-Benzopyryldiol
(no carcinogenicity)

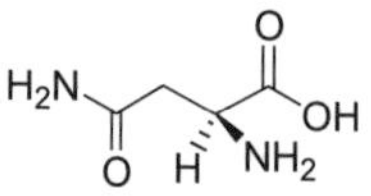

L-asparagine (bitter)

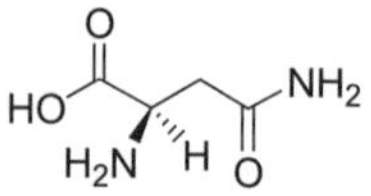

D-asparagine (sweet)

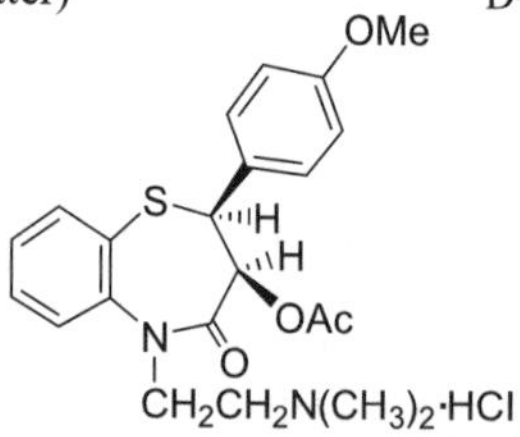

diltiazem
(the (*S*, *S*)-form is effective in relieving myocardial infarction[4])

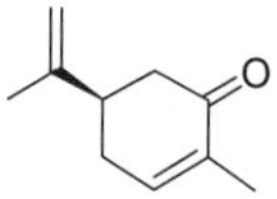

(*R*)-carvone (spearmint odor)

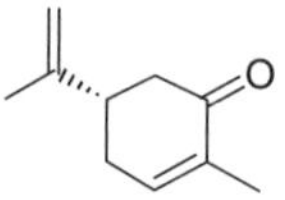

(*S*)-carvone (caraway odor)

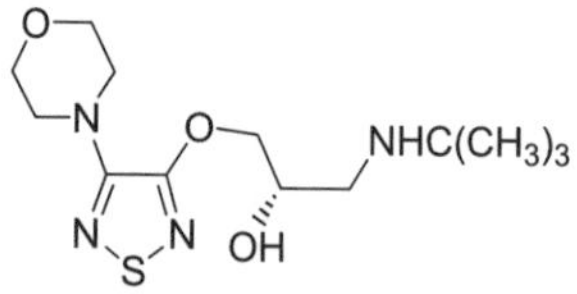

(*R*)-timolol (adrenergic blocker)

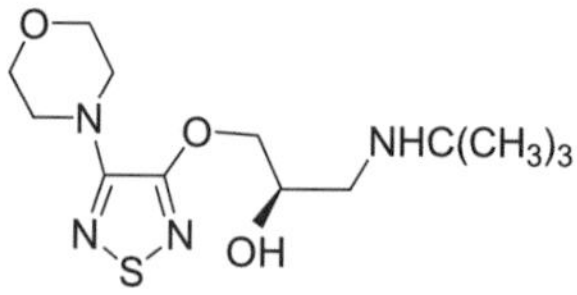

(*S*)-timolol (ineffective)

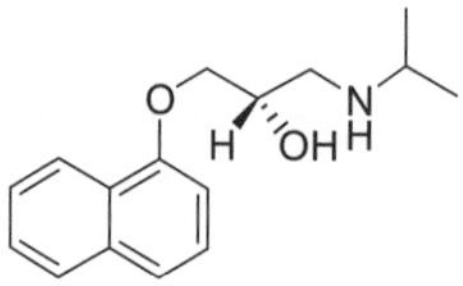

(*S*)-propranolol
[98 times the activity of its (*R*)-isomer]

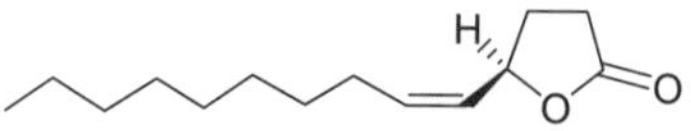

Sex pheromone of the Japanese beetle
(its enantiomer is inactive)

especially appropriate for use during early pregnancy. Unfortunately, it was soon found that this drug was a very potent teratogen and thus had serious harmful effects on the fetus. Further study showed that this teratogenicity was caused by the (S)-isomer (which had little sedative effect), but the drug was sold in racemic form. The (R)-isomer (the active sedative) was found not to cause deformities in animals even in high doses.[5] Similarly, the toxicity of naturally occurring (−)-nicotine is much greater than that of unnatural (+)-nicotine. Chiral herbicides, pesticides, and plant growth regulators widely used in agriculture also show strong biodiscriminations.

In fact, stereodiscrimination has been a crucial factor in designing enantiomerically pure drugs that will achieve better interaction with their receptors. The administration of enantiomerically pure drugs can have the following advantages: (1) decreased dosage, lowering the load on metabolism; (2) increased latitude in dosage; (3) increased confidence in dose selection; (4) fewer interactions with other drugs; and (5) enhanced activity, increased specificity, and less risk of possible side effects caused by the enantiomer.

Now it is quite clear that asymmetry (or chirality) plays an important role in life sciences. The next few sections give a brief introduction to the conventions of the study of asymmetric (or chiral) systems.

1.2 ASYMMETRY

1.2.1 Conditions for Asymmetry

Various chiral centers, such as the chiral carbon center, chiral nitrogen center, chiral phosphorous center, and chiral sulfur center are depicted in Figure 1–2.

Amines with three different substituents are potentially chiral because of the pseudotetrahedral arrangement of the three groups and the lone-pair electrons. Under normal conditions, however, these enantiomers are not separable because of the rapid inversion at the nitrogen center. As soon as the lone-pair electrons are fixed by the formation of quaternary ammonium salts, tertiary amide N-oxide, or any other fixed bonding, the inversion is prohibited, and consequently the enantiomers of chiral nitrogen compounds can be separated.

In contrast to the amines, inversion of configuration for phosphines is generally negligibly slow at ambient temperature. This property has made it possible for chiral phosphines to be highly useful as ligands in transition metal-catalyzed asymmetric syntheses.

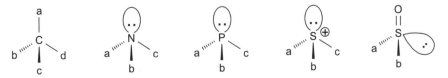

Figure 1–2. Formation of asymmetry.

Figure 1–3. Solution stable three-membered heterocyclic ring systems.

As a result of the presence of lone-pair electrons, the configuration of orga-nosulfur species is pyramidal, and the pyramidal reversion is normally slow at ambient temperature. Thus two enantiomers of chiral sulfoxides are possible and separable.

As a general rule, asymmetry may be created by one of the following three conditions:

1. Compounds with an asymmetric carbon atom: When the four groups connected to a carbon center are different from one another, the central carbon is called a *chiral center*. (However, we must remember that the presence of an asymmetric carbon is neither a necessary nor a sufficient condition for optical activity.)

2. Compounds with another quaternary covalent chiral center binding to four different groups that occupy the four corners of a tetrahedron:
   Si, Ge, N (in quaternary salts or *N*-oxides)
   Mn, Cu, Bi and Zn—when in tetrahedral coordination.

3. Compounds with trivalent asymmetric atoms: In atoms with pyramidal bonding to three different groups, the unshared pair of electrons is ana-logous to a fourth group. In the case of nitrogen compounds, if the inversion at the nitrogen center is prevented by a rigid structural ar-rangement, chirality also arises. The following examples illustrate this phenomenon.

   a. In a three-membered heterocyclic ring, the energy barrier for inversion at the nitrogen center is substantially raised (Fig. 1–3).

   b. The bridgehead structure completely prevents inversion.

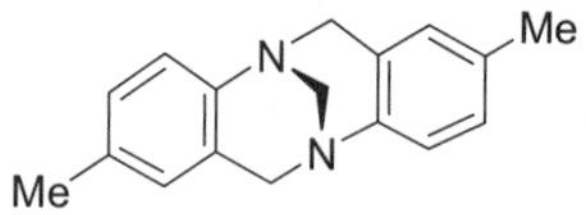

Irreversible

1.2.2 Nomenclature

If a molecule contains more than one chiral center, there are other forms of stereoisomerism. As mentioned in Section 1.1, nonsuperimposable mirror images are called *enantiomers*. However, substances with the same chemical constitution may not be mirror images and may instead differ from one another

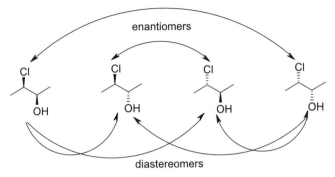

Figure 1–4. Enantiomers and diastereomers.

in having different configurations at one or more chiral centers in the molecule. These substances are called *diastereomers*. Thus, for 2-chloro-3-hydroxylbutane, one can draw four different structures, among which one can find two pairs of enantiomeric and four pairs of diastereomeric relations (Fig. 1–4).

For the unambiguous description of the various isomers, it is clearly necessary to have formal rules to define the structural configurations. These rules are explained in the following sections.

1.2.2.1 *Fischer's Convention.*

Initially, the absolute configurations of optical isomers were unknown to chemists working with optically active compounds. Emil Fischer, the father of carbohydrate chemistry, decided to relate the possible configurations of compounds to that of glyceraldehyde of which the absolute configuration was yet unknown but was defined arbitrarily.

In Fischer's projection of glyceraldehyde, the carbon chain is drawn vertically with only the asymmetric carbon in the plane of the paper. Both the carbonyl and the hydroxylmethyl groups are drawn as if they are behind the plane, with the carbonyl group on the top and the hydroxylmethyl group at the bottom of the projection. The hydroxyl group and the hydrogen atom attached to the asymmetric carbon atom are drawn in front of the plane, the hydroxyl group to the right and the hydrogen atom to the left. This configuration was arbitrarily assigned as the D-configuration of glyceraldehyde and is identified by a small capital letter D. Its mirror image enantiomer with the opposite configuration is identified by a small capital letter L.

The structure of any other optically active compound of the type R–CHX–R′ is drawn with the carbon chain

$$\begin{array}{c} R \\ | \\ \overline{} \\ R' \end{array}$$

in the vertical direction with the higher oxidative state atom (R or R′) on the top. If the X group (usually –OH, –NH$_2$, or a halogen) is on the right side, the relative configuration is designated D; otherwise the configuration is designated L.

Although the D-form of glyceraldehyde was arbitrarily chosen as the dextro-rotatory isomer without any knowledge of its absolute configuration, the choice was a fortuitous one. In 1951, with the aid of modern analytical methods, the D-configuration of the dextrorotatory isomer was unambiguously established.

The merit of Fischer's convention is that it enables the systematic stereochemical presentation of a large number of natural products, and this convention is still useful for carbohydrates or amino acids today. Its limitations, however, become obvious with compounds that do not resemble the model reference compound glyceraldehyde. For example, it is very difficult to correlate the terpene compounds with glyceraldehyde. Furthermore, selection of the correct orientation of the main chain may also be ambiguous. Sometimes different configurations may even be assigned to the same compound when the main chain is arranged in a different way.

1.2.2.2 The Cahn-Ingold-Prelog Convention.

The limitations of Fischer's convention made it clear that in order to assign the exact orientation of the four connecting groups around a chiral center it was necessary to establish a systematic nomenclature for stereoisomers. This move started in the 1950s with Cahn, Ingold, and Prelog establishing a new system called the Cahn-Ingold-Prelog (CIP) convention[6] for describing stereoisomers. The CIP convention is based on a set of sequence rules, following which the name describing the constitution of a compound is accorded a prefix that defines the absolute configuration of a molecule unambiguously. These prefixes also enable the preparation of a stereodrawing that represents the real structure of the molecule.

In the nomenclature system, atoms or groups bonded to the chiral center are prioritized first, based on the sequence rules. The rules can be simplified as follows: (1) An atom having a higher atomic number has priority over one with a lower atomic number; for isotopic atoms, the isotope with a higher mass precedes the one with the lower mass. (2) If two or more of the atoms directly bonded to the asymmetric atom are identical, the atoms attached to them will be compared, according to the same sequence rule. Thus, if there is no heteroatom involved, alkyl groups can be sequenced as tertiary > secondary > primary. When two groups have different substituents, the substituent bearing the highest atomic number on each group must be compared first. The sequence decision for these groups will be made based on the sequence of the substituents, and the one containing prior substituents has a higher precedence. A similar rule is applicable in the case of groups with heteroatoms. (3) For multiple bonds, a doubly or triply bonded atom is duplicated or triplicated with the atom to which it is connected. This rule is also applicable to aromatic systems. For example,

(4) For vinyl groups, a group having the (Z)-configuration precedes the same group having the (E)-configuration, and an (R)-group has precedence over an (S)-group for pseudochiral centers.

Based on these sequence rules, configurations can be easily assigned to chiral molecules, which are classified into different types according to spatial orientation. The detailed assignments are as follows.

Central Chirality. The system Cxyzw (**5**) has no symmetry when x, y, z, and w are different groups, and this system is referred to as a *central chiral system.*

5

Imagine that an asymmetric carbon atom C is connected to w, x, y, and z and that these four substituents are placed in priority sequence x > y > z > w according to the CIP sequencing rule. If we observe the chiral center from a position opposite to group w and from this viewpoint groups x → y → z are in clockwise sequence, then this chiral center is defined as having an (R)-configuration.* Otherwise the configuration is defined as (S).† For example, the configuration of molecule **5** is specified as (R).

Following these rules, D-glyceraldehyde **6** in Fischer's convention can be assigned an (R)-configuration.

CHO
H——OH
CH₂OH

6

D-Glyceraldehyde in Fischer's convention
(R)-glyceraldehyde in the Cahn-Ingold-Prelog convention

For an adamantane-type compound, it is possible to substitute the four tertiary hydrogen atoms and make four quaternary carbon atoms. These carbon atoms can be asymmetric if the four substituents are chosen properly. It is possible to specify these chiral centers separately, but their chiralities can also be so interlinked that they collectively produce one pair of enantiomers with only one chiral center. Usually it is more convenient to collectively specify the chirality with reference to a center of chirality taken as the unoccupied centroid of the adamantane frame.

---

*Originating from the Latin word *rectus*, which means *right* in English.
† Originating from the Latin word *sinister*, which means *left* in English.

adamantane-type compounds
(*R* if a>b>c>d)

Axial Chirality. For a system with four groups arranged out of the plane in pairs about an axis, the system is asymmetric when the groups on each side of the axis are different. Such a system is referred to as an *axial chiral system.* This structure can be considered a variant of central chirality. Some axial chiral molecules are allenes, alkylidene cyclohexanes, spiranes, and biaryls (along with their respective isomorphs). For example, compound **7a** (binaphthol), which belongs to the class of biaryl-type axial chiral compounds, is extensively used in asymmetric synthesis. Examples of axial chiral compounds are given in Figure 1–5.

The nomenclature for biaryl, allene, or cyclohexane-type compounds follows a similar rule. Viewed along the axis, the nearer pair of ligands receives the first two positions in the order of preference, and the farther ligands take the third and fourth position. The nomination follows a set of rules similar to those applied in the central chiral system. In this nomination, the end from which the molecule is viewed makes no difference. From whichever end it is viewed, the positions remain the same. Thus, compound **7a** has an (*R*)-configuration irrespective of which end it is viewed from.

It is important to note that the method for naming chiral spirocyclic compounds has been revised from the original proposal.[6] In the original nomenclature system, these compounds were treated on the basis of axial chirality like biaryls, allenes, and so forth. According to the old nomenclature, the first and second priorities are given to the prior groups in one cycle, and the third and fourth priorities are given to that in the other one. Taking the above spiro-

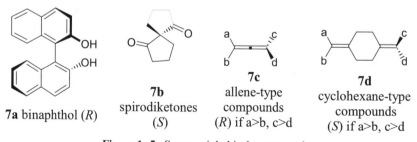

7a binaphthol (*R*)

7b
spirodiketones
(*S*)

7c
allene-type
compounds
(*R*) if a>b, c>d

7d
cyclohexane-type
compounds
(*S*) if a>b, c>d

Figure 1–5. Some axial chiral compounds.

old, obsolete method
(*R*)-configuration

currently used method
(*S*)-configuration

Figure 1–6. Examples of the old and new nomenclatures of spirocyclic compounds.

diketone **7b** as an example, the chiral center, the spiro atom, is bonded to two equivalent carbonyl carbon atoms and two equivalent methylene carbon atoms (Fig. 1–6). In the new nomenclature, the first member of the sequence is given to either one of the carbonyl atoms, and the second priority is given to the other carbonyl carbon (in the old nomenclature, the second priority is given to the methylene atom staying on the same side of the first carbonyl group); the third priority is given to the methylene carbon atom on the same ring side with the first carbonyl group. Thus, the chiral center (the spiro atom of **7b**) has configuration (*S*). If the obsolete, original method were used, the configuration of **7b** would have been designated (*R*).

Planar Chirality. Planar chirality arises from the desymmetrization of a symmetric plane in such a way that chirality depends on a distinction between the two sides of the plane and on the pattern of the three determining groups. In the definition of this chiral system, the first step is the selection of a chiral plane; the second step is to identify a preferred side of the plane. The chiral plane is the plane that contains the highest number of atoms in the molecule.

After the designation of the chiral plane, one then needs to find a descriptor or "pilot" atom. To find this atom, one views from the out-of-plane atom closest to the chiral plane. If there are two such atoms, the one closest to the atom of higher precedence in the chiral plane is selected. The leading atom, or "pilot" atom, marks the preferred side of the plane. The higher priority atom of the set bonded to the pilot atom is marked as No. 1 as in **8a**. The second priority (marked as No. 2) is given to the atom on the chiral plane directly bonded to group No. 1, and so on. Viewing from the preferred side, the designation p*R* is given to a clockwise orientation of $1 \rightarrow 2 \rightarrow 3$, and p*S* represents a counterclockwise orientation of these three atoms/groups. Thus, examples **8a** and **8b** depict a p*S*-configuration. The letter "p" indicates the planar chirality. In example **8c**, a metallocene compound, the compound can be treated as having chiral centers by replacing the η^6–π bond by six σ single bonds (**8d**). According to the CIP rules, the chirality of this molecule can then be assigned by examining the most preferred atom on the ring (marked by an arrow). Such a molecule can then be treated as a central chiral system. Thus, according to the rule for central chirality, compound **8c** can be assigned an (*S*)-configuration.

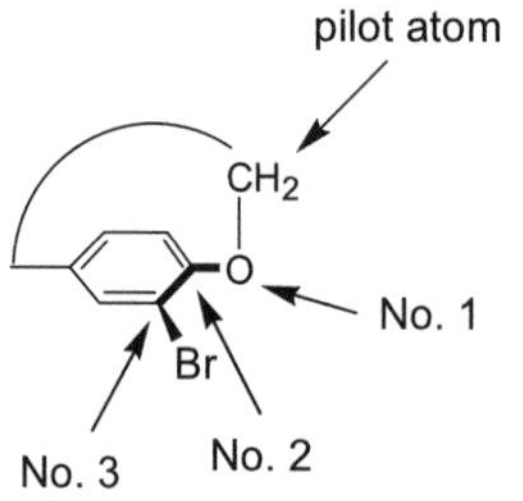

pilot atom

pilot atom

CH₂

O

No. 1

Br

No. 3   No. 2

8a

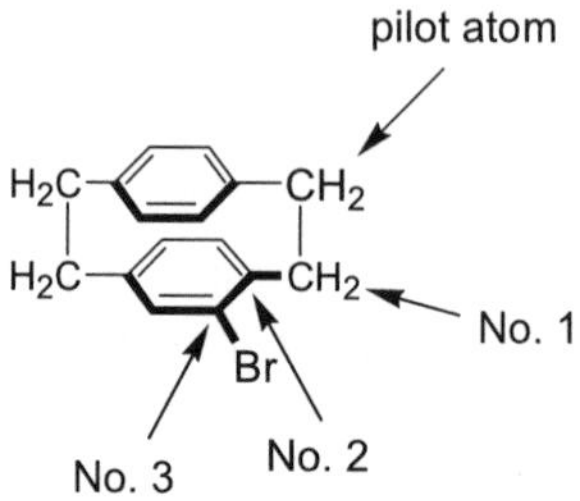

H₂C—⟨ ⟩—CH₂

H₂C—⟨ ⟩—CH₂

No. 1

Br

No. 3   No. 2

8b

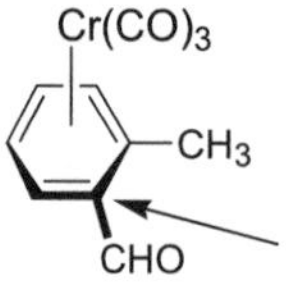

Cr(CO)₃

CH₃

CHO

8c

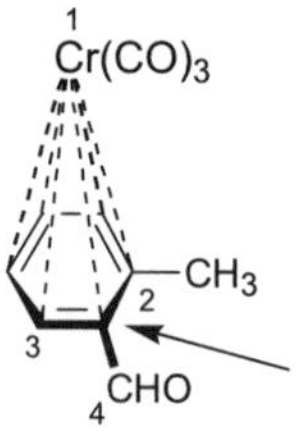

1
Cr(CO)₃

CH₃
2

3

CHO
4

8d

Helical Chirality. Helicity is a special case of chirality in which molecules are shaped as a right- or left-handed spiral like a screw or spiral stairs. The configurations are designed *M* and *P*, respectively, according to the helical direction. Viewed from the top of the axis, a clockwise helix is defined as *P*, whereas a counterclockwise orientation is defined as *M*. Thus, the configuration of example **9** is defined as *M*.

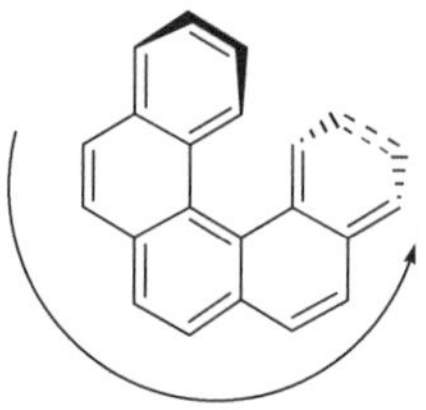

9

Octahedral Structures. Extension of the sequence rule makes it possible to arrange an octahedral structure in such a way that the ligands are placed octahedrally in an order of preference.

Special sequencing rules are applied for assigning the six substituents. Number 1 is given to the group with the highest priority according to the general CIP rule. Number 6 is then located *trans* to this group regardless of its precedence. (If the choice for No. 1 is open, No. 6 is given to the group with lowest priority, and No. 1 is the one *trans* to No. 6). The 2, 3, 4, and 5 are located in a plane and form a cyclic sequence. Number 2 will normally be assigned to the more prior group among the four.

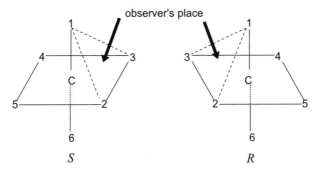

Figure 1–7. Octahedral structures.

The observer looks at the face formed by the first three preferred atoms/ groups (1, 2, and 3) from a direction opposite to the face of 4, 5, and 6. (*R*)- configuration is then defined as a clockwise arrangement of the groups 1, 2, and 3, and (*S*)-configuration is defined as a counterclockwise arrangement of the first three preferred groups (Fig. 1–7).

Pseudochiral Centers. A Cabcd system is called a *pseudochiral center* when a/b are one pair of enantiomeric groups and c/d are different from a/b as well as different from each other. Molecules with a pseudochiral center can be either achiral or chiral, depending on the properties of c and d. If both c and d are achiral, the whole molecule is also achiral; if either or both of them is chiral, the molecule is also chiral. As for the sequence rule, $R > S$ is applied when naming the pseudochiral center. The pseudochiral center is noted in italic lowercase *r* or *s*. For example, compounds **10a** and **10b** are the reduction products of D-(−)-ribose and D-(+)-xylose, respectively (Fig. 1–8). The C2 atom in these two compounds has an (*R*)-configuration, and the C4 in these two compounds has an (*S*)-configuration. The C3 atoms in these compounds can be considered as pseudochiral centers. C3 in compound **10a** is defined as *s*, and C3 in compound **10b** is defined as *r*.

Molecules that belong to C_n or D_n point groups are also chiral. For instance, *trans*-2,5-dimethylpyrrolidine (Fig. 1–9), containing a twofold rotation axis, belongs to the point group C_2 and is chiral.[7]

10a   **10b**

Figure 1–8. Pseudochiral centers.

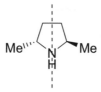

Figure 1–9. *trans*-2,5-Dimethylpyrrolidine.

1.3 DETERMINING ENANTIOMER COMPOSITION

As mentioned in Section 1.2, the presence of an asymmetric carbon is neither a necessary nor a sufficient condition for optical activity. Each enantiomer of a chiral molecule rotates the plane of polarized light to an equal degree but in opposite directions. A chiral compound is optically active only if the amount of one enantiomer is in excess of the other. Measuring the enantiomer composition is very important in asymmetric synthesis, as chemists working in this area need the information to evaluate the asymmetric induction efficiency* of asymmetric reactions.

The enantiomer composition of a sample may be described by the enantiomer excess (ee), which describes the excess of one enantiomer over the other:

$$\text{ee} = \left| \frac{[S] - [R]}{[S] + [R]} \right| \times 100\%$$

where $[R]$ and $[S]$ are the composition of R and S enantiomers, respectively.

Correspondingly, the diastereomer composition of a sample can be described by the diastereomer excess (de), which refers to the excess of one diastereomer over the other:

$$\text{de} = \left| \frac{[S^*S] - [S^*R]}{[S^*S] + [S^*R]} \right| \times 100\%$$

where $[S^*S]$ and $[S^*R]$ are the composition of the diastereomers, respectively.

Different methods have been developed for determining the enantiomer compositions of a pair of enantiomers. Some apply measurements of the original molecules, while others use derivatives of the corresponding compounds.

To determine how much one isomer is in excess over the other, analytical methods based on high-performance liquid chromatography (HPLC) or gas chromatography (GC) on a chiral column have proved to be most reliable.

---

*The goal of an asymmetric reaction is to obtain one enantiomer in high excess of the other. For this reason, after the reaction one has to measure the enantiomer excess. The larger the excess of one enantiomer over the other, the better the result of the asymmetric reaction or the higher efficiency of the asymmetric induction.

Chiral chemical shift reagents for NMR analysis are also useful, and so are optical methods.

A variety of methods are also available when the compound under investigation can be converted with a chiral reagent to diastereomeric products, which have readily detectable differences in physical properties. If a derivatizing agent is employed, it must be ensured that the reaction with the subject molecule is quantitative and that the derivatization reaction is carried out to completion. This will ensure that unintentional kinetic resolution does not occur before the analysis. The derivatizing agent itself must be enantiomerically pure, and epimerization should not occur during the entire process of analysis.

1.3.1 Measuring Specific Rotation

One of the terms for describing enantiomer composition is *optical purity*. It refers to the ratio of observed specific rotation to the maximum or absolute specific rotation of a pure enantiomer sample. For any compound for which the optical rotation of its pure enantiomer is known, the ee value may be determined directly from the observed optical rotation.

$$[\alpha]_D^{20} = \frac{[\alpha]}{L \times c} \times 100$$

where $[\alpha]$ is the measured rotation; L is the path length of cell (dm); c is concentration (g/100 ml); D is the D line of sodium, the wave length of light used for measurement (5983 Å); and 20 is the temperature in degrees (Celsius).

$$\text{Optical purity}(\%) = [\alpha]_{\text{obs.}} / [\alpha]_{\text{max}} \times 100\%$$

The classic method of determining the optical purity of a sample is to use a polarimeter. However, this method can be used to determine enantiomeric purity only when the readings are taken carefully with a homogenous sample under specific conditions. The method provides comparatively fast but, in many cases, not very precise results. There are several drawbacks to this method: (1) One must have knowledge of the specific rotation of the pure enantiomer under the experimental conditions in order to compare it with the measured result from the sample. (2) The measurement of optical rotation may be affected by numerous factors, such as the wavelength of the polarized light, the presence or absence of solvent, the solvent used for the measurement, the concentration of the solution, the temperature of measurement, and so forth. Most importantly, the measurement may be affected significantly by the presence of impurities that have large specific rotations. (3) Usually a large quantity of sample is needed, and the optical rotation of the product must be large enough for accurate measurement. (This problem, however, has somewhat been alleviated by advances in instrumentation, such as the availability of the capillary cell.) (4) In

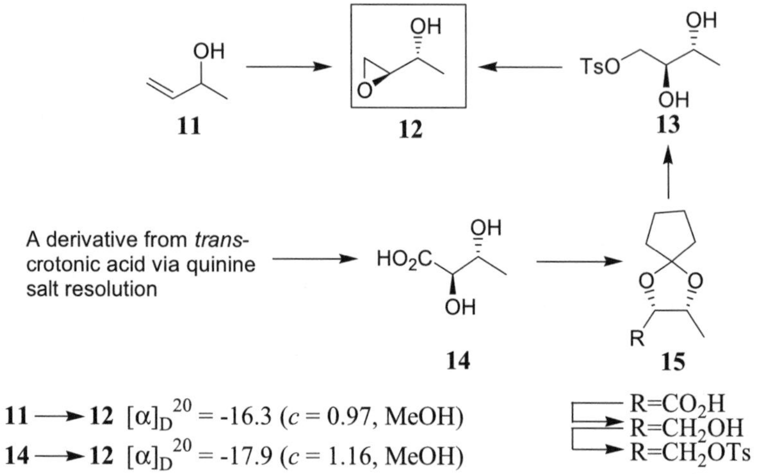

Scheme 1–2

the process of obtaining a chemically pure sample for measurement, an enrichment of the major enantiomer may occur and cause substantial errors.

An example of the application of this method is given in Scheme 1–2. White et al.[8] reported the enantioselective epoxidation of 3-buten-2-ol (**11**) using Sharpless reagent (TBHP/Ti(OPr^i)$_4$/DET, used for the asymmetric epoxidation of allyl alcohols), giving (2S,3R)-1,2-epoxy-3-butanol (−)-**12** ($[\alpha]_D^{20} = -16.3$, $c = 0.97$, MeOH), which was employed in the chiral synthesis of 2,5-dideoxyribose, a segment of the ionophoric antibiotic boromycin. Although the (3R)-enantiomer of **12** was the expected product, an unambiguous proof of the stereochemistry was still necessary. To this end, (±)-erythro-2,3-dihydroxybutyric acid (**14**), which has been prepared by the hydroxylation of trans-crotonic acid, was resolved via its quinine salt.[9] Comparing the specific rotations confirmed that (−)-**14** possesses (2S,3R)-configuration. The protection of (−)-**14** as its ketal derivative with cyclopentanone, followed by LAH reduction and tosylation, produced compound **15**, which, upon the removal of the cyclopentylidene residue, gave the diol **13**. Treatment of **13** with sodium hydride produced **12** ($[\alpha]_D^{20} = -17.9$, $c = 1.16$, MeOH), which had the same direction of optical rotation as the compound obtained from **11**. Based on this result, it was ascertained that the asymmetric epoxidation of **11** afforded (2S,3R)-epoxy alcohol **12** with an enantiomer excess of 91% (16.3/17.9 × 100%).

Example showing the potential for errors in using the optical rotation method that was found in the reduction of enantiomerically pure L-leucine to leucinol using different reducing agents.[10] When borane-dimethylsulfide was used, the product obtained had a specific rotation of $[\alpha]_D^{20} = +4.89$ (neat). When NaBH$_4$ or LiAlH$_4$ was used in the reduction of leucine ethyl ester hydrochloride, the leucinol obtained had a specific rotation of $[\alpha]_D^{20} = +1.22$ to 1.23. At first, it was thought that racemization had occurred during the reaction when $NaBH_4$ or $LiAlH_4$ was used. It was found later that the wrong value of +4.89 for the specific rotation was caused by trace amounts of a highly dextrorotatory impurity in the product. For this and other reasons, many enantiomer compositions determined by this method in earlier years have now been found to be incorrect.

1.3.2   The Nuclear Magnetic Resonance Method

NMR spectroscopy cannot normally be used directly for discriminating enantiomers in solution. The NMR signals for most enantiomers are isochronic under achiral conditions. However, NMR techniques can be used for the determination of enantiomer compositions when diastereomeric interactions are introduced to the system.

1.3.2.1   Nuclear Magnetic Resonance Spectroscopy Measured in a Chiral Solvent or with a Chiral Solvating Agent. One method of NMR analysis for enantiomer composition is to record the spectra in a chiral environment, such as a chiral solvent or a chiral solvating agent. This method is based on the diastereomeric interaction between the substrate and the chiral environment applied in the analysis.

The first example found in the literature was the use of this method in distinguishing the enantiomers of 2,2,2-trifluoro-1-phenylethanol. This was realized by recording the ^{19}F NMR of the compound in $(-)$-α-phenethylamine.[11] Burlingame and Pirkle[12] found that the ee values could also be determined by studying the ^1H NMR. Later it was found[13] that the determination can also be achieved in achiral solvents in the presence of certain chiral compounds, namely, chiral solvating agents. In these cases, the determination was achieved based on the diastereomeric interaction between the substrate and the chiral solvating agent. Sometimes, the observed chemical shift difference is very small, making the analysis difficult. This problem may be overcome by using a higher field NMR spectrometer or recording the spectra at lower temperature.

1.3.2.2   Nuclear Magnetic Resonance with a Chiral Chemical Shift Reagent. Lanthanide complexes can serve as weak Lewis acids. In nonpolar solvents (e.g., $CDCl_3$, CCl_4, or CS_2) these paramagnetic salts are able to bind Lewis bases, such as amides, amines, esters, ketones, and sulfoxides. As a result, protons, carbons, and other nuclei are usually deshielded relative to their positions in the uncomplexed substrates, and the chemical shifts of those nuclei are altered. The extent of this alteration depends on the strength of the complex and the distance of the nuclei from the paramagnetic metal ion. Therefore, the NMR signals of different types of nuclei are shifted to different extents, and this leads to spectral simplification. The spectral nonequivalence observed in the presence of chiral chemical shift reagents (CSR) can be explained by the difference in geometry of the diastereomeric CSR–chiral substrate complexes, as

well as the different magnetic environment of the coordinated enantiomers that causes the anisochrony.[14]

Achiral lanthanide shifting reagents may be used to enhance the anisochrony of diastereomeric mixtures to facilitate their quantitative analysis. Chiral lanthanide shift reagents are much more commonly used to quantitatively analyze enantiomer compositions. Sometimes it may be necessary to chemically convert the enantiomer mixtures to their derivatives in order to get reasonable peak separation with chiral chemical shift reagents.

Sometimes the enantiomer composition of a compound cannot be directly determined using a chiral CSR. In this case, another compound that can be related to the target compound will be chosen for the determination of enantiomer composition.

Disparlure (*cis*-7,8-epoxy-2-methyloctadecane **17**), as shown in Scheme 1–3, has been identified as the sex pheromone of the gypsy moth. Because the two alkyl substituents of disparlure are very similar, the molecule is effectively *meso* from an experimental viewpoint. The optical rotation of disparlure is extremely small. Estimates from +0.2° to +0.7° have been cited for the optically pure material.[15] Therefore, it is difficult to determine the optical purity of synthetic samples by the optical rotation method. Furthermore, attempts to determine the enantiomer excess using chiral solvating agents and chiral lanthanide shift agents in conjunction with ^{1}H or ^{13}C NMR failed to give satisfactory results. Pirkle and Rinaldi[16] succeeded in determining the enantiomeric purity of **17** by utilizing a chiral chemical shift reagent, tris[3-(heptafluoropropylhydroxymethlene)-*d*-camphorato]europium (III) (**18**) in the ^{13}C NMR measurement of compound **16**, an immediate precursor of disparlure (**17**). Examination of the ^{13}C NMR spectrum of racemic disparlure precursor **16** in the presence of the chiral lanthanide reagent revealed the nonequivalent resonance signals for the aromatic *ipso*- or *ortho*-carbons of the enantiomers. Because the subsequent ring closure is stereospecific, the enantiomer composition of the product **17** should correspond to that of its precursor **16**. From its ^{13}C NMR, the synthesized precursor

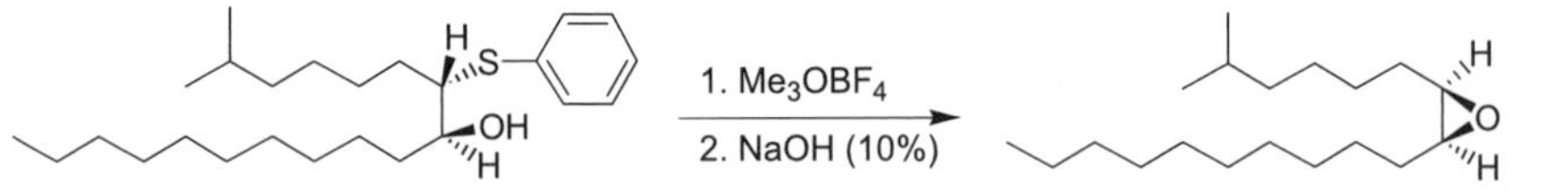

16 **17**

sex pheromone of the gypsy moth
its optical rotation is extremely small

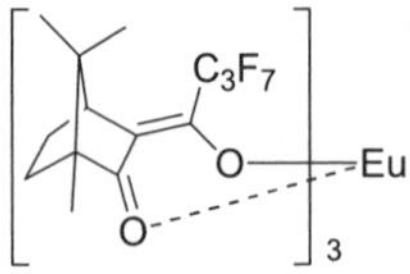

18

Scheme 1–3. Determining enantiomer composition with chiral chemical shift reagent **18**.

16 was found to have such an enantiomeric purity that the minor enantiomer could not be detected. It was thus concluded that the synthetic disparlure **17** was enantiomerically pure.

The synthesis of lanthanide chemical shift reagents has been the objective of many groups owing to their effect on NMR spectra simplification. A drawback of the commonly used reagents is their sensitivity to water or acids. Tris(tetraphenylimido diphosphinato)praseodymium [Pr(tpip)$_3$] has been developed as a CSR for the analysis of carboxylic acids.[17] Furthermore, it has been found that dinuclear dicarboxylate complexes can be obtained through reactions with ammonium or potassium salts of carboxylic acids, and these compounds can be used to determine the enantiomer composition of carboxylic acids.[18]

1.3.2.3 *Chiral Derivatizing Agents for Nuclear Magnetic Resonance Analysis.*

Chiral derivatizing agents are enantiomerically pure reagents that are used to convert test samples to diastereomers in order to determine their enantiomeric purity by NMR spectroscopy. The earliest NMR technique for the determination of enantiomer composition involved the derivatization and analysis of covalent diastereomer mixtures of esters and amides. The alcohols and amines were first converted to the corresponding ester and amide derivatives via reaction with chiral derivatizing agents. The NMR spectra of these derivatives gave some easily identifiable signals for the diastereotopic nuclei, and the enantiomer compositions were calculated from the integrated areas of these signals.[19] One of these first-generation chiral derivatizing agents was (*R*)-(−)-methylmandelyl chloride.[20] Later it was found that the derivative of this reagent had a tendency to epimerize at the α-position of the carbonyl group or to undergo kinetic resolution.[21]

In 1973, Dale and Mosher[22] proposed a reagent, α-methoxy-α-phenyl-α-trifluoromethyl acetic acid (**19**), in both the (*R*)- and (*S*)-form. This is now known as *Mosher's acid*. The chloride of the acid reacts with chiral alcohols (mostly secondary alcohols) to form diastereomeric mixtures called *MTPA esters* or *Mosher's esters*. This acid was initially designed to minimize the epimerization problem.[23] There are two advantages in using this compound: (1) The epimerization of the chiral α-C is avoided because of the absence of the α-proton; and (2) the introduction of a CF$_3$ group makes it possible to analyze the derivatives by means of ^{19}F NMR, which simplifies the analysis process. Peak overlapping is generally not observed, and the ^{19}F NMR signals are far better separated than are the ^{1}H NMR peaks. In most cases, purification of the reaction mixture is not necessary. This compound is also used in the chromatographic determination of enantiomer compositions, as well as in the determination of absolute configurations.

On account of the magnetic nonequivalence of the α-trifluoromethyl group and the α-methoxy group in diastereomeric MTPA esters, the enantiomer compositions of alcohols can be determined by observing the NMR signals of the CH$_3$O or CF$_3$ group in their corresponding MTPA esters (Scheme 1–4). Similarly, due to the different retention times of diastereomeric MTPA esters in

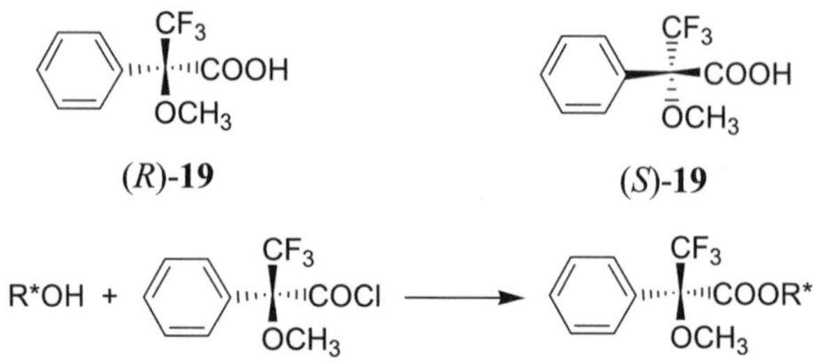

(R)-19   (S)-19

Scheme 1–4. Application of Mosher's acid.

GC or HPLC, the diastereomeric derivatives may be separated by chromatographic means.

Following Mosher's report, several publications appeared showing the preparation of Mosher's acid. One example is the chemoenzymatic preparation of Mosher's acid using *Aspergillus oryzae* protease (Scheme 1–5)[24]:

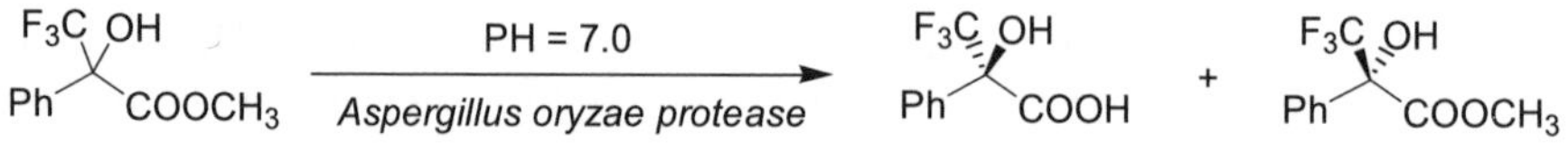

Scheme 1–5. Chemoenzymatic preparation of Mosher's acid.

Another new and simple synthesis of Mosher's acid was reported by Goldberg and Alper[25] (Scheme 1–6):

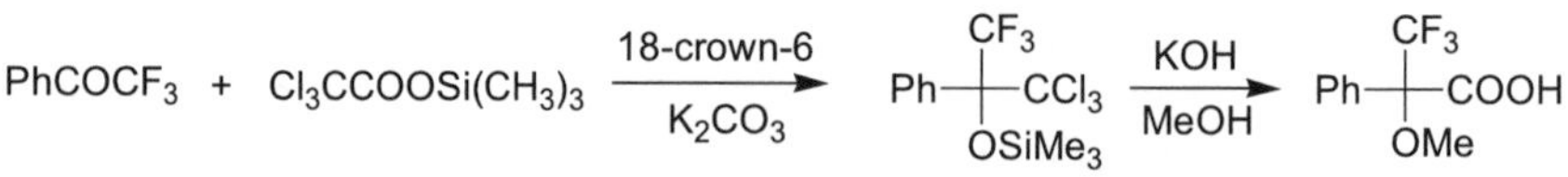

Scheme 1–6. New synthesis of Mosher's acid.

Bennani et al.[26] also reported a short route to Mosher's acid precursors via catalytic asymmetric dihydroxylation (Scheme 1–7):

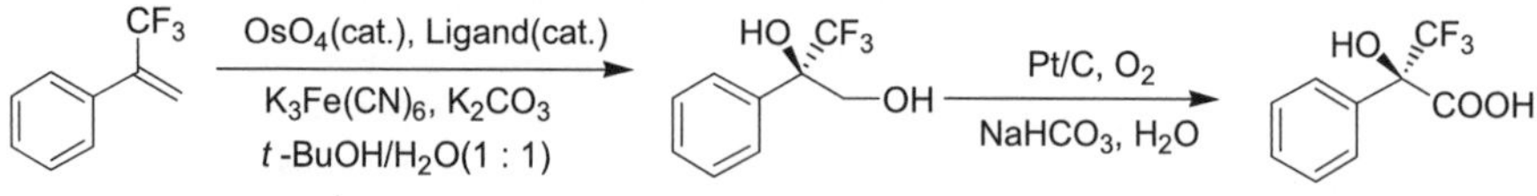

Ligand	ee (%)
(DHQD)$_2$PHAL	83
(DHQD)$_2$DPP	91

Scheme 1–7. Synthesis of Mosher's acid precursors.

Similarly, Mosher-type amines have been introduced for determining the enantiomer composition of chiral carboxylic acids (Fig. 1–10)[27]:

Figure 1–10. Mosher-type amines.

1.3.3 Some Other Reagents for Nuclear Magnetic Resonance Analysis

Various chiral derivatizing agents have been reported for the determination of enantiomer compositions. One example is determining the enantiomeric purity of alcohols using ^{31}P NMR.[28] As shown in Scheme 1–8, reagent **20** can be readily prepared and conveniently stored in tetrahydrofuran (THF) for long periods. This compound shows excellent activity toward primary, secondary, and tertiary alcohols. To evaluate the utility of compound **20** for determining enantiomer composition, some racemic alcohols were chosen and allowed to react with **20**. The diastereomeric pairs of derivative **21** exhibit clear differences in their ^{31}P NMR spectra, and the enantiomer composition of a compound can then be easily measured (Scheme 1–8).

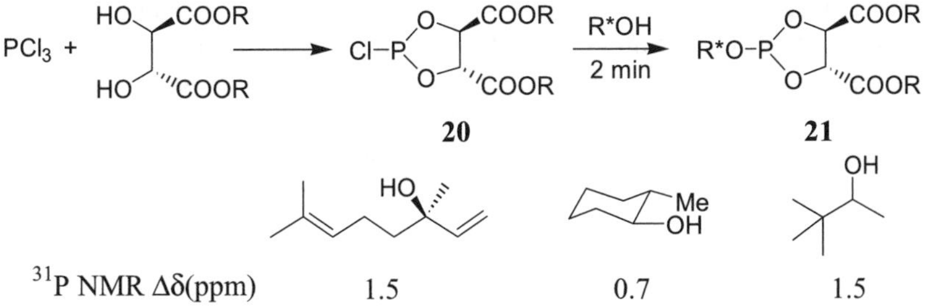

Scheme 1–8. Chemical shift differences in ^{31}P NMR ($\Delta\delta$[ppm]) of some alcohol derivatives with **20**.

Other derivatizing reagents that can be used as simple and efficient reagents for determining the enantiomer composition of chiral alcohols using the ^{31}P NMR method are shown below (Scheme 1–9 and Fig. 1–11)[29–32]:

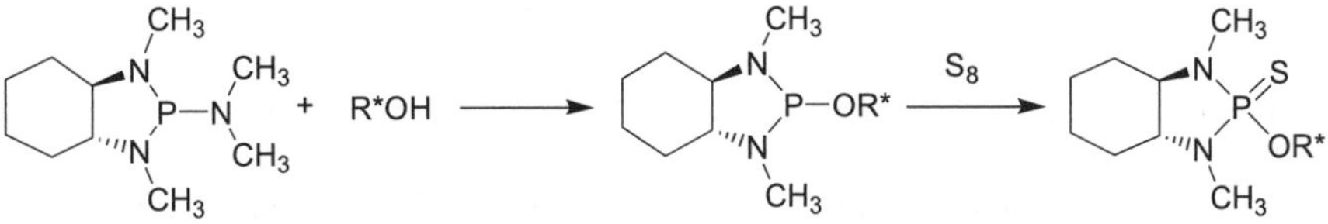

Scheme 1–9. Chiral derivatizing agents used in ^{31}P NMR analysis.

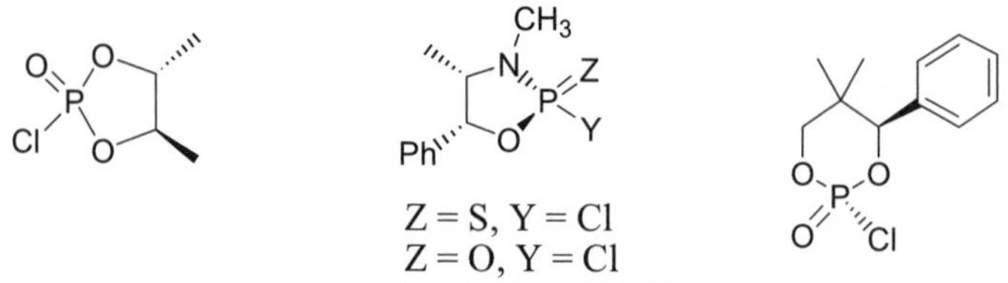

Z = S, Y = Cl
Z = O, Y = Cl
Z = S, Y = NHR*
Z = O, Y = NHR*
Z = S, Y = OR*
Z = O, Y = OR*

Figure 1–11. Some new compounds used as derivatizing agents.

α-Methoxylphenyl acetic acid can be used as an NMR chiral CSR for determining the enantiomer composition of sulfoxides.[33]

1.3.4 Determining the Enantiomer Composition of Chiral Glycols or Cyclic Ketones

Hiemstra and Wynberg reported[34] the determination of the enantiomer composition of 3-substituted cyclohexanones by observing the ^{13}C NMR signals of C-2 and C-6 in the corresponding cyclic ketals, which were prepared via the reaction of the ketones with enantiomerically pure 2,3-butanediol. This method has also been applied in determining enantiomeric composition of chiral aldehydes via the formation of acetals.[35] Similarly, chiral 2-substituted cyclohexanone **22** has been used for determining the enantiomer composition of chiral 2-substituted-1,2-glycols via ^{13}C NMR or HPLC analysis (Scheme 1–10).[36]

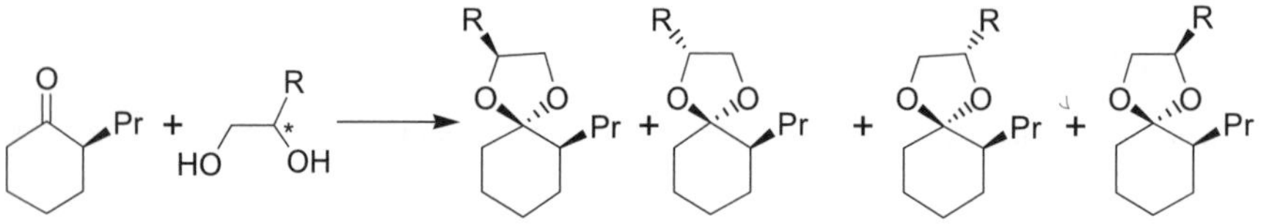

22		**23a**	**23b**	**23c**	**23d**

Scheme 1–10. Formation of ketals from glycols and 2-substituted cyclohexanone **22**.

Compound **22** can be conveniently prepared in multigram quantities and has been found to be useful for assessing the enantiomeric purity of 1,2-glycols. Because the ketal carbon represents a new chiral center, the formation of four diastereomers is possible. However, the diastereomeric pair **23a** and **23b** (or **23c** and **23d**) shows 1:1 peak height in ^{13}C NMR or equal peak areas in HPLC; the diastereomer composition measured by the ratio of **23a** to **23b** or **23c** to **23d** reflects the enantiomer composition of the original 1,2-glycol.

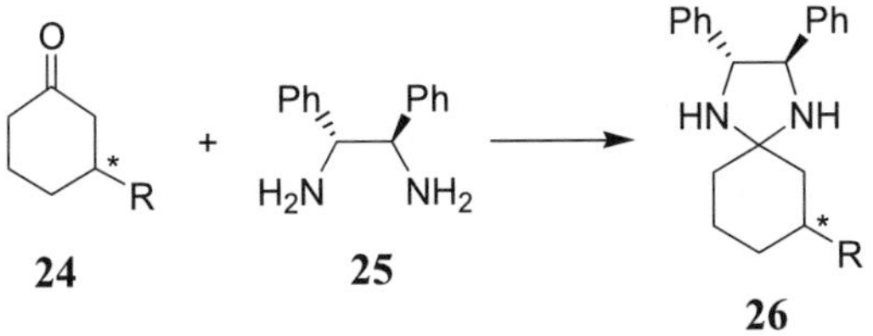

Scheme 1–11. Conversion of ketone to aminal.

Similarly, the enantiomer compositions of ketones or aldehydes can be determined using a chiral 1,2-glycol by converting the ketones or aldehydes to the corresponding ketals or acetals. The derivatization of chiral cyclic ketones or aldehydes to diastereomeric aminals by reacting the ketones or aldehydes with an enantiomerically pure diamine is also an efficient and fast method for determining their enantiomer composition. Enantiomerically pure (R,R)-1,2-diphenylethylene-diamine **25** can react readily with 3-substituted cyclohexanone **24** to form the diastereomeric aminal **26** (Scheme 1–11). The NMR spectrum of **26** in either $CDCl_3$ or C_6D_6 shows a better signal separation than that of the ketals.[37] The main advantage lies in the ease of manipulation of the sample. When ketone **24** and diamine **25** (normally in slight excess) are mixed directly in an NMR tube, the reaction is completed in a few seconds.

In the case of 3-substituted cyclopentanones or cycloheptanones, derivatization with diamine is slower, and the reaction time ranges from a few minutes to several hours. This method is not applicable to acyclic ketones and enones.

The general pattern of the spectra of aminals is similar to that of the corresponding ketals, and the measurement of enantiomer composition can be done on the same carbon nuclei. In addition, the signals are clearly distinguishable in the aminals, giving more accurate results.[38]

1.3.5 Chromatographic Methods Using Chiral Columns

One of the most powerful methods for determining enantiomer composition is gas or liquid chromatography, as it allows direct separation of the enantiomers of a chiral substance. Early chromatographic methods required the conversion of an enantiomeric mixture to a diastereomeric mixture, followed by analysis of the mixture by either GC or HPLC. A more convenient chromatographic approach for determining enantiomer compositions involves the application of a chiral environment without derivatization of the enantiomer mixture. Such a separation may be achieved using a chiral solvent as the mobile phase, but applications are limited because the method consumes large quantities of costly chiral solvents. The direct separation of enantiomers on a chiral stationary phase has been used extensively for the determination of enantiomer composition. Materials for the chiral stationary phase are commercially available for both GC and HPLC.

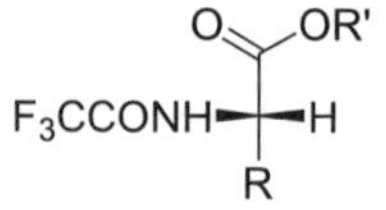

27 R = *s*-Bu, R' = dodecyl

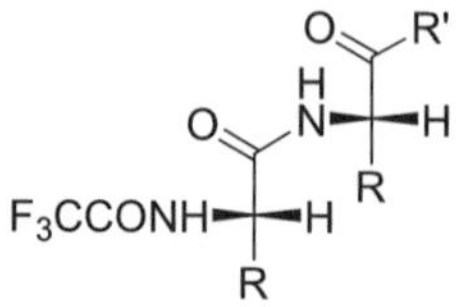

28 R = *i*-Pr, R' = *c*-Hexyl

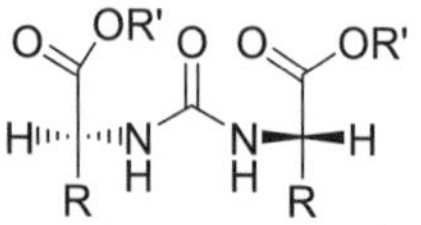

29
R = R' = *i*-Pr

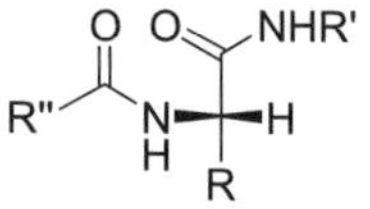

30 R = *i*-Pr
R' = *t*-Bu, R" = undecyl

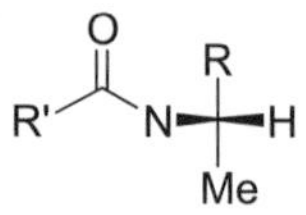

31
R = naphthyl, R' = undecyl

Figure 1–12. Basic structures of chiral materials used as the stationary phase in gas chromatographic resolution via hydrogen bonding.

1.3.5.1 Gas Chromatography. A very commonly used method for the analysis of mixtures of enantiomers is chiral GC.[39–41] In addition to being quick and simple, this sensitive method is normally unaffected by trace impurities. The method is based on the principle that molecular association between the chiral stationary phase and the sample may lead to some chiral recognition and sufficient resolution of the enantiomers. The chiral stationary phase contains an auxiliary resolving agent of high enantiomeric purity. The enantiomers to be analyzed undergo rapid and reversible diastereomeric interactions with the stationary phase and hence may be eluted at different rates (indicated as t_R, the retention time). Two examples of chiral stationary phases used for gas chromatography are illustrated below.

Hydrogen Bonding of the Substrates with the Stationary Phase. In this category (Fig. 1–12), the chiral stationary phase normally contains amide bonds that can provide hydrogen bonding sites for the substrates.[42] Such chiral stationary phases were initially designed for amino acid analysis based on the assumption that hydrogen bonding between the amino acid substrate and the chiral stationary phase can provide a small degree of enantioselectivity sufficient for the quantitative analysis of the enantiomer compositions of chiral amino acids.[43] This separation can be amplified by using long capillary columns.

Complexation with Chiral Metal Complexes. This idea was first suggested by Feibush et al.[44] The separation is realized by the dynamic formation of diastereomeric complexes between gaseous chiral molecules and the chiral stationary phase in the coordination sphere of metal complexes. A few typical examples of metal complexes used in chiral stationary phase chromatography are presented in Figure 1–13.[45]

Separation of enantiomeric or diastereomeric mixtures by GC is a good

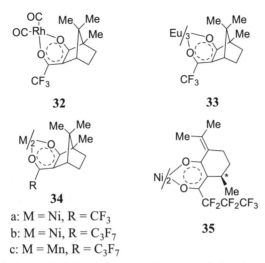

Figure 1–13. Chiral metal chelates for enantiomer resolution by complexation gas chromatography.

method for determining enantiomer compositions. However, this method is limited to samples that are both volatile and thermally stable. Normally, if the compound to be separated has a low boiling point (lower than 260°C, for example), or it can be converted to a low boiling substance, and no racemization occurs during the analysis, it is possible to analyze it by GC. In general, the lower the temperature at which the compound is eluted, the greater the opportunity for a clean separation. If the compound has a high boiling point, or the compound tends to decompose or racemize at high temperature, HPLC using either a chiral stationary phase or a chiral mobile phase would be the choice of separation.

1.3.5.2 Liquid Chromatography. The development of rapid, simple liquid chromatographic methods for determining the enantiomeric purity of chiral compounds is probably one of the most important developments in the study of asymmetric synthesis in the last 10 years. Several books have been published providing thorough evaluations of various enantiomeric separation techniques and their practical applications.[46]

Initially, chiral stationary phases for chiral liquid chromatography were designed for preparative purposes, mostly based on the concept of "three-point recognition".[47] Pirkle and other scientists[48] developed a series of chiral stationary phases that usually contain an aryl-substituted chiral compound connected to silica gel through a spacer. Figure 1–14 depicts the general concept and an actual example of such a chiral stationary phase.

Another chiral stationary phase is modified cyclodextrin. Cyclodextrins are cyclic chiral carbohydrates composed of six, seven, or eight glucopyranose

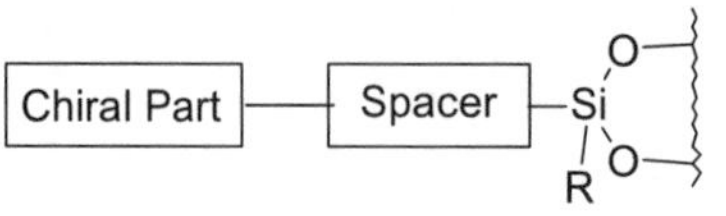

36a (general structure)

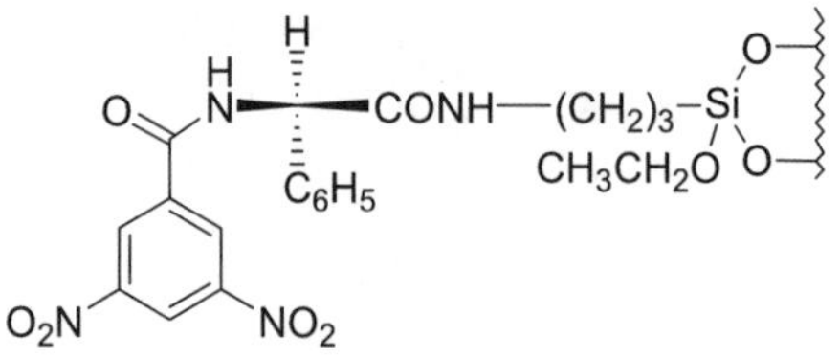

36b (used for the separation of amino acids)

Figure 1–14. Chiral stationary phase for high-performance liquid chromatography.

units designated as α-, β-, and γ-cyclodextrin, respectively. Cyclodextrins are cylinder-shaped molecules with an axial void cavity. Their outer surface is hydrophilic, and therefore they are soluble in water. The cavity is nonpolar and can include other nonpolar molecules of appropriate dimensions and bind them through hydrophobic interactions.[49]

The complexation of cyclodextrin is highly selective. The inclusion processes are influenced mainly by the hydrophobicity and shape of the guest molecules. Specifically, the guest molecules must fit the cyclodextrin cavity. Complexation processes occurring in solution are reversible, and the equilibration in solution is relatively fast. For these reasons, cyclodextrin immobilized on silica gel is also used for chromatographic separation of chiral compounds, especially for compounds containing aromatic groups.[50] An aromatic group on the substrate is essential for getting enantioselective binding through interaction with the glycosidic oxygen atoms. A substrate without an aromatic group will occupy random positions within the cavity and consequently lose enantioselectivity.[51]

1.3.6 Capillary Electrophoresis with Enantioselective Supporting Electrolytes

Electrophoresis is based on the transport of electrically charged compounds in a gel or a buffer solution under the influence of an electric field. The instrumentation involves a capillary tube filled with buffer solution and placed between two buffer reservoirs. The electric field is applied by means of a high-voltage power supply. This is similar to a chromatographic method in which the enantiomer mixture forms diastereomer complexes with a chiral mobile phase to accomplish the separation. In chromatographic separation, the driving force comes from the mobile phase, whereas in electrophoresis the driving force is the electroosmotic and electrophoretic action. Differences in complexation constants cause these transient charged species to acquire different mobilities under the influence of the applied electric field. It should be noted that in electrophoresis no mobile phase is used. The method depends on the different migration rates of charged enantiomers in a chiral supporting electrolyte. The method is fast and highly sensitive, which permits the rapid (about 10 minutes) and accurate analysis of samples in femtomolar concentration.[52]

Capillary electrophoresis (CE) was originally developed as a microanalytical technique for analysis and purification of biopolymers. The separation of biopolymers can be achieved according to their different electrophoretic mobilities. Capillary gel electrophoresis is based on the distribution of analytes in a carrier electrolyte, and this method has been extensively used in analysis and separation of proteins and nucleic acids.

Compared with GC and HPLC, the most important advantage of CE is its high peak efficiency. It can give a baseline resolution of peaks even when the separation factor is low. Volatile chiral samples are best analyzed by GC, whereas HPLC and CE are more suitable for nonvolatile samples. CE is the best choice for a charged compound or for a high-molecular-weight sample.

As the running medium in electrophoresis, the buffer solution should have a high capacity in the selected pH range and should not give a strong background signal in the detector. Furthermore, to minimize the electric current, the buffer should also have a low mobility under the voltage applied and under the experimental conditions.

The applied voltage has a significant effect on the separation. Excess voltage may degrade the analysis for two reasons: first by speeding up the mobility of the analyte and second by causing Joule heating, which changes the separation conditions.

Several modes of capillary electrophoretic separation are available: ordinary CE, capillary zone electrophoresis, capillary electrokinetic chromatography, capillary gel electrophoresis, capillary electrochromatography, capillary isotachophoresis, and capillary isoelectric focusing. The different separation mechanisms make it possible to separate a wide variety of substances depending on their mass, charge, and chemical nature.[53]

In a solution without chiral selectors, enantiomers cannot be distinguished from each other through their electrophoretic mobility. Separation can, however, be achieved when the buffer solution contains certain chiral compounds. The chiral compounds used to distinguish enantiomers are referred to as selectors.

When a sample is loaded into the capillary, a transient diastereomer complex may be formed between the sample and the selector. The differing mobilities of the diastereomers in the buffer solution in the presence of an electric field is the reason for the separation. The differences of mobility between the diastereomers are the result of different effective charge sensitivities caused by the different spatial orientations of diastereomers or the specific intermolecular interactions between them.

Many chiral compounds can be used as selectors, for example, chiral metal complexes, native and modified cyclodextrins, crown ethers, macrocyclic antibiotics, noncyclic oligosaccharides, and polysaccharides all have been shown to be useful for efficient separation of different types of compounds.

1.4 DETERMINING ABSOLUTE CONFIGURATION

Thus far, we have discussed the nomenclature of different types of chiral systems as well as techniques for determining enantiomer composition. Currently, the most commonly used nomenclature for chiral systems follows the CIP rules or sequence rules, although Fischer's convention is still applied for carbohydrates and amino acids. In the area of asymmetric synthesis, one of the most important parameters one has to know in order to evaluate the efficiency of asymmetric induction is the enantiomer composition. Another important parameter is the configuration of the major product of an asymmetric reaction. Thus, in an asymmetric reaction, there are two important elements. One is to know the predominant configuration, and the other is to determine the extent to which this configuration is in excess of the other.

It is very important to define the absolute configuration of a chiral molecule in order to understand its function in a biosystem. First, definite chirality is involved in most biological processes; second, only one of the enantiomeric forms is involved in most of the building blocks for proteins, nucleotides, and carbohydrates, as well as terpenes and other natural products. Many biological activities are exclusive to one specific absolute configuration. Without a good understanding of the absolute configuration of a molecule, we often cannot understand its chemical and biological behavior.

Under normal conditions, the two enantiomers of a chiral compound have exactly the same boiling and melting points and the same solubility in normal achiral solvents. Their chemical reactions are also identical under achiral conditions. However, under chiral conditions, the enantiomers may behave very differently. For example, physical property or chemical reactivity may change significantly under chiral conditions.

Determining the absolute configuration of a chiral center involves assigning spatial orientation to the molecule and then correlating this orientation with the negative or positive rotation of polarized light caused by this substance under given conditions. Several methods are available to determine the absolute configuration of chiral compounds.

1.4.1 X-Ray Diffraction Methods

Normal X-ray diffraction cannot distinguish between enantiomers. The amplitude of a given reflection depends on the scattering power of the atoms and phase differences in the wavelets scattered by them. When the diffraction involves light nuclei (e.g., C, H, N, O, F), the interference pattern is determined only by the internuclear separations, and the phase coincidence is independent of the spatial orientation of these nuclei. Thus, from the diffraction pattern it is possible to calculate various internuclear distances and constitutions in the molecule and to deduce the relative positions of these nuclei in space. One can build the relative configuration of a compound, but it is normally difficult to distinguish enantiomers or to get the absolute configurations for chiral compounds containing only light atoms.

When molecules containing only light nuclei are subjected to X-ray analysis, only diffraction occurs and no significant absorption can be observed. During the experiment, the phase change in the radiation is almost the same for both enantiomers. Nuclei of heavy atoms absorb X-rays over a particular range of the absorption curve. If the wavelength of the radiation coincides with the absorption edge of the heavy atom, there will be absorption, and both diffraction and phase lag can be observed. Because of this phase lag or anomalous scattering, the interference pattern will depend not only on the distance between atoms but also on their relative positions in space, thus making it possible to determine the absolute configuration of molecules containing heavy atoms.

For example, K_α radiation of zirconium is on the edge of rubidium, and L_α radiation of uranium is on that of bromine. Therefore, for a molecule containing rubidium, the absolute configuration can be determined by using Zr-K_α as the X-ray source, and the absolute configuration can be determined by using U-L_α as the X-ray source for molecules containing bromine. This is called the *anomalous X-ray scattering method*. In 1930, Coster and his co-workers used this method to determine the sequence of planes of zinc and sulfur atoms in a crystal of zincblende. In this experiment, an X-ray wavelength was chosen near the absorption edge of zinc, and this resulted in a small phase change of the X-rays scattered by zinc atoms related to sulfur.

Normally, a relatively heavy atom is chosen because the phase change generally increases with increasing atom mass. This principle was first applied[54] to determine the absolute configuration of a sodium rubidium salt of natural tartaric acid by using Zr-K_α X-rays in an X-ray crystallographic study. This method is now referred to as the *Bijvoet method*. With the absolute configuration of sodium rubidium tartrate as a starting point, the absolute configuration of other compounds has been determined in a step by step fashion through correlation based on either physical–chemical comparison or transformation by chemical reactions.

In general, the result of an individual determination of absolute configuration by this method is more prone to error than are results from other methods of structure determination because it depends on a difference in the intensity of related diffraction pairs. The heavy atom method is suitable for determining the absolute configuration of organic acids or bases, because it is easy to introduce heavy metals into these molecules by means of salt formation. Currently, X-ray analysis by this heavy atom method is a standard technique for resolving the structure of organic molecules. It is almost always possible to attach a heavy atom to the molecule. The probability of error increases in the absence of a heavy metal, but this can be offset by applying a neutron diffraction method. As a variation of X-ray diffraction, neutron diffraction analysis can also be used to determine the absolute configuration of chiral compounds that do not contain heavy atoms.

For a molecule without a heavy atom, the absolute configuration can also be determined by attaching another chiral moiety of known configuration to the sample. The absolute configuration can then be determined by comparison with this known configuration. For example, the absolute configuration of compound **37** or **38** cannot be determined by X-ray diffraction because of the lack of heavy atoms in the molecules. But the configuration can be determined by introducing chiral groups of known configuration. Thus, the absolute configuration of the phosphor atoms in the quinoline salt of $(+)$-(R)-**37**[55] and the brucine salt of $(-)$-(S)-**38**[56] has been determined by X-ray single crystal diffraction analysis.

$$
\begin{array}{cc}
\underset{\text{EtO}}{\overset{\displaystyle S}{\underset{\displaystyle \parallel}{\underset{\displaystyle \text{P}}{\quad}}}}\text{''''OH} & \underset{\text{Ph}}{\overset{\displaystyle S}{\underset{\displaystyle \parallel}{\underset{\displaystyle \text{P}}{\quad}}}}\text{''''OH} \\
\text{OPh} & \text{OEt}
\end{array}
$$

S   S
‖   ‖
EtO–P‴OH   Ph–P‴OH
　OPh   OEt

(R)-$(+)$-**37** (S)-$(-)$-**38**

1.4.2 Chiroptical Methods

The electric vectors of a beam of normal light are oriented in all planes, whereas in polarized light the electric vectors lie in the same plane perpendicular to the direction of propagation. Materials capable of rotating the plane of polarized light are termed *optically active*.

Optical activity comes from the different refractions of right and left circularly polarized light by chiral molecules. The difference in refractive indices in a dissymmetric medium corresponds to the slowing down of one beam in relation to the other. This can cause a rotation of the plane of polarization or optical rotation. The value of specific rotation varies with wavelength of the incident polarized light. This is called *optical rotatory dispersion* (ORD).

Optical activity also manifests itself in small differences in the molar extinction coefficients ε_L and ε_R of an enantiomer toward the right and left circularly polarized light. The small differences in ε are expressed by the term *molecular ellipticity* $[\theta]_\lambda^T = 3300(\varepsilon_L - \varepsilon_R)$. As a result of the differences in molar extinction coefficients, a circularly polarized beam in one direction is absorbed more than the other. Molecular ellipticity is dependent on temperature, solvent, and wavelength. The wavelength dependence of ellipticity is called *circular dichroism* (CD). CD spectroscopy is a powerful method for studying the three-dimensional structures of optically active chiral compounds, for example, for studying their absolute configurations or preferred conformations.[57]

CD spectra are usually measured in solution, and these spectra result from the interaction of the individual chromophores of a single molecule with the electromagnetic field of light. The interaction with neighboring molecules is often negligible. Moreover, because molecules in solution are tumbling and randomly oriented, the mutual interaction between two molecules, which is approximated by a dipole–dipole interaction, is negligible.

Organic molecules with π-electron systems interact with the electromagnetic field of ultraviolet or visible light to absorb resonance energy. The ultraviolet and visible absorption spectra of a variety of π-electron systems have been applied extensively in structural studies. Measuring the CD of optically active compounds is a powerful method for studying the three-dimensional structure of organic molecules, and, most importantly, this method is being used for the structural study of biopolymers.

The wavelength dependence of specific rotation and/or molecular ellipticity is called the *Cotton effect*. The Cotton effect can provide a wealth of information on relative or absolute configurations. The sign of the Cotton effect reflects the stereochemistry of the environment of the chromophore. By comparing the Cotton effect of a compound of known absolute configuration with that of a structurally similar compound, it is possible to deduce the absolute configuration or conformation of the latter.

In a plot of molecular specific rotation or molecular ellipticity vesus wavelength, the extremum on the side of the longer wavelength is called the *first extremum*, and the extremum on the side of the shorter wavelength is called the *second extremum*. If the first extremum is positive and the second one is negative, this is called a *positive* Cotton effect; the first extremum is called a *peak*, and the second extremum is called a *trough*. Conversely, in a *negative* Cotton effect curve, the first extremum is a trough and the second one is a peak.

Comparing the signs of the Cotton effect is applicable to substances with suitable chromophores connected to a rigid cyclic substructure. With the aid of an empirical rule, or "octant rule," it is possible using this comparison to predict the absolute configurations of certain five-, six-, and seven-membered cyclic ketones.[58] According to this empirical rule, three planes A, B, and C divide the space around the carbonyl group into octants. Plane A bisects the carbonyl group, plane B is perpendicular to A and resides on the carbonyl oxygen, and plane C is perpendicular to both A and B (Fig. 1–15):

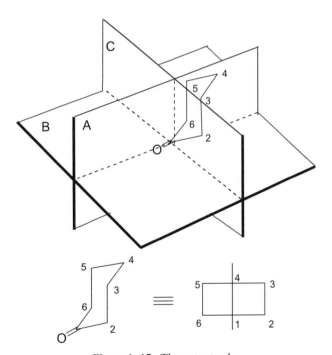

Figure 1–15. The octant rule.

Viewed from the carbonyl side, the four octants behind plane C are the *rear octants*; the four octants in front of plane C are referred to as the *forward octants*. The empirical octant rule establishes that the Cotton effect in a molecule can be correlated with its substituents. The substituents in rear octants are the most important because there are rarely substituents on the cycloalkane ring pointing forward. Substituents that reside on plane A, B, or C make no contribution to the n-π^* Cotton effect in the CD of the cycloalkanones. However, rear substituents in the lower-left octant and the upper-right octant contribute a negative Cotton effect. Rear substituents in the upper-left and lower-right octants have a positive contribution. For multisubstituted systems, the sign of the n-π^* Cotton effect can be estimated from the sum of the contributions made by the substituents in each of the eight octants.

For a system containing two chromophores *i* and *j*, the exciton chirality (positive or negative) governing the sign and amplitude of the split Cotton effect can be theoretically defined as below[59]:

$$\vec{R} \cdot (\vec{\mu}_{ioa} \times \vec{\mu}_{joa}) V_{ij}$$

where $\vec{R}$ is an interchromophore distance vector, $\vec{\mu}_{ioa}$ and $\vec{\mu}_{joa}$ are the electric transition dipole moments of excitation o $\rightarrow$ a for groups i and j, and V_{ij} is the interaction energy between the two chromophores i and j.

In the case of a molecule having two identical chromophores connected by σ-bonds in some orientation, it is probable that the state of one excited chromophore is the same as that of the other, as the excited state (exciton) delocalizes between the two chromophores.

A molecule containing two chromophores oriented in chiral positions can be defined to have either negative or positive chirality as depicted in Figure 1–16:

Positive chirality   Negative chirality

Figure 1–16. Positive and negative chirality.

In the case of a positive chirality, a Cotton effect with positive first and negative second is observed, whereas a Cotton effect with negative first is found for negative chirality. As this method is based on theoretical calculations, the absolute configuration of organic compounds can be deduced unambiguously from their corresponding CD curves.

There are several criteria for chromophores that are to be used for CD chirality studies:

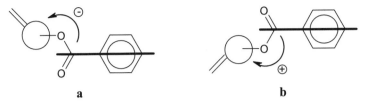

a   b

Figure 1–17. Exciton chirality of acyclic allylic benzoates and the sign of the predicted benzoate Cotton effects. The thick line denotes the electric transition moment of the benzoate group. Reprinted with permission by Am. Chem. Soc., Ref. 61.

1. The chromophore must have strong $\pi–\pi^*$ transition bands.
2. The chromophore should have as high symmetry as possible so that polarization of the transition bands is established in the geometry of the chromophores.

para-Substituted benzoate is suitable for determining the absolute configuration in glycol systems. The intramolecular charge transfer band of the chromophore undergoes a red shift when electron-donating or electron-withdrawing groups are substituted in *para*-positions. The stronger the electron donating or withdrawing property in the *para*-position, the more significant the red shift. Benzamido chromophore can be used for chiral amino alcohol or diamine systems.

Considering an olefinic functionality as a chromophore, the absolute configuration of cyclic allylic alcohols can be determined using a method that involves the conversion of the alcohol to the corresponding benzoate.[60] This can also be extended to acyclic alcohols where the conformations are dynamic (see Fig. 1–17). Interested readers may consult the literature for details.[61]

1.4.3 The Chemical Interrelation Method

The chemical interrelation method for determining the absolute configuration of a compound involves the conversion of this compound to a compound with a known configuration, and then the absolute configuration is deduced from the resulting physical properties, such as optical rotation or GC behavior. An example is shown in Scheme 1–12.

Alkylation of the configurationally unknown compound (+)-**39** followed by chlorination in Scheme 1–12 afforded product (*R*)-(−)-**41** with retained configurations. This was then converted to (*S*)-(−)-**42** with an inversion of configuration.[62] In this manner, the correlation between compounds (+)-**39** and (*S*)-(−)-**42** in the sense of absolute configuration has been established, and the starting (+)-**39** is determined to have an absolute configuration of (*R*) (Scheme 1–12).

This method of determining the absolute configuration is commonly used, as it is convenient, economical, and does not need expensive instruments. For ex-

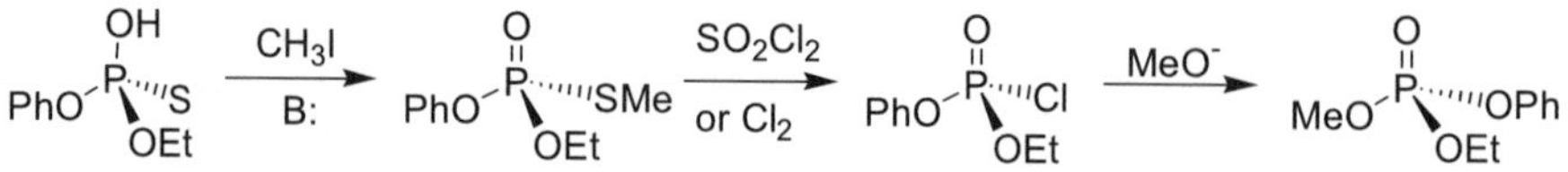

(R)-$(+)$-**39** (R)-$(-)$-**40** (R)-$(-)$-**41** (S)-$(-)$-**42**

Scheme 1–12. Chemical interrelation method.

ample, manicone, $(4E)$-4,6-dimethyl-4-octen-3-one (**43**), has been identified as an active pheromone present in the mandibular glands of two North American species of ants: *Manica mutica* and *M. bradleyi*, as well as in the mandibular gland secretion of the *Eurasiatic manica* species *M. rubida latr*. Samples of *rac*-**43** could not be separated by complexation GC on a chiral stationary phase. To determine the absolute configuration of this chiral natural product, hydroge-nation of the C=C bond of synthetic *rac*-**43** with Pd/C was carried out to give *rac*-**44**, which is composed of two pairs of diastereomers. These four isomers can be separated by complexation GC on nickel(II)-bis[3-heptafluoro-butyryl-$(1R)$-camphorate] (Fig. 1–18a). In the same manner, natural **43** was subjected to hydrogenation, and the two isomers thus formed were also separated under the same conditions. Only two diastereomeric hydrogenation products **44** appeared in the gas chromatogram (Fig. 1–18b). Compound **44** with an (S)-configura-tion on C-6 was then synthesized starting from a commercially available (S)-$(-)$-2-methylbutanol, giving a mixture of diastereomers $(4R,6S)$- and $(4S,6S)$-**44**. Chromatograms of the mixture of these diastereomers are shown in Figure 1–18d. By co-injecting two samples of natural-**44**/*rac*-**44** and $(4RS,6S)$-**44**/*rac*-**44** (Fig. 1–18c,e), the natural pheromone was finally confirmed to have an (S)-configuration at C-6.[63]

1.4.4 Prelog's Method

In 1953, Prelog[64] put forward an empirical rule from which the absolute con-figuration of an optically active secondary alcohol can be deduced. According to this rule, nucleophilic attack on an α-keto carboxylic acid ester of a chiral secondary alcohol can give a chiral α-hydroxyl carboxylic acid. From the pre-dominant absolute configuration of the resulting chiral acid, the absolute con-figuration of the original alcohol can be deduced. This rule is outlined in Scheme 1–13. The abbreviations R_L and R_M refer to large- and medium-sized substituents, respectively, on the asymmetric carbon atom in the alcohol.

Esterification of an α-keto acid or its chloride (phenylglyoxyl chloride **45**) with an optically active alcohol **46** gives an optically active α-keto ester **47**. Treatment of this ester with an achiral reagent such as a methyl Grignard re-agent results in the formation of the diastereomeric α-hydroxyl acid ester, and **48** can be obtained by hydrolysis. There is a correlation between asymmetric induction and the relative size of the substituent groups, which are located nearest to the trigonal atom undergoing reaction.

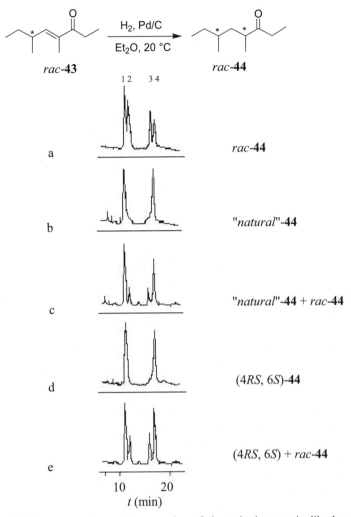

Figure 1–18. Gas chromatographic separation of a) synthetic racemic dihydromanicone *rac*-**44**; b) "natural" **44**, obtained by hydrogenation of material from the heads of *M. rubida*; c) co-injected natural-**44** and *rac*-**44**; d) synthetic (4*RS*,6*S*)-**44**; and e) co-injected synthetic (4*RS*,6*S*)-**44** and *rac*-**44**. Chiral GC phase: nickel(II)-bis[3-heptafluorobutyryl-(1*R*)-camphorate]. Signals 1 and 4 correspond to the pair of diastereomers (4*RS*,6*S*)-**44**; signals 2 and 3 correspond to (4*RS*,6*R*)-**44**. Reprinted, with permission, by VCH, Ref. 63.

Conformational analysis reveals that the ester of type RCOOR′ (**47**) adopts a planar conformation in which CO and OR′ groups are cisoid and the two carbonyl groups antiparallel, as shown in **47** (Scheme 1–13). Three conformations for the alcohol substituents might be considered (**49–51**). Examination of the attacking mode of the reagent suggests that **49** and **50**, the favorable

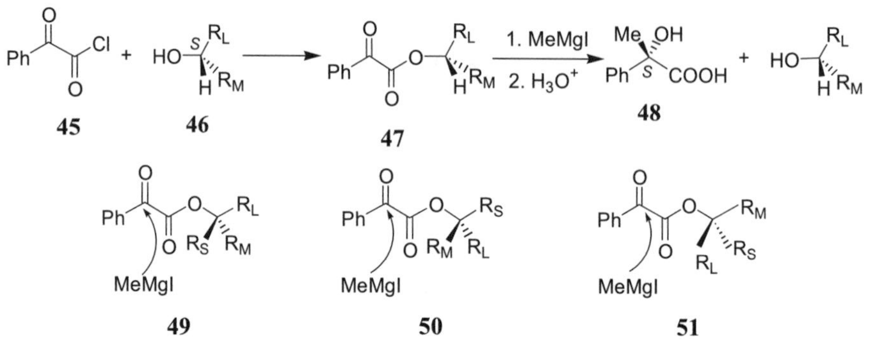

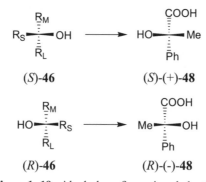

Scheme 1–13. Prelog's rule.

transition states for the subsequent reaction, will lead to the same enantiomer (α-hydroxyl acid) as indicated by Prelog. The third conformation **51** will give the antipode. Conformations **49** and **50** are preferred over **51** for the front end attack by the Grignard reagent for two reasons. First, there is considerable steric interaction between the large substituent on the alcohol alkoxyl group in **47** and the attacking methyl Grignard reagent. Second, the reagent approaches from the least sterically hindered side. It is thus possible to predict which configuration of the asymmetric addition product will be formed preferentially.

As a result, the configuration of the alcohol can be deduced as (S) or (R) depending on which form, (S)-(+)- or (R)-(−)-**48**, of the hydrolyzed product is isolated in excess at the end of the reaction (Fig. 1–19).

It is always advisable to examine the complete molecular topology in the neighborhood of the chiral carbon atom and to confirm the results by employing another analytical method before the final assignment. In conclusion, Prelog's rule does predict the steric course of an asymmetric synthesis carried out with a chiral α-keto ester, and the predictions have been found to be correct in most cases. Indeed, this method has been widely used for determining the absolute configuration of secondary alcohols.

(S)-**46** (S)-(+)-**48**

(R)-**46** (R)-(-)-**48**

Figure 1–19. Alcohol configuration deduction.

1.4.5 Horeau's Method

Another method for determining the absolute configurations of secondary alcohols is Horeau's method, which is based on kinetic resolution. As shown in Scheme 1–14, an optically active alcohol reacts with racemic 2-phenylbutanoic anhydride (**54**), and an optically active 2-phenylbutanoic acid (**52**) is obtained after hydrolysis of the half-reacted anhydride.

It has been found experimentally that, as with Prelog's rule, there is a relationship between the sign of optical rotation of the isolated 2-phenylbutyric acid and the absolute configuration of the alcohol involved. If the 2-phenylbutanoic acid isolated is levorotatory, the secondary alcohol **53** will be configured such that, in a Fischer projection, the hydroxy group is down, the hydrogen atom is up, and the larger group of the two remaining substituents is on the right. If the isolated acid is dextrorotatory, the secondary alcohol will arrange with the larger substituent on the left. Accordingly, the absolute configuration of the secondary alcohol being studied can be defined as (R) or (S) according to the CIP rule (Scheme 1–14).[6] Interested readers may consult the review written by Horeau.[65]

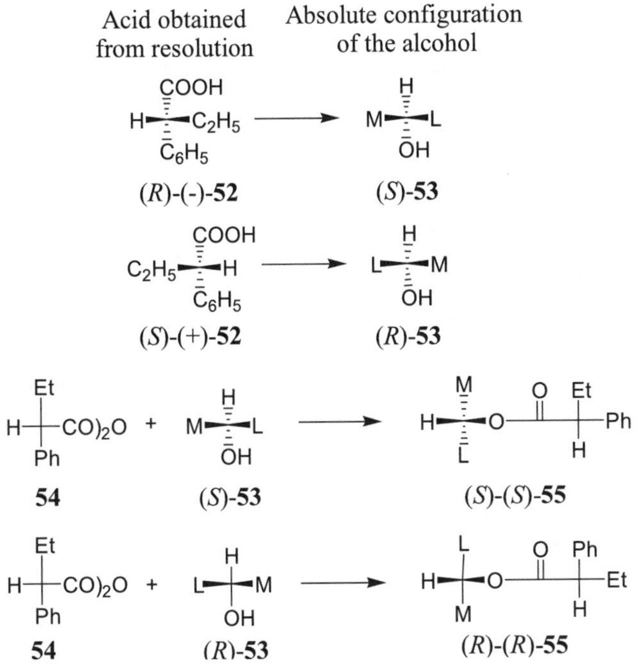

Acid obtained from resolution   Absolute configuration of the alcohol

Scheme 1–14. Horeau's method.

2-Phenylbutanoic anhydride can be easily prepared as shown in Scheme 1–15[66]:

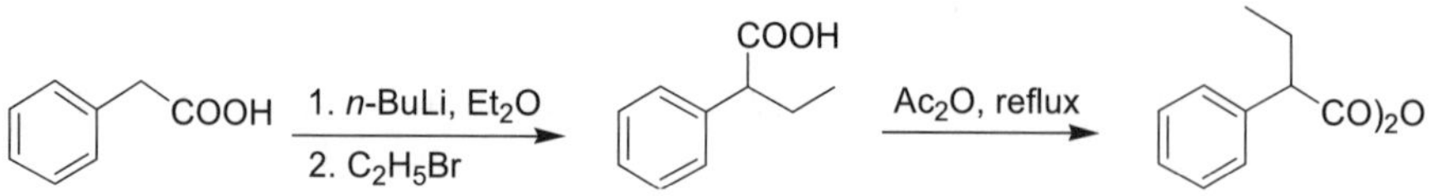

Scheme 1–15. Preparation of 2-phenylbutanoic anhydride.

2-Phenylbutanoic anhydride can be used for assigning the absolute configuration of most chiral secondary alcohols. Unfortunately this reagent fails to give a satisfactory result when the alcohol contains very bulky groups.[67] A modification of Horeau's method, using $PhCH(C_2H_5)COCl$ as the resolving reagent, has been shown to be effective for alcohols with bulky groups. In fact, with this modification following a similar esterification procedure, even very hindered neopentyl alcohols react with the acid chloride. (Acid anhydride did not react in this case.) The result is consistent with that achieved through the use of phenylbutanoic acid anhydride.[68]

1.4.6   Nuclear Magnetic Resonance Method for Relative Configuration Determination

1.4.6.1   Nuclear Overhauser Effect for Configuration Determination.
When one resonance in an NMR spectrum is perturbed by saturation or inversion, the net intensities of other resonances in the spectrum may change. This phenomenon is called the *nuclear Overhauser effect* (NOE). The change in resonance intensities is caused by spins close in space to those directly affected by the perturbation. In an ideal NOE experiment, the target resonance is completely saturated by selected irradiation, while all other signals are completely unaffected. An NOE study of a rigid molecule or molecular residue often gives both structural and conformational information, whereas for highly flexible molecules or residues NOE studies are less useful.

The NOE is a product of double magnetic resonance. The intensity of an NMR signal depends on the rate of spin relaxation (i.e., the rate at which the nuclei return from a high magnetic energy level to a low one). The process of relaxation of one nucleus can be significantly influenced by the magnetic events occurring in the other nuclei in the vicinity.

Because the NOE falls off with the inverse sixth power of distance, an observed NOE between two protons implies that the two protons are reasonably close to each other. Thus, it is possible to determine their relative position using the NOE. The following examples illustrate the use of NOE for stereochemical assignments.

In Figure 1–20, the protected amino acids **56a** and **56b** are synthetic precursors of the two diastereomers of aminostatine,[69] and they can be distinguished by the NOE spectra. The spatial interaction between H-2 and H-3 causes the difference in the NOE in the cyclized derivatives **56a′** and **56b′**. For compound **56a′**, no NOE is observed between these two vicinal protons, while for **56b′** a significant NOE between H-3 and H-2 can be observed (Fig. 1–20).

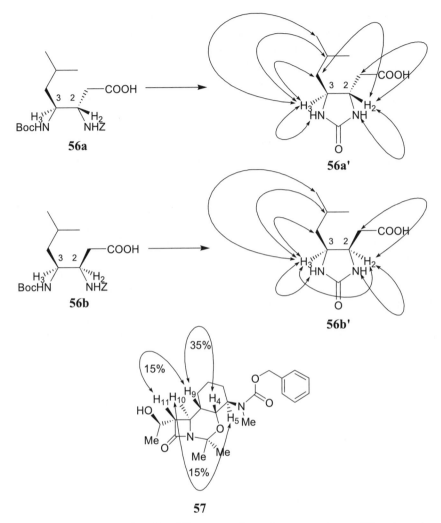

Figure 1–20. Compounds with different configurations may have different nuclear Overhauser effects.

Another example is in the determination of the absolute configuration of the antibiotic analogue GV 129606X (**57**).[70] From the NOE, the relative positions of H-4, H-5, H-9, H-10, and H-11 can easily be established, and the configurations of the corresponding chiral carbon centers to which the protons are attached can also be deduced unambiguously by relating them to the known absolute configuration.

1.4.6.2 Modified Mosher's Method for Determining the Absolute Configuration. Among a number of methods used to determine the absolute configuration of organic compounds, Mosher's method[71] using 2-methoxy-2-

trifluoromethyl-2-phenylacetic acid (MTPA) has been the most frequently used. This method involves converting chiral alcohols (or amines) to their corresponding MTPA esters (or amides), followed by NMR analysis of the resulting derivatives. With a high-field FT-NMR technique, the absolute configuration of these chiral alcohols (or amines) can be deduced from the corresponding chemical shift difference, and this method is now known as *Mosher's method*.

Mosher proposed that, in solution, the carbinyl proton and ester carbonyl, as well as the trifluoromethyl group of an MTPA moiety, lie in the same plane (Fig. 1–21). Calculations on this MTPA ester demonstrate that the proposed conformation is just one of two stable conformations.[72] X-ray studies on the (*R*)-MTPA ester of 4-*trans-t*-butylcyclohexanol[73] and (1*R*)-hydroxy-(2*R*)-bromo-1,2,3,4-tetrahydronaphthalene[74] reveal that the position of the MTPA moiety is almost identical with that proposed by Mosher, as shown in Figure 1–21A. As a result of the diamagnetic effect of the benzene ring, the NMR signals of H_A, H_B, H_C of the (*R*)-MTPA ester should appear upfield relative to those of the (*S*)-MTPA ester. The reverse should be true for the NMR signals of H_X, H_Y, H_Z. Therefore, for $\Delta\delta = \delta_S - \delta_R$, protons on the right side of the MTPA plane A must have positive values ($\Delta\delta > 0$), and protons on the left side of the plane must have negative values ($\Delta\delta < 0$), as illustrated in Figure 1–21.

Procedures for using this method to determine the absolute configuration of secondary alcohols can be outlined as follows (Fig. 1–21B):

1. Assign as many proton signals as possible with respect to each of the (*R*)- and (*S*)-MTPA esters.

2. Calculate the $\Delta\delta$ ($\delta_S - \delta_R$) values for these signals of (*R*)- and (*S*)-MTPA esters.

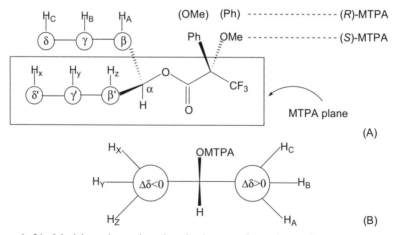

Figure 1–21. Model to determine the absolute configurations of secondary alcohols. Reprinted with permission by Am. Chem. Soc., Ref. 75.

3. Put the protons with positive $\Delta\delta$ on the right side and those with negative $\Delta\delta$ on the left side of the model.

4. Construct a molecular model of the compound in question and confirm that all the assigned protons with positive and negative $\Delta\delta$ values are actually found on the right and left sides of the MTPA plane, respectively. The absolute values of $\Delta\delta$ must be proportional to the distance from the MTPA moiety. When these conditions have all been met, the model will indicate the correct absolute configuration of the compound.[75]

The following are some examples of the application of Mosher's method.

Combretastatins D-1 (**58**) and D-2 (**59**) are two 15-membered macrocyclic lactone compounds isolated from the South African tree *Cambretum caffrum* and have been shown to inhibit PS cell line growth with ED_{50} values of 3.3 and 5.2 μg/ml, respectively.[76] Compound (+)-**58** can be prepared via asymmetric epoxidation of the acetate of compound **59** catalyzed by the (S,S)-Mn-salen complex (Mn-salen catalyzed epoxidation is discussed in Chapter 4). The absolute configuration of this compound was determined according to Mosher's method. At first, the epoxide was hydrogenated to give 3-ol **60** as the major product. It was then esterified to give the *di*-MTPA ester **61** and was subjected to NMR analysis. The $\Delta\delta_H$ indicates that C-3 of the major product has an (S)-configuration; thus the synthetic combretastatin (+)-**58** is concluded to have a $(3R,4S)$-configuration, and this is consistent with the expected result of epoxidation catalyzed by the (S,S)-Mn-salen complex (Scheme 1–16, note the change of priority for the two carbon atoms attached to C-3 in **58** and **60**.).

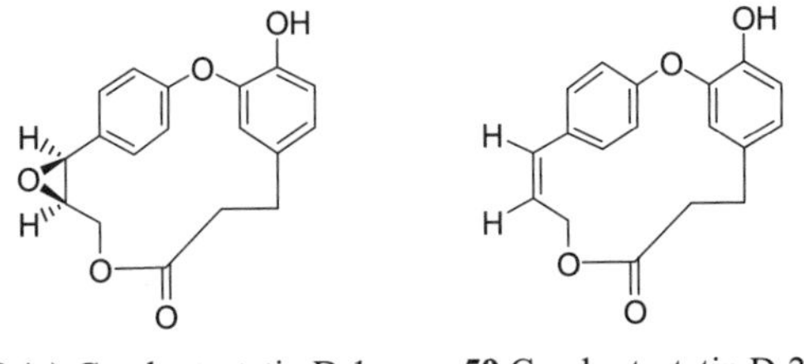

58 (+)-Combretastatin D-1   **59** Combretastatin D-2

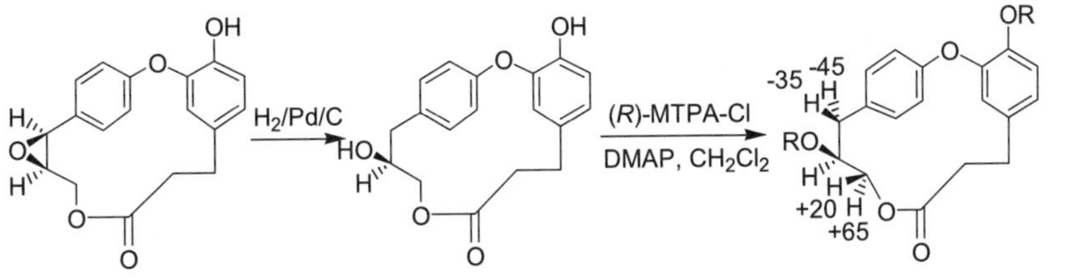

61 R = MTPA

$$\Delta\delta_H = \delta_S - \delta_R \text{ (in Hz at 500 MHz)}$$

Scheme 1–16. Deduction of the absolute configuration of **58**.

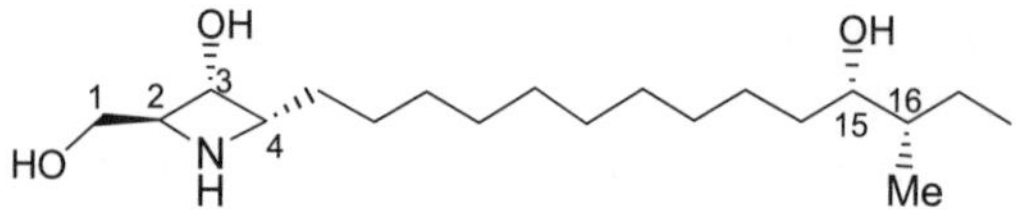

2*S*, 3*R*, 4*S*, 15*S*, 16*S*

62

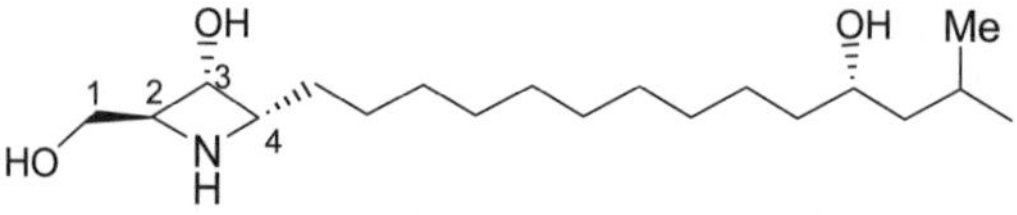

63

$$\Delta\delta_H = (\delta_S - \delta_R) \text{ in ppm}$$

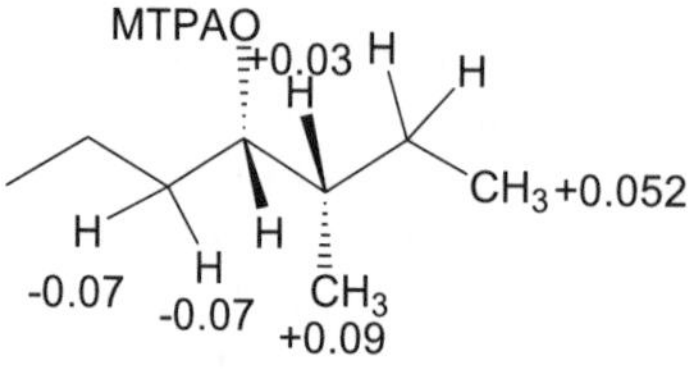

part of **62** 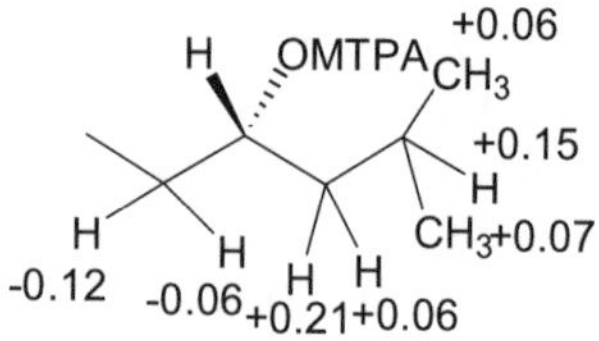   part of **63**

Figure 1–22. Determining the absolute configuration of chiral centers in compounds **62** and **63** by measuring $\Delta\delta$ values.

Because the natural **58** exhibited a negative value of optical rotation, the stereochemistry of natural $(-)$-combrestatin can be assigned as $(3S,4R)$.[77]

Penaresidins A **62** and B **63** are sphingosine-related compounds isolated from the Okinawa marine sponge *Penares* sp. that have shown potent actomyosin ATPase-activating activity.[78] Through detailed analysis of the ^{1}H–^{1}H COSY and HOHAHA spectra, the ^{1}H NMR spectrum for the MTPA ester (acetamide) can be assigned.[78] As shown in Figure 1–22, the chemical shift differences $\Delta\delta = \delta_S - \delta_R$ for the (S)- and (R)-MTPA esters of *N*-acetyl penaresidins A and B reveal that the absolute configurations of these compounds are as follows[79]: C-15: *S*, C-2: *S*, C-3: *R*, C-4: *S*, C-16: *S*. [C-16-(S) is only for penaresidin A.]

Similarly, 2-anthrylmethoxyacetic acid (2ATMA **64**) can also be used as a chiral anisotropic reagent (Fig. 1–23).[80] The advantage of using this compound is that the absolute configuration assignment can be accomplished from only one isomer of the ester without calculating the $\Delta\delta$ value $(\delta_S - \delta_R)$:

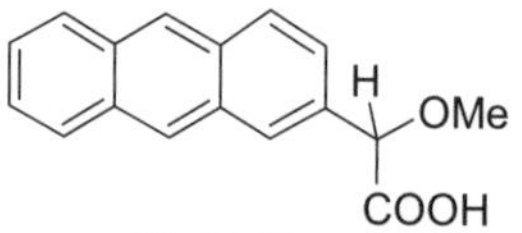

64 (2ATMA)

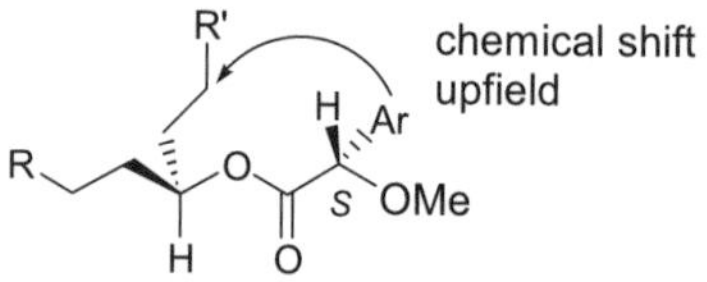

Figure 1–23. 2-Anthrylmethoxyacetic acid in absolute configuration deduction.

This is illustrated by the determination of the absolute configurations of C-7 and C-13 in taurolipid B (**65**). The absolute configuration can be deduced by analyzing the ^{1}H NMR spectrum of either the (*R*)- or the (*S*)-2ATMA ester without calculating the corresponding $\Delta\delta$ values. Taurolipids A, B, and C were isolated from the freshwater protozoan *Tetradymena thermophila*. Taurolipid B (**65**) had shown growth-inhibitory activity against HL-60.[81] Mild hydrolysis of **65** gave a tetrahydroxy compound that possesses four chiral centers. Subsequent derivatization gave an acetonide with two free hydroxyl groups at C-7 and C-13 for the formation of 2ATMA ester. Their 2ATMA esters exist in a conformation similar to that of its MTPA esters, in which the carbinyl proton, carbonyl oxygen, and methoxy groups are oriented on the same plane. Upfield shifts are observed for protons shielded by the anthryl group, and the shifted signals are wide ranging. Thus, the C-7 can be assigned an (*R*)-configuration and C-13 an (*S*)-configuration (Fig. 1–24):

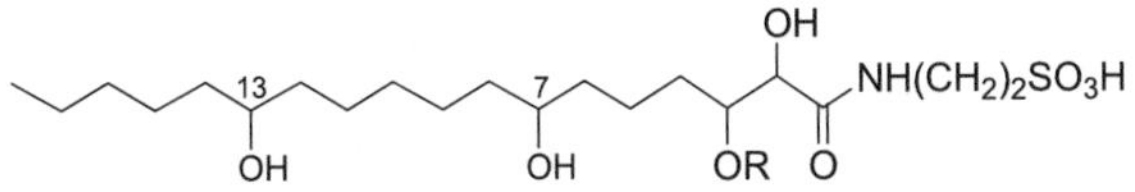

65

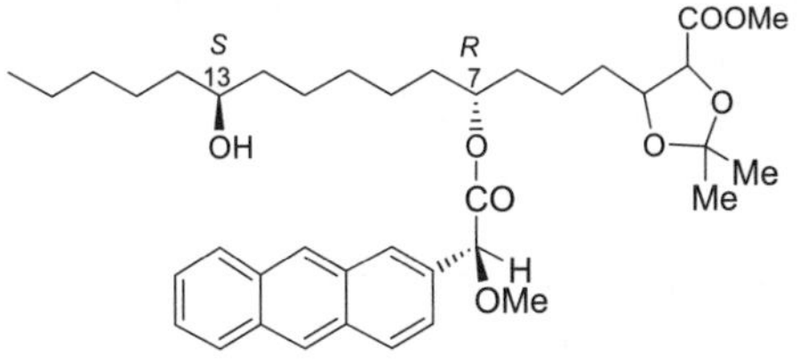

$\Delta\delta = \delta_{OH} - \delta_{2ATMAester} > 0.05$

upfield signals assigned to H8-12: $\delta = 1.36$ (H-8), 0.82(H-9), 0.90(H-10), 0.72(H-11), 0.93(H-12).

(*R*) configuration for C-7

65a

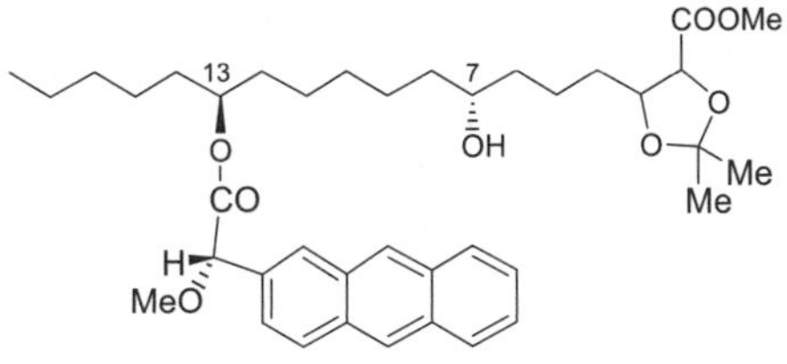

H6-12 shielded. (*S*) for C-13

65b

Figure 1–24. Deduction of absolute configuration of **65**.

The result of this method is consistent with that obtained by using the corresponding Mosher's ester method ($\Delta\delta_H = \delta_{(R)\text{-ATMA}} - \delta_{(S)\text{-ATMA}}$) (Fig. 1–25). Note that in the method of Mosher's ester, ($\delta_S - \delta_R$) was applied to calculate $\Delta\delta_H$, the difference in the 2ATMA method is only due to the configuration nomenclature difference caused by the CIP rule.

The absolute configuration of a secondary alcohol can also be determined through the NMR spectra of a single methoxyphenylacetic ester derivative

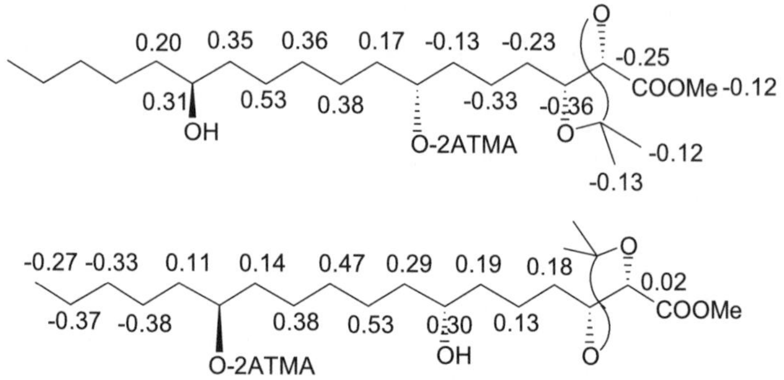

Figure 1–25. Mosher's method for determining the absolute configuration of compound **65**.

(MPA, either the [R]- or the [S]-form) recorded at two different temperatures. This approach, requiring just one derivatizing reaction, also simplifies the NMR-based methodologies for absolute configuration determination.[82]

For a given chiral secondary alcohol L^1L^2CHOH ($L^1 \neq L^2$), its (R)- or (S)-methoxyphenylacetic acid ester (or MPA ester) may exist in two conformations, sp and ap, in equilibrium around the CR–CO bond, with the $CO-O-C_1'HL^1L^2$ fragment virtually rigid (Fig. 1–26).[83] Experimental and theoretical data[83] indicate that the sp conformation is the more stable one in both the (R)- and the (S)-MPA esters. In addition, the relative population of the sp conformer is practically unaffected by the nature of the alcohol.

Figure 1–27 depicts the application of this method. According to Boltzman's law, a selective increase of the sp conformation population over the ap is expected at lower temperatures. Thus, an upfield shift should be observed for the L^1 group of an (R)-MPA ester (positive $\Delta\delta^{T_1T_2}$) with decreasing temperature, because the fraction of molecules having L^1 under the shielding cone of the phenyl ring is being increased. Analogously, a downfield shift for L^2 should be expected (negative $\Delta\delta^{T_1T_2}$). Similar analysis of the (S)-MPA ester leads to op-

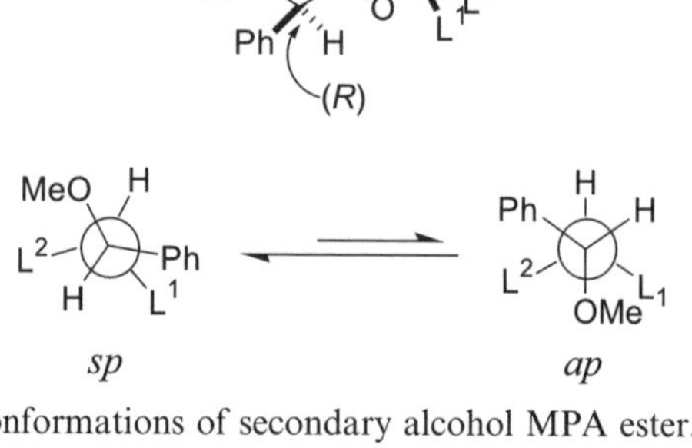

Figure 1–26. Two conformations of secondary alcohol MPA ester. Reprinted with permission by Am. Chem. Soc., Ref. 82.

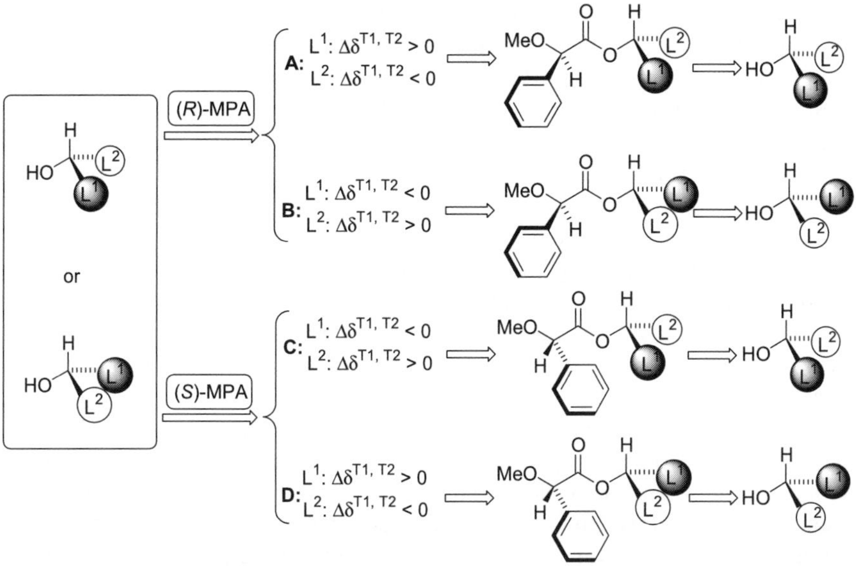

Figure 1–27. Absolute configuration deduction using MPA. Reprinted with permission by Am. Chem. Soc., Ref. 82.

posite shifts, which are equally useful. L^1 should be shifted downfield (negative $\Delta\delta^{T_1, T_2}$) at decreased temperatures, while L^2 should be moved upfield, and thus the corresponding $\Delta\delta^{T_1 T_2}$ values must become positive.

As a result, the relative position of L^1 or L^2 in relation to the aromatic ring in the *sp* conformation can be established from the sign of the variations in the chemical shifts of substituents L^1 and L^2 with temperature (positive or negative $\Delta\delta^{T_1 T_2}$). Assigning the configuration of the chiral center is then straightforward.

1.5 GENERAL STRATEGIES FOR ASYMMETRIC SYNTHESIS

Asymmetric organic reactions have proved to be very valuable in the study of reaction mechanisms, in the determination of relative and absolute configurations, and in the practical synthesis of optically active compounds. The pharmaceutical industry, in particular, has shown markedly increased interest in asymmetric organic reactions. Currently, an expanding number of drugs, food additives, and flavoring agents are being prepared by synthetic methods. Most often, the desired compound is obtained through resolution of the corresponding racemic species performed at the end of the synthetic sequence. Because only one optical antipode is useful, half of the synthetic product is often discarded. Obviously, this is a wasteful procedure from the preparative point of view. Even if the wrong isomer can be converted to the active form via racemization and resolution, extensive work is required. Also, resolution is usually a tedious, repetitive, and laborious process. It is economically appealing to exclude the unwanted optical isomers at the earliest possible stage through the asymmetric creation of chiral centers. In the interest of effective use of raw material, it is wise to choose an early step in the synthetic sequence for the asymmetric operation and to consider carefully the principles of convergent synthesis.

Asymmetric synthesis refers to the conversion of an achiral starting material to a chiral product in a chiral environment. It is presently the most powerful and commonly used method for chiral molecule preparation. Thus far, most of the best asymmetric syntheses are catalyzed by enzymes, and the challenge before us today is to develop chemical systems as efficient as the enzymatic ones.

The resolution of racemates has been an important technique for obtaining enantiomerically pure compounds. Other methods involve the conversion or derivatization of readily available natural chiral compounds (chiral pools) such as amino acids, tartaric and lactic acids, terpenes, carbohydrates, and alkaloids. Biological transformations using enzymes, cell cultures, or whole microorganisms are also practical and powerful means of access to enantiomerically pure compounds from prochiral precursors, even though the scope of such reactions is limited due to the highly specific action of enzymes. Organic synthesis is characterized by generality and flexibility. During the last three decades, chemists have made tremendous progress in discovering a variety of versatile stereoselective reactions that complement biological processes.

In an asymmetric reaction, substrate and reagent combine to form diastereomeric transition states. One of the two reactants must have a chiral element to induce asymmetry at the reaction site. Most often, asymmetry is created upon conversion of trigonal carbons to tetrahedral ones at the site of the functionality. Such asymmetry at carbon is currently a major area of interest for the synthetic organic chemists.

1.5.1   "Chiron" Approaches

Naturally occurring chiral compounds provide an enormous range and diversity of possible starting materials. To be useful in asymmetric synthesis, they should be readily available in high enantiomeric purity. For many applications, the availability of both enantiomers is desirable. Many chiral molecules can be synthesized from natural carbohydrates or amino acids. The syntheses of (+)-exo-brevicomin (**66**) and negamycin (**67**) illustrate the application of such naturally occurring materials.

6,8-Dioxalicyclo[3.2.1]octane, or (+)-*exo*-brevicomin (**66**), is the aggregating pheromone of the western pine beetle. It has been prepared from glucose using a procedure based on the retro synthesis design shown in Figure 1–28[84]:

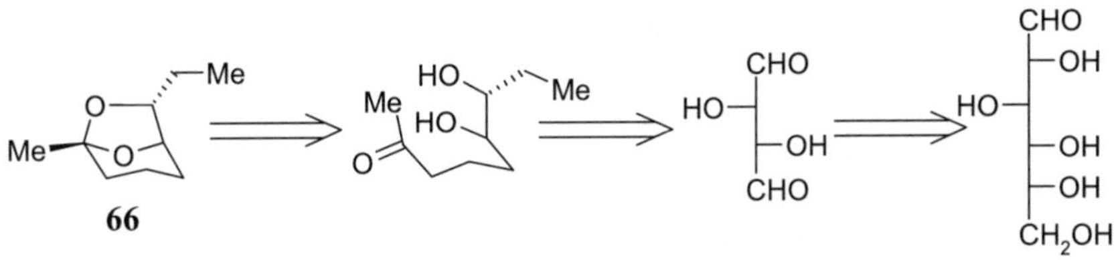

Figure 1–28. Retro synthesis of (+)-*exo*-brevicomin (**66**).

Negamycin (**67**), a broad-spectrum antibiotic produced naturally by *Streptomyces purpeofuscus*, has also been synthesized from glucose (Fig. 1–29)[85]:

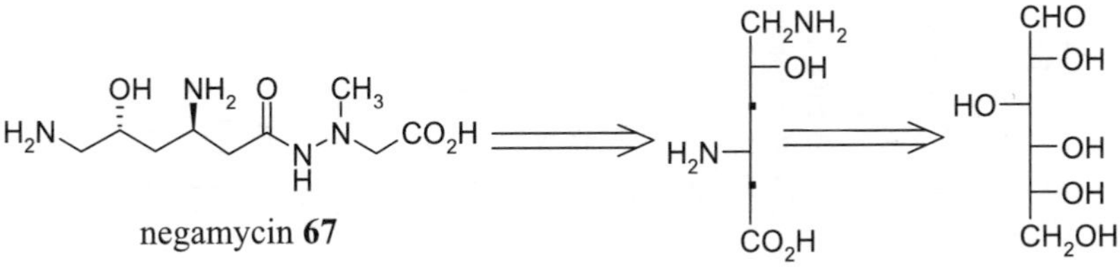

Figure 1–29. Retro synthesis of negamycin **67**.

A great number of natural compounds have been employed as chiral starting materials for asymmetric syntheses. Table 1–2 classifies such inexpensive reagents.

1.5.2 Acyclic Diastereoselective Approaches

In principle, asymmetric synthesis involves the formation of a new stereogenic unit in the substrate under the influence of a chiral group ultimately derived from a naturally occurring chiral compound. These methods can be divided into four major classes, depending on how this influence is exerted: (1) substrate-controlled methods; (2) auxiliary-controlled methods; (3) reagent-controlled methods, and (4) catalyst-controlled methods.

The substrate-controlled reaction is often called the first generation of asymmetric synthesis (Fig. 1–30, 1). It is based on intramolecular contact with a stereogenic unit that already exists in the chiral substrate. Formation of the new stereogenic unit most often occurs by reaction of the substrate with an achiral reagent at a diastereotopic site controlled by a nearby stereogenic unit.

The auxiliary-controlled reaction (Fig. 1–30, 2) is referred to as the second generation of asymmetric synthesis. This approach is similar to the first-generation method in which the asymmetric control is achieved intramolecularly by a chiral group in the substrate. The difference is that the directing

TABLE 1–2. Inexpensive Chiral Starting Materials and Resolving Agents[86]

Amino Acids	Hydroxy Acids	Carbohydrates	Terpenes	Alkaloids
L-alanine	L-lactic acid	D-arabinose	l-Borneol	Cinchonidine
L-arginine	D-lactic acid	L-arabinose	endo-3-Bromo-d-camphor	Cinchonine
D-asparagine	(S)-malic acid	L-ascorbic acid	d-Camphene	D-(+)-ephedrine
L-asparagine	(Poly)-3(R)-hydroxybutyrate	α-chloralose	d-Camphor	l-Nicotine
L-aspartic acid	L-tartaric acid	Diacetone-D-glucose	D-(+)-camphoric acid	Quinidine
L-cysteine	D-tartaric acid	D-fructose	d-10-Camphor-sulfonic acid	Quinine
L-glutamic acid	D-threonine	D-galactonic acid	d-3-Carene	D-(+)-pseudoephedrine
L-isoleucine	L-threonine	D-galactonic acid	l-Carvone	L-(−)-pseudoephedrine
L-glutamine		γ-Lactone		
L-leucine		D-galactose	d-Citronellal	
L-lysine		D-glucoheptonic acid	d-Fenchone	
L-methionine		α-D-glucoheptonic	l-Fenchone	
		Acid γ-lactone		
L-ornithine		D-gluconic acid	d-Isomenthol	
L-phenylalanine		D-gluconic acid	d-Limonene	
		δ-lactone		
D-phenylglycine		L-gluconic acid	l-Limonene	
		γ-lactone		
L-proline		D-glucosamine	l-Menthol	
L-pyroglutamic acid		D-glucose	d-Menthol	
L-serine		D-glucurone	l-Menthone	
L-tryptophan		D-gluconic acid	Nopol	
L-tyrosine		L-glutamine	(−)-α-Phellandrene	

L-valine

D-isoascorbic acid
D-mannitol
D-mannose
D-quinic acid
D-ribolactone
D-ribose
D-saccharic acid
D-sorbitol
L-sorbose
D-xylose

(−)-α-Pinene
(+)-α-Pinene
(−)-β-Pinene
(R)-(+)-pulegone

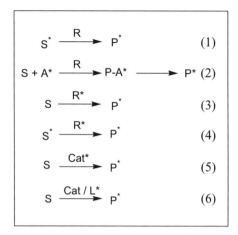

Figure 1–30. Development of asymmetric synthesis.

group, the "chiral auxiliary", is deliberately attached to the original achiral substrate in order to direct the enantioselective reaction. The chiral auxiliary will be removed once the enantioselective transformation is completed.

Although second-generation methods have proved useful, the requirement for two extra steps, namely, the attachment and the removal of the chiral auxiliary, is a cumbersome feature. This is avoided in the third-generation method in which an achiral substrate is directly converted to the chiral product using a chiral reagent (Fig. 1–30, 3). In contrast to the first- and second-generation methods, the stereocontrol is now achieved intermolecularly.

In all three of the above-mentioned chiral transformations, stoichiometric amounts of enantiomerically pure compounds are required. An important development in recent years has been the introduction of more sophisticated methods that combine the elements of the first-, second-, and third-generation methods and involve the reaction of a chiral substrate with a chiral reagent. The method is particularly valuable in reactions in which two new stereogenic units are formed stereoselectively in one step (Fig. 1–30, 4).

The most significant advance in asymmetric synthesis in the past three decades has been the application of chiral catalysts to induce the conversion of achiral substrates to chiral products (Fig. 1–30, 5 and 6). In ligand-accelerated catalysis (Fig. 1–30, 6), the addition of a ligand increases the reaction rate of an already existing catalytic transformation. Both the ligand-accelerated and the basic catalytic process operate simultaneously and complement each other. The nature of the ligand and its interaction with other components in the metal complex always affect the selectivity and rate of the organic transformation catalyzed by such a species. The obvious benefit of catalytic asymmetric synthesis is that only small amounts of chiral catalysts are needed to generate large quantities of chiral products. The enormous economic potential of asymmetric catalysis has made it one of the most extensively explored areas of research in recent years.

1.5.3 Double Asymmetric Synthesis

Double asymmetric synthesis was pioneered by Horeau et al.,[87] and the subject was reviewed by Masamune et al.[88] in 1985. The idea involves the asymmetric reaction of an enantiomerically pure substrate and an enantiomerically pure reagent. There are also reagent-controlled reactions and substrate-controlled reactions in this category. Double asymmetric reaction is of practical significance in the synthesis of acyclic compounds.

Figure 1–31 formulates this transformation: Chiral substrate *A-C(x) is converted to A*-(*Cn)-C(z) by process I, where both C(x) and C(z) denote appropriate functional groups for the chemical operation. To achieve this task, a chiral reagent *B-C(y) is allowed to react with *A-C(x) to provide a mixture of stereoisomers—*A-*C-*C-*B (process II). The reagent *B-C(y) is chosen in such a manner that high stereoselectivity at *C is achieved in the reaction process.

In selecting the right reagent B*-C(y), the following observations are important:

1. When the desired *A-*C-*C-*B is the major product in the matched pair reaction, the resultant stereoselectivity should be higher than the diastereofacial selectivity of *A-C(x).
2. If the product *A-*C-*C-*B occurs as the minor product, this presents a mismatched pair reaction, and the reagent with the opposite chirality should be used. The diastereofacial selectivity of the reagent must be large enough to outweigh that of *A-C(x) in order to create the desired *C-*C stereochemistry with high selectivity.

The above strategy can be illustrated by the following two examples of reactions (Schemes 1–17 and 1–18)[88,89]:

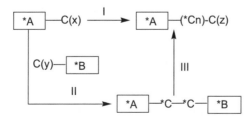

Figure 1–31. Strategy for generation of new chiral centers on a chiral substrate. A and B must be homochiral. *A-C(x), chiral substrate; *B-C(y), chiral reagent; I, desired transformation; II, double asymmetric induction; III, removal of the chiral auxiliary. Reprinted with permission by VCH, Ref. 88.

68

(*R*)-69

68

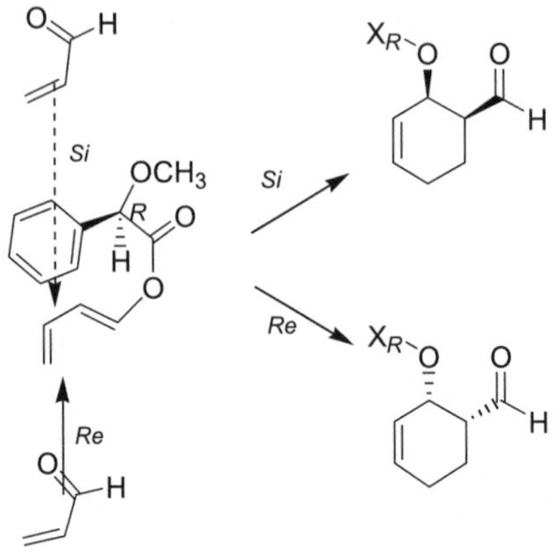

The ratio of the *Si* attack product to the *Re* attack product:

1 : 4.5

Scheme 1–17

The first reaction is the pericyclic reaction of chiral diene (*R*)-**69** with achiral acrolein **68**. In the presence of $BF_3 \cdot OEt_2$, a mixture with a diastereoisomeric ratio of 1:4.5 results. The phenyl group of **69** covers one face of the butadiene π-system, while the *Si*-face of (*R*)-**69** is more shielded from the attack of **68** than is the *Re*-face (Scheme 1–17).

In a similar manner, butadienyl phenylacetate **71**, an achiral diene, is expected to approach the chiral dienophile (*R*)-**70** from its *Re*-prochiral face. The two faces of the chelate ring are differentiated by the small hydrogen and large benzyl groups attached to the chiral center of (*R*)-**70** (Scheme 1–18); the ratio of the *Si* attack product to the *Re* attack product is 1:8.[88]

The interaction of these two chiral reagents (*R*)-**69** and (*R*)-**70** can be evaluated as in Schemes 1–19 and 1–20. The diastereofacial selectivity of (*R*)-**70**

(*R*)-70

71

(*S*)-70

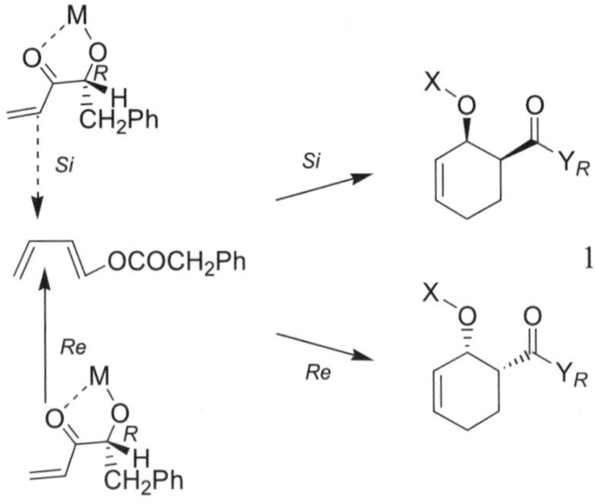

1 : 8

Scheme 1–18

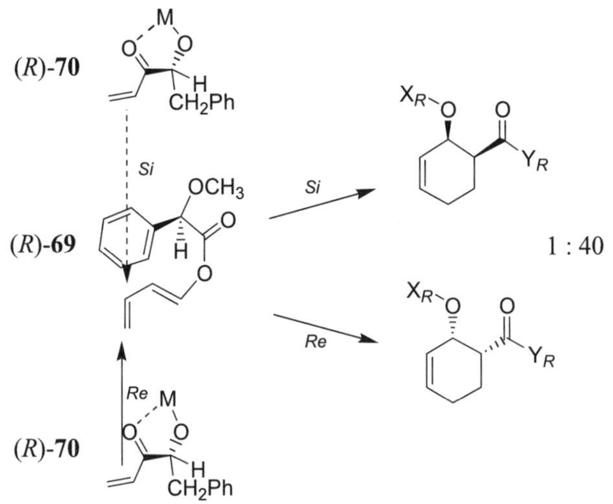

Scheme 1–19. Reaction of (R)-**69** and (R)-**70**: matched pair.

and (R)-**69** is in concert, and this pair is called a *matched pair* (Scheme 1–19); the ratio of the *Si* attack product to the *Re* attack product is 1:40.[88]

On the other hand, the diastereofacial selectivity of (R)-**69** and (S)-**70** counteract each other, as depicted in Scheme 1–20, and these are referred to as a *mismatched pair*. The ratio of the *Si* attack product to the *Re* attack product is 1:2.

Another example is the aldol reaction of benzaldehyde with the chiral enolate (S)-**72**, from which a 3.5:1 mixture of diastereoisomers is obtained. When

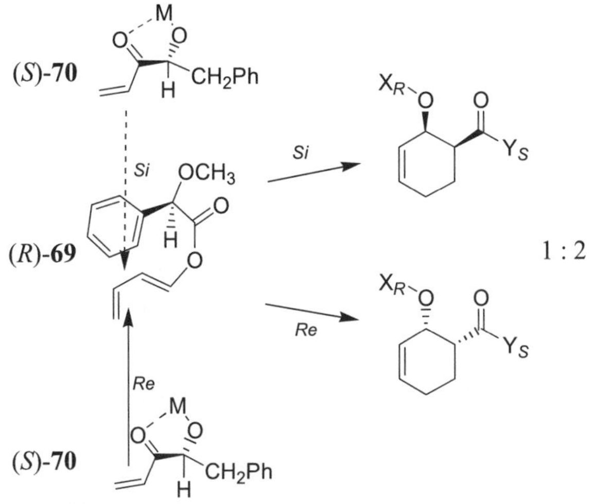

Scheme 1–20. (R)-**69** and (S)-**70**: mismatched pair.

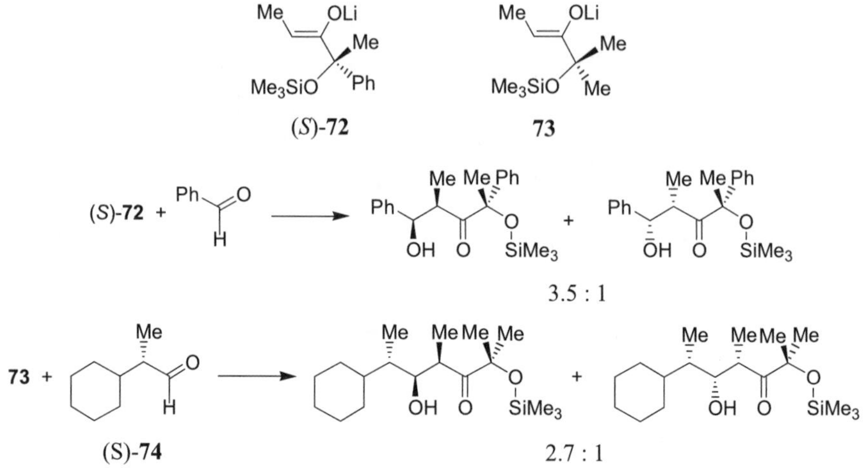

Scheme 1–21. Matched pair and mismatched pair.

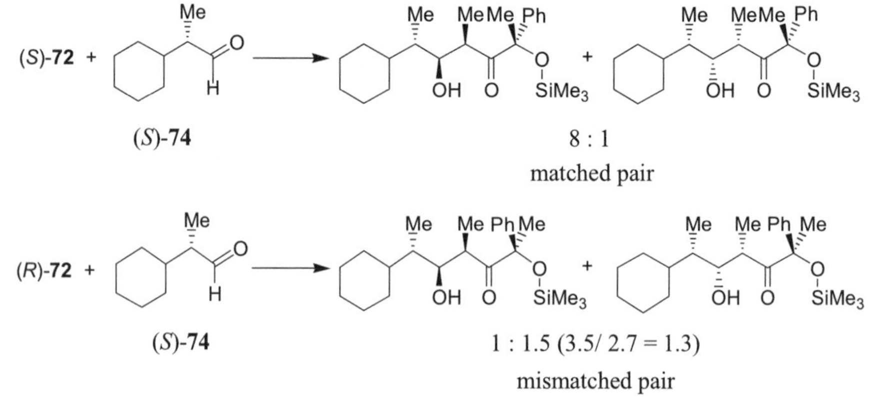

Scheme 1–22. Matched pair and mismatched pair.

the chiral aldehyde (*S*)-**74** is treated with **73**, two diastereomers are formed in a similar manner with a ratio of 2.7:1 (Scheme 1–21).[88,90]

In the following example, a matched pair is found in (*S*)-**72** and (*S*)-**74**. In contrast, (*S*)-**74** and (*R*)-**72** constitute a mismatched pair (Scheme 1–22).[88]

1.6   EXAMPLES OF SOME COMPLICATED COMPOUNDS

There are a number of complicated molecules whose synthesis without using asymmetric methods would be extremely difficult. This section introduces some of these compounds.

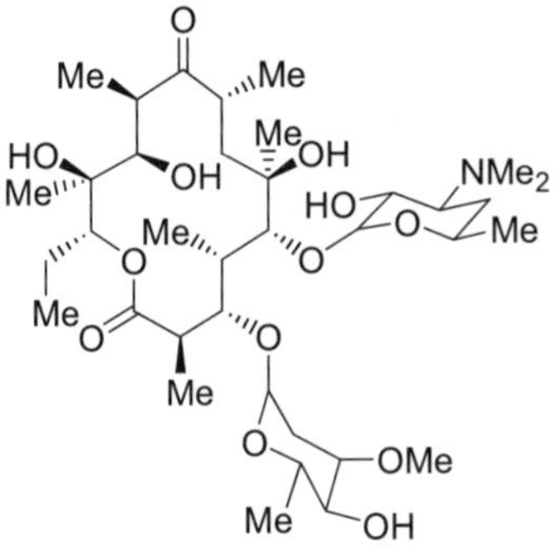

Erythromycin A

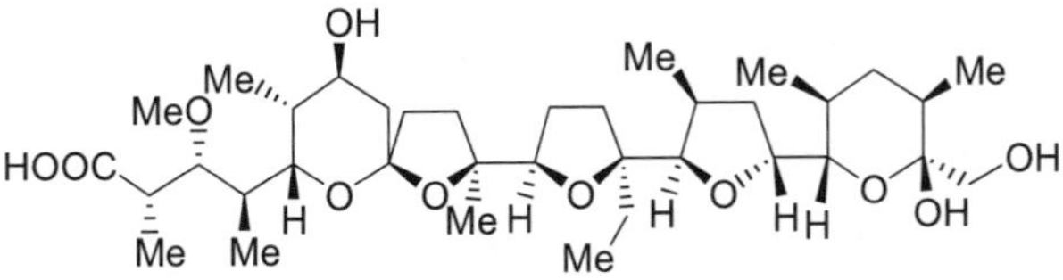

Monensin, $2^{17} = 131,072$ stereoisomers

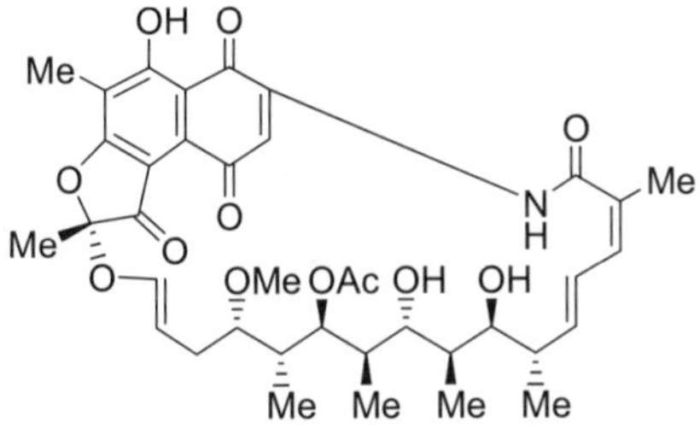

Rifamycin S[91]

Erythromycin A and rifamycin S are representatives of two classes of antibiotics. Due to their large number of chiral centers, constructing the aglycone part of these molecules was considered a major synthetic challenge in the late 1970s and early 1980s when suitable asymmetric synthesis methods had not yet been developed. This challenge was met by several groups whose approaches depended on different synthetic strategies. The total synthesis is evaluated from a methodological point of view in Chapter 7.

Forskolin was isolated in 1977 from the Indian medicinal plant *Coleus forskohlii Brig* by Hoechst Pharmaceutical Research in Bombay as the result of a screening program for the discovery of new pharmaceuticals. The structure and absolute configuration of forskolin were determined by extensive spectroscopic, chemical, and X-ray crystallographic studies. Pharmacologically, forskolin has blood pressure–lowering and cardioactive properties. The unique structural features and biological properties of forskolin have aroused the interest of syn-

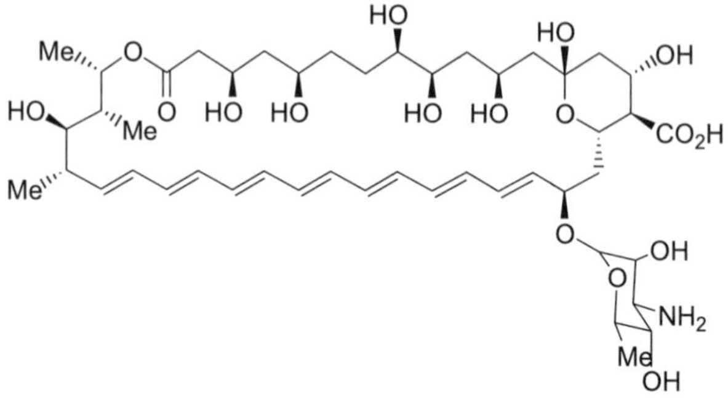

Amphotericin B

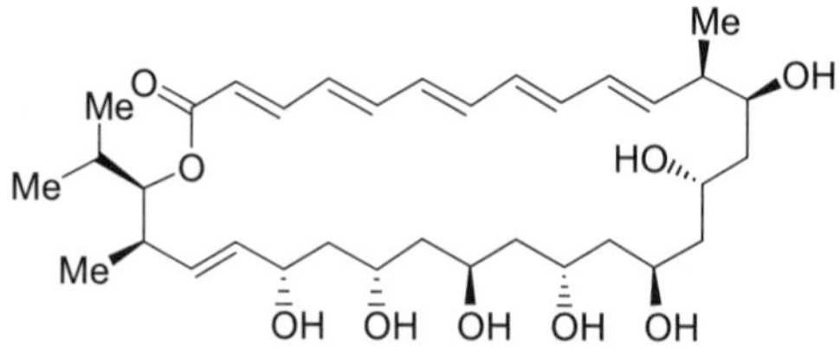

Roxaticin[92]

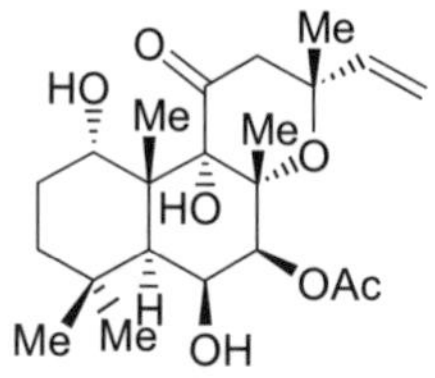

Forskolin

thetic organic chemists and have resulted in enormous activity directed toward the synthesis of this challenging target. There are now several synthetic routes available to this compound.[93]

Since the discovery of the anticancer potential of Taxol™, a complex compound isolated from the bark extract of the Pacific yew tree, more than 20 years ago, there has been an increasing demand for the clinical application of this compound. First, the promising results of the 1991 clinical trials in breast cancer patients were announced, and soon after Bristol-Myers-Squibb trademarked the name Taxol™ and used it as an anticancer drug. At that point, the only source of the drug was the bark of the endangered yew tree. Fortunately, it was soon discovered that a precursor of Taxol™ could be obtained from an extract of the tree needles instead of the bark.

In the meantime, the race toward the total synthesis of Taxol™ and the synthesis of Taxol-like compounds started. It was announced in February 1994 that two groups, led by K. C. Nicolaou in the Scripps Research Institute and by Robert Holton at Florida State University, had independently completed the total synthesis of Taxol™. At present, the nucleus of Taxol™ can be obtained from the needles of the tree, and the C-13 side chain can be prepared on large scale.[94]

Taxol™, a compound with powerful
antileukemic and tumor-inhibiting activity

Palytoxin,[95] 68 stereogenic units

Several novel natural products with an intriguing system containing the *cis*-endiyne moiety have attracted considerable attention from chemists in recent years. Several derivatives with this characteristic skeleton have now been isolated: neocarzinostatin,[96] esperamicin,[97] calicheamicin γ_1^I,[98] and dynemicin A_1.[99] The high antitumor activity of these compounds is based on an elegant

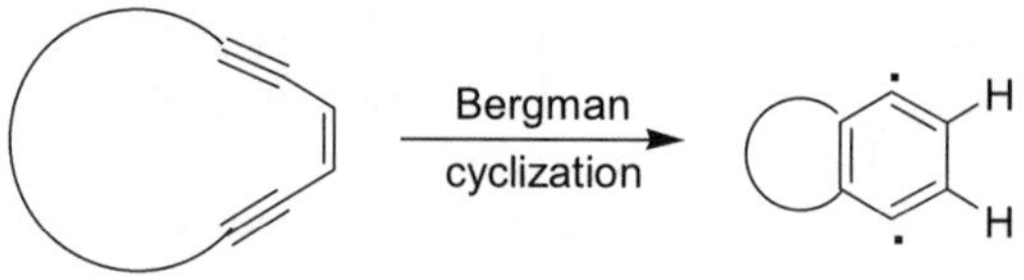

Figure 1–32. Ring-closure reaction of endiyne anticancer antibiotics.

initiation of the ring-closure reaction illustrated in Figure 1–32. The resulting aromatic di-radical reacts with a nucleotide unit of DNA to cause chain cleavage and thus cause the antitumor activity of the compounds.[100]

The following compounds contain a great number of chiral centers that must be built up in asymmetric synthesis. They exemplify the significance of asymmetric synthesis.

• Esperamicin A_1[101]:

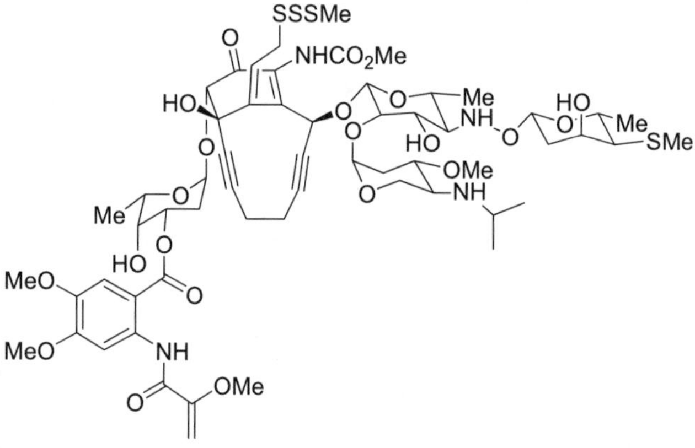

• Kedarcidin, a new chromoprotein antitumor antibiotic[102]:

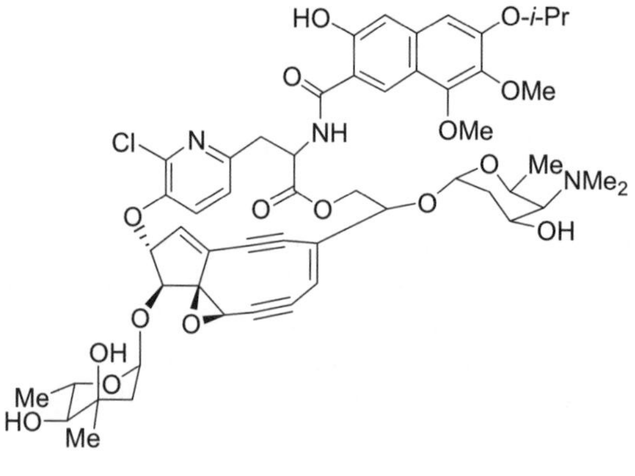

- Calicheamicin γ_1^I [103]:

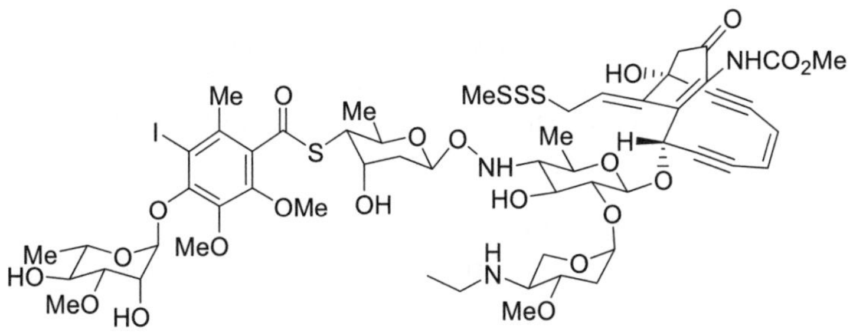

- Rapamycin, [104] an antiproliferative agent:

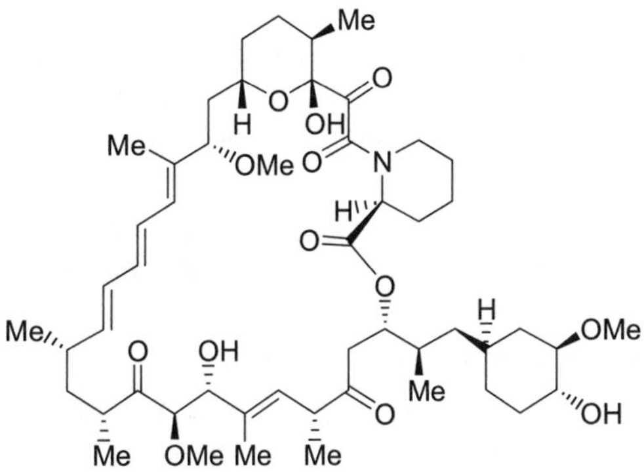

FK-506 was isolated[105] from *Streptomyces trukubaensis*, possessing a unique 21-membered macrolide, in particular an unusual α,β-diketoamide hemiketo system. It shows immunosuppressive activity superior to that of cyclosporin in the inhibition of delayed hypersensitivity response in a variety of allograft transplantation and autoimmunity models.

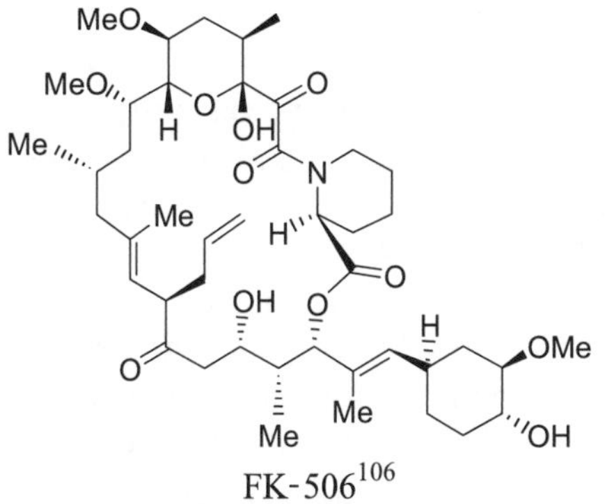

FK-506[106]

1.7 SOME COMMON DEFINITIONS IN ASYMMETRIC SYNTHESIS AND STEREOCHEMISTRY

In this chapter a number of common terms in the field of stereochemistry have been introduced. These terms appear repeatedly throughout this book. Therefore, it is essential that we establish common definitions for these frequently used terms.

Asymmetric and **dissymmetric compounds**

> **Asymmetric**: Lack of symmetry. Some asymmetric molecules may exist not only as enantiomers; they can exist as diastereomers as well.
>
> **Dissymmetric**: Compounds lacking an alternating axis of symmetry and usually existing as enantiomers. Some people prefer this to the term *asymmetric*.

D/L and *d/l*

> D or L: Absolute configurations assigned to a molecule through experimental chemical correlation with the configuration of D- or L-glyceraldehyde; often applied to amino acids and sugars, although (R) and (S) are preferred.
>
> *d* or *l*: Dextrorotatory or levorotatory according to the experimentally determined rotation of the plane of monochromatic plane-polarized light to the right or left.

Diastereomer or **diastereoisomer** and **enantiomer**

> **Stereoisomer**: Molecules consisting of the same types and same number of atoms with the same connections but different configurations.
>
> **Diastereoisomer**: Stereoisomers with two or more chiral centers and where the molecules are not mirror images of one another, for example, D-erythrose and D-threose; often contracted to *diastereomer*.
>
> **Enantiomer**: Two stereoisomers that are nonsuperimposable mirror images of each other.

Enantiomer excess

> **Enantiomer excess (ee)**: Percentage by which one enantiomer is in excess over the other in a mixture of the two, ee $\equiv |(E_1 - E_2)/(E_1 + E_2)| \times 100\%$.

Optical activity, optical isomer, and optical purity

> **Optical activity**: Experimentally observed rotation of the plane of monochromatic plane-polarized light to the observer's right or left. Optical activity can be observed with a polarimeter.
>
> **Optical isomer**: Synonym for enantiomer, now disfavored, because most enantiomers lack optical activity at some wavelengths of light.
>
> **Optical purity**: The optical purity of a sample is expressed as the magnitudes of its optical rotation as a percentage of that of its pure enantiomer (which has maximum rotation).

Racemic, *meso*, and racemization

Racemic: Compounds existing as a racemate, or a 50–50 mixture of two enantiomers; also denoted as *dl* or ($\pm$). Racemates are also called *racemic mixtures*.

***Meso* compounds**: Compounds whose molecules not only have two or more centers of dissymmetry but also have plane(s) of symmetry. They do not exist as enantiomers, for example, *meso*-tartaric acid:

$$
\begin{array}{c}
\text{COOH} \\
\text{H} \!-\!\!-\! \text{OH} \\
\hline
\text{H} \!-\!\!-\! \text{OH} \\
\text{COOH}
\end{array}
$$

meso-Tartaric acid

Racemization: The process of converting one enantiomer to a 50–50 mixture of the two.

Scalemic: Compounds existing as a mixture of two enantiomers in which one is in excess. The term was coined in recognition of the fact that most syntheses or resolutions do not yield 100% of one enantiomer.

Prochirality: Refers to the existence of stereoheterotopic ligands or faces in a molecule that, upon appropriate replacement of one such ligand or addition to one such face in an achiral precursor, gives rise to chiral products.

Pro-*R* and Pro-*S*: Refer to heterotopic ligands present in the system. It is arbitrarily assumed that the ligand to be introduced has the highest priority, and replacement of a given ligand by this newly introduced ligand creates a new chiral center. If the newly created chiral center has the (R)-configuration, that ligand is referred to as pro-*R*; while pro-*S* refers to the ligand replacement that creates an (S)-configuration. For example, as shown in Figure 1–33, H_A in ethanol is pro-*R* and H_B in the molecule is pro-*S*.

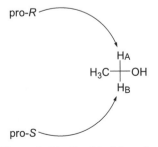

Figure 1–33. Prochiral ligands.

Re* and *Si: Labels used in stereochemical descriptions of heterotopic faces. If the CIP priority of the three ligands a, b, and c is assigned as a > b > c, the face that is oriented clockwise toward the viewer is called *Re*, while the face with a counterclockwise orientation of a → b → c is called *Si*, as shown in Figure 1–34:

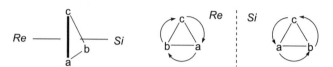

Figure 1–34. Prochiral faces.

Syn/anti and *Erythro/threo*

Syn/anti: Prefixes that describe the relative positions of substituents with respect to the defined plane of a ring: *syn* for the same side and *anti* for the opposite side (Fig. 1–35).

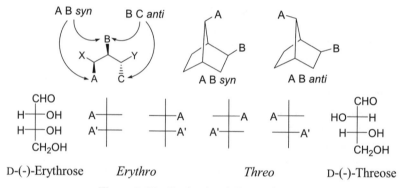

Figure 1–35. *Syn/anti* and *Erythro/threo*.

Erythro/threo: Terms derived from carbohydrate nomenclature used to describe the relative configuration at adjacent stereocenters. *Erythro* refers to a configuration with identical or similar substituents on the same side of the vertical chain in Fischer projection. Conversely, a *threo* isomer has these substituents on opposite sides. These terms came from the nomenclature of two carbohydrate compounds, threose and erythrose (see Fig. 1–35).

This chapter has provided a general introduction to stereochemistry, the nomenclature for chiral systems, the determination of enantiomer composition and the determination of absolute configuration. As the focus of this volume is asymmetric synthesis, the coming chapters provide details of the asymmetric syntheses of different chiral molecules.

1.8 REFERENCES

1. (a) Stinson, S. C. *Chem. Eng. News* **1997**, *75(42)*, 38. (b) Stinson, S. C. *Chem. Eng. News* **1995**, *73(41)*, 44. (c) Stinson, S. C. *Chem. Eng. News* **1994**, *72(38)*, 38. (d) Stinson, S. C. *Chem. Eng. News* **1993**, *71(39)*, 28. (e) Stinson, S. C. *Chem. Eng. News* **1992**, *70(39)*, 46.

2. Veloo, R. A.; Koomen, G. J. *Tetrahedron Asymmetry* **1993**, *4*, 2401.

3. Baker, R.; Rao, V. B. *J. Chem. Soc. Perkin Trans. 1* **1982**, 69.

4. Schwartz, A.; Madan, P. B.; Mohacsi, E.; O'Brien, J. P.; Tadaro, L. J.; Coffen, D. L. *J. Org. Chem.* **1992**, *57*, 851.

5. Blaschke, G.; Kraft, H. P.; Markgraf, H. *Chem. Ber.* **1980**, *113*, 2318.

6. Cahn, R. S.; Ingold, C.; Prelog, V. *Angew. Chem. Int. Ed. Engl.* **1966**, *5*, 385.

7. An introduction to point groups containing chiral molecules can be found in any of the leading text books. For example, in Eliel, E. L; Wilen, S. H.; Mander, L. N. *Stereochemistry of Organic Compounds*, John Wiley & Sons, Inc., New York, **1994**.

8. White, J. D.; Kang, M.; Sheldon, B. G. *Tetrahedron Lett.* **1983**, *24*, 4539.

9. Bachelor, F. W.; Miana, G. A. *Can. J. Chem.* **1969**, *47*, 4089.

10. Poindexter, G. S.; Meyers, A. I. *Tetrahedron Lett.* **1977**, 3527.

11. Pirkle, W. H. *J. Am. Chem. Soc.* **1966**, *88*, 1837.

12. Burlingame, T. G.; Pirkle, W. H. *J. Am. Chem. Soc.* **1966**, *88*, 4294.

13. For reviews, see, for example: (a) Pirkle, W. H.; Hoover, D. J. "NMR Chiral Solvating Agents" in *Top. Stereochem.* **1982**, *13*, 263. (b) Weisman, G. R. "Nuclear Magnetic Resonance Analysis Using Chiral Solvating Agents" in Morrison, J. D., ed. *Asymmetric Synthesis*, vol. 1, Chap. 8, Academic Press, New York, **1983**. (c) Parker, D. *Chem. Rev.* **1991**, *91*, 1441.

14. (a) Goering, H. L.; Eikenberry, J. N.; Koermer, G. S. *J. Am. Chem. Soc.* **1971**, *93*, 5913. (b) McCreary, M. D.; Lewis, D. W.; Wernick, D. L.; Whitesides, G. M. *J. Am. Chem. Soc.* **1974**, *96*, 1038. (c) Peterson, P. E.; Stepanian, M. *J. Org. Chem.* **1988**, *53*, 1907.

15. (a) Iwaki, S.; Marumo, S.; Saito, T.; Yamada, M.; Katagiri, K. *J. Am. Chem. Soc.* **1974**, *96*, 7842. (b) Mori, K.; Takigawa, T.; Matsui, M. *Tetrahedron Lett.* **1976**, 3953. (c) Beames, D. J.; Mander, L. N. *Chem. Commun.* **1969**, 498. (d) Stork, G.; Malhotra, S.; Thompson, H.; Uchibayashi, M. *J. Am. Chem. Soc.* **1965**, *87*, 1148.

16. Pirkle, W. H.; Rinaldi, P. L. *J. Org. Chem.* **1979**, *44*, 1025.

17. (a) Rodriguez, I.; Alvarez, C.; Gomez-Lara, J.; Toscano, R. A.; Platzer, N.; Mulheim C.; Rudler, H. *J. Chem. Soc. Chem. Commun.* **1987**, 1502. (b) Alvarez, C.; Goasdoue, N.; Platzer, N.; Rodriguez, I.; Rudler, H. *J. Chem. Soc. Chem. Commun.* **1988**, 1002.

18. Alvarez, C.; Barkaoui, L.; Goasdoue; N.; Daran, J. C.; Platzer, N.; Rudler, H.; Vaissermann, J. *J. Chem. Soc. Chem. Commun.* **1989**, 1507.

19. (a) Raban, M.; Mislow, K. *Tetrahedron Lett.* **1965**, 4249. (b) Jacobus, J.; Raban, M.; Mislow, K. *J. Org. Chem.* **1968**, *33*, 1142.

20. (a) Jacobus, J.; Raban, M. *J. Chem. Educ.* **1969**, *46*, 351. (b) Jacobus, J.; Jones, T. B. *J. Am. Chem. Soc.* **1970**, *92*, 4583.

21. Dale, J. A.; Mosher, H. S. *J. Am. Chem. Soc.* **1968**, *90*, 3732.

22. Dale, J. A.; Mosher, H. S. *J. Am. Chem. Soc.* **1973**, *95*, 512.

23. Dale, J. A.; Dull, D. L; Mosher, H. S. *J. Org. Chem.* **1969**, *34*, 2543.

24. Feichter, C.; Faber, K.; Griengl, H. *J. Chem. Soc. Perkins Trans. 1* **1991**, 653.

25. Goldberg, Y.; Alper, H. *J. Org. Chem.* **1992**, *57*, 3731.

26. Bennani, Y. L.; Vanhessche, K. P. M.; Sharpless, K. B. *Tetrahedron Asymmetry* **1994**, *5*, 1473.

27. You, T. B.; Mosher, H. S.; Okamoto, K.; Wang, Y. *Youji Huaxue* **1990**, *10*, 498.

28. Brunel, J. M.; Pardigon, O.; Maffei, M.; Buono, G. *Tetrahedron Asymmetry* **1992**, *3*, 1243.

29. Alexakis, A.; Mutti, S.; Normant, J. F.; Mangeney, P. *Tetrahedron Asymmetry* **1990**, *1*, 437.

30. Anderson, R. C.; Shapiro, M. J. *J. Org. Chem.* **1984**, *49*, 1304.

31. Johnson, C. R.; Elliott, R. C.; Penning, T. D. *J. Am. Chem. Soc.* **1984**, *106*, 5019.

32. Hulst, R.; Zijlstra, R. W. J.; Feringa, B. L.; de Vries, N. K.; ten Hoeve, W.; Wynberg, H. *Tetrahedron Lett.* **1993**, *34*, 1339.

33. Buist, P. H.; Marecak, D.; Holland, H. L; Brown, F. M. *Tetrahedron Asymmetry* **1995**, *6*, 7.

34. Hiemstra, H.; Wynberg, H. *Tetrahedron Lett.* **1977**, 2183.

35. Alexakis, A.; Mangeney, P. *Tetrahedron Asymmetry* **1990**, *1*, 477.

36. Meyers, A. I.; White, S. K.; Fuentes, L. M. *Tetrahedron Lett.* **1983**, *24*, 3551.

37. (a) Mangeney, P.; Alexakis, A.; Normant, J. F. *Tetrahedron Lett.* **1988**, *29*, 2677. (b) Cuvinot, D.; Mangeney, P.; Alexakis, A.; Normant, J. F.; Lellouche, J. P. *J. Org. Chem.* **1989**, *54*, 2420.

38. Alexakis, A.; Frutos, J. C.; Mangeney, P. *Tetrahedron Asymmetry* **1993**, *4*, 2431.

39. Schurig, V.; Wistuba, D. *Tetrahedron Lett.* **1984**, *25*, 5633.

40. Schurig, V.; Nowotny, H.-P. *Angew. Chem. Int. Ed. Engl.* **1990**, *29*, 939.

41. Dougherty, W.; Liotta, F.; Mondimore, D.; Shum, W. *Tetrahedron Lett.* **1990**, *31*, 4389.

42. König, W. A.; Benecke, I.; Sievers, S. *J. Chromatogr.* **1982**, *238*, 427.

43. (a) Gil-Av, E.; Feibush, B.; Charles-Sigler, R. *Tetrahedron Lett.* **1966**, 1009. (b) Gil-Av, E.; Feibush, B.; Charles-Sigler, R. in Littlewood, A. B. ed. *Gas Chromatography 1967* Inst. Petroleum, London, **1967**. (c) Beitler, U.; Feibush, B. *J. Chromatogr.* **1976**, *123*, 149. (d) Feibush, B. *J. Chem. Soc. Chem. Commun.* **1971**, 544. (e) Feibush, B.; Gil-Av, E. *J. Gas Chromatogr.* **1967**, *5*, 257. (f) Feibush, B.; Gil-Av, E. *Tetrahedron* **1970**, *26*, 1361. (g) Gil-Av, E.; Feibush, B. *Tetrahedron Lett.* **1967**, 3345. (h) Nakaparksin, S.; Birell, P.; Gil-Av, E.; Oró, J. *J. Chromatogr. Sci.* **1970**, *8*, 177. (i) Parr, W.; Yang, C.; Bayer, E.; Gil-Av, E. *J. Chromatogr Sci.* **1970**, *8*, 591. (j) Weinstein, S.; Feibush, B.; Gil-Av, E. *J. Chromatogr.* **1976**, *126*, 97. (k) Chang, S. C.; Charles, R.; Gil-Av, E. *J. Chromatogr.* **1982**, *238*, 29.

44. Feibush, B.; Richardson, M. F.; Sievers, R. E.; Springer, C. S. *J. Am. Chem. Soc.* **1972**, *94*, 6717.

45. (a) Schurig, V.; Gil-Av, E. *J. Chem. Soc. Chem. Commun.* **1971**, 650. (b) Schurig, V. *Inorg. Chem.* **1972**, *11*, 736. (c) Schurig, V. *Angew. Chem. Int. Ed. Engl.* **1977**, *16*, 110. (d) Golding, B. T.; Sellars, P. J.; Wong, A. K. *J. Chem. Soc. Chem. Commun.* **1977**, 570. (e) Schurig, V.; Bürkle, W. *J. Am. Chem. Soc.* **1982**, *104*,

7573. (f) Weber, R.; Hintzer, K.; Schurig, V. *Naturwissenschaften* **1980**, *67*, 453. (g) Weber, R.; Schurig, V. *Naturwissenschaften* **1981**, *68*, 330.

46. (a) Lough, W. J., ed. *Chiral Liquid Chromatography*. Blackie/Chapman and Hall, New York, **1989**. (b) Ahuja, S., ed. *Chiral Separations: Application and Technology*. American Chemical Society, Washington, D.C., **1997**. (c) Zief, M.; Cranes, L. J., ed. *Chromatographic Chiral Separations*, vol. 40, Marcel Dekker, New York, **1988**.

47. For a review of liquid chromatographic separation based on chiral recognition, see Pirkle, W. H.; Pochapsky, T. C. *Chem. Rev.* **1989**, *89*, 347.

48. Pirkle, W. H.; House, D. W.; Finn, J. M. *J. Chromatogr.* **1980**, *192*, 143.

49. For inclusion of cyclodextrins, interested readers may see Bender, M. L.; Komiyama, M. *Cyclodextrin Chemistry*, Springer-Verlag, Berlin, **1978**.

50. For initial work on cyclodextrin-mediated chromatographic separation of enantiomers, see Hinze, W. L.; Armstrong, D. W. ed. *Ordered Media in Chemical Separations*, ACS Symposium Series 342, **1986**.

51. Cyclodextrin-mediated chromatographic separation of enantiomers is also discussed in Braithwaite, A.; Smith, F. J. *Chromatographic Methods*, 5th Edition, Blackie Academic & Professional, London, New York, **1996**.

52. (a) Gassmann, E.; Kuo, J. E.; Zare, R. N. *Science* **1985**, *230*, 813. (b) Gozel, P.; Gassmann, E.; Michelsen, H.; Zare, R. N. *Anal. Chem.* **1987**, *59*, 44.

53. For a general introduction to the application of capillary electrophoresis in chiral analysis, see Chankvetadze, B. *Capillary Electrophoresis in Chiral Analysis*, John Wiley & Sons, Ltd, Chichester, **1997**.

54. Bijvoet, J. M.; Peerdeman, A. F.; Van Bommel, J. A. *Nature* **1951**, *168*, 271.

55. Dou, S. Q.; Zheng, Q. T.; Dai, J. B.; Tang, C. C.; Wu, G. P. *Wuli Xuebao* **1982**, *31*, 554 (in Chinese).

56. Dou, S. Q.; Yao, J. X.; Tang, C. C. *Kexue Tongbao* **1987**, *32*, 141 (in Chinese).

57. Harada, N.; Nakanishi, K. *Circular Dichroic Spectroscopy—Exciton Coupling in Organic Stereochemistry*, Oxford University Press, Oxford, **1983**.

58. (a) Moffitt, W.; Woodward, R. B.; Moscowitz, A.; Klyne, W.; Djerassi, C. *J. Am. Chem. Soc.* **1961**, *83*, 4013. (b) Eliel, E. L. *Stereochemistry of Carbon Compounds*, McGraw-Hill, New York, **1962**, pp 412–427.

59. Harada, N.; Chen, S. L.; Nakanishi, K. *J. Am. Chem. Soc.* **1975**, *97*, 5345.

60. Harada, N.; Iwabuchi, J.; Yokota, Y.; Uda, H.; Nakanishi, K. *J. Am. Chem. Soc.* **1981**, *103*, 5590.

61. Gonnella, N. C.; Nakanishi, K.; Martin, V. S.; Sharpless, K. B. *J. Am. Chem. Soc.* **1982**, *104*, 3775.

62. Tang, C. C.; Wu, G. P. *Chem. J. Chin. Univ.* **1983**, *4*, 87.

63. Bestmann, H. J.; Attygalle, A. B.; Glasbrenner, J.; Riemer, R.; Vostrowsky, O. *Angew. Chem. Int. Ed. Engl.* **1987**, *26*, 784.

64. Prelog, V. *Helv. Chim. Acta* **1953**, *36*, 308.

65. Horeau, A. "Determintaion of the Configuration of Secondary Alcohols by Partial Resolution," in Kagan, H. B., ed. *Stereochemistry: Fundamentals and Methods*, vol. 3 George Thieme, Stuttgart, **1977**, p 51.

66. Wang, Z. M. *Youji Huaxue* **1986**, *2*, 167.

67. (a) Cardellina II, J. H.; Barnekow, D. E. *J. Org. Chem.* **1988**, *53*, 882. (b) Barnekow, D. E.; Cardellina II, J. H.; Zektzer, A. S.; Martin, G. E. *J. Am. Chem. Soc.* **1989**, *111*, 3511.

68. Barnekow, D. E.; Cardellina II, J. H. *Tetrahedron Lett.* **1989**, *30*, 3629.

69. Arrowsmith, R. J.; Carter, K.; Dann, J. G.; Davies, D. E.; Harris, C. J.; Morton, J. A.; Lister, P.; Robinson, J. A.; Williams, D. J. *J. Chem. Soc. Chem. Commun.* **1986**, 755.

70. Pecunioso, A.; Maffeis, M.; Marchioro, C. *Tetrahedron Asymmetry* **1998**, *9*, 2787.

71. Sullivan, G. R.; Dale, J. A.; Mosher, H. S. *J. Org. Chem.* **1973**, *38*, 2143.

72. Merckx, E. M.; Vanhoeck, L.; Lepoivre, J. A.; Alderweireldt, F. C.; Van der Veken, B. J.; Tollenaere, J. P.; Raymaekers, L. A. *Spectros Int. J.* **1983**, *2*, 30.

73. Doesburg, H. M.; Petit, G. H.; Merckx, E. M. *Acta Crystallogr.* **1982**, *B38*, 1181.

74. Oh, S. S.; Butler, W. M.; Koreeda, M. *J. Org. Chem.* **1989**, *54*, 4499.

75. Ohtani, I.; Kusumi, T.; Kashman, Y.; Kakisawa, H. *J. Am. Chem. Soc.* **1991**, *113*, 4092.

76. (a) Pettit, G. R.; Singh, S. B.; Niven, M. L. *J. Am. Chem. Soc.* **1988**, *110*, 8539. (b) Singh, S. B.; Pettit, G. R. *J. Org. Chem.* **1990**, *55*, 2797.

77. Rychnovsky, S. D.; Hwang, K. *Tetrahedron Lett.* **1994**, *35*, 8927.

78. Kobayashi. J.; Cheng, J. F; Ishibashi, M.; Wälchli, M. R.; Yamamura, S.; Ohizumi, Y. *J. Chem. Soc. Perkin Trans. 1* **1991**, 1135.

79. Kobayashi, J.; Tsuda, M.; Cheng, J.; Ishibashi, M.; Takikawa, H.; Mori, K. *Tetrahedron Lett.* **1996**, *37*, 6775.

80. Kouda, K.; Ooi, T.; Kaya, K.; Kusumi, T. *Tetrahedron Lett.* **1996**, *37*, 6347.

81. Kaya, K.; Uchida, K.; Kusumi, T. *Biochim. Biophys. Acta* **1986**, *875*, 97.

82. Latypov, Sh. K.; Seco, J. M.; Quiñoá, E.; Riguera, R. *J. Am. Chem. Soc.* **1998**, *120*, 877.

83. (a) Seco, J. M.; Latypov, Sh. K.; Quiñoá, E.; Riguera, R. *Tetrahedron Lett.* **1994**, *35*, 2921. (b) Latypov, Sh. K.; Seco, J. M.; Quiñoá, E.; Riguera, R. *J. Org. Chem.* **1995**, *60*, 504.

84. Sherk, A. E.; Fraser-Reid, B. *J. Org. Chem.* **1982**, *47*, 932.

85. Bernardo, S. D.; Tengi, J. P.; Sasso, G.; Weigele, M. *Tetrahedron Lett.* **1988**, *29*, 4077.

86. For details: (a) Crosby, J. *Tetrahedron* **1991**, *47*, 4789. (b) Scott, J. S. in Morrison, J. D., ed. *Asymmetric Synthesis*, Academic Press, Orlando, Vol 4, **1984**, 1.

87. Horeau, A.; Kagan, H. B.; Vigneron, J. P. *Bull. Soc. Chim. Fr.* **1968**, 3795.

88. Masamune, S.; Choy, W.; Petersen, J. S.; Sita, L. R. *Angew. Chem. Int. Ed. Engl.* **1985**, *24*, 1.

89. Trost, B. M.; O'Krongly, D.; Belletire, J. L. *J. Am. Chem. Soc.* **1980**, *102*, 7595.

90. (a) Masamune, S.; Ali, A.; Snitman, D. L.; Garvey, D. S. *Angew. Chem. Int. Ed. Engl.* **1980**, *19*, 557. (b) Buse, C. T.; Heathcock, C. H. *J. Am. Chem. Soc.* **1977**, *99*, 8109.

91. (a) Rinehart, K. L.; Shield, L. S. in Herz, W.; Grisebach, H.; Kirby, G. W. eds. *Progress in the Chemistry of Organic Natural Products*, Springer-Verlag, New York, **1976**, vol. 33, p 231. (b) Wehrli, W. *Top. Curr. Chem.* **1977**, *72*, 22. (c) Brufani, M.

in Sammes, P. G. ed. *Top. Antibiotic Chem.*, Wiley, New York, **1977**, vol. 1, p 91. (d) Harada, T.; Kagamihara, Y.; Tanaka, S.; Sakamoto, K.; Oku, A. *J. Org. Chem.* **1992**, *57*, 1637.

92. Rychnovsky, S. D.; Rodriguez, C. *J. Org. Chem.* **1992**, *57*, 4793.

93. (a) Nicolaou, K. C.; Kubota, S.; Li, W. S. *J. Chem. Soc. Chem. Commun.* **1989** 512. (b) Colombo, M. I.; Zinczuk, J.; Ruvéda, E. A. *Tetrahedron* **1992**, *48*, 963.

94. (a) Bormans, B. *Chem. Eng. News* **1991**, *69*, Sep 2, 11. (b) Guénard, D.; Guéritte-Voegelein, F.; Potier, P. *Acc. Chem. Res.* **1993**, *26*, 160. (c) Denis, J.; Greene, A. E.; Guénard, D.; Guéritte-Voegelein, F.; Mangatal, L.; Potier, P. *J. Am. Chem. Soc.* **1988**, *110*, 5917. (d) Flam, F. *Science* **1994**, *263*, 911. (e) Nicolaou, K. C.; Dai, W. M.; Guy, R. K. *Angew. Chem. Int. Ed. Engl.* **1994**, *33*, 15. (f) Wang, Z. M.; Kolb, H. C.; Sharpless, K. B. *J. Org. Chem.* **1994**, *59*, 5104.

95. (a) Armstrong, R. W.; Beau, J. M.; Cheon, S. H.; Christ, W. J.; Fujioka, H.; Ham, W. H.; Hawkins, L. D.; Jin, H.; Kang, S. H.; Kishi, Y.; Martinelli, M. J.; McWhorter, W. W.; Mizuno, M.; Nakata, M.; Stutz, A. E.; Talamas, F. X.; Taniguchi, M.; Tino, J. A.; Ueda, K.; Uenishi, J.; White, J. B.; Yonaga, M. *J. Am. Chem. Soc.* **1989**, *111*, 7530. (b) Cha, J. K.; Christ, W. J.; Finan, J. M; Kishi, Y.; Klein, L. L.; Ko, S. S.; Leder, J.; McWhorter, W. W.; Pfaff, K.-P.; Yonaga, M. *J. Am. Chem. Soc.* **1982**, *104*, 7369.

96. Napier, M. A.; Holmquist, B.; Strydom, D. J.; Goldberg, I. H. *Biochem. Biophys. Res. Commun.* **1979**, *89*, 635.

97. Konishi, M.; Ohkuma, H.; Saitoh, K.; Kawaguchi, H.; Golik, J.; Dubay, G.; Groenewold, G.; Krishnan, B.; Doyle, T. W. *J. Antibiot.* **1985**, *38*, 1605.

98. Lee, M. D.; Dunne, T. S.; Siegel, M. M.; Chang, C. C.; Morton, G. O.; Borders, D. B. *J. Am. Chem. Soc.* **1987**, *109*, 3464.

99. Konishi, M.; Ohkuma, H.; Matsumoto, K.; Tsuno, T.; Kamei, H.; Miyaki, T.; Oki, T.; Kawaguchi, H.; VanDuyne, G. D.; Clardy, J. *J. Antibiot.* **1989**, *42*, 1449.

100. For a review, see Nicolau, N. C.; Dai, W. M. *Angew. Chem. Int. Ed. Engl.* **1991**, *30*, 1387.

101. Nicolaou, K. C.; Clark, D. *Angew. Chem. Int. Ed. Engl.* **1992**, *31*, 855.

102. Leet, J. E.; Schroeder, D. R.; Hofstead, S. J.; Golik, J.; Colson, K. L.; Huang, S.; Klohr, S. E.; Doyle, T. W.; Matson, J. A. *J. Am. Chem. Soc.* **1992**, *114*, 7946.

103. Nicolaou, K. C.; Hummel, C. W.; Pitsinos, E. N.; Nakada, M., Smith, A. L.; Shibayama, K.; Saimoto, H. *J. Am. Chem. Soc.* **1992**, *114*, 10082.

104. (a) Romo, D.; Johnson, D. D.; Plamondon, L.; Miwa, T.; Schreiber, S. L. *J. Org. Chem.* **1992**, *57*, 5060. (b) Romo, D.; Meyer, S. D.; Johnson, D. D. Schreiber, S. L. *J. Am. Chem. Soc.* **1993**, *115*, 7906.

105. Tanaka, H.; Kuroda, A.; Marusawa, H.; Hatanaka, H.; Kino, T.; Goto, T.; Hashimoto, M.; Taga, T. *J. Am. Chem. Soc.* **1987**, *109*, 5031.

106. (a) Villalobos, A.; Danishefsky, S. J. *J. Org. Chem.* **1990**, *55*, 2776. (b) Jones, T. K.; Mills, S. G.; Reamer, R. A.; Askin, D.; Desmond, R.; Volante, R. P.; Shinkai, I. *J. Am. Chem. Soc.* **1989**, *111*, 1157. (c) Jones, T. K.; Reamer, R. A.; Desmond, R.; Mills, S. G. *J. Am. Chem. Soc.* **1990**, *112*, 2998. (d) Nakatsuka, M.; Ragan, J. A.; Sammakia, T.; Smith, D. B.; Uehling, D. E.; Schreiber, S. L. *J. Am. Chem. Soc.* **1990**, *112*, 5583.

α-Alkylation and Catalytic Alkylation of Carbonyl Compounds

Chapter 1 introduced the nomenclature for chiral systems, the determination of enantiomer composition, and the determination of absolute configuration. This chapter discusses different types of asymmetric reactions with a focus on asymmetric carbon–carbon bond formation. The asymmetric alkylation reaction constitutes an important method for carbon–carbon bond formation.

2.1 INTRODUCTION

The carbonyl group in a ketone or aldehyde is an extremely versatile vehicle for the introduction of functionality. Reaction can occur at the carbonyl carbon atom using the carbonyl group as an electrophile or through enolate formation upon removal of an acidic proton at the adjacent carbon atom. Although the carbonyl group is an integral part of the nucleophile, a carbonyl compound can also be considered as an enophile when involved in an asymmetric "carbonyl-ene" reaction or dienophile in an asymmetric hetero Diels-Alder reaction. These two types of reaction are discussed in the next three chapters.

The prime functional group for constructing C–C bonds may be the carbonyl group, functioning as either an electrophile (Eq. 1) or via its enolate derivative as a nucleophile (Eqs. 2 and 3). The objective of this chapter is to survey the issue of asymmetric inductions involving the reaction between enolates derived from carbonyl compounds and alkyl halide electrophiles. The addition of a nucleophile toward a carbonyl group, especially in the catalytic manner, is presented as well. Asymmetric aldol reactions and the related allylation reactions (Eq. 3) are the topics of Chapter 3. Reduction of carbonyl groups is discussed in Chapter 4.

Carbonyl compounds including ketones, aldehydes and carboxylic acid derivatives constitute a class of carbon acids, the acidity of which falls in the pK_a range of 25 to 35 in dimethylsulfoxide (DMSO). Representative values for selected carbonyl substrates are summarized in Table 2–1.[1] Different methods may be invoked for generating the enolates according to the pK_a value of their parent compounds.

To generate an enolate from a carbonyl substrate, a suitable base should be chosen to meet two criteria:

1. Adequate basicity to ensure the selective deprotonation process for enolate generation
2. A sterically hindered structure so that nucleophilic attack of this base on the carbonyl centers can be prevented.

TABLE 2–1. pK_a **Data for Representative Carbonyl Compounds and Related Substances in DMSO**

Substrate	pK_a (DMSO)	Substrate	pK_a (DMSO)
H_3CCOCH_3	26.5	$NCCH_3$	31.3
$PhCOCH_3$	24.6	$EtOCOCH_3$	30–31
$PhCOCH_2CH_3$	24.4	$EtOCOCH_2Ph$	22.7
$PhCOCH_2OMe$	22.9	$EtOCOCH_2SPh$	21.4
$PhCOCH_2Ph$	17.7	Me_2NCOCH_3	34–35
$PhCOCH_2SPh$	17.1	CH_3SOCH_3	35.1
HOH	27.5	NH_3	41
CH_3OH	27.9	$HN(CH_3)_2$	44
$(CH_3)_2CHOH$	29.3	$(CH_3)_3COH$	29.4

Reprinted with permission by Am. Chem. Soc., Ref. 1(a).

The metal amide bases had enjoyed much popularity since the introduction of sterically more hindered bases **1–4**. The introduction of sterically hindered bases **1–4** has been a particularly important innovation in this field, and these reagents have now been accepted as the most suitable and commonly used bases for carbonyl deprotonation. The metal dialkylamides are all quite soluble in ethereal solvent systems. Lithium diisopropylamide (LDA, **1**)[2] has been recognized as the most important strong base in organic chemistry. Both LDA (**1**) and lithium isopropylcyclohexyl amide (LICA **2**)[3] exhibit similarly high kinetic deprotonation selectivity. Furthermore, silylamides **4a–c** have been found to exhibit good solubility in aromatic hydrocarbon solvents,[4] and these silyl amides are comparably effective in enolate generation. In fact, lithium tetramethylpiperidine (LTMP **3**)[5] is probably the most sterically hindered amide base at this time.

$(Me_2CH)_2NLi$ $(Me_2CH)(c\text{-}C_6H_{11})NLi$

LDA **1**   LICA **2**

LTMP **3**

$(Me_3Si)_2N\text{-}M$

4a M = Li, LHMDS
4b M = Na, NHMDS
4c M = K, KHMDS
4d $(Et_3Si)_2N\text{-}Li$
4e $(Me_2PhSi)_2N\text{-}Li$

Before the emergence in the mid-1980s of the asymmetric deprotonation of *cis*–dimethyl cyclohexanone using enantiomerically pure lithium amide bases, few reports pertaining to the chemistry of these chiral reagents appeared. Although it is not the focus of this chapter, the optically active metal amide bases are still considered to be useful tools in organic synthesis. Readers are advised to consult the appropriate literature on the application of enantiomerically pure lithium amides in asymmetric synthesis.[6]

2.2 CHIRALITY TRANSFER

The asymmetric alkylation of a carbonyl group is one of the most commonly used chirality transfer reactions. The chirality of a substrate can be transferred to the newly formed asymmetric carbon atom through this process. In surveying chiral enolate systems as a class of nucleophile, three general subdivisions can be made in such asymmetric nucleophilic addition reactions: intra-annular, extra-annular, and chelation enforced intra-annular.

Considering enolate **5**, where X, Y, and Z stand for three connective points for cyclic enolate formation, the formation can occur between any two of such points, leading to three possible oxygen enolates: *endo*-cyclic **6**, **7**, and *exo*-cyclic enolate **8**. In such cases, the resident asymmetric center (*) may be positioned on any connective atom in the substrate.

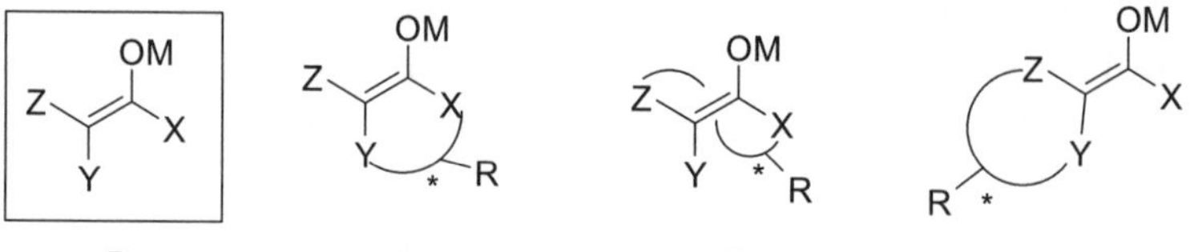

5   **6**   **7**   **8**

endo-cyclic   *exo*-cyclic

These classes of enolates are central to our discussion of chirality transfer and thus are briefly discussed in the following sections.

2.2.1 Intra-annular Chirality Transfer

The definition of intra-annular chirality transfer can be best established by the following examples (Scheme 2–1)[7]:

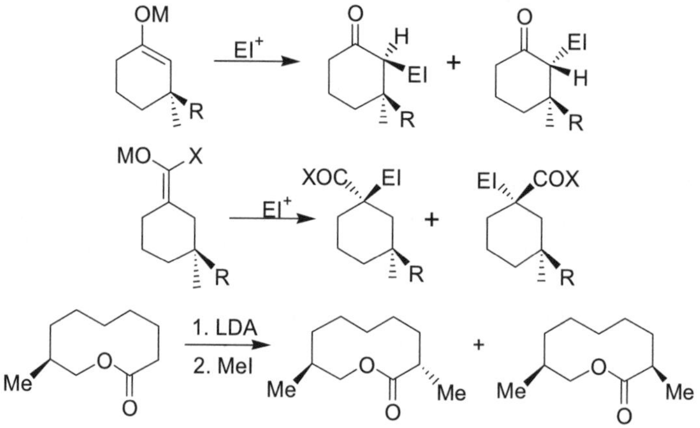

Scheme 2–1. Intra-annular chirality transfer.

The examples in Scheme 2–1 show that in intra-annular chirality transfer the resident asymmetric center connects to the enolate through annular covalent bonds. The geometric configuration of the enolate can be either immobile or irrelevant to the sense of asymmetric induction.

2.2.1.1 *Six-Membered Ring (exo-Cyclic)*.
The diastereoselective alkylation reactions of *exo*-cyclic enolates involving 1,2-asymmetric inductions are *anti*-inductions. In Scheme 2–2, there are two possible enolate chair conformations in which the two possible transition-state geometries lead to the major diastereomer **9e** (where the substituent takes the equatorial orientation). However, for the case in which R = methyl and X = alkoxyl or alkyl, one would expect the

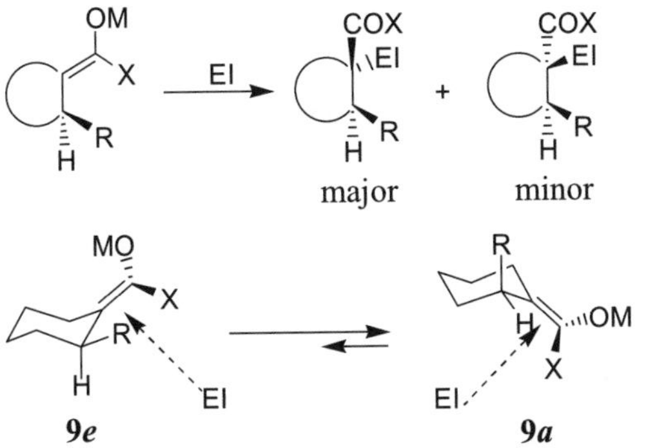

Scheme 2–2. Two possible enolate chair conformations.

axial conformation **9a** to be favored over the equatorial conformation **9e**. As shown in Scheme 2–2, **9e** suffers from the steric strain R ↔ X(OM), and **9a** is more likely to be favored to receive axial attack by the approaching electrophile rather than the less stable enolate **9e** by the equatorial attack.[8]

Examples are given in Scheme 2–3. Enolate **10** exhibits selective alkylation *anti* to the alkyl substituent, and the same *anti*-induction occurs in the case of 1,4-asymmetric induction of **11**.[9]

2.2.1.2 Six-Membered Ring (endo-Cyclic).

All the previous discussion of stereo-attack is based on steric hindrance, but in the case of a six-membered ring (*endo*-cyclic) enolate, the direction is affected simultaneously by stereo-electronic effects (Scheme 2–4).[10] In the transition state, the attacking electrophiles must obey the principle of maximum overlap of participating orbitals by perpendicularly approaching the plane of atoms that constitute the enolate functional group. Electrophile attacks take place on the two diastereotopic

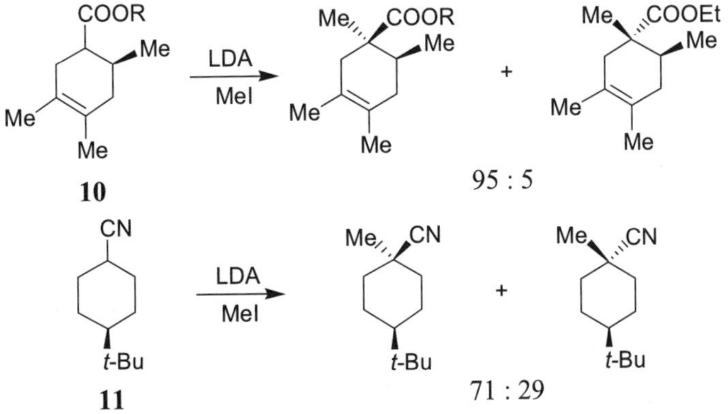

Scheme 2–3. Alkylation of *exo*-cyclic six-membered ring substrates.

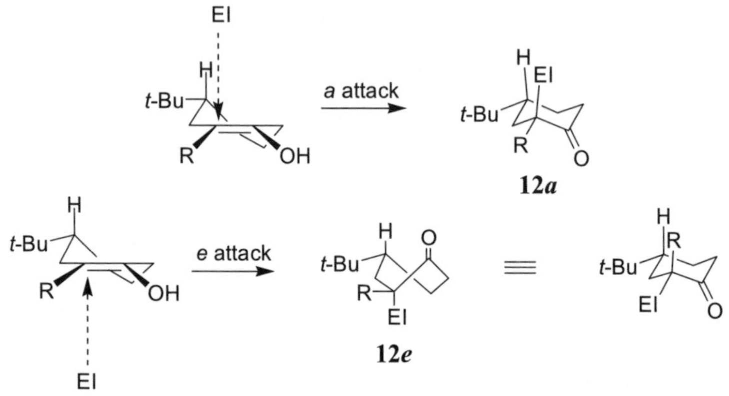

Scheme 2–4. Transition states of the *endo*-cyclic six-membered ring.

faces of the enolate, the *a* attack and the *e* attack, which lead to the ketone products **12a** and **12e**. Ketone **12a**, obtained via a postulated chair-like transition state, could be presumed to be formed in preference to ketone **12e**, which resulted from a boat-like transition state. The energetic bias for "axial alkylation" via a chair-like transition state to form **12a** is relatively small.[11] Therefore, for a given enolate, the ratio **12a:12e** is insensitive to the alkylating agent employed.

Scheme 2–5 is one of such examples in which stereoelectronic control has to be taken into account in diastereoselective alkylation of substituted cyclohexanone enolates.[12]

The diastereoselective alkylation reaction of *endo*-cyclic five-membered ring enolates exhibits good potential for both 1,3- and 1,2-asymmetric induction. In Scheme 2–6, the factor controlling the alkylation transition state is steric rather than stereoelectronic, leading to an *anti*-induction.[13]

The reaction shown in Scheme 2–7 is an example of 1,3-asymmetric induction. The oxidative hydroxylation of a five-membered lactone led to an α-hydroxyl product **14**.[14] The α-hydroxylation of carbonyl compounds is further discussed in Chapter 4.

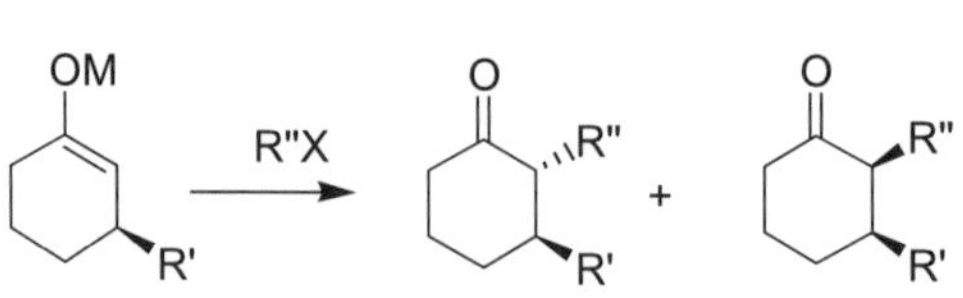

R'	R"X	Ratio (*anti/syn*)
n-C$_4$H$_9$	MeI	88 : 12
CH$_2$=CH	MeI	75 : 25
CH$_3$	CH$_2$=CHCH$_2$Br	89 : 11

Scheme 2–5. Alkylation of *endo*-cyclic six-membered ring systems.

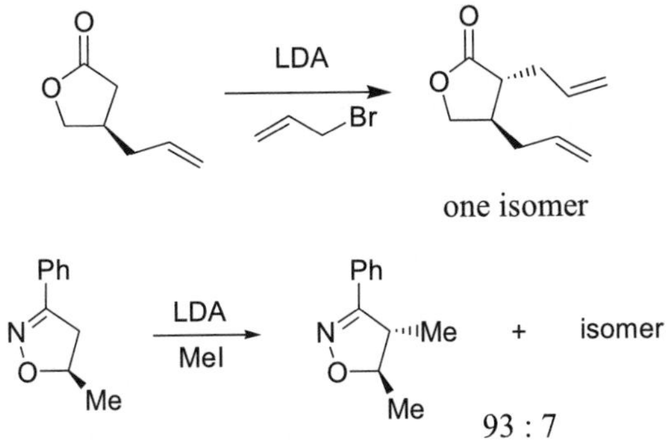

Scheme 2–6. Reaction for a five-membered ring system.

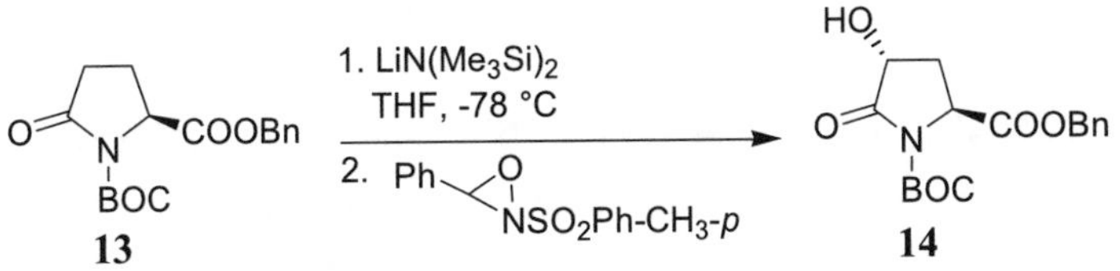

Scheme 2–7. Oxidative hydroxylation.

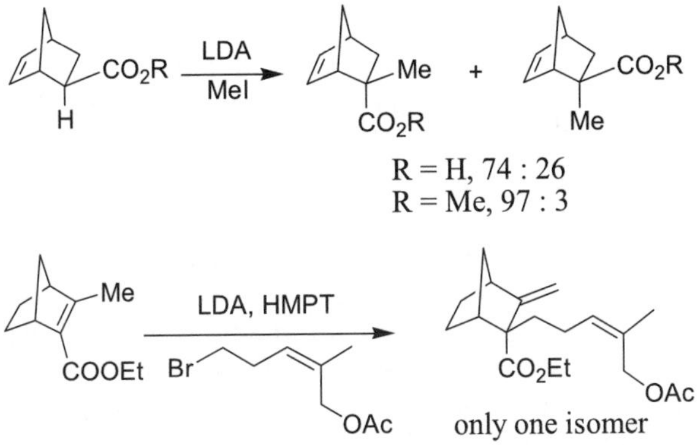

R = H, 74 : 26
R = Me, 97 : 3

only one isomer

Scheme 2–8. Diastereoselective alkylation reactions in the norbornyl ring system.

There are also many examples of alkylation reactions involving the norbornyl ring system in which the enolate can be either *endo*- or *exo*-cyclic. Both the *endo*-cyclic (**6, 7**) and *exo*-cyclic (**8**) enolates exhibit high levels of asymmetric induction due to the rigid ring system. Scheme 2–8 presents some examples for alkylation involving the norbornyl ring system.[15]

2.2.2 Extra-annular Chirality Transfer

The cases illustrated here are typical examples of extra-annular chirality transfer via the alkylation process (Scheme 2–9)[16]:

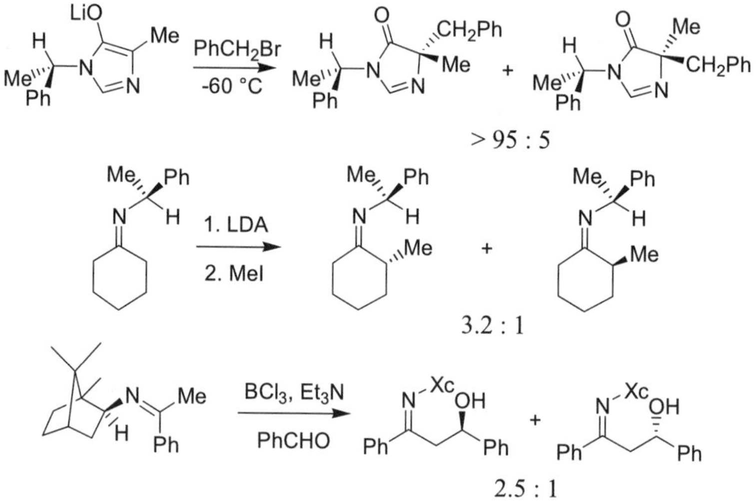

Scheme 2–9. Extra-annular chirality transfer. Xc stands for the chiral auxiliary.

One can see from the examples in Scheme 2–9 that, although the asymmetric center thus formed connects to the enolate through a covalent bond, the stereo-relationship between the chirality transfer and the enolate cannot be established because the resident chiral moiety is not conformationally fixed at two or more contact points via covalent bonds to the trigonal center undergoing substitution. As a consequence of this conformational flexibility, it is frequently difficult to make the desired prediction for stereoselectivity. Nonetheless, with an increasing understanding of acyclic conformational analysis, particularly the implications of ring strain or steric hindrance, greater acyclic diastereoselectivity is achievable.[17] Scheme 2–10 provides two further examples of extra-annular chirality transfer.[18,19]

In Eq. 1, lithium enolate generated from tetronic acid–derived vinylogous urethanes undergoes alkylation with an electrophile to give a product with a diastereoselectivity of 97:3. Eq. 2 in Scheme 2–10 is a diastereoselective conjugate addition reaction. When treated with organocopper reagents, compound **15** undergoes reactions with high diastereoselectivity.[19] Asymmetric conjugate addition is further discussed in Chapter 8.

Examination of different molecular models shows that the bicyclic nature of **15** creates a rigid molecule where steric hindrance plays a predominant role. This can help explain the high enantioselectivity for the product.

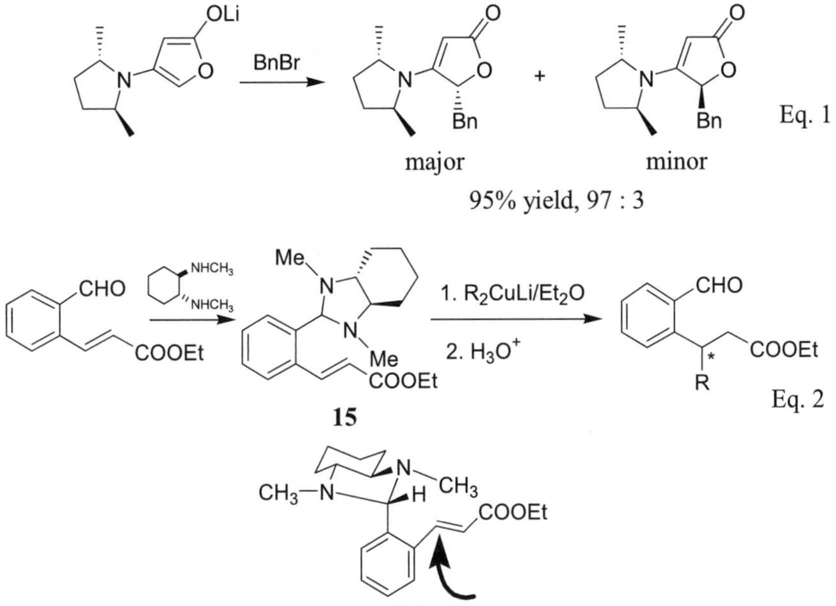

Eq. 1

major   minor

95% yield, 97 : 3

Eq. 2

15

15

Scheme 2–10. Two examples of extra-annular chirality transfer.

TABLE 2–2. Reaction of 15 With Various Reagents

	Me$_2$CuLi	Bu$_2$CuLi	Ph$_2$CuLi	(EtCH=CH)$_2$CuLi
Yield	85	90	84	80
ee	94(S)	95(S)	96(R)	90(R)*

*This inversion of configuration is only due to the CIP selection rules and does not correspond to the steric course of the reaction.

2.2.3 Chelation-Enforced Intra-annular Chirality Transfer

As a result of the combination of intra- and extra-annular chirality transfer, a productive approach has been developed for the design of chiral enolate systems in which a structurally organized diastereofacial bias is established, as illustrated in the equations in Scheme 2–11. A lithium-coordinated five-membered or six-membered ring fixes the orientation between the inducing asymmetric center and the enolate[20]:

Thus, the postulated chelated enolates and their alkylation reaction make the intra-annular chirality transformation possible. This method for enolate formation is the focal point of this chapter, as this is by far the most effective approach to alkylation or other asymmetric synthesis involving carbonyl are compounds.

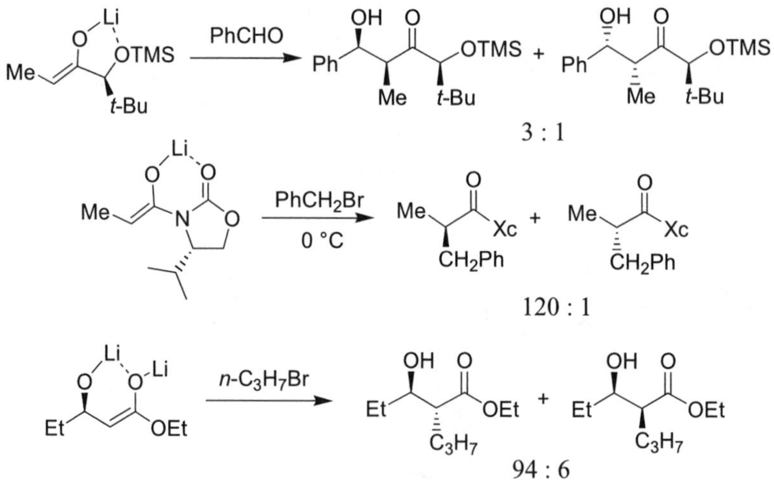

Scheme 2–11. Chelation-enforced intra-annular chirality transfer.

Much attention has been devoted to the examination of chiral enolate systems in which metal ion chelation may play an important role in establishing a fixed stereochemical relationship between the resident chirality and the enolate moiety. This has resulted in the conclusion that enolate geometry is critical in the definition of π-facial selection. The following sections discuss this effort in several different chemical systems.

2.2.3.1 β-Hydroxy Acid Systems.

The alkylation of β-hydroxy ester enolates is an excellent example of a reaction in which metal chelation plays a critical role.[21] In this system, two points should be taken into consideration. First, in the enolization process these substrates may form either (E)- or (Z)-enolates; second, from either of these enolate systems, chelation could be involved in determining the enolate-facial selection. As shown in Scheme 2–12, the major component of the reaction product is formed as a result of Re-face attack on the (Z)- or (E)-enolate.

The enolization product depends on the structure of the carbonyl substrate: When β-OH is present in the carbonyl compound, (Z)-enolate is the major product due to the metal ion chelation,[20c] whereas (E)-enolate is the major product in the absence of a β-OH group.[22] It is worth noting that the yield is normally low for β-OH carbonyl substrates because of the tendency for β-elimination.

Generally speaking, if care is taken, alkylation of β-hydroxyl esters can be successful (Scheme 2–13).[21a,f]

2.2.3.2 Prolinol-Type Chiral Auxiliaries.

In this section, applications of chelation-enforced chirality transfers with nitrogen derivatives are discussed

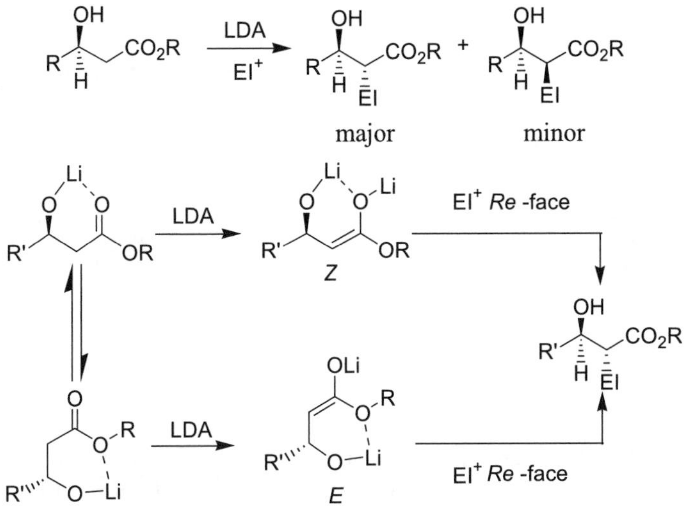

Scheme 2–12

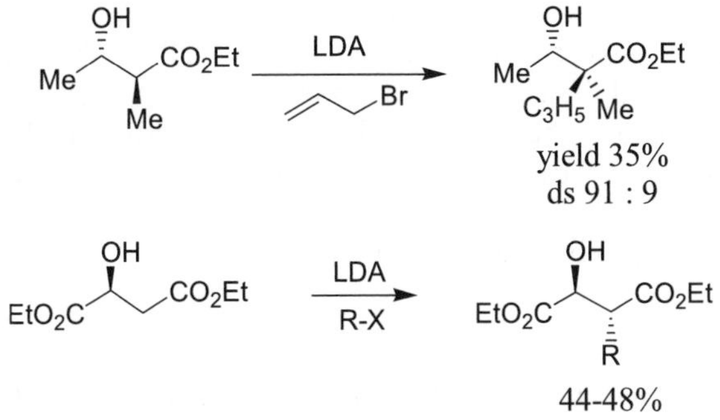

yield 35%
ds 91 : 9

44-48%

R-X	anti : syn
MeI	> 91 : 9
CH$_2$=CHCH$_2$Br	> 95 : 5
PhCH$_2$Br	> 91 : 1

Scheme 2–13. Diastereoselective alkylation of β-hydroxy enolates.

briefly. The formation of nitrogen derivatives, such as chiral imines, imides, amides, and sultams or other nitrogen analogs, from ketones or acids allows the incorporation of a chiral auxiliary that can be removed through hydrolysis or reduction after alkylation. These chiral enolates provide the entry into α-substituted ketones, carboxylic acids, and other related compounds.

Evans and Takacs[23] demonstrated a diastereoselective alkylation based on metal ion chelation of a lithium enolate derived from a prolinol-type chiral auxiliary. This method can provide effective syntheses of α-substituted carbox-

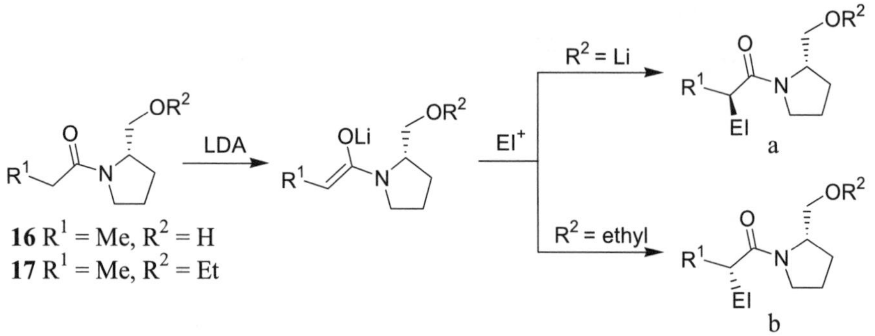

Scheme 2–14

ylic acids. The alkylation occurs preferentially from the *Si*-face of the enolate system **16** (R^1 = Me, R^2 = H) or *Re*-face of **17** (R^1 = Me, R^2 = Et). The sense of asymmetric induction is strongly influenced by the nature of the pendent oxygen substituent R^2. When R^2 is lithium, preferential alkylation from the *Si*-face of the enolate can be found, whereas the analogous alkylation reaction of the derived ethers exhibited a reversal in π-selection. Thus, starting from either substrate **16** or **17**, a pair of enantiomers of the final α-substituted carboxylic acids can be obtained after acidic hydrolysis of the alkylated product (Scheme 2–14). Table 2–3 shows the results of the enantioselective alkylation of **16**, indicating that α-alkylated carboxylic acid can be obtained upon hydrolysis of the reaction product.

In this reaction, prolinol serves as a chiral auxiliary, but it cannot be easily

TABLE 2–3. Enantioselective Alkylations and Conversions of 16 to Carboxylic Acids[23]

Entry	Electrophile	a:b	Hydrolysis products	Yield (%)
1	CH₃CH₂I	92:8	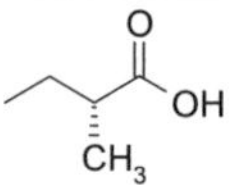	84
2	n-C₄H₉I	94:6	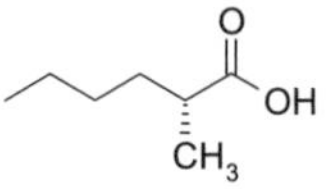	78
3	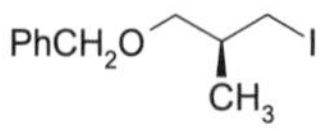	97:3	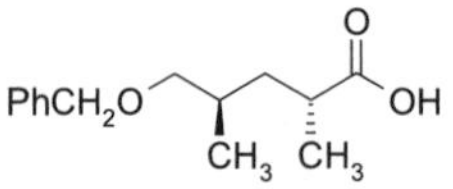	91
4	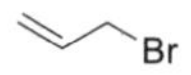	96:4	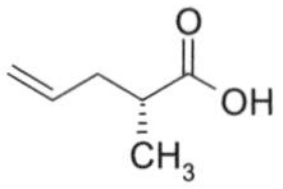	87

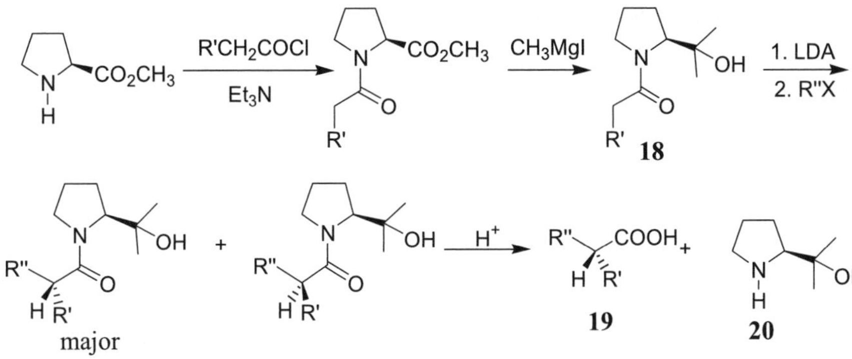

Scheme 2–15

Entry	R'	R''	Yield (%)	ee (%)	Config. of **19**
1	CH_3	C_2H_5	86	84	R
2	CH_3	n-C_4H_9	77	87	R
3	n-C_4H_9	CH_3	96	75	S
4	CH_3	n-C_8H_{17}	75	90	R
5	n-C_8H_{17}	CH_3	82	78	S
6	CH_3	$C_6H_5CH_2$	87	87	R
7	$C_6H_5CH_2$	CH_3	72	> 99	S

recovered from the reaction mixture due to its high water solubility. To overcome this problem, Lin et al.[24] modified Evans' reagent by introducing two methyl groups to prolinol to form a tertiary alcohol **18**, which can easily be recovered after the workup. By changing the reaction sequence of R and R' in the acyl and alkyl groups, both of the enantiomers of the carboxylic acid can be obtained (Scheme 2–15).

In the development of asymmetric synthesis methodology, the advantage of a chiral auxiliary having C_2 asymmetry has been realized and applied to the pyrrolidines, a prolinol structure–based derivative.[25] The asymmetric alkylation of the corresponding carboxylamide enolates developed by Kawanami et al.[26] has proved to be highly successful in providing good chemical yield and high enantioselectivity.

Racemic *trans-N*-benzyl-2,5-bis-(ethoxycarbonyl)pyrrolidine has been resolved via its dicarboxylic acid, followed by subsequent transformation to offer $(2R,5R)$-**21** or $(2S,5S)$-**21**. The absolute configuration of the alkylated carboxylic acids indicates that the approach of alkyl halides is directed to one of the diastereotopic faces of the enolate thus formed. In the following case, the approached face is the *Si*-face of the (Z)-enolate. By employing the chiral auxiliary $(2R,5R)$-**21** or its enantiomer $(2S,5S)$-**21**, the (R)- or (S)-form of carboxylic acids can be obtained with considerably high enantioselectivity (Table 2–4).

TABLE 2–4. Asymmetric Alkylation Using (2R,5R)-21 in THF at −78°C

Entry	R in **21** Y = CH$_3$	R'X	Yield (%)	de (%)	Configuration*
1	CH$_3$	C$_2$H$_5$I	87	>95	R
2	CH$_3^\dagger$	C$_2$H$_5$I	78	>95	S
3	C$_2$H$_5$	CH$_3$I	91	>95	S
4	CH$_3$	C$_4$H$_9$I	81	>95	(R)
5	C$_4$H$_9$	CH$_3$I	81	>95	(S)
6	CH$_3$	PhCH$_2$Br	80	>95	R
7	PhCH$_2$	CH$_3$I	76	>95	S
8	C$_{16}$H$_{33}$	CH$_3$I	61	>95	(S)
9	CH$_3$	CH$_2$=CHCH$_2$Br	81	>95	(R)
10	CH$_3$	PhCH$_2$OCH$_2$Cl	74	>95	(R)
11	CH$_3$	R''OCH$_2$CH$_2$CH$_2$Br‡	78	>95	(R)

*Tentative assignment in parentheses. Reprinted with permission by Pergamon Press Ltd., Ref. 26.
† (2S,5S)-enantiomer of **21** was used.
‡ R'' = TBS.
de = diastereometric excess.

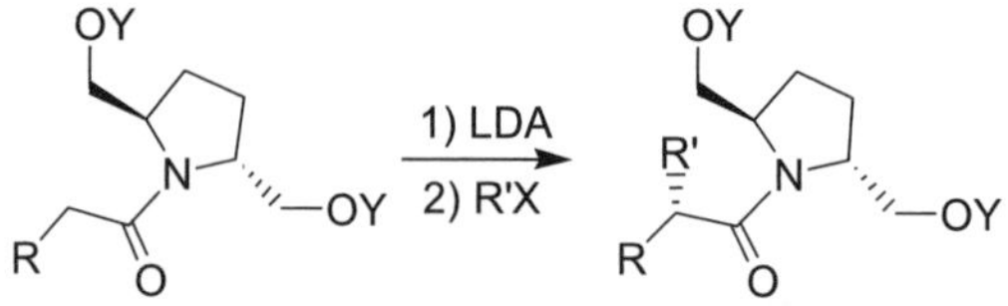

21 Y = CH$_3$ or CH$_2$OCH$_3$

The chiral auxiliary *trans*-(2R,5R)-*bis*-(benzyloxymethyl)pyrrolidine can be prepared from mannitol as shown in Scheme 2–16[27]:

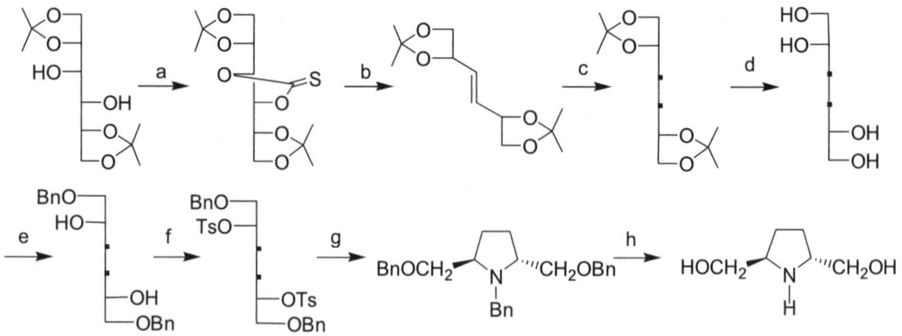

Scheme 2–16. Synthesis of pyrrolidine. *Reagents and conditions:* a: TCDI (1,1-thiono-carbonyldiimidazole), THF; b: P(OEt)$_3$, DEAD; c: H$_2$, Rh/Al$_2$O$_3$, EtOH; d: TsOH, aq. MeOH; e: Bu$_2$SnO, toluene, reflux; BnBr, Bu$_4$N$^+$Br$^-$; f: TsCl, Py, 0°C; g: BnNH$_2$, Δ; h: H$_2$, Pd(OH)$_2$/C, EtOH.

TABLE 2–5. Diastereoselective Alkylation Reaction of the Lithium Enolates Derived from Imides 22 and 23

Entry	Imide	EI$^+$	Ratio	Yield (%)
1	**22** (R = CH$_3$)	PhCH$_2$Br	99:1	92
2	**23** (R = CH$_3$)	PhCH$_2$Br	2:98	78
3	**22** (R = C$_2$H$_5$)	CH$_3$I	89:11	79
4	**23** (R = C$_2$H$_5$)	CH$_3$I	13:87	82
5	**22** (R = CH$_3$)	C$_2$H$_5$I	94:6	36
6	**23** (R = CH$_3$)	C$_2$H$_5$I	12:88	53
7	**22** (R = CH$_3$)	CH$_2$=CHCH$_2$Br	98:2	71
8	**23** (R = CH$_3$)	CH$_2$=CHCH$_2$Br	2:98	65

EI$^+$ = electrophiles in Scheme 2–17.

Reprinted with permission by Am. Chem. Soc., Ref. 28.

2.2.3.3 Imide Systems. Imide compounds **22** and **23**, or Evans' reagents, derived from the corresponding oxazolidines are chiral auxiliaries for effective asymmetric alkylation or aldol condensation and have been widely used in the synthesis of a variety of substances.

Table 2–5 summarizes the results of the asymmetric alkylation (Scheme 2–17) of the lithium enolates derived from **22** or **23**.[28] When chiral auxiliary **22** or **23** is involved in the alkylation reactions, the substituent at C-4 of the oxazolidine ring determines the stereoselectivity and therefore controls the stereogenic outcome of the alkylation reaction.

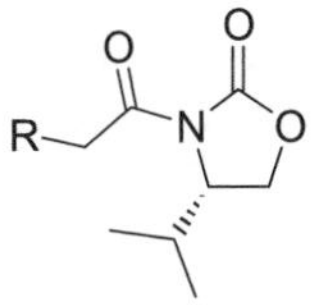

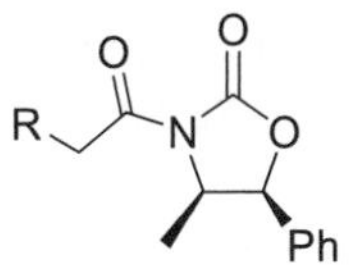

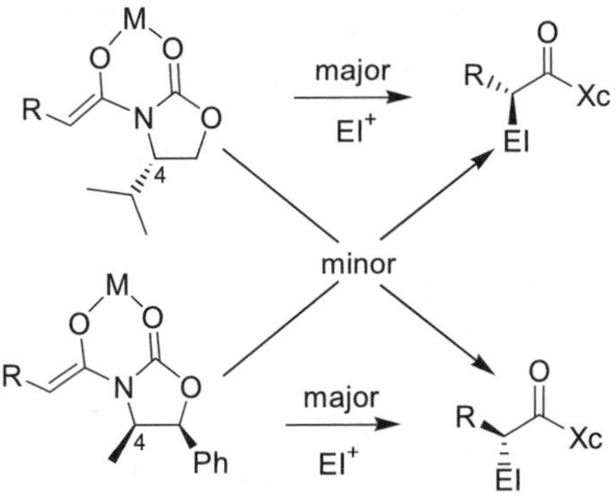

Scheme 2–17

The application of Evans' imides in the preparation of various alkyl acids or the corresponding derivatives can be depicted as in Scheme 2–18[29]:

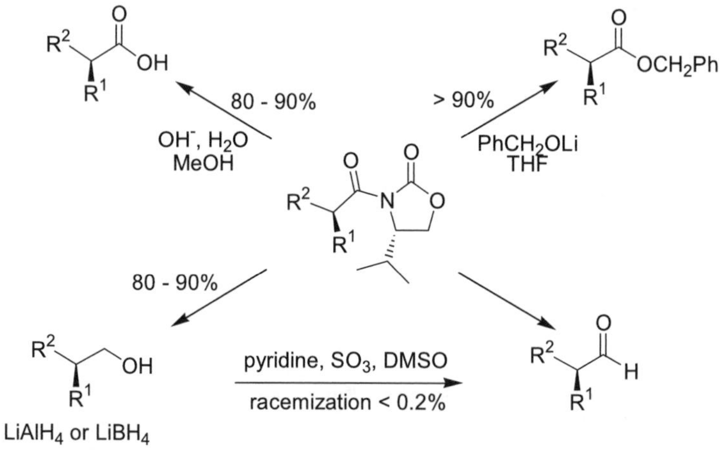

Scheme 2–18

The main disadvantages of Evans' auxiliaries **22** and **23** are that they are expensive to purchase and inconvenient to prepare, as the preparation involves the reduction of (S)-valine **24** to water-soluble (S)-valinol, which cannot be readily extracted to the organic phase. The isolation of this water-soluble valinol is difficult and requires a high vacuum distillation, which is not always practical, especially on an industrial scale. Therefore, an efficient synthesis of Evans' chiral auxiliary **25** has been developed, as depicted in Scheme 2–19[30]:

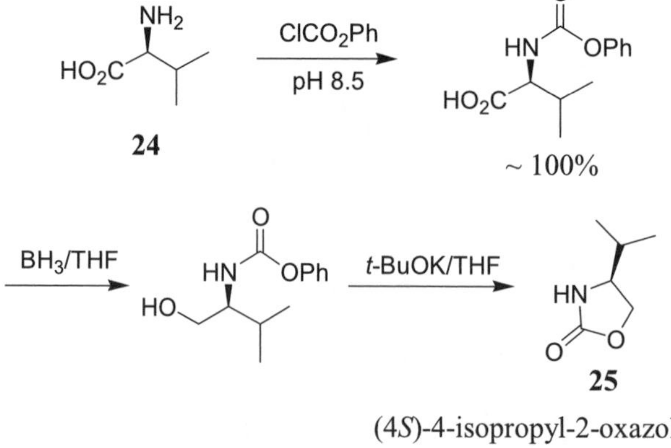

(4S)-4-isopropyl-2-oxazolidinone, 82%

Scheme 2–19. Synthesis of Evans' chiral auxiliary **25**.

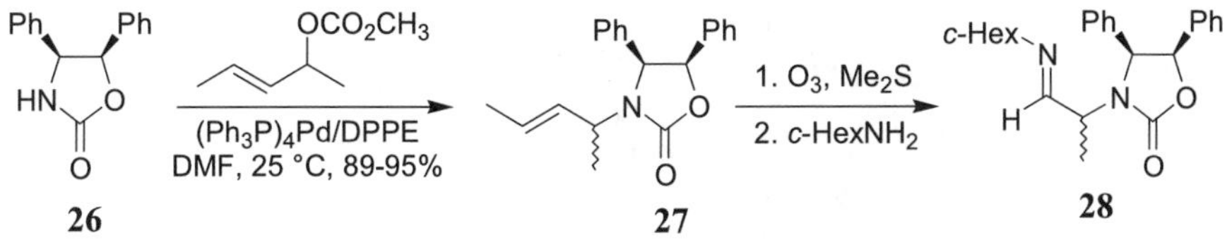

Scheme 2–20

This imide system can also be used for the asymmetric synthesis of optically pure α,α-disubstituted amino aldehydes, which can be used in many synthetic applications.[31] These optically active α-amino aldehydes were originally obtained from naturally occurring amino acids, which limited their availability. Thus, Wenglowsky and Hegedus[32] reported a more practical route to α-amino aldehydes via an oxazolidinone method. As shown in Scheme 2–20, chiral diphenyl oxazolidinone **26** is first converted to allylic oxazolidinone **27**; subsequent ozonolysis and imine formation lead to compound **28**, which is ready for the α-alkylation using the oxazolidinone method. The results are shown in Table 2–6.

2.2.3.4 Chiral Enamine Systems. At this stage it is appropriate to introduce some important studies in the field of metalloenamines. To start with, note that metalloenamine generated from chiral cyclohexanone imine **29** or **31** is highly diastereoselective in alkylation (Scheme 2–21 and the results therein).[33] This can be explained by the possible transition states. If we take **29** as an example, there is an equilibrium between two possible transition states as shown in Figure 2–1. It appears that the left structure is more stable than the right one, thus favoring the formation of the product **30** in (R)-configuration.

TABLE 2–6. Synthesis of α-Amino Aldehydes

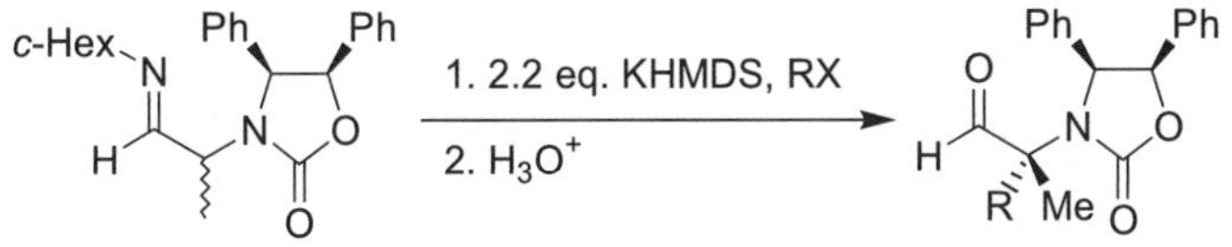

Entry	RX	de	Yield (%)
1	PhCH$_2$Br	94:6	62
2	3,4-di-MeOPhCH$_2$Br	90:10	47
3	CH$_2$=CHCH$_2$Br	92.5:7.5	62
4	(CH$_3$)$_2$CHCH$_2$I	93.5:6.5	62
5	(CH$_3$)$_2$CHI	97.5:2.5	48
6	n-BuI	92:8	75

de = diastereomeric excess; RX = electrophiles in the reaction.

Reprinted with permission by Am. Chem. Soc., Ref. 32.

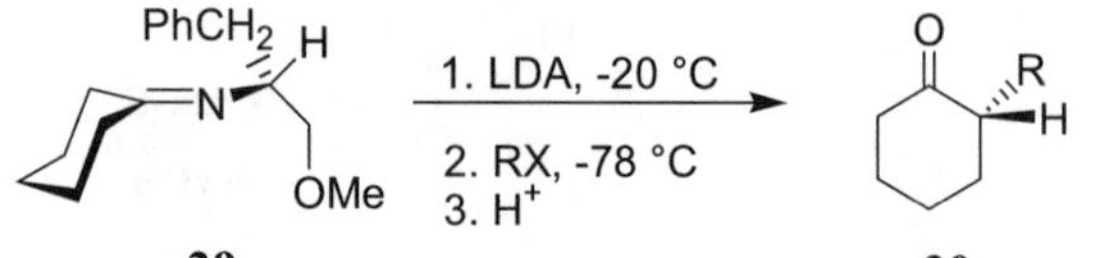

29 **30**

R-X	ee (%)
Me_2SO_4	82 (*R*)
n-C_3H_7I	> 95 (*R*)
$CH_2{=}CHCH_2Br$	> 90 (*S*)

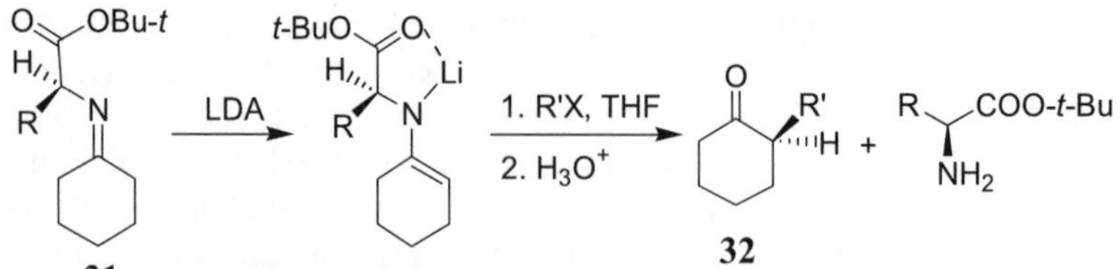

31 **32**

R	R'-X	ee (%)
i-C_3H_7	Me_2SO_4	84 (*S*)
t-C_4H_9	Me_2SO_4	98 (*S*)
t-C_4H_9	MeI	97 (*S*)
t-C_4H_9	*n*-C_3H_7I	97 (*S*)

Scheme 2–21

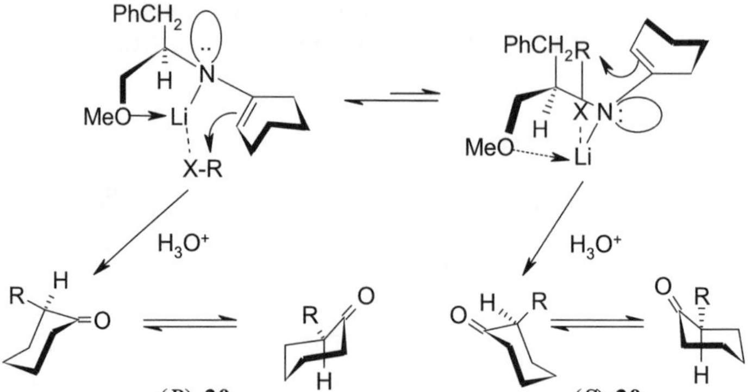

(*R*)-**30** (*S*)-**30**

Figure 2–1. Transition state for α-alkylation of enamines.

2.2.3.5 Chiral Hydrazone Systems. In 1976, Corey and Enders[34] demonstrated the great synthetic potential of metalated dimethylhydrazones as highly reactive intermediates in regio- and diastereoselective C–C bond formation reactions. The procedure for carrying out the electrophilic substitution reaction

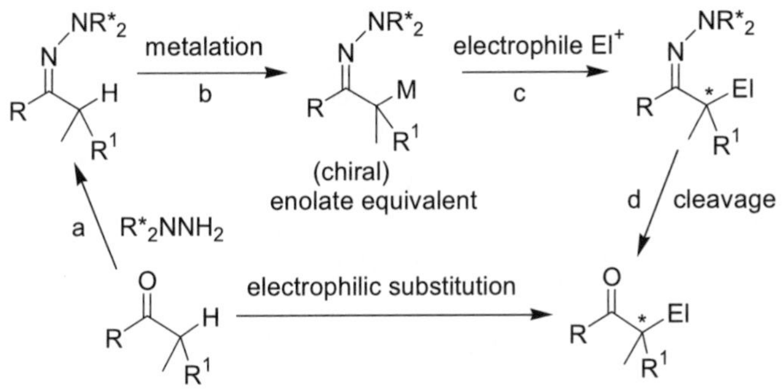

Figure 2–2. Electrophilic substitution to the carbonyl group of aldehydes and ketones via metalated (chiral) hydrazones.

at the α-carbon of the carbonyl group is shown in Figure 2–2. The carbonyl compounds are metalated to give enolate equivalents, which can be trapped with electrophiles. As indicated in Figure 2–2, if one uses an achiral ketone and chiral hydrazine in step a, this will be a chiral version of the hydrazone method, and chiral substituted ketone will be the final product.

Alkylation of chiral hydrazones has several advantages. The starting hydrazones can be conveniently prepared, even for sterically hindered ketones. The product thus formed is highly stable, and its metalated derivative has very high reactivity. The subsequent electrophilic substitution reaction will give very good yield, and a variety of procedures are available to remove the hydrazine moiety and to release the final alkylated product.[35] For example, the hydrazine moiety can be removed through a very mild oxidation of the alkylation product under neutral conditions (pH = 7). The reaction can be carried out with the corresponding cuprates, which are readily available as well.

Enders developed the "hydrazone methods" by choosing SAMP and its enantiomer RAMP. The application and scope of SAMP/RAMP are summarized in Figure 2–3. SAMP and RAMP can be prepared on a large scale from (S)-proline[36] and (R)-glutamic acid,[37] respectively.

A good example of applying the hydrazone method is the preparation of the optically active pheromone **34** (Scheme 2–22).[38] Further study of the crude product prepared from SAMP-hydrazone and 3-pentanone **33** shows that, among the four possible stereoisomers, (Z,S,S)-isomer **35** predominates along with the minor (E,S,S)-isomer, the geometric isomer of **35**. The final product **34** was obtained with over 97% enantiomeric excess (ee).

It has been reported that the cleavage of SAMP hydrazones can proceed smoothly with a saturated aqueous oxalic acid, and this allows the efficient recovery of the expensive and acid-sensitive chiral auxiliaries SAMP and RAMP. No racemization of the chiral ketones occurs during the weak acid oxalic acid treatment, so this method is essential for compounds sensitive to oxidative cleavage.[39a]

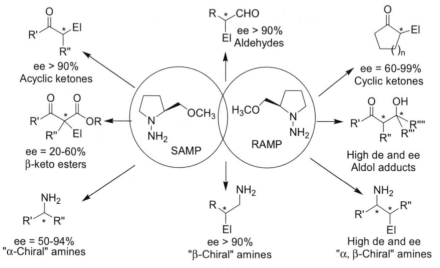

Figure 2–3. Synthetic scope of the SAMP/RAMP-hydrazone method.

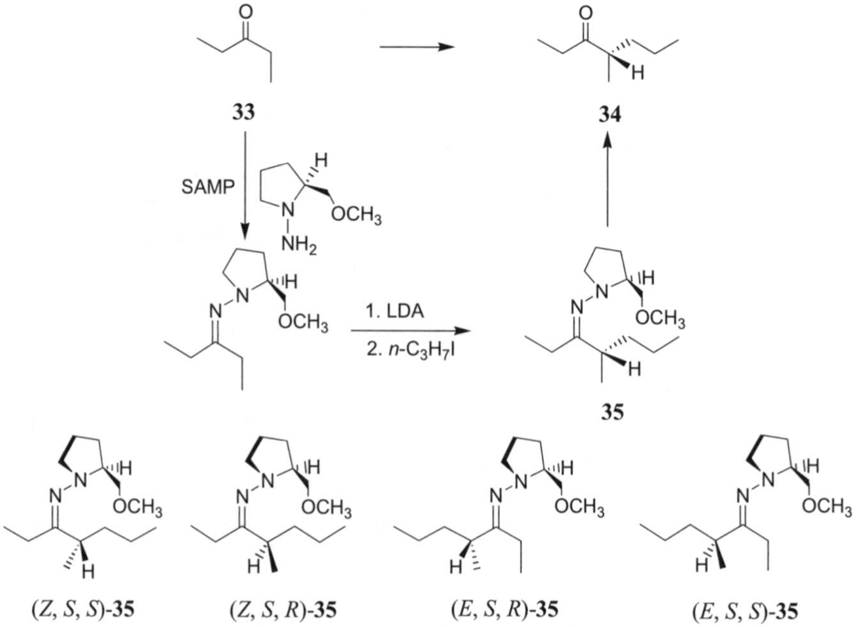

(Z, S, S)-**35**   (Z, S, R)-**35**   (E, S, R)-**35**   (E, S, S)-**35**

Scheme 2–22

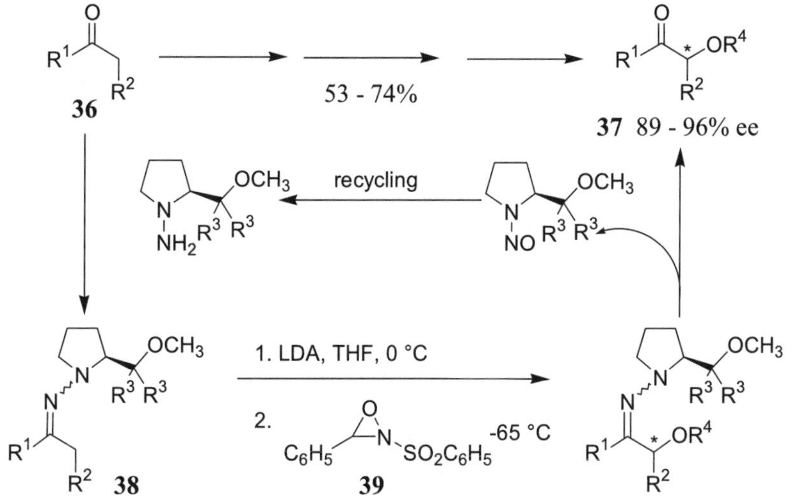

Scheme 2–23

Another application of the hydrazone method is the preparation of α-hydroxy carbonyl compounds (R^4 = H in **37**). The aldehydes/ketones **36** are first transformed into their corresponding SAMP-hydrazones **38**, followed by deprotonation with *t*-butyllithium or LDA in THF. The resulting anion undergoes facile oxidation by treatment with 2-phenylsulfonyl-3-phenyloxaziridine (**39**), and the product can be obtained with good to excellent enantioselectivity (Scheme 2–23).[39b]

Several reviews and research papers discussing the application and extension of this method have appeared.[40] For example, Weber et al.[41] reported an interesting result in which cerium acted as a counterion in the modified proline auxiliary (SAMEMP **40**) for selective addition of organocerium reagents to hydrazones. The initial adduct was trapped with either methyl or benzyl chloroformate to afford the stable *N*-aminocarbonate **41** (Scheme 2–24). From this example readers can see that this proline chiral auxiliary can be used not only for α-alkylation but also for nucleophilic addition, which is discussed in detail later.

In the study of Weber et al.,[41] a series of proline-derived hydrazones were prepared, and the reactions of the hydrazones with organocerium reagents were examined. It is clear from the table in Scheme 2–24 that the diastereoselectivity of the examined reactions depends on the nature of the side chain. (*S*)-1-amino-2-(2-methoxyethoxymethyl) pyrrolidine (**40**) gave the highest selectivity for various nucleophiles.

2.2.3.6 Oxazoline Systems.

The 2-oxazoline system has long been known[42] and can readily be prepared[43] from 2-aminoethanol derivatives and carboxylic acids. This compound has served as a potential precursor for elaborated car-

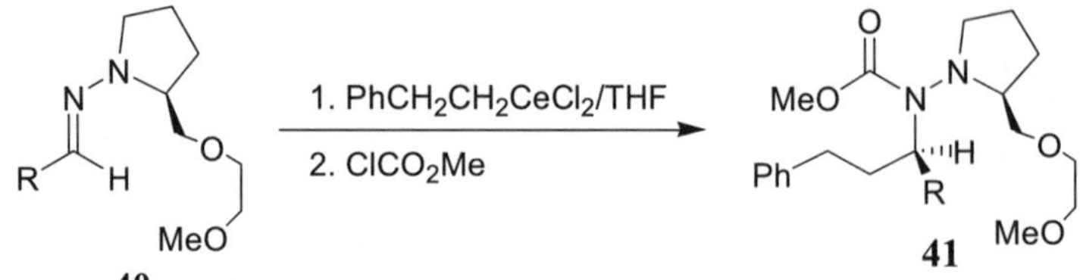

40

(S)-1-amino-2-(2-methoxyethoxy
methyl)pyrrolidine (**SAMEMP**)

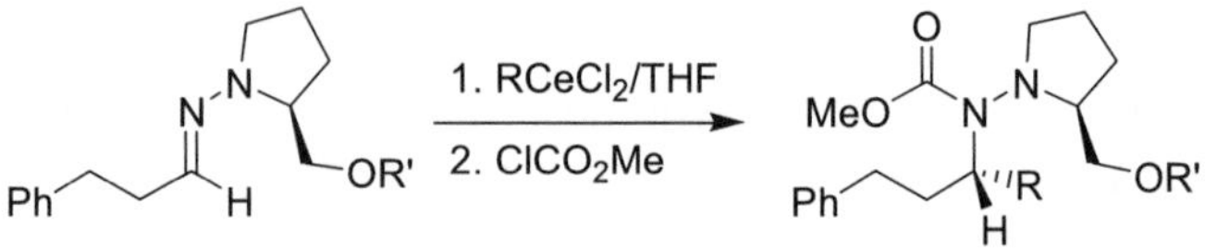

R	ds (R' = CH₂OCH₃)	ds (R' = CH₃)
Bu	97 : 3	93 : 7
Me	97 : 3	98 : 2
t-Bu	>99 : 1	96 : 4
Ph	95 : 5	96 : 4

Scheme 2–24. SAMEMP mediated reactions.

boxylic acids because of its ease of formation,[44] the availability of the starting material, and the stability of the compounds in the face of a wide range of temperatures and reagents. The 2-position of this oxazoline compound can easily be metalated with butyl lithium (THF, −78°C), and the resulting lithio derivative can be alkylated with various electrophiles (alkyl halides, carbonyl compounds, and epoxides). Oxazoline-mediated asymmetric synthesis was first reported in 1974.[45] Since then, great effort has been expended in the development and application of these versatile substrates.

Metalation of **42** using n-butyllithium or LDA provides an azaenolate that exists as a mixture of (Z)- and (E)-**43**. Alkylation followed by hydrolysis yields the optically active α,α-disubstituted carboxylic acid (S)-**44** in 72–80% ee (Scheme 2–25 and Table 2–7).

Lithiated chiral oxazolines have been shown to react with various electrophiles, generating a new asymmetric center with considerable bias. This process has led to the synthesis of optically active α-alkylalkanoic acids,[47] α-hydroxy(methoxy)alkanoic acids,[48] β-hydroxy(methoxy)alkanoic acids,[49] α-substituted γ-butyrolactones,[50] and 2-substituted-1,4-butanediols (Fig. 2–4).[50]

The oxazoline methodology can be applied in the total synthesis of natural products. For example, in the course of the total synthesis of European pine-saw fly pheromone **47**, the key intermediate, chiral α-methyl carboxylic acid **46**, was prepared via the reaction of α-lithioethyloxazoline with n-octyl iodide. The product 2-methyl decanoic acid **46** was obtained, after hydrolysis, in 72% ee (Scheme 2–26).[51]

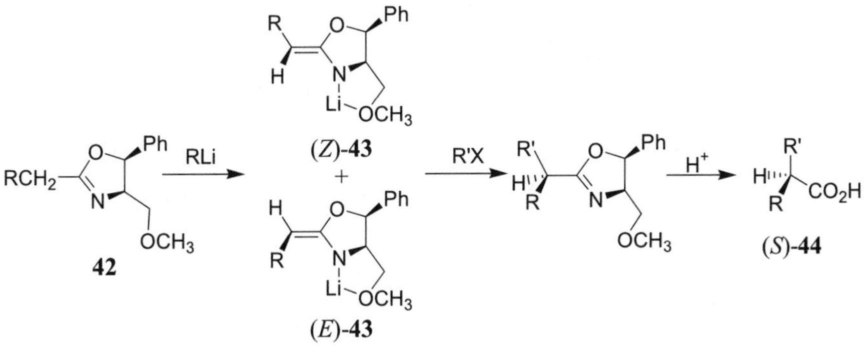

Scheme 2–25. Alkylation of chiral oxazolines to carbonyl acids **44**.

TABLE 2–7. Alkylation of Chiral Oxazolines to Carbonyl Acids 44[46]

Entry	R	R′X	ee (%)	Config.	Overall Yield (%)
1	Me	EtI	78	S	84
2	Et	Me$_2$SO$_4$	79	R	83
3	Me	n-PrI	72	S	79
4	n-Pr	Me$_2$SO$_4$	72	R	74
5	Me	PhCH$_2$Cl	74	S	62
6	PhCH$_2$	Me$_2$SO$_4$	78	R	75

ee = Enantiomeric excess; R = R in **42**; R′X = R′X in Scheme 2–25.
Reprinted with permission by Am. Chem. Soc., Ref. 46.

In addition to the reactions discussed above, α,β-unsaturated oxazolines can also act as chiral electrophiles to undergo conjugated addition of organolithium reagent to give optically active β,β-disubstituted carboxylic acids.[52] The vinyl oxazolines **48** are prepared using the two methods outlined in Scheme 2–27.

After treating **48** with various organolithium reagents and the subsequent hydrolysis of the thus formed products, a variety of β,β-disubstituted propionic acids **49** can be obtained in good yield with high enantioselectivity (Table 2–8).

2.2.3.7 Acylsultam Systems.
Oppolzer et al.[53] developed a general route to enantiomerically pure crystalline α,α-disubstituted carboxylic acid derivatives by asymmetric alkylation of N-acylsultams. Acylsultam **50** can be readily prepared from the inexpensive chiral auxiliary sultam **53**.[54]

Successive treatments of chiral acylsultam **50** with n-BuLi or NaHMDS and primary alkyl halides, followed by crystallization, give the pure α,α-alkylation product **52** (Scheme 2–28). Under these conditions, the formation of C-10–alkylated by-product is inevitable. It is worth mentioning, however, that product **52** can readily be separated from the C(α)-epimers by crystallization. In fact,

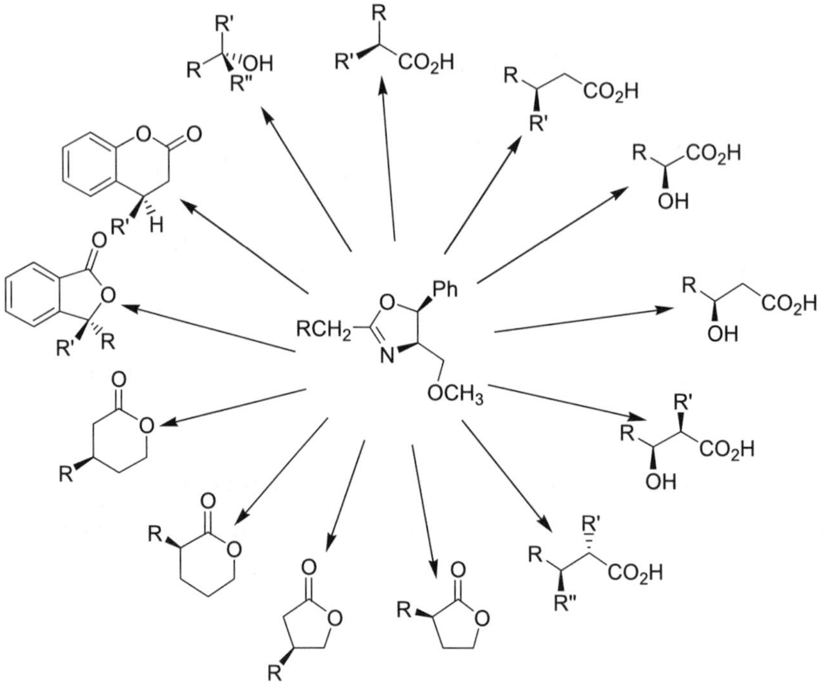

Figure 2–4. Oxazoline methodology as a tool in organic synthesis.

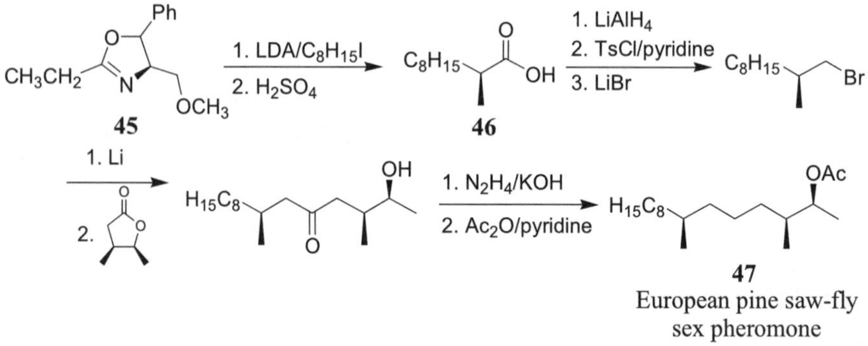

Scheme 2–26. Synthesis of European pine-saw fly pheromone.

via appropriate cleavage, enantiomerically pure alcohol **55** or carboxylic acid **54** can be obtained and the sultam can be recovered. The observed topicity is consistent with the kinetically controlled formation of chelated (*Z*)-enolate **51** (Scheme 2–28). In the process of alkylation of **50** to **52**, the alkylating reagent attacks from the *Re*-face, which is opposite to the lone pair electrons on the

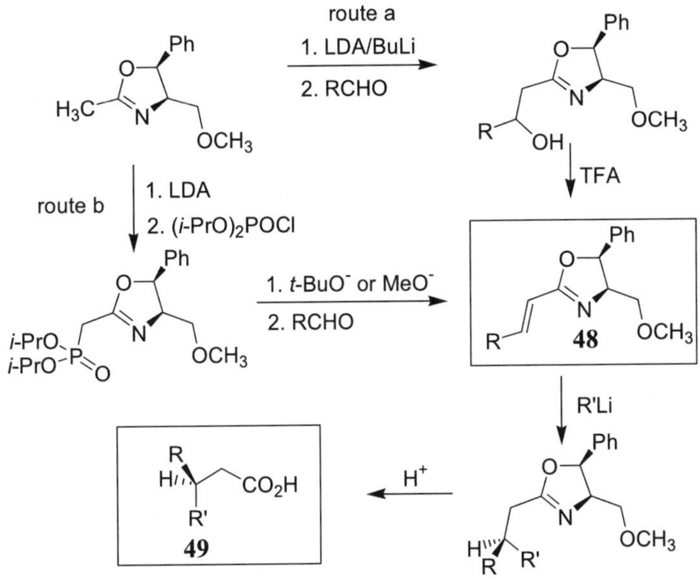

Scheme 2–27. Preparation of β,β'-substituted carboxylic acids.

TABLE 2–8. Preparation of β, β-Disubstituted Carboxylic Acids 49[52]

Entry	R	R'	ee (%)	Config.	Yield (%)
1	Me	Et	92	R	30
2	Me	Ph	98	S	34
3	i-Pr	n-Bu	99	R	53
4	t-Bu	n-Bu	98	R	50
5	c-hexyl	Et	99	R	73
6	MeOCH$_2$CH$_2$	n-Pr	99	S	50
7	o-MeOPh	n-Bu	95	R	75

ee = Enantiomeric excess; R = R in Scheme 2–27; R' = R' in Scheme 2–27.

nitrogen atom. Table 2–9 illustrates the alkylation results using various halides as electrophiles.

In addition to the asymmetric induction mentioned above, sultam **53** can also be used to prepare enantiomerically pure amino acids (Scheme 2–29 and Table 2–10).[55] Me$_3$Al-mediated acylation of **53** with methyl N-[bis(methylthio)-methylene]glycinate **56** provided, after crystallization, glycinate **57**, which can serve as a common precursor for various α-amino acids. In agreement with a kinetically controlled formation of chelated (Z)-enolates, alkylation happened from the Si-face of the α-C, opposite to the lone pair electrons on the sultam nitrogen atom. High overall yield for both the free amino acid **58** and the

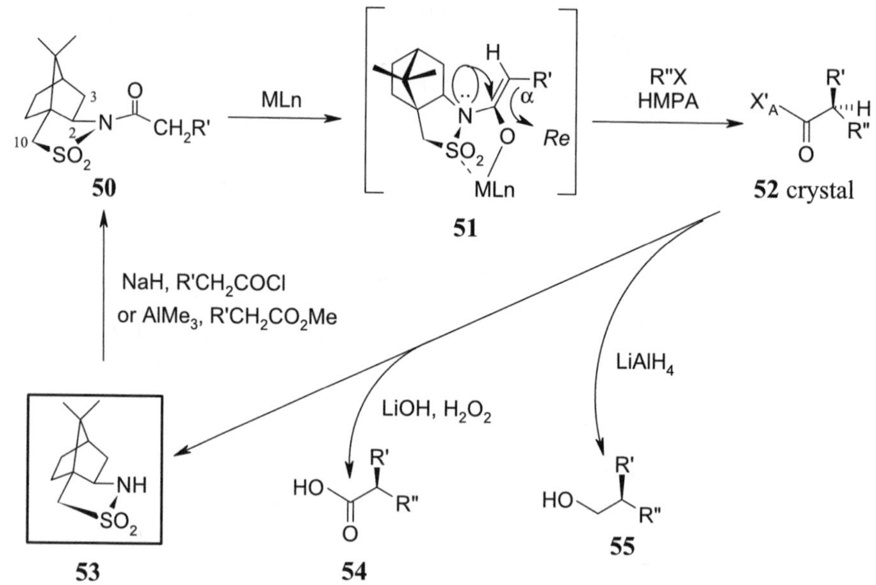

Scheme 2–28. Reprinted with permission by Pergamon-Elsevier, Ref. 53.

TABLE 2–9. Asymmetric Alkylation of 50

R′	R″X	ML	de (%, crude)	de (%, crystal)
Me	PhCH$_2$I	NHDMS	96.5	98.4
Me	PhCH$_2$I	KHDMS	92.9	
Me	PhCH$_2$I	BuLi	96.9	98.5
Me	CH$_2$=CHCH$_2$I	NHDMS	94.2	94.5
Me	CH$_2$=CHCH$_2$I	BuLi	96.6	96.6
Me	CH$_2$=CHCH$_2$Br	BuLi	98.8	>99
Me	HC≡CCH$_2$Br	BuLi	98.3	>99

de = Diastereomeric excess; ML = ML$_n$ in Scheme 2–28; R′ = R′ in **50**; R″X = R″X in Scheme 2–28.

readily separable sultam **53** can be obtained via mild acidic N-deprotection of **59** and subsequent gentle saponification. Analogous alkylation of the glycinate equivalent affords a variety of α-amino acids (Table 2–10).

Sultam **53** has proved to be an excellent chiral auxiliary in various asymmetric C–C bond formation reactions. One more example of using sultam **53** is the asymmetric induction of copper(I) chloride-catalyzed 1,4-addition of alkyl magnesium chlorides to α,β-disubstituted (E)-enesultams **60**. Subsequent protonation of the reaction product gives compound **61c** as the major product (Scheme 2–30 and Table 2–11).[56]

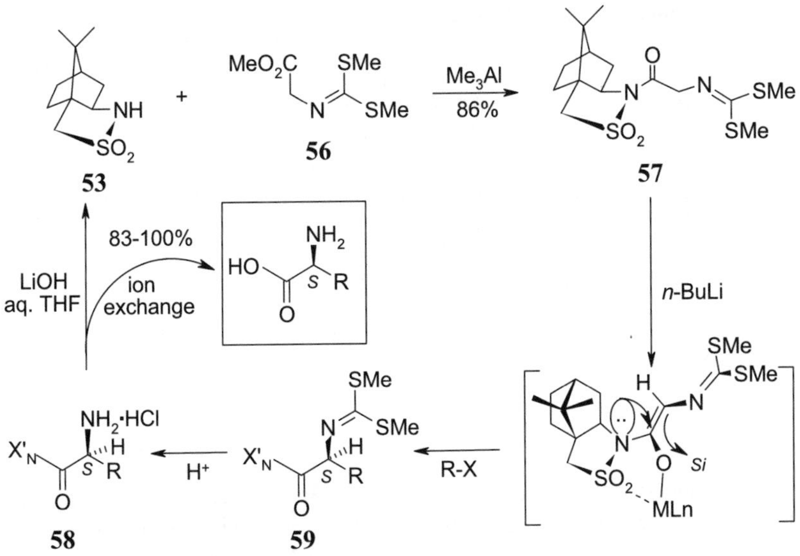

Scheme 2-29

TABLE 2-10. Alkylation of Glycinate Equivalents

RX	ee (%)
MeI	>99.8
PhCH$_2$I	>99.8
CH$_2$=CHCH$_2$I	>99.8
t-BuOOCCH$_2$Br	>99.8
(CH$_3$)$_2$CHCH$_2$I	>99.8
(CH$_3$)$_2$CHI	99.5

ee = Enantiomeric excess; RX = R-X in Scheme 2-29.

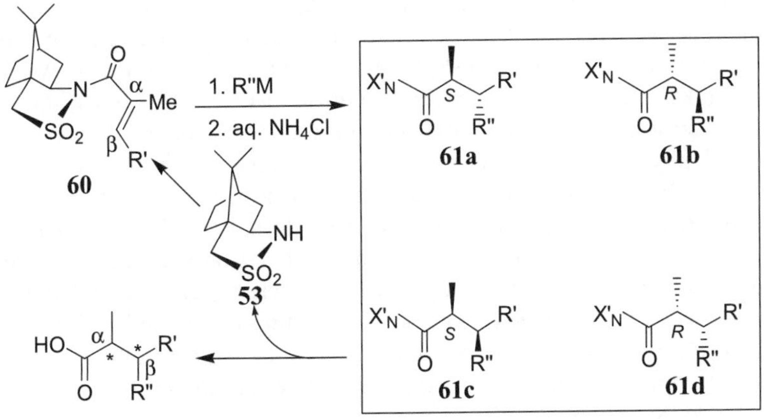

Scheme 2-30

TABLE 2–11. Sultam 53 in the Preparation of 60c

R′	R″	Cu(I) salt	Ratio (a:b:c:d)	Crystal purity (%)	Config.
Me	Bu	CuCl	2.1:0:86.3:11.6	97.7	2S,3S
Me	Bu	CuCN	2.5:0:83.7:14.4	—	2S,3S
Me	Et	CuCl	2.3:0:85.4:12.3	98.6	2S,3S
Et	Bu	CuCl	0:0:91.5:8.5	99.8	2S,3S
Bu	Et	CuCl	0:0:97.3:2.7	99	2S,3R
TBSOCH₂	Bu	CuCl	0:0:97.0:3.0	99.4	2S,3S
Bu	Me	CuCl	10.5:8.2:68.6:12.7	—	2S,3R
Me	Ph	CuCl	2.7:3.2:72.5:21.6	—	2S,3R

R′ = R′ in Scheme 2–30; R″ = R″ in Scheme 2–30.

For a review of sultam chemistry, interested readers can refer to Oppolzer's article[57] on "Camphor as a Natural Source of Chirality in Asymmetric Synthesis."

2.3 PREPARATION OF QUATERNARY CARBON CENTERS

The previous section discussed chelation enforced intra-annular chirality transfer in the asymmetric synthesis of substituted carbonyl compounds. These compounds can be used as building blocks in the asymmetric synthesis of important chiral ligands or biologically active natural compounds. Asymmetric synthesis of chiral quaternary carbon centers has been of significant interest because several types of natural products with bioactivity possess a quaternary stereocenter, so the synthesis of such compounds raises the challenge of enantiomer construction. This applies especially to the asymmetric synthesis of amino group–substituted carboxylic acids with quaternary chiral centers.

A new method for the stereoselective introduction of a quaternary asymmetric carbon atom was developed by Meyers, based on the interactive lithiation and alkylation of chiral bicyclic lactam 62–64 derived from γ-keto acids and (S)-valinol. Although the initial step proceeds with poor diastereocontrol, the second alkylation can proceed with excellent *endo*-selectivity. Chiral bicyclic lactams have now proved to be useful compounds for synthesizing a variety of chiral, nonracemic compounds containing quaternary carbons at the stereocenter. The substrates 62–64 undergo double alkylation with lithium base and two alkyl halides to yield the products with quaternary carbon centers in high diastereoselectivity. Acid treatment of the resulting compound yields enantiomerically pure γ-keto acid 66,[58] while reduction of the resulting compound followed by base-catalyzed aldol condensation yields the cyclic pentenone 68 with high ee (Scheme 2–31).[59,60]

Meyers et al. also studied the stereoelectronic and steric effects of the π-facial

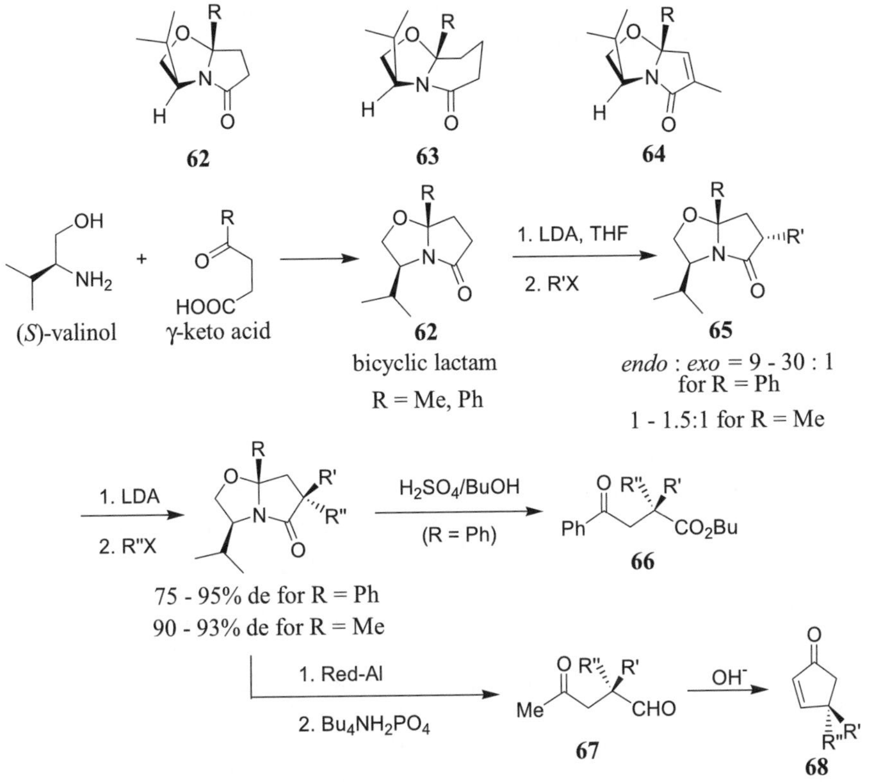

Scheme 2–31

addition of electrophiles to lactam enolates in order to explain the observed stereoselectivity. In previous studies, Romo and Meyers[61] found that angularly placed *exo*-substituents imparted steric bias for *endo*-alkylation. Systematic replacement of the *exo*-alkyl and aryl substituents on bicyclic lactam **69** with hydrogen results in a drop in *endo*-alkylation selectivity from 98:2 to 69:31 (Scheme 2–32 and Table 2–12). However, the *endo*-alkylation is still preferred, even when both the A and B substituents are hydrogen. Thus, it is presumed

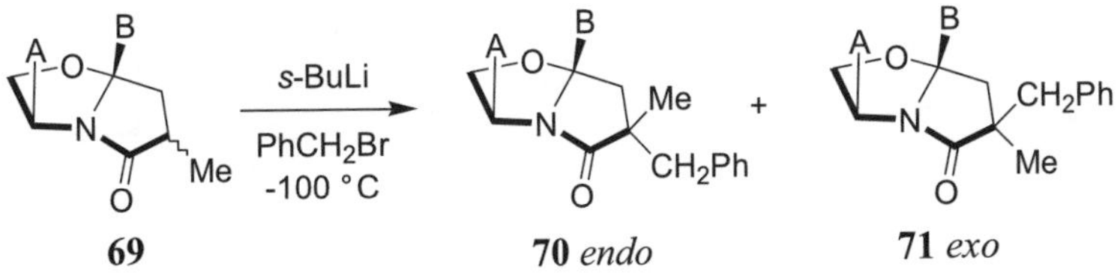

Scheme 2–32

**TABLE 2–12. Effect of Substituents A and B on the
Diastereoselective Alkylation of 69.**

Entry	Substituents		71 (*endo*)	72 (*exo*)
	A	B		
1	i-Pr	Me	97	3
2	t-Bu	Me	98	2
3	i-Pr	Ph	98	2
4	i-Pr	H	80	20
5	H	Me	70	30
6	H	H	69	31

that the steric effects of substituents A and B may be the only factor in the determination of diastereofacial alkylation selectivity.

Application of Meyers' method can be extended to the synthesis of some other functionalized compounds. The wide varieties of natural products[62] that contain the cyclopropane ring in a chiral environment provide further impetus for having broadly applicable synthetic routes for introducing a cyclopropane ring. A novel asymmetric synthesis of substituted cyclopropane[63] uses this bicyclic lactam chemistry. In Scheme 2–33, the starting bicyclic lactam is first transformed to the α,β-unsaturated bicyclic lactam **72** through metalation, selenation, and oxidative elimination (LDA, PhSeBr, and H_2O_2). Compound **72** (R = Ph) can also be prepared by treating α-substituted 4-oxo-2-phenyl-2-pentenoate with (S)-valinol in toluene with the removal of water. Dimethyl sulfonium methylide reacts with this chiral unsaturated lactam, yielding the cyclopropanated compound **73** in more than 93% de (Scheme 2–33).[64]

Fuji et al.[65] reported on the asymmetric induction via an addition–elimination process of nitro-olefination of α-substituted lactone to the formation of chiral quaternary carbon centers. This is an interesting method for asymmetric synthesis of quaternary carbon centers involving the addition and elimination of a chiral leaving group. The main advantage of asymmetric induction by a chiral leaving group is that it provides the direct formation of chiral products, without the need for a later step removing the chiral auxiliary. Nitroenamines[66] have been known to react with a variety of nucleophiles giving addition–elimination products.[67] Both the chemical yield and the ee

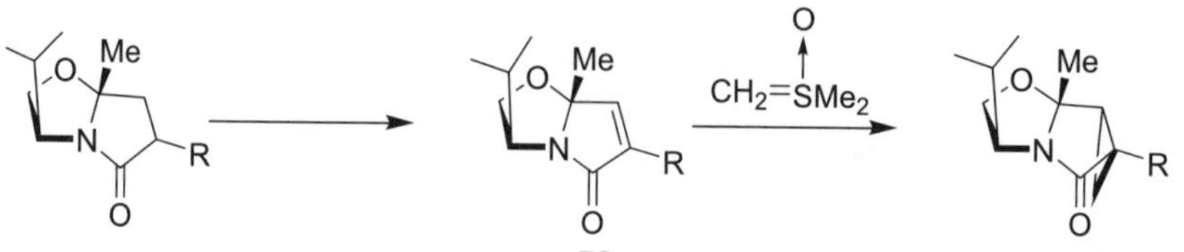

Scheme 2–33

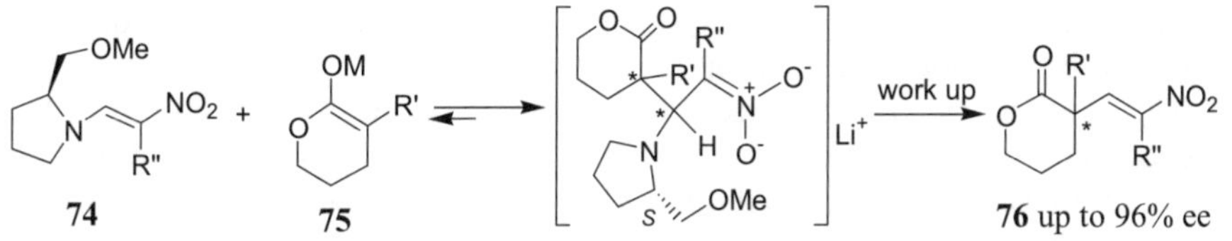

Scheme 2–34

increase when Zn^{2+} is used as a countercation. Moreover, the resulting α,β-unsaturated nitro function in **76** is a versatile moiety for further transformation.

The study of Fuji et al. shows that the addition of lithium enolate **75** to nitroamine **74** is readily reversible; quenching conditions are thus essential for getting a good yield of product **76**. An equilibrium mixture of the adducts exists in the reaction mixture, and the elimination of either the prolinol or lactone moiety can take place depending on the workup condition (Scheme 2–34). A feature of this asymmetric synthesis is the direct one pot formation of the enantiomer with a high ee value. One application of this reaction is the asymmetric synthesis of a key intermediate for indole type *Aspidosperma* and *Hunteria* alkaloids.[68] Fuji[69] has reviewed the asymmetric creation of quaternary carbon atoms.

Seebach et al.[70] introduced another interesting idea for creating chiral quaternary carbon centers, namely, the self-regeneration of stereocenters (SRS). To replace a substituent at a single stereogenic center of a chiral molecule without racemization, a temporary center of chirality is first generated diastereoselectively, such as *t*-BuCH in **78**. The original tetragonal center is then trigonalized by removal of a substituent, such as forming enolate **79** (Scheme 2–35). A new ligand is then introduced diastereoselectively, such as the introduction of group R′ in **80**. Finally, the temporary chiral center is removed to provide product **81** with a chiral quaternary carbon center. By means of these four steps, 2- and 3-amino, hydroxy, and sulfonyl carboxylic acids have been successfully alkylated with the formation of tertiary alkylated carbon centers without using a chiral

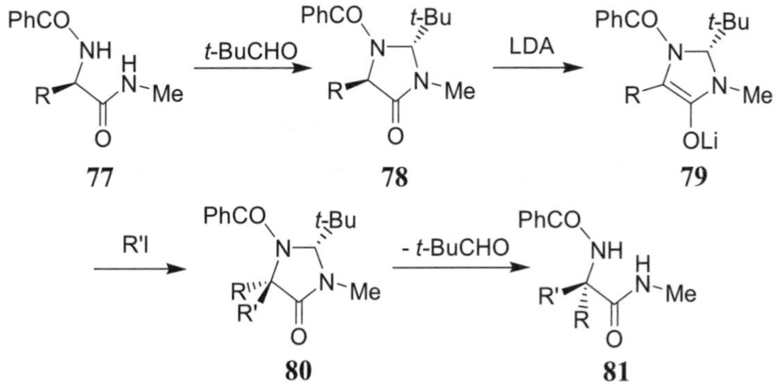

Scheme 2–35

auxiliary. This method allows the potential of these inexpensive chiral building blocks to be extended considerably.

This method can be regarded as an example of memory of chirality,[71] a phenomenon in which the chirality of the starting material is preserved in a reactive intermediate for a limited time. The example in Scheme 2–35 can also be explained by the temporary transfer of chirality from the α-carbon to the *t*-BuCH moiety so that the newly formed chiral center *t*-BuCH* acts as a memory of the previous chiral center. The original chirality can then be restored upon completion of the reaction.

Very recently, Matsushita et al.[72] reported an efficient route for the synthesis of α,α-disubstituted α-amino acid derivatives **82–84** starting from some readily available expoxy silyl ethers such as **86**. The key step involves an MABR (**85**)–catalyzed rearrangement[73] for converting **86** to **87** and a Curtis rearrangement for introducing the isocyanate and the subsequent build up of an amino group (Scheme 2–36).[74] This method complements the currently applied methods for α,α-disubstituted amino acid synthesis that are based on the stereoselective alkylation of cyclic compounds (e.g., Schöllkopf's bislactim method,[75] Seebach and Aebi's oxazolidine method,[76] and William's oxazinone method[77]). Interested readers may consult the recent review by E. J. Corey[78] on the catalytic enantioselective construction of carbon stereocenters.

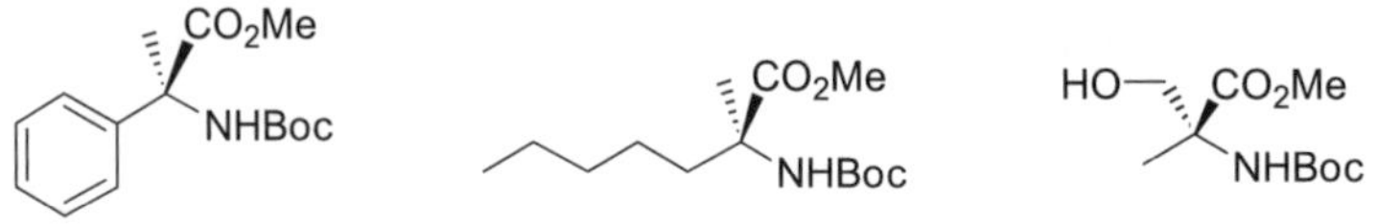

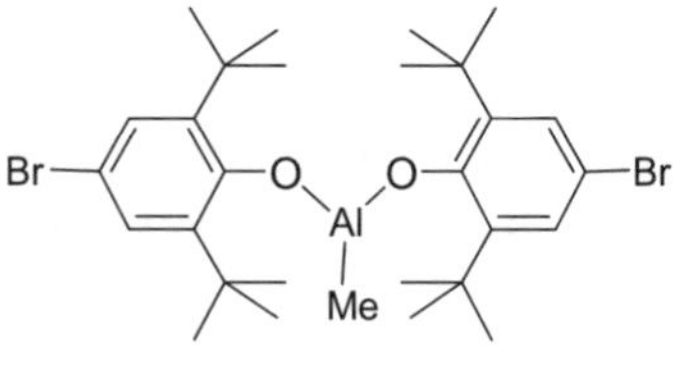

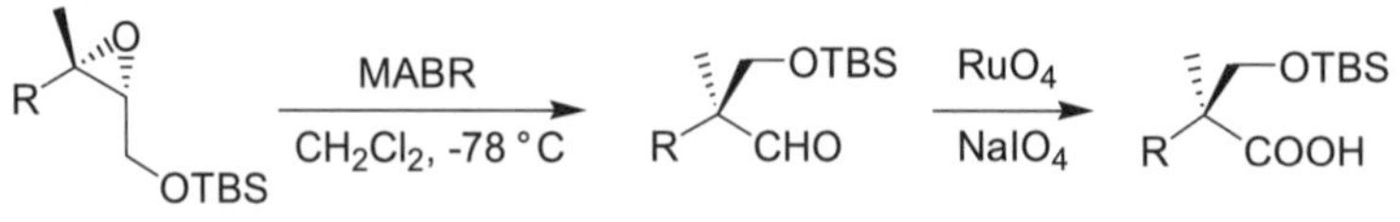

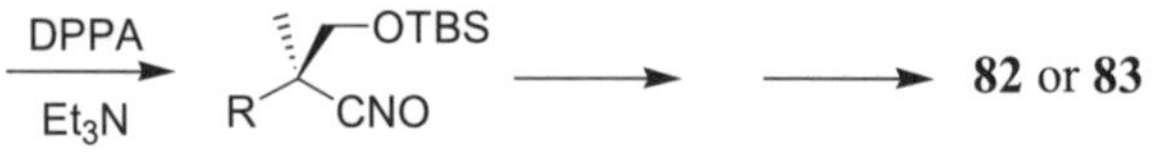

Scheme 2–36

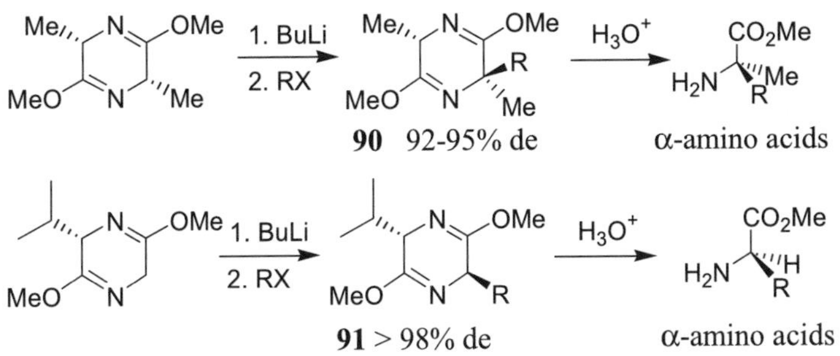

Scheme 2–37

There are several recent publications regarding the syntheses of α,α-disubstituted amino acids.[79]

2.4   PREPARATION OF α-AMINO ACIDS

Owing to their possible biological activity, enantiomerically pure nonprotein α-amino acids have become increasingly important. α-Alkylation of a chiral glycine derivative is among the most attractive methods for the asymmetric synthesis of these nonprotein α-amino acids. Good results have been obtained using the bislactim system,[80] which is conceptually very similar to the SRS proposed by Seebach. Six-membered heterocyclic products (e.g., **90** and **91**) are obtained from glycine and other amino acids via diketopiperazine, followed by O-methylation with Meerwein salt (Scheme 2–37). Finally, α-methyl amino acids with high enantiomeric excess can be obtained through acidic hydrolysis of **90** and **91**.

Table 2–13 summarizes some useful chiral auxiliaries for α-alkylation of a carbonyl compound.

2.5   NUCLEOPHILIC SUBSTITUTION OF CHIRAL ACETAL

Acetals/ketals are among the most widely used protecting groups for aldehydes/ketones and can be used as important tools in the synthesis of enantiomerically pure compounds. Under neutral condition, acetals are inert toward nucleophiles. However, in the presence of a Lewis acid, the acetal functional group becomes a powerful electrophile, which is capable of undergoing reactions with electron-rich double bonds or nucleophiles. The origin of their selectivity is believed to be the preferential complexation of the Lewis acid with the less-hindered oxygen as shown in **93** (Scheme 2–38). The reaction takes place by means of an S_N2 displacement with the inversion at the electrophilic carbon to give **94**. Cleavage of the chiral auxiliary leads to the asymmetric hydroxy

TABLE 2–13. A Summary of Chiral Auxiliaries Reported To Be Useful in the α-Alkylation of Carbonyl Compounds

Chiral Auxiliary	Reference	Chiral Auxiliary	Reference
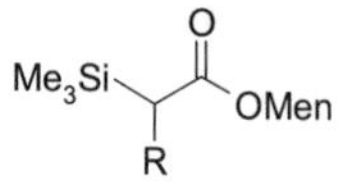	81	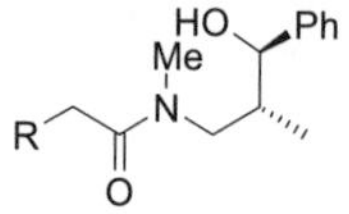	82
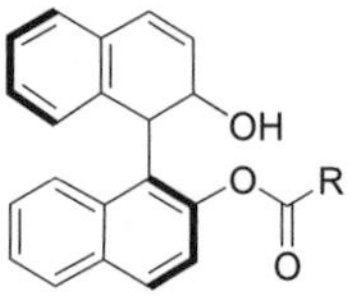	83	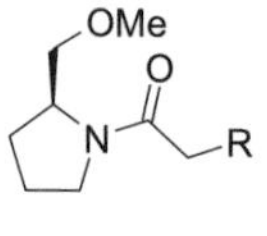	84
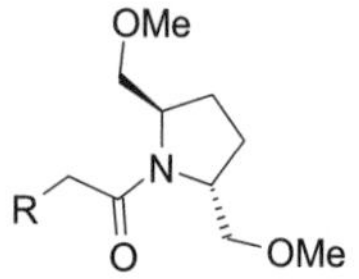	26	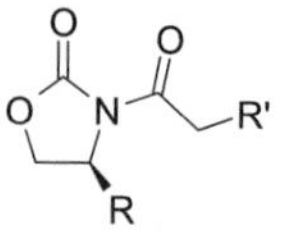 R = Bn, *i*-Pr	85
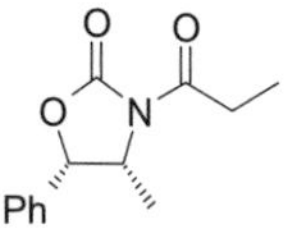	85a	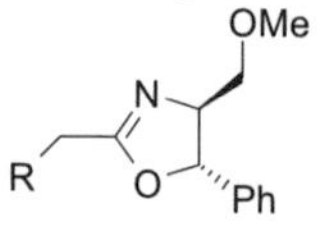	86
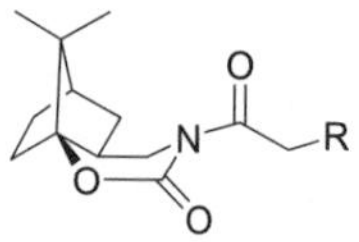	87	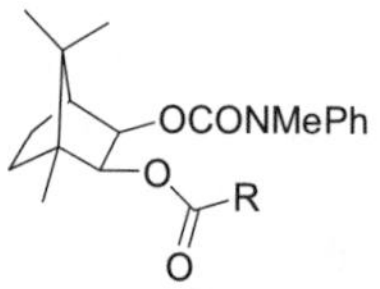	88
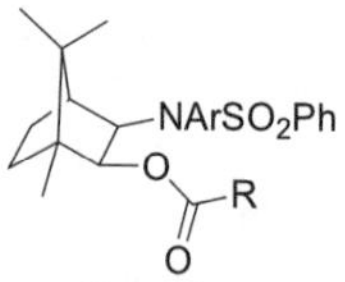	88	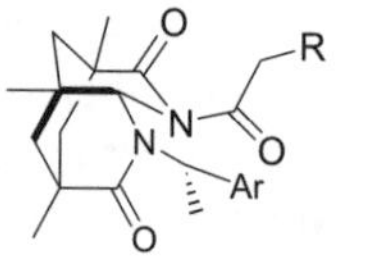	89
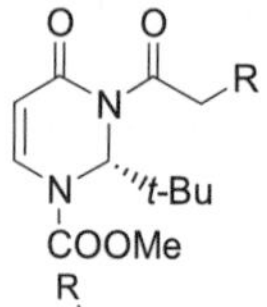	90	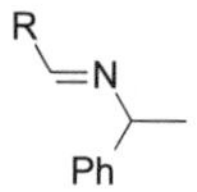	91
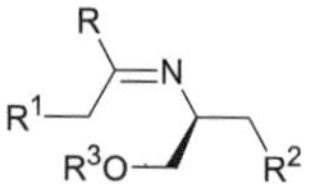	86a, 92	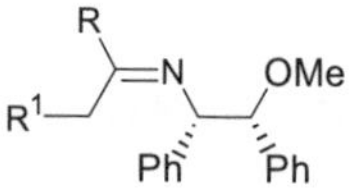	93
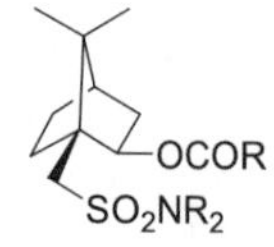	88	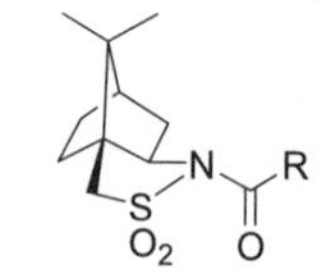	53, 88
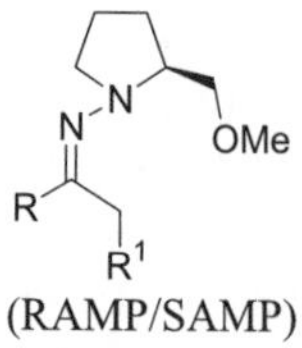 (RAMP/SAMP)	94	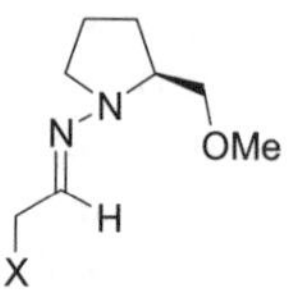	95

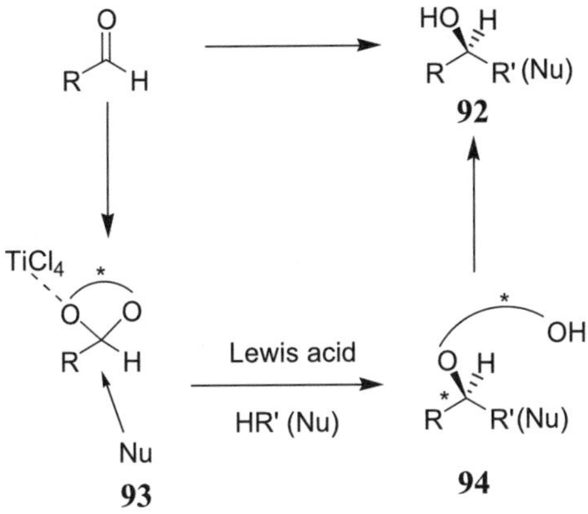

Scheme 2–38

molecule **92**. Thus, the chirality is transferred from the diol to the newly formed carbon center in **92**.[96]

The following auxiliaries and nucleophiles are often employed for this purpose:

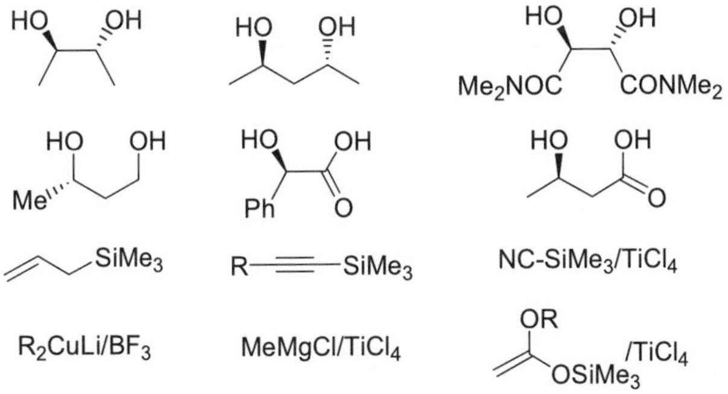

Chiral acetals/ketals derived from either (R,R)- or (S,S)-pentanediol have been shown to offer considerable advantages in the synthesis of secondary alcohols with high enantiomeric purity. The reaction of these acetals with a wide variety of carbon nucleophiles in the presence of a Lewis acid results in a highly diastereoselective cleavage of the acetal C–O bond to give a β-hydroxy ether, and the desired alcohols can then be obtained by subsequent degradation through simple oxidation elimination. Scheme 2–39 is an example in which H⁻ is used as a nucleophile.[97]

TiCl₄-induced cleavage of chiral acetal can be used to prepare β-adrenergic blocking agents **95** bearing the glycerol structure (Scheme 2–40).[98]

β-Hydroxy carboxylic acid can be used as a 1,3-diol analog in a similar reaction. Subsequent Lewis acid–mediated electrophilic attack takes place with excellent diastereoselectivity.[99]

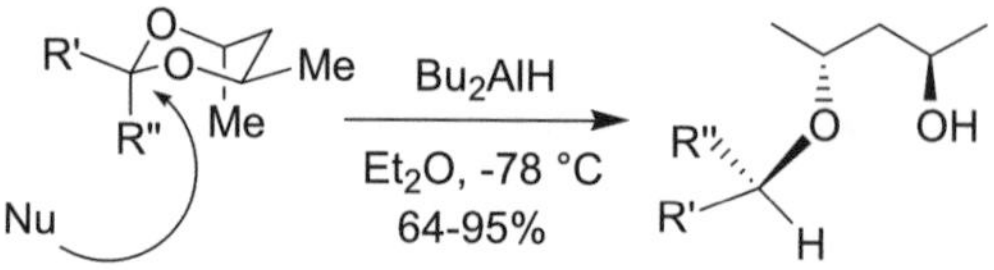

R'	R"	de (%)
c-C_6H_{11}	Me	92
Ph	Me	96
n-C_6H_{13}	Me	77

Scheme 2–39

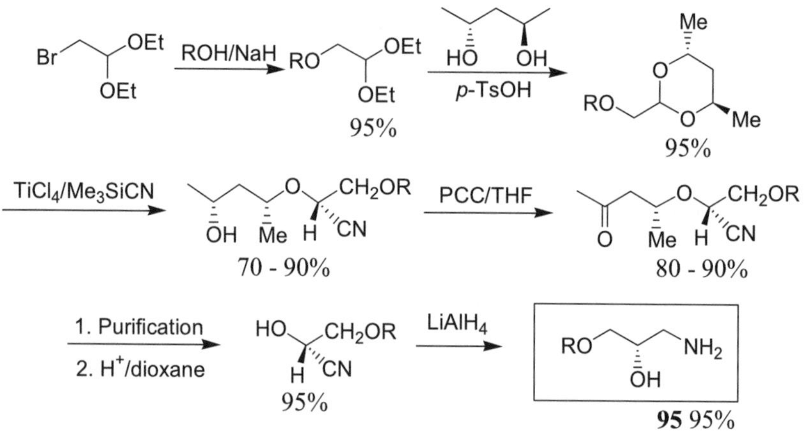

95 95%

Scheme 2–40. A route to amino alcohol.

On the other hand, acetal cleavage in the presence of a chiral Lewis acid could also be a route to chiral alcohols. Recently, Harada et al.[100] reported the kinetic resolution of cyclic acetals derived from 1,3-alkanediols in ring-cleavage reactions mediated by *N*-mesyloxazaborolidine **96** (Scheme 2–41). The enantiotopic C–O groups in racemic acetals *rac*-**97** were differentiated by the ring-cleavage reaction using allylmethylsilane **98** as a nucleophile.

These reactions were carried out using *N*-mesyloxazaborolidine **96** (0.5 eq.) and allylsilane **98** (1.5 eq.) in CH_2Cl_2 at −50°C. Conversion of **97** as high as 63% was observed, and the remaining (2*S*,4*R*)-**97** was recovered in 92% ee. Modification of the electronic nature of the aryl substituent attached to the acetal carbon at the *para* position of **97** did not affect the enantioselectivity of ring cleavage.

Harada et al.[101] extended this oxazaborolidine-mediated ring-cleavage method to biacetals, that is, a desymmetrization of *meso*-1,3-tetrol derivatives (Scheme 2–42). Ring cleavage of **100** was examined, and the mono-cleavage products **101a** and **101b** were obtained in 95% and 82% yields, respectively.

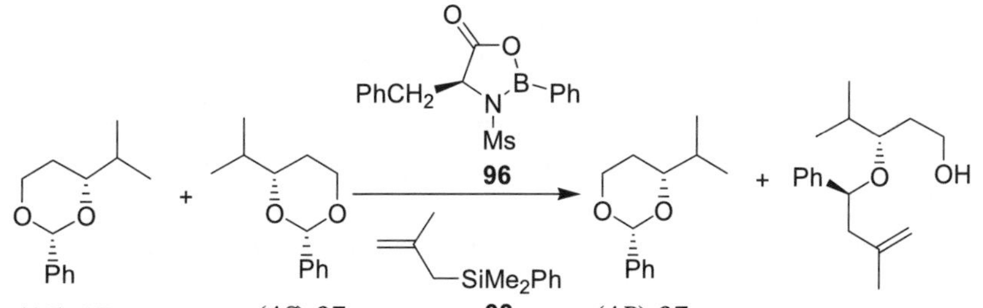

(4*R*)-**97** (4*S*)-**97** **98** (4*R*)-**97** **99**

Scheme 2–41

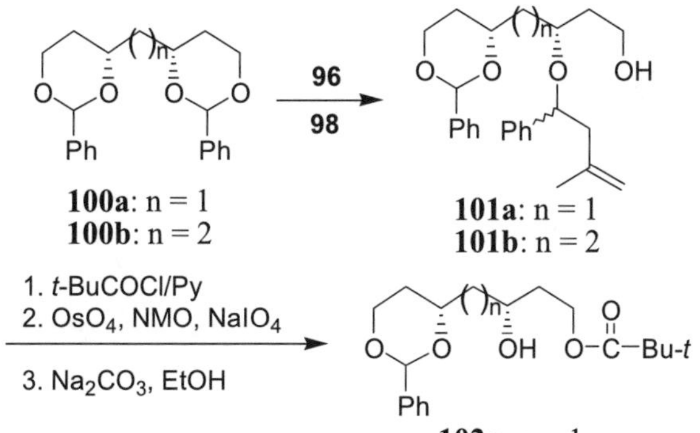

100a: n = 1
100b: n = 2

101a: n = 1
101b: n = 2

102a: n = 1
102b: n = 2

Scheme 2–42

After three steps of transformation, the (*S*)-products **102a** and **102b** were obtained in 88% and 95% ee, respectively.

A review of chiral acetals in asymmetric synthesis is available.[102]

2.6 CHIRAL CATALYST-INDUCED ALDEHYDE ALKYLATION: ASYMMETRIC NUCLEOPHILIC ADDITION

Nucleophilic addition of metal alkyls to carbonyl compounds in the presence of a chiral catalyst has been one of the most extensively explored reactions in asymmetric synthesis. Various chiral amino alcohols as well as diamines with C_2 symmetry have been developed as excellent chiral ligands in the enantioselective catalytic alkylation of aldehydes with organozincs. Although dialkylzinc compounds are inert to ordinary carbonyl substrates, certain additives can be used to enhance their reactivity. Particularly noteworthy is the finding by Oguni and Omi[103] that a small amount of (*S*)-leucinol catalyzes the reaction of diethylzinc to form (*R*)-1-phenyl-1-propanol in 49% ee. This is a case where the ligand accelerates the catalytic reaction. In the following sections of this chapter, stereoselective addition of dialkylzinc to aldehyde, promoted by amino alcohols or titanium derivatives bearing chiral ligands such as ditriflamides **103**,[104] TADDOL **104**,[105] binaphthol **105**,[106] norephedrine **106a**, **106b**,[107] and camphor sulfonamide derivatives **107**,[108] are discussed. It will become evident that development of a metal-complex system that can activate both nucleophiles and electrophiles is an efficient way to reach the high enantioselectivity.

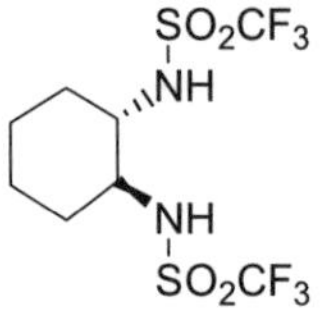

103

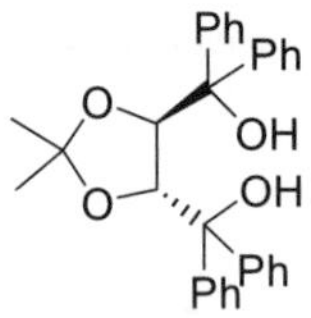

104

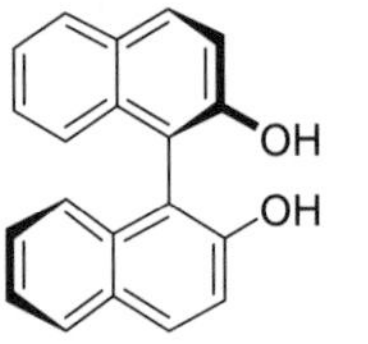

105

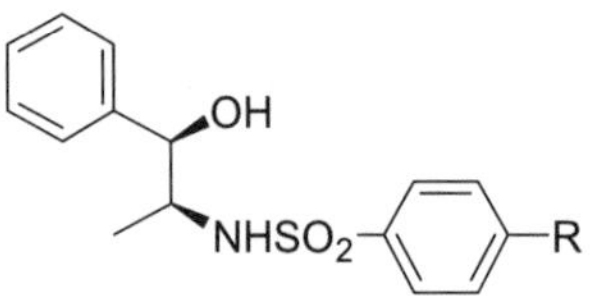

106a: R = CH_3
106b: R = OCH_3

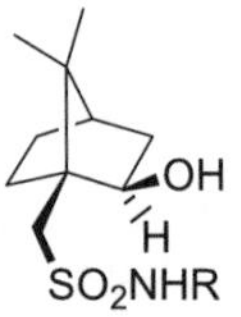

107

Figure 2–5 presents a possible pathway for catalytic asymmetric alkylation using a protonic auxiliary. The metallic compounds **108** are not simple monomers, but usually exist as aggregates. To obtain high enantioselectivity, the ligand X* must possess a suitable three-dimensional structure that is able to differentiate the diastereomeric transition states during the alkyl delivery step **108** → **109**. The key issue is that at first the rate of alkylation by RMX* (**108**) should substantially exceed that of the original achiral nucleophile R_2M; then, chiral ligand X* must be quickly detached from the initially formed metal alkoxide **109** by the action of the alkyl donor or carbonyl substrate to complete the catalytic cycle.

The reaction between dialkylzinc and several chiral amino alcohol ligands satisfies these two key factors. Since the discovery by Oguni that various addi-

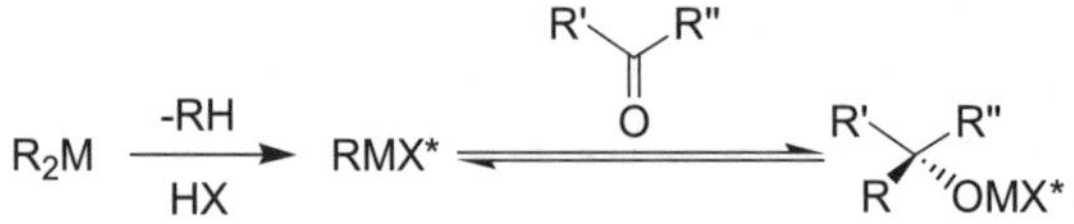

108 **109**

Figure 2–5. Enantioselective alkylation catalyzed by protonic auxiliary HX*. M = Metallic species; X* = chiral heteroatom ligand.

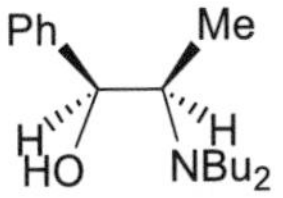

110 (1*S*, 2*R*)-DBNE
N,*N*-di-*n*-butylnorephedrine

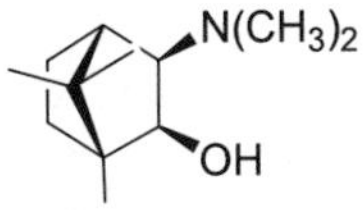

111 (-)-DAIB
3-*exo*-(dimethylamino)isoborneol

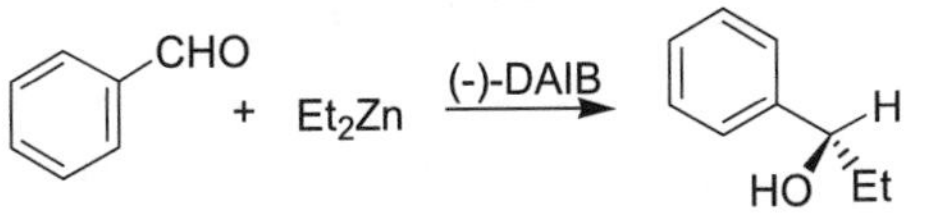

yield 97%, ee 98%

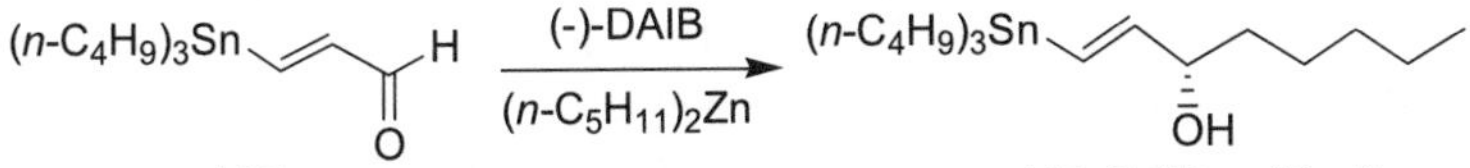

112

113 S / R = 93 : 7

Scheme 2–43

tives catalyze the addition of dialkylzinc reagents to aldehydes, there has been a rapid growth of research in this area. Most of these efforts have been directed toward the design of new chiral ligands, most of them being β-amino alcohols. Perhaps the best examples are DBNE (*N*,*N*-di-*n*-butylnorephedrine) (**110**)[109] and DAIB (**111**).[110]

Treating benzaldehyde with diethylzinc in the presence of 2 mol% (−)-DAIB gives (*S*)-alcohol in 98% ee (Scheme 2–43). When compound **112** is treated in the same manner, compound **113**, a chiral building block in the three-component coupling prostaglandin synthesis, is also obtained with high ee (Scheme 2–43).

The optically active reagent (*S*)-1-methyl-2-(diphenylhydroxymethyl)-azitidine [(*S*)-**114**] has also been reported to catalyze the enantioselective addition of diethylzinc to various aldehydes. The resulting chiral secondary alcohols **115** are obtained in up to 100% ee under mild conditions (Scheme 2–44).[111] Furthermore, most of the **114**-type ligands have also been used in the ox-

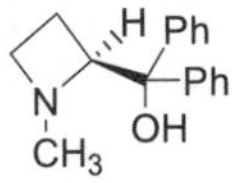

(*S*)-**114**

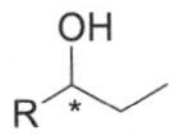

115

R	Ph	4-Cl-Ph	2-MeO-Ph	4-MeO-Ph	4-Me-Ph	(*E*)-PhCH=CH
ee (%)	98%	100%	94%	100%	99%	80%
config.	*S*	*S*	*S*	*S*	*S*	*S*

Scheme 2–44

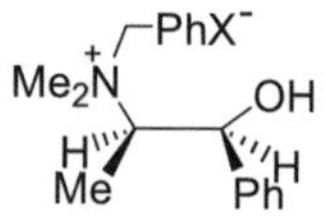

$$\underset{\textbf{116 X = Cl, Br}}{\overset{\overset{\displaystyle \overset{+}{Me_2N} \quad \overset{\frown}{PhX^-}}{\underset{\underset{Me}{\overset{|}{H}}}{\quad} \underset{\underset{Ph}{\overset{|}{H}}}{OH}}}{}}$$

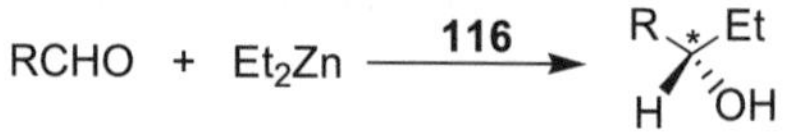

$$RCHO \; + \; Et_2Zn \xrightarrow{\quad \textbf{116} \quad} \underset{H}{\overset{R}{\diagdown}}\underset{OH}{\overset{Et}{\diagup}}$$

Substrate and Catalyst		Solvent	ee (%)	Config.
R	X			
Ph	Cl	Hexane	74	*S*
Ph	Cl	Toluene–Hexane	64	*S*
Ph	Cl	Benzene–Hexane	73	*S*
4-MePh	Cl	Hexane	61	*S*
Ph	Br	Hexane	62	*S*

Scheme 2–45. Reprinted with permission by Royal Chem. Soc., Ref. 112.

azaborolidine catalytic reduction of carbonyl compounds, which is discussed in detail in Chapter 6.

Chiral quaternary ammonium salts in solid state have also been used as catalysts for the enantioselective addition of diethylzinc to aldehydes (Scheme 2–45).[112] In most cases, homogeneous chiral catalysts afford higher enantio-selectivities than heterogeneous ones. Scheme 2–45 presents an unusual asymmetric reaction in which chiral catalysts in the solid state afford much higher enantioselectivities than its homogeneous counterpart.[112]

Most organometallic reagents, such as alkyllithium and Grignard reagents, are such strong nucleophiles that they usually fail to react chemoselectively with only aldehydes in the presence of ketones. Scheme 2–46 depicts the advantage of catalytic asymmetric synthesis of hydroxyketone **118** by the chemo- and enantioselective alkylation of **117** with dialkylzinc reagents using **119** or **120** as the chiral catalyst. In these reactions, optically active hydroxyketones can be obtained with high chemo- and enantioselectivity (up to 93% ee).[113]

The optically active β-amino alcohol (1R,3R,5R)-3-(diphenylhydroxymethyl)-2-azabicyclo[3.3.0]octane [(1R,3R,5R)-**121**], can be derived from a bicyclic proline analog. It catalyzes the enantioselective addition of diethylzinc to various aldehydes. Under mild conditions, the resulting chiral secondary alcohols are obtained in optical yields up to 100%. The bicyclic catalyst gives much better results than the corresponding (S)-proline derivative (S)-**122** (Scheme 2–47).[114]

Wally et al.[115] report a homoannularly bridged hydroxyamino ferrocene (+)-**123** as an efficient catalyst for enantioselective ethylation of aromatic or aliphatic aldehydes.

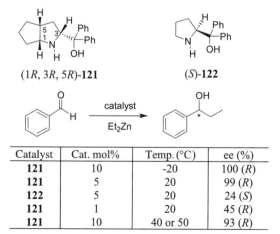

Scheme 2–46. Chemo- and enatioselective alkylation of ketoaldehydes.

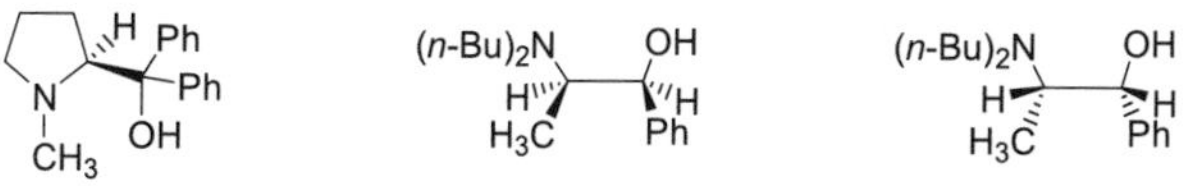

Entry	R$_2$Zn	Substrate	Catalyst	ee (%)
1	Et$_2$Zn	Ph—⬡—CHO (O)	(S)-**119**	93
2	Et$_2$Zn	Ph—⬡—CHO (O)	(1S,2R)-**120**	91
3	(n-Bu)$_2$Zn	Ph—⬡—CHO (O)	(S)-**119**	92
4	Et$_2$Zn	Ph—(O)—CHO	(1S,2R)-**120**	87
5	Et$_2$Zn	Ph—(O)—CHO	(1R,2S)-**120**	85
6	Et$_2$Zn	PhCH$_2$—(O)—CHO	(1S,2R)-**120**	81
7	Et$_2$Zn	Cl, CHO Ph—(O)—Cl	(S)-**119**	88

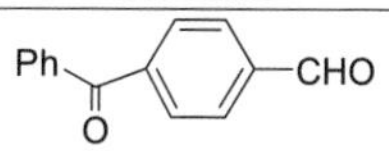

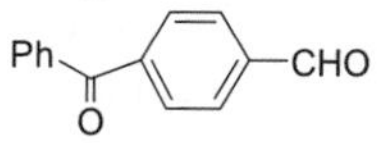

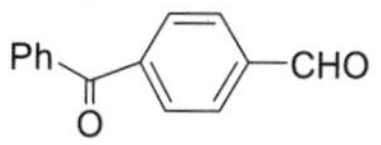

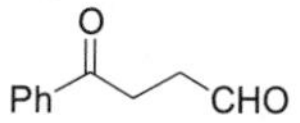

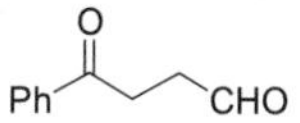

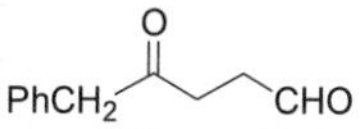

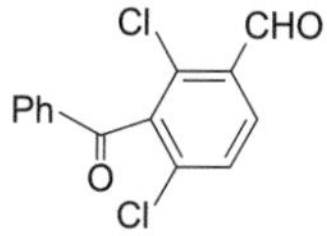

(S)-(+)-**119**   (1S, 2R)-(-)-**120**   (1R, 2S)-(+)-**120**

(1R, 3R, 5R)-**121**   (S)-**122**

Catalyst	Cat. mol%	Temp. (°C)	ee (%)
121	10	-20	100 (R)
121	5	20	99 (R)
122	5	20	24 (S)
121	1	20	45 (R)
121	10	40 or 50	93 (R)

Scheme 2–47. Application of a new bicyclic catalyst. Reprinted with permission by Pergamon Press Ltd., Ref. 114.

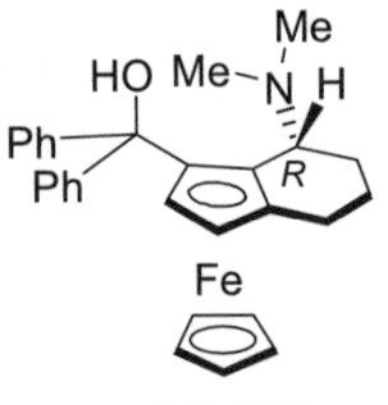

(+)-**123** 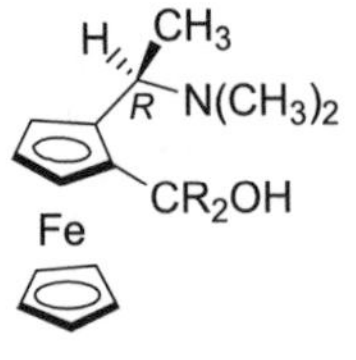   **124**

Eleven aromatic and aliphatic aldehydes have been alkylated with Et_2Zn in the presence of homoannularyl bridged hydroxyamino ferrocene (−)-**123**. The resulting carbinols have ee values varying from 66% to 97%. This new ferrocenyl catalyst has been used successfully to alkylate aromatic and linear or branched chain aliphatic aldehydes to secondary alcohols with up to 97% ee. This ligand is effective even for β-branched aliphatic substrate.

The transition state for the configuration-determining step has been presented by Kitamura et al.[116] and Watanabe et al.[117] (Fig. 2–6). Both groups propose the participation of two molecules of Et_2Zn and the formation of a seven-membered ring, which can be considered as a two-center catalysis system or bimetallic catalyst. This cyclic system adopts a chair-like conformation in which Zn bonded covalently to O and coordinated to N. The ethyl groups attached to Zn are arranged in equatorial positions. The Zn in the seven-membered ring is coordinated with the substrate. The second Zn in the Et_2Zn molecule (attached to the O atom) possesses the minimum energy in steric repulsion, thus favoring an Si-side approach.

With the knowledge that the presence of $Ti(OPr^i)_4$ promotes the alkylation of diethylzinc to benzaldehyde, Ho et al.[118] demonstrated that the chiral tetradentate sulfonamide ligand **125** catalyzes the addition of diethylzinc to aldehyde in the presence of $Ti(OPr^i)_4$ with good yield and enantioselectivity (Scheme 2–48).

Pritchett et al.[119] found that $Ti(OPr^i)_4$ did not react with the bis(sulfonamide) ligand itself, so they postulated that a chiral ligand initially reacted with the diethylzinc and was subsequently transferred to the titanium in the next step. Based on this assumption, they presented an improved procedure for the asymmetric alkylation of aldehyde to overcome the poor solubility of the li-

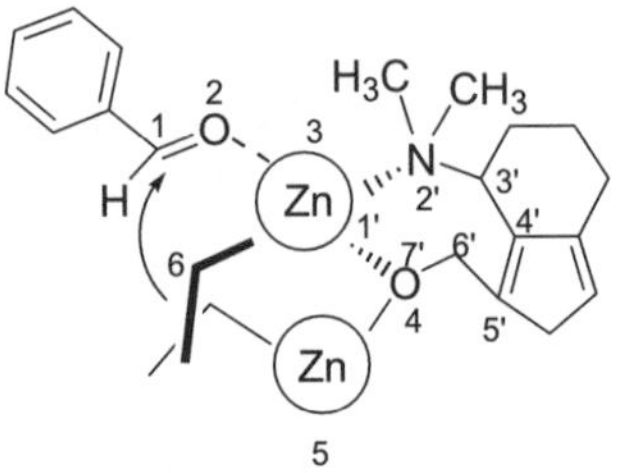

Figure 2–6. Transition state of the reaction.

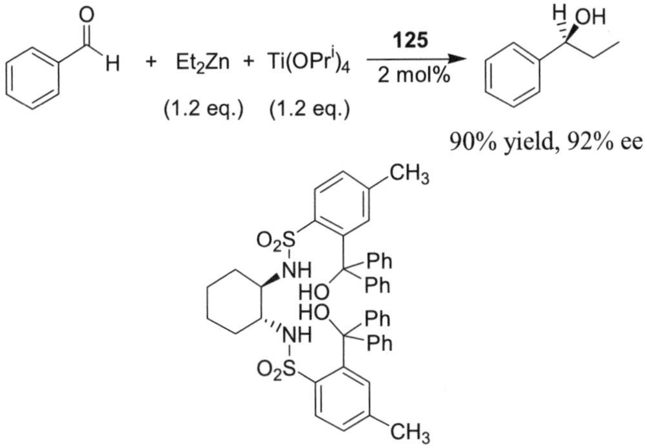

Scheme 2-48

gands in the nonpolar reaction mixture. The reaction was carried out by initial reaction of the bis(sulfonamide) with the zinc species, followed by the addition of titanium and subsequent addition of aldehyde.

Ito et al.[120] reported TADDOL **126** as a new type of chiral ligand in place of amino alcohol and examined the catalytic ligand effect of using various chiral diols in the presence of $Ti(OPr^i)_4$.

TADDOL **104** and **126** afford 95–99% ee in the asymmetric addition of organozinc reagents to a variety of aldehydes. The best enantioselectivities are observed when a mixture of the chiral titanium TADDOL compound **127** and excess $[Ti(OPr^i)_4]$ are employed (Scheme 2–49). The mechanism of the alkylzinc addition involves acceleration of the asymmetric catalytic process by the

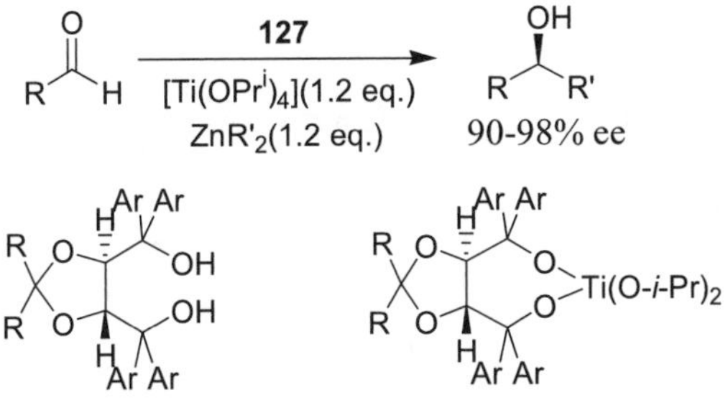

TADDOL
104: R = Me, Ar = Ph,
126: R = Me, Ar = 2-naphthyl

Scheme 2–49. TADDOL and its analogs as titanium ligands in enantioselective addition of diethylzinc reagents to benzaldehyde.

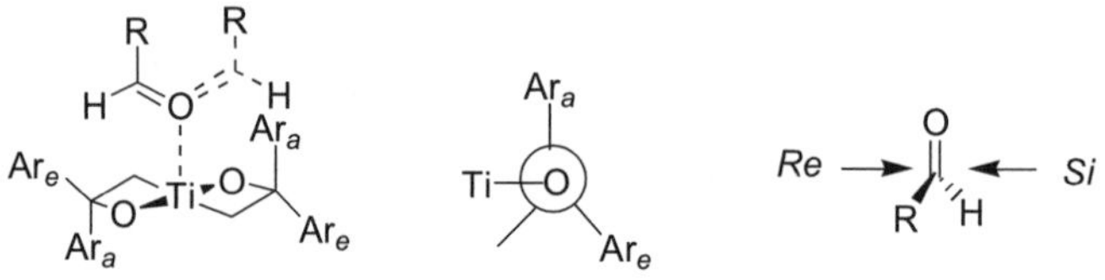

Figure 2–7. The role of [Ti(OPr^i)$_4$] in dialkylzinc addition reactions. The dioxolane in the rear is deleted for clarity.

TADDOL ligand over the competing (achiral) catalyst [Ti(OPr^i)$_4$]. The rate enhancement by the TADDOL ligands is due to an increase in the rate of ligand exchange in the TADDOL complex over the *iso*-propoxyl complex because of the steric bulk of the TADDOL compared with two *iso*-propoxides.

The role of Ti(OPr^i)$_4$ in this process is shown in Figure 2–7. The aldehyde is illustrated in two conformations, the solid lines indicating the more favorable orientation. The conformation represented by the dashed line is disfavored by a steric interaction with a pseudo-axial aryl group. Assuming that the attack of a nucleophile comes from the direction of the viewer, this hypothesis accounts for the *Si*-face selectivity in all known Ti-TADDOLate–mediated nucleophilic additions to aldehydes.

Prasad and Joshi[121] presented a conceptually different catalyst system—zinc amides of oxazolidine. Because the addition of dialkylzinc to aldehyde is known to involve a chiral zinc alkoxide with a coordinately unsaturated tricoordinated center, they anticipated that a zinc amide with dicoordinate zinc should be a better Lewis acid. Examining three different zinc species **128–130**, zinc amide derived from the corresponding oxazolidine **130** was found to lead to a very fast reaction (4 hours, 0°C) and 100% ee (Scheme 2–50). The reaction proceeds even faster at room temperature (completed within 1 hour) without significant loss of stereoselectivity. This reaction can provide excellent ee for aromatic aldehydes,

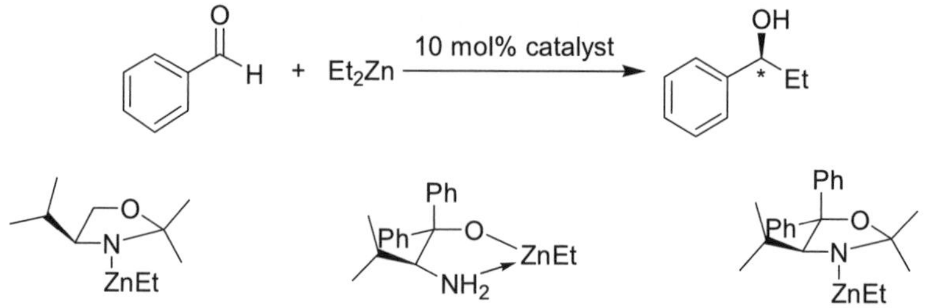

(*R*)-product, 38% ee **128** | (*S*)-product, < 5% ee **129** | (*S*)-product, 100% **130**

Scheme 2–50

though not for aliphatic ones. For this catalyst system, aliphatic aldehyde normally fails to give a good enantioselectivity.

A model explaining the stereochemical outcome of this catalytic system is based on the following transition state **131**:

Transition state:

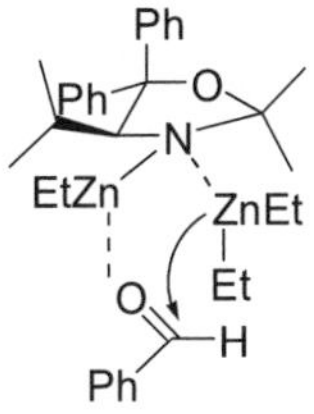

131

Both the aldehyde and diethylzinc are activated by the zinc amide, and the ethyl group transformation from diethylzinc to aldehyde furnishes the highly enantioselective alkylation of aromatic aldehydes.

BINOL and related compounds have proved to be effective catalysts for a variety of reactions. Zhang et al.[106a] and Mori and Nakai[106b] used an (R)-BINOL-Ti(OPr^i)$_4$ catalyst system in the enantioselective diethylzinc alkylation of aldehydes, and the corresponding secondary alcohols were obtained with high enantioselectivity. This catalytic system works well even for aliphatic aldehydes. Dialkylzinc addition promoted by Ti(OPr^i)$_4$ in the presence of (R)- or (S)-BINOL can give excellent results under very mild conditions. Both conversion of the aldehyde and the ee of the product can be over 90% in most cases. The results are summarized in Table 2–14.

TABLE 2–14. Asymmetric Alkylation of Aromatic and Aliphatic Aldehydes

Entry	Aldehyde	BINOL	Condition	Yield (%)	ee (%)
1	PhCHO	0.2	0°C, 20 min	100 (conversion)	91.9 (S)[106a]
2	2-Naph-CHO	0.2	0°C, 20 min	100 (conversion)	93.6 (S)[106a]
3	m-MeOPhCHO	0.2	0°C, 20 min	100 (conversion)	94 (S)[106a]
4	m-ClPhCHO	0.2	0°C, 20 min	98.7 (conversion)	88.2 (S)[106a]
5	n-C$_8$H$_{17}$CHO	0.2	−30°C, 40 h	94	86 (S)[106b]
6	n-C$_6$H$_{13}$CHO	0.2	−30°C, 40 h	75	85 (S)[106b]
7	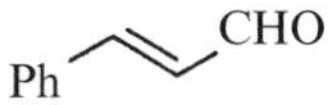 Ph⁀CHO	0.2	0°C, 1 h	97	82 (S)[106b]
8	TMS—≡—CHO	0.2	0°C, 1 h	>98	56 (S)[106b]
9	TBS —≡—CHO	0.2	0°C, 1 h	>98	79 (S)[106b]

ee = Enantiomeric excess.

Reprinted with permission by Elsevier Science Ltd., Ref. 106.

The chiral complex **132** (X = OPr^i) is easily available by mixing Ti(OPr^i)$_4$ and (R)- or (S)-BINOL. The ratio of BINOL to Ti(OPr^i)$_4$ is a key factor for inducing enantioselectivity. A large excess of Ti(OPr^i)$_4$ over BINOL is required to make the reaction efficient, and excess Et$_2$Zn (over aldehyde) is needed to get high yields.

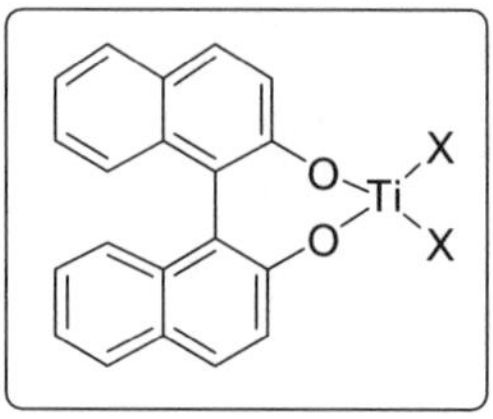

(R)- or (S)-**132** (X = O-i-Pr or CN)

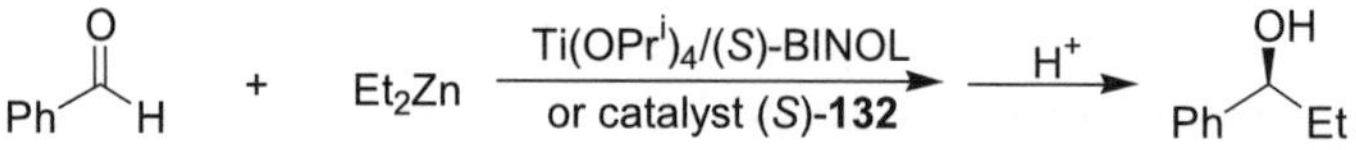

Nakai has proposed that the involved asymmetric catalyst was not **132** itself, but the following complex **133**:

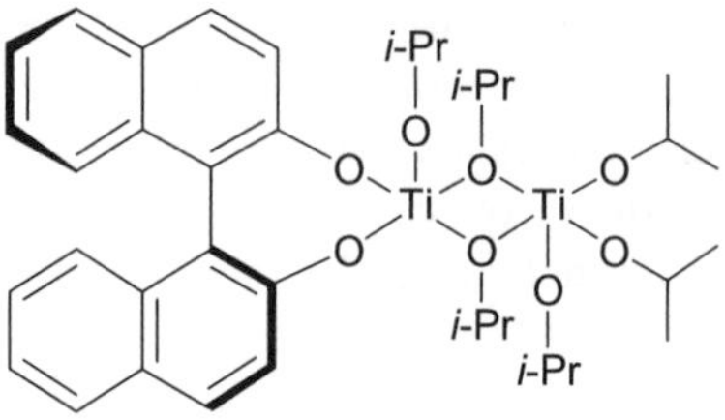

133

In Scheme 2–51, species **133** is formed from the precatalyst **132** and Ti(OPr^i)$_4$. It is then converted to complex G upon addition of diethylzinc. Reaction between species G and an aldehyde furnishes intermediate E, which accomplishes the enantioselective addition of the nucleophile to the carbonyl group. Intervention of two molecules of Ti(OPr^i)$_4$ releases the alkylated product, regenerates the active catalyst **133**, and also completes the catalytic cycle. This cycle explains the fact that at least one equivalent of Ti(OPr^i)$_4$ is required for an effective reaction.

Zhang and Chan[122] found that H$_8$-BINOL, (R)- or (S)-**134**, in which the naphthyl rings in the BINOL were partially hydrogenated,[123] can give even better results in the diethylzinc reactions. Using (R)- or (S)-**134** as the chiral ligand, addition of diethylzinc to aromatic aldehydes proceeds smoothly with over 95% ee and, in most cases, quantitative conversion.[122]

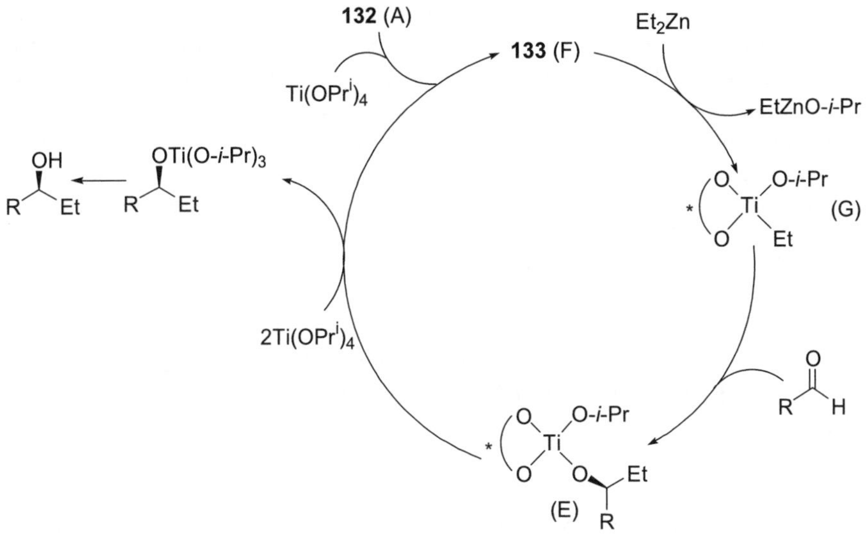

Scheme 2–51. Reprinted with permission by Elsevier Science Ltd., Ref. 106b.

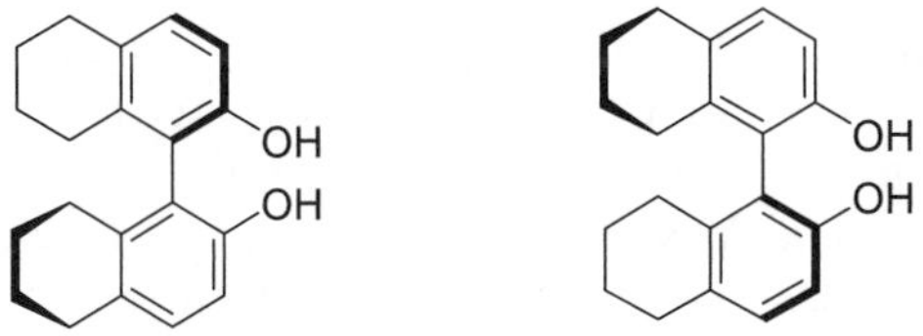

(R)-**134** (S)-**134**

Triethylaluminum can be economically prepared on an industrial scale from aluminum hydride and ethylene,[124] so a successful alkylation using organoaluminum compound will certainly open up a new area for active research. Asymmetric alkylation of aromatic aldehydes with triethylaluminum was carried out by Chan et al.[125] In the presence of (R)- or (S)-**134** and $Ti(OPr^i)_4$, alkylation proceeded readily, yielding the alcohol with high ee (Scheme 2–52).

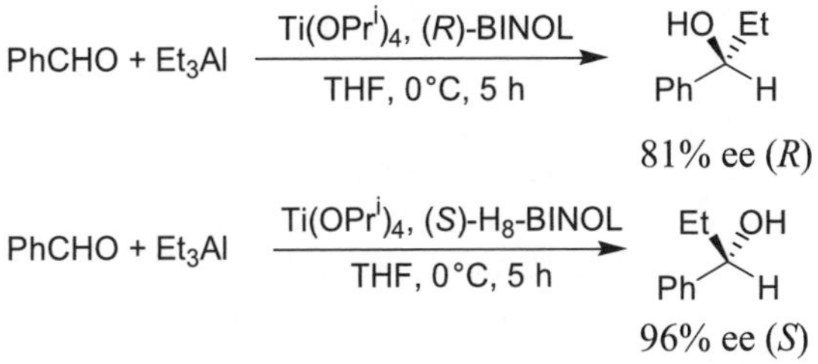

Scheme 2–52

Since the discovery of amino alcohol–induced dialkylzinc addition to aldehydes, many new ligands have been developed. It has recently been reported that chiral amino thiols and amino disulfides can form complexes or structurally strained derivatives with diethylzinc more favorably than chiral amino alcohols and thus enhance the asymmetric induction. Table 2–15 is a brief summary of such chiral catalysts.

For more information on diethylzinc addition reactions, see Ito et al.,[120] Wirth,[129] and others.[138] For a detailed discussion of the nonlinear stereochemical effects in diethylzinc addition, see Chapter 8.

2.7 CATALYTIC ASYMMETRIC ADDITIONS OF DIALKYLZINC TO KETONES: ENANTIOSELECTIVE FORMATION OF TERTIARY ALCOHOLS

As mentioned in Section 2.3, a large number of biologically active natural products contain quaternary carbon atoms, and the addition of carbon nucleophiles to ketones has attracted increasing attention for the construction of quaternary carbon centers.

Fu and Dosa[139] report the enantioselective addition of diphenylzinc to a range of aryl-alkyl and dialkyl ketones with good to excellent stereocontrol. Addition of 1.5 eq. of MeOH in the presence of a catalytic amount of (+)-DAIB **135** results in enhanced enantioselectivity and improved yield (Scheme 2–53). Table 2–16 gives the results of this reaction.

Similarly, Ramón and Yus[140] reported the enantioselective addition of diethylzinc and dimethylzinc to prochiral ketones catalyzed by camphorsulfonamide-titanium alkoxide derivatives as shown in Scheme 2–54.

The reaction of diethylzinc or dimethylzinc with prochiral ketones, in the presence of a stoichiometric amount of $Ti(OPr^i)_4$ and a catalytic amount (20%) of camphor-sulfonamide derivative **136**, leads to the formation of the corresponding tertiary alcohols with enantiomeric ratios of up to 94.5:5.5.

Nakamura et al.[141] reported a closely related reaction, that is, the enantioselective addition of allylzinc reagent to alkynyl ketones catalyzed by a bisoxazoline catalyst **137**. High ee values were obtained in most cases (Scheme 2–55).

2.8 ASYMMETRIC CYANOHYDRINATION

Cyanohydrination (addition of a cyano group to an aldehyde or ketone) is another classic reaction in organic synthesis. Enantioselective addition of TMSCN to aldehyde, catalyzed by chiral metal complexes, has also been an active area of research for more than a decade. The first successful synthesis using an (S)-binaphthol–based complex came from Reetz's group[142] in 1986. Their best result, involving Ti complex, gave 82% ee. Better results were reported shortly thereafter by Narasaka and co-workers.[143] They showed that by

TABLE 2–15. Newly Developed Ligands for Alkylation Reactions

Chiral Auxiliary	Reference	Chiral Auxiliary	Reference
	126		127
	128		129
	130		130
	131	 R = H, R' = Me R = R' = Me R = R' = n-Bu	132
	133		134
	135		135
 R' = R" = Me R' = SO₃CF₃ R" = H	136	X = H , Cl, OCH₃	120
	137		

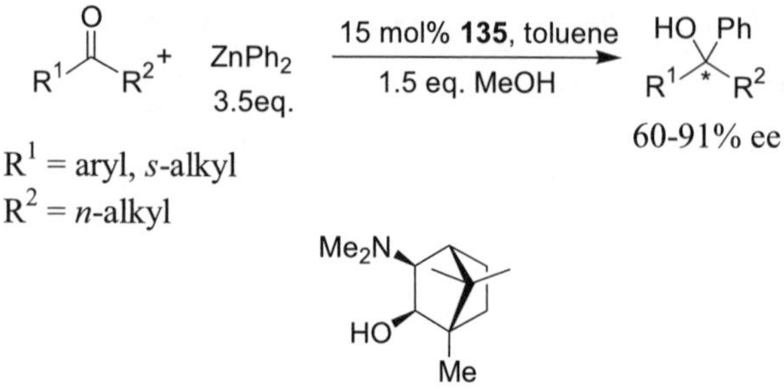

Scheme 2–53

TABLE 2–16. Enantioselective Alkylation of Ketones

Entry	Substrate	ee (%)	Yield (%)	Entry	Substrate	ee (%)	Yield (%)
1	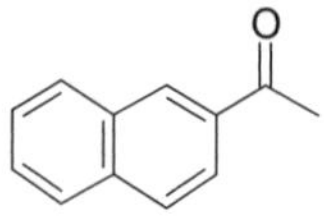	72 (+)-(R)-	58	5	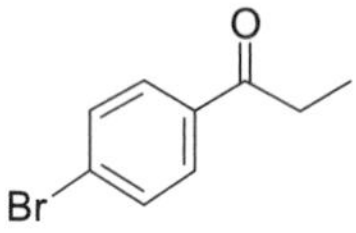	90 (−)-	83
2	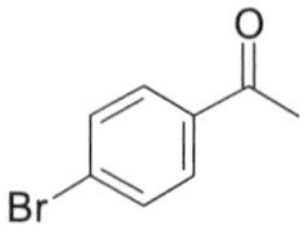	80 (−)-	53	6	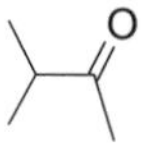	60 (+)-	63
3	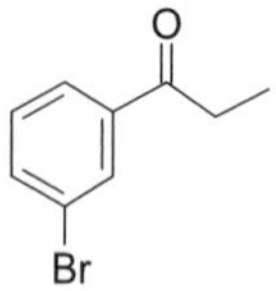	91 (−)-	91	7	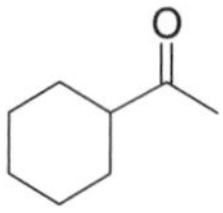	75 (+)-	76
4	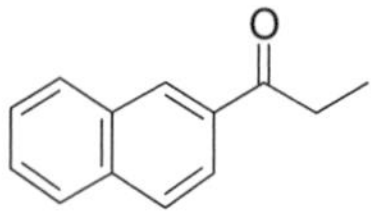	86 (−)-(R)-	79				

ee = Enantiomeric excess.
Reprinted with permission by Am. Chem. Soc., Ref. 139.

using highly substituted chiral 1,4-diol as ligand, both aromatic and aliphatic aldehyde could be converted to the corresponding cyanohydrin with more than 85% yield and over 90% ee. While the Narasaka method was effective in preparing optically active cyanohydrins, it required a stoichiometric amount of titanium and tartaric acid derivatives. Hayashi et al.[144] reported that a similar catalytic system based on the modified Sharpless catalyst was also effective as an asymmetric catalyst for the addition of TMSCN to aromatic aldehydes. The

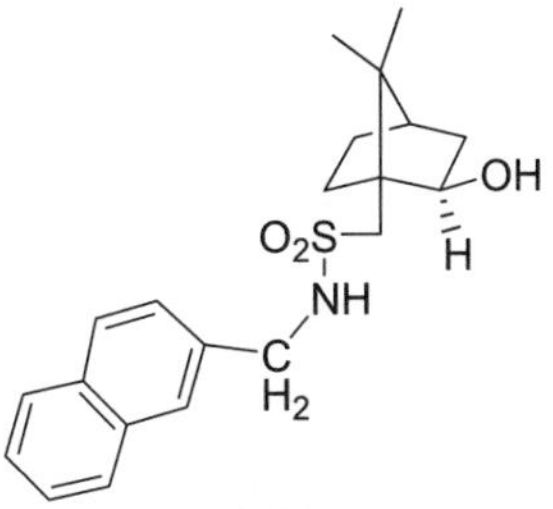

136

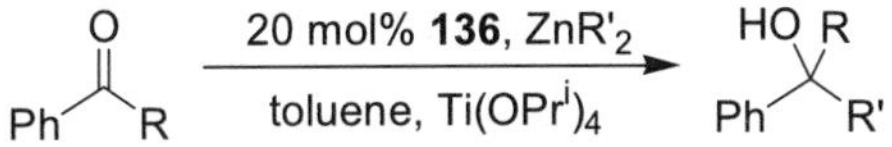

Scheme 2–54

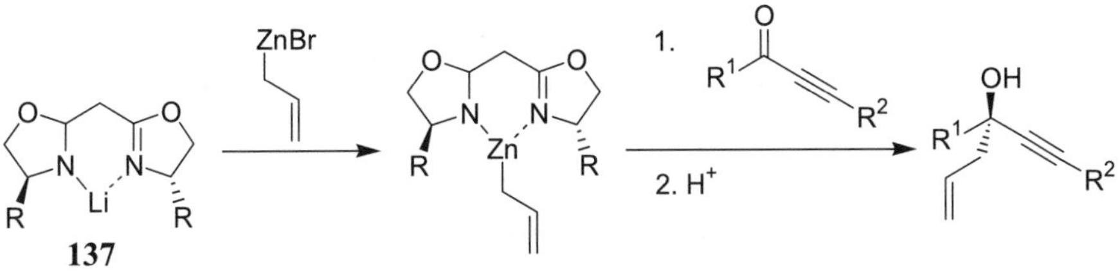

Scheme 2–55

use of cyclic dipeptides, described by Mori et al.,[145] worked satisfactorily as well.

The best results for the asymmetric cyanohydrination reactions are obtained through biocatalysis, using the readily available enzyme oxynitrilase. This provides cyanohydrins from a number of substances with over 98% ee.[146]

Hayashi et al.[147] reported another highly enantioselective cyanohydrination catalyzed by compound **138**. In this reaction, a Schiff base derived from β-amino alcohol and a substituted salicylic aldehyde were used as the chiral ligand, and the asymmetric addition of trimethylsilylcyanide to aldehyde gave the corresponding cyanohydrin with up to 91% ee (Scheme 2–56).

Bolm and Müller[148] reported that a chiral titanium reagent generated from optically active sulfoximine (R)-**139** and Ti(OPr^i)$_4$ promotes the asymmetric addition of trimethylsilyl cyanide to aldehydes, affording cyanohydrins in high yields with good enantioselectivities (up to 91% ee) (Scheme 2–57). The aldehydes can be either aromatic or aliphatic. For example, in the presence of a stoichiometric amount of Ti(OPr^i)$_4$ and 1.1 eq. of (R)-**139**, trimethylsilylcyanation of benzaldehyde at $-50°C$, followed by acidic cleavage of the trimethylsilyl group gave (S)-mandelonitrile with 72% yield and 91% ee. Lowering the reaction temperature did not significantly improve the ee values.

The proposed reaction mechanism is shown in Figure 2–8. First, a chiral titanium complex (R)-**140** is formed by the exchange of two titanium alkoxides.

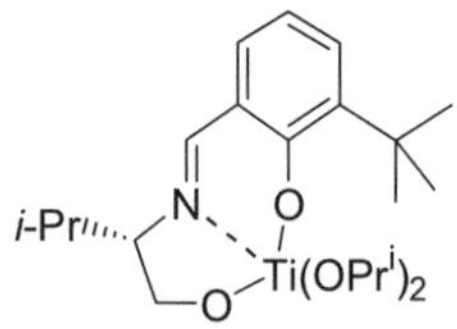

138

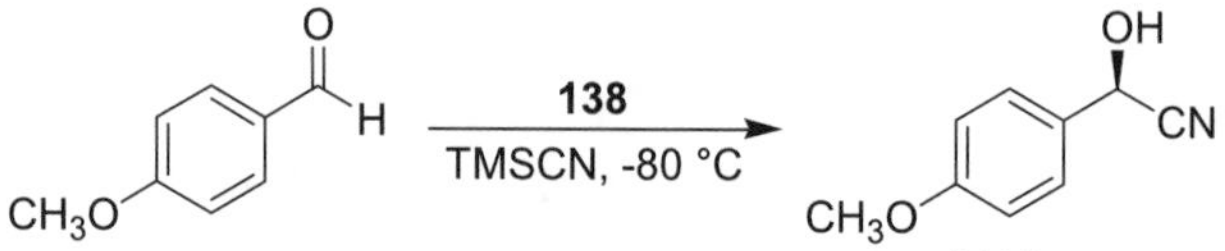

91% ee

Scheme 2–56

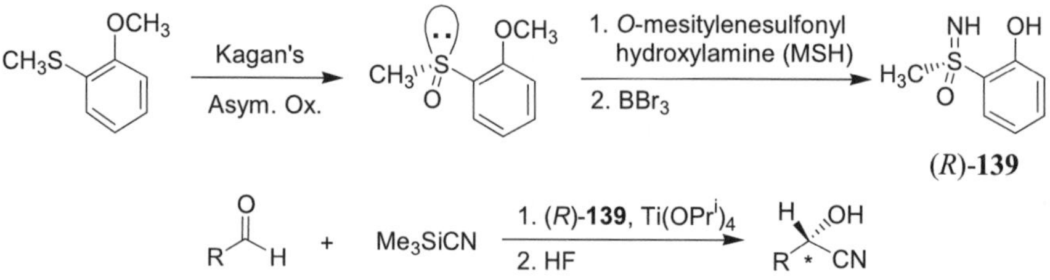

(R)-**139**

up to 91% ee

Scheme 2–57

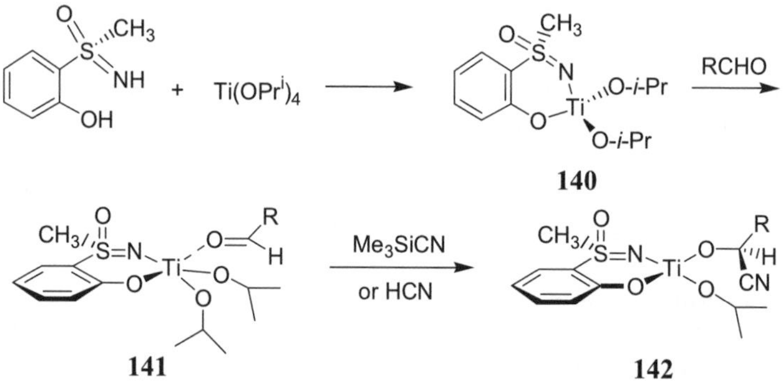

140

141   **142**

Figure 2–8. Proposed reaction mechanism for Ti(OPr^i)$_4$-mediated asymmetric silylcyanation.

Complex (R)-**140** serves as a chiral Lewis acid and coordinates to the aldehyde at the less hindered β-face of **141**. *Re*-side cyanation of (R)-**141** and the subsequent cleavage of the alkoxide group give the product **142**. Because at this stage the catalyst turnover is blocked, the reaction cannot be carried out in a catalytic manner.

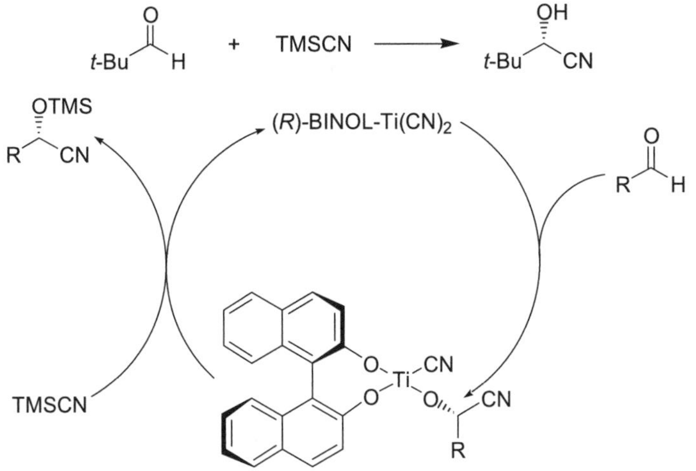

Scheme 2–58. Reprinted with permission by Elsevier Science Ltd., Ref. 149.

Mori et al.[149] also reported the asymmetric cyanosilylation of aldehyde with TMSCN using **132** (X = CN) as the precatalyst. The chiral dicyano complex was generated in situ, and the asymmetric cyanosilylation gave ee values of up to 75%. Scheme 2–58 depicts the proposed reaction process.

The addition of cyanide to imines, the Strecker reaction, constitutes an interesting strategy for the asymmetric synthesis of α-amino acid derivatives. Sigman and Jacobsen[150] reported the first example of a metal-catalyzed enantioselective Strecker reaction using chiral salen Al(III) complexes **143** as the catalyst (see Scheme 2–59).

Among the complexes of Ti, Cr, Mn, Co, Ru, and Al, which catalyzed the reaction with varying degrees of conversion and enantioselectivity, complex **143** was found to give the best result, and it was found that the uncatalyzed reaction between HCN and **144** could be completely suppressed at −70°C. For example, in Scheme 2–59, the reaction for an aromatic substrate **144**, such as R = Ph, can be completed within 15 hours, providing product **145** (R = Ph, without a trifluoroacetyl group) with 91% isolated yield and 95% ee. Because the cyano addition product has been observed to undergo racemization upon exposure to silica gel during the isolation procedure, the product is transformed to the corresponding stable trifluoroacetamide derivative. This is the first example in which a main group metal–salen complex has been identified as a highly effective asymmetric catalyst.

In contrast, testing substrates in Scheme 2–59 demonstrates that alkyl-substituted imines undergo the addition of HCN with considerably lower ee. (For R = cyclohexyl, 57% ee; and 37% ee for R = t-butyl.) The N-substituent does not exert a significant influence on the enantioselectivity of the reaction.

For more information about the asymmetric addition of trimethylsilyl cyanide to aldehydes, see Belokon et al.[151]

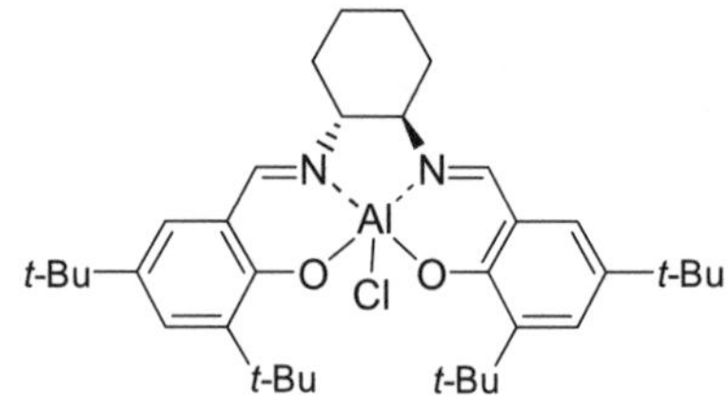

143

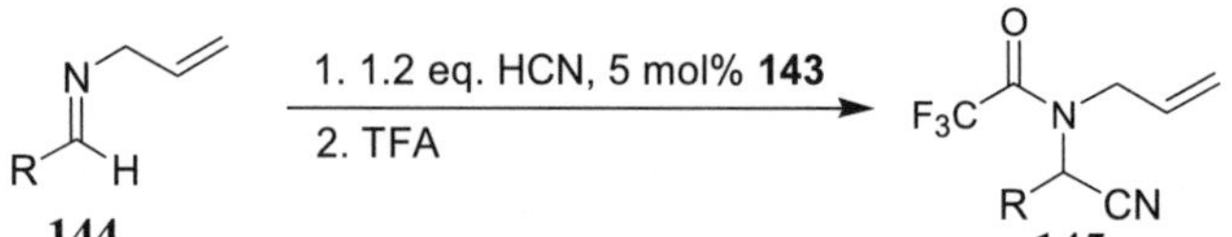

Scheme 2–59. Chiral Al-salen–catalyzed Strecker reaction.

2.9   ASYMMETRIC α-HYDROXYPHOSPHONYLATION

α-Hydroxyphosphonyl compounds (phosphonates and phosphonic acids) are biologically active and can be used for enzyme inhibitors (e.g., renin synthase inhibitor[152] and HIV protease inhibitor[153]). Although the biological activities of α-substituted phosphonyl compounds depend on their absolute configuration,[154] it is only recently that detailed studies on the synthesis of optically active phosphonyl compounds have begun to emerge. The most efficient and economic route to chiral hydroxylphosphonate involves asymmetric α-hydroxyphosphonylation.

One common approach incorporates an oxazaborolidine-mediated catecholborane reduction starting from α-ketophosphonates (**146**).[155] The reaction proceeds with good yield and gives excellent ee (up to 99%).

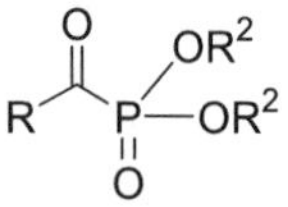

146

Enantioselective synthesis of α-hydroxy phosphonates can also be achieved by asymmetric oxidation with camphorsulfonyl oxaziridines (Scheme 2–60).[156] Reasonable yields can usually be obtained. (+)-**147a** or (+)-**147b** favors formation of the (S)-product, as would be expected, because these oxidations proceed via a transition state that parallels that previously discussed for the stereoselectivity observed with ketones.[157]

Attempts have also been made to explore chiral catalysis in the Pudovik reaction (the addition of dialkylphosphites to aldehydes). Rath and Spilling[158]

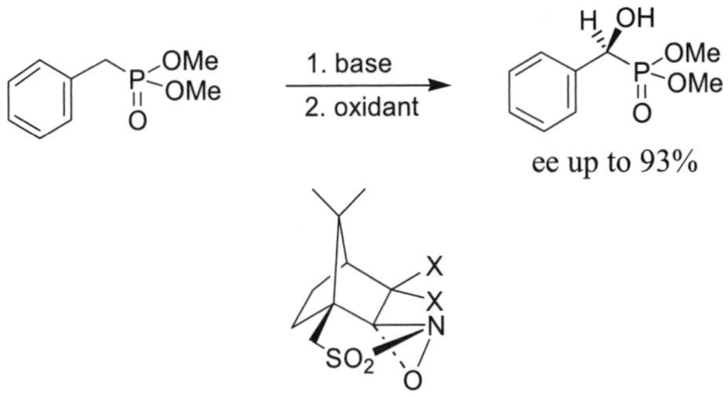

ee up to 93%

(+)-**147a** X = H
(+)-**147b** X = Cl

Scheme 2–60

and Yokomatsu et al.[159] independently published the lanthanum binaphth-oxide complex catalyzed addition of diethylphosphite to aromatic aldehydes. Lanthanum (R)-binaphthoxide complex gives (S)-hydroxyphosphonates in good yield with modest enantioselectivity. The catalyst LaLi$_3$(BINOL)$_3$ (LLB) was prepared from lanthanum trichloride by the method reported by Sasai et al.[160] for catalytic enantioselective nitroaldol reaction.

Sasai et al.[161] revealed an improved condition for the preparation of LLB, which involves the reaction of a mixture of LaCl$_3$ · 7H$_2$O (1 eq.), (R)- or (S)-BINOL dilithium salt (2.7 eq.), and t-BuONa (0.3 eq.) in THF at 50°C. The LLB obtained is effective for the hydrophosphonylation of various aldehydes, and the desired α-hydroxyphosphonates can be obtained in up to 95% ee (89% yield). With slow addition of the aldehyde, the ee of the product can be further increased (Scheme 2–61).

LLB, a so-called heterobimetallic catalyst, is believed to activate both nucleophiles and electrophiles.[162] For the hydrophosphonylation of comparatively unreactive aldehydes, the activated phosphite can react with only the molecules precoordinated to lanthanum (route A). The less favored route (B) is a competing reaction between Li-activated phosphite and unactivated aldehyde, and this unfavored reaction can be minimized if aldehydes are introduced slowly to the reaction mixture, thus maximizing the ratio of activated to inactivated aldehyde present in solution. Route A regenerates the catalyst and completes the catalysis cycle (Fig. 2–9).

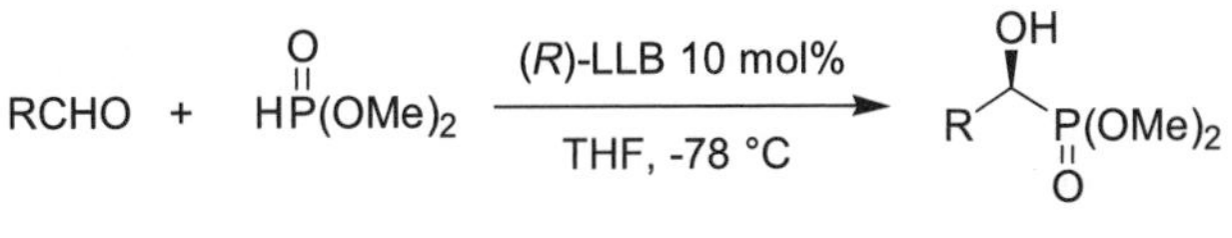

$$RCHO + HP(OMe)_2 \xrightarrow[\text{THF, -78 °C}]{(R)\text{-LLB 10 mol\%}} R\overset{OH}{\underset{O}{\wedge}}P(OMe)_2$$

Scheme 2–61

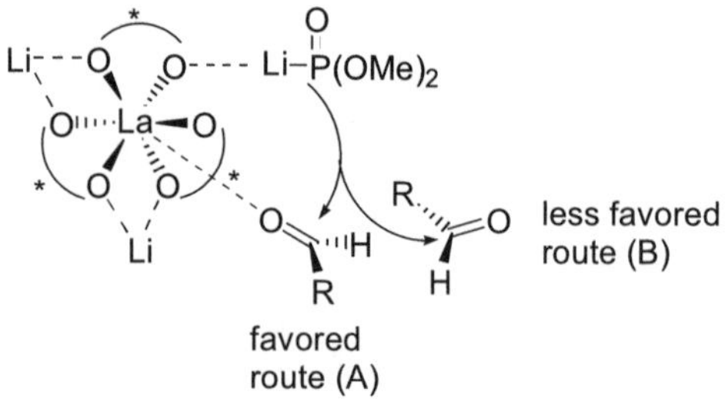

Figure 2–9. Proposed mechanism for the asymmetric hydroxyphosphonylation catalyzed by LLB.

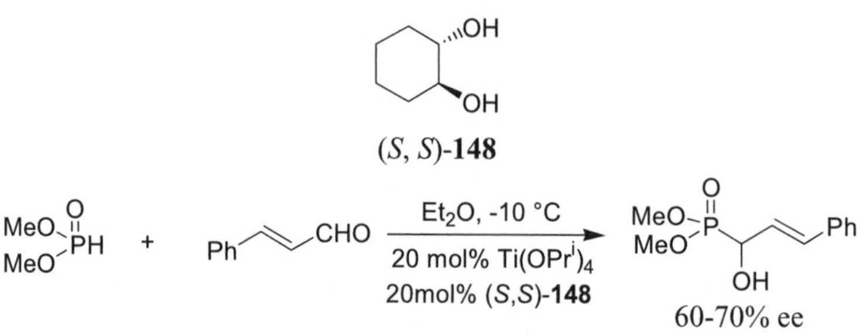

(S, S)-**148**

Scheme 2–62

Other homochiral cyclic diol ligands such as (S,S)-**148** for titanium alkoxide have also been tested for catalyzing phosphonylation of aldehydes, but it has been found that these diols are a poor choice of ligand for asymmetric phosphonylation.[163] For most of the aldehydes studied (substituted benzaldehydes, α,β-unsaturated aldehydes, and cyclohexanecarboxaldehyde), only moderate enantioselectivity was obtained (Scheme 2–62).

To complement the above information, a highly enantioselective synthesis of α-amino phosphonate diesters should be mentioned.[164] Addition of lithium diethyl phosphite to a variety of chiral imines gives α-amino phosphonate with good to excellent diastereoselectivity (de ranges from 76% to over 98%). The stereoselective addition of the nucleophile can be governed by the preexisting chirality of the chiral auxiliaries (Scheme 2–63).

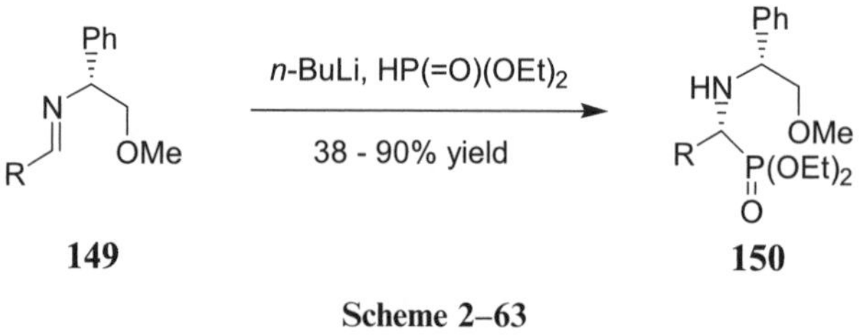

Scheme 2–63

The diastereofacial selectivity is explained by the proposed chelated intermediate **151**. Internal delivery of the nucleophile takes place from the less hindered side. Removal of the chiral directing moiety with a catalytic amount of palladium hydroxide on carbon in absolute ethanol then furnishes the final product. This process yields the amino ester in 83–100% yield without observable racemization.

151

2.10 SUMMARY

This chapter has given a general introduction to the α-alkylation of carbonyl compounds, as well as the enantioselective nucleophilic addition to carbonyl compounds. Chiral auxiliary aided α-alkylation of a carbonyl group can provide high enantioselectivity for most substrates, and the hydrazone method can provide routes to a large variety of α-substituted carbonyl compounds. Chiral sultam and chiral oxazoline are also useful chiral auxiliaries for the asymmetric synthesis of such carbonyl compounds. The SRS method (self-regeneration of stereocenters), starting from inexpensive chiral compounds, provides a convenient synthesis for chiral compounds with quaternary chiral centers. Perhaps the most important method developed in this area is the enantioselective addition of dialkylzinc to carbonyl groups. The reaction is normally carried out under very mild conditions, giving excellent results in both conversion and enantioselectivity.

2.11 REFERENCES

1. (a) Arnett, E. M.; Small, L. E. *J. Am. Chem. Soc.* **1977**, *99*, 808. (b) Novak, M.; Loudon, G. M. *J. Org. Chem.* **1977**, *42*, 2494. (c) Haspra, P.; Sutter, A.; Wirz, J. *Angew. Chem. Int. Ed. Engl.* **1979**, *18*, 617. (d) Bordwell, F. G.; Fried, H. E. *J. Org. Chem.* **1981**, *46*, 4327. (e) Bordwell, F. G.; Drucker, G. E.; Fried, H. E. *J. Org. Chem.* **1981**, *46*, 632.
2. (a) House, H. O.; Czuba, L. J.; Gall, M.; Olmstead, H. D. *J. Org. Chem.* **1969**, *34*, 2324. (b) Creger, P. L. *Org. Syn.* **1970**, *50*, 58. (c) Wittig, G.; Hesse, A. *Org. Syn.* **1970**, *50*, 66. (d) Cregge, R. J.; Herrmann, J. L.; Lee, C. S.; Richman, J. E.; Schlessinger, R. H. *Tetrahedron Lett.* **1973**, 2425. (e) Herrmann, J. L.; Schlessinger, R. H. *Tetrahedron Lett.* **1973**, 2429. (f) Herrmann, J. L.; Kieczykowski, G. R.; Schlessinger, R. H. *Tetrahedron Lett.* **1973**, 2433.

3. Rathke, M. W.; Lindert, A. *J. Am. Chem. Soc.* **1971**, *93*, 2318.

4. (a) Wannagat, U.; Niederprum, H. *Chem. Ber.* **1961**, *94*, 1540. (b) Kruger, C. R.; Rochow, E. G. *J. Organomet. Chem.* **1964**, *1*, 476. (c) Barton, D. H. R.; Hesse, R. H.; Tarzia, G.; Pechet, M. M. *J. Chem. Soc. Chem. Commun.* **1969**, 1497. (d) Masamune, S.; Ellingboe, J. W.; Choy, W. *J. Am. Chem. Soc.* **1982**, *104*, 5526.

5. (a) Olofson, R. A.; Dougherty, C. M. *J. Am. Chem. Soc.* **1973**, *95*, 582. (b) Stowell, J. C.; Padegimas, S. J. *J. Org. Chem.* **1974**, *39*, 2448.

6. Simpkins, N. S. *Pure Appl. Chem.* **1996**, *68*, 691.

7. Still, W. C.; Galynker, I. *Tetrahedron* **1981**, *37*, 3981.

8. (a) Johnson, F.; Malhotra, S. K. *J. Am. Chem. Soc.* **1965**, *87*, 5492. (b) Chow, Y. L.; Colon, C. J.; Tam, J. N. S. *Can. J. Chem.* **1968**, *46*, 2821.

9. (a) House, H. O.; Bare, T. M. *J. Org. Chem.* **1968**, *33*, 943. (b) Ziegler, F. E.; Wender, P. A. *J. Am. Chem. Soc.* **1971**, *93*, 4318.

10. Velluz, L. Valls, J.; Nomine, G. *Angew. Chem. Int. Ed. Engl.* **1965**, *4*, 181.

11. Allinger, N. L.; Blatter, H. M.; Freiberg, L. A.; Karkowski, F. M. *J. Am. Chem. Soc.* **1966**, *88*, 2999.

12. (a) Posner, G. H.; Chapdelaine, M. J.; Sterling, J. J.; Whitten, C. E.; Lenz, C. M. *J. Org. Chem.* **1979**, *44*, 3661. (b) Coates, R. M.; Sandefur, L. O. *J. Org. Chem.* **1974**, *39*, 275.

13. (a) Takano, S.; Tamura, N.; Ogasawara, K. *J. Chem. Soc. Chem. Commun.* **1981**, 1155. (b) Jager, V.; Schwab, W. *Tetrahedron Lett.* **1978**, 3129.

14. (a) Ohta, T.; Hosoi, A.; Nozoe, S. *Tetrahedron Lett.* **1988**, *29*, 329. For preparation of the optically active α-hydroxycarboxylic acid derivatives, see (b) Gamboni, R.; Mohr, P.; Waespe-Sarcevic, N.; Tamm, C. *Tetrahedron Lett.* **1985**, *26*, 203. (c) Davis, F. A.; Vishwakarma, L. C. *Tetrahedron Lett.* **1985**, *26*, 3539. (d) Enders, D.; Bhushan, V. *Tetrahedron Lett.* **1988**, *29*, 2437. (e) Davis, F. A.; Sheppard, A. C.; Lal, G. S. *Tetrahedron Lett.* **1989**, *30*, 779. (f) Gamboni, R.; Tamm, C. *Helv. Chim. Acta* **1986**, *69*, 615. (g) Evans, D. A.; Morrissey, M. M.; Dorow, R. L. *J. Am. Chem. Soc.* **1985**, *107*, 4346. (h) Brown, H. C.; Pai, G. G.; Jadahav, P. K. *J. Am. Chem. Soc.* **1984**, *106*, 1531. (i) Davis, F. A.; Haque, M. S.; Ulatowski, T. G.; Towson, J. C. *J. Org. Chem.* **1986**, *51*, 2402. (j) Davis, F. A.; Haque, M. S. *J. Org. Chem.* **1986**, *51*, 4083. (k) Gore, M. P.; Vederas, J. C. *J. Org. Chem.* **1986**, *51*, 3700. (l) Smith III, A. B.; Dorsey, B. D.; Ohba, M.; Lupo, A. T.; Malamas, M. S. *J. Org. Chem.* **1988**, *53*, 4314. (m) Davis, F. A.; Ulatowski, T. G.; Haque, M. S. *J. Org. Chem.* **1987**, *52*, 5288. (n) Askin, D.; Volante, R. P.; Reamer, R. A.; Ryan, K. M.; Shinkai, I. *Tetrahedron Lett.* **1988**, *29*, 277.

15. (a) Krapcho, A. P.; Dundulis, E. A. *J. Org. Chem.* **1980**, *45*, 3236. (b) Sato, K.; Miyamoto, O.; Inoue, S.; Honda, K. *Chem. Lett.* **1981**, 1183.

16. (a) Schöllkopf, U.; Hausberg, H. H.; Segal, M.; Reiter, U.; Hoppe, I.; Saenger, W.; Lindner, K. *Liebigs Ann. Chem.* **1981**, 439. (b) Fraser, R. R.; Akiyama, F.; Banville, J. *Tetrahedron Lett.* **1979**, 3929 (c) Sugasawa, T.; Toyoda, T. *Tetrahedron Lett.* **1979**, 1423.

17. Johnson, F. *Chem. Rev.* **1968**, *68*, 375.

18. Schlessinger, R. H.; Iwanowicz, E. J.; Springer, J. P. *Tetrahedron Lett.* **1988**, *29*, 1489.

19. Alexakis, A.; Sedrani, R.; Mangeney, P.; Normant, J. F. *Tetrahedron Lett.* **1988**, *29*, 4411.

20. (a) Heathcock, C. H.; Pirrung, M. C.; Lampe, J.; Buse, C. T.; Young, S. D. *J. Org. Chem.* **1981**, *46*, 2290. (b) Evans, D. A.; Ennis, M. D.; Mathre, D. J. *J. Am. Chem. Soc.* **1982**, *104*, 1737. (c) Kraus, G. A.; Taschner, M. J. *Tetrahedron Lett.* **1977**, 4575.

21. (a) Frater, G. *Helv. Chim. Acta* **1979**, *62*, 2825. (b) Frater, G. *Helv. Chim. Acta* **1979**, *62*, 2829. (c) Frater, G. *Helv. Chim. Acta* **1980**, *63*, 1383. (d) Frater, G. *Tetrahedron Lett.* **1981**, *22*, 425. (e) Zuger, M.; Welle, T.; Seebach, D. *Helv. Chim. Acta* **1980**, *63*, 2005. (f) Seebach, D.; Wasmuth, D. *Helv. Chim. Acta* **1980**, *63*, 197. (g) Seebach, D. Wasmuth, D. *Angew. Chem. Int. Ed. Engl.* **1981**, *20*, 971.

22. (a) Ireland, R. E.; Mueller, R. H.; Willard, A. K. *J. Am. Chem. Soc.* **1976**, *98*, 2868. (b) Heathcock, C. H.; Buse, C. T.; Kleschick, W. A.; Pirrung, M. C.; Sohn, J. E.; Lampe, J. *J. Org. Chem.* **1980**, *45*, 1066. (c) Meyers, A. I.; Reider, P. J. *J. Am. Chem. Soc.* **1979**, *101*, 2501.

23. (a) Evans, D. A.; Takacs, J. M. *Tetrahedron Lett.* **1980**, *21*, 4233. (b) Sonnet, P. E.; Heath, R. R. *J. Org. Chem.* **1980**, *45*, 3137.

24. Lin, G. Q.; Hjalmarsson, M.; Högberg, H. E.; Jernstedt, K.; Norin, T. *Acta Chem. Scand. B* **1984**, *38*, 795.

25. (a) ApSimon, J. W.; Seguin, R. P. *Tetrahedron* **1979**, *35*, 2797. (b) Mukaiyama, T. *Tetrahedron* **1981**, *37*, 4111. (c) Evans, D. A. *Aldrichimica Acta* **1982**, *15*, 23. (d) Knowles, W. S. *Acc. Chem. Res.* **1983**, *16*, 106.

26. Kawanami, Y.; Ito, Y.; Kitagawa, T.; Taniguchi, Y.; Katsuki T.; Yamaguchi, M. *Tetrahedron Lett.* **1984**, *25*, 857.

27. Marzi, M.; Misiti, D. *Tetrahedron Lett.* **1989**, *30*, 6075.

28. Evans, D. A.; Ennis, M. D.; Mathre, D. J. *J. Am. Chem. Soc.* **1982**, *104*, 1737.

29. Evans, D. A.; Takacs, J. M.; McGee, L. R.; Ennis, M. D.; Mathre, D.; Bartroli, J. *Pure Appl. Chem.* **1981**, *53*, 1109.

30. Wuts, P. G. M.; Pruitt, L. E. *Synthesis* **1989**, 622.

31. For examples of the synthetic application of α-amino aldehydes, see (a) Jurczak, J.; Golebiowski, A. *Chem. Rev.* **1989**, *89*, 149. (b) Reetz, M. T. *Angew. Chem. Int. Ed. Engl.* **1991**, *30*, 1531. (c) Sardina, F. J.; Rapoport, H. *Chem. Rev.* **1996**, *96*, 1825.

32. Wenglowsky, S.; Hegedus, L. S. *J. Am. Chem. Soc.* **1998**, *120*, 12468.

33. (a) Meyers, A. I.; Williams, D. R.; Druelinger, M. *J. Am. Chem. Soc.* **1976**, *98*, 3032. (b) Hashimoto, S.; Koga, K. *Tetrahedron Lett.* **1978**, 573. (c) Hashimoto, S.; Koga, K. *Chem. Pharm. Bull.* **1979**, *27*, 2760.

34. (a) Corey, E. J.; Enders, D. *Tetrahedron Lett.* **1976**, 3. (b) Corey, E. J.; Enders, D.; Bock, M. G. *Tetrahedron Lett.* **1976**, 7. (c) Corey, E. J.; Enders, D. *Tetrahedron Lett.* **1976**, 11. (d) Enders, D.; Weuster, P. *Tetrahedron Lett.* **1978**, 2853.

35. (a) Enders, D.; Eichenauer, H. *Chem. Ber.* **1979**, *112*, 2933. (b) Kitamoto, M.; Hiroi, K.; Terashima, S. Yamada, S. *Chem. Pharm. Bull.* **1974**, *22*, 459.

36. (a) Enders, D.; Eichenauer, H. *Angew. Chem. Int. Ed. Engl.* **1976**, *15*, 549. (b) Enders, D.; Eichenauer, H. *Chem. Ber.* **1979**, *112*, 2933.

37. (a) Enders, D.; Eishenauer, H.; Pieter, R. *Chem. Ber.* **1979**, *112*, 3703. (b) Hardy, P. M. *Synthesis* **1978**, 290.

38. Enders, D., Eichenauer, H. *Angew. Chem. Int. Ed. Engl.* **1979**, *18*, 397.

39. (a) Enders, D.; Hundertmark, T.; Lazny, R. *Synlett* **1998**, 721. (b) Enders, D.; Schäfer, T.; Mies, W. *Tetrahedron* **1998**, *54*, 10232.

40. Related reviews and publications: (a) Enders, D.; Bhushan, V. *Tetrahedron Lett.* **1988**, *29*, 2437. (b) Enders, D.; Lohray, B. B. *Angew. Chem. Int. Ed. Engl.* **1987**, *26*, 351. (c) Synthesis of (*R*, *R*)-Statin: Enders, D.; Reinhold, U. *Angew. Chem. Int. Ed. Engl.* **1995**, *34*, 1219. (d) Synthesis of β-amino acids: Enders, D.; Wahl, H.; Bettary, W. *Angew. Chem. Int. Ed. Engl.* **1995**, *34*, 455.

41. Weber, T.; Edwards, J. P.; Denmark, S. E. *Synlett* **1989**, 20.

42. Cornforth, J. W. *Heterocycl. Compounds* **1957**, *5*, 386.

43. Wehrmeister, H. L. *J. Org. Chem.* **1962**, *27*, 4418.

44. Allen, P.; Ginos, J. *J. Org. Chem.* **1963**, *28*, 2759.

45. Meyers, A. I.; Knaus, G.; Kamata, K. *J. Am. Chem. Soc.* **1974**, *96*, 268.

46. Meyers, A. I.; Knaus, G.; Kamata, K.; Ford, M. E. *J. Am. Chem. Soc.* **1976**, *98*, 567.

47. Meyers, A. I.; Knaus, G. *J. Am. Chem. Soc.* **1974**, *96*, 6508.

48. Meyers, A. I.; Knaus, G.; Kendall, P. M. *Tetrahedron Lett.* **1974**, 3495.

49. Meyers, A. I.; Knaus, G. *Tetrahedron Lett.* **1974**, 1333.

50. Meyers, A. I.; Mihelich, E. D. *J. Org. Chem.* **1975**, *40*, 1186.

51. Byström, S.; Högberg, H.; Norin, T. *Tetrahedron* **1981**, *37*, 2249.

52. (a) Meyers, A. I.; Whitten, C. E. *J. Am. Chem. Soc.* **1975**, *97*, 6266. (b) Meyers, A. I.; Smith, R. K.; Whitten, C. E. *J. Org. Chem.* **1979**, *44*, 2250.

53. Oppolzer, W.; Moretti, R.; Thomi, S. *Tetrahedron Lett.* **1989**, *30*, 5603.

54. For reviews, see (a) Oppolzer, W.; Chapuis, S.; Bernardinelli, G. *Helv. Chim. Acta* **1984**, *67*, 1397. (b) Vandewalle, M.; Van der Eycken, J.; Oppolzer, W.; Vullioud, C.; *Tetrahedron* **1986**, *42*, 4035. (c) Davis, F. A; Towson, J. C.; Weismiller, M. C.; Lal, S.; Carroll, P. J. *J. Am. Chem. Soc.* **1988**, *110*, 8477.

55. Oppolzer, W.; Moretti, R.; Thomi, S. *Tetrahedron Lett.* **1989**, *30*, 6009.

56. Oppolzer, W.; Kingma, A. J. *Helv. Chim. Acta* **1989**, *72*, 1337.

57. Oppolzer, W. *Pure Appl. Chem.* **1990**, *62*, 1241.

58. Meyer, A. I.; Harre, M.; Garland, R. *J. Am. Chem. Soc.* **1984**, *106*, 1146.

59. Meyer, A. I.; Wanner, K. T. *Tetrahedron Lett.* **1985**, *26*, 2047.

60. Meyers, A. I.; Lefker, B. A. *Tetrahedron Lett.* **1987**, *28*, 1745.

61. (a) Romo, D.; Meyers, A. I. *Tetrahedron* **1991**, *47*, 9503. (b) Meyers, A. I.; Brengel, G. P. *Chem. Commun.* **1997**, 1.

62. (a) Wender, P. A.; Keenan, R. M.; Lee, H. Y. *J. Am. Chem. Soc.* **1987**, *109*, 4390. (b) Arlt, D.; Jautelat, M.; Lantzsch, R. *Angew. Chem. Int. Ed. Engl.* **1981**, *20*, 703.

63. Meyers, A. I.; Romine, J. L.; Fleming, S. A. *J. Am. Chem. Soc.* **1988**, *110*, 7245.

64. Corey, E. J.; Chaykovsky, M. *J. Am. Chem. Soc.* **1965**, *87*, 1353.

65. Fuji, K.; Node, M.; Nagasawa, H.; Naniwa, Y.; Terada, S. *J. Am. Chem. Soc.* **1986**, *108*, 3855.

66. Rajappa, S. *Tetrahedron* **1981**, *37*, 1453.

67. (a) Severin, T.; Pehr, H. *Chem. Ber.* **1979**, *112*, 3559. (b) Corey, E. J.; Estreicher, H. *J. Am. Chem. Soc.* **1978**, *100*, 6294. (c) Miyashita, M.; Yanami, T.; Kumazawa, T.; Yoshikoshi, A. *J. Am. Chem. Soc.* **1984**, *106*, 2149.

68. Fuji, K.; Node, M.; Nagasawa, H.; Naniwa, Y.; Terada, S. *J. Am. Chem. Soc.* **1986**, *108*, 3855.

69. Fuji, K. *Chem. Rev.* **1993**, *93*, 2037.

70. Seebach, D.; Sting, A. R.; Hoffmann, M. *Angew. Chem. Int. Ed. Engl.* **1996**, *35*, 2708.

71. Fuji, K.; Kawabata, T. *Chem. Eur. J.* **1998**, *4*, 373.

72. Matsushita, M.; Maeda, H.; Kodama, M. *Tetrahedron Lett.* **1998**, *39*, 3749.

73. Maruoka, K.; Ooi, T.; Yamamoto, H. *J. Am. Chem. Soc.* **1989**, *111*, 6431.

74. Shioiri, T.; Ninomiya, K.; Yamada, S. *J. Am. Chem. Soc.* **1972**, *94*, 6203.

75. Schöllkopf, U. *Pure Appl. Chem.* **1983**, *55*, 1799.

76. Seebach, D.; Aebi, J. D. *Tetrahedron Lett.* **1983**, *24*, 3311.

77. William, R. M. in Hasser, A. ed. *Advances in Asymmetric Synthesis*, vol. 1, JAI Press, London, **1995**, pp 45–94.

78. Corey, E. J.; Guzman-Perez, A. *Angew. Chem. Int. Ed. Engl.* **1998**, *37*, 388.

79. (a) Chinchilla, R.; Galindo, N.; Nájera, C. *Tetrahedron Asymmetry* **1998**, *9*, 2769. (b) Carloni, A.; Porzi, G.; Sandri, S. *Tetrahedron Asymmetry* **1998**, *9*, 2987.

80. Fitzi, R.; Seebach, D. *Angew. Chem. Int. Ed. Engl.* **1986**, *25*, 345.

81. Gilday, J. P.; Gallucci, J. C.; Paquette, L. A. *J. Org. Chem.* **1989**, *54*, 1399.

82. Larcheveque, M.; Ignatova, E.; Cuvigny, T. *Tetrahedron Lett.* **1978**, 3961.

83. (a) Fuji, K.; Node, M.; Tanaka, F. *Tetrahedron Lett.* **1990**, *31*, 6553. (b) Fuji, K.; Node, M.; Tanaka, F.; Hosoi, S. *Tetrahedron Lett.* **1989**, *30*, 2825.

84. Evans, D. A.; Takacs, J. M. *Tetrahedron Lett.* **1980**, *21*, 4233.

85. (a) Evans, D. A.; Ennis, M. D.; Mathre, D. J. *J. Am. Chem. Soc.* **1982**, *104*, 1737. (b) Evans, D. A.; Urpi, F.; Somers, T. C.; Clark, J. S.; Bilodeau, M. T. *J. Am. Chem. Soc.* **1990**, *112*, 8215.

86. (a) Meyers, A. I. *Pure Appl. Chem.* **1979**, *51*, 1255. (b) Meyers, A. I.; Knaus, G.; Kamata, K.; Ford, M. E. *J. Am. Chem. Soc.* **1976**, *98*, 567.

87. Ahn, K. H.; Lim, A.; Lee, S. *Tetrahedron Asymmetry* **1993**, *4*, 2435.

88. Oppolzer, W. *Tetrahedron* **1987**, *43*, 1969.

89. Jeong, K.; Parris, K.; Ballester, P.; Rebek, J. *Angew. Chem. Int. Ed. Engl.* **1990**, *29*, 555.

90. Negrete, G. R.; Konopelski, J. P. *Tetrahedron Asymmetry* **1991**, *2*, 105.

91. Fraser, R. R.; Akiyama, F.; Banville, J. *Tetrahedron Lett.* **1979**, 3929.

92. Meyers, A. J.; Williams, D. R.; Erickson, G. W.; White, S.; Druelinger, M. *J. Am. Chem. Soc.* **1981**, *103*, 3081.

93. Saigo, K.; Kasahara, A.; Ogawa, S.; Nohira, H. *Tetrahedron Lett.* **1983**, *24*, 511.

94. Enders, D.; Gatzweiler, W.; Dederichs, E. *Tetrahedron* **1990**, *46*, 4757.

95. Enders, D.; Zamponi, A.; Raabe, G. *Synlett* **1992**, 897.

96. For formation of acetals, see Ott, J.; Ramos Tombo, G. M.; Schimd, B.; Venanzi, L. M.; Wang, G.; Ward, T. R. *Tetrahedron Lett.* **1989**, *30*, 6151.

97. Mori, A.; Fujiwara, J.; Maruoka, K.; Yamamoto, H. *Tetrahedron Lett.* **1983**, *24*, 4581.

98. Solladie-Cavallo, A.; Suffert, J.; Gordon, M. *Tetrahedron Lett.* **1988**, *29*, 2955.

99. (a) Seebach, D.; Imwinkelried, R.; Stucky, G. *Angew. Chem. Int. Ed. Engl.* **1986**, *25*, 178. (b) Schreiber, S. L.; Reagan, J. *Tetrahedron Lett.* **1986**, *27*, 2945.

100. Harada, T.; Egusa, T.; Kinugasa, M.; Oku, A. *Tetrahedron Lett.* **1998**, *39*, 5531.

101. Harada, T.; Egusa, T.; Oku, A. *Tetrahedron Lett.* **1998**, *39*, 5535.

102. Alexakis, A.; Mangeney, P. *Tetrahedron Asymmetry* **1990**, *1*, 477.

103. Oguni, N.; Omi, T. *Tetrahedron Lett.* **1984**, *25*, 2823.

104. Berger, S.; Langer, F.; Lutz, C.; Knochel, P.; Mobley, T. A.; Reddy, C. K. *Angew. Chem. Int. Ed. Engl.* **1997**, *36*, 1496.

105. Seebach, D.; Beck, A. K. *Chimia* **1997**, *51*, 293.

106. (a) Zhang, F.; Yip, C.; Cao, R.; Chan, A. S. C. *Tetrahedron Asymmetry* **1997**, *8*, 585. (b) Mori, M.; Nakai, T. *Tetrahedron Lett.* **1997**, *38*, 6233.

107. Ito, K.; Kimura, Y.; Okamura, H.; Katsuki, T. *Synlett* **1992**, 573.

108. Ramón, D. J.; Yus, M. *Tetrahedron Asymmetry* **1997**, *8*, 2479.

109. Soai, K.; Yokoyama, S.; Hayasaka, T. *J. Org. Chem.* **1991**, *56*, 4264.

110. Kitamura, M.; Suga, S.; Kawai, K.; Noyori, R. *J. Am. Chem. Soc.* **1986**, *108*, 6071.

111. Behnen, W.; Mehler, T.; Martens, J. *Tetrahedron Asymmetry* **1993**, *4*, 1413.

112. Soai, K. Watanabe, M. *J. Chem. Soc. Chem. Commun.* **1990**, 43.

113. Soai, K.; Watanabe, M.; Koyano, M. *J. Chem. Soc. Chem. Commun.* **1989**, 534.

114. Wallbaum, S.; Martens, J. *Tetrahedron Asymmetry* **1993**, *4*, 637.

115. Wally, H.; Widhalm, M.; Weissensteiner, W.; Schlögl, K. *Tetrahedron Asymmetry* **1993**, *4*, 285.

116. Kitamura, M.; Okata, S.; Suga, S.; Noyori, R. *J. Am. Chem. Soc.* **1989**, *111*, 4028.

117. Watanabe, M.; Araki, S.; Butsugan, Y. *J. Org. Chem.* **1991**, *56*, 2218.

118. Ho, D. E.; Betancort, J. M.; Woodmansee, D. H.; Larter, M. L.; Walsh, P. J. *Tetrahedron Lett.* **1997**, *38*, 3867.

119. Pritchett, S.; Woodmansee, D. H.; Davis, T. J.; Walsh, P. J. *Tetrahedron Lett.* **1998**, *39*, 5941.

120. Ito, Y. N.; Ariza, X.; Beck, A. K.; Boháe, A.; Ganter, C.; Gawley, R. E.; Kühnle, F. N. M.; Tuleja, J.; Wang, Y. M.; Seebach, D. *Helv. Chim. Acta* **1994**, *77*, 2071.

121. Prasad, K. R. K.; Joshi, N. N. *J. Org. Chem.* **1997**, *62*, 3770.

122. Zhang, F.; Chan, A. S. C. *Tetrahedron Asymmetry* **1997**, *8*, 3651.

123. Cram, D. J.; Helgeson, R. C.; Peacock, S. C.; Kaplan, L. J.; Domeier, L. A.; Moreau, P.; Koga, K.; Mayer, J. M.; Chao, Y.; Siegel, M. G.; Hoffman, D. H.; Sogah, G. D. Y. *J. Org. Chem.* **1978**, *43*, 1930.

124. Cotton, F. A.; Wilkinson, G. *Advanced Inorganic Chemistry*, 4th ed., Wiley, New York, **1980**, 342.

125. Chan, A. S. C.; Zhang, F.; Yip, C. *J. Am. Chem. Soc.* **1997**, *119*, 4080.

126. Gibson, C. L. *J. Chem. Soc. Chem. Commun.* **1996**, 645.

127. Kang, J.; Lee, J.; Kin, J. *J. Chem. Soc. Chem. Commun.* **1994**, 2009.

128. Soai, K.; Suzuki, T.; Shono, T. *J. Chem. Soc. Chem. Commun.* **1994**, 317.

129. Wirth, T. *Tetrahedron Lett.* **1995**, *36*, 7849.

130. Jin, M.; Ahn, S.; Lee, K. *Tetrahedron Lett.* **1996**, *37*, 8767.

131. Williams, D. R.; Fromhold, M. G. *Synlett* **1997**, 523.

132. Bringmann, G.; Breuning, M. *Tetrahedron Asymmetry* **1998**, *9*, 667.

133. Brunel, J.; Constantieux, T.; Legrand, O.; Buono, G. *Tetrahedron Lett.* **1998**, *39*, 2961.

134. Kossenjans, M.; Martens, J. *Tetrahedron Asymmetry* **1998**, *9*, 1409.

135. Watanabe, M.; Hashimoto, N.; Araki, S.; Butsugan, Y. *J. Org. Chem.* **1992**, *57*, 742.

136. Kimura, K.; Sugiyama, E.; Ishizuka, T.; Kunieka, T. *Tetrahedron Lett.* **1992**, *33*, 3147.

137. Cho, B. T.; Chun, Y. S. *Tetrahedron Asymmetry* **1998**, *9*, 1489.

138. (a) Tomioka, K. *Synthesis* **1990**, 541. (b) Oguni, N.; Matsuda, Y.; Kaneko, T. *J. Am. Chem. Soc.* **1988**, *110*, 7877. (c) Kitamura, M.; Okada, S.; Suga, S.; Noyori, R. *J. Am. Chem. Soc.* **1989**, *111*, 4028. (d) Oppolzer, W.; Radinov, R. N. *Tetrahedron Lett.* **1988**, *29*, 5645. (e) Yoshioka, M.; Kawakita, T.; Ohno, M. *Tetrahedron Lett.* **1989**, *30*, 1657. (f) Zhang, X.; Guo, C. *Tetrahedron Lett.* **1995**, *36*, 4947. (g) Noyori, R.; Suga, S.; Kawai, K.; Okada, S.; Kitamura, M. *Pure Appl. Chem.* **1988**, *60*, 1597. (h) Smaardijk, A. A.; Wynberg, H. *J. Org. Chem.* **1987**, 52, 135. (i) Corey, E. J.; Yuen, P. W.; Hannon, F. J.; Wierda, D. A. *J. Org. Chem.* **1990**, *55*, 784. (j) Giffels, G.; Dreisbach, C.; Kragl, U.; Weigerding, M.; Waldmann, H.; Wandrey, C. *Angew. Chem. Int. Ed. Engl.* **1995**, *34*, 2005.

139. Dosa, P. I.; Fu, G. C. *J. Am. Chem. Soc.* **1998**, *120*, 445.

140. (a) Ramón, D. J.; Yus, M. *Tetrahedron* **1998**, *54*, 5651. (b) Ramón, D. J.; Yus, M. *Tetrahedron Lett.* **1998**, *39*, 1239.

141. Nakamura, M.; Hirai, A.; Sogi, M.; Nakamura, E. *J. Am. Chem. Soc.* **1998**, *120*, 5846.

142. Reetz, M. T.; Kyung, S.; Bolm, C.; Zierke, T. *Chem. Ind.* **1986**, 824.

143. (a) Narasaka, K.; Yamada, T.; Minamikawa, H. *Chem. Lett.* **1987**, 2073. (b) Minamikawa, H.; Hayakawa, S.; Yamada, T.; Iwasawa, N.; Narasaka, K. *Bull. Chem. Soc. Jpn.* **1988**, *61*, 4379.

144. Hayashi, M.; Matsuda, T.; Oguni, N. *J. Chem. Soc. Chem. Commun.* **1990**, 1364.

145. Mori, A.; Ohno, H.; Nitta, H.; Tanaka, K.; Inoue, S. *Synlett* **1991**, 563.

146. (a) Hayashi, M.; Inoune, T.; Miyamoto, Y.; Oguni, N. *Tetrahedron* **1994**, *50*, 4385. (b) Hayashi, M.; Miyamoto, Y.; Inoue, T.; Oguni, N. *J. Org. Chem.* **1993**, *58*, 1515. (c) Hayashi, M.; Tamura, M.; Oguchi, N. *Synlett* **1992**, 663. (d) Effenberger, F.; Heid, S. *Tetrahedron Asymmetry* **1995**, *6*, 2945. (e) Cainelli, G.; Giacomini, D.; Treré, A.; Galletti, P. *Tetrahedron Asymmetry* **1995**, *6*, 1593. (f) Klempier, N.; Pichler, U.; Griengl, H. *Tetrahedron Asymmetry* **1995**, *6*, 845. (g) Belokon, Y.; Ikonnikov, N.; Moscalenko, M.; North, M.; Orlova, S.; Tararov, V.; Yashkina, L. *Tetrahedron Asymmetry* **1996**, *7*, 851. (h) North, M. *Synlett* **1993**, 807.

147. Hayashi, M.; Miyamoto, Y.; Inoue, T.; Oguni, N. *J. Chem. Soc. Chem. Commun.* **1991**, 1752.

148. Bolm, C.; Müller, P. *Tetrahedron Lett.* **1995**, *36*, 1625.

149. Mori, M.; Imma, H.; Nakai, T. *Tetrahedron Lett.* **1997**, *38*, 6229.

150. Sigman, M. S.; Jacobsen, E. N. *J. Am. Chem. Soc.* **1998**, *120*, 5315.

151. Belokon', Y. N.; Caveda-Cepas, S.; Green, B.; Ikonnikov, N. S.; Khrustalev, V. N.; Larichev, V. S.; Moscalenko, M. A.; North, M.; Orizu, C.; Tararov, V. I.; Tasinazzo, M.; Timofeeva, G. I.; Yashkina, L. V. *J. Am. Chem. Soc.* **1999**, *121*, 3968.

152. Sikorshi, J. A.; Miller, M. J.; Braccolino, D. S.; Cleary, D. G.; Corey, S. D.; Font, J. L.; Gruys, K. J.; Han, C. Y.; Lin, K. C.; Pansegrau, Ream, J. E.; Schnur, D.; Shah, A.; Walker, M. C. *Phosphorous Sulfur Silicon* **1993**, *76*, 115.

153. Stowasser, B.; Budt, K.; Li, J.; Peyman, A.; Ruppert, D. *Tetrahedron Lett.* **1992**, *33*, 6625.

154. Kametani, T.; Kigasawa, K.; Hiiragi, M.; Wakisaka, K.; Haga, S.; Sugi, H.; Tanigawa, K.; Suzuki, Y.; Fukawa, K.; Irino, O.; Saita, O.; Yamabe, S. *Hetereocycles* **1981**, *16*, 1205.

155. Meier, C.; Laux, W. H. G. *Tetrahedron Asymmetry* **1995**, *6*, 1089.

156. Pogatchnik, D. M.; Wiemer, D. F. *Tetrahedron Lett.* **1997**, *38*, 3495.

157. Davis, F. A.; Chen, B. C. *Chem. Rev.* **1992**, *92*, 919.

158. Rath, N. P.; Spilling, C. D. *Tetrahedron Lett.* **1994**, *35*, 227.

159. (a) Yokomatsu, T.; Yamagishi, T.; Shibuya, S. *Tetrahedron Asymmetry* **1993**, *4*, 1779. (b) Yokomatsu, T.; Yamagishi, T.; Shibuya, S. *Tetrahedron Asymmetry* **1993**, *4*, 1783.

160. Sasai, H.; Suzuki, T.; Arai, S.; Arai, T.; Shibasaki, M. *J. Am. Chem. Soc.* **1992**, *114*, 4418.

161. Sasai, H.; Bougauchi, M.; Arai, T.; Shibasaki, M. *Tetrahedron Lett.* **1997**, *38*, 2717.

162. Sasai, H.; Arai, T.; Satow, Y.; Houk, K. N.; Shibasaki, M. *J. Am. Chem. Soc.* **1995**, *117*, 6194.

163. Groaning, M. D.; Rowe, B. J.; Spilling, C. D. *Tetrahedron Lett.* **1998**, *38*, 5485.

164. Smith III, A. B.; Yager, K. M.; Taylor, C. M. *J. Am. Chem. Soc.* **1995**, *117*, 10879.

■■■■■■ CHAPTER 3

Aldol and Related Reactions

3.1 INTRODUCTION

Chapter 2 provided a general introduction to the α-alkylation of carbonyl compounds, as well as the enantioselective nucleophilic addition on carbonyl compounds. Chiral auxiliary aided α-alkylation of a carbonyl group can provide high enantioselectivity for most substrates, and the hydrazone method can provide routes to a large variety of α-substituted carbonyl compounds. While α-alkylation of carbonyl compounds involves the reaction of an enolate, the well known aldol reaction also involves enolates.

Aldol reactions refer to the condensation of a nucleophilic enolate species with an electrophilic carbonyl moiety along with its analogs. These reactions are among those transformations that have greatly simplified the construction of asymmetric C–C bonds and, thus, satisfied the most stringent requirements for asymmetric organic synthesis methodology. Numerous examples of asymmetric aldol reactions can be found for syntheses of both complex molecules and small optically active building blocks.[1]

Acyclic stereocontrol has been a striking concern in modern organic chemistry, and a number of useful methods have been developed for stereoregulated synthesis of conformationally nonrigid complex molecules such as macrolide and polyether antibiotics. Special attention has therefore been paid to the aldol reaction because it constitutes one of the fundamental bond constructions in biosynthesis.

In the synthesis of complex natural products, one is frequently confronted with the task of creating intermediates possessing multiple contiguous stereogenic centers. The most efficient synthetic strategies for such compounds are those in which the joining of two subunits results in the simultaneous creation of adjacent stereocenters. To have a better understanding of the aldol and related reactions, it is essential to be familiar with the foregoing strategies. When using them it is desirable to exert control over relative (*syn/anti*) as well as absolute (*R/S*) stereochemistry. Many studies have focused on the diastereoselective (enantioselective) aldol reactions. The major control variables in these asymmetric aldol reactions are the metal counterions, the ligands binding to these metals, and the reaction conditions. Several approaches are available for imposing asymmetric control in aldol reactions:

1. Substrate control: This refers to the addition of an achiral enolate (or allyl metal reagent) to a chiral aldehyde (generally bearing a chiral center at the α-position). In this case, diastereoselectivity is determined by transition state preference according to Cram-Felkin-Ahn considerations.[2]
2. Reagent control: This involves the addition of a chiral enolate or allyl metal reagent to an achiral aldehyde. Chiral enolates are most commonly formed through the incorporation of chiral auxiliaries in the form of esters, acyl amides (oxazolines), imides (oxazolidinones) or boron enolates. Chiral allyl metal reagents are also typically joined with chiral ligands.
3. Double stereodifferentiation: This refers to the addition of a chiral enolate or allyl metal reagent to a chiral aldehyde. Enhanced stereoselectivity can be obtained when the aldehyde and reagent exhibit complementary facile preference (matched case). Conversely, diminished results might be observed when their facial preference is opposed (mismatched pair).

When chelated with proper chiral ligands, enolates of many metals (such as Li, Mg, Zr, B, Al, Sb, Si, and Ti) can afford good stereoselectivity in asymmetric aldol reactions. Lithium and magnesium form chelates that can offer selectivity through Cram-Felkin-Ahn or chelation-controlled additions. The applications of titanium in particular are marvelous and diverse, and titanium enolates containing chiral ligands present an important area of enantioselective transformations. Similarly, boron enolates are widely used because of their high enantioselectivity. Heterobimetallic catalysts and/or two carbon-center catalysts activate both nucleophiles and electrophiles, thus being very good catalysts for asymmetric aldol reactions. These issues are further discussed in subsequent sections of this chapter.

It is only since the early 1980s that significant progress has been made with aldol reactions. This chapter introduces some of the most important developments on the addition of metallic enolates and the more important of the related allylic metal derivatives to carbonyl compounds. These processes are depicted as paths A and B in Scheme 3–1.

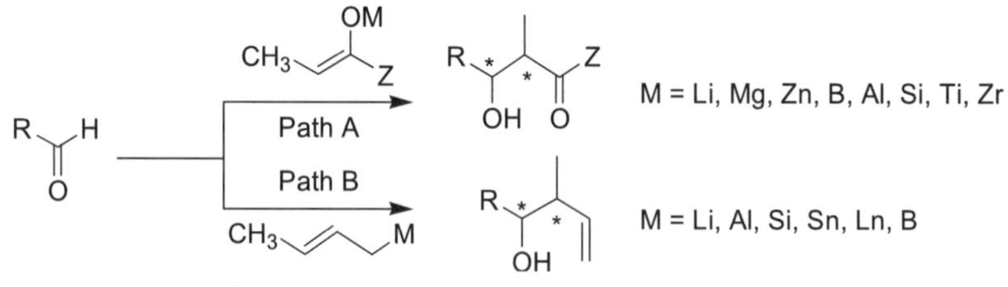

Scheme 3–1

In general, the aldol reaction of an aldehyde with metal enolate creates two new chiral centers in the product molecule, and this may lead to four possible stereoisomers **2a**, **2b**, **2c**, and **2d** (Scheme 3–2 and Fig. 3–1).

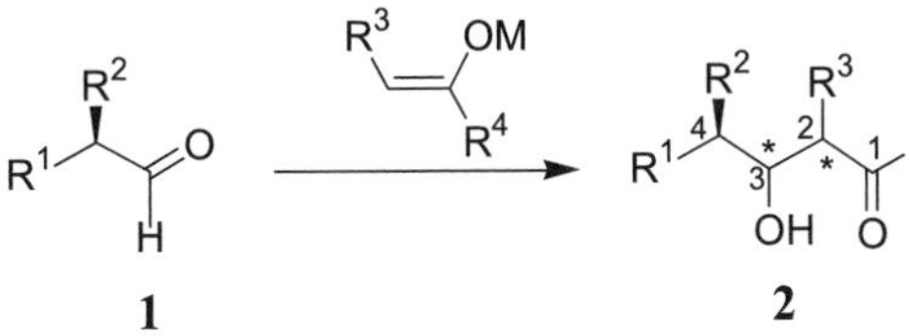

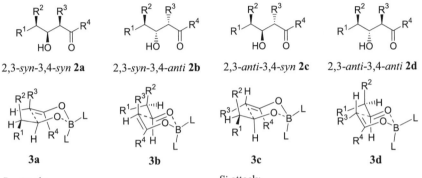

Scheme 3–2

2,3-*syn*-3,4-*syn* **2a**   2,3-*syn*-3,4-*anti* **2b**   2,3-*anti*-3,4-*syn* **2c**   2,3-*anti*-3,4-*anti* **2d**

3a   **3b**   **3c**   **3d**

Re-attack: *Si*-attack:
(*Z*)-enolate + **1** via **3a** to **2a**: 2,3-*syn*, 3,4-*syn*   (*Z*)-enolate + **1** via **3b** to **2b**: 2,3-*syn*, 3,4-*anti*
(*E*)-enolate + **1** via **3c** to **2c**: 2,3-*anti*, 3,4-*syn*   (*E*)-enolate + **1** via **3d** to **2d**: 2,3-*anti*, 3,4-*anti*

Figure 3–1. Routes to the aldol products **2a–d**.

Taking the boron-mediated aldol reaction as an example, one can conclude from the chair-like cyclic transition states **3a–d** (Zimmerman-Traxler model[3] as depicted in Fig. 3–1) that the enolate geometry can be translated into 2,3-stereochemistry in the product. One can see from Figure 3–1 that (*Z*)-enolate tends to give 2,3-*syn*-product, whereas (*E*)-enolate gives the 2,3-*anti*-one. The rationales of the high stereo-outcome are that the dialkylboron enolates have relatively short metal–oxygen bonds, and this is essential for maximizing 1,3-diaxial interactions in the transition states. The $R^1R^2CH–$ moiety occupies a more stable transition state, a pseudo-equatorial position, which leads to aldol products in high stereoselectivity.

The following parameters are critically important for stereochemical control:

1. The size of the substituent moiety in the enolate
2. The proper choice of reagents
3. The conditions chosen for enolization

Accordingly, Liu et al.[4] have designed two types of aldol reagents that can lead to opposite stereochemistry in aldol condensation reactions. In the following structures, compound **4** can be used for obtaining *anti*-aldol products, and compound **5** can be employed for synthesizing *syn*-aldol products (Scheme 3–3).

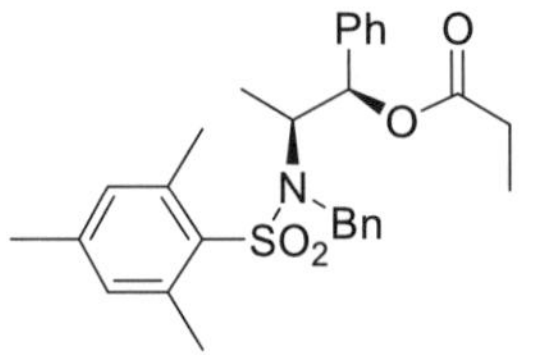

4

anti- aldol reagent

c-Hex$_2$BOTf/Et$_3$N

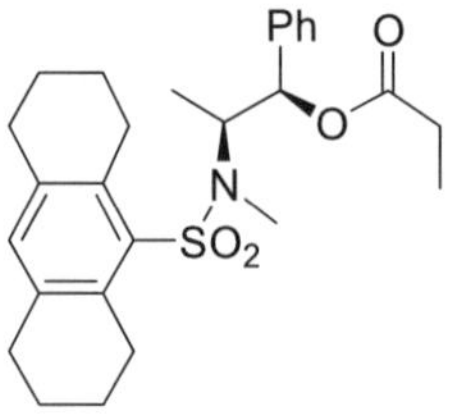

5

syn- aldol reagent

n-Bu$_2$BOTf/*i*-Pr$_2$NEt

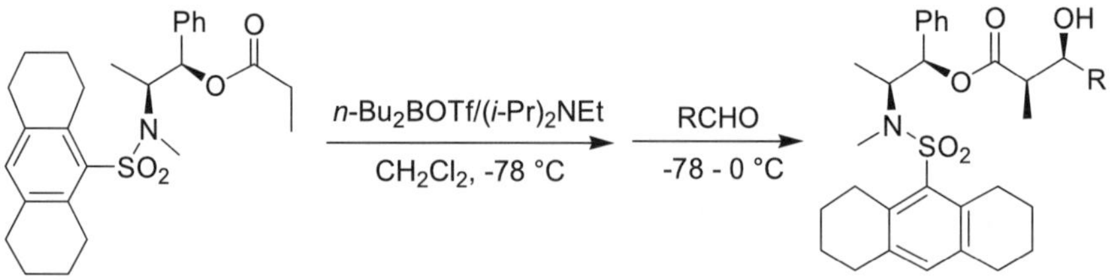

5

$$\xrightarrow[\text{CH}_2\text{Cl}_2, -78\ °\text{C}]{n\text{-Bu}_2\text{BOTf}/(i\text{-Pr})_2\text{NEt}} \xrightarrow[-78 - 0\ °\text{C}]{\text{RCHO}}$$

6

Scheme 3–3

When aldol reagent **5** is treated with aldehyde in the presence of *n*-Bu$_2$BOTf and Et$_3$N, *syn*-aldol product **6** can be produced with high diastereoselectivity (Table 3–1).

3.2   SUBSTRATE-CONTROLLED ALDOL REACTION

3.2.1   Oxazolidones as Chiral Auxiliaries: Chiral Auxiliary-Mediated Aldol-Type Reactions

In 1964, Mitsui et al.[5] used a chiral auxiliary to achieve asymmetric aldol condensation, although the stereoselectivity was not high (58%) at that time. Significant improvement came in the early 1980s when Evans et al.[6] and Masamune et al.[7] introduced a series of chiral auxiliaries that led to high stereo-

TABLE 3–1. Diastereoselective Aldol Reaction Using Chiral Reagent 5

Aldehyde	*syn:anti*	ds for *syn*	Yield (%)
EtCHO	93:7	97:3	95
PrCHO	94:6	>97:3	93
(*E*)-CH$_3$CH=CHCHO	93:7	>97:3	98
PhCHO	94:6	95:5	97

ds = diastereoselectivity.

Reprinted with permission by Pergamon-Elsevier Science Ltd., Ref. 4.

selectivity. When bonded to dialkylboron enolates, these chiral auxiliaries induced aldol reactions with high selectivity.

The chiral boron enolates generated from N-acyl oxazolidones such as **7** and **8** (which were named *Evans' auxiliaries* and have been extensively used in the α-alkylation reactions discussed in Chapter 2) have proved to be among the most popular boron enolates due to the ease of their preparation, removal, and recycling and to their excellent stereoselectivity.[8]

Usually, (Z)-boron enolates can be prepared by treating N-acyl oxazolidones with di-n-butylboron triflate and triethylamine in CH_2Cl_2 at $-78°C$, and the enolate then prepared can easily undergo aldol reaction at this temperature to give a *syn*-aldol product with more than 99% diastereoselectivity (Scheme 3–4). In this example, the boron counterion plays an important role in the stereoselective aldol reaction. Triethylamine is more effective than di-*iso*-propylethyl amine in the enolization step. Changing boron to lithium leads to a drop in stereoselectivity.

The stereoselectivity probably results from bidentate chelation of the metal (such as boron) with the oxazolidone carbonyl and the enolate oxygen via a chair-type transition state **9** (Scheme 3–4).[1a,9]

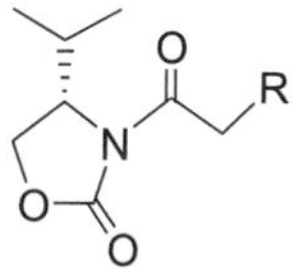

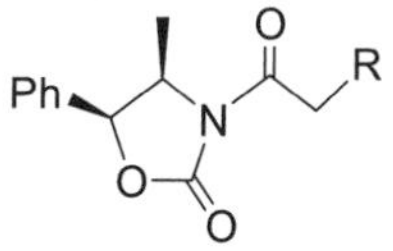

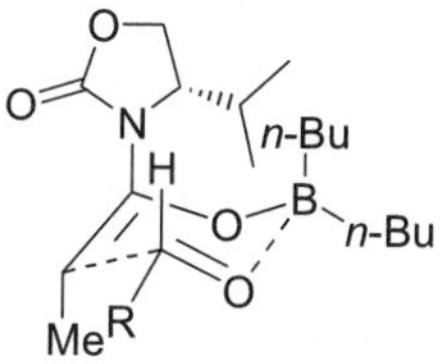

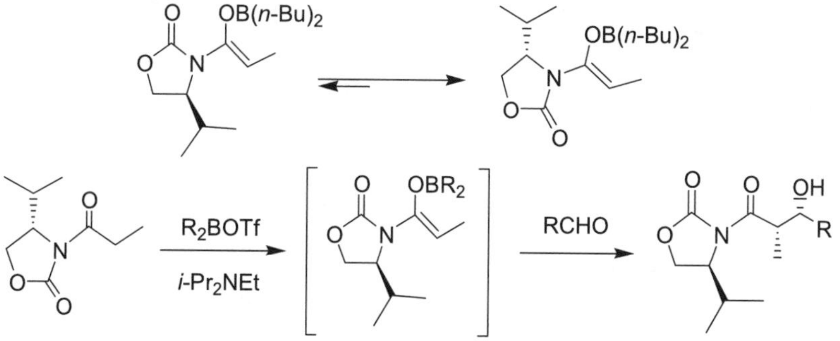

Scheme 3–4

When amide derivatives **10** and **12** are used in the reaction, a pair of enantiomers **11** and **13** (R = CH_3) can be obtained (Scheme 3–5).[6]

Double asymmetric induction (See section 1.5.3) can also be employed in aldol reactions. When chiral aldehyde **15** is treated with achiral boron-mediated enolate **14**, a mixture of diastereomers is obtained in a ratio of 1.75:1. However, when the same aldehyde **15** is allowed to react with enolates derived from Evans' auxiliary **8**, a *syn*-aldol product **16** is obtained with very high stereo-

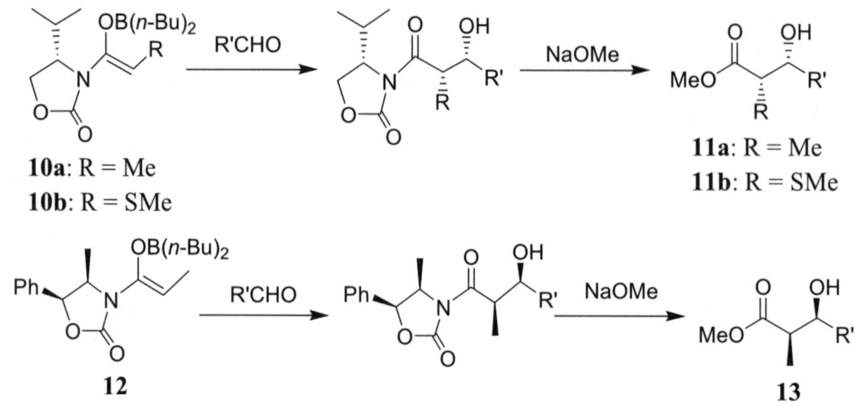

Scheme 3–5

selectivity. A diastereofacial ratio of 600:1 for the matched pair was obtained. Compound **16** can be easily transformed to the Prelog-Djerrassi lactone **17** via standard well-established procedures. Even in the case of a mismatched pair, for example, treatment of aldehyde **15** with another Evans' auxiliary **7**, which exerts the opposite function in terms of stereoselectivity in comparison with **8**, product **18** can still be obtained with highly satisfactory diastereoselectivity (400:1) (Scheme 3–6).[10]

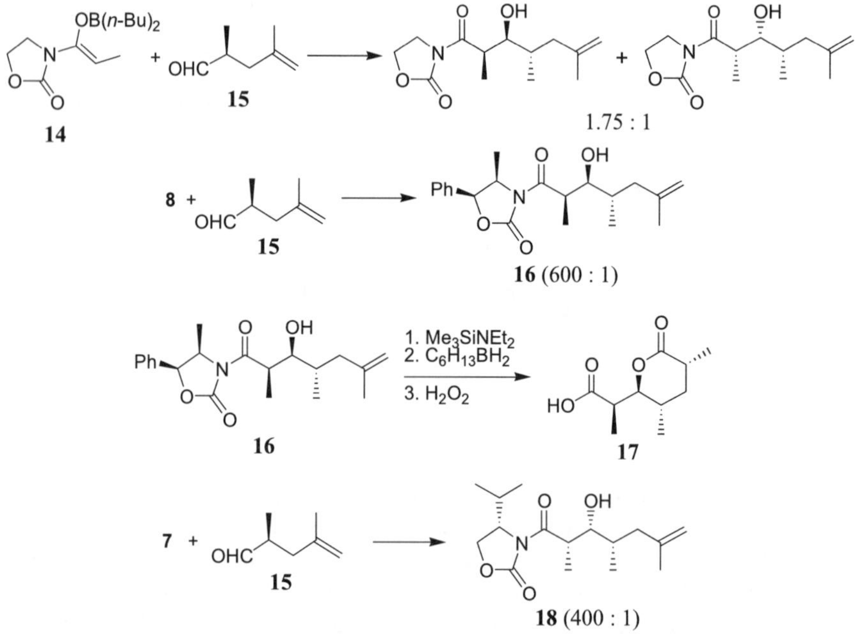

Scheme 3–6

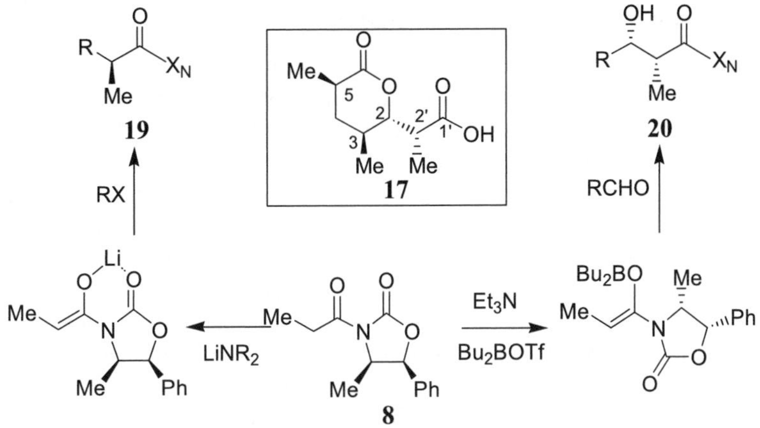

Scheme 3–7. X_N = chiral auxiliary.

Compound **17** is the so-called (+)-Prelog-Djerassi lactonic acid derived via the degradation of either methymycin or narbomycin. This compound embodies important architectural features common to a series of macrolide antibiotics and has served as a focal point for the development of a variety of new stereoselective syntheses. Another preparation of compound **17** is shown in Scheme 3–7.[11] Starting from **8**, by treating the boron enolate with an aldehyde, **20** can be synthesized via an asymmetric aldol reaction with the expected stereochemistry at C-2 and C-2′. Treating the lithium enolate of **8** with an electrophile affords **19** with the expected stereochemistry at C-5. Note that the stereochemistries in the aldol reaction and in α-alkylation are opposite each other. The combination of **19** and **20** gives the final product **17**.

Compound α-vinyl-β-hydroxyimide **21′**, which can be used in the total synthesis of natural products, can be prepared through aldol reaction. In most cases of aldol reactions mediated by **21**, products of more than 98% de can be obtained (Scheme 3–8).[12]

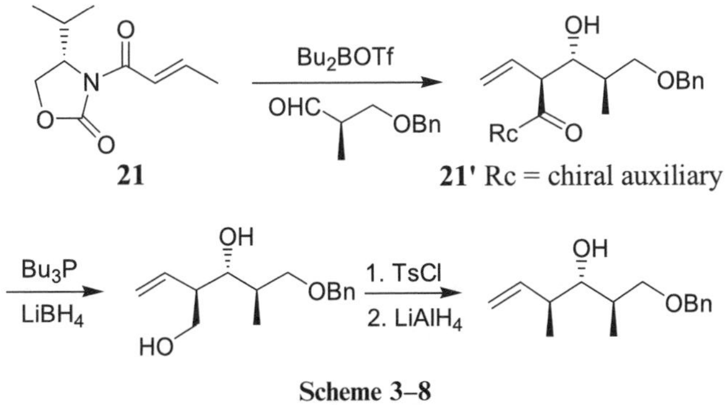

Scheme 3–8

3.2.2   Pyrrolidines as Chiral Auxiliaries

The frequent occurrence of β-hydroxy carbonyl moiety in a variety of natural products (such as macrolide or ionophore antibiotics or other acetogenics) has stimulated the development of stereocontrolled synthetic methods for these compounds. Indeed, the most successful methods have involved aldol reactions.[13]

Amide enolate **22** bearing a *trans*-2,5-disubstituted pyrrolidine moiety as the amine component has proved to be an excellent substrate in asymmetric alkylation[14] and acylation[15] reactions. In contrast to these successes, using its lithium enolate in an aldol reaction fails to give good stereoselectivity (entry 1 in Table 3–2). On the other hand, zirconium enolate[16] prepared from the corresponding lithium enolate and bis(cyclopentadienyl)zirconium dichloride exhibits a remarkably high stereoselectivity (entries 2–5 in Table 3–2).

Studies show that the Zr-bearing bulky ligand is exclusively located in the bottom hemisphere with respect to the plane of the (Z)-enolate. The aldehyde molecule coordinates with the Zr atom and approaches from the same side, adopting a chair-like transition state. This leads to the formation of *erythro*-aldols (Scheme 3–9 and **23**). For lithium enolate, the attack of alkyl or acyl halides in alkylation or acylation occurs directly on the top face of the enolate.

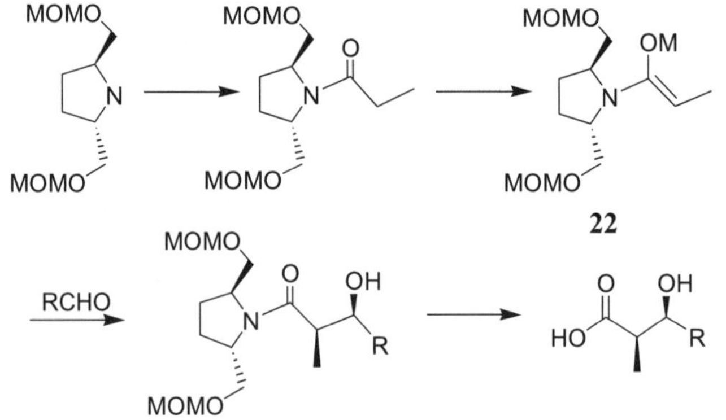

> 97% de

Scheme 3–9

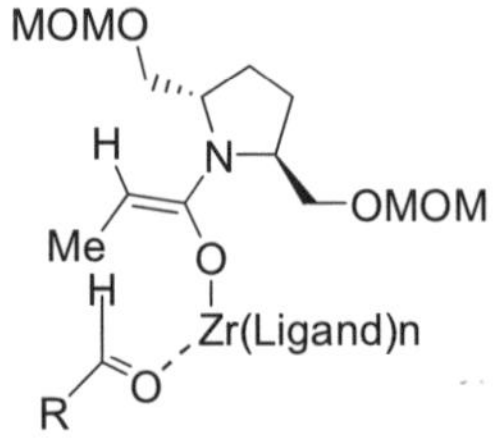

23

TABLE 3–2. Aldol Reaction of Propionamide Enolate with Aldehydes

			Ratio of diastereomers in product				Yield (%)
Entry	R in RCHO	Metal	erythro	erythro	threo	threo	
1	C_6H_5	Li	30	50	20	<1	82
			$(2S^*,5R^*,2'R^*,3'R^*)$			$(2S^*,5R^*,2'S^*,3'R^*)$	
2	C_6H_5	Zr	60	1	<1 (other two isomers)		88
			$(2S^*,5S^*,2'R^*,3'R^*)$				
3	$(CH_3)_2CH$	Zr	100		<1 (other three isomers)		85
			$(2S^*,5S^*,2'R^*,3'S^*)$				
4	CH_3CH_2	Zr	100		<1 (other three isomers)		92
			$(2S^*,5S^*,2'R^*,3'S^*)$				
5	$CH_3CH=CH$	Zr	100	1	<1 (other two isomers)		98
			$(2S^*,5S^*,2'R^*,3'R^*)$				

The importance of the sterically demanding metal centers in aldol regulations is thus apparent.

The stereochemical assignment of the condensation products reveals that asymmetric induction in this reaction is opposite to that observed in the previously discussed alkylation or acylation.

As with the above pyrrolidine, proline-type chiral auxiliaries also show different behaviors toward zirconium or lithium enolate–mediated aldol reactions. Evans found that lithium enolates derived from prolinol amides exhibit excellent diastereofacial selectivities in alkylation reactions (see Section 2.2.3.2), while the lithium enolates of proline amides are unsuccessful in aldol condensations. Effective chiral reagents were zirconium enolates, which can be obtained from the corresponding lithium enolates via metal exchange with Cp_2ZrCl_2. For example, excellent levels of asymmetric induction in the aldol process with *syn/ anti* selectivity of 96–98% and diastereofacial selectivity of 50–200:1[16a] can be achieved in the Zr-enolate–mediated aldol reaction (see Scheme 3–10).

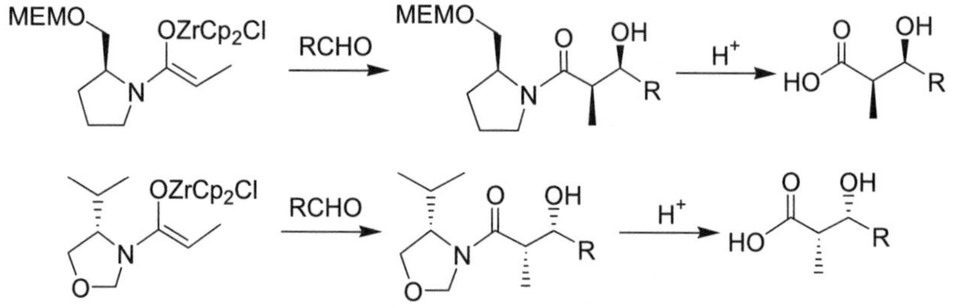

Scheme 3–10. MEM = methoxyethoxymethyl; Cp = cyclopentadienyl.

Acylated product **25** can be obtained by reacting the enolate of **24** with acyl chloride. Interestingly, *syn-* or *anti-***26** can be obtained upon treating the acylated enolate **25** with $Zn(BH_4)_2$ and KEt_3H, respectively (Scheme 3–11 and

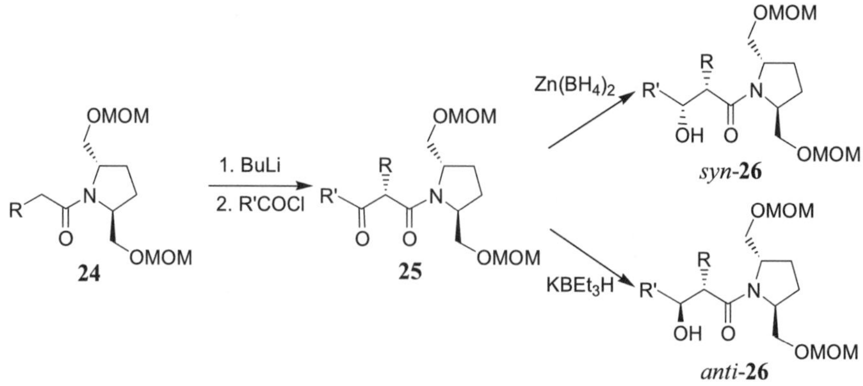

Scheme 3–11

TABLE 3–3. Asymmetric Acylation Using (2S,5S)-24 and Subsequent Stereoselective Reduction with Zinc Borohydride in THF at −78°C

			Acylation Product 25		Reduction Product 26	
Entry	R'	R	Yield (%)	de (%)	Yield (%)	syn:anti
1	C_2H_5	CH_3	74	98	96	99:1
2	$(CH_3)_2CH$	CH_3	95	98	99	>99:1
3	$(CH_3)_3C$	CH_3	80	98	99	96:4
4	C_9H_{15}	CH_3	90	98	98	>97:3
5	Ph	CH_3	90	98	93	>99:1
6	$PhCH_2$	CH_3	76	98	98	99:1
7	$(CH_3)_2CH$	$PhCH_2$	96	98	98	99:1
8	$CH_3CH=CH$	CH_3	47	98	97	97:3
9	$PhCH=CH$	CH_3	61	98	95	>99:1

de = Diastereomeric excess.

Table 3–3). Although the initial step is an α-acylation reaction, the final resultant compound can still be considered an aldol reaction product.

3.2.3 Aminoalcohols as the Chiral Auxiliaries

The aldol reactions introduced thus far have been performed under basic conditions where *enolate species* are involved as the reactive intermediate. In contrast to the commonly accepted carbon-anion chemistry, Mukaiyama developed another practical method in which *enol species* can be used as the key intermediates. He is the first chemist to successfully demonstrate that acid-catalyzed aldol reactions using Lewis acid (such as $TiCl_4$) and silyl enol ether as a stable enol equivalent can work as well.[17] Furthermore, he developed the boron trifluoromethane sulfonate (triflate)–mediated aldol reactions via the formation of formyl enol ethers.

(Silyloxy)alkenes were first reported by Mukaiyama as the requisite latent enolate equivalent to react with aldehydes in the presence of Lewis acid activators. This process is now referred to as the *Mukaiyama aldol reaction* (Scheme 3–12). In the presence of Lewis acid, *anti*-aldol condensation products can be obtained in most cases via the reaction of aldehydes and silyl ketene acetals generated from propionates under kinetic control.

Another chiral auxiliary for controlling the absolute stereochemistry in Mukaiyama aldol reactions of chiral silyl ketene acetals has been derived from *N*-methyl ephedrine.[18] This has been successfully applied to the enantioselective synthesis of various natural products[19] such as α-methyl-β-hydroxy esters (ee 91–94%),[18,20] α-methyl-β-hydroxy aldehydes (91% ee),[21] α-hydrazino and α-amino acids (78–91% ee),[22] α-methyl-δ-oxoesters (72–75% ee),[20b] *cis*- and *trans*-β-lactams (70–96% ee),[23] and carbapenem antibiotics.[24]

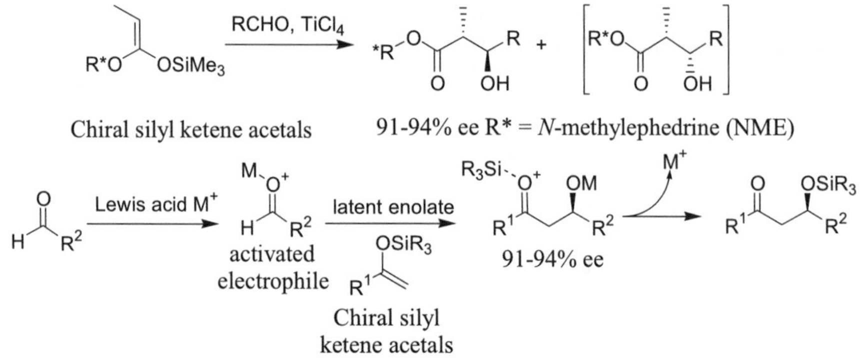

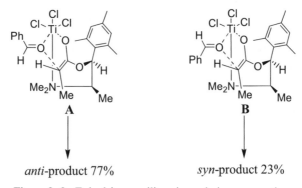

Scheme 3–12. Mukaiyama aldol reactions.

With chiral enol species (*E*)-silylketene acetal derived from (1*R*,2*S*)-*N*-methyl ephedrine-*O*-propionate, both the aldehyde carbonyl and the ephedrine NMe$_2$ group are expected to bind to TiCl$_4$, which usually chelates two electron-donating molecules to form *cis*-octahedral six-coordinated complexes.[25] Conformational freedom is therefore reduced, and the C–C bond formation occurs on the six-coordinated metal in a highly stereoselective manner.[18]

As shown in Figure 3–2, titanium is coordinated with the oxygen from both the aldehyde and the alkene enol silyl ether. When aldehyde approaches the enol species, intermediate **A** is favored to **B**, and *anti*-aldol is obtained as the major product. Table 3–4 presents some results of these reactions.

Ligands for catalytic Mukaiyama aldol addition have primarily included bidentate chelates derived from optically active diols,[26] diamines,[27] amino acid derivatives,[28] and tartrates.[29] Enantioselective reactions induced by chiral Ti(IV) complex have proved to be one of the most powerful stereoselective transformations for synthetic chemists. The catalytic asymmetric aldol reaction introduced by Mukaiyama is discussed in Section 3.4.1.

Catalyst **29**,[30] obtained by treating **28** with Ti(OPr^i)$_4$, does not give a good

anti-product 77%   *syn*-product 23%

Figure 3–2. Ephedrine auxiliary in mukaiyama reaction.

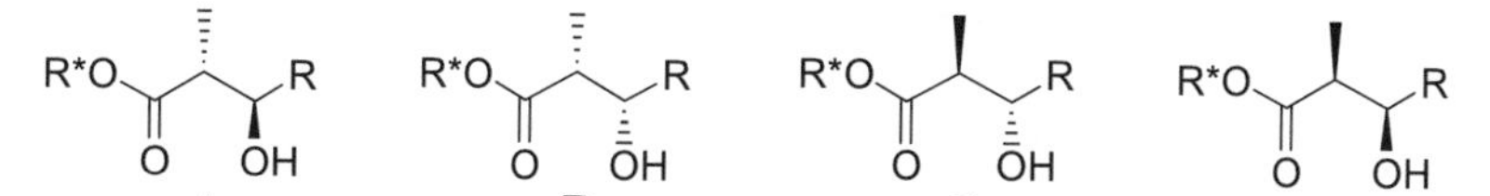

A **B** **C** **D**

TABLE 3–4. Reaction of Silyl Ketene Acetals with Aldehydes

Entry	R*OH	RCHO	*anti*/*syn*(**A** + **C**)/(**B** + **D**)	**A**/**C**
1	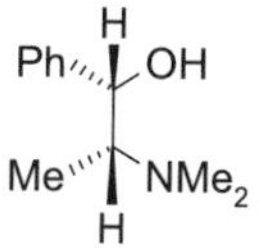	PhCHO	85:15	97:3
2	"	*n*-C_3H_7CHO	80:20	95.5:4.5
3	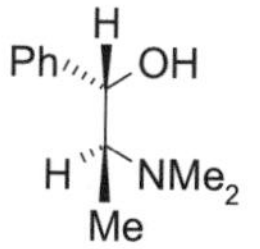	PhCHO	77:23	93:7

Reprinted with permission by Am. Chem. Soc., Ref. 18.

result. Both the yield and ee are low. This can be explained by reference to Figure 3–3, which indicates the nonspecific transfer of the TMS group to either the isopropoxide or the aldolate in the following transition state.

As indicated in Scheme 3–13, replacing the isopropoxide counterions with less basic oxyanions, namely, commercially available 3,5-di-*tert*-butylsalicylic acid, will lead to new catalyst **30**, which shows very good results for the asymmetric aldol addition of alkyl acetate to ketene acetals (Scheme 3–13).

This catalyst is readily soluble in ether, and the aldol reaction proceeds with high yield (>95% in most cases) and enantioselectivity (>90% in most cases) in the presence of a catalytic amount of **30**. The salicylate counterion probably

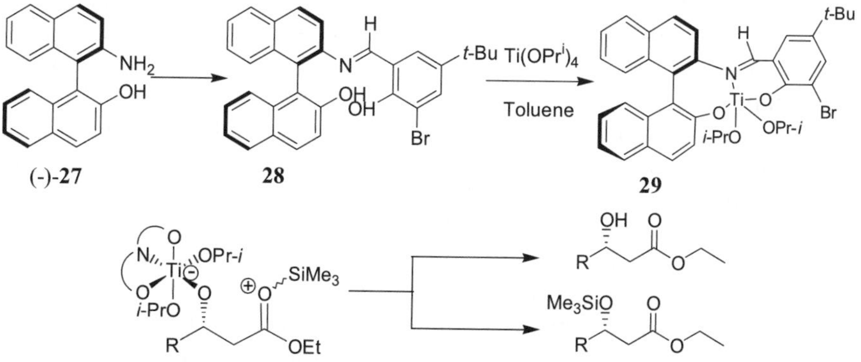

Figure 3–3. Poor result in **29** mediated aldol reaction.

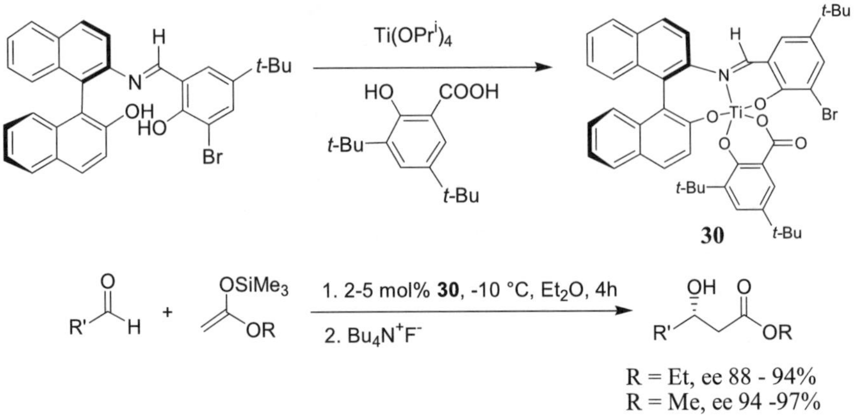

Scheme 3–13

serves as a shuttle for the TMS group between the catalyst and the silylated aldol product, thereby facilitating the regeneration of the catalyst.

3.2.4   Acylsultam Systems as the Chiral Auxiliaries

Besides their application in asymmetric alkylation, sultams can also be used as good chiral auxiliaries for asymmetric aldol reactions, and *anti*-product can be obtained with good selectivity. As can be seen in Scheme 3–14, reaction of the propionates derived from chiral auxiliary R*-OH with LICA in THF affords the lithium enolates. Subsequent reaction with TBSCl furnishes the *O*-silyl ketene acetals **31**, **33**, and **35** with good yields.[31] Upon reaction with $TiCl_4$ complexes of an aldehyde, product β-hydroxy carboxylates **32**, **34**, and **36** are obtained with high diastereoselectivity and good yield. Products from direct aldol reaction of the lithium enolate without conversion to the corresponding silyl ethers show no stereoselectivity.[32]

In Scheme 3–15, boryl enolates can be obtained by treating *N*-propionyl-sultam with Et_2BOTf and i-Pr_2NEt. No reaction is observed in the presence of $SnCl_4$ or $BF_3 \cdot OEt_2$. However, in the presence of $TiCl_4$, the aldol reaction proceeds smoothly, yielding *anti*-product with good stereoselectivity. Stereoselectivity decreased slightly when the amount of $TiCl_4$ was lowered. The optimized procedure involves the addition of a mixture of aldehyde/$TiCl_4$ in CH_2Cl_2 to a stirred solution of boryl enolate prepared in situ at $-78°C$, stirring at $-78°C$ for 0.5 to 4 hours, and aqueous workup.[33]

In summary, boryl enolate **38** can be obtained via in situ *O*-borylation of *N*-propionylsultam **37** and converted to aldol product **40** upon treatment with aliphatic, aromatic, or α,β-unsaturated aldehdyes at $-78°C$ in the presence of $TiCl_4$. As aldol product **40** can normally be obtained in crystalline form, in most cases diastereomerically pure *anti*-aldol **40** can also be obtained after the recrystallization.

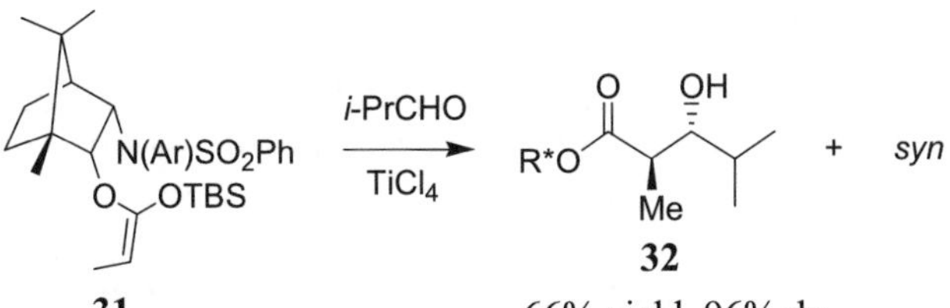

31 → 32

66% yield, 96% de
anti : *syn* = 93 : 7

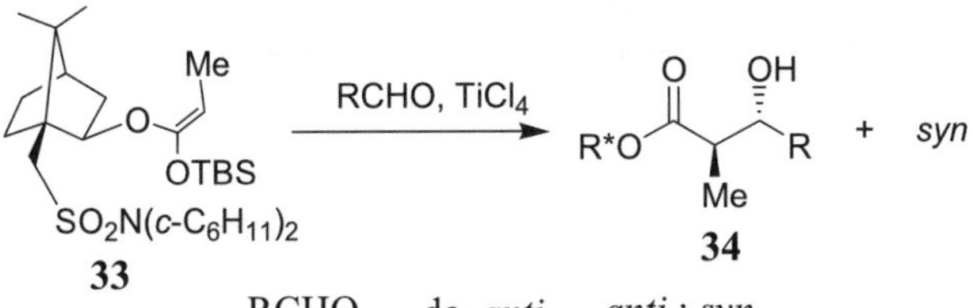

33 → 34

RCHO	de, *anti*	*anti* : *syn*
PhCHO	90	81 : 19
n-PrCHO	85	94 : 6
i-PrCHO	85	98 : 2

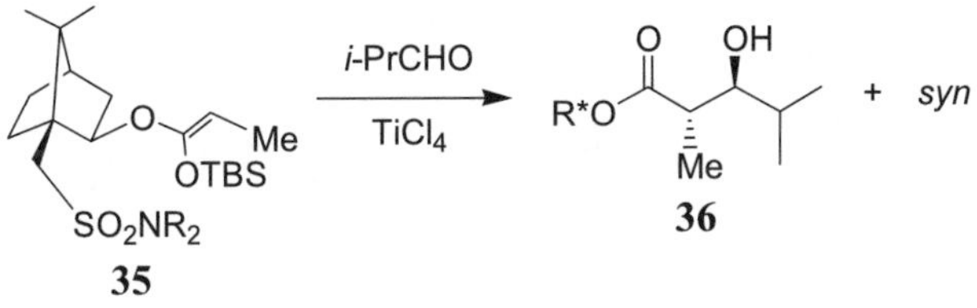

35 → 36

Scheme 3–14

37 → 38 → 39 + 40

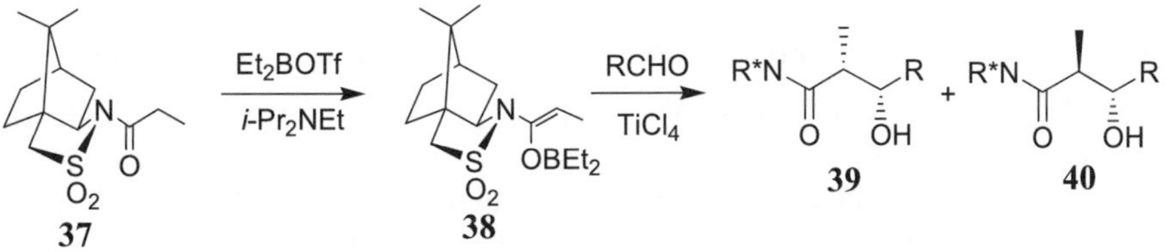

Lewis acid	mol. equiv. of Lewis acid/EtCHO	**40/39**/others
Et$_2$AlCl	2	27/37/36
Et$_2$BOTf	2	78/0/22
TiCl$_4$	2	98/0/2
TiCl$_4$	1	97/0/3
TiCl$_4$	0.5	93/0/7

Scheme 3–15. Reprinted with permission by Pergamon Science Ltd., Ref. 33.

3.2.5   α-Silyl Ketones

α-Silyl ketones of type **41** can be employed in aldol reactions, and *syn*-configurated β-hydroxy ketone **42** can be obtained in high de and ee. Direct aldol reaction for the synthesis of β-hydroxy ketones is less widely used because of its low ee value. Thus, (*R*)-**41** can be converted to the corresponding boron enolate via the reaction with *n*-Bu₂BOTf in CH₂Cl₂. The resulting compound is then reacted with the aldehyde at −78°C, giving the aldol product **44** with 92–98% de and more than 98% ee. After desilylation with 60% aqueous tetra-fluoroboric acid, the *syn*-aldol **42** can be obtained with high de and ee (Scheme 3–16).[34]

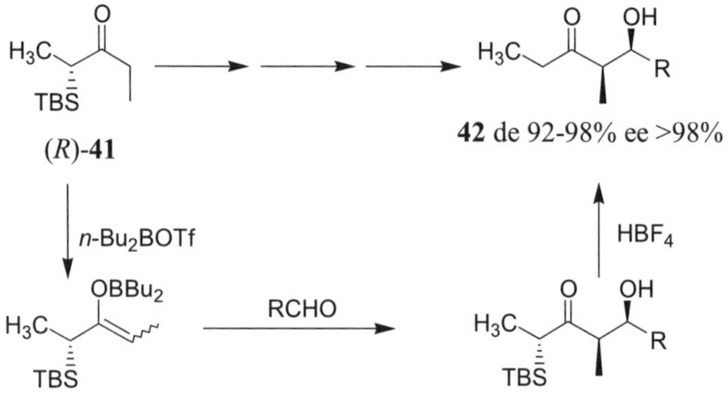

Scheme 3–16

3.3   REAGENT-CONTROLLED ALDOL REACTIONS

3.3.1   Aldol Condensations Induced by Chiral Boron Compounds

Boron triflates **45a** and **45b** are very useful chiral auxiliaries. Boron azaenolate derived from achiral[35] and chiral[36] oxazolines gives good stereoselectivity in the synthesis of acyclic aldol products, particularly for the rarely reached *threo*-isomers. By changing the chiral auxiliary, the stereochemistry of the reaction can be altered.[37]

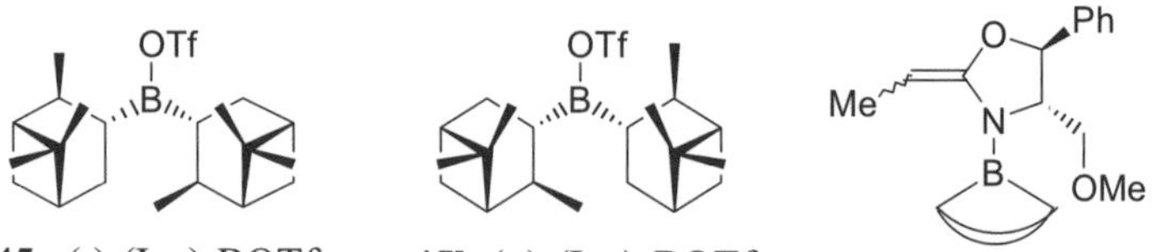

45a (-)-(Ipc)₂BOTf   **45b** (+)-(Ipc)₂BOTf   **46**

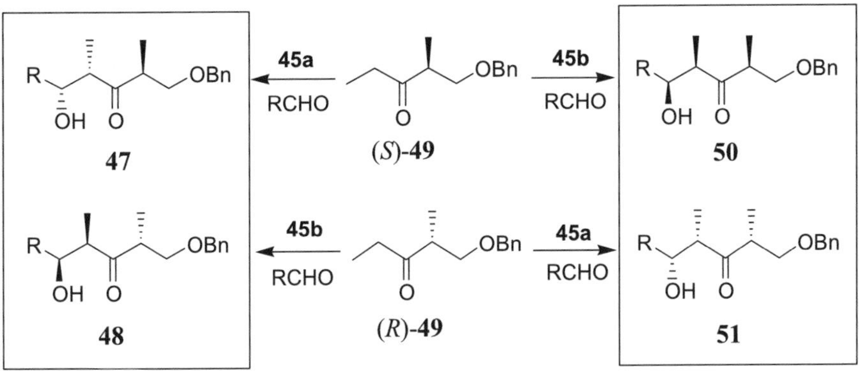

Scheme 3–17

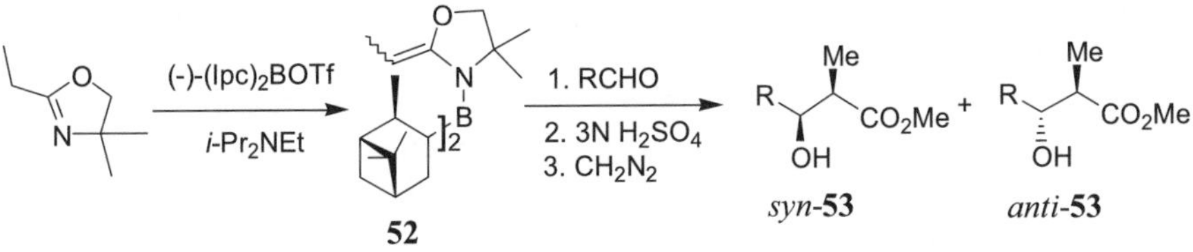

RCHO	ee (%) for *anti*	*anti:syn*
n-PrCHO	77	91:9
c-HexCHO	84	95:5
t-BuCHO	79	94:6

Scheme 3–18

Starting from ketone(*R*)-/(*S*)-**49**, the asymmetric aldol reaction with aldehyde in the presence of **45a** or **45b** affords all four isomers of β-hydroxyl ketone **47**, **48**, **50**, and **51** with high yields and stereoselectivities (Scheme 3–17).

Treating boron reagent **45a** with an oxazoline compound gives the azaenolate **52**. Subsequent aldol reaction of **52** with aldehyde yields mainly *threo*-product (*anti*-**53**) with good selectivities (Scheme 3–18).[38]

When a chiral auxiliary is present in the oxazoline ring and the boron part is replaced with an achiral bicyclic system (**46** bearing 9-BBN), *erythro*-β-hydroxy esters (*syn*-**53**) can be obtained as the major product upon reaction of the enolate with several aldehydes.[37]

3.3.2 Aldol Reactions Controlled by Corey's Reagents

Corey et al.[39] introduced a chiral controller system, chiral auxiliary **55**, which has shown excellent practical potential because of its availability, recoverability, and high enantioselectivity. Furthermore, using conformation analysis,

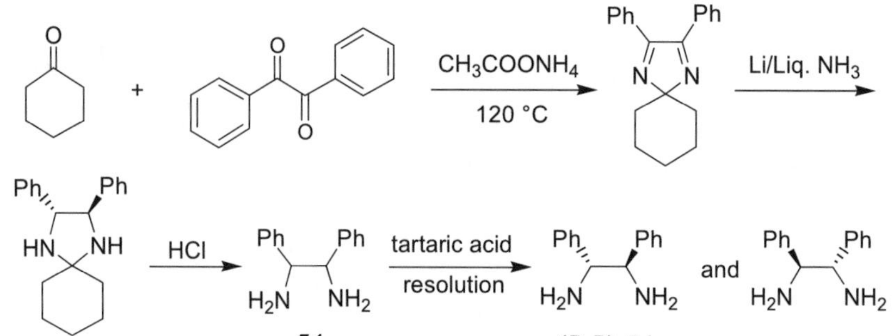

Scheme 3–19. Synthesis and resolution of chiral compound **54**.

the absolute configuration of the product can be predicted when **55** is involved in the reaction (see Scheme 3–20 and Table 3–5, later, for some examples). Starting from benzil, compound 1,2-diamino-1,2-diphenylethane (stilbenediamine) is synthesized. Using tartaric acid as the resolving agent, stilbenediamine in both (R,R)- and (S,S)-forms can be obtained.[40] Scheme 3–19 depicts the synthesis of both isomers of **54**. Compound **54** is a key compound for synthesizing **55** and related compounds.

Conversion of diamine **54** to bis-sulfonaminde, followed by treatment with BBr$_3$, yields the corresponding catalyst **55**. Application of this controller in enantioselective aldol reactions has led to striking success. The reaction can be used not only in aldol reactions but also in allylation reactions. Treating a solution of **55a** with allyl tributyltin results in chiral allyl borane, and reaction of this allyl borane with a variety of aldehydes gives the corresponding homoallyl alcohols with high enantiomeric excess ($>90\%$).[41] The allylation reaction is discussed in detail in a later section.

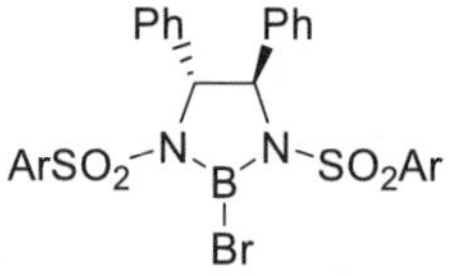

55
55a Ar = p-CH$_3$C$_6$H$_4$
55b Ar = p-NO$_2$C$_6$H$_4$

As illustrated in Scheme 3–20 and Table 3–5, using **55a** or **55b** as the chiral auxiliary, *syn*-aldol adduct **56** can be obtained with high stereoselectivity via aldol reaction of diethyl ketone with various aldehydes.[39]

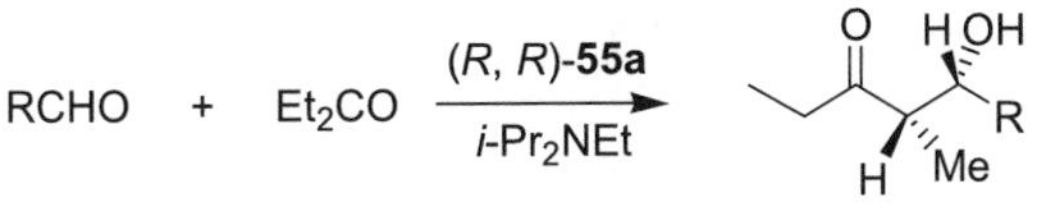

Scheme 3–20

TABLE 3–5. Reaction of Aldehydes with the Enolate from Diethyl Ketone and Bromoborane (*R,R*)-55b

RCHO	Yield (%)	*syn:anti*	ee (%)
PhCHO	95	94.3:5.7	97
Me$_2$CHCHO	85	98:2	95
EtCHO	91	>98:2	>98

ee = Enantiomeric excess.

Reprinted with permission by Am. Chem. Soc., Ref. 39.

With **55** as the chiral auxiliary, good enantioselectivity can be obtained in reactions between aldehydes and esters. In the case of phenyl thioacetate, as depicted in Scheme 3–21, an aldol reaction induced by **55a** or **55b** can also give acceptable stereoselectivity.

It is proposed that stereochemistry of the controlled aldol addition originates when the phenyl groups of the stien ligand force the vicinal *N*-sulfonyl substituents to occupy the face of the five-membered ring opposite the face where they are linked. The optimized stereoelectronic and steric arrangements of the favored transition state for the formation of the major product are depicted in Figure 3–4.

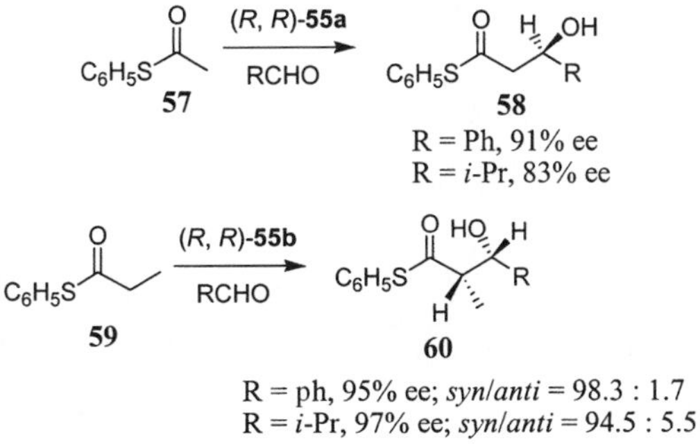

R = Ph, 91% ee
R = *i*-Pr, 83% ee

R = ph, 95% ee; *syn/anti* = 98.3 : 1.7
R = *i*-Pr, 97% ee; *syn/anti* = 94.5 : 5.5

Scheme 3–21

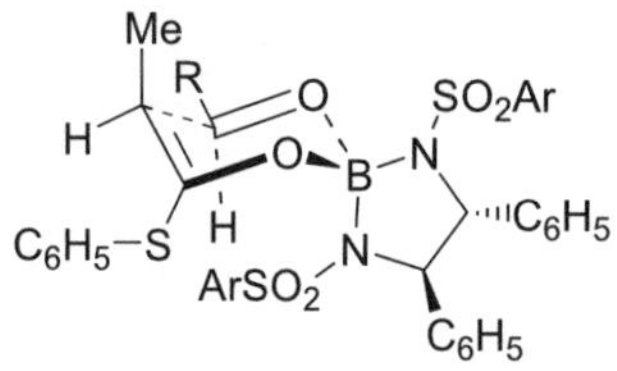

Figure 3–4. Transition state.

3.3.3   Aldol Condensations Controlled by Miscellaneous Reagents

Both *anti-* and *syn*-3-hydroxy-2-methylcarbonyl units (e.g., **61** and **62**) are frequently embedded in natural products of propionate origin, such as macrolide antibiotics.

Double asymmetric aldol reaction has been widely used for the efficient construction of the *syn*-unit.[10b,42] With above-described organoboron compounds (Section 3.3.1), *anti*-selectivity can be obtained.

Because *anti/syn* ratios in the product can be correlated to the $E(O)/Z(O)$ ratio of the involved boron enolate mixture,[10b] initial experiments were aimed at the preparation of highly $E(O)$-enriched boron enolate. The $E(O)/Z(O)$ ratio increases with the bulk of the alkanethiol moiety, whereas the formation of $Z(O)$ enolates prevails with (*S*)-aryl thioates. ($E/Z = 7{:}93$ for benzenethiol and 5:95 for 2-naphthalenethiol esters). $E(O)$ reagent can be formed almost exclusively by reaction of (*S*)-3,3-diethyl-3-pentyl propanethioate **64** with the chiral boron triflate. High reactivity toward aldehydes can be retained in spite of the apparent steric demand (Scheme 3–22).[43]

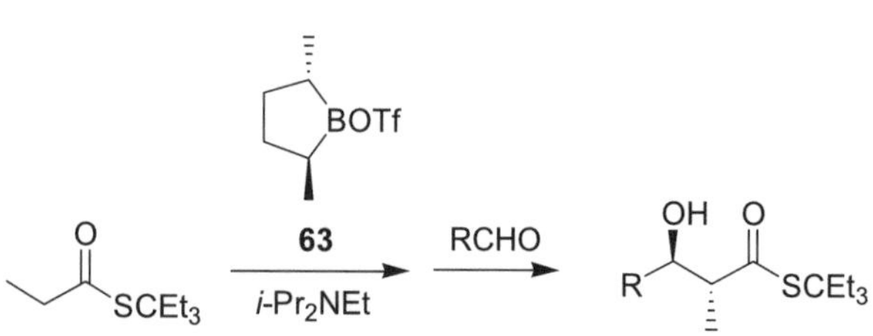

Scheme 3–22

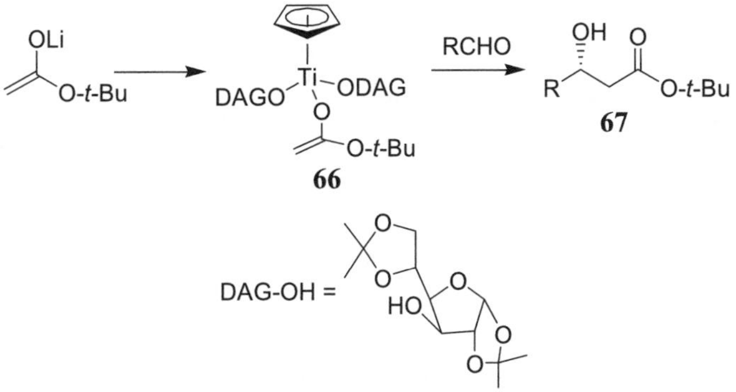

Scheme 3–23

Covalently bonded chiral auxiliaries readily induce high stereoselectivity for propionate enolates, while the case of acetate enolates has proved to be difficult. Alkylation of carbonyl compound with a novel cyclopentadienyl titanium carbohydrate complex has been found to give high stereoselectivity,[44] and a variety of β-hydroxyl carboxylic acids are accessible with 90–95% optical yields. This compound was also tested in enantioselective aldol reactions. Transmetalation of the relatively stable lithium enolate of t-butyl acetate with chloro(cyclopentadienyl)-bis(1,2:5,6-di-O-isopropylidene-α-D-glucofuranose-3-O-yl)titanate provided the titanium enolate **66**. Reaction of **66** with aldehydes gave β-hydroxy esters in high ee (Scheme 3–23).

As the t-butyl group can readily be removed upon acidic or basic hydrolysis, this method can also be used for β-hydroxyl acid synthesis. In analogy with allylation reactions, the enolate added preferentially to the Re-face of the aldehydes in aldol reactions. Titanium enolate **66** tolerates elevated temperatures, while the enantioselectivity of the reaction is almost temperature independent. The reaction can be carried out even at room temperature without significant loss of stereoselectivity. We can thus conclude that this reaction has the following notable advantages: High enantiomeric excess can be obtained (ee > 90%); the reaction can be carried out at relatively high temperature; the chiral auxiliary is readily available; and the chiral auxiliary can easily be recovered.[44]

3.4 CHIRAL CATALYST-CONTROLLED ASYMMETRIC ALDOL REACTION

3.4.1 Mukaiyama's System

Thus far, most of the stereoselective approaches to aldol reactions mentioned have depended on substrate-based asymmetric induction by employing chiral

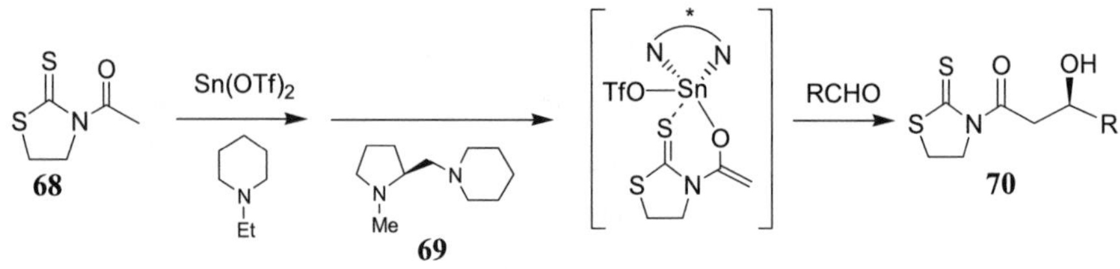

Scheme 3–24

enolates or chiral aldehydes. In addition to their discovery of the usefulness of enol equivalents in aldol reactions, Mukaiyama et al.[45] also demonstrated that tributyltin fluoride, stannous triflate, or trityl triflates are even more effective reagents for aldol reactions. Based on the concept of "enol species", they developed various highly selective and practical asymmetric reactions, particularly stannous triflate-mediated aldol reactions. Divalent tin has vacant d orbitals that enable it to form complexes with amines. The tin(II) metal is bonded with two nitrogen atoms, leaving one vacant orbital coordinatable to an aldehyde without losing the favorable chiral environment. Almost complete stereochemical control is therefore achieved starting from achiral aldehyde and silyl enol ethers with a catalytic amount of diamine.[46] Thus, compound **68** is first treated with Sn(OTf)$_2$ and a chiral amine **69**, followed by an aldehyde, yielding β-cyclohexyl (S)-imide **70** in about 90% ee (Scheme 3–24).[47]

In the presence of a chiral promoter, the asymmetric aldol reaction of prochiral silyl enol ethers **71** with prochiral aldehydes will also be possible (Table 3–6). In this section, a chiral promoter, a combination of chiral diamine-coordinated tin(II) triflate and tributyl fluoride, is introduced. In fact, this is the first successful example of the asymmetric reactions between prochiral silyl enol ethers and prochiral aldehyde using a chiral ligand as promoter.

As depicted in Scheme 3–25, the aldol reaction carried out at −78°C can give the corresponding aldol adduct **72** in 78% yield with 82% ee. The combination of chiral diamine-coordinated tin(II) triflate and tributyltin fluoride is so essential that the enantioselectivity cannot be obtained without tributyltin flu-

TABLE 3–6. The Reaction of Silyl Enol Ether of S-Ethyl Ethanethiate with Benzaldehyde

Promoter	Yield (%)	ee (%)
Sn(OTf)$_2$ + chiral diamine A	74	0
Sn(OTf)$_2$ + chiral diamine A + n-Bu$_3$SnF	78	82
Sn(OTf)$_2$ + chiral diamine B + n-Bu$_3$SnF	52	92
Sn(OTf)$_2$ + chiral diamine C + n-Bu$_3$SnF	74	78

ee = Enantiomeric excess.

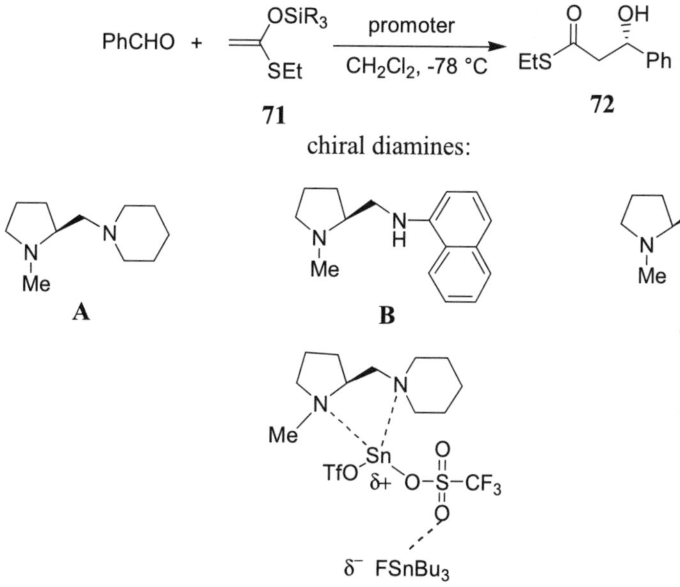

73

Scheme 3–25

oride in the reaction system. When Bu_3SnF was present, diamine A induced good yield for the product. The reaction induced by diamine B gives a higher ee but a moderate yield (Table 3–6).

It has been proposed that the reaction is promoted by the formation of the active species **73**, in which, the cationic center of the tin(II) triflate activates an aldehyde, and, at the same time, the electronegative fluoride is able to interact with a silicon atom of the silyl enol ether to make the enol ether more reactive. This dual process results in the formation of the entropically favored intermediate (Scheme 3–25).[46]

Perfect stereochemical control in the synthesis of *syn*-α-methyl-β-hydroxy thioesters has been achieved by asymmetric aldol reaction between the silyl enol ether of *S*-ethyl propanethioate (1-trimethylsiloxy-1-ethylthiopropene) and aldehydes using a stoichiometric amount of chiral diamine-coordinated tin(II)

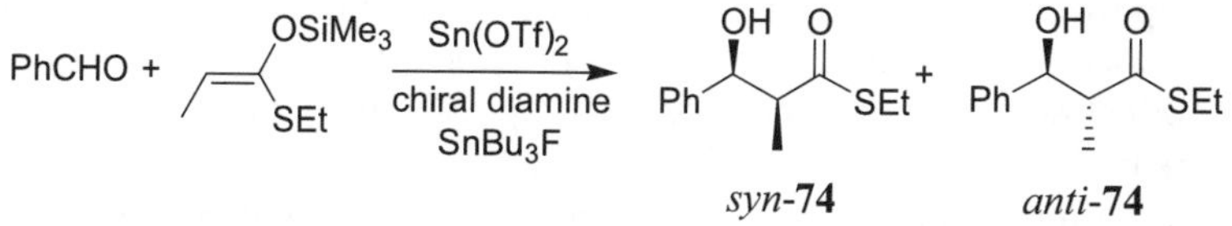

Scheme 3–26

TABLE 3–7. Effect of Chiral Diamine

Chiral Diamine	Time (h)	Yield (%)	*syn:anti*	ee (%)
A	20	80	93:7	80
B	3	86	100:0	>98
C	20	77	88:12	44

ee = Enantiomeric excess.

triflate and tributyltin fluoride or n-Bu$_2$Sn(OAc)$_2$ as chiral promoters (Scheme 3–26). The effect of chiral diamine on the stereo-outcome of the product is shown in Table 3–7.[45]

This method has been applied in the enantioselective synthesis of D-*erythro*-sphingosine and phytosphingosine. Sphingosine became an important substance for studying signal transduction since the discovery of protein kinase C inhibition by this compound.[48] Many efforts have been made to synthesize sphingosine and its derivatives.[49] Kobayashi et al. reported another route to this type of compound in which a Lewis acid–catalyzed asymmetric aldol reaction was a key step.

In the synthesis of D-*erythro*-sphingosine (**78** without BOC protection), the key step is the asymmetric aldol reaction of trimethylsilylpropynal **75** with ketene silyl acetal **76** derived from α-benzyloxy acetate. The reaction was carried out with 20 mol% of tin(II) triflate chiral diamine and tin(II) oxide. Slow addition of substrates to the catalyst in propionitrile furnishes the desired aldol adduct **77** with high diastereo- and enantioselectivity (*syn/anti* = 97:3, 91% ee for *syn*). In the synthesis of protected phytosphingosine (**80**, OH and NH$_2$ protected as OAc and NHAc, respectively), the asymmetric aldol reaction is again employed as the key step. As depicted in Scheme 3–27, the reaction between acrolein and ketene silyl aectal **76** proceeds smoothly, affording the desired product **80** with 96% diastereoselectivity (*syn/anti* = 98:2) and 96% ee for *syn* (Scheme 3–27).[50]

For a comprehensive review of Sn(II) catalyst systems, see Nelson's review article, *Catalyzed Enantioselective Aldol Additions of Latent Enolate Equivalents.*[51]

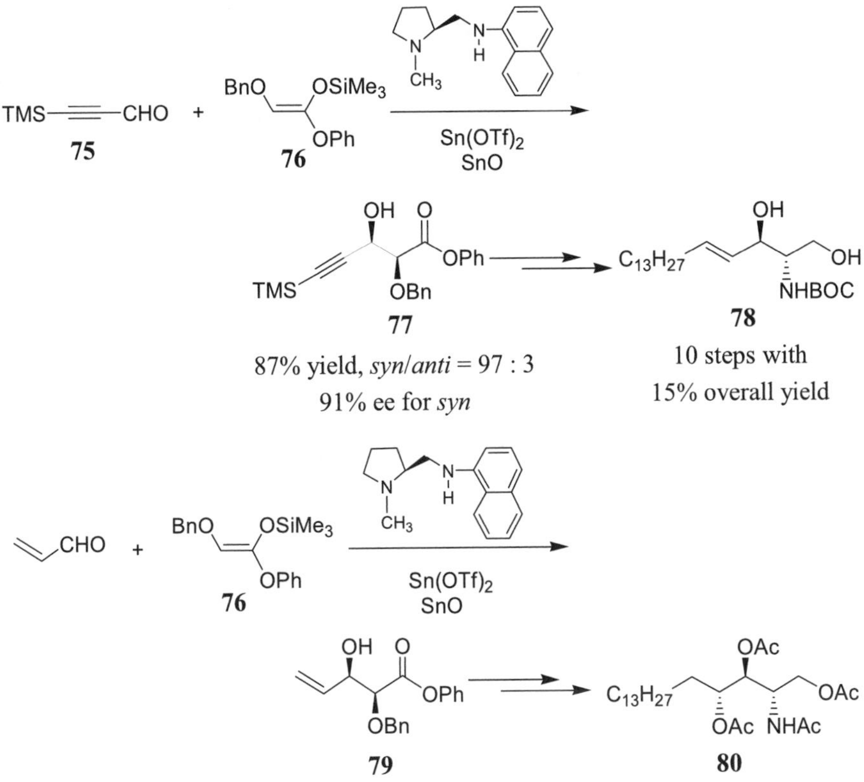

Scheme 3–27

3.4.2 Asymmetric Aldol Reactions with a Chiral Ferrocenylphosphine–Gold(I) Complex

Ito et al.[52] found that gold(I) complex containing an optically active ferrocenylphosphine ligand bearing a 2-dialkylaminoethylamino side chain was an effective catalyst for an asymmetric aldol-type reaction of α-isocyanocarboxylate (CNCHRCO$_2$CH$_3$), α-isocyanocarboxyamides (CNCH$_2$CONR$_2$), or α-isocyanophosphonates [CNCH$_2$PO(OR)$_2$] with aldehydes. This reaction can be used for β-hydroxy-α-amino acid synthesis (Scheme 3–28).

The *trans*-selectivity is due to steric repulsion between the alkyl substituent of the aldehydes and the large carboxylate or dialkoxy phosphinyl moiety (Scheme 3–29).[53]

As shown in Scheme 3–30, α,β-unsaturated aldehyde is treated with α-isocyanocarboxylate in the presence of gold complex (*S*)-(*R*)-**82** to afford, after several steps, D-*erythro*-sphingosine derivative.[53b]

The asymmetric aldol reaction of α-ketoesters (RCOCOOMe: R=Me, *i*-Bu, Ph) with methyl isocyanoacetate or *N,N*-dimethyl-isocyanoacetamide in the

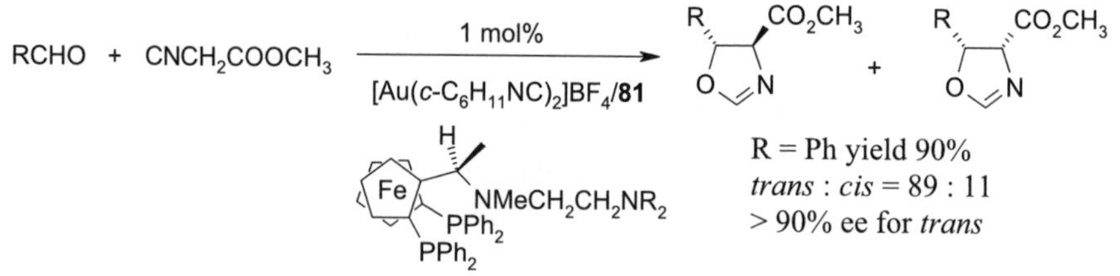

RCHO + CNCH₂COOCH₃ $\xrightarrow[\text{[Au}(c\text{-}C_6H_{11}NC)_2]BF_4/\mathbf{81}]{\text{1 mol\%}}$

R = Ph yield 90%
trans : *cis* = 89 : 11
> 90% ee for *trans*

81 R = Me, Et

Scheme 3–28

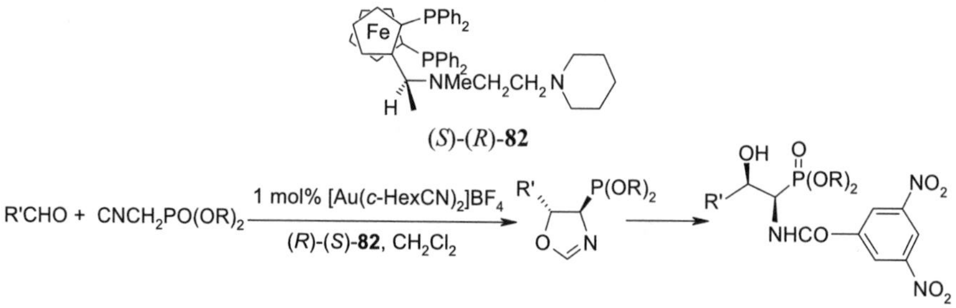

(S)-(R)-**82**

R'CHO + CNCH₂PO(OR)₂ $\xrightarrow[\text{(R)-(S)-\textbf{82}, CH}_2\text{Cl}_2]{\text{1 mol\% [Au}(c\text{-HexCN})_2]BF_4}$

Scheme 3–29

R = H, Ac

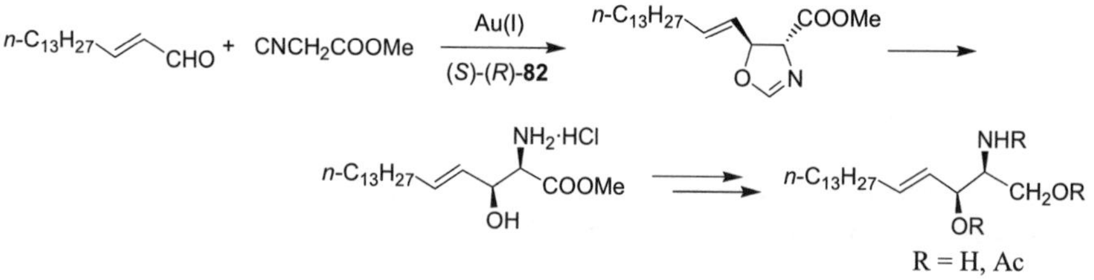

Scheme 3–30. Gold(I) complex–catalyzed asymmetric aldol reactions.

presence of 1 mol% of chiral aminoalkylferrocenylphosphine gold(I) (L* = **85**) catalyst proceeds with high enantioselectivity, giving the corresponding oxazolines **83** in up to 90% ee. This compound can be further converted to optically active α-alkyl-β-hydroxyaspartic acid derivative **84** (Scheme 3–31).[54]

3.4.3 Asymmetric Aldol Reactions Catalyzed by Chiral Lewis Acids

The aldol reaction between enolsilanes and aldehydes mediated by chiral Lewis acids may be considered the most notable achievement in the area of asymmetric aldol reactions. However, the design of new catalyst systems to tolerate

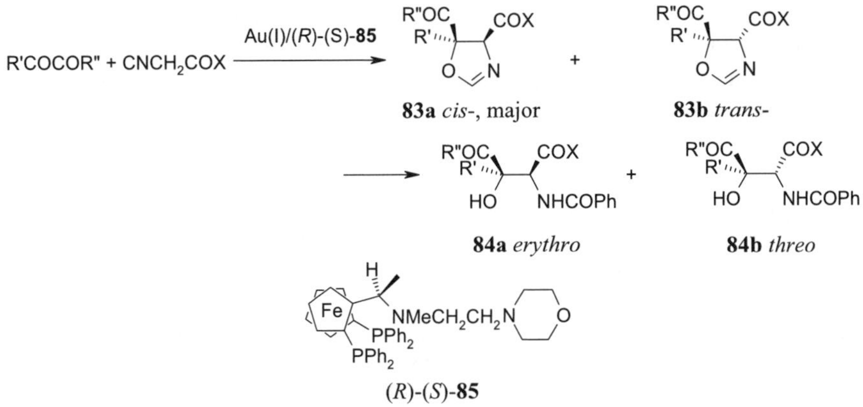

R'COCOR" + CNCH$_2$COX $\xrightarrow{\text{Au(I)/(R)-(S)-85}}$

R"OC$\cdots$R'$_{\prime\prime\prime}$ COX
O$\diagdown$N

83a *cis-*, major

+

R"OC$\cdots$R'$_{\prime\prime\prime}$ COX
O$\diagdown$N

83b *trans-*

$\longrightarrow$

R"OC$\cdots$R'$_{\prime\prime\prime}$ COX
HO NHCOPh

84a *erythro*

+

R"OC$\cdots$R'$_{\prime\prime\prime}$ COX
HO NHCOPh

84b *threo*

Fe — H, NMeCH$_2$CH$_2$N$\diagdown$O
PPh$_2$
PPh$_2$

(R)-(S)-**85**

Scheme 3–31

substantial variation in both nucleophilic and electrophilic components at low catalyst loading still remains problematic.

Realizing high enantioselectivity for the catalyzed aldol reaction necessarily relies on effective channeling of the reactants through a transition state that is substantially lower in energy than other competing diastereomeric transition states. Several factors must be considered when designing the catalyst. These are (A) mode of binding of the carbonyl group to the Lewis acid; (B) the regiochemistry of complexation of the two available C=O lone pairs; and (C) establishing a fixed diastereofacial bias, therefore directing the enol/enolate addition to one of the two carbonyl π-faces. Recent investigations have incorporated additional stabilizing interactions such as H-bonding, π-stacking, or chelation into the catalyst–aldehyde complexes to give a highly defined carbonyl facial differentiation.[55,56] Developing chiral catalysts that exhibit a strong tendency toward substrate chelation and also meet other criteria necessary for asymmetric aldol reaction is likely to be the focal point of the research (Fig. 3–5).

Evans' group has demonstrated that bidentate bis(oxazolinyl)-Cu(II) (**86** and **87**) and tridentate bis(oxazolinyl)pyridine-Cu(II) (**88** and **89**) can be used

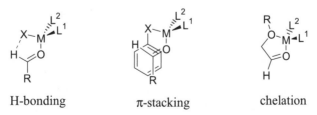

H-bonding   π-stacking   chelation

Figure 3–5. Additional stabilizing interactions between substrate and catalyst.

as effective catalysts for asymmetric Diels-Alder reactions (see Chapter 5 for asymmetric Diels-Alder reactions). They also found that these complexes gave excellent reactivity in aldol reactions of Mukaiyama reagents with a series of aldehydes.[57]

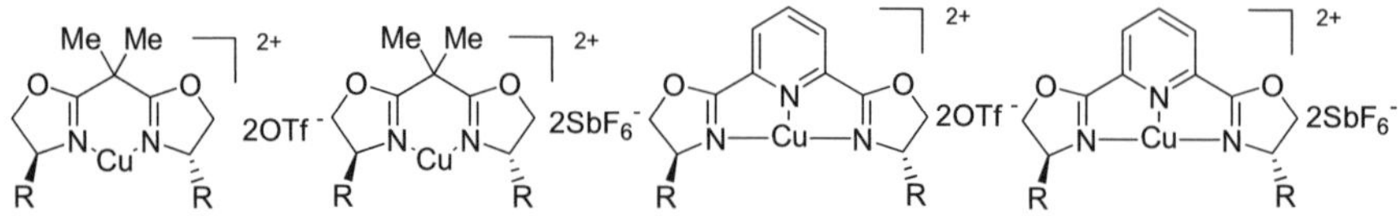

86a R = CMe$_3$
86b R = CHMe$_2$
86c R = Ph
86d R = Bn

87a R = CMe$_3$
87b R = CHMe$_2$
87c R = Ph
87d R = Bn

88a R = CMe$_3$
88b R = CHMe$_2$
88c R = Ph
88d R = Bn

89a R = CMe$_3$
89b R = CHMe$_2$
89c R = Ph
89d R = Bn

The efficiency of catalysts **86–89** for the asymmetric aldol reaction of a series of nucleophiles toward benzyloxyacetaldehyde was studied. For example, compound **89c** was found to be an excellent catalyst for the asymmetric aldol reaction of silylketene acetal derivatives of *t*-butyl thioacetate, ethyl thioacetate, and ethyl acetate with benzyloxyacetaldehyde. In the presence of 0.5 mol% of the catalyst, the asymmetric aldol reaction took place at $-78°C$ in CH$_2$Cl$_2$, affording the respective β-hydroxy esters with excellent enantioselectivity (Scheme 3–32).

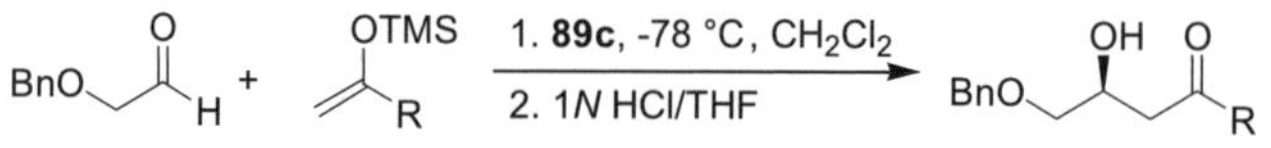

Entry	R	Catalyst (mol%)	Time (h)	ee (%)	Yield (%)
1	S-*t*-Bu	0.5	12–24	99	99
2	SEt	0.5	12–24	98	95
3	OEt	0.5	12	98	99

Scheme 3–32. Asymmetric aldol reaction catalyzed by chiral bis(oxazolinyl)pyridine compound **89c**. Reprinted with permission by Am. Chem. Soc., Ref. 57.

Catalyst **86a** also catalyzed the enantioselective aldol reaction between α-keto esters and silylketene acetals or enolsilanes with high ee (ranging from 93% to 99%).[58]

Several other chiral Lewis acids have also been reported to effect asymmetric aldol reactions. Krüger and Carreira[59] reported a catalytic aldol addition of silyl dienolate to a range of aldehydes in the presence of a bisphosphanyl–Cu(II) fluoride complex generated in situ from (*S*)-Tol-BINAP, Cu(OTf)$_2$, and (Bu$_4$N)Ph$_3$SiF$_2$. Aromatic, heteroaromatic, and α,β-unsaturated aldehydes provided the aldol adducts with up to 95% ee and 98% yield (Scheme 3–33).

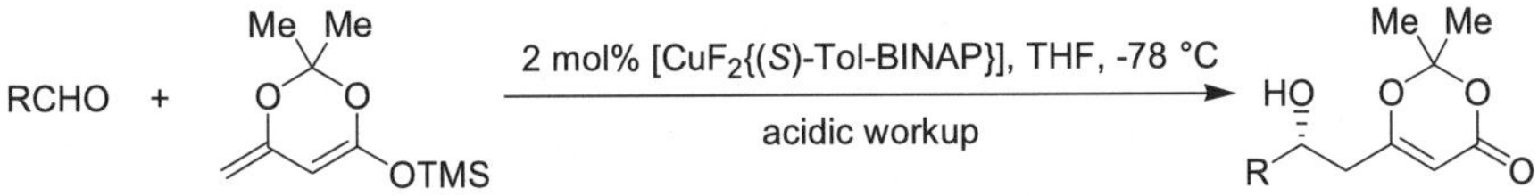

Scheme 3–33

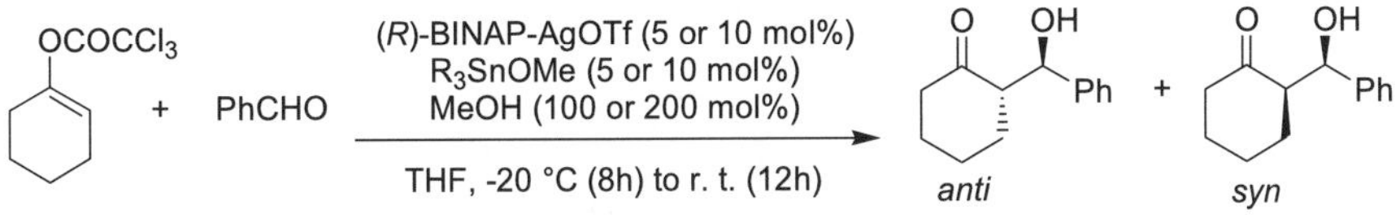

Scheme 3–34

TABLE 3–8. BINAP-Ag Complex Catalyzed Aldol Reaction

Entry	(R)-BINAP-AgOTf (mol%)	R$_3$SnOMe (mol%)	MeOH (mol%)	Yield (%)	anti (ee)	syn (ee)
1	10	10 (R = Bu)	100	62	84 (93)	16 (18)
2	10	10 (R = Bu)	200	94	92 (95)	8 (16)
3	5	5 (R = Bu)	200	82	92 (95)	8 (17)
4	5	5 (R = Me)	200	88	93 (94)	7 (6)
5	5	5 (R = Me)	200	86	94 (96)	6 (18)

ee = Enantiomeric excess.
Reprinted with permission by Am. Chem. Soc., Ref. 61.

Besides the silyl enolate-mediated aldol reactions, organotin(IV) enolates are also versatile nucleophiles toward various aldehydes in the absence or presence of Lewis acid.[60] However, this reaction requires a stoichiometric amount of the toxic trialkyl tin compound, which may limit its application. Yanagisawa et al.[61] found that in the presence of one equivalent of methanol, the aldol reaction of an aldehyde with a cyclohexenol trichloroacetate proceeds readily at −20°C, providing the aldol product with more than 70% yield. They thus carried out the asymmetric version of this reaction using a BINAP–silver(I) complex as chiral catalyst (Scheme 3–34). As shown in Table 3–8, the Sn(IV)-mediated aldol reaction results in a good diastereoselectivity (*anti/syn* ratio) and also high enantioselectivity for the major component.

3.4.4 Catalytic Asymmetric Aldol Reaction Promoted by Bimetallic Catalysts: Shibasaki's System

Most of the asymmetric aldol reactions discussed thus far deal with the nucleophilic addition of a chiral or achiral enolate onto a chiral or achiral aldehyde, and the preparation of an enol derivative is necessary. In the course of developing direct catalytic asymmetric aldol reactions starting from aldehydes and unmodified ketones without the involvement of enolate, Yamada et al.[62] developed a bimetallic compound (binaphthoxide) LLB **90** from binaphthol and LaCl$_3$. This complex has been shown to exhibit properties of both Lewis acid and base and may be considered to be an ideal catalyst for simple asymmetric aldol reactions.

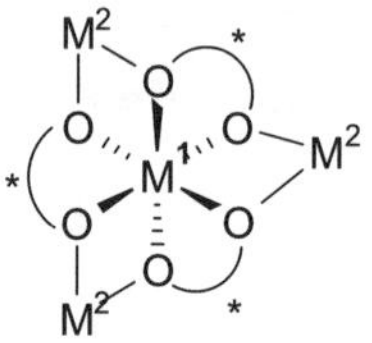

90

The LLB catalyst system needs a rather long reaction time and the presence of excess ketone to get a reasonable yield. Yamada and Shibasaki[63] found that another complex, BaBM (**91**), was a far superior catalyst. Complex **91** also contains a Lewis acidic center to activate and control the orientation of the aldehyde, but it has stronger Brønsted basic properties than LLB. The preparation of BaBM is shown in Scheme 3–35.

Scheme 3–35. Preparation of BaBM **91**.

Reaction of aldehydes and 2 equivalents of ketone in the presence of 5 mol% of (*R*)-BaBM gives good yield of aldol product (77–99%) with moderate enantioselectivity (54–70% ee) after a 2-day reaction (Scheme 3–36 and Table 3–9). Although the enantioselectivity is not very high, this is one of the first examples of direct aldol condensation using barium catalyst as the promoter.

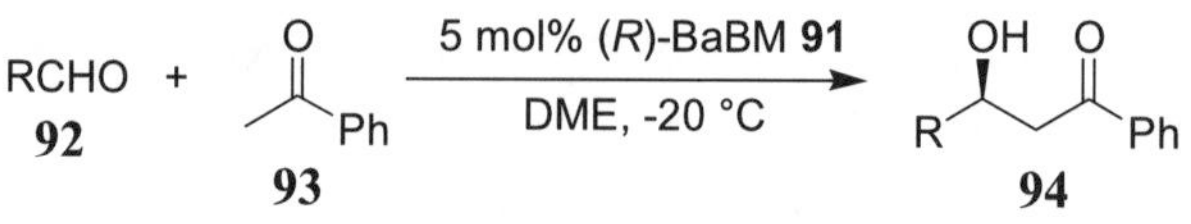

Scheme 3–36

TABLE 3–9. Direct Aldol Reaction Promoted by 91

Entry	R in **92**	Yield for **94** (%)	ee for **94** (%)
1	t-Bu	77	67
2	$PhCH_2(CH_3)_2C$	77	55
3	c-C_6H_{11}	87	54
4	i-C_3H_7	91	50
5	$BnOCH_2(CH_3)_2C$	83	69
6	$BnO(CH_3)_2C$	99	70

ee = Enantiomeric excess.

Reprinted with permission by Pergamon-Elsevier Science Ltd., Ref. 63.

This catalyst can also be applied in the reaction of aldehydes bearing a bulky α-group, and moderate to good yields can be obtained. The advantage of this reagent, bearing both Lewis acidic and Lewis basic properties, are further discussed in Chapter 8.

3.5 DOUBLE ASYMMETRIC ALDOL REACTIONS

As shown in Scheme 3–37, reaction of (−)-**96** with achiral (Z)-O-enolate **95** provides a mixture of **97** and **98** in an approximate ratio of 3:2.[64] After screening a variety of chiral enolate reagents, (S)-**100c** has been found to provide good asymmetric induction. Aldol reaction of achiral aldehyde **99** with (S)-**100c** (the most stereoselective boron enolate among the three compounds **100a–c**; see Table 3–10) provides a mixture of diastereoisomers **101** and **102** in a ratio of 100:1.[65] Successive treatment of a mixture of **101**/**102** with hydrogen fluoride followed by oxidation with sodium metaperiodate provides the corresponding 2,3-syn-3-hydroxy-2-methylcarboxylic acid **103** with an enantiomeric excess of over 98%. This three-step process is now referred to as the *three-carbon atom extension*.

Now, we examine the interaction of chiral aldehyde (−)-**96** with chiral enolate (S)-**100b**. This aldol reaction gives **104** and **105** in a ratio of **104**:**105** > 100:1. Changing the chirality of the enolate reverses the result: Compound **104** and **105** are synthesized in a ratio of 1:30 (Scheme 3–38).[66] The two reactions (−)-**96** + (S)-**100b** and (−)-**96** + (R)-**100b** are referred to as the *matched* and *mismatched pairs*, respectively. Even in the mismatched pair, stereoselectivity is still acceptable for synthetic purposes. Not only is the stereochemical course of the aldol reaction fully under control, but also the power of double asymmetric induction is clearly illustrated.

Scheme 3–39 shows the reaction of boron enolates **106** and **108** with chiral aldehydes (2R,4S)- and (2S,4R)-**96**. In the matched case, lactone 3,4-*anti*-**107** is obtained with very high ee.[67]

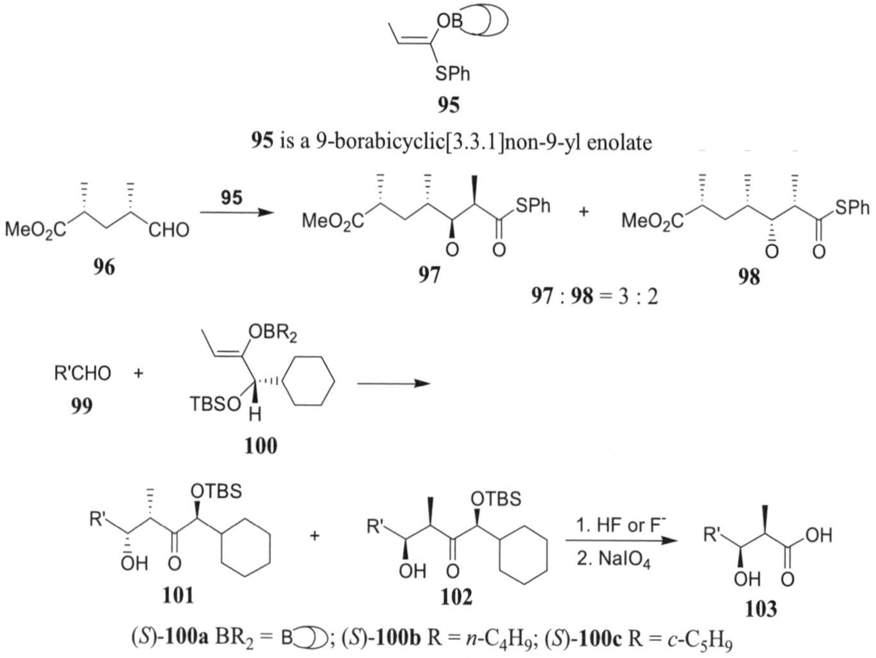

95

95 is a 9-borabicyclic[3.3.1]non-9-yl enolate

97 : 98 = 3 : 2

(S)-**100a** BR$_2$ = B⬡; (S)-**100b** R = n-C$_4$H$_9$; (S)-**100c** R = c-C$_5$H$_9$

Scheme 3–37

TABLE 3–10. Reaction of Aldehydes with Boron Enolates

Aldehyde	R'	Boron Enolate	98:97
99a	PhCH$_2$OCH$_2$CH$_2$	(S)-**100a**	16:1
		(S)-**100b**	28:1
		(S)-**100c**	100:1
99b	(CH$_3$)$_2$CH	(S)-**100a**	>100:1
		(S)-**100b**	>100:1
		(S)-**100c**	No reaction

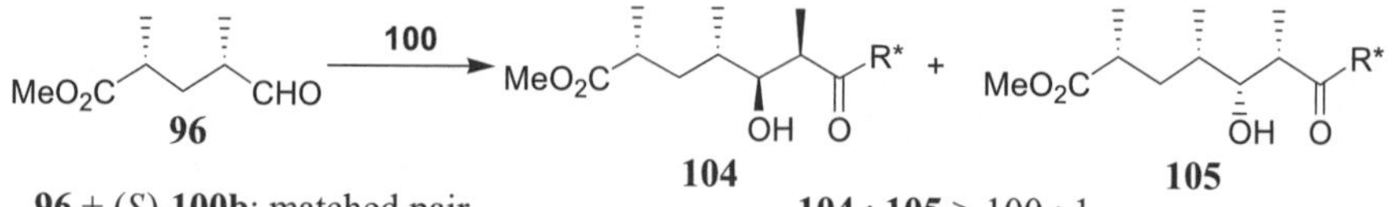

96 + (S)-**100b**: matched pair   104 : 105 > 100 : 1
96 + (R)-**100b**: mismatched pair   104 : 105 = 1 : 30

Scheme 3–38

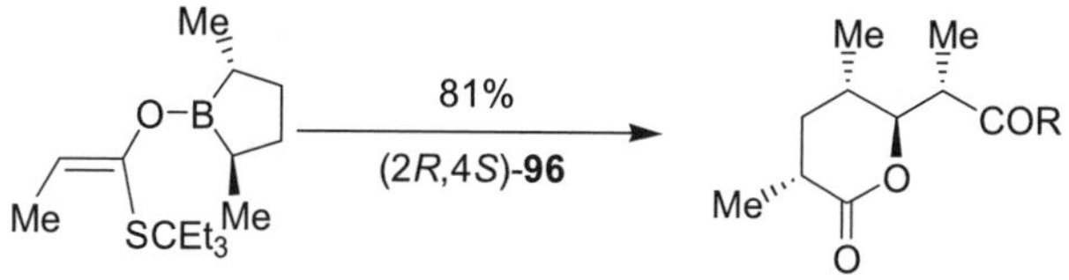

106 → 81% (2R,4S)-**96** → 3,4-*anti*-**107**
matched case: 200 : 1

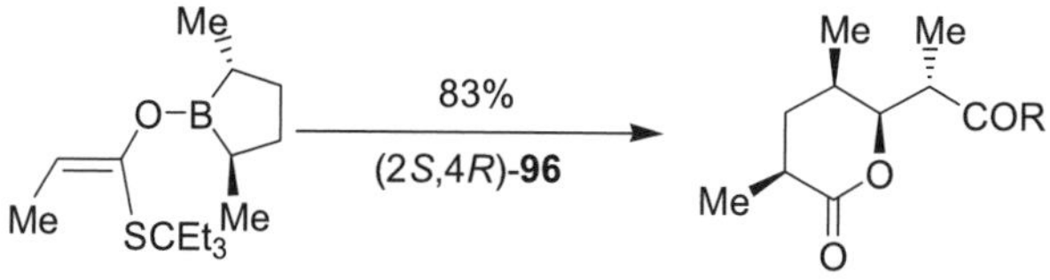

106 → 83% (2S,4R)-**96** → 3,4-*syn*-**107**
mismatched case: 55 : 1

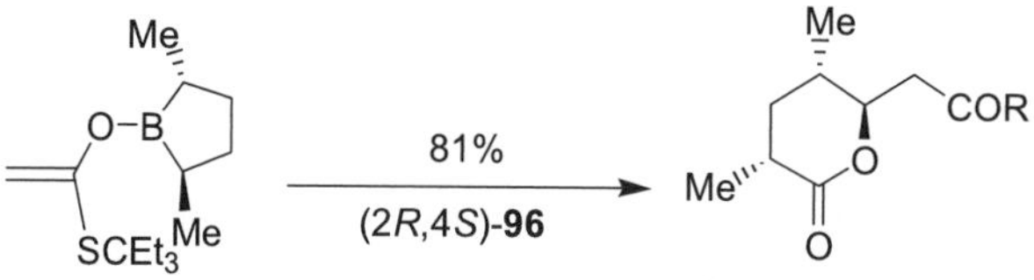

108 → 81% (2R,4S)-**96** → 3,4-*anti*-**109**
matched case: 13 : 1

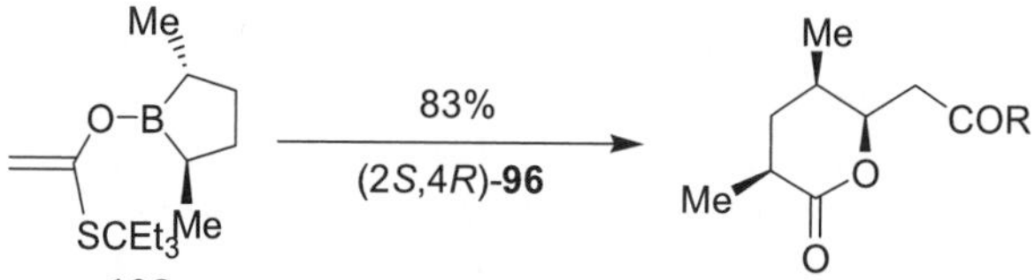

108 → 83% (2S,4R)-**96** → 3,4-*syn*-**109**
mismatched case: 7 : 1

Scheme 3–39

3.6 ASYMMETRIC ALLYLATION REACTIONS

Given the fact that homoallyl alcohols can be easily converted to the corresponding aldol compounds, allylation of aldehydes with allylic and crotyl organometallic reagents is synthetically analogous to the aldol addition of metal enolates. This reaction has significant advantages over aldol condensations because the produced alkenes may not only be readily converted into aldehydes but also undergo a facile one-carbon homologation to δ-lactones via hydroformylation or be selectively epoxidized to introduce a third chiral center.

As a result, allylic organometallic reactions have attracted much attention, and the allylation or crotylation reactions via the corresponding organometallic compounds have been extensively studied. This method has become one of the most useful procedures for controlling the stereochemistry in acyclic systems. Except for allyl silanes, most of the allylic organometallic or organometalloid systems are reactive enough toward aldehyde carbonyl groups, even if Lewis acid is not applied.

The general synthetic equation can be expressed by Eq. 3–1. "M" represents various metals. These conversions generate two new stereocenters and four possible diastereomeric products. The product *syn/anti* ratio reflects the $(Z):(E)$ ratio in the crotyl moiety, which may include B-, Al-, Sn-, Si-, or Ti-based reagents.

$$(3-1)$$

where M = SiMe₃; SnBu₃; BR₂; AlR₂; MgX; Li; CrX₂; TiCp₂X; ZrCp₂X, and so forth.

Before the late 1970s, allylic organometallic compounds were studied primarily by a limited number of organometallic chemists who were interested only in the structural determination of metal-allyl species. A number of studies on the reactions of metal-allyl with electrophiles have focused solely on the regioselectivity of the allylic unit. In the late 1970s, significant synthetic interest began to emerge in controlling the stereochemistry of C–C bond formation in reactions of metal-allyl compounds with aldehydes or ketones.

Over the past two decades, chiral allyl- and crotyl-boron reagents have proved to be extremely valuable in the context of acyclic stereoselection. The development of superior allyl-boron reagents, which can give enantio- and diastereoselectivities approaching 100%, has become both challenging and desirable.[68]

3.6.1   The Roush Reaction

The use of tartrates as chiral auxiliaries in asymmetric reactions of allenyl boronic acid was first reported by Haruta et al.[69] in 1982. However, it was not for several years that Roush et al.,[70] after extensive study, achieved excellent results in the asymmetric aldol reactions induced by a new class of tartrate ester based allyl boronates.

(R, R)- or (S, S)-**110** (R, R)- or (S, S)-**111** (R, R)- or (S, S)-**112**

Roush first reported that good yield and stereoselectivity could be obtained by reacting allyl-boronates (R,R)- or (S,S)-**110** with achiral or chiral aldehydes. The auxiliary system was then extended to the (E)- and (Z)-crotyl family (Scheme 3–40) in which reagent **111** and **112** served as (E)- or (Z)-propionate enolate equivalents. Compound **111/112** can be easily prepared by modified Schlosser's crotyl-boronate synthesis[71] from (E)- or (Z)-butene. *Syn-* or *anti-* **113** can be obtained via the corresponding allylation reaction. Molecular sieves are used to help maximize the enantioselectivity by maintaining an anhydrous reaction condition and to prevent adventitious hydrolysis of the chiral boronic reagents to achiral allyl-boronic acid, which may lead to unselective products.

In the reaction of (R,R)-tartrate allyl-boronate with aldehydes, *Si* attack of the nucleophile on the carbonyl group has been observed, while *Re* attack occurs in (S,S)-tartrate allyl-boronate reactions. Thus, an (S)-alcohol is produced preferentially when an (R,R)-allyl reagent is used, and the (R)-product can be obtained from an (S,S)-reagent, assuming that the "R" substituent in the aldehyde substrate takes priority over the allyl group to be transferred. In fact, no exceptions to this generalization have yet been found in over 40 well-characterized cases where the tartrate auxiliary controls the stereochemical outcome of the allyl or crotyl transfer.[72]

The asymmetric induction cannot be explained simply by steric interaction because the R group in the aldehyde is far too remote to interact with the tartrate ester. In addition, the alkyl group present in the tartrate ligand seems to have a relatively minor effect on the overall stereoselectivity. It has thus been proposed that stereoelectronic interaction may play an important role. A more likely explanation is that transition state **A** is favored over transition state **B**, in which an n–n electronic repulsion involving the aldehyde oxygen atom and the β-face ester group causes destabilization (Fig. 3–6). This description can help explain the stereo-outcome of this type of allylation reaction.

Systematic studies of the reactions of tartrate allyl-boronates with a series of chiral and achiral alkoxy-substituted aldehydes show that conformationally unrestricted α- and β-alkoxy aldehyde substrates have a significant negative impact on the stereoselectivity of asymmetric allyl-boration. In contrast, con-

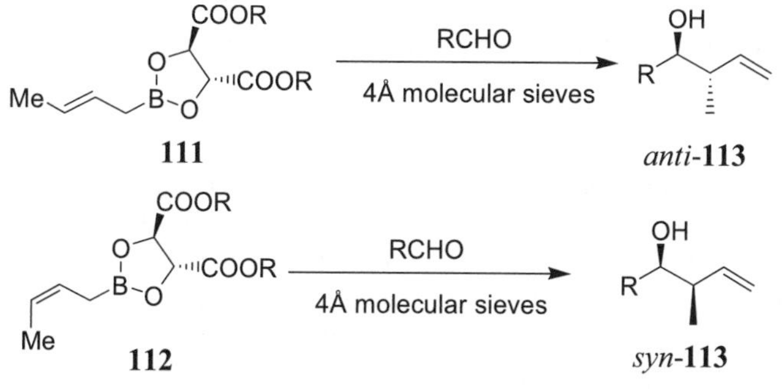

Scheme 3–40

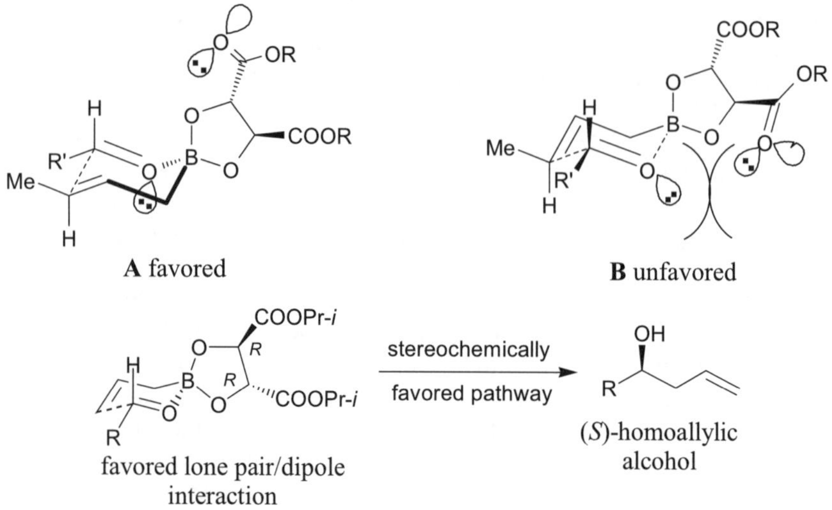

A favored

B unfavored

favored lone pair/dipole interaction

stereochemically favored pathway

(*S*)-homoallylic alcohol

Figure 3–6. Favored and unfavored transition states.

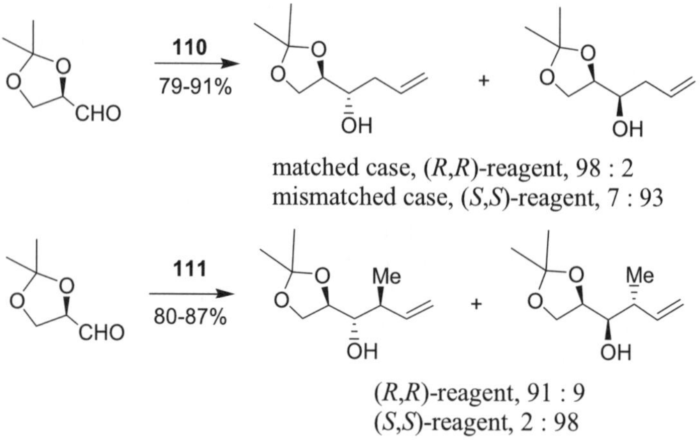

matched case, (*R,R*)-reagent, 98 : 2
mismatched case, (*S,S*)-reagent, 7 : 93

(*R,R*)-reagent, 91 : 9
(*S,S*)-reagent, 2 : 98

Scheme 3–41. Reaction with protected strained aldehyde (glyceraldehyde).

formationally constrained α- or β-alkoxyl–substituted aldehydes are excellent allyl-boration substrates. The diminished stereoselectivity is introduced not by the steric effect but by the unfavorable lone pair/lone pair interaction between the tartrate carbonyl and the alkoxy substituents. This is true especially for conformationally unconstrained aldehyde substrates.[73] In fact, Roush reagents (**110**, **111**, and **112**) exhibit a useful level of matched and mismatched diastereoselectivity in reactions with both chiral strained (Scheme 3–41) and unstrained aldehydes (Scheme 3–42).[74]

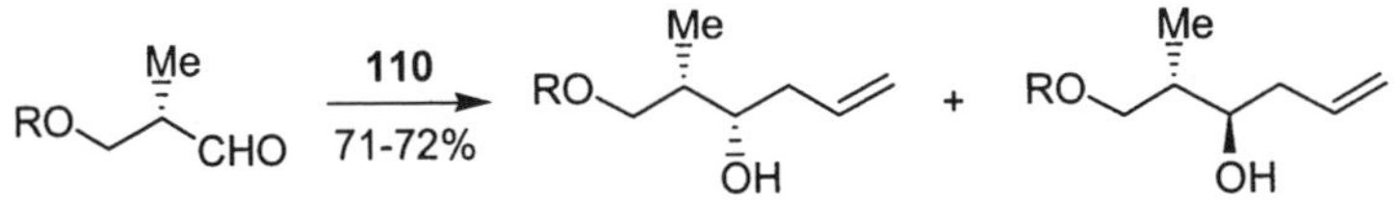

Aldehyde	Reagent	Selectivity
R = TBS	(R,R)-**110** (matched)	89 : 11
	(S,S)-**110** (mismatched)	19 : 81
R = TBDPS	(R,R)-**110**	79 : 21
	(S,S)-**110**	13 : 87
R = Bn	(R,R)-**110**	83 : 17
	(S,S)-**110**	20 : 80

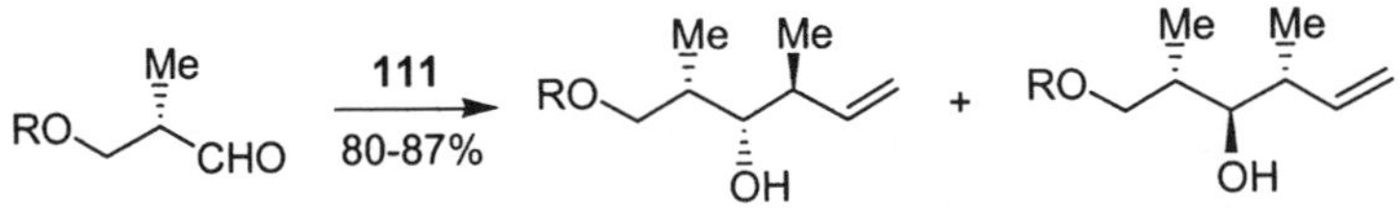

R = TBS	(R,R)-reagent (matched)	97 : 3
	(S,S)-reagent (mismatched)	16 : 84

Scheme 3–42. Example of allylation of unstrained aldehyde.

Garcia et al.[75] have introduced another boron reagent **114** that can also be used in asymmetric allylation reactions.

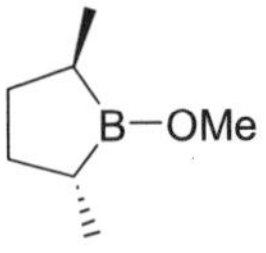

114

With the aid of $BF_3 \cdot OEt_2$, methoxyborolane (R,R)-**114** reacts with (E)- or (Z)-crotylpotassium to provide (E,R,R)-**115** and (Z,R,R)-**115**, respectively. After adding the aldehyde to a solution of crotyl-borolane in THF at $-78°C$ for 4 hours, 2-aminoethanol is added. The solution is warmed to room temperature, and oxidative cleavage at this point gives the homoallylic alcohols with high stereoselectivity. The borolane moiety can be recovered by precipitating it as an amino alcohol complex and can be reused without any loss of enantiomeric purity. As shown in Scheme 3–43, the (E)- and (Z)-crotyl compounds lead to anti- and syn-products **116**, respectively. The diastereoselectivity is about 20:1, and the ee for most cases is over 95% (Table 3–11).

Kijanolide **117**,[76] tetronolide **118**,[76] and chlorothricolide **119**,[77] the aglycones of the structurally novel antitumor antibiotics kijimicin, tetrocaricin A, and chlorothrimicin, are highly valued targets for total synthesis. All three structures share a similar octahydronaphthalene fragment **121**, which can be obtained by cyclization of **120**. Compound **120**, appropriately functionalized 2,8,10,12-tetradecatetraene acid, can be constructed via aldol reactions. Two

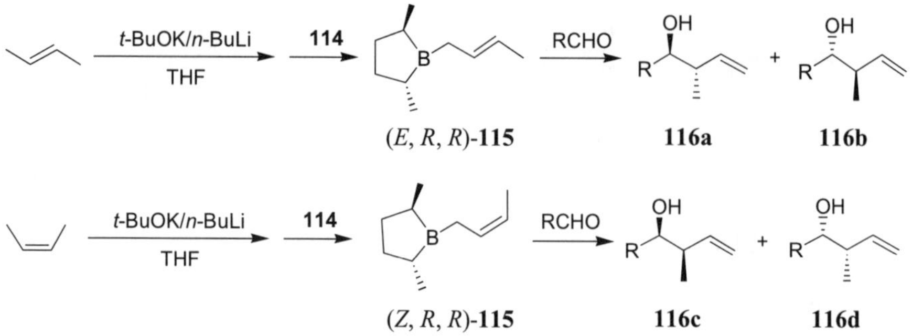

(E, R, R)-**115** **116a** **116b**

(Z, R, R)-**115** **116c** **116d**

Scheme 3–43

TABLE 3–11. Reaction of Crotylboranes (E,R,R)-115 and (Z,R,R)-115 with Representative Achiral Aldehydes

Entry	Crotylborane	Aldehyde	Yield (%)	*anti/syn* Ratio	Major Product ee (%)
1	(E)-**115**	C_2H_5CHO	81	93/7	96
2	(E)-**115**	i-C_3H_7CHO	76	96/4	97
3	(E)-**115**	i-C_4H_9CHO	72	96/4	95
4	(Z)-**115**	C_2H_5CHO	73	7/93	86
5	(Z)-**115**	i-C_3H_7CHO	70	4/96	93
6	(Z)-**115**	i-C_4H_9CHO	75	5/95	97

ee = Enantiomeric excess.

Reprinted with permission by Am. Chem. Soc., Ref. 75.

pairs of chiral centers, C-4/C-5 and C-6/C-7, can be regarded as aldol products, and the Roush reaction provides excellent access to these asymmetric centers.

Treatment of **122** with (R,R)-tartrate crotyl-boronate (E,R,R)-**111** provides the alcohol corresponding to **123** with 96% stereoselectivity. Benzylation of this alcohol yields **123** with 64% overall yield. The crude aldehyde intermediate obtained by ozonolysis of **123** is again treated with (Z,R,R)-**111** (the second Roush reaction), and a 94:5:1 mixture of three diastereoisomers is produced, from which **124** can be isolated with 73% yield. A routine procedure completes the synthesis of compound **120**, as shown in Scheme 3–44. Heating a toluene solution of **120** in a sealed tube at 145°C under argon for 7 hours provides the cyclization product **127**. Subsequent debromination, deacylation, and Barton deoxygenation accomplishes the stereoselective synthesis of **121** (Scheme 3–44).

With this method, that is, the reaction of tartaric acid ester–modified crotyl-boronates with chiral 2-methyl aldehydes, the C-19 to C-29 fragment of rifamycin has been constructed similarly.[76]

Roush reported another tartrate boronate, (E)-γ-[(menthofuryl)-dimethyl silyl]-allylboronate **130**, for *anti*-α-hydroxyallylation of aldehydes. Reagent **130** can be obtained from commercially available menthofuran, which was selected

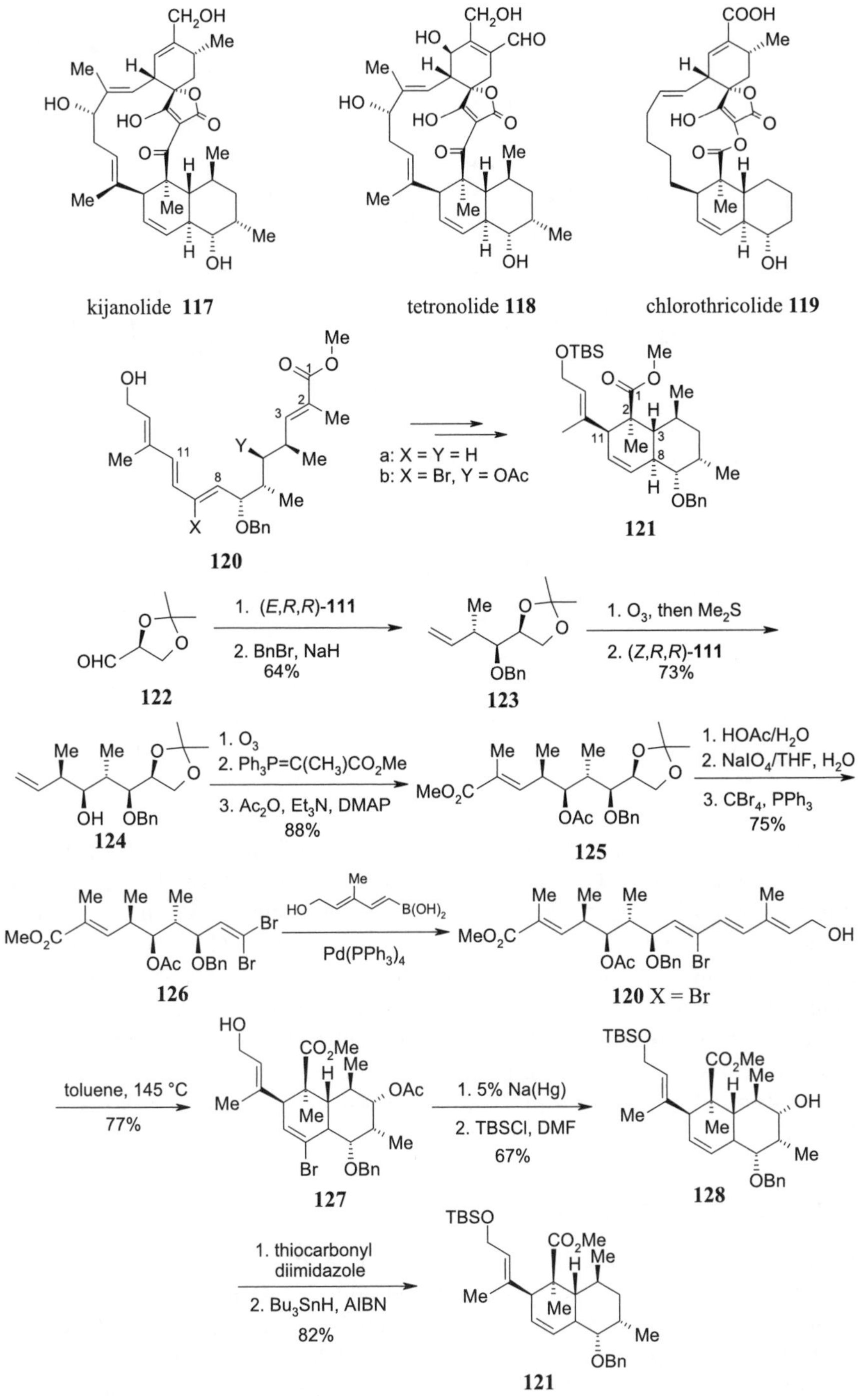

kijanolide **117**

tetronolide **118**

chlorothricolide **119**

Scheme 3–44. Asymmetric synthesis of octahydronaphthalene fragment **121**.

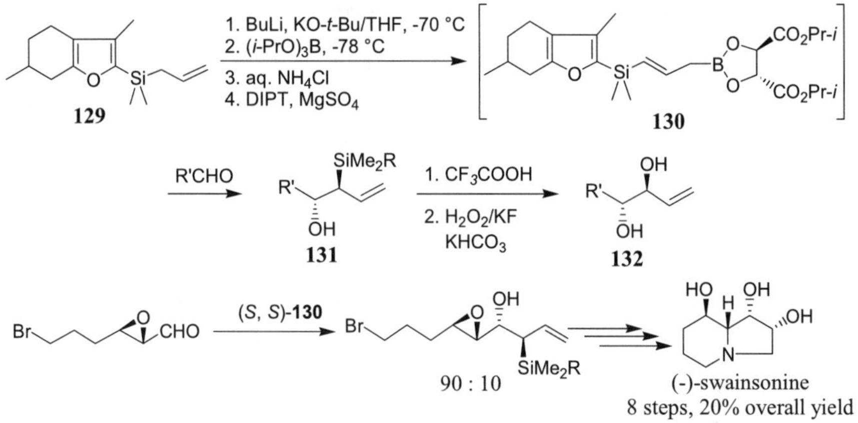

Scheme 3–45. Application of the boronate.

Scheme 3–46. Synthesis of α,β-disubstituted tetrahydropyrans and tetrahydrofurans.

because of the significantly easier protodesilylation of the resulting intermediate 2-furyl dimethylsilanes. On treatment with aldehyde, transformation of the allyl group accomplishes the synthesis of **131**. The silyl group in **131** can be removed by the Fleming method, resulting in the free alcohol **132**. This reagent provides an excellent method for *anti*-diol synthesis. Application of this compound is exemplified by the synthesis of (−)-swainsonine (Scheme 3–45).[78]

Starting from substituted allyl bis-(2,4-dimethyl-3-pentyl)-L-tartrate boronic acid, synthesis of α,β-disubstituted tetrahydrofurans (**134**, n = 1) or tetrahydropyrans (**134**, n = 2) can be accomplished with high enantioselectivity (Scheme 3–46).[79]

3.6.2   The Corey Reaction

As discussed in Section 3.3.2, Corey demonstrated the utility of compound **55**, prepared from 1,2-diphenyl-1,2-diamino ethane **54**, as a chiral auxiliary for asymmetric aldol reaction. In a similar manner, his group utilized this compound **55** in both (R,R)- and (S,S)-forms for allylation reactions. Treatment of **55** with allyltributyltin in dry CH_2Cl_2 at 0°C and then 23°C for 2 hours gives chiral allyl-borane **135**. In this process, both the (R,R)- and (S,S)-forms can be obtained and applied in asymmetric allylation reactions. Thus, treatment of

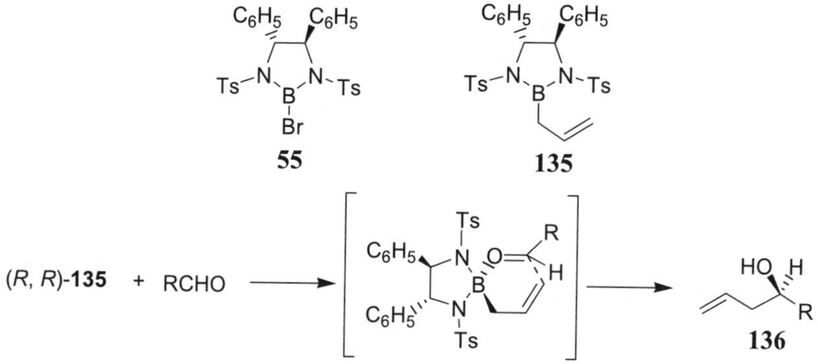

Scheme 3–47. Reaction of aldehydes with chiral allyl boranes **135**. Reprinted with permission by Am. Chem. Soc., Ref. 41.

R of RCHO	Solvent	ee (%) of **136**	Config.
Ph	Toluene	95	R
(E)-PhCH=CH-	CH_2Cl_2	98	R
c-C_6H_{11}	Toluene	97	R
n-C_5H_{11}	Toluene	95	S

various aldehydes with (R,R)-**135** in toluene or CH_2Cl_2 at $-78°C$ furnishes the homoallylic alcohol **136** with high optical purity (Scheme 3–47). The absolute configuration of the product **136** can be predicted based on a chair-like transition state with optimized stereoelectronic and steric interaction between the substituents on the five-membered ring.[41]

2-Haloallyl reagents **137** have been produced by treating enantiomers of **55** with the corresponding 2-haloallyl-n-butyltin; and 2-haloallyl carbinol **138** is obtained in high yield and predictable diastereoselectivity by reacting aldehydes with **137** (Scheme 3–48).[41]

Product 2-haloallyl carbinol **138** has wide synthetic utility in a number of transformations, as shown in Figure 3–7.[41]

3.6.3   Other Catalytic Asymmetric Allylation Reactions

The most commonly used method for achieving asymmetric allylation is to use an organometallic reagent in which the metal is ligated by chiral modifiers. Excellent results can be obtained for boron-containing[80] and titanium-containing[81] allyl moieties; but only low or modest results are obtained with silicon[82] or tin[83] compounds under similar conditions. The modest results with certain compounds suggest that the key issue is a difference in the reaction mechanism: Allyl-boron or titanium reagents react through an associative cyclic transition structure,[84] while allyl silanes or stannanes react through a less rigid, open chain transition structure. Good results have been achieved with asymmetric addition of allyl silane and/or allyl stannane (Sakurai reaction[85] to carbonyl compouds in the presence of a Lewis acid).

137a X = Br
137b X = Cl

138a X = Br
138b X = Cl

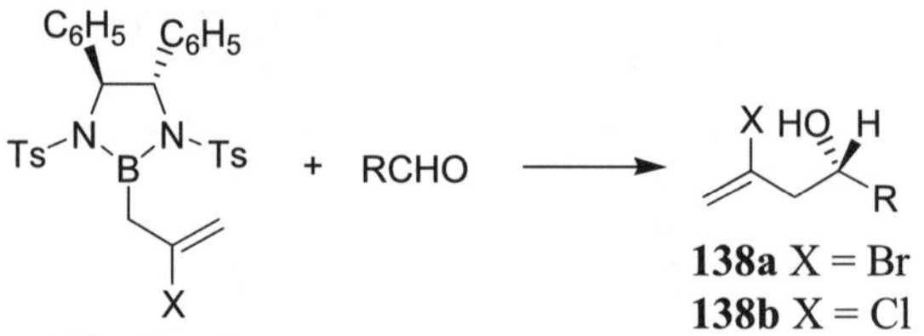

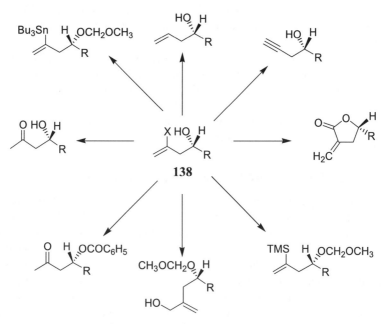

R of RCHO	Reagent	Yield (%)	ee (%) of **138**	Config.
Ph	**137a**	73	79	*S*
Ph	**137b**	79	84	*S*
(*E*)-PhCH=CH	**137a**	79	87	*S*
(*E*)-PhCH=CH	**137b**	84	92	*S*
c-C_6H_{11}	**137a**	75	94	*S*
c-C_6H_{11}	**137b**	81	99	*S*
n-C_5H_{11}	**137a**	71	94	*R*
n-C_5H_{11}	**137b**	77	99	*R*

Scheme 3–48. Reaction of aldehydes with chiral 2-haloallylboranes. Reprinted with permission by Am. Chem. Soc., Ref. 41.

Figure 3–7. Applications of compound **138** in organic synthesis. Reprinted with permission by Am. Chem. Soc., Ref. 41.

Denmark and Almstead[86] reported that Lewis acid promoted asymmetric allylation and crotylation of aldehydes with allyltrichlorosilanes. Kobayashi and Nishio[87] demonstrated the allyl and crotyl trichlorosilane addition in which good yield (>85%) and diastereoselecvitity [anti/syn > 93:7 for (E)-crotyl and syn/anti > 96:4 for (Z)-crotyl] were obtained when DMF was used as the solvent. When one equivalent of DMF was used, the reaction proceeded very slowly, and a rather long reaction time was required for a reasonable conversion (e.g., 70 hours were required for 80% conversion). Better results were obtained by changing DMF to another Lewis base, phosphoramide. For example, with HMPA as a promotor, rapid reaction was observed ($t_{1/2} = 18$ minutes). Using chiral phosphoramide as a promotor for asymmetric allylation, however, failed to enhance the stereoselectivity.

As an alternative approach, chiral Lewis base has been tested for catalytic allylation. Compound **139**, reported by Iseki et al.,[88] was the first example of a chiral Lewis base that effectively serves as a catalyst in asymmetric allylation in combination with HMPA. Allylation of aliphatic aldehydes with allyl- and crotyltrichlorosilanes in the presence of **139** provides up to 98% ee (Scheme 3–49).

Allylation of aromatic aldehydes with allyl trimethylsilane catalyzed by chiral acyloxyborane gives good results. In contrast, the results are normally poor for aliphatic aldehydes.[89] Costa et al.[90] introduced another enantioselective allylation procedure aiming to overcome this problem. In the following example, the enantioselective allylation of aldehyde octanal was carried out using

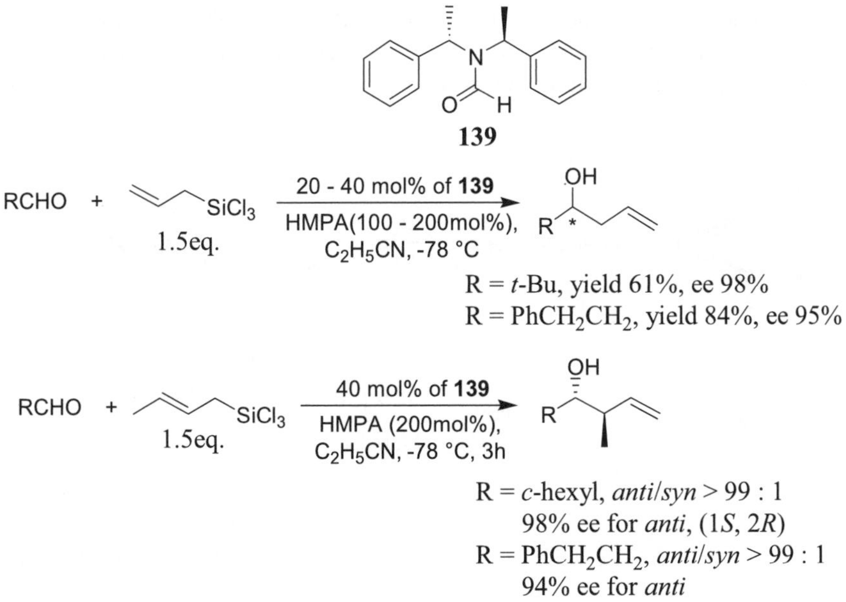

139

RCHO + (allyl)SiCl₃
1.5eq.

20 - 40 mol% of **139**
HMPA(100 - 200mol%),
C_2H_5CN, -78 °C

R = t-Bu, yield 61%, ee 98%
R = PhCH₂CH₂, yield 84%, ee 95%

RCHO + (crotyl)SiCl₃
1.5eq.

40 mol% of **139**
HMPA (200mol%),
C_2H_5CN, -78 °C, 3h

R = c-hexyl, anti/syn > 99 : 1
98% ee for anti, (1S, 2R)
R = PhCH₂CH₂, anti/syn > 99 : 1
94% ee for anti

Scheme 3–49

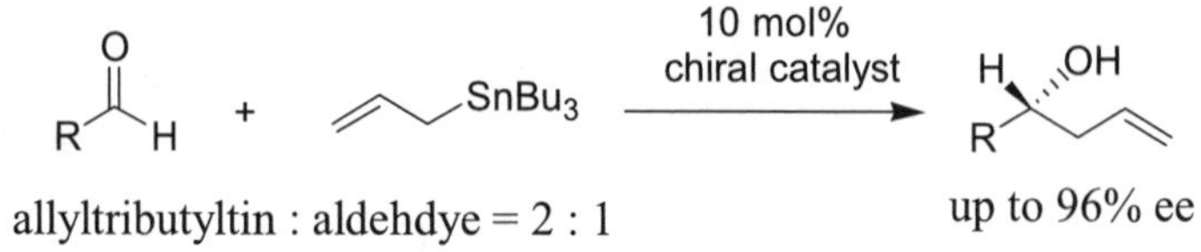

allyltributyltin : aldehdye = 2 : 1

up to 96% ee

Scheme 3–50

allyl-tributyltin as the allyl donor. The enantiomeric excess obtained normally can be over 90% when the reaction is carried out in the presence of a catalytic amount of (20 mol%) BINOL-TiCl$_2$ complex and activated 4 Å MS (Scheme 3–50).[90]

Preparation of the catalyst can be accomplished under mild conditions without stirring, heating, or cooling, and allyl addition can also be conducted more conveniently using 10 mol% of a 2:1 BINOL/Ti catalyst system at room temperature.[91]

In the last few years, asymmetric catalysis by means of chiral Lewis acids has led to highly enantioselective protocols for a variety of synthetic transformations, including important C–C bond formation processes. The most successful chiral Lewis acids for catalytic enantioselective C–C bond formation contain B(III), Al(III), Ti(IV), Sn(II), and rare earth metals.

Ti-BINOL–catalyzed reactions have been well established. When the Ti is replaced by Zr,[92] the resulting complex **140** can also catalyze the addition of allyl-tributyltin to aldehydes (aldehydes:allyl-tributyltin:**140** = 1:2:0.2 mol ratio) in the presence of 4 Å MS. Product 1-alken-4-ols are obtained in good yield and high ee. The *Si*-face of the aldehyde is attacked if (*S*)-BINOL is used, and *Re*-face attack takes place when (*R*)-BINOL is used as the chiral ligand. For Zr complex–catalyzed reactions, the reaction proceeds much faster, although the

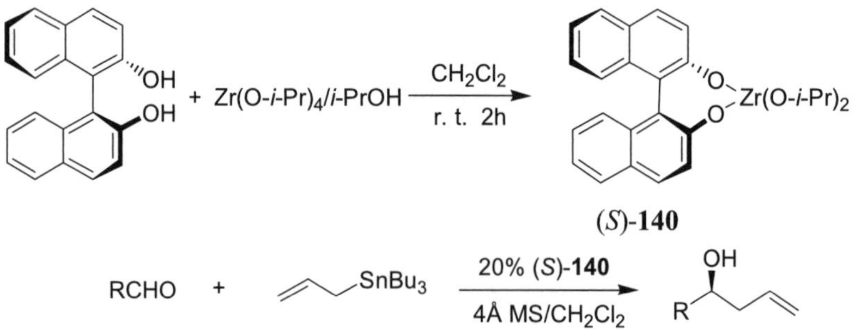

(*S*)-**140**

R =		
n-C$_7$H$_{15}$	87% ee	
n-C$_5$H$_{11}$	89% ee	
(*E*)-PhCH=CH	91% ee	
Ph	92% ee	

Scheme 3–51. Zirconate-catalyzed asymmetric allylation reactions.

TABLE 3–12. Some Commonly Used Allylating Agents

Chiral Auxiliary	Reference	Chiral Auxiliary	Reference
(structure, O–BAll, Ph)	80a	(structure, SO₂Me, N, O–BAll, Ph)	93
(structure, Ph, SO₂Tol, N, BAll, N, SO₂Tol, Ph)	41	(structure, i-PrO, i-PrO, BAll)	70a
(structure, Bn, N, N, Bn, O, O, BAll)	94	(structure, BAll)	75
(structure, BAll, SiMe₃)	95	(structure,]₂BAll)	96
(structure,]₂BAll)	96	(structure,]₂BAll)	97
(structure, Ph, Ph, Ph, Ph, O, O, O, O, Ti, All)	81		

stereofacial selectivity is not improved. These two catalysts can complement each other in the way that the Ti catalyst is suitable for aliphatic aldehydes, while Zr catalyst can be used for aromatic ones.

For convenient reference, some commonly used allylating agents are listed in Table 3–12.

3.7 ASYMMETRIC ALLYLATION AND ALKYLATION OF IMINES

Stereoselective addition of organometallics to C=N bond is not fully understood due to a number of difficulties. First, imines are not as electrophilic as carbonyl compounds, and they are less susceptible to nucleophilic attack. Second, imine compounds tend to undergo deprotonation rather than additions. Usually, the presence of a Lewis acid is required to promote the nucleophilic attack of an allyl group on a carbonyl group. Imines, however, are susceptible to E/Z isomerization under such conditions, which complicates the stereochemical outcome of the reaction. In addition, even if the alkylation or allylation is successful, the amine thus formed will strongly bind to the Lewis acid catalyst, rendering the catalyst inactive.

To circumvent these problems, several strategies have been adopted. One is to increase the electrophilicity of the carbon atom of the C=N substrate by N-alkylation, N-oxidation, N-acylation, or N-sulfonylation to give reactive iminium salts, nitrones, acylimines, or sulfonimines. The problem associated with this method arises from the subsequent removal of the activating agent. Another method involves the activation of the C=N bond by coordination with Lewis acid or addition of an external promoter. In this method, reactive alkyl organometallic reagents can be used in imine additions. The less basic reagents such as allyl-boranes, allyl-stannanes, alkyl-coppers, or alkyl-cuprates and organocerium reagents can all be employed for this purpose.

Increasing interest is expressed in diastereoselective addition of organometallic reagents to the C=N bond of chiral imines or their derivatives, as well as chiral catalyst-facilitated enantioselective addition of nucleophiles to prochiral imines.[98] The imines frequently selected for investigation include N-masked imines such as oxime ethers, sulfenimines, and N-trimethylsilylimines (**150–153**). A variety of chiral modifiers, including chiral boron compounds, chiral diols, chiral hydroxy acids, N-sulfonyl amino acids, and N-sulfonyl amido alcohols **141–149**, have been evaluated for their efficiency in enantioselective allylboration reactions.[68c]

Substrates:

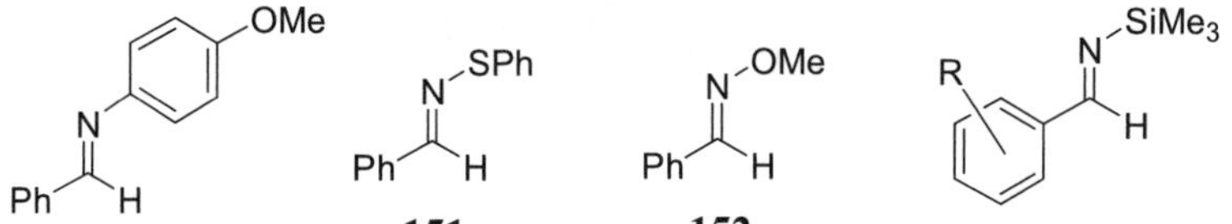

150   **151**   **152**   **153a** R = H
153b R = 2-Cl
153c R = 4-MeO

The first example of allylation of imines using dialkyl 2-allyl-1,3-dioxaborolane-4,5-dicarboxylates **110** or B-allyldiisopinocamphenyl borane **154** was reported in 1995.[99] After hydrolysis of the reaction product, homoallyl primary amine was obtained in 54–90% yield and up to 73% ee (Scheme 3–52).

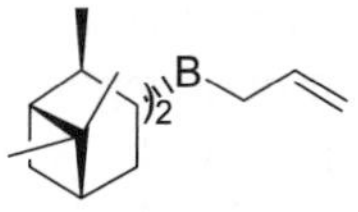

154

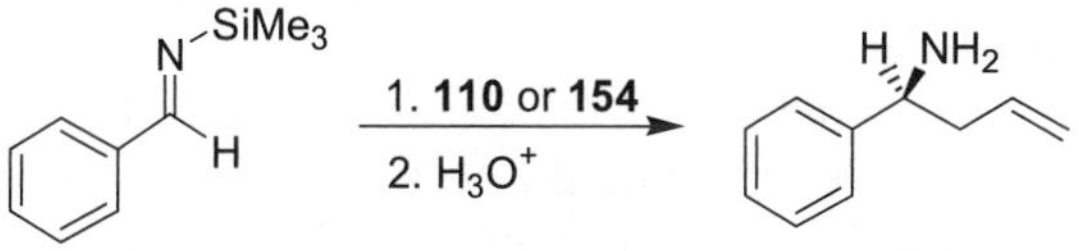

Scheme 3–52

Itsuno et al.[68c] found that among the imines **150–153**, N-trimethylsilylimine derivative **153a** has the highest reactivity. When it is treated with β-allyloxazaborolidine **156**, a compound derived from toluenesulfonyl norephedrine **155**, a product with high selectivity (92% ee) is obtained. Scheme 3–53 depicts the

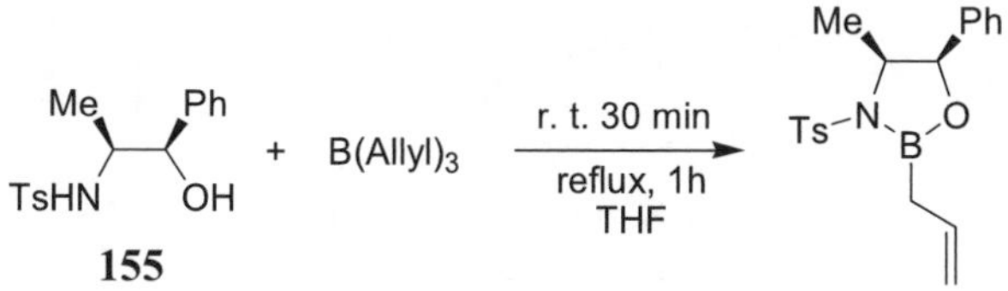

155

156

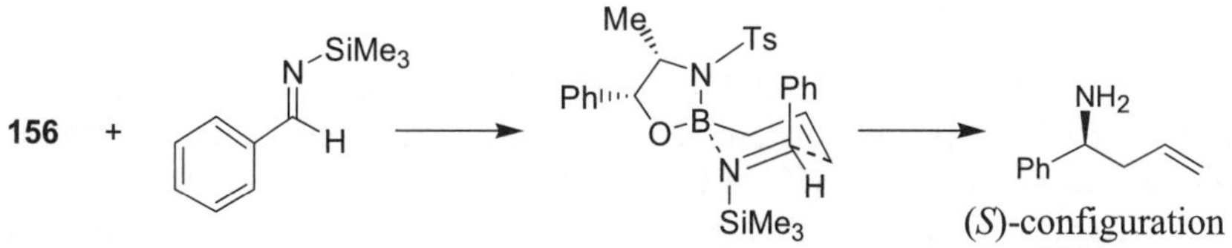

(S)-configuration
92% ee

Scheme 3–53

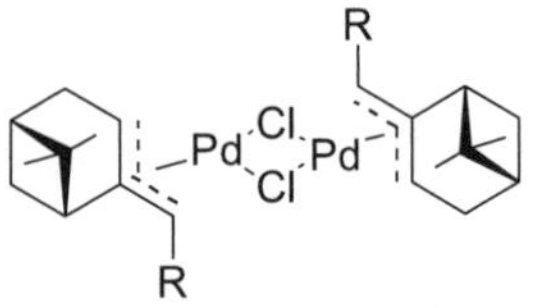

157

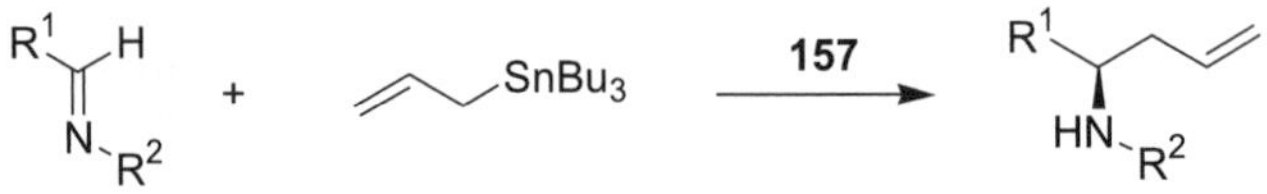

Scheme 3–54

reaction transition state. From the transition state, it seems likely that the addition of **156** proceeds with an optimized stereoelectronic interaction and minimized steric repulsion, providing the (*S*)-homoallylamine as the major product. In this transition state the steric repulsion between the bulky trimethyl silyl and the toluenesulfonyl group seems to be avoided.

Allylation of imines using this type of reagent has been extensively studied, and this transformation has become important for the synthesis of acyclic and cyclic amine derivatives.

Nakamura et al.[100] found that in the presence of palladium catalysts imines undergo allylation readily, providing the corresponding homoallylamines with high yields. Thus, chiral palladium complex **157** has been synthesized and applied in the asymmetric allylation of imines using allyl tributyltin as the allylation reagent. In Scheme 3–54, moderate yield and up to 82% ee have been obtained with **157** as the chiral catalyst.[101]

Park et al.[102] demonstrated an interesting intramolecular asymmetric allylation of imine based on a substrate-controlled mode.

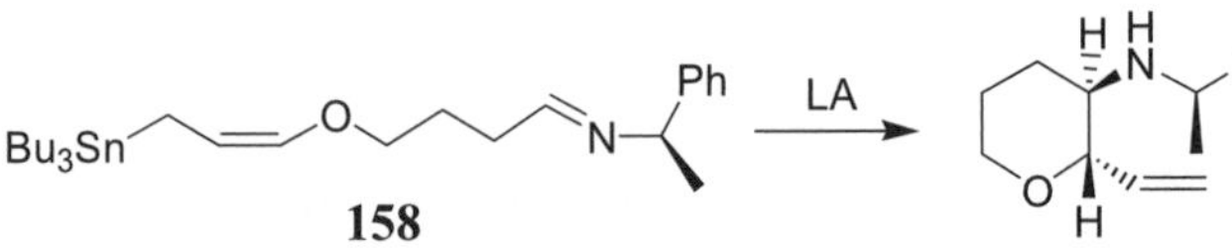

158   **159**

LA = ZrCl$_4$, ZnCl$_2$ + aq. HCl,
AlCl$_3$, EtAlCl$_2$, Et$_2$AlCl

In the presence of ZrCl$_4$ or HCl, cyclization of γ-alkoxyallylstannane **158** bearing (*R*)-(+)-1-phenylethylamine as a chiral auxiliary occurs to produce *trans-β*-aminocyclic ether **159** with high de (91%). As shown in Scheme 3–55, asymmetric addition of an allyl group to the imine carbon can be explained by the modified Cram model **160**. The attack of the allylic γ-carbon approaches

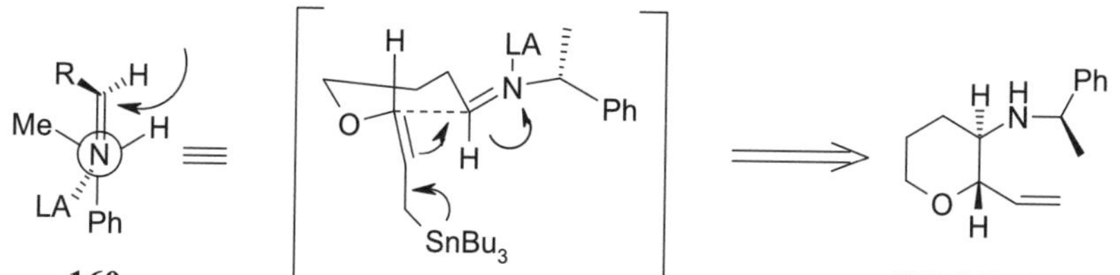

Scheme 3–55

the imine carbon from a direction opposite to the Lewis acid chelating site. This orientation of the nucleophile to the Lewis acid chelating site produces the C-3 chirality of **159**, giving the (2S,3R)-chirality of compound **159**. The allylation of imines has been discussed elsewhere.[103]

Besides the allylation reactions, imines can also undergo enol silyl ether addition as with carbonyl compounds. Carbon–carbon bond formation involving the addition of resonance-stabilized nucleophiles such as enols and enolates or enol ethers to iminium salt or imine can be referred to as a *Mannich reaction,* and this is one of the most important classes of reactions in organic synthesis.[104]

Scheme 3–56 shows an example of the generation of chiral amines via nucleophilic attack onto an imine substrate in the presence of an external homochiral auxiliary. Moderate ee can be obtained from **161**-induced reactions, and moderate to high ee can be expected from **162**-induced reactions. For instance, when **161** (R^1 = Et, R^2 = *t*-Bu) is involved in the reaction, nucleophilic attack of RLi (R = Me, *n*-Bu, and vinyl) on imine **163** gives product **164** with 81–92%

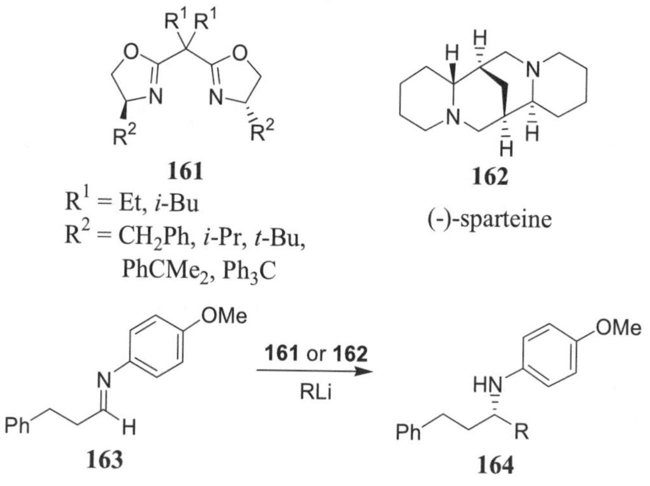

161

R^1 = Et, *i*-Bu
R^2 = CH$_2$Ph, *i*-Pr, *t*-Bu,
PhCMe$_2$, Ph$_3$C

162

(-)-sparteine

163　　　　　**164**

Scheme 3–56

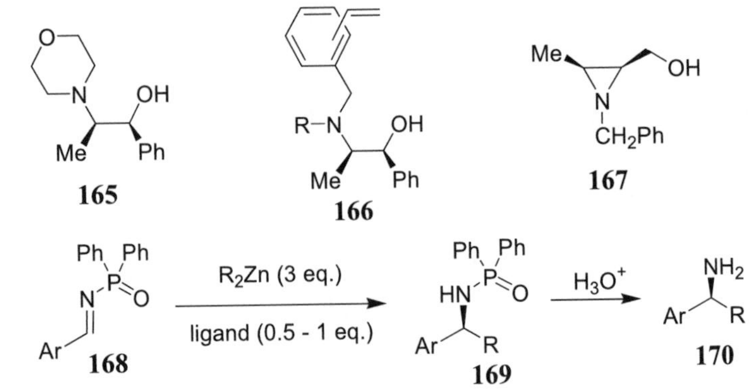

Scheme 3–57

yield and 51–82% ee. The **162**-induced reaction (R = Me, n-Bu, Ph) gives product **164** with 71–91% yield and over 91% ee. Thus, in the presence of **161** or **162**, various substrates derived from both aryl and alkyl aldehydes can undergo feasible organolithium nucleophilic addition reactions, and the chiral ligand can be used even in only catalytic amounts.[105]

The reactivity of N-diphenylphosphinyl imines toward dialkylzinc addition in the presence of a stoichiometric or catalytic amount of chiral ligand **165**, **166**, or **167** has also been meticulously investigated. The reaction in Scheme 3–57 gives good yield with up to 95% ee.[106]

Hagiwara et al.[107] reported the chiral Pd(II) complex–catalyzed asymmetric addition of enol silyl ethers to imines, based on the belief that Pd(II) enolate was involved in the reaction. They found that with compound **171a** as the catalyst, very low enantioselectivity was obtained in the asymmetric reactions between silyl enol ether and imine compounds (Scheme 3–58). However, in the

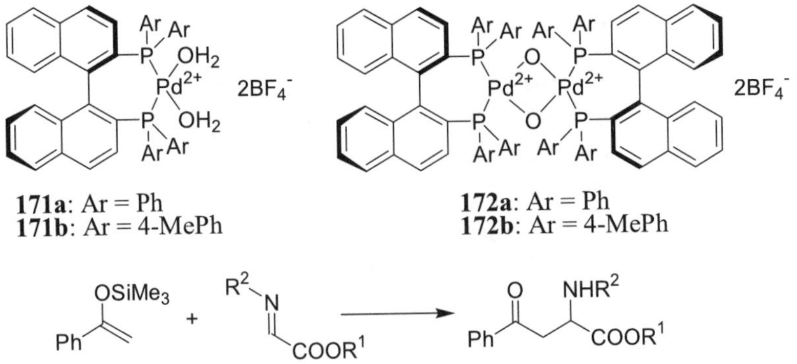

171a: Ar = Ph
171b: Ar = 4-MePh

172a: Ar = Ph
172b: Ar = 4-MePh

Scheme 3–58

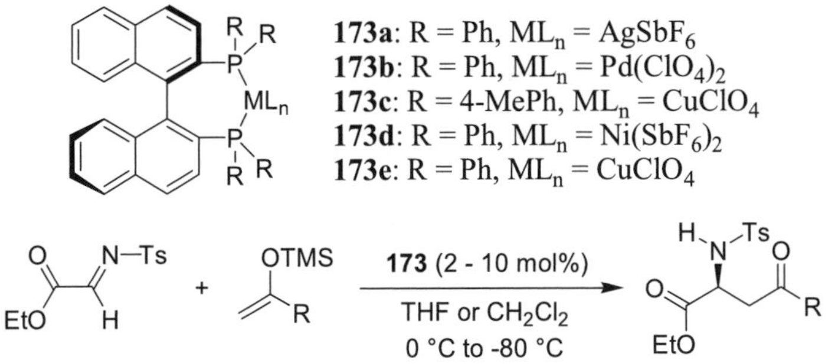
173a: R = Ph, MLn = AgSbF6
173b: R = Ph, MLn = Pd(ClO4)2
173c: R = 4-MePh, MLn = CuClO4
173d: R = Ph, MLn = Ni(SbF6)2
173e: R = Ph, MLn = CuClO4
173 (2 - 10 mol%)
THF or CH2Cl2
0 °C to -80 °C

Scheme 3–59

presence of catalyst **172**, a series of enantioselective Mannich-type reactions proceeded smoothly, providing the product with up to 95% yield and up to 90% ee (Scheme 3–58, $R^1 = i$-Pr, $R^2 =$ 4-MeOPh).

Ferraris et al.[108] demonstrated an asymmetric Mannich-type reaction using chiral late-transition metal phosphine complexes as the catalyst. As shown in Scheme 3–59, the enantioselective addition of enol silyl ether to α-imino esters proceeds at −80°C, providing the product with moderate yield but very high enantioselectivity (over 99%).

Ferraris et al.[109] also studied the diastereoalkylation of α-imino esters catalyzed by Cu complexes. As depicted in Scheme 3–60, the reaction results in both high diastereoselectivity (*syn*/*anti* up to 21/1) and high enantioselectivity for the major product (ee up to 99%).

β-Amino alcohol can be found as important subunits of many bioactive compounds such as *α*/*β*-adrenergic agents or antagonists,[110] HIV protease inhibitor,[111] and antifungal or antibacterial peptides.[112] For this reason, many attempts have been made at the asymmetric construction of *β*-amino alcohol subunits.[113] Kobayashi et al.[113] have reported a route to both *syn*- and *anti*-*β*-amino alcohols via catalytic diastereoselective or enantioselective Mannich-type reactions of α-alkoxy enolates with aldimines. By varying the protective groups on the α-alkoxy part and the ester group of the enolate part, both *syn*- and *anti*-selectivity can be achieved. As shown in Scheme 3–61 and Table 3–13, the expected product can be obtained in high diastereoselectivity as well as high enantioselectivity.

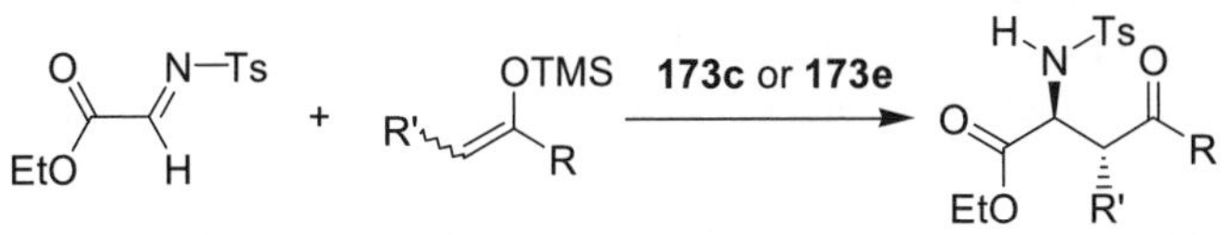
173c or 173e

Scheme 3–60

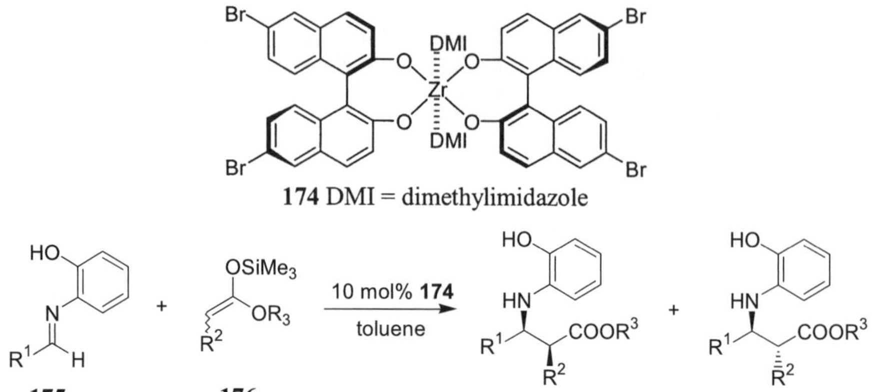

174 DMI = dimethylimidazole

Scheme 3–61

TABLE 3–13. BINOL-Zr Catalyzed Mannich-Type Reaction

Entry	R^1	R^2	R^3	Temp. (°C)	Yield (%)	syn/anti	ee (%)
1	Ph	(E)-TBSO	i-Pr	−78	Quant.	96:4	95
2	1-Nap	(E)-TBSO	i-Pr	−78	65	>99:1	91
3	2-Furyl	(E)-TBSO	i-Pr	−45	68	82:18	92
4	4-ClPh	(E)-TBSO	i-Pr	−78	73	92:8	98
5	Ph	(Z)-BnO	i-Pr	−45	Quant.	32:68	95
6	Ph	(Z)-BnO	PMP	−45	91	6:94	80
7	1-Nap	(Z)-BnO	c-C$_6$H$_{11}$	−45	80	8:92	96
8	2-Furyl	(Z)-BnO	PMP	−45	68	13:87	80
9	4-ClPh	(Z)-BnO	i-Pr	−45	Quant.	43:57	91
10	4-ClPh	(Z)-BnO	PMP	−45	72	8:92	76
11	c-C$_6$H$_{11}$	(Z)-BnO	c-C$_6$H$_{11}$	−45	41	18:82	92

ee = Enantiomeric excess.

Reprinted with permission by Am. Chem. Soc., Ref. 113.

3.8 OTHER TYPES OF ADDITION REACTIONS: HENRY REACTION

Nitroalkane can be used as a convenient reagent for alkyl anion synthesis. Aliphatic nitro compounds can also be considered as versatile building blocks and intermediates in organic synthesis. They are readily available, and there are a wide variety of methods for converting the nitro group to other functional groups.

Acting as a strong electron-withdrawing group, the nitro group can activate a neighboring carbon–hydrogen bond for alkylation, as well as acylation or sulfonation. Through the reaction with saturated or conjugated carbonyl compounds, 1,2- or 1,4-difunctionalized derivatives can be obtained, and other novel compounds can be synthesized upon oxidation, reduction, hydrolysis, and dehydration on both nitro and hydroxy groups.

Nitroalkanes can be reduced to amines, amides, hydroxylamines, nitrones, oximes, or nitriles or converted into azidosulfones and azidosulfides. More importantly, the nitro group can be converted into a carbonyl group. This allows a wide range of transformations, because it effectively reverses the polarity of the neighboring carbon from nucleophilic to electrophilic. With the wide range of methods available for converting a nitro group to a carbonyl group,[114] primary nitroalkanes can be considered useful and convenient intermediates in organic synthesis, particularly because they can be efficiently used as acyl anion synthons.

The ease of converting a nitro group to a carbonyl or other functional group has significantly increased the synthetic potential of nitroalkane derivatives as reagents for the nucleophilic introduction of functionalized alkyl groups in the synthesis of natural products.

The nucleophilic addition of nitroalkane to carbonyl groups is known as the Henry reaction. The products of the Henry reaction are 2-nitroalkanols,[115] which are useful intermediates for nitroalkenes, 2-amino alcohols, and 2-nitroketones. However, this does not always give high yields because of the possible *O*-alkylation in preference to *C*-alkylation during the Henry reaction.

Scheme 3–62 is an example of an asymmetric Henry reaction reported by Sasai et al.[116] in 1993. The catalyst acts in a bimetallic manner. This multifunctional effect is further discussed in Chapter 8.

Scheme 3–63 gives another example in which an asymmetric Henry reaction is involved in the synthesis of the β-receptor–blocking drug (*S*)-propranolol.

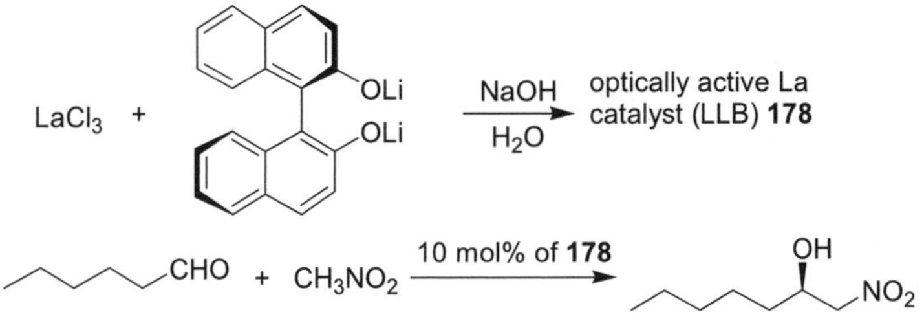

Scheme 3–62

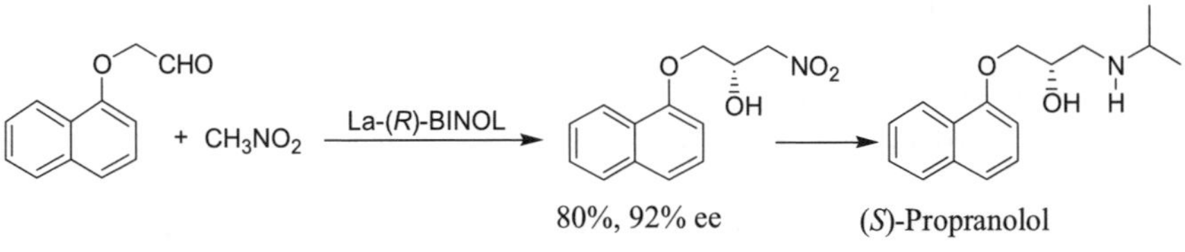

80%, 92% ee   (*S*)-Propranolol

Scheme 3–63

Similarly, this reaction has also been applied in the synthesis of other β-receptor–blocking drugs such as (S)-metoprolol,[117] (S)-pindolol,[118] and the HIV protease inhibitors KNI-227 and KNI-272.[119]

For a review about nitroaldol and the Henry reaction, see Rosini and Ballini.[120]

3.9 SUMMARY

This chapter has introduced the aldol and related allylation reactions of carbonyl compounds, the allylation of imine compounds, and Mannich-type reactions. Double asymmetric synthesis creates two chiral centers in one step and is regarded as one of the most efficient synthetic strategies in organic synthesis. The aldol and related reactions discussed in this chapter are very important reactions in organic synthesis because the reaction products constitute the backbone of many important antibiotics, anticancer drugs, and other bioactive molecules. Indeed, study of the aldol reaction is still actively pursued in order to improve reaction conditions, enhance stereoselectivity, and widen the scope of applicability of this type of reaction.

3.10 REFERENCES

1. For references on enantioselective aldol methodologies, see (a) Heathcock, C. H., in Trost, B. M., Fleming I. eds. *Comprehensive Organic Synthesis*, Pergamon Press., New York, **1991**, vol. 2, Chapter 1.6, p 181. (b) Kim, B. M.; Williams, S. F.; Masamune, S., in Trost, B. M.; Fleming I. eds. *Comprehensive Organic Synthesis*, Pergamon Press, New York, **1991**, vol. 2, Chapter 1.7, p 239. (c) Paterson, I., in Trost, B. M.; Fleming I. eds. *Comprehensive Organic Synthesis*, Pergamon Press, New York, **1991**, vol. 2, Chapter 1.9, p 301. (d) Caine, D., in Trost, B. M.; Fleming I. eds. *Comprehensive Organic Synthesis*, Pergamon Press, New York, **1991**, vol. 3, Chapter 1.1, p 1. (e) Braun, M.; Devant, R. *Tetrahedron Lett.* **1984**, *25*, 5031. (f) Ambler, P. W.; Davies, S. G. *Tetrahedron Lett.* **1985**, *26*, 2129. (g) Katsuki, T.; Yamaguchi, M. *Tetrahedron Lett.* **1985**, *26*, 5807. (h) Evans, D. A.; Sjogren, E. B.; Weber, A. E.; Conn, R. E. *Tetrahedron Lett.* **1987**, *28*, 39. (i) Sasai, H.; Suzuki, T.; Itoh, N.; Shibasaki, M. *Tetrahedron Lett.* **1993**, *34*, 851. (j) Kodota, I.; Sakaihara, T.; Yamamoto, Y. *Tetrahedron Lett.* **1996**, *37*, 3195. (k) Ito, Y.; Sawamura, M.; Hayashi, T. *J. Am. Chem. Soc.* **1986**, *108*, 6405. (l) Abiko, A.; Liu, J.; Masamune, S. *J. Am. Chem. Soc.* **1997**, *119*, 2586.

2. Eliel, E. in Morrison, J. D. ed. *Asymmetric Synthesis*, Academic Press, Orlando, **1983**, Vol. 2, p 125.

3. Zimmerman, H. E.; Traxler, M. D. *J. Am. Chem. Soc.* **1957**, *79*, 1920.

4. Liu, J. F.; Abiko, A.; Pei, Z. H.; Buske, D. C.; Masamune, S. *Tetrahedron Lett.* **1998**, *39*, 1873.

5. Mitsui, S.; Konno, K.; Onuma, I.; Shimizu, K. *J. Chem. Soc. Jpn.* **1964**, *85*, 437.

6. Evans, D. A.; Bartroli, J.; Shih, T. L. *J. Am. Chem. Soc.* **1981**, *103*, 2127.

7. Masamune, S.; Choy, W.; Kerdesky, F. A. J.; Imperiali, B. *J. Am. Chem. Soc.* **1981**, *103*, 1566.

8. For a new and racemization-free reductive removal of *N*-acyloxazolidinones, see Prashad, M.; Har, D.; Kim, H.; Repic, O. *Tetrahedron Lett.* **1998**, *39*, 7067.

9. Nerz-Stormes, M.; Thornton, E. R. *Tetrahedron Lett.* **1986**, *27*, 897.

10. (a) Heathcock, C. H. in Morrison, J. D. ed. *Asymmetric Synthesis*, Academic Press, New York, **1984**, vol. 3, p 111. (b) Masamune, S.; Choy, W.; Petersen, J. S.; Sita, L. R. *Angew. Chem. Int. Ed. Engl.* **1985**, *24*, 1.

11. Evans, D. A.; Bartroli, J. *Tetrahedron Lett.* **1982**, *23*, 807.

12. Evans, D. A.; Sjogren, E. B.; Bartroli, J.; Dow, R. L. *Tetrahedron Lett.* **1986**, *27*, 4957.

13. Evans, D. A.; Nelson, J. V.; Taber, T. R. in Allinger, N. L.; Eliel, E. L.; Silen, S. H. ed. *Top. Stereochem.*, John Wiley & Sons, New York, **1982**, vol. 13, p 1.

14. Kawanami, Y.; Ito, Y.; Kitagawa, T.; Taniguchi, Y.; Katsuki, T.; Yamaguchi, M. *Tetrahedron Lett.* **1984**, *25*, 857.

15. Ito, Y.; Katsuki, T.; Yamaguchi, M. *Tetrahedron Lett.* **1984**, *25*, 6015.

16. (a) Evans, D. A.; McGee, L. R. *J. Am. Chem. Soc.* **1981**, *103*, 2876. (b) Yamamoto, Y.; Maruyama, K. *Tetrahedron Lett.* **1980**, *21*, 4607.

17. Mukaiyama, T.; Narasaka, K.; Banno, K. *Chem. Lett.* **1973**, 1011.

18. Gennari, C.; Bernardi, A.; Colombo, L.; Scolastico, C. *J. Am. Chem. Soc.* **1985**, *107*, 5812.

19. (a) Zelle, R. E.; DeNinno, M. P.; Selnick, H. G.; Danishefsky, S. J. *J. Org. Chem.* **1986**, *51*, 5032. (b) Danishefsky, S. J.; Selnick, H. G.; DeNinno, M. P.; Zelle, R. E. *J. Am. Chem. Soc.* **1987**, *109*, 1572. (c) Barbier, P.; Schneider, F.; Widmer, U. *Helv. Chim. Acta* **1987**, *70*, 1412.

20. (a) Palazzi, C.; Colombo, L.; Gennari, C. *Tetrahedron Lett.* **1986**, *27*, 1735. (b) Gennari, C.; Colombo, L.; Bertolini, G.; Schimperna, G. *J. Org. Chem.* **1987**, *52*, 2754.

21. Gennari, C.; Bernardi, A.; Scolastico, C.; Potenza, D. *Tetrahedron Lett.* **1985**, *26*, 4129.

22. Gennari, C.; Colombo, L.; Bertolini, G. *J. Am. Chem. Soc.* **1986**, *108*, 6394.

23. (a) Gennari, C.; Venturini, I.; Gislon, G.; Schimperna, G. *Tetrahedron Lett.* **1987**, *28*, 227. (b) Gennari, C.; Schimperna, G.; Venturini, I. *Tetrahedron* **1988**, *44*, 4221.

24. (a) Gennari, C.; Cozzi, P. G. *J. Org. Chem.* **1988**, *53*, 4015. (b) Gennari, C.; Cozzi, P. G. *Tetrahedron* **1988**, *44*, 5965.

25. Poll, T.; Metter, J. O.; Helmchen, G. *Angew. Chem. Int. Ed. Engl.* **1985**, *24*, 112.

26. (a) Mikami, K.; Matsukawa, S. *J. Am. Chem. Soc.* **1994**, *116*, 4077. (b) Reetz, M. T.; Kyung, S.; Bolm, C.; Zierke, T. *Chem. Ind.* **1986**, 824.

27. (a) Kobayashi, S.; Uchiro, H.; Shiina, I.; Mukaiyama, T. *Tetrahedron* **1993**, *49*, 1761. (b) Kobayashi, S.; Furuya, M.; Ohtsubo, A.; Mukaiyama, T. *Tetrahedron Asymmetry* **1991**, *2*, 635.

28. (a) Corey, E. J.; Cywin, C. L.; Roper, T. D. *Tetrahedron Lett.* **1992**, *33*, 6907. (b) Parmee, E. R.; Hong, Y.; Tempkin, O.; Masamune, S. *Tetrahedron Lett.* **1992**, *33*, 1729. (c) Kiyooka, S.; Kaneko, Y.; Kume, K. *Tetrahedron Lett.* **1992**, *33*, 4927. (d) Parmee, E. R.; Tempkin, O.; Masamune, S.; Abiko, A. *J. Am. Chem. Soc.*

1991, *113*, 9365. (e) Kiyooka, S.; Kaneko, Y.; Komura, M.; Matsuo, H.; Nakano, M. *J. Org. Chem.* **1991**, *56*, 2276.

29. (a) Furuta, K.; Maruyama, T.; Yamamoto, H. *J. Am. Chem. Soc.* **1991**, *113*, 1041. (b) Furuta, K.; Maruyama, T.; Yamamoto, H. *Synlett* **1991**, 439.

30. Carreira, E. M.; Singer, R. A.; Lee, W. *J. Am. Chem. Soc.* **1994**, *116*, 8837.

31. Mukaiyama, T. *Org. React.* **1982**, *28*, 203.

32. (a) Helmchen, G.; Leikauf, U.; Taufer-Knöpfel, I. *Angew. Chem. Int. Ed. Engl.* **1985**, *24*, 874. (b) Oppolzer, W.; Marco-Contelles, J. *Helv. Chim. Acta* **1986**, *69*, 1699.

33. Oppolzer, W.; Lienard, P. *Tetrahedron Lett.* **1993**, *34*, 4321.

34. Enders, D.; Lohray, B. B. *Angew. Chem. Int. Ed. Engl.* **1988**, *27*, 581.

35. Meyers, A. I.; Temple, D. L.; Nolen, R. L.; Mihelich, E. D. *J. Org. Chem.* **1974**, *39*, 2778.

36. Meyers, A. I.; Knaus, G.; Kamata, K.; Ford, M. E. *J. Am. Chem. Soc.* **1976**, *98*, 567.

37. Meyers, A. I.; Yamamoto, Y. *J. Am. Chem. Soc.* **1981**, *103*, 4278.

38. Meyers, A. I.; Yamamoto, Y. *Tetrahedron* **1984**, *40*, 2309.

39. Corey, E. J.; Imwinkelried, R.; Pikul, S.; Xiang, Y. B. *J. Am. Chem. Soc.* **1989**, *111*, 5493.

40. Corey, E. J.; Lee, D.; Sarshar, S. *Tetrahedron Asymmetry* **1995**, *6*, 3.

41. Corey, E. J.; Yu, C. M.; Kim, S. S. *J. Am. Chem. Soc.* **1989**, *111*, 5495.

42. Paterson, I.; Mansuri, M. M. *Tetrahedron* **1985**, *41*, 3569.

43. Masamune, S.; Sato, T.; Kim, B.; Wollmann, T. A. *J. Am. Chem. Soc.* **1986**, *108*, 8279.

44. Duthaler, R. O.; Herold, P.; Lottenbach, W.; Oertle, K.; Riediker, M. *Angew. Chem. Int. Ed. Engl.* **1989**, *28*, 494.

45. Mukaiyama, T.; Uchiro, H.; Kobayashi, S. *Chem. Lett.* **1989**, 1001.

46. (a) Kobayashi, S.; Mukaiyama, T. *Chem. Lett.* **1989**, 297. (b) Kobayashi, S.; Uchiro, H.; Shiina, I.; Mukaiyama, T. *Tetrahedron* **1993**, *49*, 1761.

47. (a) Iwasawa. N.; Mukaiyama, T. *Chem. Lett.* **1982**, 1441. (b) Mukaiyama, T.; Iwasawa, N.; Stevens, R. W.; Haga, T. *Tetrahedron* **1984**, *40*, 1381.

48. Hannun, Y. A.; Loomis, C. R.; Merrill, A. H.; Bell, R. M. *J. Biol. Chem.* **1986**, *261*, 12604.

49. (a) Kiso, M.; Nakamura, A.; Tomita, Y.; Hasegawa, A. *Carbohydr. Res.* **1986**, *158*, 101. (b) Herold, P. *Helv. Chim. Acta* **1988**, *71*, 354. (c) Garner, P.; Park; J. M.; Malecki, E. *J. Org. Chem.* **1988**, *53*, 4395. (d) Julina, R.; Herzig, T.; Bernet, B.; Vasella, A. *Helv. Chim. Acta* **1986**, *69*, 368. (e) Ito, Y.; Sawamura, M.; Hayashi, T. *Tetrahedron Lett.* **1988**, *29*, 239. (f) Sugawara, T.; Narisada, M. *Carbohydr. Res.* **1989**, *194*, 125. (g) Fujita, S.; Sugimoto, M.; Tomita, K.; Nakahara, Y.; Ogawa, T. *Agric. Biol. Chem.* **1991**, *55*, 2561. (h) Marukami, T.; Minamikawa, H.; Hato, M. *Tetrahedron Lett.* **1994**, *35*, 745.

50. (a) Kobayashi, S.; Hayashi, T.; Kawasuji, T. *Tetrahedron Lett.* **1994**, *35*, 9573. (b) For more on the synthesis of sphingosine, sphingofungins B and F see: Kobayashi S.; Furuta, T. *Tetrahedron* **1998**, *54*, 10275.

51. Nelson, S. G. *Tetrahedron Asymmetry* **1998**, *9*, 357.

52. Ito, Y.; Sawamura, M.; Hayashi, T. *J. Am. Chem. Soc.* **1986**, *108*, 6405.

53. (a) Sawamura, M.; Ito, Y.; Hayashi, T. *Tetrahedron Lett.* **1989**, *30*, 2247. (b) Ito, Y.; Sawamura, M.; Hayashi, T. *Tetrahedron Lett.* **1988**, *29*, 239.

54. Ito, Y.; Sawamura, M.; Hamashima, H.; Emura, T.; Hayashi, T. *Tetrahedron Lett.* **1989**, *30*, 4681.

55. (a) Corey, E. J.; Barnes-Seeman, D.; Lee, T. W. *Tetrahedron Lett.* **1997**, *38*, 1699. (b) Corey, E. J.; Rohde, J. J. *Tetrahedron Lett.* **1997**, *38*, 37.

56. (a) Keck, G. E.; Krishnamurthy, D. *J. Am. Chem. Soc.* **1995**, *117*, 2363. (b) Mikami, K.; Matsukawa, S. J. *J. Am. Chem. Soc.* **1994**, *116*, 4077. (c) Corey E. J.; Cywin, C. L.; Roper, T. D. *Tetrahedron Lett.* **1992**, *33*, 6907. (d) Parmee, E. R.; Hong, Y.; Tempkin, O.; Masamune, S. *Tetrahedron Lett.* **1992**, *33*, 1729. (e) Kiyooka, S.; Kaneko, Y.; Kume, K. *Tetrahedron Lett.* **1992**, *33*, 4927. (f) Yanagisawa, A.; Matsumoto, Y.; Nakashima, H.; Asakawa, K.; Yamamoto, H. *J. Am. Chem. Soc.* **1997**, *119*, 9319, and references therein.

57. Evans, D. A.; Kozlowski, M. C.; Murry, J. A.; Burgey, C. S.; Campos, K. R.; Connell, B. T.; Staples, R. J. *J. Am. Chem. Soc.* **1999**, *121*, 669.

58. Evans, D. A.; Burgey, C. S.; Kozlowski, M. C.; Tregay S. W. *J. Am. Chem. Soc.* **1999**, *121*, 686.

59. (a) Krüger, J.; Carreira, E. M. *J. Am. Chem. Soc.* **1998**, *120*, 837. (b) Pagenkopf, B.; Krüger, J.; Stojanovic, A.; Carreira, E. M. *Angew. Chem. Int. Ed. Engl.* **1998**, *37*, 3124.

60. (a) Pereyre, M.; Quintard, J.; Rahm, A. *Tin in Organic Synthesis*, Butterworths, London, **1987**, p 286. (b) Davies, A. G. *Organotin Chemistry*, VCH, Weinheim, **1997**, p 185.

61. Yanagisawa, A.; Matsumoto, Y.; Asakawa, K.; Yamamoto, H. *J. Am. Chem. Soc.* **1999**, *121*, 892.

62. Yamada, Y. M. A.; Yoshikawa, N.; Sasai, H.; Shibasaki, M. *Angew. Chem. Int. Ed. Engl.* **1997**, *36*, 1871.

63. Yamada, Y. M. A.; Shibasaki, M. *Tetrahedron Lett.* **1998**, *39*, 5561.

64. Masamune, S.; Kaiho, T.; Garvey, D. S. *J. Am. Chem. Soc.* **1982**, *104*, 5521.

65. Masamune, S.; Choy, W.; Kerdesky, F. A. J.; Imperiali, B. *J. Am. Chem. Soc.* **1981**, *103*, 1566.

66. Masamune, S.; Ali, S. A.; Snitman, D. L.; Garvey, D. S. *Angew. Chem. Int. Ed. Engl.* **1980**, *19*, 557.

67. Short, R. P.; Masamune, S. *Tetrahedron Lett.* **1987**, *28*, 2841.

68. (a) For a review about allylation of carbonyl groups by allylic metals as a highly efficient tool for selective functionalization, see Yamamoto, Y.; Asao, N. *Chem. Rev.* **1993**, *93*, 2207. (b) For a review about the addition of allylic tin reagents to carbonyl compounds, see Marshall, J. A. *Chem. Rev.* **1996**, *96*, 31. (c) For a review about the nucleophilic addition of chiral modified allylboron reagents to imines, see Itsuno, S.; Watanabe, W.; Ito, K.; EI-Shehawy, A. A.; Sarhan, A. A. *Angew. Chem. Int. Ed. Engl.* **1997**, *36*, 109.

69. Haruta, R.; Ishiguro, M.; Iketa, N.; Yamamoto, H. *J. Am. Chem. Soc.* **1982**, *104*, 7667.

70. (a) Roush, W. R.; Walts, A. E.; Hoong, L. K. *J. Am. Chem. Soc.* **1985**, *107*, 8186. (b) Roush, W. R.; Halterman, R. L. *J. Am. Chem. Soc.* **1986**, *108*, 294.

71. Fujita, K.; Schlosser, M. *Helv. Chim. Acta* **1982**, *65*, 1258.

72. Roush, W. R.; Hoong, L. K.; Palmer, M. A. J.; Park, J. *J. Org. Chem.* **1990**, *55*, 4109.

73. Roush, W. R.; Hoong, L. K.; Palmer, M. A. J.; Straub, J. A.; Palkowitz, A. D. *J. Org. Chem.* **1990**, *55*, 4117.

74. Roush, W. R.; Palkowitz, A. D.; Palmer, M. A. J. *J. Org. Chem.* **1987**, *52*, 316.

75. Garcia, J.; Kim, B. M.; Masamune, S. *J. Org. Chem.* **1987**, *52*, 4831.

76. Roush, W. R.; Brown, B. B.; Drozda, S. E. *Tetrahedron Lett.* **1988**, *29*, 3541.

77. Roush, W. R.; Sciotti, R. J. *Tetrahedron Lett.* **1992**, *33*, 4691.

78. (a) Hunt, J. A.; Roush, W. R. *Tetrahedron Lett.* **1995**, *36*, 501. (b) Hunt, J. A.; Roush, W. R. *J. Org. Chem.* **1997**, *62*, 1112.

79. Brown, H. C.; Phadke, A. S. *Synlett* **1993**, 927.

80. (a) Hoffmann, R. W.; Herold, T. *Chem. Ber.* **1981**, *114*, 375. (b) Racherla, U. S.; Brown, H. C. *J. Org. Chem.* **1991**, *56*, 401.

81. Hafner, A.; Duthaler, R. O.; Marti, R.; Rihs, G.; Rothe-Streit, P.; Schwarzenbach, F. *J. Am. Chem. Soc.* **1992**, *114*, 2321.

82. (a) Hathaway, S. J.; Paquette, L. A. *J. Org. Chem.* **1983**, *48*, 3351. (b) Coppi, L.; Mordini, A.; Taddei, M. *Tetrahedron Lett.* **1987**, *28*, 969. (c) Nativi, C.; Ravida, N.; Ricci, A.; Seconi, G.; Taddei, M. *J. Org. Chem.* **1991**, *56*, 1951. (d) Chan, T. H.; Wang, D. *Tetrahedron Lett.* **1989**, *30*, 3041.

83. (a) Otera, J.; Kawasaki, Y.; Mizuno, H.; Shimizu, Y. *Chem. Lett.* **1983**, 1529. (b) Otera, J. Yoshinaga, Y. Yamaji, T.; Yoshioka, T.; Kawasaki, Y. *Organometallics* **1985**, *4*, 1213. (c) Boldrini, G. P.; Tagliavini, E.; Trombini, C.; Umandi-Ronchi, A. *J. Chem. Soc. Chem. Commun.* **1986**, 685. (d) Augé, J.; Bourleaux, G. *J. Organomet. Chem.* **1989**, *377*, 205.

84. Denmark, S. E.; Weber, E. J. *Helv. Chim. Acta* **1983**, *66*, 1655.

85. Hosomi, A.; Sakurai, H. *Tetrahedron Lett.* **1976**, 1295.

86. (a) Denmark, S. E.; Almstead, N. G. *J. Org. Chem.* **1994**, *59*, 5130. (b) Denmark, S. E.; Hosoi, S. *J. Org. Chem.* **1994**, *59*, 5133.

87. (a) Kobayashi, S.; Nishio, K. *Tetrahedron Lett.* **1993**, *34*, 3453. (b) Kobayashi, S.; Nishio, K; *Synthesis* **1994**, 457.

88. Iseki, K.; Mizuno, S.; Kuroki, Y.; Kobayashi, Y. *Tetrahedron Lett.* **1998**, *39*, 2767.

89. Furuta, K.; Mouri, M.; Yamamoto, H. *Synlett* **1991**, 561.

90. Costa, A. L.; Piazza, M. G.; Tagliavini, E.; Trombini, C.; Umani-Ronchi, A. *J. Am. Chem. Soc.* **1993**, *115*, 7001.

91. Keck, G. E.; Geraci, L. S. *Tetrahedron Lett.* **1993**, *34*, 7827.

92. Bedeschi, P.; Casolari, S.; Costa, A. L.; Tagliavini, E.; Umani-Ronchi, A. *Tetrahedron Lett.* **1995**, *36*, 7897.

93. Reetz, M. T.; Zierke, T. *Chem. Ind.* **1988**, 663.

94. Roush, W. R.; Banfi, L. *J. Am. Chem. Soc.* **1988**, *110*, 3979.

95. Short, R. P.; Masamune, S. *J. Am. Chem. Soc.* **1989**, *111*, 1892.

96. Jadhav, P. K.; Bhat, K. S.; Perumal, P. T.; Brown, H. C. *J. Org. Chem.* **1986**, *51*, 432.

97. Brown, H. C.; Randad, R. S.; Bhat, K. S.; Zaidlewicz, M.; Racherla, U. S. *J. Am. Chem. Soc.* **1990**, *112*, 2389.

98. For a recent review on addition of organometallic reagent to C=N bonds, see Bloch, R. *Chem. Rev.* **1998**, *98*, 1407.

99. Watanabe, K.; Ito, K.; Itsuno, S. *Tetrahedron Asymmetry* **1995**, *6*, 1531.

100. (a) Nakamura, H.; Iwama, H.; Yamamoto, Y. *J. Chem. Soc. Chem. Commun.* **1996**, *1459*. (b) Nakamura, H.; Iwama, H.; Yamamoto, Y. *J. Am. Chem. Soc.* **1996**, *118*, 6641.

101. Nakamura, H.; Nakamura, K.; Yamamoto, Y. *J. Am. Chem. Soc.* **1998**, *120*, 4242.

102. Park, J. Y.; Park, C. H.; Kadota, I.; Yamamoto, Y. *Tetrahedron Lett.* **1998**, *39*, 1791.

103. For other references on the asymmetric allylation of imines, see (a) Yamamoto, Y.; Nishii, S.; Maruyama, K.; Komatsu, T.; Ito, W. *J. Am. Chem. Soc.* **1986**, *108*, 7778. (b) Basile, T.; Bocoum, A.; Savoia, D.; Umani-Ronchi, A. *J. Org. Chem.* **1994**, *59*, 7766.

104. For review, see, for example, Kleinman, E. F., in Trost, B. M.; Fleming, I. eds. *Comprehensive Organic Synthesis*, Pergamon Press, New York, **1991**, vol. 2, p 893, and references therein.

105. Denmark, S. E.; Nakajima, N.; Nicaise, O. J. *J. Am. Chem. Soc.* **1994**, *116*, 8797.

106. (a) Soai, K.; Hatanaka, T.; Miyazawa, T. *J. Chem. Soc., Chem. Commun.* **1992**, 1097. (b) Suzuki, T.; Narisata, N.; Shibata, T.; Soai, K. *Tetrahedron Asymmetry* **1996**, *7*, 2519. (c) Anderson, P. G.; Guijarro, D.; Tanner, D. *Synlett* **1996**, 727.

107. Hagiwara, E.; Fujii, A.; Sodeoka, M. *J. Am. Chem. Soc.* **1998**, *120*, 2474.

108. Ferraris, D.; Young, B.; Dudding, T.; Lectka, T. *J. Am. Chem. Soc.* **1998**, *120*, 4548.

109. Ferraris, D.; Young, B.; Cox, C.; Drury, W. J.; Dudding, T.; Lectka, T. *J. Org. Chem.* **1998**, *63*, 6090.

110. (a) Howe, R.; Rao, B. S.; Holloway, B. R.; Stribling, D. *J. Med. Chem.* **1992**, *35*, 1751. (b) Bloom, J. D.; Dutia, M. D.; Johnson, B. D.; Wissner, A.; Burns, M. G.; Largis, E. E.; Dolan, J. A.; Claus, J. H. *J. Med. Chem.* **1992**, *35*, 3081.

111. Askin, D.; Wallace, M. A.; Vacca, J. P.; Reamer, R. A.; Volante, R. P.; Shinkai, I. *J. Org. Chem.* **1992**, *57*, 2771.

112. Ohfune, Y. *Acc. Chem. Res.* **1992**, *25*, 360.

113. Kobayashi, S.; Ishitani, H.; Ueno, M. *J. Am. Chem. Soc.* **1998**, *120*, 431, and the references cited therein.

114. Nef, J. U. *Liebigs Ann. Chem.* **1894**, *280*, 263.

115. For an introduction to Henry reactions, see Coombes, R. G., in Barton, D.; Ollis, W. D. eds. *Comprehensive Organic Chemistry*, Pergamon, Oxford, **1979**, Vol. 2, p 303.

116. Sasai, H.; Itoh, N.; Suzuki, T.; Shibasaki, M. *Tetrahedron Lett.* **1993**, *34*, 855.

117. Sasai, H.; Suzuki, T.; Ito, N.; Arai, S.; Shibasaki, M. *Tetrahedron Lett.* **1993**, *34*, 2657.

118. Sasai, H.; Yamada, Y. M. A.; Suzuki, T.; Shibasaki, M. *Tetrahedron* **1994**, *50*, 12313.

119. Sasai, H.; Kim, W.; Suzuki, T.; Shibasaki, M. *Tetrahedron Lett.* **1994**, *35*, 6123.

120. Rosini, G.; Ballini, R. *Synthesis* **1988**, 833.

CHAPTER 4

Asymmetric Oxidations

The asymmetric oxidation of organic compounds, especially the epoxidation, dihydroxylation, aminohydroxylation, aziridination, and related reactions have been extensively studied and found widespread applications in the asymmetric synthesis of many important compounds. Like many other asymmetric reactions discussed in other chapters of this book, oxidation systems have been developed and extended steadily over the years in order to attain high stereoselectivity. This chapter on oxidation is organized into several key topics. The first section covers the formation of epoxides from allylic alcohols or their derivatives and the corresponding ring-opening reactions of the thus formed 2,3-epoxy alcohols. The second part deals with dihydroxylation reactions, which can provide diols from olefins. The third section delineates the recently discovered aminohydroxylation of olefins. The fourth topic involves the oxidation of unfunctionalized olefins. The chapter ends with a discussion of the oxidation of enolates and asymmetric aziridination reactions.

4.1 ASYMMETRIC EPOXIDATION OF ALLYLIC ALCOHOLS: SHARPLESS EPOXIDATION

Asymmetric epoxidation of allylic alcohols was once one of the leading areas of investigation in synthetic organic chemistry, mainly due to the fact that very high enantioselective induction for a wide range of substrates is possible using several classes of reagents. In terms of both chemical and optical yields, this procedure allows a chemical reaction to compete with an enzymatic process. Among the reagents serving as an essential element in epoxidation, the Sharpless titanium method needs to be introduced first.

In studies of the asymmetric epoxidation of olefins, chiral peroxycarboxylic acid–induced epoxidation seldom gives enantiomeric excess over 20%.[1] Presumably, this is due to the fact that the controlling stereocenters in peroxycarboxylic acids are too remote from the reaction site. An enantiomeric excess of over 90% has been reported for the poly-(S)-alanine–catalyzed epoxidation of chalcone.[2] The most successful nonmetallic reagents for asymmetric epoxidation have been the chiral N-sulfonyloxaziridines[3] until asymmetric epoxidation reactions mediated by chiral ketones were reported. Today, the

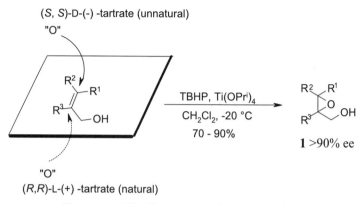

Figure 4–1. The Sharpless epoxidation reaction.

most successful asymmetric epoxidation reaction is the titanate-mediated epoxidation of allylic alcohols, or Sharpless epoxidation, which enables the achievement of an enantiomeric excess of over 90% in most cases.

The development of transition metal–mediated asymmetric epoxidation started from the dioxomolybdenum-N-ethylephedrine complex,[4] progressed to a peroxomolybdenum complex,[5] then vanadium complexes substituted with various hydroxamic acid ligands,[6] and the most successful procedure may now prove to be the tetraisopropoxyltitanium-tartrate–mediated asymmetric epoxidation of allylic alcohols.

The Sharpless epoxidation is a popular laboratory process that is both enantioselective and catalytic in nature. Not only does it employ inexpensive reagents and involve various important substrates (allylic alcohols) and products (epoxides) in organic synthesis, but it also demonstrates unusually wide applicability because of its insensitivity to many aspects of substrate structure. Selection of the proper chirality in the starting tartrate esters and proper geometry of the allylic alcohols allows one to establish both the chirality and relative configuration of the product (Fig. 4–1).

Since its discovery in 1980,[7] the Sharpless expoxidation of allylic alcohols has become a benchmark classic method in asymmetric synthesis. A wide variety of primary allylic alcohols have been epoxidized with over 90% optical yield and 70–90% chemical yield using TBHP (t-BuOOH) as the oxygen donor and titanium isopropoxide-diethyl tartrate (DET, the most frequently used dialkyl tartrate) as the catalyst. One factor that simplifies the standard epoxidation reaction is that the active chiral catalyst is generated in situ, which means that the pre-preparation of the active catalyst is not required.

The wide scope application of this transformation arises not only from the utility of epoxide compounds but also from the subsequent regiocontrolled and stereocontrolled nucleophilic substitution (ring-opening) reactions of the derived epoxy alcohol. These, through further functionalization, allow access to an impressive array of target molecules in enantiomerically pure form.

Like the vanadium-based epoxidation reaction, the Sharpless reaction in-

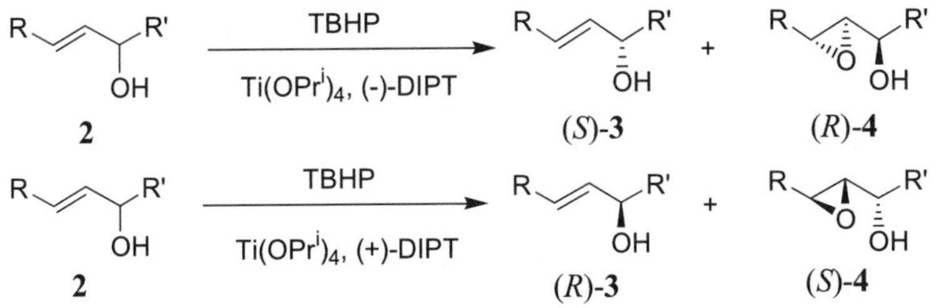

Scheme 4–1. Kinetic resolution of secondary allylic alcohols.

trinsically favors 1,2-*anti* products. With a racemic allylic alcohol, one of the enantiomers reacts faster, and this rate differentiation step can be used to selectively epoxidize the more reactive enantiomer in the presence of its antipode. In general, by reducing the amount of TBHP to 0.6 equivalent in the reaction system, the same reaction can be used to kinetically resolve secondary allylic alcohols (Scheme 4–1).[8]

4.1.1 The Characteristics of Sharpless Epoxidation

For the asymmetric epoxidation of achiral allyl alcohols, high ee can normally be obtained. For example, in Scheme 4–2, asymmetric epoxidation of the achiral allylic alcohol **5** provides epoxyl alcohol (**6**) with a selectivity of 99:1.

The idea of double asymmetric induction is also applicable to asymmetric epoxidation (see Chapter 1 for double asymmetric induction). In the case of asymmetric epoxidation involving double asymmetric induction, the enantioselectivity depends on whether the configurations of the substrate and the chiral ligand are matched or mismatched. For example, treating **7** with titanium tetraisopropoxide and *t*-butyl hydroperoxide without (+)- or (−)-diethyl tartrate yields a mixture of epoxy alcohols **8** and **9** in a ratio of 2.3:1 (Scheme 4–3). In a

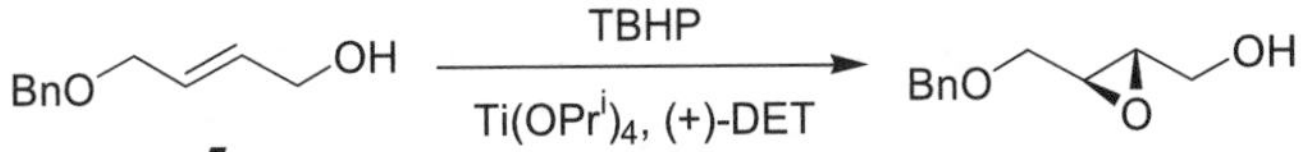

Scheme 4–3

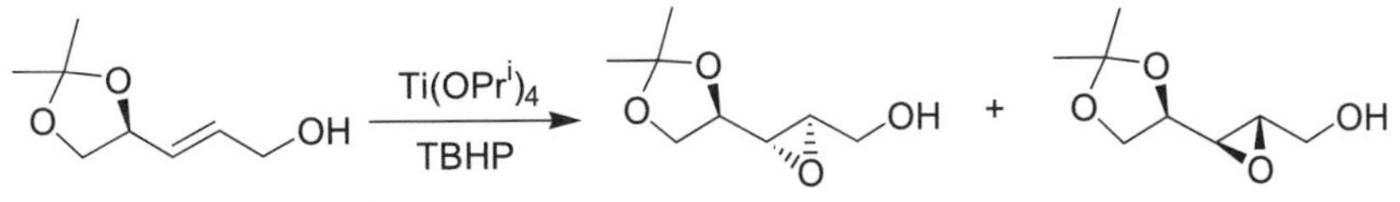

in the absence of tartrate: **8** : **9** = 2.3 : 1
in the presence of (+)-DET, mismatched, **8** : **9** = 1 : 22
in the presence of (−)-DET: matched, **8** : **9** = 90 : 1

Scheme 4–2

double asymmetric reaction, asymmetric epoxidation reaction of **7** with (+)- or (−)-diethyl tartrate proceeds smoothly to provide the epoxides **8** and **9** in ratios of 1:22 and 90:1, referring to the mismatched and matched cases, respectively.

In Sharpless epoxidation reactions, (*Z*)-substituted allylic alcohols react much more slowly than the corresponding (*E*)-substituted substrates, and sometimes the reaction is sensitive to the position of preexisting chirality in the selected substrate. For instance, in the presence of (+)-DET, chiral (*E*)-allylic alcohol **10** undergoes epoxidation in 15 hours to give product **11** as the major product with a diastereomeric ratio of >20:1. As for reaction with (−)-DET, **12** is then obtained, also with a diastereoselectivity of >20:1 (Scheme 4–4).

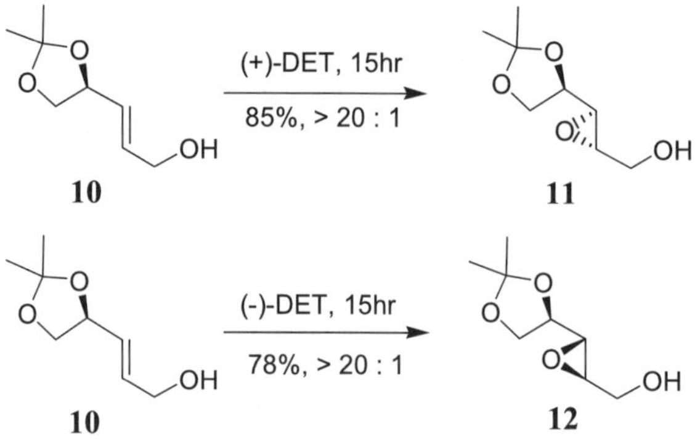

Scheme 4–4

In the case of (*Z*)-allylic alcohol **13**, however, it takes 2 weeks to get product **14** in a ratio of **14:15** = 30:1 for matched pairs, while the epoxide **14** is obtained in the much lower ratio of **14:15** = 3:2 for mismatched pairs (Scheme 4–5).

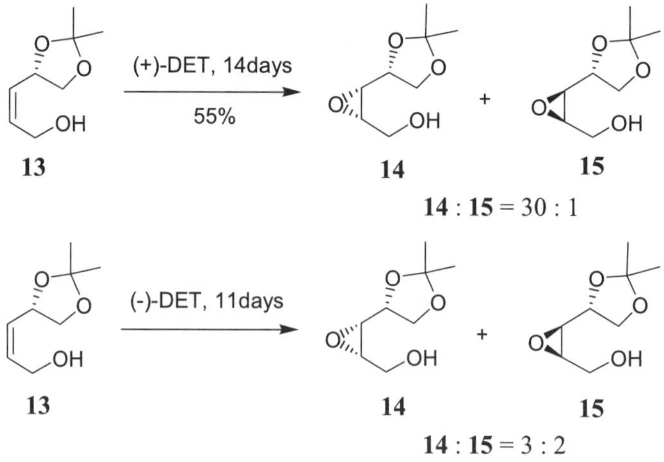

Scheme 4–5

In summary, the following characteristics describe the effectiveness of this epoxidation reaction.

- Simplicity: All the ingredients are inexpensive and commercially available.
- Reliability: It succeeds with most allylic alcohols, although bulky substituents at R are deleterious (Fig. 4–1).
- High optical purity: Optical purity of the product is generally >90% ee and usually >95% (99.5% ee is the highest measured accurately to date).
- Predictable absolute stereochemistry: Thus far, when dealing with a prochiral allylic alcohol substrate, no exception to the rules laid down in Figure 4–1 has been observed.
- Relative insensitivity to preexisting chiral centers: In allylic alcohols with preexisting chiral centers, the diastereofacial preference of the chiral titanium-tartrate catalyst is often strong enough to override diastereofacial preferences inherent in the chiral olefinic substrate.
- Versatility of 2,3-epoxy alcohols as intermediates: New selective transformations widen the utility and significance of the reaction.

4.1.2 Mechanism

There are several Ti–tartrate complexes present in the reaction system. It is believed that the species containing equal moles of Ti and tartrate is the most active catalyst. It promotes the reaction much faster than Ti(IV) tetraalkoxide alone and exhibits selective ligand-accelerated reaction.[9]

Sharpless suggested that epoxidation was catalyzed by a single Ti center in a dimeric complex with a C_2 symmetric axis. Molecular weight measurement, infrared spectroscopy, and ^1H, ^{13}C, and ^{17}O NMR spectrometry all suggest that such a dinuclear structure is dominant in the solution phase (Fig. 4–2).[10]

As shown in Scheme 4–6, the reaction proceeds via a Ti(IV) mixed-ligand complex **A** bearing allyl alkoxide and TBHP anions as ligands. The alkyl peroxide is electrophilically activated by bidentate coordination to the Ti(IV) center. Oxygen transfer to the olefinic bond occurs to provide the complex **B**, in which Ti(IV) is coordinated by epoxy alkoxide and *t*-butoxide. In complex **B**,

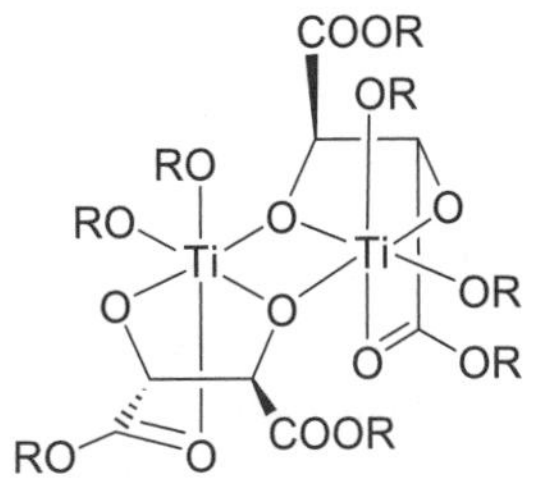

Figure 4–2. Structure of dinuclear Ti–tartrate complexes.

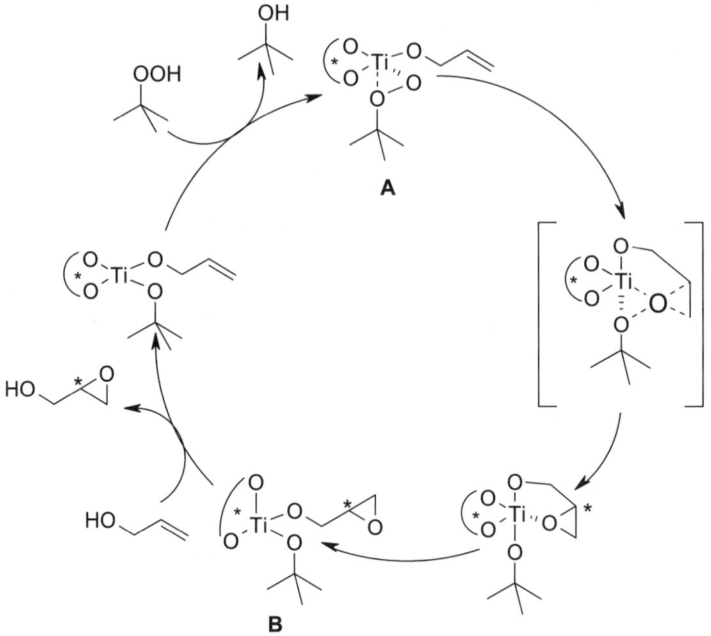

Scheme 4–6. Mechanism of Ti-catalyzed Sharpless epoxidation.

alkoxide products are replaced by allylic alcohol and TBHP to regenerate **A** and complete the catalytic cycle. It seems clear that enantioselectivity is controlled by the chiral ligands on Ti(IV), which determines the conformation of the coordinated allylic alcohol. The exact nature of the catalytic species remains only partially understood.

Corey[11] also proposed another mechanism for the origin of enantioselectivity in the reaction by suggesting the presence of an ion pair in the reaction pathway. Interested readers are advised to consult the original papers.

4.1.3 Modifications and Improvements of Sharpless Epoxidation

4.1.3.1 The CaH₂/SiO₂ System. Almost by chance, Zhou and colleagues found that the reaction time in Sharpless epoxidation could be reduced dramatically by adding a catalytic amount of calcium hydride and silica gel to the reaction system, although the mechanism is not yet clarified (Table 4–1).[12]

Using this modification, Zhou et al.[13] succeeded in the kinetic resolution of α-furfuryl amide **16a–f** (Scheme 4–7).

Under these oxidation conditions, (*S*)-**16a–f** and (*R*)-**16a–f** remain as part of the slow reacting enantiomers and can be obtained in high enantiomeric purity (90–100% ee) and 40–50% chemical yield when using the corresponding L-(+)- and D-(−)-DIPT. It should be pointed out that 2.0–2.5 equivalents of TBHP is required to get ~50% conversion. Otherwise, the reaction proceeds extremely

TABLE 4–1. Asymmetric Epoxidation of Allylic Alcohols by Sharpless Reagents (Method A) and by the Modified Sharpless Reagents (Method B)

Substrate	Method	Time (h)	Yield (%)	$[\alpha]_D$	ee (%)	Config.
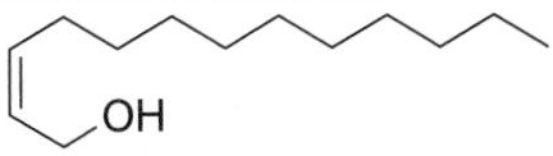	A	96	76–80	−7.6	95	2R,3S
	B	8	76	−7.8		
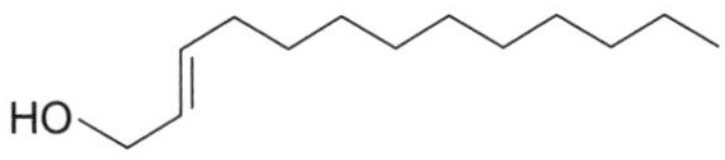	A	72	76–80	+26.5	96	2S,3S
	B	6	76.4	+25.9		
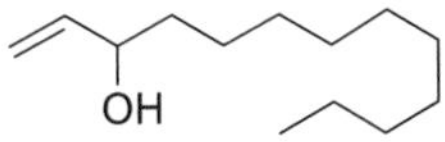	A	360	81	+16.2	91	2S,3S
	B	25	84	+15.2		

Conditions: Method A: Epoxidation using Sharpless reagent; method B: addition of 0.05–0.1 equivalent of calcium hydride and 0.1–0.15 equivalent of silica gel to the Sharpless reagent.

ee = Enantiomeric excess.

slowly. For example, no reaction occurs when 0.6 equivalent of TBHP is used. This reaction is important because the furan ring compounds (S)-**16a–f** can be converted to α-amino acids by ozonolysis or oxidation using RuCl$_3$/NaIO$_4$.

Zhou and Wei[14] then further extended the modified epoxidation to kinetic resolution of α-pyrroyl carbinol (Scheme 4–8).

The reaction was carried out in CH$_2$Cl$_2$ using Ti(OPr^i)$_4$ (1.0 equivalent), L-(+)-DIPT or D-(−)-DIPT (1.2 equivalents), 5–10 mol% of CaH$_2$, 10–20% of silica gel, and TBHP (1.0 equivalent) at −10°C. The substituent R can be a primary or secondary alkyl group. The presence of a tosyl substituent on the

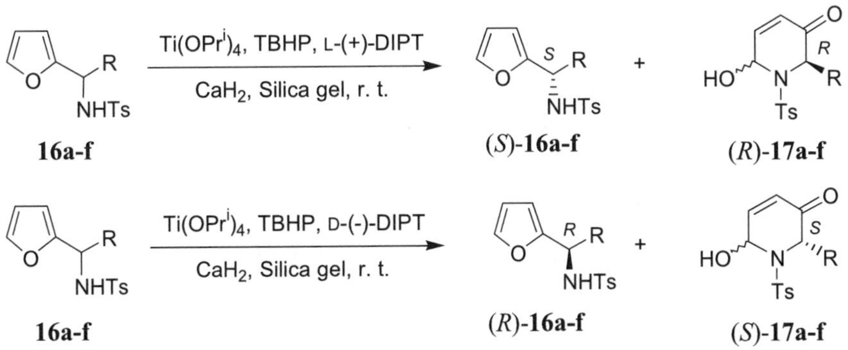

a. R = Me   b. R = Et
c. R = Pr   d. R = Bu
e. R = i-Bu   f. R = n-hexyl

Scheme 4–7

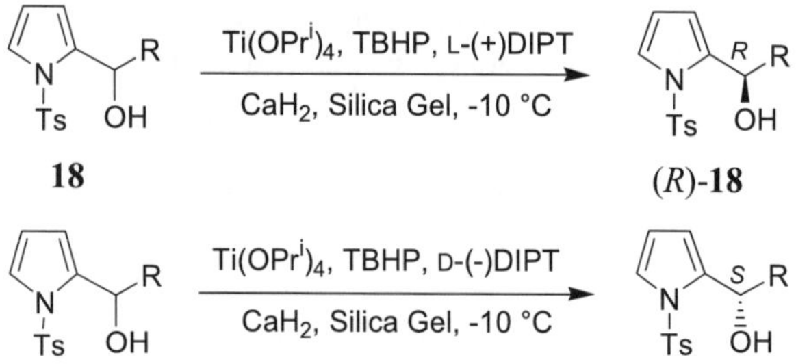

Scheme 4–8

nitrogen atom avoids N-oxide formation, and kinetic resolution takes place with high enantioselectivity (90–95% ee). When L-(+)-DIPT is used, the slow reacting enantiomer is the one related to (R)-**18**, while (S)-**18** remains as the slow reacting enantiomer when D-(−)-DIPT is employed (Scheme 4–8).

4.1.3.2 *The 4 Å Molecular Sieves System.* The initial procedure for the Sharpless reaction required a stoichiometric amount of the tartrate–Ti complex promoter. In the presence of 4 Å molecular sieves, the asymmetric reaction can be achieved with a catalytic amount of titanium tetraisopropoxide and DET (Table 4–2).[15] This can be explained by the fact that the molecular sieves may remove the co-existing water in the reaction system and thus avoid catalyst de-activation. Similar results may be observed in kinetic resolution (Table 4–3).[15]

TABLE 4–2. Catalytic Asymmetric Epoxidation With (+)-DET

Product	Ti–Tartrate	Temp. (°C)	Time (h)	Yield (%)	ee (%)
R〜〜OH R = C₃H₇, Ph	5/6.0	−20	2.5	85	94
	5/7.5	−20	3	89	>98
R〜〜OH R = C₇H₁₅, BnOCH₂	10/14	−10	29	74	86
	10/14	−20	43		85
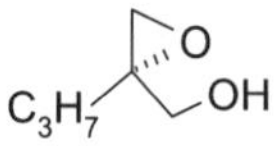 C₃H₇〜OH R = C₃H₇	4.7/5.9	−12	11	88	95
 Ph〜〜OH	5/7.5	−35	2	79	>98

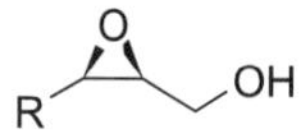

ee = Enantiomeric excess.

TABLE 4–3. Kinetic Resolution with (+)-DIPT

Product	Yield (%)	Conversion (%)	ee (%)
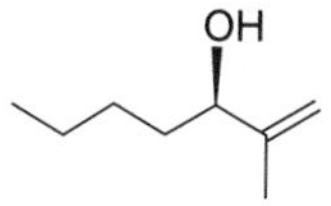	93	53	94
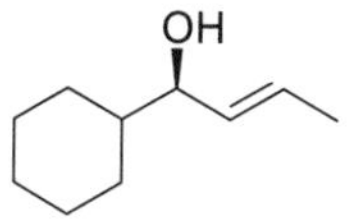	96	54	94
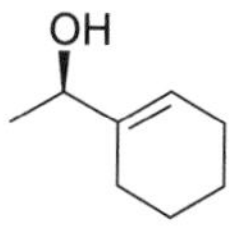	93	63	>98
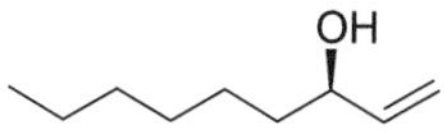	92	51	86

ee = Enantiomeric excess.
Reprinted with permission by Am. Chem. Soc., Ref. 15.

4.1.3.3 Asymmetric Epoxidation Using Polymer-Supported Ti(IV) Catalysts.

The advantages of polymer-supported reactive species are now widely recognized by organic chemists. The strategy often affords several advantages over the use of homogeneous catalysts. One advantage of these polymer-supported catalysts is their ease of separation from the reaction system, which allows their efficient recovery and potential reuse. Using polymer-supported catalysts also makes it possible to carry out the reactions in flow reactors or flow membrane reactors for continuous production.[16]

Canali et al.[17] reported the use a linear poly(tartrate) ligand in the asymmetric epoxidation of allylic alcohols. Moderate results were obtained. They also reported the use of branched/crosslinked poly(tartrate), which gave moderate to good results in the asymmetric epoxidation of allylic alcohols. As shown in Scheme 4–9, when L-(+)-tartaric acid and 1,8-octanediol are heated

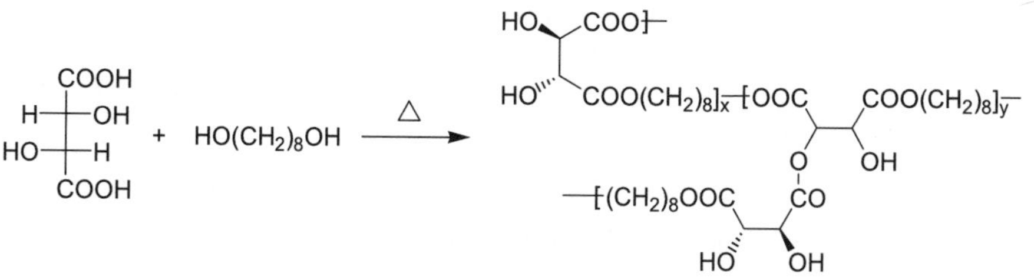

Scheme 4–9

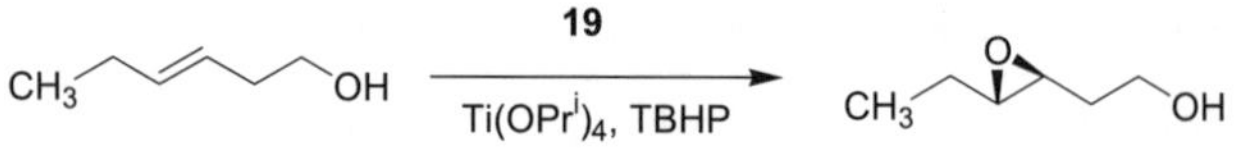

$$CH_3 \diagup\!\!\!\diagdown\!\!\!\diagup OH \xrightarrow[\text{Ti(OPr}^i)_4,\ \text{TBHP}]{\textbf{19}} CH_3 \diagup\!\!\!\triangle\!\!\!\diagup OH$$

Entry	Ligands [(%), branching, crosslinking]	Molar ratio substrate: Ti: tartrate	Reaction time (days)	Yield (%)	ee (%)
1	DMT	100 : 100 : 120	10	58	29
2	3 (0%)	100 : 200 : 400	21	26	54
3	3 (8%)	100 : 200 : 400	5	45	54
4	3 (13%)	100 : 200 : 400	1	75	38

Scheme 4–10

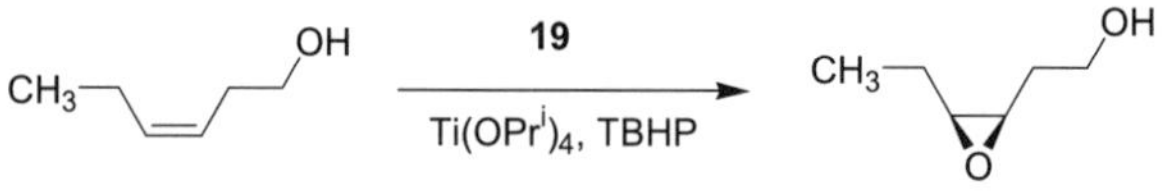

$$CH_3 \diagdown\!\!\!=\!\!\!\diagup OH \xrightarrow[\text{Ti(OPr}^i)_4,\ \text{TBHP}]{\textbf{19}} CH_3 \diagdown\!\!\!\triangle\!\!\!\diagup OH$$

Entry	Ligands [(%), branching, crosslinking]	Molar ratio substrate: Ti: tartrate	Reaction time (days)	Yield (%)	ee (%)
1	DMT	100 : 100 : 120	6	15	28
2	3 (10%)	100 : 200 : 400	21	20	51
3	3 (13%)	100 : 200 : 400	1	52	41
4	3 (16%)	100 : 100 : 200	13	33	41

Scheme 4–11

together, condensation polymerization takes place, yielding the branched/crosslinked polyester **19**, which shows moderate to high asymmetric induction capability for the asymmetric epoxidation of allylic alcohols.[18]

As discussed thus far, Sharpless epoxidation deals with allylic alcohol substrates, giving 2,3-epoxy alcohols as the reaction product. 3,4-Epoxy alcohols are also important building blocks in organic synthesis. However, only a few reports about the enantioselective synthesis of 3,4-epoxy alcohols have been published,[19] and the results were not as satisfactory as those with 2,3-epoxy alcohols. Karjalainen et al.[20] found that, in the presence of the branched/crosslinked poly(tartrate), epoxidation of 3-en-1-ol substrates proceeded readily, producing the corresponding 3,4-epoxy alcohols with moderate yields but higher enantiomeric excess as compared with that obtained from a reaction catalyzed by monomeric tartaric acid esters. As shown in Schemes 4–10 and 4–11, polymer catalyst **19** facilitates the asymmetric epoxidation of 3-en-1-ol compounds, providing product with much higher ee than that from monomer DMT (dimethyl tartrate)–mediated reactions.

4.2 SELECTIVE OPENING OF 2,3-EPOXY ALCOHOLS

Sharpless epoxidation is considered highly valuable because it combines the powerful nature of the reaction with the capacity of the resultant epoxy alcohols to undergo regioselective and stereoselective reactions with various nucleophiles. The regiochemistry is often determined by the functional group in the substrate or by chelation between substrates and the reagents.

4.2.1 External Nucleophilic Opening of 2,3-Epoxy Alcohols

4.2.1.1 Ti(OPr^i)$_4$-Mediated Nucleophilic Opening of 2,3-Epoxy Alcohols.

Caron and Sharpless[21] have demonstrated that in the presence of 1.5 equivalents of Ti(OPr^i)$_4^{-1}$, nucleophiles such as secondary amine, azide, thiol, and free alcohol preferentially attack the C-3 atom in chiral 2,3-epoxy alcohols with configuration inversion at the C-3 position. Excess nucleophile can be used to overcome the regiochemical problems associated with the inherently low reaction rate. In the absence of Ti(IV) compounds, ring opening does not occur (Table 4–4). For a wide variety of nucleophiles, Ti(OPr^i)$_4$ not only enhances the rate of nucleophilic ring opening of 2,3-epoxy alcohols, but also leads to an increase in regioselectivity for C-3 attack in the reaction.

Ti(OPr^i)$_4$-mediated nucleophilic ring opening of 2,3-epoxy-alcohol with primary amine requires more rigorous conditions, and the product is a complex mixture. Lin and Zeng[22] found that this problem could be overcome and moderate to good yields could be obtained under weak base conditions by in situ N-acylation of the aminolysis product with benzoyl chloride.

20   **21** C-3 opening   **22** C-2 opening

TABLE 4–4. Nucleophilic Opening of 3-Propyloxiranemethanol[21]

Nucleophile	Ti(OPr^i)$_4$ (eq.)	Reaction Conditions	Regioselectivity (C-3/C-2)	Yield (%)
Et$_2$NH	0	Et$_2$NH (excess), reflux, 18 h	3.7/1	4
Et$_2$NH	1.5	Et$_2$NH (excess), r.t., 5 h	20/1	90
i-PrOH	0	i-PrOH (excess), reflux, 18 h		0
i-PrOH	1.5	i-PrOH (excess), reflux, 18 h	100/1	88
PhSH	0	PhSH (5.0 eq.), benzene, r.t., 22 h		0
PhSH	1.5	PhSH (1.6 eq.), benzene, r.t., 5 min	6.4/1	95
Me$_3$SiN$_3$	1.5	Me$_3$SiN$_3$ (3.0 eq.), benzene, reflux, 3 h	14/1	74

r.t. = Room temperature.

Reprinted with permission by Am. Chem. Soc., Ref. 21.

4.2.1.2 Regioselective Azide Opening of 2,3-Epoxy Alcohols by [Ti(OPr^i)$_2$(N$_3$)$_2$] and Other Azidic Compounds.

Similarly, an azide can be introduced into 2,3-epoxy alcohols like **23** and those shown in Table 4–5. Using Ti(OPr^i)$_2$(N$_3$)$_2$ as a nucleophile, ring opening proceeds readily in the presence of Ti(OPr^i)$_4$, yielding the C-3 ring opening product as the major product. The C-3 ring opening products can be used for preparing various α-amino acids.[23]

C-3 opening product   C-2 opening product

TABLE 4–5. Preparation of the Azide Compounds

Entry	Substrate	Conditions	Regioselectivity C-3/C-2	Yield (%)
1		7 h^a	5.8:1	95
2		0.08 h^b	36:1	88
3	Ph	3.5 h^c	1.4:1	71
4		0.16 h^b	27:1	96
5		10 h^c	1.7:1	93
6		0.25 h^b	20:1	94
7		12 h^c	1:100	47
8		0.75 h^b	2:1	96
9		2.75 h^a	100:1	100
10	Ph	0.08 h^d	100:1	76

^{a} NaN$_3$/NH$_4$Cl, 65°C, MeOH/H$_2$O = 8:1.
^{b} Ti(OPr^i)$_2$(N$_3$)$_2$, benzene, 70°C.
^{c} NaN$_3$/NH$_4$Cl, CH$_3$OCH$_2$CH$_2$OH:H$_2$O = 8:1; 124°C.
^{d} Ti(OPr^i)$_2$(N$_3$)$_2$, ether, 25°C.
Reprinted with permission by Am. Chem. Soc., Ref. 23.

The preparation of Ti(OPr^i)$_2$(N$_3$)$_2$, is described elsewhere.[24]

Besides Ti(OPr^i)$_2$(N$_3$)$_2$, other azide compounds are also effective for C-3 ring opening with 2,3-epoxy alcohols. Benedetti et al.[25] have demonstrated another regioselective and stereoselective ring-opening reaction using diethyl aluminum azide as the nucleophile. High regioselectivity (C-3 ring opening over C-2 ring opening) has been observed for both *cis*- and *trans*-substituted epoxides, and the C-3 attack is not affected by bulky substituents at C-3.

4.2.1.3 Ring-Opening Reactions of Epoxy Alcohols with X_2-Ti(OPr^i)$_4$.

Treating allylic and homoallylic epoxy alcohols with an equivalent amount of halogen (Br$_2$, I$_2$) in the presence of a stoichiometric amount of Ti(OPr^i)$_4$ provides halohydrins under mild conditions with a high degree of generality and with good regioselectivity (Scheme 4–12).[26]

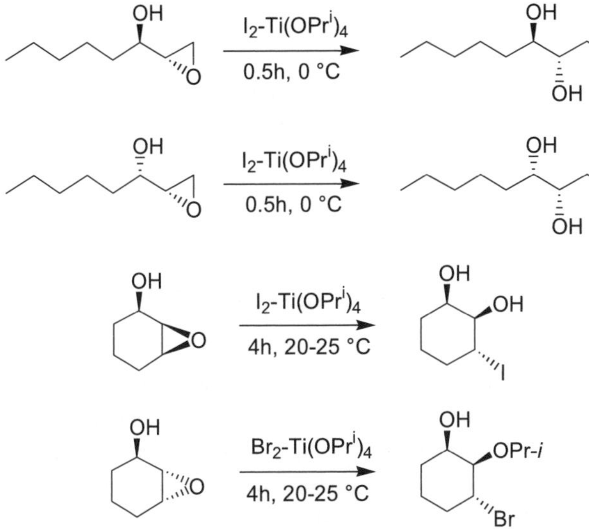

Scheme 4–12

4.2.2 Opening by Intramolecular Nucleophiles

Another approach that produces regioselective ring opening in 2,3-epoxy alcohols is to take advantage of intramolecular nucleophiles by attaching a potential N-nucleophile or O-nucleophile to the hydroxy group.[27] The application of this intramolecular ring opening is exemplified in the synthesis of sphingosine isomers. In the enantioselective synthesis of D- and L-*erythro*-sphingosine,[28] 2,3-epoxy alcohol **26** is treated with excess CCl$_3$CN in the presence of DBU. The resulting trichloroimidate **27** can serve as an N-nucleophile. Treating **27** with triethyl aluminum leads to a single product **28** in which the C-3 configuration has been reversed. After acid hydrolysis and Li/NH$_3$ reduction, L-*erythro*-sphingosine is obtained and can be characterized as its triacetate (**30b**) (Scheme 4–13).

Cerebrosides are major constituents of the membrane of brain cells. They are the simplest glycosphingolipids, serving as model substances for more complex lipids of this kind. Furthermore, they are credited with important properties as receptors for hormones and toxins.[29] Schemes 4–13 and 4–14 provide a method for preparing sphingosine and its analogs that can be used for the synthesis of cerebroside compounds.

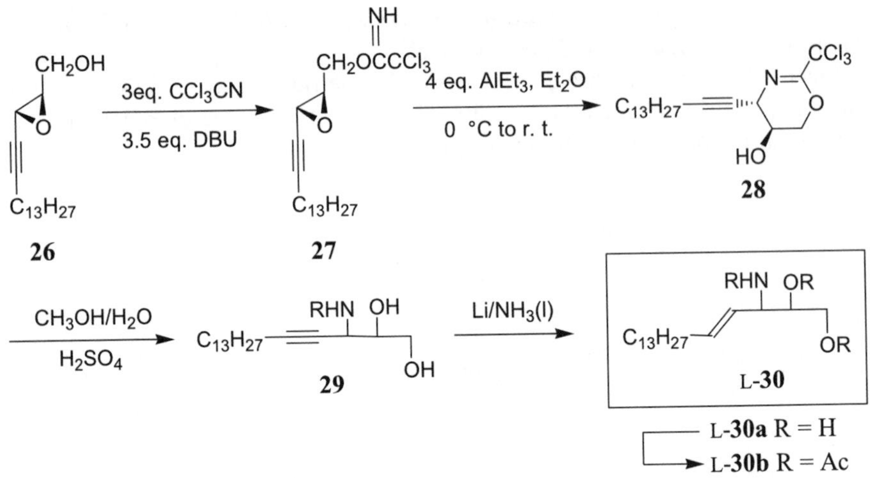

Scheme 4–13. Enantioselective synthesis of L-*erythro*-sphingosine.

To prepare the desired D-*erythro*-isomer D-**30b**, benzyl urethane **32** is prepared via the *p*-nitrophenylcarbonate **31** in a one pot reaction. Treating this *N*-nucleophile **32** with 5 equivalents of NaN(SiMe$_3$)$_2$ leads to the desired oxazolidine **33**. Li/NH$_3$ cleavage affords D-*erythro*-sphingosine, which may also be characterized as its triacetate D-**30** (Scheme 4–14).

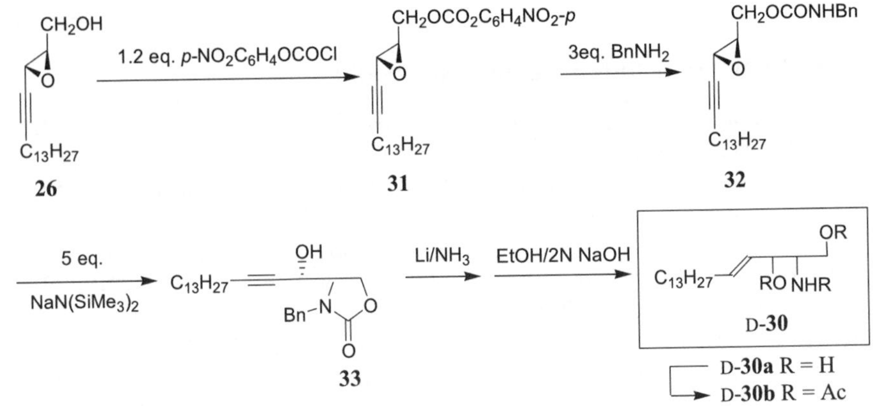

Scheme 4–14. Enantioselective synthesis of D-*erythro*-sphingosine (D-**30**).

Intramolecular ring opening is also exemplified in the synthesis of (4R)-4-[(E)-2-butenyl]-4-*N*-dimethyl-L-threonine (MeBmt **34**), an unusual β-hydroxy-α-amino acid of cyclosporine (**35**),[30] which is a clinically used immunosuppressing agent.

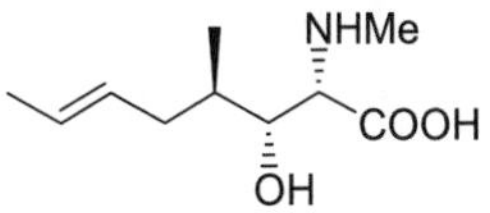

34 (MeBmt)
an unusual amino acid of cyclosporine

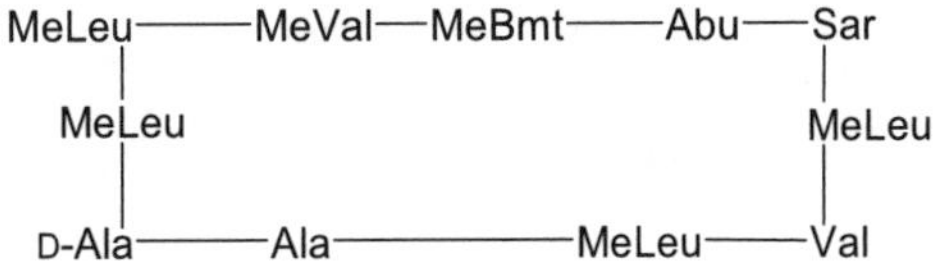

35
cyclosporine

Key steps, as shown in Scheme 4–15, involve the formation of a urethane intermediate **37** by treating epoxide **36** with methyl isocyanate in the presence of sodium hydride. Intramolecular *N*-nucleophilic ring opening of oxirane affords oxazolidine **38**. Subsequent treatment furnishes product **34**.

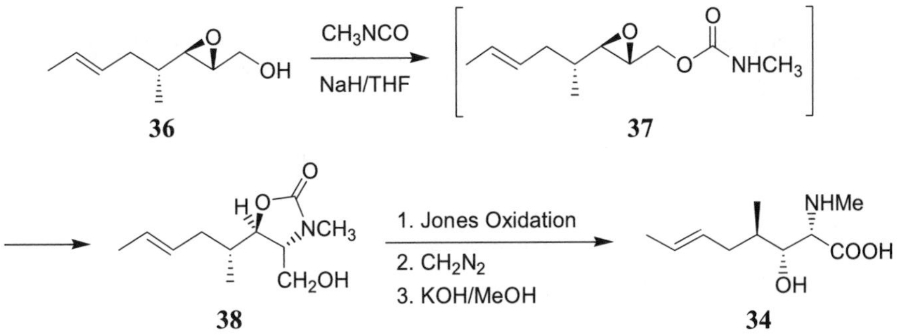

Scheme 4–15. Key steps in the synthesis of MeBmt.

4.2.3 Opening by Metallic Hydride Reagents

Using different reagents or under various conditions, 2,3-epoxy alcohols can undergo ring-opening reactions with metallic hydrides, giving 1,3-diols or 1,2-diols. As shown in Scheme 4–16, reduction of 3-substituted 2,3-epoxy alcohols with Red-Al leads to the exclusive formation of 1,3-diols, and this can be applied in the preparation of 1,3-diol compounds.[31]

In contrast to Red-Al reductions, DIBAL-H or $LiBH_4/Ti(OPr^i)_4$ reduction of epoxides yields 1,2-diols as the major products.[32] When treated with DIBAL-H, ratios of 1,3- to 1,2-diol ranging from 1:6 to 1:13 have been observed.

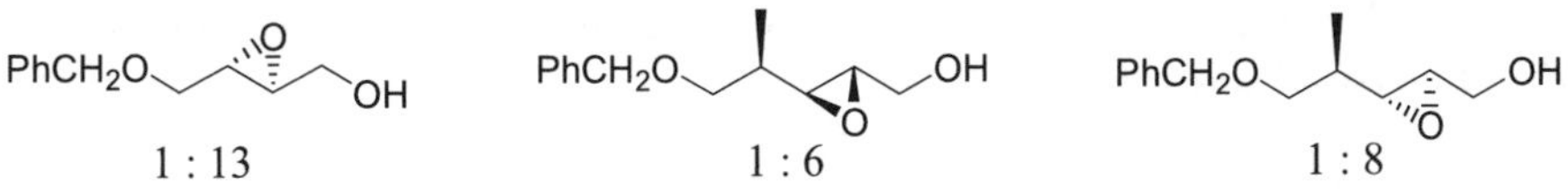

1 : 13   1 : 6   1 : 8

An alternative method for reducing 2,3-epoxy alcohols to 1,2-diols through regioselective delivery of hydride to C-3 is realized by treating the correspond-

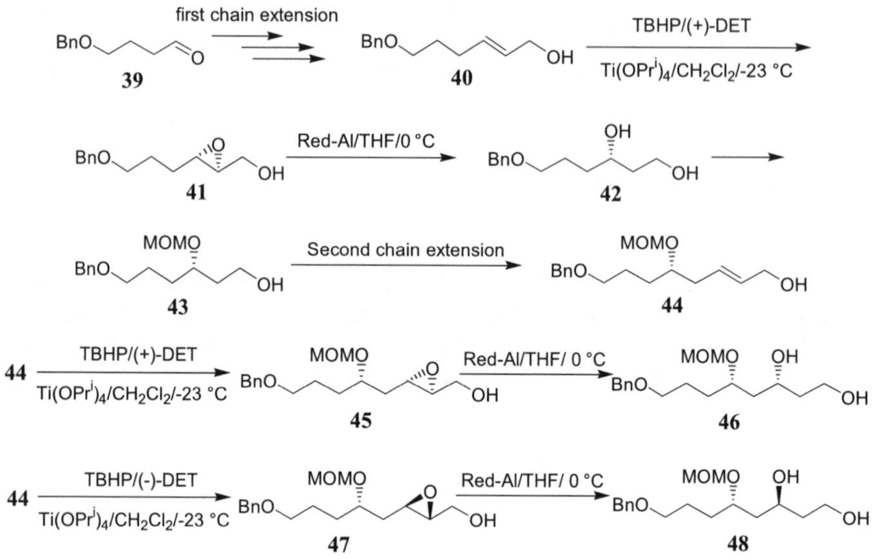

Scheme 4–16. Synthesis of 1,3-diol.

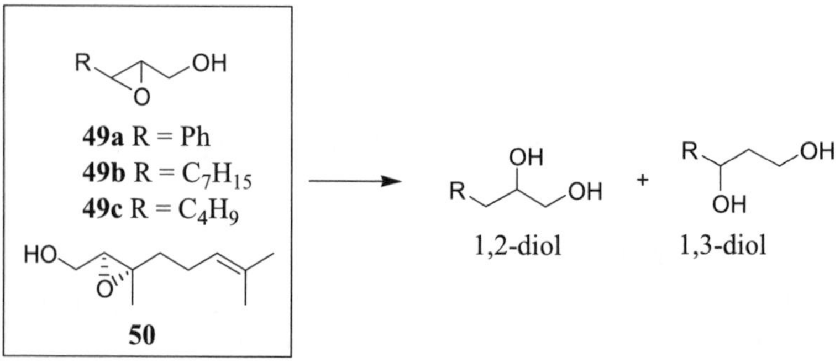

Scheme 4–17. Reduction of 2,3-epoxy alcohols by LiBH$_4$/Ti(OPr^i)$_4$.

ing epoxide compound with excess LiBH$_4$ in the presence of Ti(OPr^i)$_4$ (Scheme 4–17 and Table 4–6). This reaction carried out at low temperature gives high stereoselectivity but rather low reaction rate. Raising the reaction temperature generally results in a drop in enantioselectivity.[33]

A similar ring opening reported by Sajiki et al.[34] involves the catalytic hydrogenolysis of terminal epoxides.

4.2.4 Opening by Organometallic Compounds

When treated with organocuprates, 2,3-epoxy alcohols can be converted to substituted 1,3-diols with high regioselectivity and stereoselectivity. Thus, as

TABLE 4–6. LiBH$_4$ Reduction of 2,3-Epoxy Alcohols

Entry	Substrate	Ti(OR)$_4$ eq.	Condition			Ratio 1,2-:1,3-	Yield (%)
			Solvent	Temp. (°C)	Time		
1	**49a**	1.5	THF	65	1 h	7.3:1	83
2	**49b**	1.9	C$_6$H$_6$	10	20 h	145:1	93.2
3	**49c**	1.7	C$_6$H$_6$	50	15 min	46:1	99
4	**49c**	1.7	C$_6$H$_6$	10	18 h	150:1	97
5	**50**	1.5	THF	65	15 min	1.7:1	84.3
6	**50**	1.6	C$_6$H$_6$	50	45 min	6.8:1	96.3

Reprinted with permission by Pergamon-Elsevier Science Ltd., Ref. 33.

Scheme 4–18 shows, epoxides **51**, **54**, and **56** can be converted to the corresponding 1,3-diols **52**, **53**, **55**, and **57** in high yield and ee through dialkylcuprate treatment.[35]

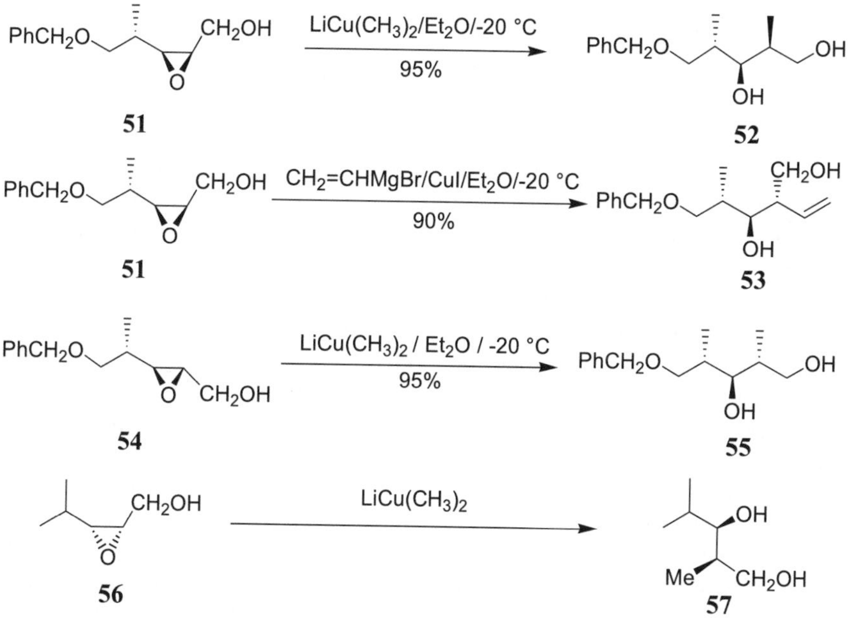

Scheme 4–18. Ring opening by dialkylcuprate.

4.2.5   Payne Rearrangement and Ring-Opening Processes

Thus far, we have discussed nucleophilic ring opening in 2,3-epoxy-1-ol taking place at the C-2 and C-3 positions (see compound **58** in Scheme 4–19). However, in the presence of a base, nucleophilic ring opening can take place at C-1 via Payne rearrangement to produce 2,3-diol.[36] For example, compound 1,2-

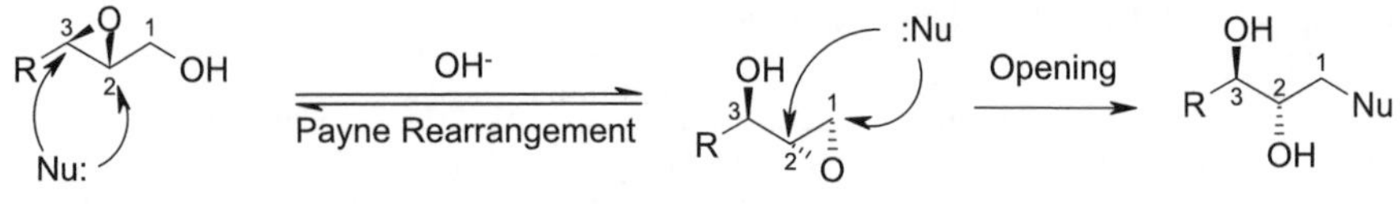

Scheme 4–19. Payne rearrangement and subsequent ring opening.

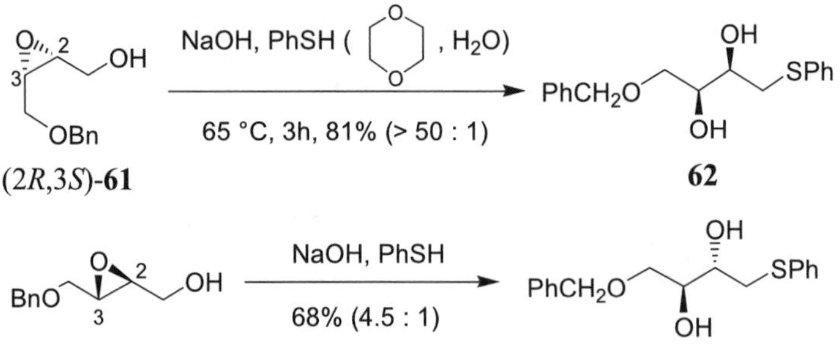

Scheme 4–20. PhS⁻ as a nucleophile in the preparation of tetritol precursors.

epoxy-3-ol **59** was first produced via this rearrangement. Subsequent ring opening at C-1 gives the corresponding 2,3-diol **60** (Scheme 4–19). Various nucleophiles such as PhS⁻, BH₄⁻, CN⁻, and TsNH⁻ can be used for this purpose.

Scheme 4–20 exemplifies PhS⁻ attack mediated by a Payne rearrangement. The selective ring-opening product can be applied to prepare tetritols.[37]

This approach provides a new method for carbohydrate synthesis. In the synthesis of tetritols, pentitols, and hexitols, for example, titanium-catalyzed asymmetric epoxidation and the subsequent ring opening of the thus formed 2,3-epoxy alcohols can play an essential role.

Scheme 4–21 shows the preparation of L-threitol and L-erythritol.[38] Epoxy alcohols (2R,3S)-**61** and (2S,3R)-**61**, generated by asymmetric epoxidation, are exposed to sodium benzenethiolate and sodium hydroxide in a protonic solvent to undergo base-catalyzed rearrangement, yielding the *threo*-diol **62** and *erythro*-diol **63**, which can then be converted to the corresponding tetraacetate of L-threitol **67** and L-erythritol **69** through subsequent transformations.

Another approach to L-threitol **67** and L-erythritol **69** is stereoselective ring opening at the C-2 position of (2R,3S)-**61** using intramolecular oxygen as the nucleophile. With the aid of an acid catalyst, compound phenylurethane **64** undergoes ring opening to give carbonate **66**, which can then be converted to the known tetraacetate **67** via routine chemistry. Similarly, compound **69** can be obtained from (2S,3R)-**61**, which furnishes a synthesis of L-threitols and L-erythritols.

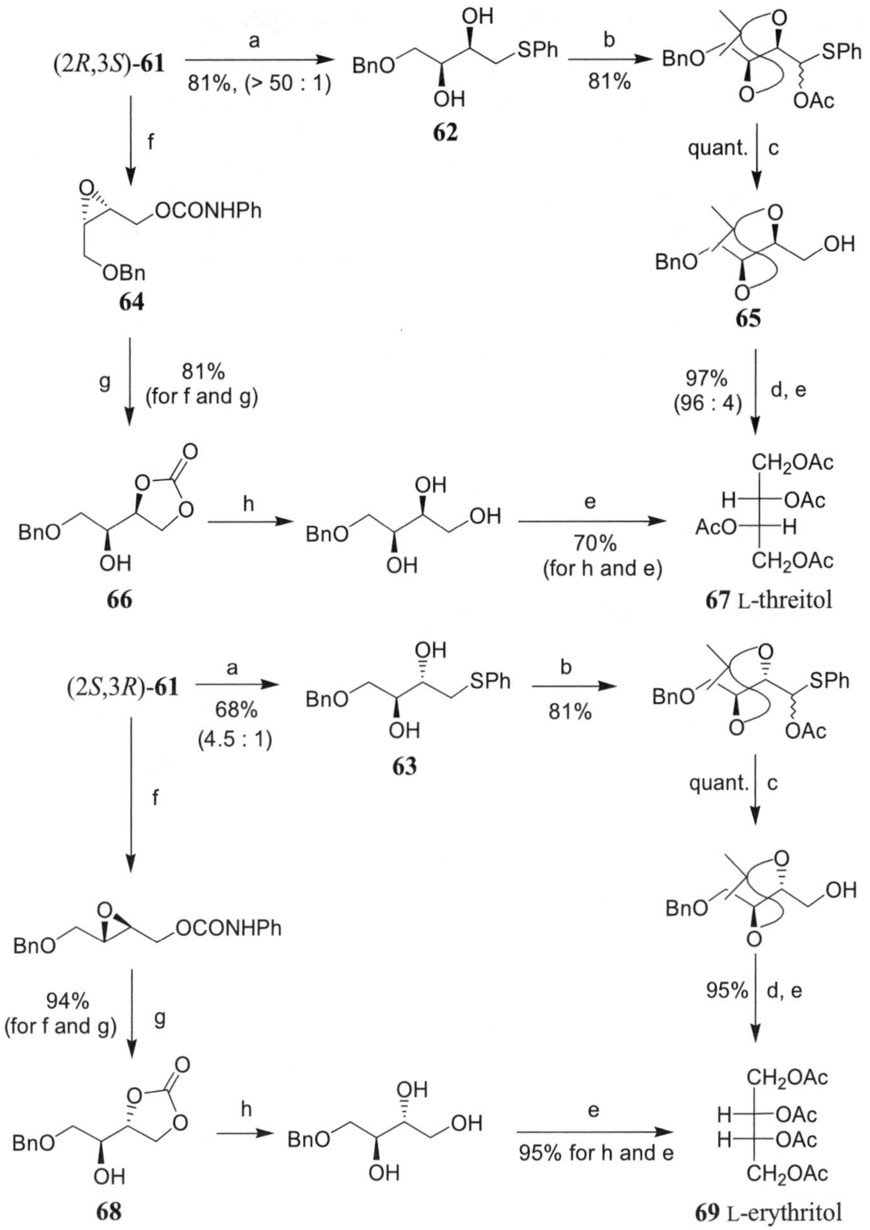

Scheme 4–21. Asymmetric synthesis of tetritol isomers **67** and **69**. Reagents and conditions: a: NaOH, PhSH (dioxane, H₂O), 65°C, 3 h. b: (1) Me₂C(OMe)₂, H⁺; (2) *m*-CPBA, CH₂Cl₂, −20°C, 1 h; (3) Ac₂O, NaOAc, reflux, 6 h. c: LAH, ether, 0°C, 1 h. d: MeOH, H⁺, 70°C, 1 h. e: (1) H₂, Pd/C, acidic MeOH, 25°C, 6 h; (2) Ac₂O, C₅H₅N. f: PhNCO, (Et)₃N, CH₂Cl₂, 25°C, 24 h. g: 5% HClO₄, CH₃CN, 25°C, 24 h. h: NaOH, aq. MeOH, 25°C, 24 h.

4.2.6 Asymmetric Desymmetrization of *meso*-Epoxides

Complementary to the regioselective and stereoselective ring opening of epoxides, desymmetrization of *meso*-epoxides by oxirane-ring opening with nucleophiles in an asymmetric manner, as shown in Scheme 4–22, should also be mentioned. In the presence of certain Lewis acids, the metal center of a catalyst or reagent is able to coordinate to the epoxide oxygen atom. The chiral environment provided by the Lewis acid will then allow an appropriate achiral nucleophile to discriminate the formal enantiotropic carbon–oxygen bond of the epoxide.

Scheme 4–23 presents several examples[39] in which the desymmetrization takes place with moderate to excellent enantioselectivity.

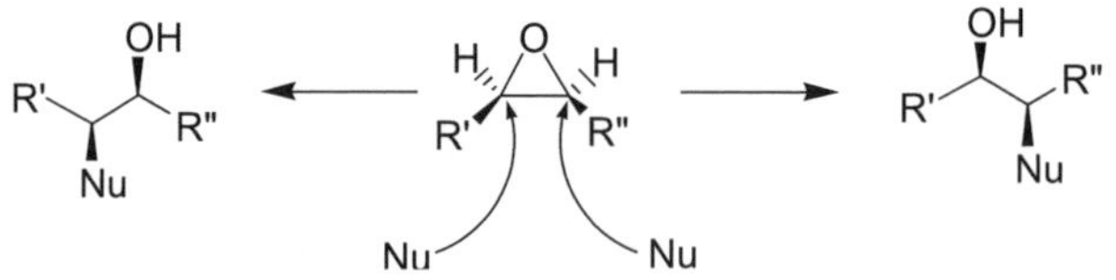

Scheme 4–22. Enantioselective ring opening of *meso*-epoxides (R′ = R″).

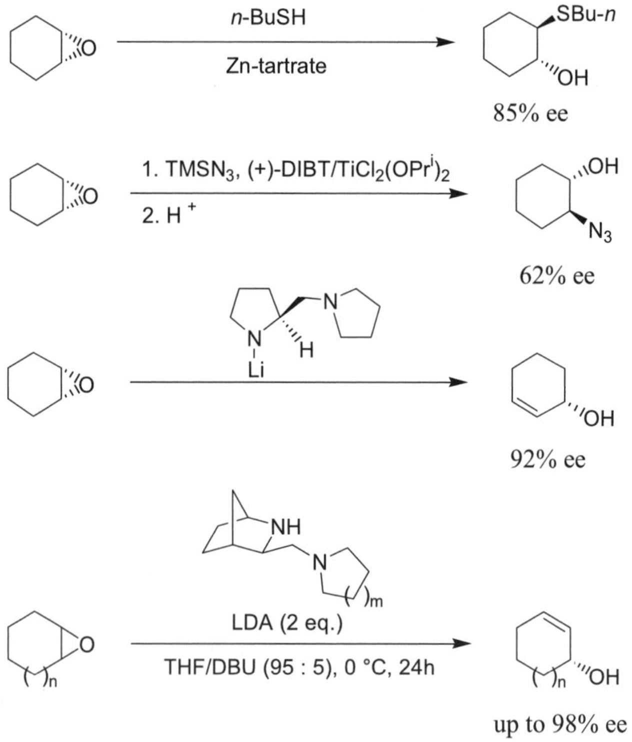

Scheme 4–23. Asymmetric ring opening of *meso*-epoxides.

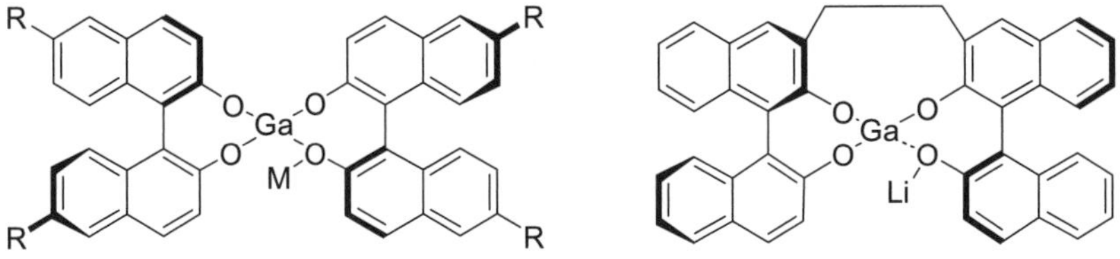

70a   **70b**

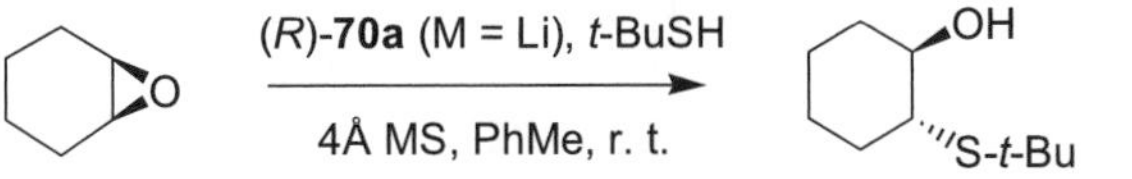

(R)-**70a** (M = Li), t-BuSH

4Å MS, PhMe, r. t.

80% yield, 97% ee

Scheme 4–24. Gallium–lithium complex-catalyzed ring opening.

Iida et al.[40] presented gallium complexes **70a** and **70b** containing chiral BINOL ligand. These complexes have been applied to catalyze the highly enantioselective ring opening of epoxides with thiols or phenols. Scheme 4–24 is an example of using compound **70a** as the catalyst for asymmetric ring opening of cyclohexene oxide by t-BuSH. The product can be obtained in good yield and excellent ee. Compound **70a** can also be used for enantioselective ring opening of epoxides with 4-methoxyphenol, providing 1,2-diol monoethers with good yields and moderate ee.[41] Following his success with this desymmetrization ring opening reaction, Shibasaki then introduced another complex, **70b**, in which two binaphthyl ligands are connected. This complex can also be used as a catalyst for phenol-mediated ring-opening reactions.[42]

The enantioselective ring opening of *meso*-epoxides with thiols can also be facilitated by chiral (salen)Ti(IV) complex.[43] As shown in Scheme 4–25, in the presence of salen compound **71** and Ti(OPr^i)$_4$, ring opening of *meso*-epoxide proceeds at −25° to −40°C, giving a product with good chemical yield and moderate ee.

Salen–transition metal complex, or Jacobsen reagent, has been found useful in a range of asymmetric reactions, such as Diels-Alder reactions or the epox-

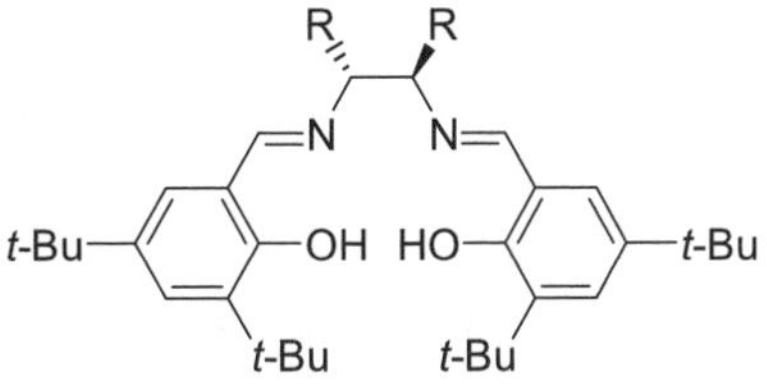

71 R = -(CH$_2$)$_4$-, Ph

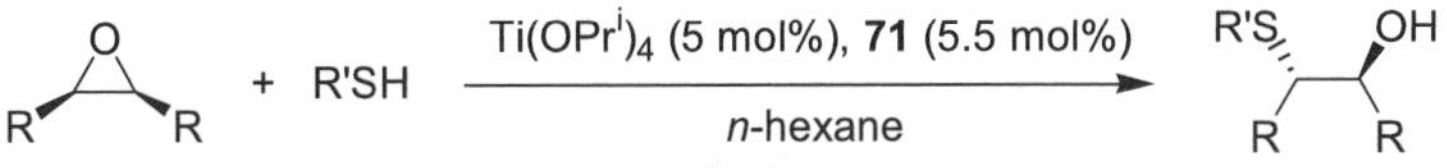

+ R'SH

Ti(OPr^i)$_4$ (5 mol%), **71** (5.5 mol%)

n-hexane

Scheme 4–25

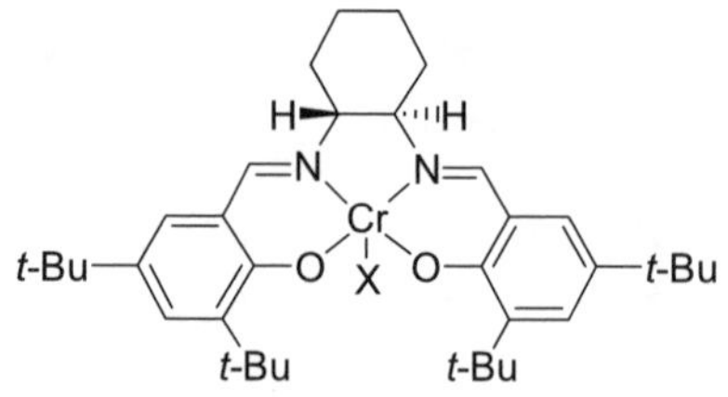

72 X = Cl, N$_3$

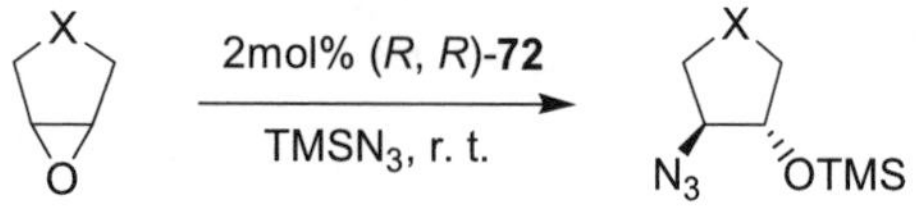

73

74 up to 97% ee

Scheme 4–26

idation of unfunctionalized olefins. (These reactions are discussed later.) Schaus et al.[44] have reported a practical synthesis of enantiomerically pure cyclic 1,2-amino alcohols via catalytic asymmetric ring opening of *meso*-epoxides. As shown in Scheme 4–26, reagent **72** catalyzes azide attack on the epoxide **73**, resulting in chiral product **74** in high ee. From **74**, a series of 1,2-amino alcohols can be prepared. For example, **72** (X = Cl)–catalyzed reactions give ee of 94%, 88%, 95%, and 97% for product **74** where X = CH$_2$, (CH$_2$)$_2$, NCOCF$_3$, and O, respectively.

Jacobsen and colleagues[45] also report applying the (salen)Cr–N$_3$-catalyzed epoxide ring-opening reaction in the kinetic resolution of racemic terminal epoxides. Not only can the remaining epoxides be recovered with high ee, but also 1-azido-2-trimethyl siloxyalkanes can be obtained in good yield and very high ee. As an example, the precursor for (*S*)-propranolol, an antihypertensive agent, can be readily prepared via asymmetric ring opening of the epoxide compound (Scheme 4–27).

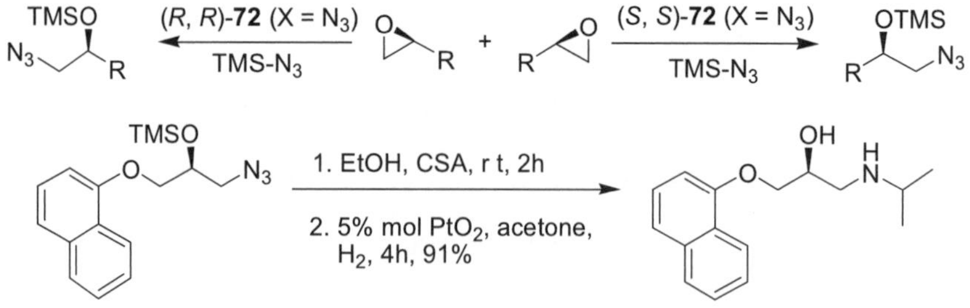

(*S*)-propranolol
anti-hypertensive agent

Scheme 4–27

Jacobsen has also designed a dimeric catalyst in which two salen complexes are connected together. A pronounced cooperative catalysis effect is observed, leading to reaction rate enhancement even at a very low catalyst concentration. For example, the rate constant for asymmetric ring opening catalyzed by **75** (n = 5) is two orders of magnitude higher than that catalyzed by the control catalyst **76**, while the enantioselectivity is comparable.[46] This reaction and the cooperative catalysis are further discussed in Chapter 8.

75 n = 2, 4, 5, 6, 7, 8, 10

76

4.3 ASYMMETRIC EPOXIDATION OF SYMMETRIC DIVINYL CARBINOLS

An important consideration in designing and performing asymmetric reactions with high regioselectivity and enantioselectivity, especially for those in which more than one stereogenic center is formed, is the need to carry out the reaction with a combination of kinetic resolution in the initial asymmetric synthesis.

Schreiber et al.[47] have described a mathematical model that combines enantiotopic group and diastereotopic face selectivity. They applied the model to a class of examples of epoxidation using several divinyl carbinols as substrates to predict the asymmetric formation of products with enhanced ee (Scheme 4–28).

Consider Sharpless epoxidation with an achiral substrate. With certain ligands, the epoxidation can take place at any one of the four stereotopic faces of the substrate, affording X^1, X^2, X^3, and X^4. In Scheme 4–28, X^1 reacts fast when A or B is OH, and the reaction is performed in an asymmetric way. When

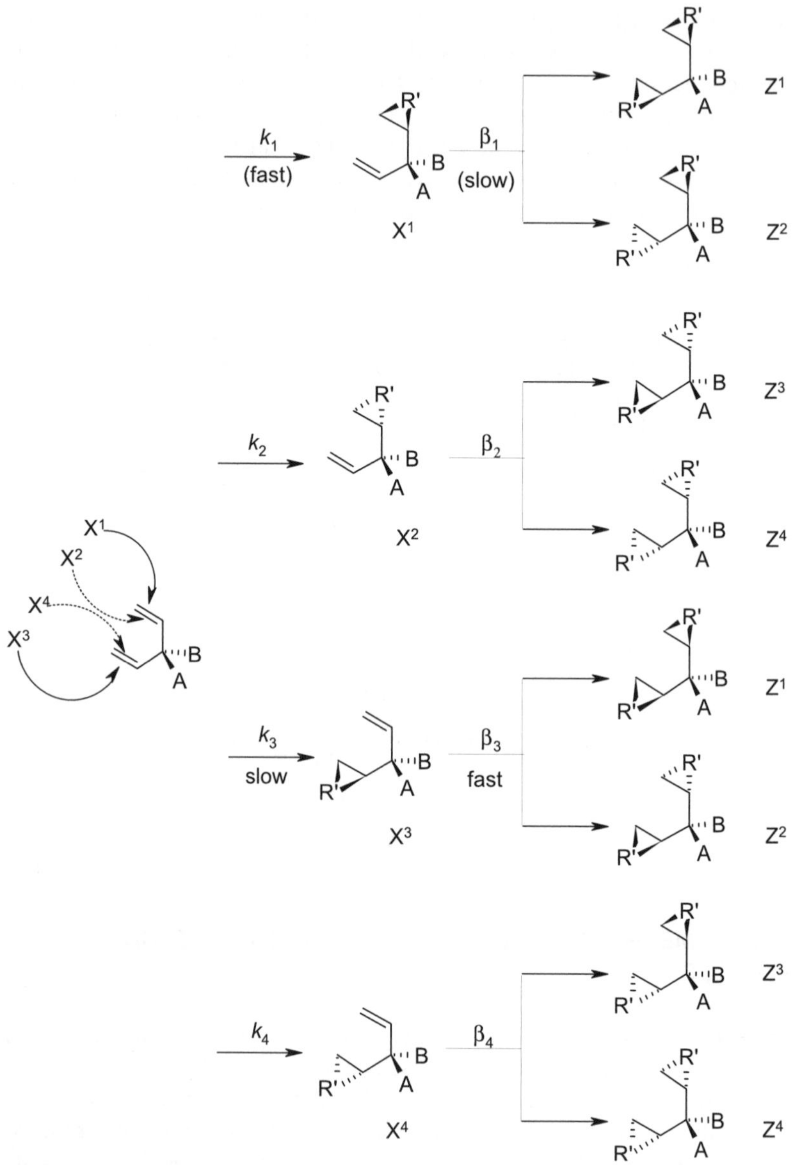

Scheme 4–28. The asymmetric epoxidation of divinyl carbinols. Reprinted with permission by Am. Chem. Soc., Ref. 47b.

a second epoxidation takes place kinetically, the minor enantiomers X^2 and X^3 proceed faster and are destroyed due to the instability of the diepoxide products Z. Thus, the ratio of X^1 to X^3 should increase as the reaction proceeds. In other words, the first reaction converts an achiral divinyl carbinol with a pro-stereogenic atom into a chiral nonracemic epoxy alcohol, and the second

TABLE 4–7. Epoxidation in the Presence of L-(+)-DIPT

Conditions	ee (%)	de (%)
3 h, −25°C	84	92
24 h, −25°C	93	99.7
140 h, −25°C	≥97	>99.7

Reprinted with permission by Am. Chem. Soc., Ref. 46b.

reaction occurring in the epoxidation system enhances the ee via a kinetically controlled process.

Three divinyl carbinol substrates have been chosen as examples. They are good substrates for examination because the vinyl carbinols are known to undergo Sharpless reaction at low reaction rates. The results presented in Table 4–7 clearly show that the ee of **79** improves as the reaction proceeds toward completion. Note that the minor enantiomer **78** can be removed through a second, faster epoxidation that converts enantiomer **78** into an easily destroyed bis-epoxide **80** (Scheme 4–29 and Table 4–7). The same trend is apparent in the second demonstration with diisopropenyl carbinol **81** (Scheme 4–30 and Table 4–8). Similarly, the third reaction is the reaction of (E,E)-divinyl carbinol **82** (Scheme 4–31 and Table 4–9).

Schreiber's model has been successfully applied to synthesize intermediates of several important natural products, such as prostaglandin intermediate[48] (Scheme 4–32) and 2,6-dideoxyhexoses, such as D-(+)-digitoxose, D-(+)-cymarose, D-(+)-olivose, and D-(−)-oleandrose[49] (see Fig. 4–3 for the structures).

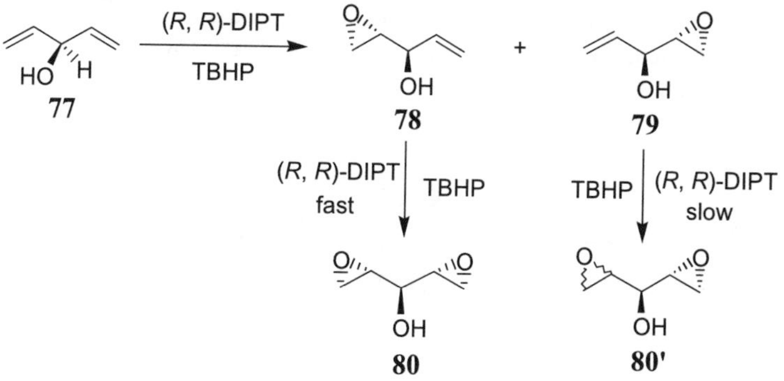

Scheme 4–29

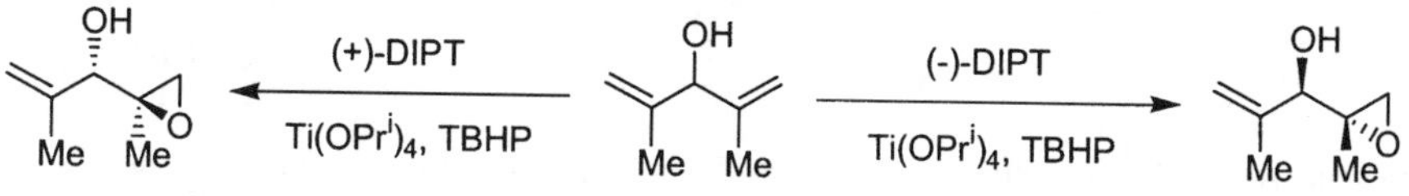

81

yield 80-85%

Scheme 4–30. Yield 80–85%.

TABLE 4–8. Epoxidation in the Presence of D-(−)-DIPT

Conditions	ee (%)	de (%)
0.5 h, −25°C	88	99
1.0 h, −25°C	94	>99
1.5 h, −25°C	>99.3	>99

de = Diastereomeric excess; ee = enantiomeric excess.
Reprinted with permission by Am. Chem. Soc., Ref. 47b.

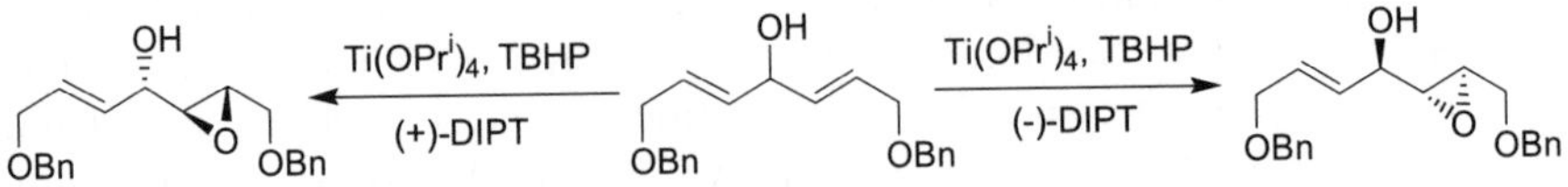

Scheme 4–31. Yield 70–80% without 4 Å molecular sieves.

TABLE 4–9. Epoxidation in the Presence of L-(+)-DIPT

Conditions	ee (%)	de (%)
1 h, −25°C	93	≥97
3 h, −25°C	95	≥97
44 h, −25°C	≥97	≥97

de = Diastereomeric excess; ee = enantiomeric excess.
Reprinted with permission by Am. Chem. Soc., Ref. 47b.

a: R = MOM
b: R = Me

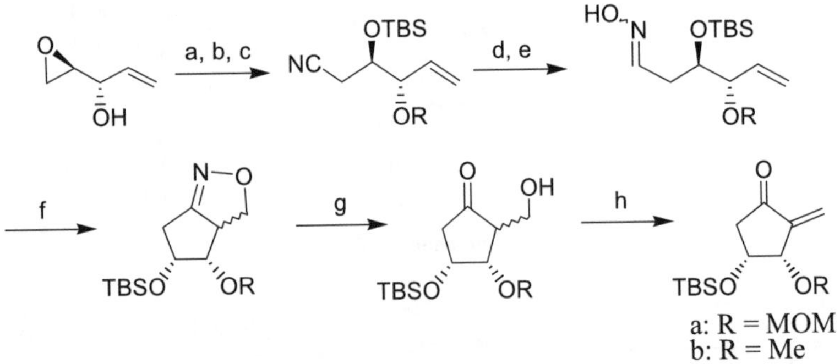

Scheme 4–32. Highly efficient synthesis of chiral prostagladin intermediate. Reagents and conditions: a: NaH/MeOCH$_2$Cl or CH$_3$I, 0°C. b: KCN/AcOH. c: TBSCl, imidazole, room temperature. d: DIBAL. e: HONH$_2$ · HCl/Py. f: 0.7 N NaOCl/CH$_2$Cl$_2$, room temperature. g: 10% Pd/C, H$_2$. h: MeSO$_2$Cl/Et$_3$N.

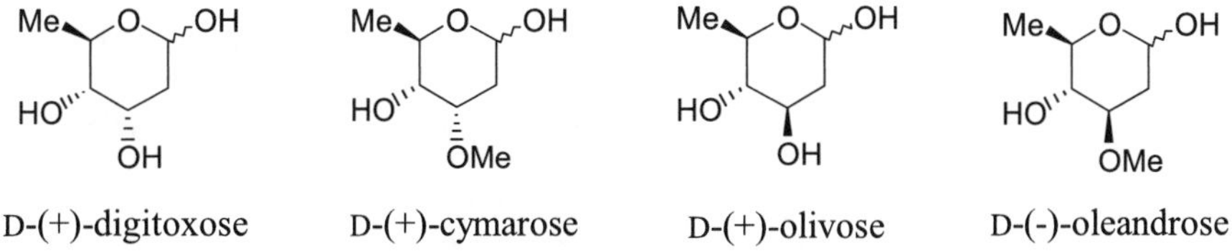

Me
O
OH
HO
OH
D-(+)-digitoxose
OMe
D-(+)-cymarose
D-(+)-olivose
D-(-)-oleandrose

Figure 4–3

Schreiber's model has also proved to be a general approach to a series of oxygenated metabolites of arachidonic acid, such as lipoxin A and lipoxin B.[50] The family of linear oxygenated metabolites of arachidonic acid has been implicated in immediate hypersensitivity reactions, inflammation, and a number of other health problems. Among these metabolites, several compounds, such as lipoxin A, lipoxin B, 5,6-diHETE, and 14,15-diHETE possess 1-substituted (*E*)-1-alken-3,4-diol **84** as a common substructural moiety. Therefore, the carbinol **83** is an ideal substrate for generating compound **84** by applying Sharpless epoxidation reaction.[50]

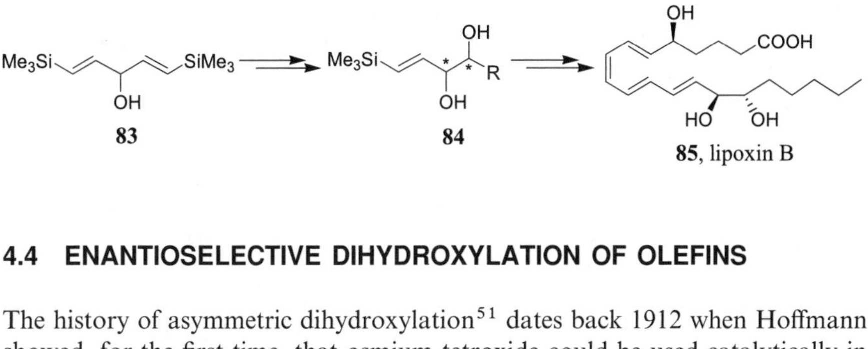

Me_3Si
$SiMe_3$
OH
R
COOH
HO

83 **84** **85**, lipoxin B

4.4 ENANTIOSELECTIVE DIHYDROXYLATION OF OLEFINS

The history of asymmetric dihydroxylation[51] dates back 1912 when Hoffmann showed, for the first time, that osmium tetroxide could be used catalytically in the presence of a secondary oxygen donor such as sodium or potassium chlorate for the *cis*-dihydroxylation of olefins.[52] About 30 years later, Criegee et al.[53] discovered a dramatic rate enhancement in the osmylation of alkene induced by tertiary amines, and this finding paved the way for asymmetric dihydroxylation of olefins.

The first attempt to effect the asymmetric *cis*-dihydroxylation of olefins with osmium tetroxide was reported in 1980 by Hentges and Sharpless.[54] Taking into consideration that the rate of osmium(VI) ester formation can be accelerated by nucleophilic ligands such as pyridine, Hentges and Sharpless used *l*-2-(2-menthyl)-pyridine as a chiral ligand. However, the diols obtained in this way were of low enantiomeric excess (3–18% ee only). The low ee was attributed to the instability of the osmium tetroxide chiral pyridine complexes. As a result, the naturally occurring cinchona alkaloids quinine and quinidine were derived to dihydroquinine and dihydroquinidine acetate and were selected as chiral

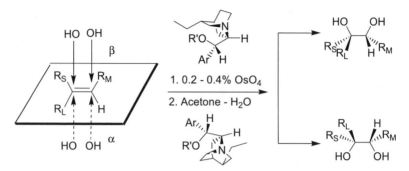

Figure 4–4. Asymmetric dihydroxylation reaction.

auxiliaries. Because the bonding of osmium tetroxide to quininuclidine nitrogen is much tighter than the bonding of chiral pyridine, the diols obtained were in reasonably high enantiomeric excess (Fig. 4–4).

The reaction mechanism is shown in Figure 4–5. This reaction is a good example of ligand-accelerated asymmetric catalysis, as the alkaloid ligands enhance the rate by one to two orders of magnitude. In the initial stages, slow addition of the olefin is essential to obtain high ee due to a competing second catalytic cycle with low enantioselectivity. In the work of Sharpless' group, improved enantioselectivities were observed with potassium ferricyanide as the primary oxidant under alkaline (K_2CO_3) conditions in aqueous t-butanol. In

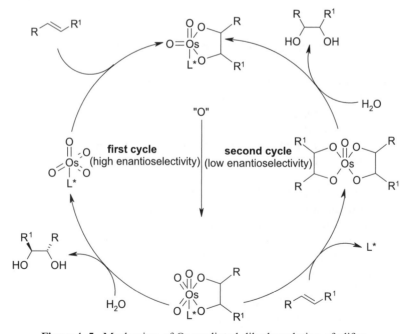

Figure 4–5. Mechanism of Os-mediated dihydroxylation of olifins.

this case, the second cycle is precluded because hydrolysis of the monoglycolate complex precedes reoxidation of the osmium under these conditions. In contrast to Sharpless epoxidation, which may be influenced by the bulky nature of the substituents on the substrates, asymmetric dihydroxylation is successful with a broad range of substituents.

Since Sharpless' discovery of asymmetric dihydroxylation reactions of alkenes mediated by osmium tetroxide–cinchona alkaloid complexes, continuous efforts have been made to improve the reaction. It has been accepted that the tighter binding of the ligand with osmium tetroxide will result in better stability for the complex and improved ee in the products, and a number of chiral auxiliaries have been examined in this effort. Table 4–11 (below) lists the chiral auxiliaries thus far used in asymmetric dihydroxylation of alkenes. In most cases, diamine auxiliaries provide moderate to good results (up to 90% ee).

The major breakthrough in the catalytic asymmetric dihydroxylation reactions of olefins was reported by Jacobsen et al.[55] in 1988. Combining 9-acetoxy dihydroquinidine as the chiral auxiliary with N-methylmorphine N-oxide as the secondary oxidant in aqueous acetone produced optically active diols in excellent yields, along with efficient catalytic turnover.

The 4-chlorobenzoate derivatives of dihydroquinidine (DHQD-CLB **91a**) and dihydroquinine (DHQ-CLB **91b**) have been found to yield optically active diols with high optical purity. Although several aryl-substituted alkenes yield the corresponding diols with high optical purity, reactions with alkyl-substituted olefins normally show lower stereoselectivity. Poor ee values are also observed for *cis* or cyclic olefins. Slow addition of the olefins to the reaction mixture improves the rate, as well as the ee with all diols, and all of the *trans*-disubstituted aromatic olefins can give products with ee in the range of 80–99%.[56] For aliphatic *cis*- and *trans*-olefins or terminal, tri-, and tetrasubstituted olefins, the ee of the products are still not high enough for the reactions to be synthetically useful.

The highest enantioselectivity in the dialkyl-substituted olefines has been obtained with the aryl ethers of DHQD **94a** and DHQ **94b**. With potassium ferricyanide as secondary oxidant, it is possible to carry out the reaction at room temperature, and slow addition of the olefins is not required. Under these conditions, the diols can be obtained in 85–90% yield and excellent enantioselectivity.

Although high asymmetric inductions have been obtained with various alkenes using cinchona alkaloids as chiral ligands, the exploration for better catalytic systems is still under way. In fact, even better ee values have been achieved using C_2-symmetric ligands **96**, **97**, **99**, and **101** introduced by Sharpless[57–59] and colleagues and Lohray and Bhushan.[60] Table 4–10 presents some results of catalytic asymmetric dihydroxylation of olefins. The high enantioselectivity in the dihydroxylation of various substituted olefins, which previously was possible only by applying stoichiometric reagents at low temperature, can now be achieved in catalytic fashion using C_2-symmetric ligands **96**, **97**, **99**, and **101** at room temperature.[58] Among these chiral ligands, **101** is the superior ligand for asymmetric dihydroxylation reactions, with most olefins bearing aliphatic substituents or heteroatoms in the alkylic position.[59]

TABLE 4–10. Enantiomeric Excess of Diols Obtained by Catalytic Asymmetric Dihydroxylation of Alkenes[61]

Entry	Substrate	(DHQD)$_2$-PHAL 96a	(DHQ)$_2$-PHAL 96b	C$_2$-DHQD 99	C$_2$-DHQ 99
1	Ph⟋⟍Ph	>99.5	>99.5	>98	98
2	Ph⟋⟍Me	—	—	>98	—
3	Ph⟍	97	97	92	85
4	⟍⟋⟍	—	—	93	—
5	Ph⟋⟍COOMe	97	95	94	—
6	⟍ (2-methylstyrene)	—	—	76	—
7	⟍ (tert-butyl alkene)	—	—	48	—
8	⟋⟍C$_8$H$_{17}$	84	80	45	—
9	n-Bu⟋⟍n-Bu	97	93	—	—
10	α-methylstyrene	94	93	—	—

Reprinted with permission by Pergamon Press Ltd., Ref. 61.

In summary, the reaction of osmium tetroxide with alkenes is a reliable and selective transformation. Chiral diamines and cinchona alkakoid are most frequently used as chiral auxiliaries. Complexes derived from osmium tetroxide with diamines do not undergo catalytic turnover, whereas dihydroquinidine and dihydroquinine derivatives have been found to be very effective catalysts for the oxidation of a variety of alkenes. OsO$_4$ can be used catalytically in the presence of a secondary oxygen donor (e.g., H$_2$O$_2$, TBHP, N-methylmorpholine-N-oxide, sodium periodate, O$_2$, sodium hypochlorite, potassium ferricyanide). Furthermore, a remarkable rate enhancement occurs with the addition of a nucleophilic ligand such as pyridine or a tertiary amine. Table 4–11 lists the preferred chiral ligands for the dihydroxylation of a variety of olefins.[61] Table 4–12 lists the recommended ligands for each class of olefins.

To give a better understanding of the scope of application for epoxidation and dihydroxylation reactions in organic synthesis, the studies by several groups on these reactions are discussed in the remainder of this section.

Corey et al.[66] have developed a bidentate chiral ligand **93** for asymmetric dihydroxylation of olefins. As shown in Table 4–13, asymmetric dihydroxylation of a series of olefins using **93** as a chiral catalyst and OsO$_4$ as the oxidant gives good to excellent yield as well as good enantioselectivity in most cases.

TABLE 4–11. Ligands Used in Asymmetric Dihydroxylations

Ligand	References
86 **87**	54
88	62
89	63
90	64
91	55
92	65

TABLE 4–11 (*Continued*)

Ligand	References
93	66
94	67
95	68
96	57
97	60

TABLE 4–11 (*Continued*)

Ligand	References

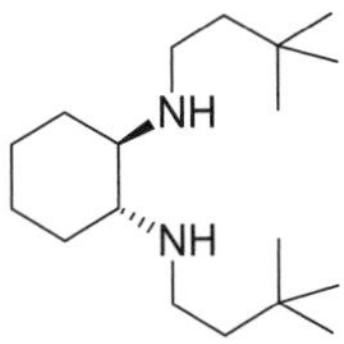

98

69

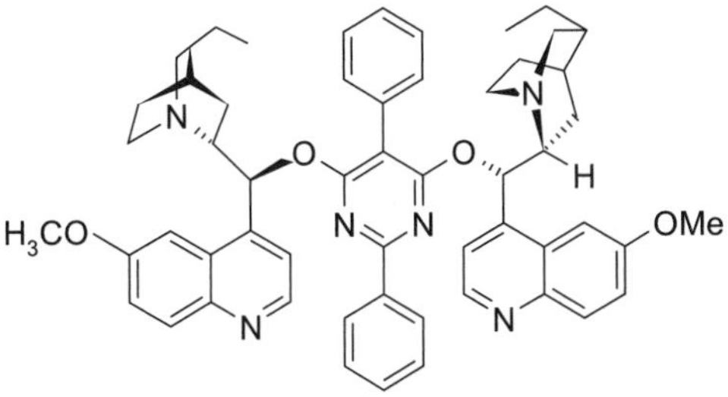

(DHQ)₂PYR **99a** and (DHQD)₂PYR **99b**

58

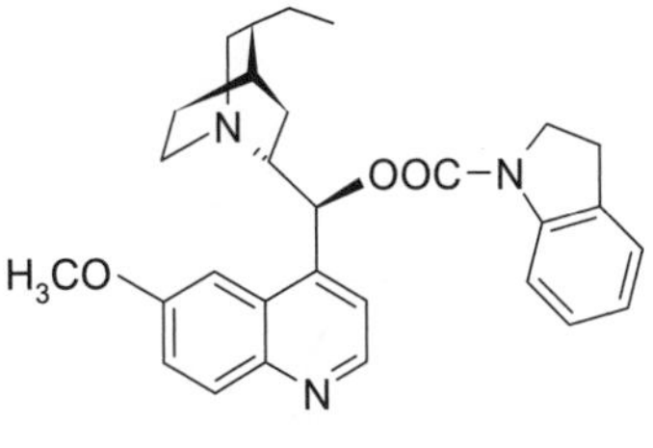

(DHQ)IND **100a** and (DHQD)IND **100b**

70

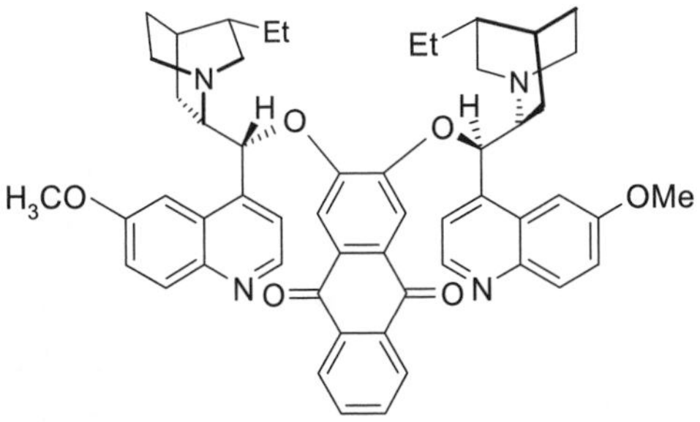

(DHQ)₂AQN **101a** and (DHQD)₂AQN **101b**

59

TABLE 4–12. Recommended Ligands for Each Class of Olefins[51]

Olefin Class	Preferred Ligand	ee Range (%)
	PYR, PHAL	30–97
	PHAL	70–97
	IND	20–80
	PHAL	90–99.8
	PHAL	90–99
	PYR, PHAL	20–97

ee = Enantiomeric excess.

TABLE 4–13. Enantioselective Hydroxylation of Olefin by OsO$_4$ · 93

$$\underset{R^1}{\overset{R^2}{\diagup}} \quad \xrightarrow[\substack{OsO_4 \,(1\ eq),\ CH_2Cl_2 \\ -90\ °C,\ 2h}]{1\ eq.\ (S,S)\text{-}\mathbf{93}} \quad HO^{\prime\prime\prime}\underset{R^1}{\overset{H}{\diagup}}\underset{H}{\overset{R^2}{\diagup}}OH$$

Olefin	Yield (%)	ee (%)	Config.
Ph⌒	81	92	S
Ph⌒⌒	95	93	S,S
Ph⌒⌒Ph	95	92	S,S
p-MeOC$_6$H$_4$⌒⌒C$_6$H$_4$OMe-p	90	82	S,S
C$_2$H$_5$⌒⌒C$_2$H$_5$	90	98	S,S
MeO$_2$C⌒⌒CO$_2$Me	75	92	2R,3R
⌒⌒CO$_2$Me	82	97	2R,3S
t-BuOCONH⌒⌒CO$_2$Me	91	97	2R,3S
Ph⌒⌒COOMe	83	92	2R,3S
⌒⌒⌒OTBDPS	87	95	2S,3S

ee = Enantiomeric excess.

Reprinted with permission by Am. Chem. Soc., Ref. 66.

TABLE 4–14. Enantioselective Dihydroxylation of Olefins Using $OsO_4 \cdot$ 92b

92a R = pentyl
92b R = neohexyl

Entry	Olefin	Solvent	ee (%)	Yield (%)	Config.
1	(E)-stilbene	Toluene	100	96	S,S
2	(E)-stilbene	Acetone	80	93	S,S
3	(E)-stilbene	CH_2Cl_2	56	87	S,S
4	Ethyl (E)-3-phenylacrylate	Toluene	99	97	
5	(E)-phenylpropene	Toluene	92	95	S,S
6	Dimethyl fumarate	CH_2Cl_2	98	79	R,R
7	Ethyl (E)-crotonate	CH_2Cl_2	98	90	
8	(E)-2-heptene	CH_2Cl_2	98	93	
9	(E)-3-hexene	CH_2Cl_2	96	82	S,S
10	(E)-3-heptene	CH_2Cl_2	93	90	
11	(E)-1-heptene	CH_2Cl_2	91	90	S
12	Styrene	Toluene	88	90	S

ee = Enantiomeric excess.
Reprinted with permission by Am. Chem. Soc., Ref. 71.

Hirama and co-workers[71] developed another chiral bidentate ligand **92** for OsO_4-mediated dihydroxylation of *trans*-disubstituted and monosubstituted olefins. As shown in Table 4–14, asymmetric dihydroxylation of olefins using (S,S)-(−)-**92b** as the chiral ligand provides excellent yield and enantioselectivity.

Chiral compounds **91a** and **91b**, as shown in Table 4–15, were first reported by Jacobsen et al.[55] for the asymmetric dihydroxylation of olefins. These catalysts can be used for asymmetric dihydroxlation of a variety of substrates.

Chiral catalysts $(DHQD)_2PHAL$ **96a** and $(DHQ)_2PHAL$ **96b**, developed by Sharpless' group, are highly effective in asymmetric dihydroxylation reactions. With an oxidant and an Os source, high ee can be obtained, and the approach of hydroxyl groups can be directed to either the α- or the β-side of the prochiral face of the substrate by choosing the appropriate catalyst, **96a** or **96b**. Sharpless and co-workers have formulated the chiral ligand, metal, and oxidant as AD mix-α and AD mix-β. For example, a mixture of $(DHQ)_2PHAL$, $K_3Fe(CN)_6$, K_2CO_3 and potassium osmate is AD mix-α. Currently, both AD mix-α and AD mix-β are commercially available.

TABLE 4–15. Asymmetric Dihydroxylation of Olefins with OsO$_4$ Induced by 91a or 91b

DHQD-CLB **91a**

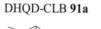

0.2 - 0.4% OsO$_4$, acetone
$\xrightarrow{\hspace{2cm}}$
H$_2$O, NMO

DHQ-CLB **91b**
R = *p*-chlorobenzoyl

Entry	Olefin	Ligand	ee (%)	Config.	Sign of $[\alpha]_D$	Time (h)
1		**91a**	62	*R*	(−)	3
		91a	60	*R*	(−)	7
		91b	53.6	*S*	(+)	7
2		**91a**	65	*R,R*	(−)	5
		91b	55.4	*S,S*	(+)	12
3		**91a**	33	*R*	(−)	1.5
4		**91a**	46	*R*	(+)	1
5		**91a**	76	*R*	(+)	7
6		**91a**	65		(−)	3
7		**91a**	20	*R,R*	(+)	17
8		**91a**	88	*R,R*	(+)	7
		91a	85	*R,R*	(+)	15
		91b	78.5	*S,S*	(−)	17

ee = Enantiomeric excess.
Reprinted with permission by Am. Chem. Soc., Ref. 55.

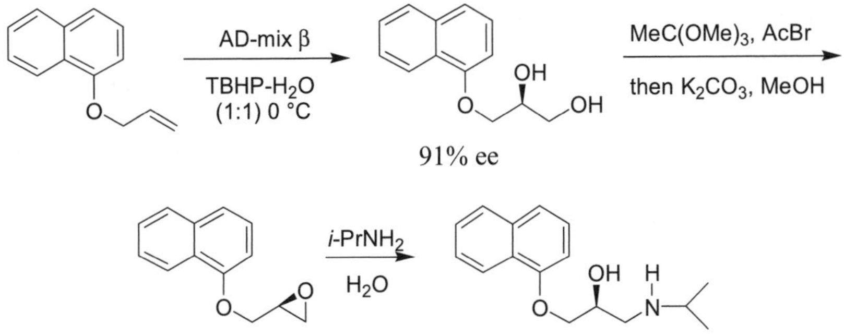

(2*S*)-propranolol

Scheme 4–33. Asymmetric dihydroxylation as a key step in the synthesis of (2*S*)-propranolol.

Using AD mix-α or AD mix-β as the dihydroxylation agent, various olefins can be dihydroxylated with high ee.[57,67b,72] As an example, in Scheme 4–33, aryl-allyl ethers undergo dihydroxylation yielding products with good ee. The procedure can be used as an alternative for the synthesis of (2*S*)-propranolol.[73]

Tomioka et al.[74] reported an interesting example of applying chiral diamine (−)-**102** in the synthesis of the chromophore part of anthracycline antibiotics (Scheme 4–34).

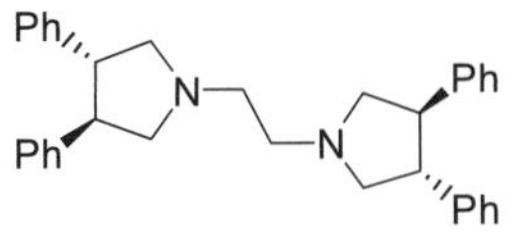

(-)-**102**

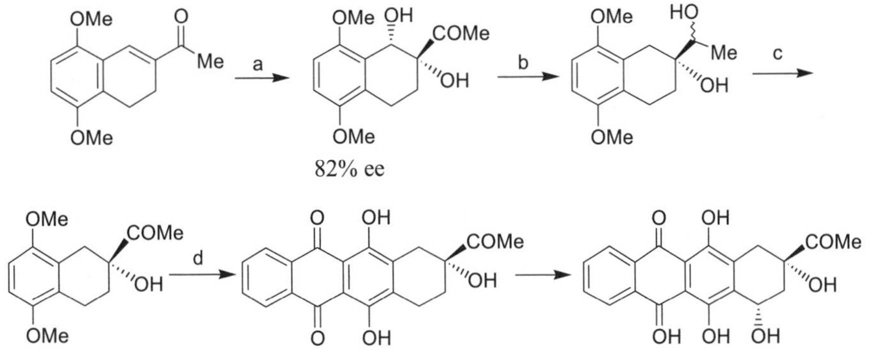

chromophore of anthracycline antibiotics

Scheme 4–34. Reagents and conditions: a: OsO_4-(−)-**102**/THF (96%). b: Et_3SiH/CF_3COOH (78%). c: Pyridine-SO_3-NEt_3/DMSO (87%). d: *o*-$C_6H_4(COCl)_2$-$AlCl_3$/$PhNO_2$, (76%, 53% after recrystallization).

4.5 ASYMMETRIC AMINOHYDROXYLATION

The β-amino alcohol structural unit is a key motif in many biologically important molecules. It is difficult to imagine a more efficient means of creating this functionality than by the direct addition of the two heteroatom substituents to an olefin, especially if this transformation could also be in regioselective and/or enantioselective fashion. Although the osmium-mediated[75] or palladium-mediated[76] aminohydroxylation of alkenes has been studied for 20 years, several problems still remain to be overcome in order to develop this reaction into a catalytic asymmetric process.

With Sharpless' recently discovered osmium-mediated asymmetric aminohydroxylation,[77] this functionality can now be obtained directly from olefins with excellent enantioselectivities and very good yields. This process first emerged as the reaction using TsNClNa (chloramine T) as the nitrogen source/oxidant. Product α-sulfonamido hydroxy compounds can be obtained when the olefin substrates are subjected to the aminohydroxylation reaction using chloramine-T as the nitrogen source and water as the oxygen source. In the presence of a chiral alkaloid ligand and a catalytic amount of $K_2OsO_2(OH)_4$, the asymmetric aminohydroxylation reaction results in the product with good yield and enantiomeric excess. This is a process that greatly benefits from ligand-accelerated catalysis, as Schemes 4–35 and 4–36 illustrate.

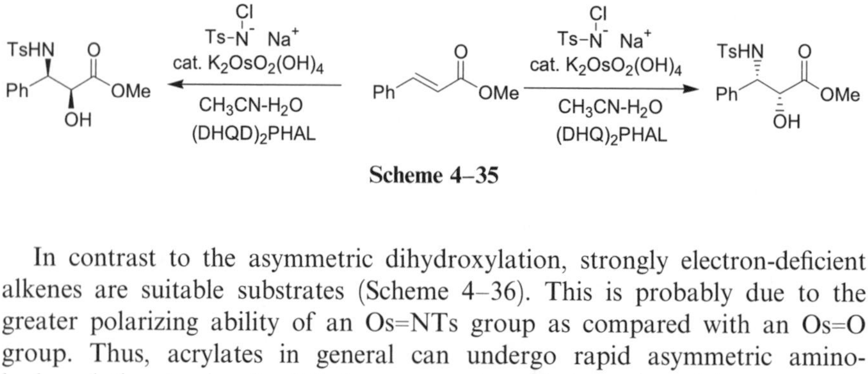

Scheme 4–35

In contrast to the asymmetric dihydroxylation, strongly electron-deficient alkenes are suitable substrates (Scheme 4–36). This is probably due to the greater polarizing ability of an Os=NTs group as compared with an Os=O group. Thus, acrylates in general can undergo rapid asymmetric aminohydroxylation to give the 2-hydroxy aspartic acid derivative with good results [see **103 → 104**, Scheme 4–36, (1)], although **103** is a very poor asymmetric dihydroxylation substrate. Less electron-deficient substrates such as stilbene (**107** and **109**) are viable substrates, but the enantioselectivities are generally lower. (E)-alkenes, in the same situation as in the dihydroxylation process, are better substrates than (Z)-alkenes (comparing the process of **107 → 108** with **109 → 110**).

With sulfonamide-derived chloramine salts bearing smaller organic substituents on the sulfur, for example, methanesulfonamide-derived chloramine salt **111** (chloramine-M) as the oxidant, better results are obtained. This reagent can be prepared separately and added to the reaction mixture as the stable anhydrous

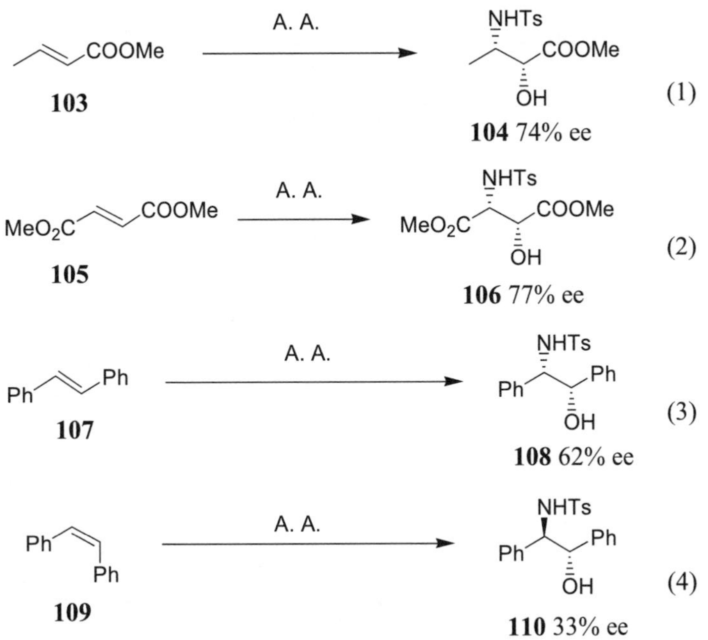

Scheme 4–36. Asymmetric aminohydroxylation.

salt, or it can be generated in situ. Thus, as illustrated in Scheme 4–37, by employing the methanesulphonamide derivative chloramine **111** as the oxidant, methyl (E)-cinnamate **112** can be converted to the corresponding α-hydroxy-β-amino product **113** with high ee (95%). Compound **113** is the taxol side chain, and this process established the shortest and the most efficient route to the side chain of this pharmaceutically important agent.[78]

As for the mechanism of asymmetric aminohydroxylation, it has been proposed that there are at least two catalytic cycles in the reaction system (Scheme 4–38).[77b] It is also suggested that both electronic and steric factors play important roles in the reaction. In the first cycle, in which the turnover occurs, effects of the ligand on selectivity are possible. For the ligand-independent

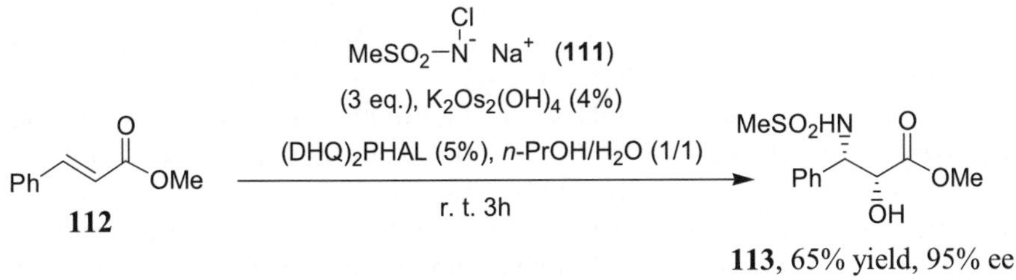

Scheme 4–37

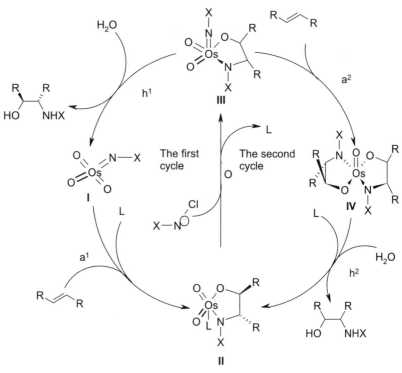

Scheme 4–38. Proposed mechanism for asymmetric aminohydroxylation. Sequence of steps in the first catalysis cycle (left): (1) addition (a^1), (2) reoxidation (O), (3) hydrolysis (h^1); in the second catalysis cycle (right): (1) addition (a^2), (2) hydrolysis (h^2), (3) reoxidation (O). The first cycle proceeds with high ee, the second with low ee. L = chiral ligand; X = CH$_3$SO$_2{}^-$.

second cycle, the ligand-mediated selectivity is lost, resulting in almost no enantioselectivity or regioselectivity. The hydrolysis step (h^1) of **III** is expected to be easy. This makes possible the suppression of the second cycle and thus accounts for most of the rate and selectivity advantages of the methyl over the *p*-tosyl substituent (X).

The original substrate-based asymmetric aminohydroxylation procedures, although very efficient, display a lack of substrate scope. For example, styrenes are not suitable substrates for these reactions. Replacement of sulfonamide with alkyl carbamates (BnO$_2$CNH$_2$, EtO$_2$CNH$_2$, and *t*-BuO$_2$CNH$_2$) or amides greatly improves the reaction in the sense of both the scope of the substrate and selectivity.[79] Carbamate-based nitrogen sources are especially useful because the resulting product can easily be converted to free amino alcohol. *t*-Butyl carbamate is superior to ethyl carbamate in terms of yield, enantioselectivity, and ease of removal of the *N*-protecting group. Application of the carbamate in asymmetric aminohydroxylation processes with a wide range of styrene substrates[79b,80] has yielded the major regioisomer in enantiomeric excess of up to 99% and yield of up to 80%. The regioselectivity is largely dependent on the substrate as well as the ligand, solvent, and ligand–solvent combination (Table 4–16).

As shown in Table 4–16, phthalazine ligand $(DHQ)_2DHAL$ in n-PrOH (alcoholic solvent) favors amine compound **115a** over **115b**. In CH_3CN, the ratio of **115a/115b** increases, and in one case the product ratio is reversed (Entry 6). The recently introduced anthraquinone (AQN) ligand appears to strongly favor this reversal of regioselectivity, especially when used in a CH_3CN/H_2O solvent system.

The observation that sterically less demanding nitrogen sources exhibit superior reactivity and give higher enantioselectivity has led to the new nitrogen source/oxidant 2-trimethylsilylethyl N-chloro-N-sodiocarbamate (TeoCNClNa).[81] This TeoCNClNa can be prepared by reacting NaOH/

TABLE 4–16. Influence of Ligand and Solvent on the Regioselectivity in Asymmetric Aminohydroxylation Reaction of Four Styrene Substrates[79b]

Entry	Substrate **114**	Ligand	Solvent	115a:115b
1		$(DHQ)_2PHAL$	n-PrOH/H_2O	88:12
2		$(DHQ)_2PHAL$	CH_3CN/H_2O	75:25
3	BnO	$(DHQ)_2AQN$	n-PrOH/H_2O	33:66
4		$(DHQ)_2AQN$	CH_3CN/H_2O	25:75
5		$(DHQ)_2PHAL$	n-PrOH/H_2O	50:50
6		$(DHQ)_2PHAL$	CH_3CN/H_2O	14:86
7	TsO	$(DHQ)_2AQN$	n-PrOH/H_2O	17:83
8		$(DHQ)_2AQN$	CH_3CN/H_2O	<1:50
9	BnO	$(DHQ)_2PHAL$	n-PrOH/H_2O	88:12
10		$(DHQ)_2PHAL$	CH_3CN/H_2O	50:50
11	BnO	$(DHQ)_2AQN$	n-PrOH/H_2O	33:66
12	BnO	$(DHQ)_2AQN$	CH_3CN/H_2O	23:77
13	MeO	$(DHQ)_2PHAL$	n-PrOH/H_2O	77:23
14		$(DHQ)_2AQN$	n-PrOH/H_2O	33:66
15	BnO	$(DHQ)_2AQN$	CH_3CN/H_2O	20:80

TABLE 4–17. Asymmetric Aminohydroxylation Using TeoCNNaCl as the Nitrogen Source

Entry	Substrate	Yield for 117 (%)	117/118	ee for (S)-117 (%)	ee for (R)-117 (%)
1	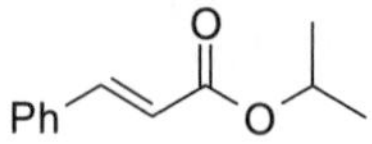 Ph~~~CO$_2$iPr	70	>98:2	99	99
2	m-O$_2$NPh~~~CO$_2$Et	74	>98:2	99	98
3	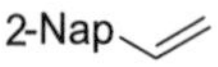 2-Nap~~~	76	86:14	97	96
4	m-MeO, p-BnOC$_6$H$_3$~~~	70	78:82	99	99
5	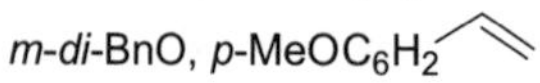 m-di-BnO, p-MeOC$_6$H$_2$~~~	81	91:9	98	97
6	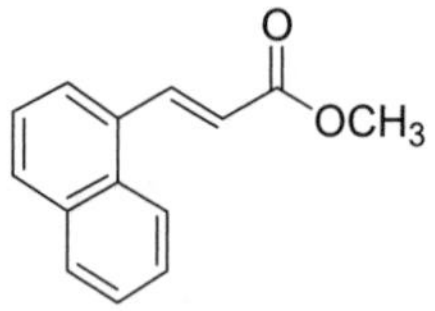 naphthyl-CH=CH-CO-OCH$_3$	86	>98:2	95	91

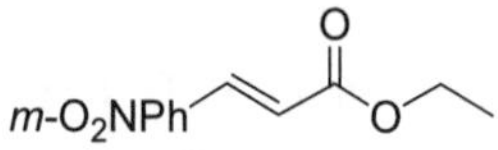

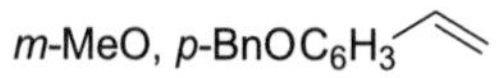

ee = Enantiomeric excess.

Reprinted with permission by Pergamon Science Ltd., Ref. 81.

t-BuOCl with 2-trimethylsilyl ethyl carbamate, which can be prepared by successively adding carbonyl diimidazole and ammonia to 2-trimethylsilylethanol in benzene. TeoCNClNa substantially extends the scope of the Os-catalyzed asymmetric aminohydroxylation of alkenes. The high reactivity of this new nitrogen source/oxidant enables the reaction to proceed effectively with generally better regioselectivity, enantioselectivity, and higher yields. The TeoC group can be cleaved by fluoride under very mild conditions, yielding the free amino alcohol with high enantiomeric purity. This result is summarized in Table 4–17.

The catalytic asymmetric aminohydroxylation of a variety of styrene derivatives, vinyl aromatics, and some other olefins using osmium tetroxide

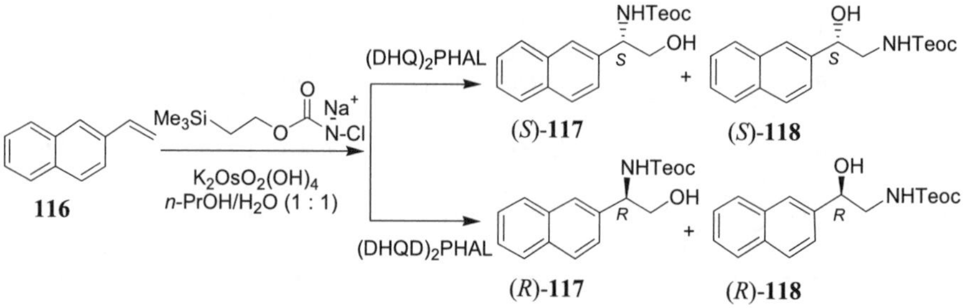

Scheme 4–39

in conjunction with alkaloid-derived ligands and haloamine salts has been extensively studied and used to complete the asymmetric synthesis of α-amino acids, taxol side chain, amino cyclitols, cyclohexylnorstatine, 2,3-diaminobutanoic acid, α-amino ketones, β-amino-α-hydroxy-phosphonic acid–derivatives, and diamines.[82]

Currently, there are seven different reagents available for carrying out asymmetric aminohydroxylations. All seven methods use a combination of osmium tetroxide [obtained from $K_2Os(OH)_4$] with alkaloid-derived ligands and the Li or Na salt of an N-halogenated sulfonamide, alkyl carbamate, or amide in an alcohol/solvent mixture. With these seven different methods in hands, chemists are in the fortunate position of being able to convert almost any olefin into an amino alcohol with high yield and high enantioselectivity.[83]

4.6 EPOXIDATION OF UNFUNCTIONALIZED OLEFINS

Following the success with the titanium-mediated asymmetric epoxidation reactions of allylic alcohols, work was intensified to seek a similar general method that does not rely on allylic alcohols for substrate recognition. A particularly interesting challenge was the development of catalysts for enantioselective oxidation of unfunctionalized olefins. These alkenes cannot form conformationally restricted chelate complexes, and consequently the differentiation of the enantiotropic sides of the substrate is considerably more difficult.

4.6.1 Catalytic Enantioselective Epoxidation of Simple Olefins by Salen Complexes

Cytochrome p-450

an oxidizing enzyme

metalloporphyrin complex   metallosalen complex

The asymmetric epoxidation of olefins catalyzed by chiral salen–Mn complexes was first reported independently in 1990 by Zhang et al.[84] and Irie et al.[85] These catalysts are related to metalloporphyrin-based epoxidation catalysts, and oxygen transfer to the alkene takes place from an oxomanganese intermediate. A good example of the catalysts for asymmetric epoxidation of unfunctionalized olefins is the biomimetic agent related to an oxidizing enzyme, cytochrome P-450. The most commonly used oxidants are iodosylbenzene in organic solvents or sodium hypochlorite in aqueous media. Hydrogen peroxide

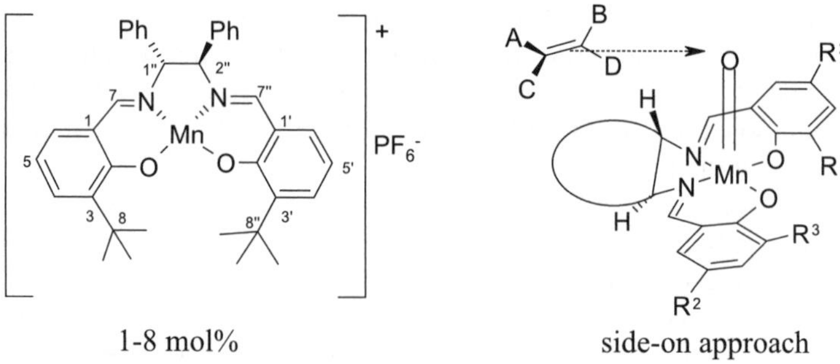

1-8 mol%   side-on approach

Figure 4–6. Side-on approach of the substrate.

in excess can also be used as an oxidant for aqueous media, but the presence of an additive ligand is essential for an effective reaction. Indeed, using molecular oxygen as the oxidant is more advisable because it is the most economical and environmentally friendly oxidant.

At first, the reaction was characterized as most effective in the epoxidation of *cis*-disubstituted olefins.[86] Later, the scope of this reaction was expanded to include the highly enantioselective synthesis of *trans*-disubstituted[87] and tri-substituted epoxides,[88] as well as certain monosubstituted epoxides.[89] The first example of nondirect asymmetric epoxidation of tetrasubstituted olefins has also appeared.[90]

The course of the enantioselectivity is interpreted in terms of a side-on approach by the substrate to the active oxomanganese(V) intermediate (Fig. 4–6). The results of the epoxidation are shown in Scheme 4–40.[84,91] To achieve higher enantioselectivities, sterically demanding *t*-butyl groups substituted at the C3 and C3′ positions in the salen complex have proved to be essential.

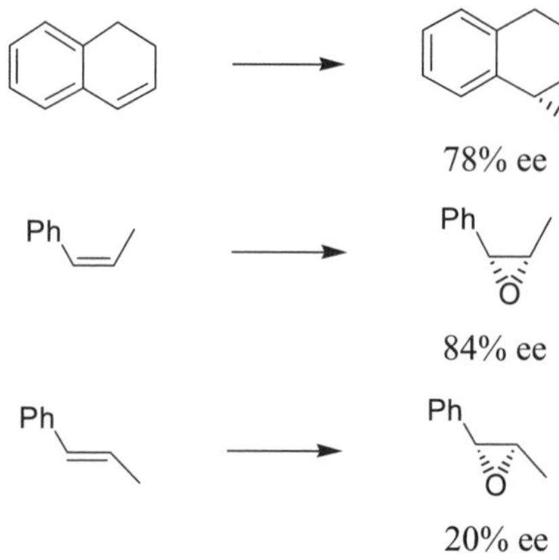

78% ee

84% ee

20% ee

Scheme 4–40

Their interactions with the more hindered side of an asymmetrical olefin determine the orientation of the substrate during its approach to the metal–oxo bond and the subsequent enantiofacial selective oxygen transfer.

Although salen complexes of chromium, nickel, iron, ruthenium, cobalt, and manganese ions are known to serve as catalysts for epoxidation of simple olefins, the cationic Mn–salen complex is the most efficient.

As pointed out by Hosoya et al.[92] the enantiofacial selection of *cis*-olefins is mainly controlled by the asymmetric centers at the C-8(8′) carbons, while that of *trans*-olefins is preferentially controlled by the asymmetric centers at the C-9(9′) carbons in **119** or **120**. Optically active Mn(III)–salen complexes have catalyzed the epoxidation of *cis*-olefins with higher ee (>90%), especially when they are conjugated with an acetylene or phenyl group. However, the epoxidation of *trans*-olefins with these salen complexes shows rather poor enantioselectivity (Table 4–18).

119   **120**

TABLE 4–18. Epoxidation of Unfunctionalized Olefins Catalyzed by 119
substrate:catalyst:iodosylbenzene = 1:0.025:1

Substrate	ee (%)
Ph~~~Ph	62
Ph~~~	9 (56 when **120** is used)
(naphthalene dihydro structure)	91
O₂N / AcNH chromene structure	96
N-oxide benzofurazan chromene structure	94

ee = Enantiomeric excess.

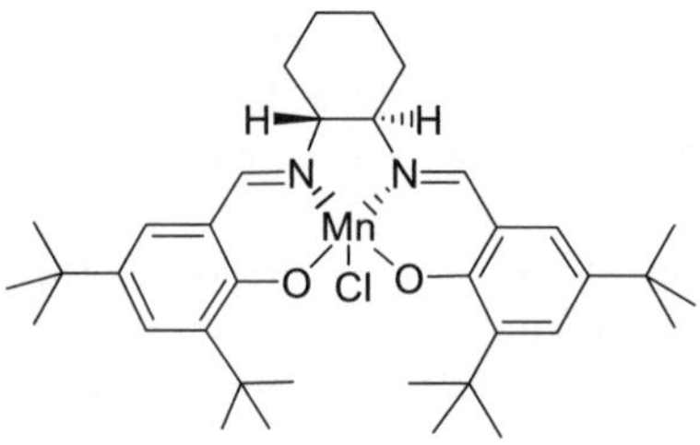

121 in both (*R,R*) and (*S,S*) form

Jacobsen applied the salen–Mn(III) complex (*R,R*)-**121** in the synthesis of two antihypertensive agents **122** and **123**[86a] (Scheme 4–41) and also the taxol side chain **124**.[86b]

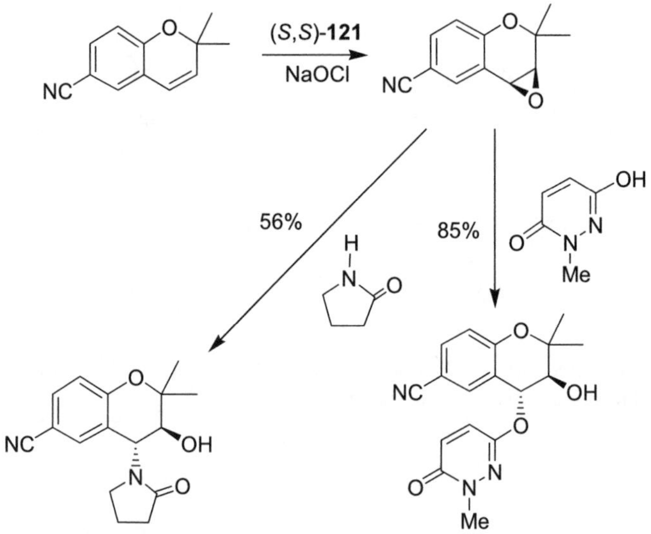

Both **122** and **123** are antihypertensive agents

Scheme 4–41. Synthesis of antihypertensive agents.

As Scheme 4–42 shows, the synthesis of **124** can be accomplished in four steps. The success of this process lies in the simplicity and low cost of all the reagents.

As an extension, this type of reagent can also be used to oxidize sulfides. Scheme 4–43 depicts the asymmetric oxidation of sulfides catalyzed by salen–Mn(III) complexes.

Iodosylarenes are impractical stoichiometric oxidants for either small-scale

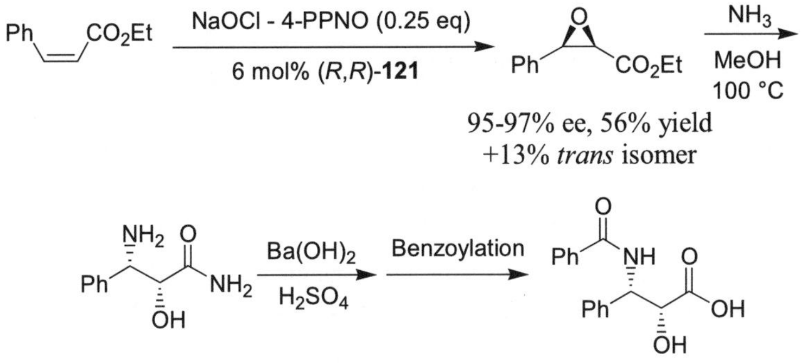

Scheme 4–42. Synthesis of the taxol side chain.

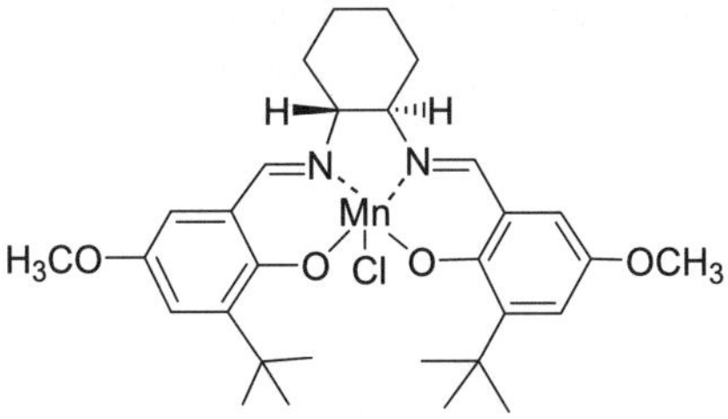

125

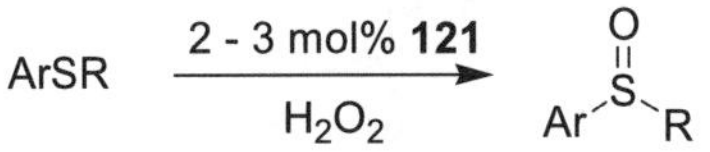

ee varies from 38% to 68%

Scheme 4–43. Extension of Salen-complex to the Oxidation of sulfides.

or large-scale reactions due to their instability in the solid state, their lack of solubility in organic solvents, their relatively high cost, and the high molecular weight of the oxygen transfer by-product. Compared with iodosylbenzene, hydrogen peroxide gives higher yields of sulfoxide, minimal overoxidation to sulfone, and enantioselectivities identical to those observed with iodosylbenzene.

Many efforts have been made to develop salen catalysts for the epoxidation of unfunctionalized olefins, and such work has been well documented.[93] Very recently, Ito and Katsuki[94] proposed that the ligand of the oxo salen species is not planar, but folded as shown in Figure 4–7 ($R' \neq H$, $R^2 = H$, L^* = achiral axial ligand). This folded chiral structure amplifies asymmetric induction by the Mn–salen complex. This transition state proposed by Ito and Katsuki is not compatible with the proposal by Palucki et al.[95] that the salen ligands of oxo species are planar.

The conformation bearing the substituent at the asymmetric center in a pseudoequatorial position (R^1) is more stable than that bearing the substituent

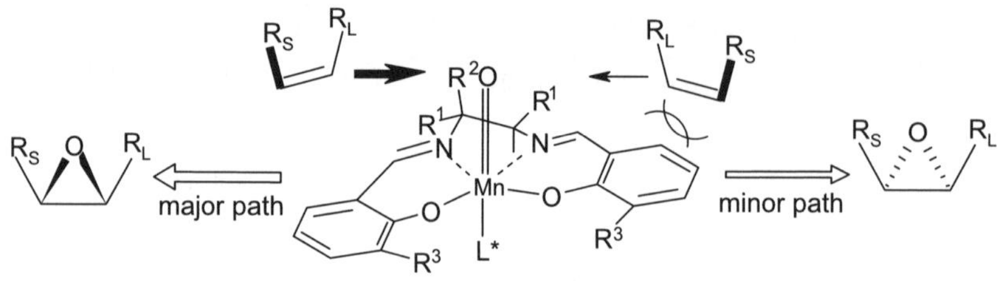

Figure 4–7. Folded conformation of the salen complex. R_S = Alkyl; RL = aryl; R^1, R^2 = H, alkyl, or aryl; R^3 = bulky group. Reprinted with permission by Elsevier Science Ltd., Ref. 94.

in a pseudoaxial position (R^2). It is this conformation that controls the asymmetric induction of the reaction. Based on the proposed nonplanarity of the salen ligand, Katsuki's group prepared a conformationally reversed optically active salen–manganese(III) complex **126** bearing a carboxylate group on the ethylenediamine moiety.[94]

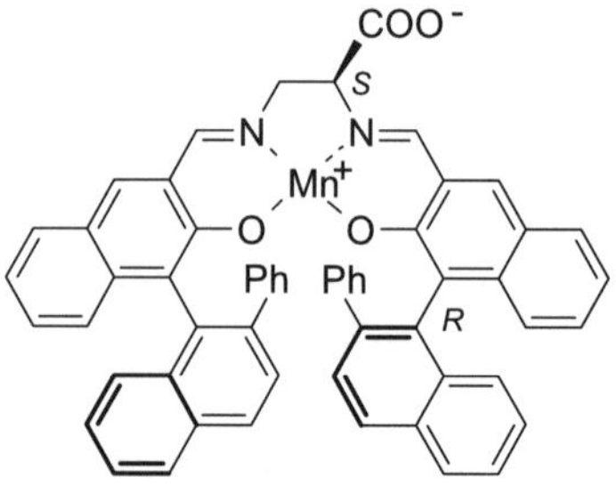

126

Reversal of the conformation of the chiral Mn–salen complex forces the substituents on the ethylenediamine moiety to take the disfavored axial position. This disfavored conformation (Fig. 4–8A) should be stabilized by the co-

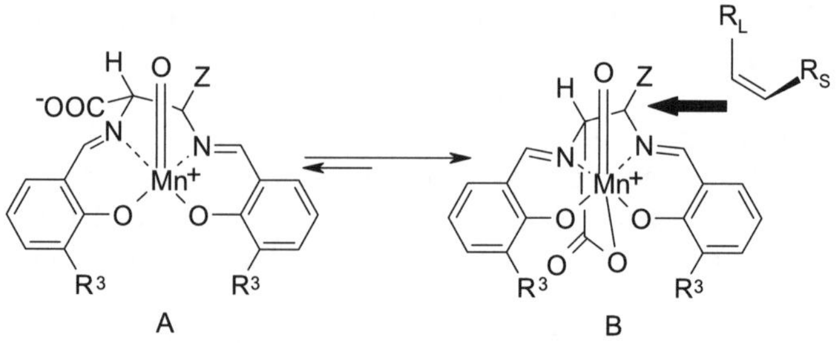

A   B

Figure 4–8. Transition state for **126** catalyzed epoxidation.

ordination of carboxylate to the Mn ion (Fig. 4–8B). Complex **126** was then found to be an efficient catalyst for the asymmetric epoxidation of several 2,2-dimethylchromene derivatives, resulting in high ee (up to 99%) and high turnover number (up to 9200).

4.6.2 Catalytic Enantioselective Epoxidation of Simple Olefins by Porphyrin Complexes

Porphyrin–metal complexes are natural mimetic substances that have attracted much attention during the past decade. The epoxidation of olefins by porphyrin complexes proceeds well, but with only modest enantioselectivity. As this area of research is growing, description of a few selected publications may be useful.[96]

Konishi et al.[97] synthesized porphyrin compound **127**. As shown in Scheme 4–44, asymmetric epoxidation of prochiral olefins such as styrene derivatives and vinyl naphthalene by iodosobenzene has been achieved by using this porphyrin complex as the catalyst in the presence of imidazole. The optically active epoxides were obtained with moderate ee.

Better results for the porphyrin complex–catalyzed asymmetric epoxidation of prochiral olefins were achieved by Naruta et al.[98] using iron complexes of chiral binaphthalene or bitetralin-linked porphyrin **128** as chiral catalysts. As shown in Scheme 4–45, asymmetric epoxidation of styrene or its analogs provided the product with good ee. Even better results were obtained with substrates bearing electron-withdrawing substituents.

Collman et al.[99] reported the asymmetric epoxidation of terminal olefins catalyzed by iron porphyrin complex **129**. The catalyst was synthesized by connecting binaphthyl moieties to a readily available $\alpha\alpha\beta\beta$-tetrakis(aminophenyl)-porphyrin (TAPP). Epoxidation of unfunctinalized olefins was carried out using iodosylbenzene as the oxidant. As shown in Scheme 4–46, excellent results were

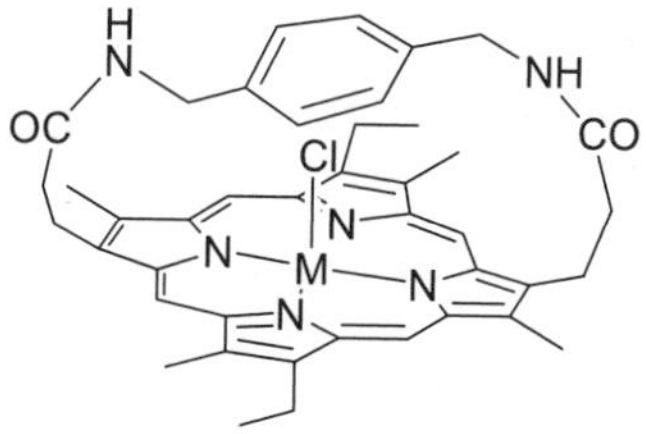

127 M = Mn, Fe

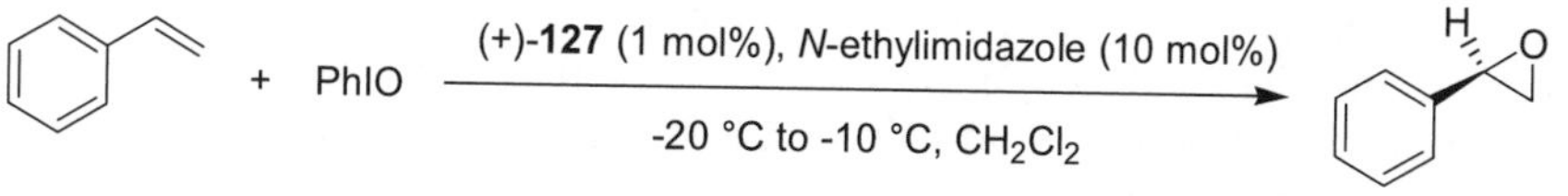

68% yield, 50% ee

Scheme 4–44

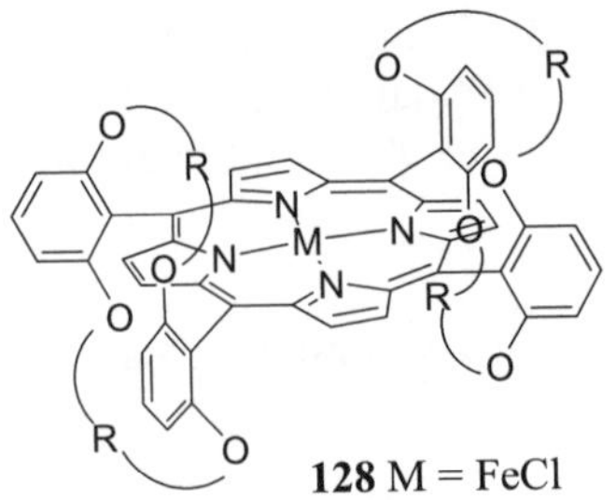

128 M = FeCl

$$Ar \diagdown \xrightarrow[\text{PhI, 0 °C, CH}_2\text{Cl}_2]{\text{Fe porphyrin}} Ar\text{—epoxide}$$

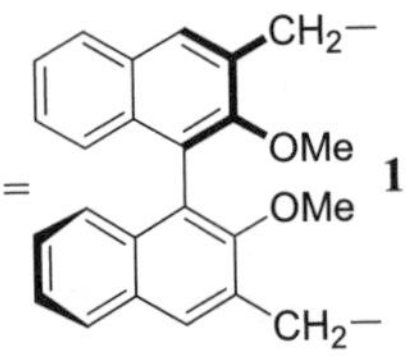

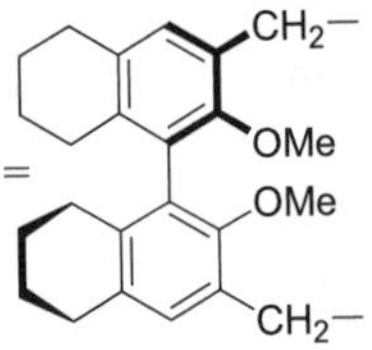

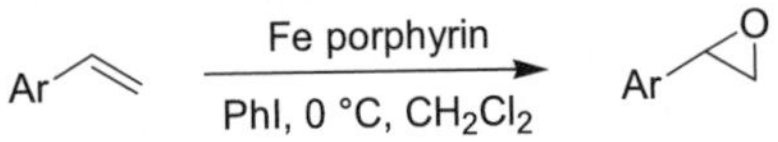

Entry	Substrate	ee (%)		Config.
		128a	**128b**	
1	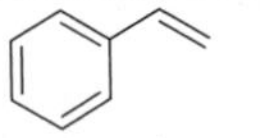 (styrene)	20	56	*S*
2	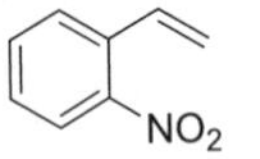 (2-nitrostyrene)	80	89	*S*
3	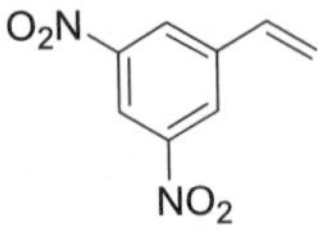 (3,5-dinitrostyrene)	74	82	*S*
4	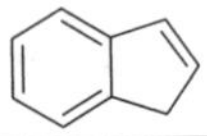 (indene)	10	70	1*S*,2*R*

Scheme 4–45

obtained for most substrates. The good ee values, high turnover number, and ready availability of the reagents make this compound a potential catalyst for many practical applications.

4.6.3 Chiral Ketone–Catalyzed Asymmetric Oxidation of Unfunctionalized Olefins

As oxiranes can be generated in situ from Oxone® (potassium peroxomonosulfate) and a ketone, dioxiranes are attractive oxidants for epoxidation reactions that may be rapid and may require only a simple workup.

The following rules should be considered when designing this type of chiral ketone compound: (1) The stereogenic centers should be close to the reaction centers in order to get efficient stereochemical communication between the substrate and the catalyst; (2) the presence of a fused ring and a quaternary center α to the carbonyl group will minimize the epimerization of the stereogenic center; and (3) one face of the catalyst should be sterically blocked in order to limit possible competing approaches from this face. In addition, it is essential to design the catalyst bearing a C$_2$-symmetric axis.

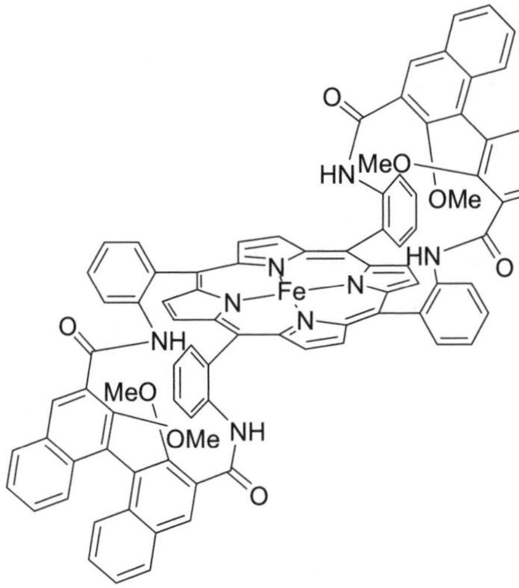

129

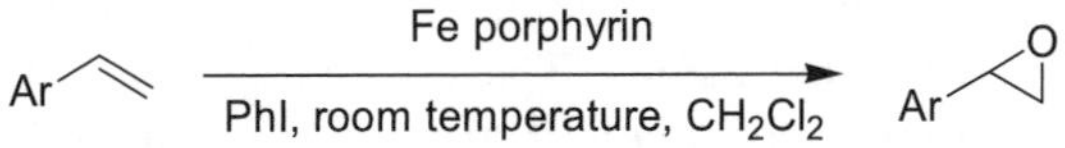

Entry	Substrate	Yield (%)	ee (%)	Config.
1	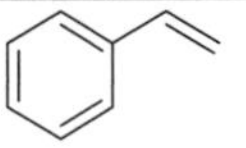	95	83	S
2		isolated yield 89	75	S
3	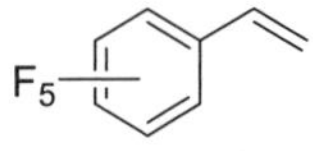 F_5—	75	88	S
4	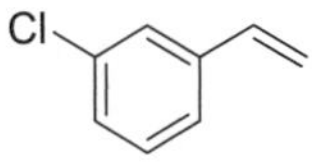 Cl—	90	82	S
5		isolated yield 85	74	S
6	NO_2	74	55	S
7	NO_2—	78	72	S S
8		isolated yield 75	68	S
9	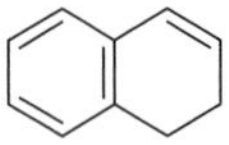	80	55	$1S,2R$
10	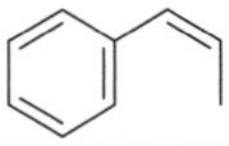	78	49	$1S,2R$

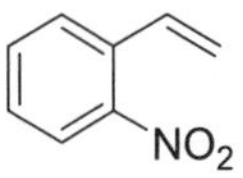

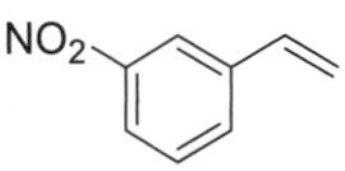

Scheme 4–46. Reprinted with permission by Am. Chem. Soc., Ref. 99.

4.6.3.1 Chiral Ketone from Carbohydrate. Tu et al.[100] reported a dioxirane-mediated asymmetric epoxidation based on the ketones derived from the low cost material D-fructose (Scheme 4–47).

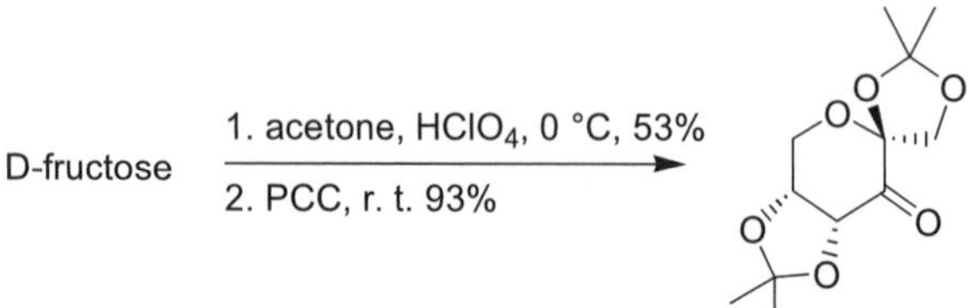

Scheme 4–47

All the reactions were carried out at 0°C, with the substrate (1 equivalent), ketone (3 equivalents), Oxone® (5 equivalents), and NaHCO₃ in CH₃CN–aqueous EDTA for 2 hours. High enantioselectivity can generally be obtained for *trans*- and trisubstituted olefins. The favored spiro and planar transition states have been proposed for ketone **130**-mediated *trans*-stilbene epoxidation (Scheme 4–48).

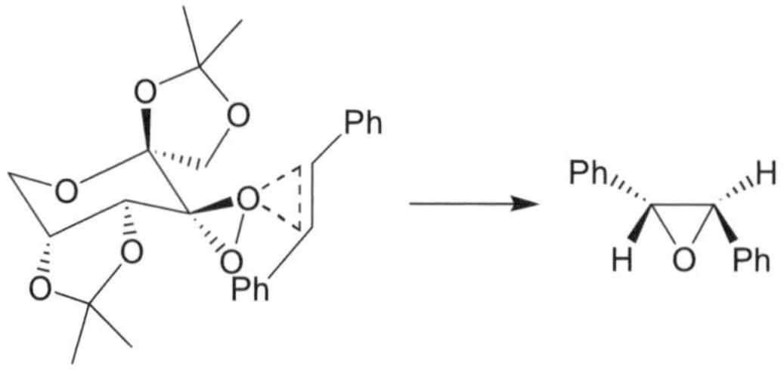

Substrate	Yield (%)	ee (%)	Config.
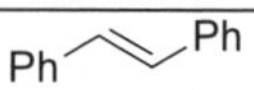 Ph⌇⌇Ph	73	>95	*R, R*
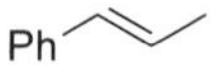 Ph⌇⌇	81	88	*R, R*
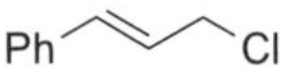 Ph⌇⌇Cl	61	93	2*S*, 3*R*
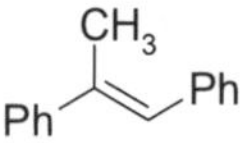 CH₃ Ph⌇⌇Ph	73	92	*R, R*
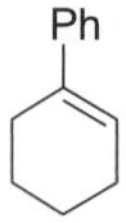 Ph (cyclohexene)	69	91	*R, R*

Scheme 4–48. Reprinted with permission by Am. Chem. Soc., Ref. 100.

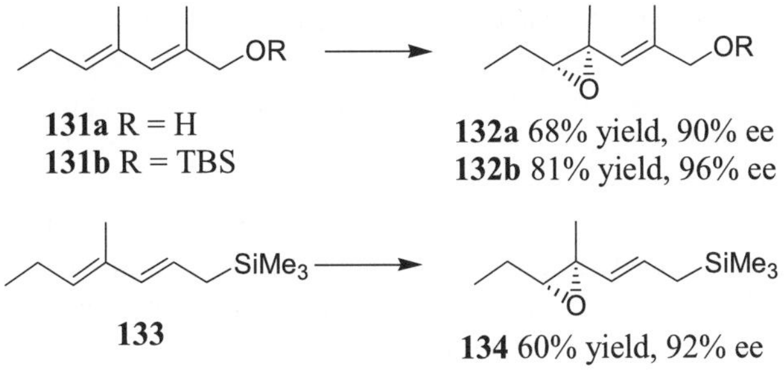
OR
OR
131a R = H
131b R = TBS
132a 68% yield, 90% ee
132b 81% yield, 96% ee
SiMe3
SiMe3
133
134 60% yield, 92% ee

Scheme 4–49

In most cases, such reactions proceed highly regioselectively.[101] Taking the epoxidation of dienol **131a**, **131b**, and **133** as an example, as shown in Scheme 4–49, under such conditions the nonallylic double bonds are epoxidized, giving the corresponding monoepoxides **132a**, **132b**, and **134**, respectively. Only trace amounts of the bis-epoxides are detected in the crude products.

Cao et al.[102] extended their discovery to the asymmetric epoxidation of enynes using ketone **130** as the catalyst and Oxone® as the oxidant (Scheme 4–50).

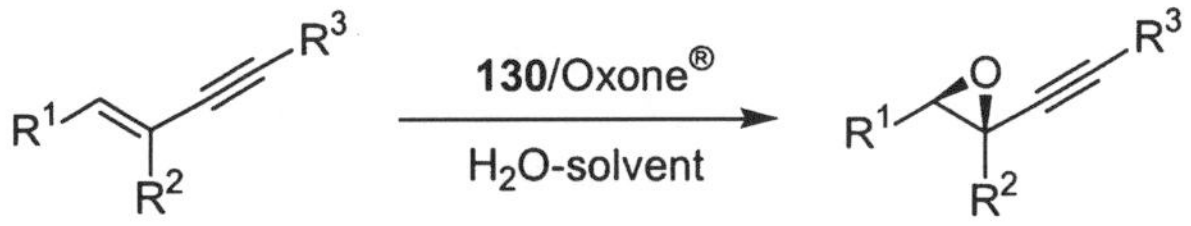
R3
R1
R2
130/Oxone®
H2O-solvent
R3
O
R1
R2

Scheme 4–50

Subsequently, high chemoselectivity and enantioselectivity have been observed in the asymmetric epoxidation of a variety of conjugated enynes using fructose-derived chiral ketone as the catalyst and Oxone® as the oxidant. Reported enantioselectivities range from 89% to 97%, and epoxidation occurs chemoselectively at the olefins. In contrast to certain isolated trisubstituted olefins, high enantioselectivity for trisubstituted enynes is noticeable. This may indicate that the alkyne group is beneficial for these substrates due to both electronic and steric effects.

Mechanistic studies[103] revealed that chiral ketone-mediated asymmetric epoxidation of hydroxyl alkenes is highly pH dependent. Lower enantioselectivity is obtained at lower pH values; at high pH, epoxidation mediated by chiral ketone out-competes the racemic epoxidation, leading to higher enantioselectivity. (For another mechanistic study on ketone-mediated epoxidation of C=C bonds, see Miaskiewicz and Smith.[104])

4.6.3.2 A C₂ Symmetric Chiral Ketone for Catalytic Asymmetric Epoxidation of Unfunctionalized Olefins.

Yang et al.[105] reported the use of C_2-symmetric chiral ketones **135a–h** for the asymmetric epoxidation of unfunctionalized olefins using Oxone® as the oxidant. Moderate to good enantioselectivities were obtained for *trans*-olefins and trisubstituted olefins (33–87% ee) (Scheme 4–51). X-ray structural analysis of **135a** shows that this chiral ketone has a rigid structure and C_2 symmetry. The keto group lies on the C_2 axis of the molecule, and the two ester groups are nearly perpendicular to the plane of the macrocyclic ring. The two naphthalenes are located on the opposite face of the ketone group, and the dihedral angle of the two naphthalene rings is about 70°.

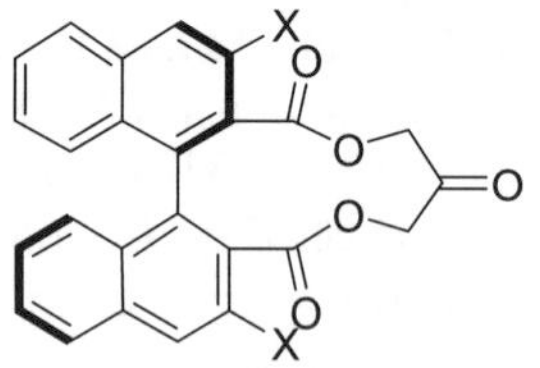

135

135a X = H	**135f** X = CH₂OCH₃
135b X = Cl	
135c X = Br	**135g** X =
135d X = I	
135e X = Me	**135h** X = TMS

10 mol% **135**, r. t.

Oxone®/NaHCO₃

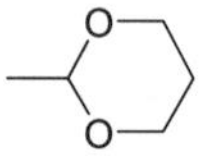

CH₃CN-H₂O

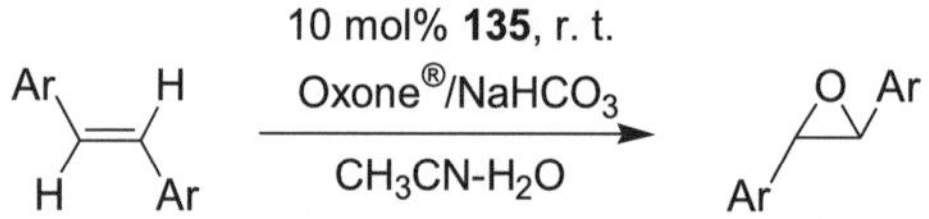

Cat.	Time (h)	Yield (%)	Config.	ee (%)
(R)-**135a**	1	91	S,S	47[a]
(R)-**135b**	2	95	S,S	76[a]
(R)-**135c**	3	92	S,S	75[a]
(R)-**135d**	22	90	S,S	32[a]
(R)-**135e**	1	93	R,R	56[a]
(R)-**135f**	1.8	92	S,S	66[a]
(R)-**135g**	0.7	95	S,S	71[a]
(S)-**135h**	20	-	R,R	44[a]
(R)-**135a**	6	82	S,S	87[b]
(S)-**135a**	6	80	R,R	87[b]

Scheme 4–51. Asymmetric epoxidation of *trans*-stilbene with catalyst **135**. [a]Substrate: (E)-stilbene. [b]Substrate: (E)-4,4'-diphenylstilbene. Reprinted with permission by Am. Chem. Soc., Ref. 105a, 106.

Yang et al.[106] found that when the *para* substituents of the *trans*-stilbenes became larger, the ee values of the *trans*-epoxides increased gradually. Their studies show that these chiral ketones are stable under the reaction conditions and can be recovered in over 80% yield and reused without loss of catalytic activity and chiral induction. The ester groups that give the rigid and C_2-symmetric structure to the cyclic ketones seem to be essential for effective asymmetric epoxidation. Lowering the reaction temperature enhances the enantioselectivity but decreases the reaction rate. The most suitable reaction temperature is $0°C$, at which optimal yields and ee values can be obtained.

4.7 CATALYTIC ASYMMETRIC EPOXIDATION OF ALDEHYDES

Thus far, the asymmetric epoxidation of olefins for preparing chiral epoxide compounds has been discussed. These are invaluable intermediates in the organic synthesis of important molecules such as pharmaceuticals or agrochemicals. The Sharpless and Jacobsen/Katsuki methods have proved to be the most powerful for preparing chiral epoxides of various types. Some drawbacks to these reactions may be that Sharpless epoxidation requires an allylic alcohol and the Jacobsen/Katsuki epoxidation generally requires a *cis*-substituted substrate bearing a π-stabilizing substituent. The chiral ketone-mediated epoxidation developed by Cao and Yang can be considered a breakthrough in asymmetric epoxidation of unfunctionalized olefins, and a wide range of olefins can now be epoxidized with high enantioselectivity.

An alternative to the synthesis of epoxides is the reaction of sulfur ylide with aldehydes and ketones.[107] This is a carbon–carbon bond formation reaction and may offer a method complementary to the oxidative processes described thus far. The formation of sulfur ylide involves a chiral sulfide and a carbene or carbenoid, and the general reaction procedure for epoxidation of aldehydes may involve the application of a sulfide, an aldehyde, or a carbene precursor as well as a copper salt. This reaction may also be considered as a thiol acetal-mediated carbene addition to carbonyl groups in the aldehyde.

In the design of chiral sulfides for sulfur ylide–mediated asymmetric epoxidation of aldehydes, two factors are important. First, a single sulfur ylide should be produced. Otherwise, the diastereomeric sulfur ylides may react with aldehydes in different ways and thus cause a drop in stereoselectivity. This may be achieved by choosing a rigid cyclic structure to make one of the lone pairs more accessible than the other. Second, the structure should be amenable to structural modification in order to study the electronic and steric effects of the sulfur on the enantioselectivity of the epoxidation reaction.

Aggarwal et al.[108] reported excellent results with the catalytic asymmetric epoxidation of aldehydes. As shown in Scheme 4–52, a series of thioacetals **137** was prepared from hydroxy thiol **136** and the corresponding carbonyl compound. Among them, compound **138**, derived from **136** and acetaldehyde, proved to be the best catalyst for asymmetric epoxidation of aldehydes.

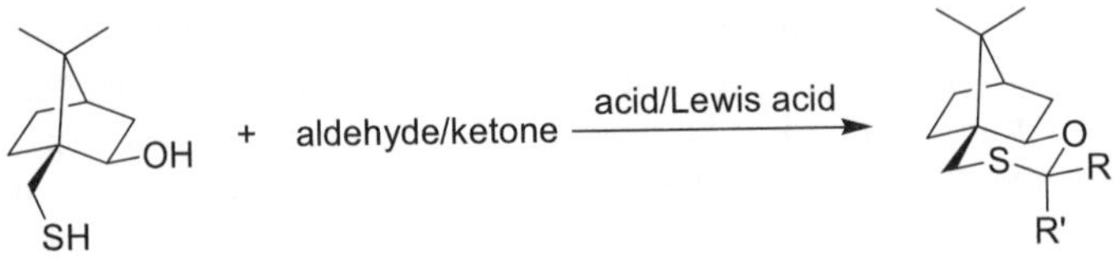

136   **137**

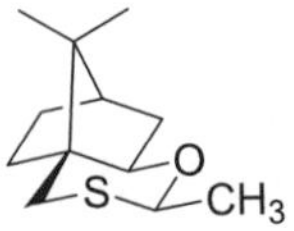

138

Scheme 4–52

This reaction is very sensitive to water because in the presence of water and a metal salt (such as copper salt) the thioacetal tends to decompose, and this may reduce the amount of thioacetal available for epoxidation. When water is excluded from all the reagents, the reaction can be carried out in the presence of a catalytic amount of thioacetal. Otherwise, a stoichiometric amount of thioacetal compound is required. Scheme 4–53 summarizes the epoxidation of aldehydes using **138** as the chiral-inducing reagent. Excellent enantioselectivities are obtained in most cases.

4.8 ASYMMETRIC OXIDATION OF ENOLATES FOR THE PREPARATION OF OPTICALLY ACTIVE α-HYDROXYL CARBONYL COMPOUNDS

In a general sense, *oxidation* refers to the introduction of oxygen or other electronegative atoms to organic molecules. In this context, α-hydroxylation of a

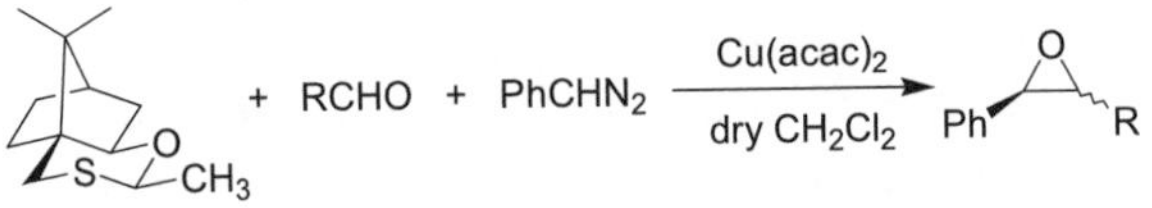

Entry	Aldehyde	Yield (%)	*trans/cis*	ee (%)
1	PhCHO	73	> 98 : 2	94 (*R,R*)
2	4-ClPhCHO	72	> 98 : 2	92 (*R,R*)
3	4-MePhCHO	64	> 98 : 2	92 (*R,R*)
4	Cinnamaldehyde	55	> 98 : 2	89
5	Valeraldehyde	35	92 : 8	68
6	Cyclohexanecarboxaldehyde	32	70 : 30	90

Scheme 4–53

carbonyl group and other related reactions can also be regarded as oxidation reactions. Numerous efforts have been made toward the development of a technique for the hydroxylation of a site adjacent to a carbonyl group. Among these, applying chiral *N*-sulphonyl oxaziridine seems to be a promising approach in a wide variety of enolate hydroxylation reactions.

4.8.1 Substrate-Controlled Reactions

In early studies, α-hydroxylation of an adjacent carbonyl group was achieved with control by a chiral auxiliary. This is the case where chiral auxiliaries are used to differentiate the two faces of an enolate, and oxygen transfer produces the corresponding α-hydroxy ketones, esters, or acids after cleavage of the chiral auxiliaries. Representative hydroxylation reactions of chiral enolates include those with substrates derived from metallopyrrolidines, metalloenamines, Oppolzer's sulfonamides, oxazolidinones, and hydrazones. All of these reactions involve a mechanism of diastereofacial discrimination similar to that applied in carbon–carbon bond formation reactions. In this manner, α-hydroxy acid can be obtained from the corresponding chiral enolates derived from pyrrolidine-type imines such as **140a** or **140b**, when *N*-sulfonyloxaziridine **141** is used as the oxidant (see Scheme 4–54).[109]

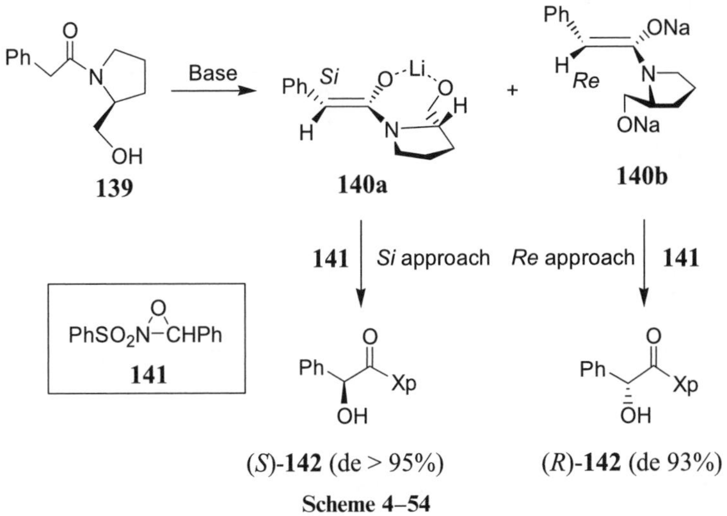

Scheme 4–54

Evans succeeded in oxidizing *N*-acyl oxazolidinone enolate **143** or **145** using oxaziridine **141** as the oxidant (Scheme 4–55).[110] Representative results are summarized in Table 4–19.

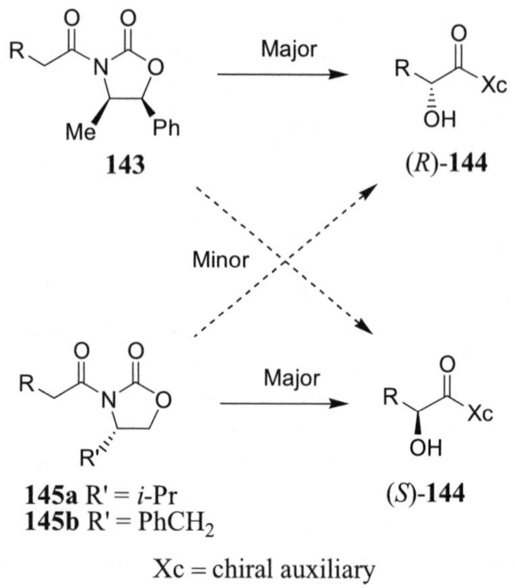

Xc = chiral auxiliary

Scheme 4–55. *N*-Acyl oxazolidinone enolate oxidation.

TABLE 4–19. Diasteroselective Hydroxylation of Chiral Carboximide Sodium Enolates Using 2-(Phenylsulfonyl)-3-Phenyl-Oxaziridine (141) in THF at $-78°C$

Imide	Yield (%)	de (%, config.)
143 (R = Bn)	86	88 (*R*)
145a (R = Bn)	85	90 (*S*)
145b (R = Bn)	83	90 (*S*)
143 (R = Ph)	77	80 (*R*)
143 (R = Et)	86	88 (*R*)
143 (R = Allyl)	91	90 (*R*)
143 (R = *t*-Bu)	94	98 (*R*)
145a (R = *i*-Pr)	86	98 (*S*)
143 (R=MeO$_2$C(CH$_2$)$_3$	68	92 (*R*)

de = Diastereomeric excess.

4.8.2 Reagent-Controlled Reactions

Davis et al.[111] developed another method for reagent-controlled asymmetric oxidation of enolates to α-hydroxy carbonyl compounds using (+)-camphorsulfonyl oxaziridine (**147**) as the oxidant. This method afforded synthetically useful ee (60–95%) for most carbonyl compounds such as acyclic keto esters, amides, and α-oxo ester enolates (Table 4–20).

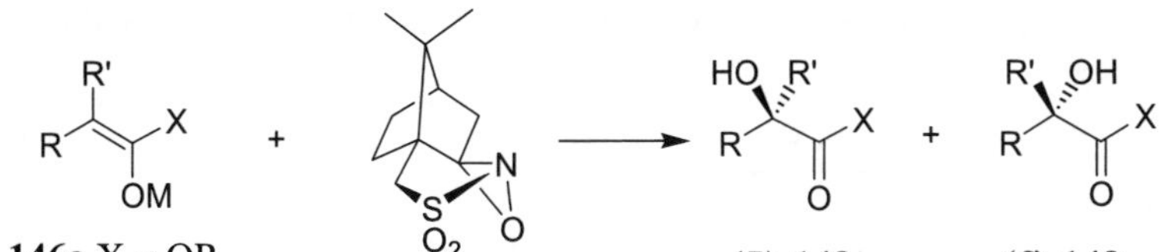

146a X = OR
146b X = NR$_2$

(+)-**147**

(R)-**148**   (S)-**148**

TABLE 4–20. Asymmetric Oxidation of Lithium Enolates and Amides Using (+)-147 as the Oxidant

Entry	Enolate **146**					Product **148**	
	R	R'	X	Co-solvent	Temp. (°C)	Yield (%)	ee (%)
1	Ph	H	t-BuO	—	−90	84	71.0(R)
2	Bn	H	OMe	—	−90	73	58.0(R)
				HMPA	−90	63	85.5(R)
3	Ph	H	OMe	—	−78	84	54.0(R)
4	Ph	H	N(C$_4$H$_8$)$_2$	—	−78	70	30.0(S)
				HMPA	−78	74	50.0(R)
5	Ph	Me	N(C$_4$H$_8$)$_2$	—	−78	77	60.0(R)
				HMPA	−78	35	20.0(S)

ee = Enantiomeric excess; HMPA = hexamethylphosphoramide.
Reprinted with permission by Am. Chem. Soc., Ref. 111.

In contrast to the oxidation of prochiral esters and amides, which induces only moderate ee, sodium enolates of ketones give high stereoselectivity with (+)-**147** or (−)-**147** as the oxidant (Scheme 4–56 and Table 4–21). The highest stereoselectivity has been observed in the oxidation of the sodium enolate of deoxybenzoin **150**, in which benzoin **149** can be obtained in over 95% optical purity.

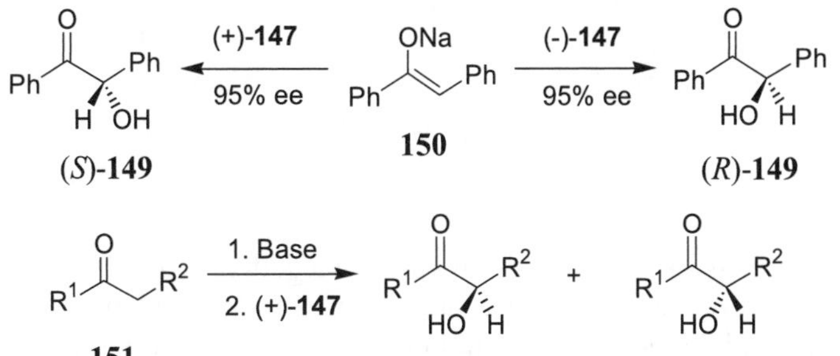

(S)-**149**   **150**   (R)-**149**

151   (R)-**152**   (S)-**152**

Scheme 4–56

TABLE 4–21. Asymmetric Oxidation of Prochiral Ketone Enolates to α-Hydroxyl Ketones Using 147

Ketone	Base	Temp. (°C)	Yield (%)	ee (%)	Config.
PhCOCH$_2$Ph	LDA	0	70	68	S
	NHMDS	−78	84	95.4	S
PhCOCH$_2$Me	LDA	0	51	43.2	S
	NHMDS	−78	77	68.5	S
t-BuCOCH$_2$Me	LDA	0	55	33	R
	NHMDS	−78	71	90	R
PhCH$_2$COMe	NHMDS	−78	70	41	S
	NHMDS/HMPA	−78	76	76	R

ee = Enantiomeric excess; LDA = lithium diisopropylamide; NHMDS = NaN(SiMe$_3$)$_2$.

α-Hydroxy ketones, which are otherwise difficult to prepare, can be obtained with high ee by applying the Davis reagent–mediated oxidation reaction. For example, as shown in Scheme 4–57, (*S*)-2-hydroxy-1-phenyl-1-propanone (*S*)-**155** can be generated with over 95% ee and 61% isolated yield by oxidation of the sodium enolate of **154** with (+)-**153** at −78°C. Oxidation of the enolate of **154** with (+)-**147**, on the other hand, yields (*S*)-**155** in only 62% ee.[112]

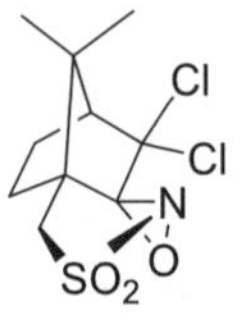

(+)-**153**

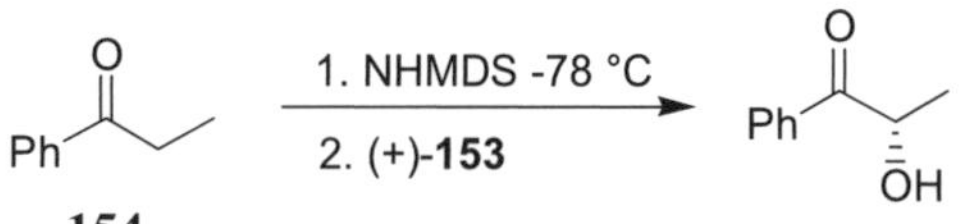

Scheme 4–57

Following their success with chiral ketone-mediated asymmetric epoxidation of unfunctionalized olefins, Zhu et al.[113] further extended this chemistry to prochiral enol silyl ethers or prochiral enol esters. As the resultant compounds can easily be converted to the corresponding α-hydroxyl ketones, this method may also be regarded as a kind of α-hydroxylation method for carbonyl substrates. Thus, as shown in Scheme 4–58, the asymmetric epoxidation of enol silyl

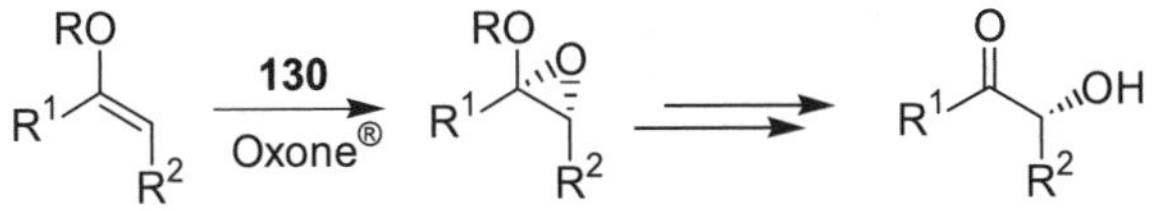

Scheme 4–58

ether or ester gives, after subsequent treatment, the corresponding α-hydroxyl ketone in up to 99% ee.

4.9 ASYMMETRIC AZIRIDINATION AND RELATED REACTIONS

Aziridines, like epoxides, have great synthetic potential and are versatile intermediates for nitrogen-containing compounds. Tanner[114] reviewed different uses of optically active aziridines as chiral synthons for the enantioselective synthesis of alkaloids, amino acids, β-lactams, and pyrrolidines as well as their potential as chiral auxiliaries or chiral ligands. These aziridine compounds can be approached through reaction of nitrene with alkene, which is analogous to epoxide formation. Indeed, this has become a burgeoning area of interest in recent years.

4.9.1 Asymmetric Aziridination

Various approaches to epoxide also show promise for the preparation of chiral aziridines. Identification of the Cu(I) complex as the most effective catalyst for this process has raised the possibility that aziridination might share fundamental mechanistic features with olefin cyclopropanation.[115] Similar to cyclopropanation, in which the generally accepted mechanism involves a discrete Cu–carbenoid intermediate, copper-catalyzed aziridation might proceed via a discrete Cu–nitrenoid intermediate as well.

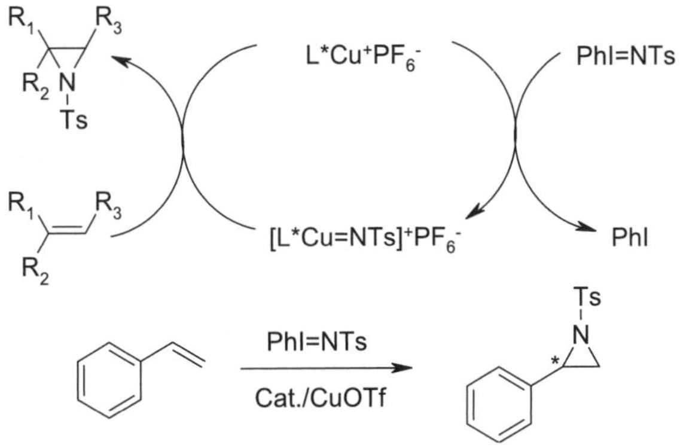

Scheme 4–59

TABLE 4–22. Asymmetric Aziridinations of Styrenes

Entry	Ligand	ee (%)	Reference
1	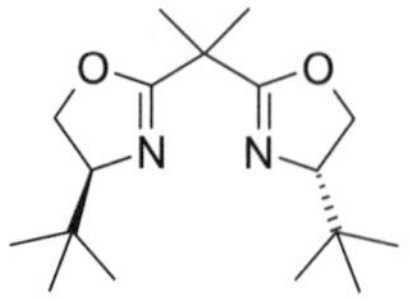	63	116
2	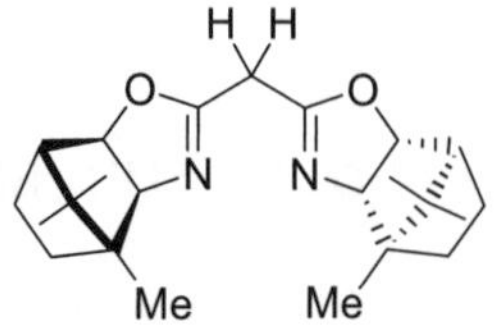	33	116
3	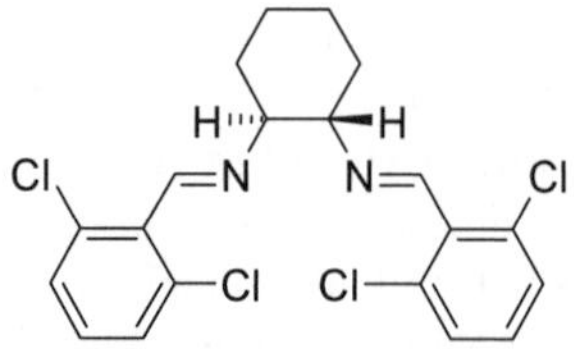	81	117

ee = Enantiomeric excess.

As the application of transition metal–salen complexes for asymmetric epoxidation has gained increasing recognition, chemists from many groups have also tried to use a salen complex for the asymmetric aziridination of alkenes.[118] For example, the chiral nitridomanganese complexes **156** and **157** were synthesized by treating the chiral ligands with $Mn(OAc)_2$, $NH_3 \cdot H_2O$ and NaOCl or reacting Mn(III) complex with gaseous NH_3 using chloramine-T as the oxidant in MeOH.

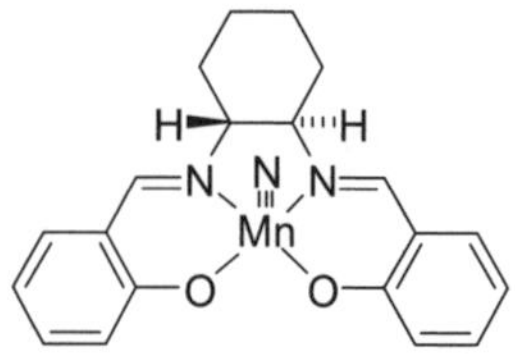

156

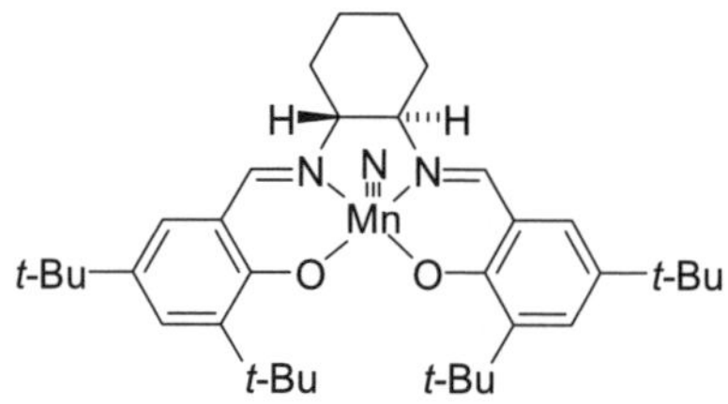

157

Both compounds were tested for their catalytic activity in asymmetric aziridination using p-toluenesulfonic anhydride (Ts_2O) to activate the nitridomanganese complex. As shown in Scheme 4–60, the aziridination generally gave poor results, while addition of pyridine N-oxide improved both the yield and the enantiomeric excess of the products.

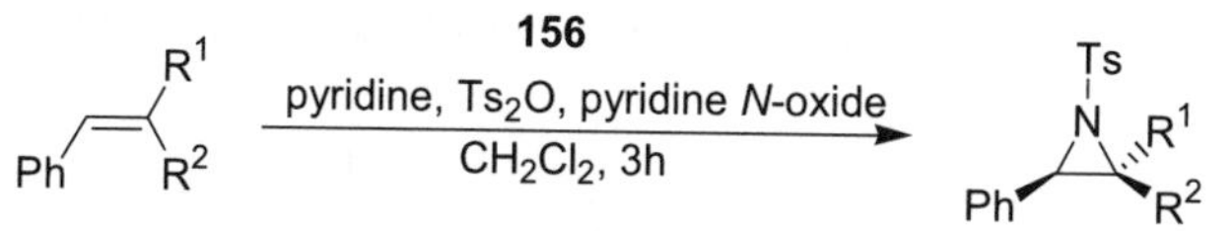

156

$$\underset{Ph}{\overset{R^1}{\diagdown}}\diagdown R^2 \xrightarrow[\text{CH}_2\text{Cl}_2,\ 3h]{\text{pyridine, Ts}_2\text{O, pyridine }N\text{-oxide}} \underset{Ph}{\overset{Ts}{\diagup}}\overset{N}{\diagdown}\underset{R^2}{\overset{R^1}{\triangle}}$$

Entry	Substrate		Additive	Temp. (°C)	Yield (%)	ee (%)
	R^1	R^2				
1	H	H	-	rt	63	31
2	H	H	pyridine N-oxide	0	78	41
3	Me	H	-	rt	60	73
4	Me	H	pyridine N-oxide	0	72	85
5	H	Me	-	rt	14	30
6	H	Me	pyridine N-oxide	rt	34	25
7	n-Pr	H	pyridine N-oxide	0	66	90
8	i-Pr	H	pyridine N-oxide	0	53	94

Scheme 4–60. Reprinted with permission by Wiley-VCH Verlag Germany, Ref. 118.

Jeong et al.[119] developed another procedure for the aziridination of olefins involving the inexpensive practical nitrogen source chloramine-T (TsNClNa). The reaction was not facilitated by any transition metal catalyst, but could proceed with good yield in the presence of inorganic bromides. They further found that using phenyltrimethylammonium bromide (PTAB) as the bromine source gave good to excellent yields of aziridines with a wide range of olefin substrates. As shown in Table 4–23, aziridination of olefins with TsNClNa gives much higher yield than the **156** catalyzed reaction.

Although the process as it stands is still not enantioselective in nature, the high yield and mild reaction conditions may attract further search for an asymmetric version of this reaction using proper chiral bromine-comtaining compounds as the catalyst.

Porphyrin complexes, which have been mentioned in the previous section as catalysts for the epoxidation of olefins, can also catalyze aziridination[120] using [N-(p-toluenesulfonyl)imino]phenyl iodinane or other nitrene precursors.

Asymmetric aziridination are further discussed elsewhere.[121]

4.9.2 Regioselective Ring Opening of Aziridines

The regiospecific reductive ring cleavage of N-substituted aziridines has been the means of many synthetic efforts for the asymmetric synthesis of amino derivatives. Just like the products from the regioselective and chemoselective ring opening of chiral epoxides, these amino derivatives can be the building blocks for biologically important compounds.

Bis(oxazoline)–copper complexes **158** have been used by Evans' group as chiral catalysts for the enantioselective aziridination of olefins.[116] Aryl-substituted olefins have been found to be particularly suitable substrates, which can be efficiently converted to N-tosylaziridines with ee of up to 97% (R = Ph

TABLE 4–23. Bromide Catalyzed Aziridination of Olefins with TsNClNa

Entry	Substrate	Product	Yield (%, isolated yield in bracket)
1			93 (90)
2			76 (62)
3			95 (88)
4			89 (72)
5			86 (80)
6			54
7			68 (65)
8			76 (60)
9			51

Reprinted with permission by Am. Chem. Soc., Ref. 119.

in **158** and $R' = $ COOPh in Eq. 1, Scheme 4–61). Reductive ring opening of **160** by transfer hydrogenation affords the corresponding 2-(R)-phenylalanine **161**, and acidic hydrolysis of **160** yields the β-hydroxy-α-amino ester **159** (Eq. 2 in Scheme 4–61).

The regioselective reductive ring opening of N-substituted aziridine-2-carboxylates gives either α- or β-amino acids.[122] Heteronucleophiles such as nitrogen,[123] oxygen,[124] sulfur,[125] and chloride preferably attack the C-3 position to provide α-amino acid derivatives. However, a mixture of products will result when carbanion or azide is used as the nucleophile (Scheme 4–62).

Lim and Lee[122] reported the regioselective reductive ring opening of N-substituted aziridine-2-carboxylates and aziridine-2-methanol via catalytic hydrogenation using Pd as a catalyst. For example, catalytic hydrogenation of **165** and **166** in AcOH with 20 mol% of Pd(OH)$_2$ proceeded with C-2 cleavage,

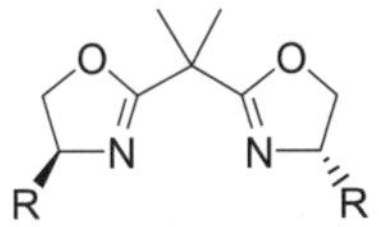

158

R = CHMe$_2$, Ph, Me$_3$C,
PhMe$_2$C, CMePh$_2$

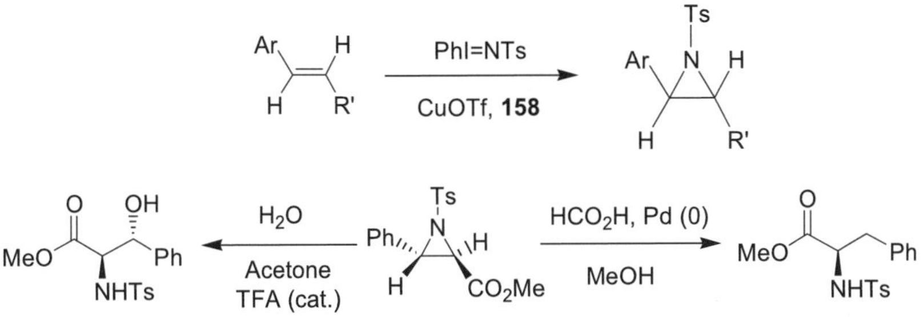

Eq. 1

Eq. 2

159   **160**   **161**

Scheme 4–61

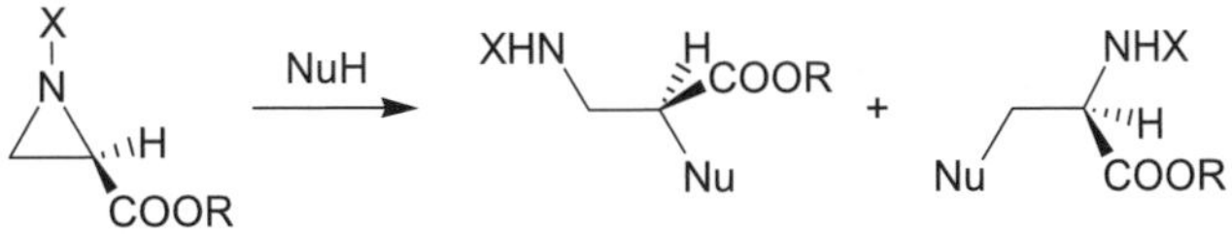

162   **163**   **164**

Scheme 4–62

yielding exclusively product **169**. The catalytic hydrogenation of **167** and **168** in EtOH with 20 mol% of Pd(OH)$_2$ gave the C-3 cleavage products **170** and **171** with yields over 95%.

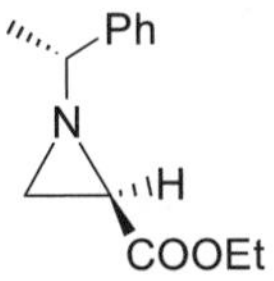

165 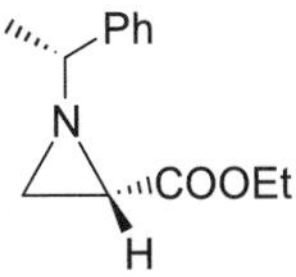   **166** 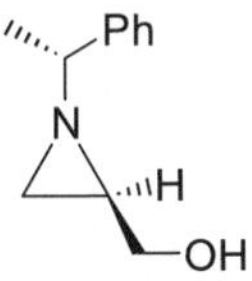   **167** 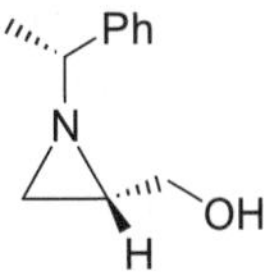   **168**

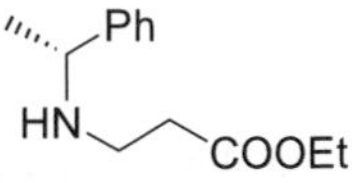

169 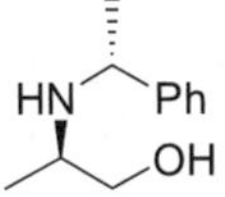   **170** 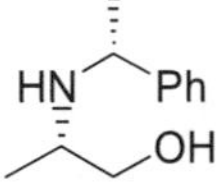   **171**

Osborn et al.[126] demonstrated the regiospecific ring opening of *N*-dimethyl phosphinoylaziridine by copper(I)-modified Grignard reagents with good to excellent yields (Scheme 4–63). The process was activated by the phosphinyl group. Either primary or secondary Grignard reagents are suitable nucleophiles, and the reaction is of general utility. When R = PhCH$_2$ (**172**), the yields of **173** for R$'$ = Me, *n*-Bu, and Ph are 67%, 89%, and 83%, respectively.

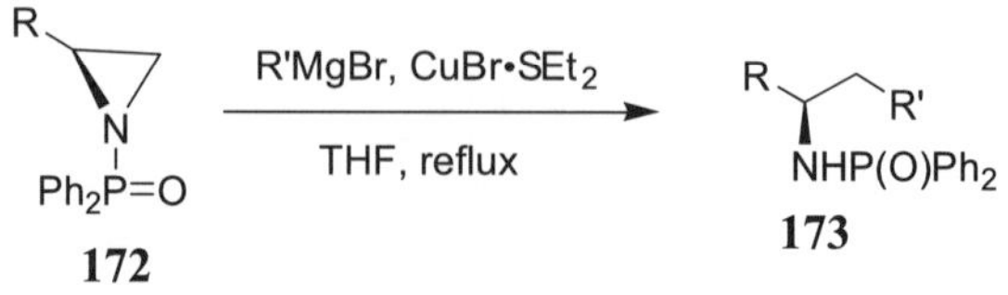

Scheme 4–63

Before closing this discussion of oxidation reactions, it is worth mentioning a demonstrated synthesis of α-aminoalkylphosphonate from vinylphosphonate via aziridinyl phosphonates, even though this reaction is thus far not asymmetric in nature.[127] In the presence of a copper catalyst, vinylphosphonates of type **174** were treated with PhI=NTs, followed by the reductive ring opening of aziridinylphosphonate **175** to afford α-aminoalkylphosphonate **176** in good yield (Scheme 4–64). α-Aminoalkylphosphonates are key substrates in the synthesis of phosphonopeptides.

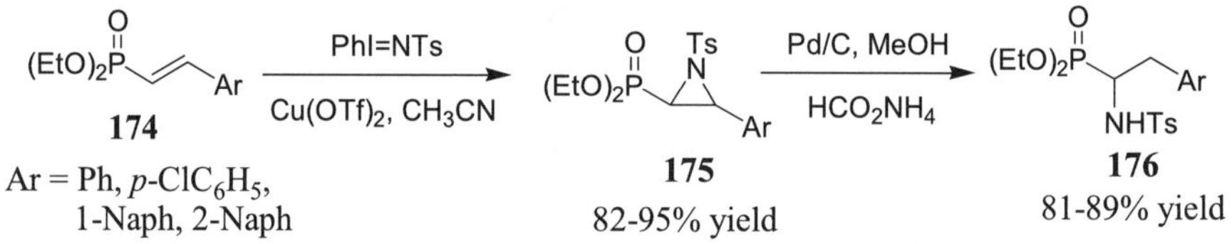

Scheme 4–64

4.10 SUMMARY

This chapter covers the asymmetric epoxidation of allylic alcohols as well as unfunctionalized olefins. Although Sharpless epoxidation is a useful tool for synthesizing 2,3-epoxy alcohols, it is not effective for the asymmetric epoxidation of unfunctionalized olefins. The Jacobsen/Katsuki metal–salen complex-promoted reaction provides a route to unfunctionalized epoxides, but this reaction generally requires a *cis*-substituted substrate bearing a π-stabilizing substituent. The chiral ketone-mediated epoxidation developed by Cao et al. and Yang et al. can be considered a breakthrough in asymmetric epoxidation of unfunctionalized olefins, and a wide range of olefins can be epoxidized with high enantioselectivity using this method. As a complement to these oxidation methods, sulfur ylide–mediated carbene addition to carbonyl groups has also been reported recently and shown some exciting results in the asymmetric synthesis of epoxide compounds from aldehydes.

Asymmetric epoxidation and the associated ring opening of epoxides, asymmetric dihydroxylation reactions, and asymmetric aminohydroxylation reactions have been extensively studied and widely used in the asymmetric synthesis of many intermediates or building blocks for important molecules. The study of its analogous aziridination reaction, however, is only at its primitive stage. Current efforts include seeking optimized reaction conditions, developing better chiral catalysts, and finding better nitrogen sources. These efforts are still ongoing, and their success will definitely open a new area for the asymmetric synthesis of amino group–containing compounds.

4.11 REFERENCES

1. (a) Ewins, R. C.; Henbest, H. B.; McKervey, M. A. *J. Chem. Soc. Chem. Commun.* **1967**, 1085. (b) Pirkle, W. H.; Rinaldi, P. L. *J. Org. Chem.* **1977**, *42*, 2080.

2. Juliá, S.; Masana, J.; Vega, J. C. *Angew. Chem. Int. Ed. Engl.* **1980**, *19*, 929.

3. (a) Davis, F. A.; Harakal, M. E.; Awad, S. B. *J. Am. Chem. Soc.* **1983**, *105*, 3123. (b) Davis, F. A.; Haque, M. S. *J. Org. Chem.* **1986**, *51*, 4083. (c) Davis, F. A.; Towson, J. C.; Weismiller, M. C.; Lal, S.; Carroll, P. J. *J. Am. Chem. Soc.* **1988**, *110*, 8477.

4. Yamada, S.; Mashiko, T.; Terashima, S. *J. Am. Chem. Soc.* **1977**, *99*, 1988.

5. Kagan, H. B.; Mimoun, H.; Mark, C.; Schurig, V. *Angew. Chem. Int. Ed. Engl.* **1979**, *18*, 485.

6. Michaelson, R. C.; Palermo, R. E.; Sharpless, K. B. *J. Am. Chem. Soc.* **1977**, *99*, 1990.

7. Katsuki, T.; Sharpless, K. B. *J. Am. Chem. Soc.* **1980**, *102*, 5974.

8. Martin, V. S.; Woodard, S. S.; Katsuki, T.; Yamada, Y.; Ikeda, M.; Sharpless, K. B. *J. Am. Chem. Soc.* **1981**, *103*, 6237.

9. Woodard, S. S.; Finn, M. G.; Sharpless, K. B. *J. Am. Chem. Soc.* **1991**, *113*, 106.

10. (a) Finn, M. G.; Sharpless, K. B. *J. Am. Chem. Soc.* **1991**, *113*, 113. (b) Potvin, P. G.; Bianchet, S. *J. Org. Chem.* **1992**, *57*, 6629.

11. Corey, E. J. *J. Org. Chem.* **1990**, *55*, 1693.

12. (a) Wang, Z. M.; Zhou, W. S.; Lin, G. Q. *Tetrahedron Lett.* **1985**, *26*, 6221. (b) Wang, Z. M.; Zhou, W. S. *Synth. Commun.* **1989**, *19*, 2627.

13. Zhou, W. S.; Lu, Z.; Wang, Z. M. *Tetrahedron Lett.* **1991**, *32*, 1467.

14. Zhou, W. Wei, D. *Tetrahedron Asymmetry* **1991**, *2*, 767.

15. Gao, Y.; Hanson, R. M.; Klunder, J. M.; Ko, S. Y.; Masamune, H.; Sharpless, K. B. *J. Am. Chem. Soc.* **1987**, *109*, 5765.

16. Kragl, U.; Dreisbach, C. *Angew. Chem. Int. Ed. Engl.* **1996**, *35*, 642.

17. Canali, L.; Karjalainen, J. K.; Sherrington, D. C.; Hormi, O. *J. Chem. Soc. Chem. Commun.* **1997**, 123.

18. (a) Karjalainen, J. K.; Hormi, O.; Sherrington, D. C. *Tetrahedron Asymmetry* **1998**, *9*, 1563. (b) Karjalainen, J. K.; Hormi, O.; Sherrington, D. C. *Tetrahedron Asymmetry* **1998**, *9*, 2019.

19. (a) Rossiter, B. E.; Sharpless, K. B. *J. Org. Chem.* **1984**, *49*, 3707. (b) Ikegami, S.; Katsuki, T.; Yamaguchi, M. *Chem. Lett.* **1987**, 83.

20. Karjalainen, J. K.; Hormi, O.; Sherrington, D. C. *Tetrahedron Asymmetry* **1998**, *9*, 3895.

21. Caron, M.; Sharpless, K. B. *J. Org. Chem.* **1985**, *50*, 1557.

22. Lin, G. Q.; Zeng, C. M. *Chin. J. Chem.* **1991**, *9*, 381.

23. Caron, M.; Carlier, P. R.; Sharpless, K. B. *J. Org. Chem.* **1988**, *53*, 5185.

24. Ti(OPr^i)$_2$(N$_3$)$_2$ can be prepared in situ by stirring a mixture of Ti(OPri)$_4$ + 2 equivalents of Me$_3$SiN$_3$ in benzene under Ar or N$_2$ for 5 hours until the solution becomes clear. A C-3 selective azide reagent consisting of NaN$_3$ supported on a calcium zeolite has also been reported. Onaka, M.; Sugita, K.; Izumi, Y. *Chem. Lett.* **1986**, 1327.

25. Benedetti, F.; Berti, F.; Norbedo, S. *Tetrahedron Lett.* **1998**, *39*, 7971.

26. Alvarez, E.; Nuñez, M. T.; Martín, V. S. *J. Org. Chem.* **1990**, *55*, 3429.

27. Jung, M. E.; Jung, Y. H. *Tetrahedron Lett.* **1989**, *30*, 6637.

28. Bernet, B.; Vasella, A. *Tetrahedron Lett.* **1983**, *24*, 5491.

29. (a) Schmidt, R. R.; Kläger, R. *Angew. Chem. Int. Ed. Engl.* **1985**, *24*, 65. (b) Tkaczuk, P.; Thornton, E. R. *J. Org. Chem.* **1981**, *46*, 4393.

30. Rama Rao, A. V.; Murali Dhar, T. G.; Chakraborty, T. K.; Gurjar, M. K. *Tetrahedron Lett.* **1988**, *29*, 2069.

31. (a) Sharpless, K. B. *J. Org. Chem.* **1982**, *47*, 1378. (b) Viti, M. S. *Tetrahedron Lett.* **1982**, *23*, 4541.

32. Finan, J. M.; Kishi, Y. *Tetrahedron Lett.* **1982**, *23*, 2719.

33. Dai, L.; Lou, B.; Zhang, Y.; Guo, G. *Tetrahedron Lett.* **1986**, *27*, 4343.

34. Sajiki, H.; Hattori, K.; Hirota, K. *Chem. Commun.* **1999**, 1041.

35. (a) Johnson, M. R.; Nakata, T.; Kishi, Y. *Tetrahedron Lett.* **1979**, 4343. (b) Wood, R. D.; Ganem, B. *Tetrahedron Lett.* **1982**, *23*, 707.

36. Payne, G. B. *J. Org. Chem.* **1962**, *27*, 3819.

37. Katsuki, T.; Lee, A. W. M.; Ma, P.; Martin, V. S.; Sharpless, K. B.; Tuddenham, D.; Walker, F. J. *J. Org. Chem.* **1982**, *47*, 1373.

38. Masamune, S. *J. Org. Chem.* **1982**, *47*, 1375.

39. (a) Paterson, I.; Berrisford, D. J. *Angew. Chem. Int. Ed. Engl.* **1992**, *31*, 1179. (b) Södergren, M. J.; Andersson, P. G. *J. Am. Chem. Soc.* **1998**, *120*, 10760.

40. Iida, T.; Yamamoto, N.; Sasai, H.; Shibasaki, M. *J. Am. Chem. Soc.* **1997**, *119*, 4783.

41. Iida, T.; Yamamoto, N.; Matsunaga, S.; Woo, H.; Shibasaki, M. *Angew. Chem. Int. Ed. Engl.* **1998**, *37*, 2223.

42. Vogl, E. M.; Matsunaga, S.; Kanai, M.; Iida, T.; Shibasaki, M. *Tetrahedron Lett.* **1998**, *39*, 7917.

43. Wu, J.; Hou, X.; Dai, L.; Xia, L.; Tang, M. *Tetrahedron Asymmetry* **1998**, *9*, 3431.

44. Schaus, S. E.; Larrow, J. F.; Jacobsen, E. N. *J. Org. Chem.* **1997**, *62*, 4197.

45. (a) Martinez, L. E.; Leighton, J. L.; Carsten, D. H.; Jacobsen, E. N. *J. Am. Chem. Soc.* **1995**, *117*, 5897. (b) Leighton, J. L.; Jacobsen, E. N. *J. Org. Chem.* **1996**, *61*, 389.

46. Konsler, R. G.; Karl, J.; Jacobsen, E. N. *J. Am. Chem. Soc.* **1998**, *120*, 10780.

47. (a) Häfele, B.; Schröter D.; Jäger, V. *Angew. Chem. Int. Ed. Engl.* **1986**, *25*, 87. (b) Schreiber, S. L.; Schreiber, T. S.; Smith, D. B. *J. Am. Chem. Soc.* **1987**, *109*, 1525.

48. Okamoto, S.; Kobayashi, Y.; Kato, H.; Hori, K.; Takahashi, T.; Tsuji J.; Sato, F. *J. Org. Chem.* **1988**, *53*, 5590.

49. Hatakeyama, S.; Sakurai, K.; Takano, S. *Tetrahedron Lett.* **1986**, *27*, 4485.

50. Kobayashi, Y.; Kato, N.; Shimazaki, T.; Sato, F. *Tetrahedron Lett.* **1988**, *29*, 6297.

51. For a review, see Kolb, H. C.; VanNieuwenhze, M. S.; Sharpless, K. B. *Chem. Rev.* **1994**, *94*, 2483.

52. Hofmann, K. A. *Chem. Ber.* **1912**, *45*, 3329.

53. Criegee, R.; Marchand, B.; Wannowius, H. *Justus Liebigs. Ann. Chem.* **1942**, *550*, 99.

54. Hentges, S. G.; Sharpless, K. B. *J. Am. Chem. Soc.* **1980**, *102*, 4263.

55. Jacobsen, E. N.; Marko, I.; Mungall, W. S.; Schröder, G.; Sharpless, K. B. *J. Am. Chem. Soc.* **1988**, *110*, 1968.

56. Lohray, B. B.; Kalantar, T. H.; Kim, B. M.; Park, C. Y.; Shibata, T.; Wai, J. S. M.; Sharpless, K. B. *Tetrahedron Lett.* **1989**, *30*, 2041.

57. Sharpless, K. B.; Amberg, W.; Bennani, Y. L.; Crispino, G. A.; Hartung, J.; Jeong, K. S.; Kwong, H.; Morikawa, K.; Wang, Z.; Xu, D.; Zhang, X. *J. Org. Chem.* **1992**, *57*, 2768.

58. Crispino, G. A.; Jeong, K.; Kolb, H. C.; Wang, Z.; Xu, D.; Sharpless, K. B. *J. Org. Chem.* **1993**, *58*, 3785.

59. Becker, H.; Sharpless, K. B. *Angew. Chem. Int. Ed. Engl.* **1996**, *35*, 448.

60. Lohray, B. B.; Bhushan, V. *Tetrahedron Lett.* **1992**, *33*, 5113.

61. Lohray, B. B. *Tetrahedron Asymmetry* **1992**, *3*, 1317.

62. Yamada, T.; Narasaka, K. *Chem. Lett.* **1986**, 131.

63. Tokles, M.; Snyder, J. K. *Tetrahedron Lett.* **1986**, *27*, 3951.

64. Tomioka, K.; Nakajima, M.; Koga, K. *J. Am. Chem. Soc.* **1987**, *109*, 6213.

65. Hirama, M.; Oishi, T.; Ito, S. *J. Chem. Soc. Chem. Commun.* **1989**, 665.

66. Corey, E. J.; Jardine, P. D.; Virgil, S.; Yuen, P. W.; Connel, R. D. *J. Am. Chem. Soc.* **1989**, *111*, 9243.

67. (a) Shibata, T.; Gilheany, D. G.; Blackburn, B. K.; Sharpless, K. B. *Tetrahedron Lett.* **1990**, *31*, 3817. (b) Sharpless, K. B.; Amberg, W.; Beller, M.; Chen, H.; Hartung, J.; Kawanami, Y.; Lübben, D.; Manoury, E.; Ogino, Y.; Shibata, T.; Ukita, T. *J. Org. Chem.* **1991**, *56*, 4585.

68. Oishi, T.; Hirama, M. *Tetrahedron Lett.* **1992**, *33*, 639.

69. Hanessian, S.; Meffre, P.; Girard, M.; Beaudoin, S.; Sanceau, J.; Bennani, Y. *J. Org. Chem.* **1993**, *58*, 1991.

70. Wang, L.; Sharpless, K. B. *J. Am. Chem. Soc.* **1992**, *114*, 7568.

71. Oishi, T.; Hirama, M. *J. Org. Chem.* **1989**, *54*, 5834.

72. Turpin, J. A.; Weigel, L. O. *Tetrahedron Lett.* **1992**, *33*, 6563.

73. Wang, Z.; Zhang, X.; Sharpless, K. B. *Tetrahedron Lett.* **1993**, *34*, 2267.

74. Tomioka, K.; Nakajima, M.; Koga, K. *J. Chem. Soc. Chem. Commun.* **1989**, 1921.

75. (a) Sharpless, K. B.; Chong, A. O.; Oshima, K. *J. Org. Chem.* **1976**, *41*, 177. (b) Herranz, E.; Sharpless, K. B. *J. Org. Chem.* **1978**, *43*, 2544.

76. Bäckvall, J. E. *Tetrahedron Lett.* **1975**, 2225.

77. (a) Li, G.; Chang, H. T.; Sharpless, K. B. *Angew Chem. Int. Ed. Engl.* **1996**, *35*, 451. (b) Rudolph, J.; Sennhenn, P. C.; Vlaar, C. P.; Sharpless, K. B. *Angew. Chem. Int. Ed. Engl.* **1996**, *35*, 2810.

78. Reiser, O. *Angew. Chem. Int. Ed. Engl.* **1996**, *35*, 1308.

79. (a) Li, G.; Angert, H. H.; Sharpless, K. B. *Angew. Chem. Int. Ed. Engl.* **1996**, *35*, 2813. (b) Reddy, K. L.; Sharpless, K. B. *J. Am. Chem. Soc.* **1998**, *120*, 1207. (c) O'Brien, P.; Osborne, S. A.: Parker, D. D. *J. Chem. Soc. Perkin Trans. 1* **1998**, 2519.

80. O'Brien, P.; Osborne, S. A.; Parker, D. D. *Tetrahedron Lett.* **1998**, *39*, 4099.

81. Reddy, K. L.; Dress, K. R.; Sharpless, K. B. *Tetrahedron Lett.* **1998**, *39*, 3667.

82. (a) Bruncko, M.; Schlingloff, G.; Sharpless, K.B. *Angew. Chem. Int. Ed. Engl.* **1997**, *36*, 1483. (b) Li, G.; Sharpless, K. B. *Acta Chem. Scand.* **1996**, *50*, 649. (c) Angelaud, R.; Landais, Y.; Schenk, K. *Tetrahedron Lett.* **1997**, *38*, 1407. (d) Upadhya, T. T.; Sudalai, A. *Tetrahedron Asymmetry* **1997**, *8*, 3685. (e) Han, H.; Yoon, J.; Janda, K. D. *J. Org. Chem.* **1998**, *63*, 2045. (f) Phukan, P.; Sudalai, A. *Tetrahedron Asymmetry* **1998**, *9*, 1001. (g) Cravotto, G.; Giovenzana, G. B.; Pagliarin, R.; Palmisano, G.; Sisti, M. *Tetrahedron Asymmetry* **1998**, *9*, 745.

83. (a) O'Brien, P. *Angew. Chem. Int. Ed. Engl.* **1999**, *38*, 326. (b) Goossen, L. J.; Liu, H.; Dress, K. R.; Sharpless, K. B. *Angew. Chem. Int. Ed. Engl.* **1999**, *38*, 1080.

84. Zhang, W.; Leobach, J. L.; Wilson, S. R.; Jacobsen E. N. *J. Am. Chem. Soc.* **1990**, *112*, 2801.

85. Irie, R.; Noda, K.; Ito, Y.; Matsumoto, N.; Katsuki, T. *Tetrahedron Lett.* **1990**, *31*, 7345.

86. (a) Lee, N. H.; Muci, A. R.; Jacobsen, E. N. *Tetrahedron Lett.* **1991**, *32*, 5055. (b) Deng, L.; Jacobsen, E. N. *J. Org. Chem.* **1992**, *57*, 4320.

87. (a) Chang, S.; Lee, N. H.; Jacobsen, E. N. *J. Org. Chem.* **1993**, *58*, 6939. (b) Chang, S.; Galvin, J. M.; Jacobsen, E. N. *J. Am. Chem. Soc.* **1994**, *116*, 6937.

88. Brandes, B. D.; Jacobsen, E. N. *J. Org. Chem.* **1994**, *59*, 4378.

89. Palucki, M.; Pospisil, P. J.; Zhang, W.; Jacobsen, E. N. *J. Am. Chem. Soc.* **1994**, *116*, 9333.

90. Brandes, B. D.; Jacobsen, E. N. *Tetrahedron Lett.* **1995**, *36*, 5123.

91. Bolm, C. *Angew. Chem. Int. Ed. Engl.* **1991**, *30*, 403.

92. Hosoya, N.; Irie, R.; Katsuki, T. *Synlett* **1993**, 261.

93. (a) Jacobsen, E. N.; Zhang, W.; Güler, M. L. *J. Am. Chem. Soc.* **1991**, *113*, 6703. (b) Jacobsen, E. N.; Zhang, W.; Muci, A. R.; Ecker, J. R.; Deng, L. *J. Am. Chem. Soc.* **1991**, *113*, 7063. (c) Irie, R.; Noda, K.; Ito, Y.; Katsuki, T. *Tetrahedron Lett.* **1991**, *32*, 1055. (d) Chang, S.; Heid, R. M.; Jacobsen, E. N. *Tetrahedron Lett.* **1994**, *35*, 669. (e) Rychnovsky, S. D.; Hwang, K. *Tetrahedron Lett.* **1994**, *35*, 8927. (f) Pietikäinen, P. *Tetrahedron Lett.* **1995**, *36*, 319. (g) Zhang, W.; Jacobsen, E. N. *J. Org. Chem.* **1991**, *56*, 2296. (h) Larrow, J. F.; Jacobsen, E. N.; Gao, Y.; Hong,

Y.; Nie, X.; Zepp, C. M. *J. Org. Chem.* **1994**, *59*, 1939. (i) Irie, R.; Noda, K.; Ito, Y.; Matsumoto, N.; Katsuki, T. *Tetrahedron Asymmetry* **1991**, *2*, 481. (j) Collman, J. P.; Zhang, X.; Lee, V. J.; Uffelman, E. S.; Brauman, J. I. *Science* **1993**, *261*, 1404. (k) Jacobsen, E. N.; Deng, L.; Furukawa, Y.; Martinez, L. E. *Tetrahedron* **1994**, *50*, 4323.

94. Ito, Y. N.; Katsuki, T. *Tetrahedron Lett.* **1998**, *39*, 4325.

95. Palucki, M.; Finney, N. S.; Pospisil, P. J.; Güler, M. L.; Ishida, T.; Jacobsen, E. N. *J. Am. Chem. Soc.* **1998**, *120*, 948.

96. (a) Groves, J. T.; Viski, P. *J. Org. Chem.* **1990**, *55*, 3628. (b) Groves, J. T.; Myers, R. S. *J. Am. Chem. Soc.* **1983**, *105*, 5971. (c) O'Malley, S.; Kodadek, T. *J. Am. Chem. Soc.* **1989**, *111*, 9116. (d) Collman, J. P.; Lee, V. J.; Zhang, X.; Ibers, J. A.; Brauman, J. I. *J. Am. Chem. Soc.* **1993**, *115*, 3834. (e) Mansuy, D.; Battioni, P.; Renaud, J.; Guerin, P. *J. Chem. Soc. Chem. Commun.* **1985**, 155. (f) Collman, J. P.; Zhang, X.; Lee, V. J.; Brauman, J. I.; *J. Chem. Soc. Chem. Commun.* **1992**, 1647. (g) Maillard, Ph.; Guerquin-Kern, J. L.; Momenteau, M. *Tetrahedron Lett.* **1991**, *32*, 4901. (h) Che, C.-M.; Yu, W.-Y. *Pure Appl. Chem.* **1999**, *71*, 281.

97. (a) Konishi, K.; Sugino, T.; Aida, T.; Inoue, S. *J. Am. Chem. Soc.* **1991**, *113*, 6487. (b) Chiang, L.; Konishi, K.; Aida, T.; Inoue, S. *J. Chem. Soc. Chem. Commun.* **1992**, 254. (c) Konishi, K.; Oda, K.; Nishida, K.; Aida, T.; Inoue, S. *J. Am. Chem. Soc.* **1992**, *114*, 1313.

98. (a) Naruta, Y.; Tani, F.; Maruyama, K. *Chem. Lett.* **1989**, 1269. (b) Naruta, Y.; Tani, F.; Ishihara, N.; Maruyama, K. *J. Am. Chem. Soc.* **1991**, *113*, 6865. (c) Naruta, Y.; Ishihara, N.; Tani, F.; Maruyama, K. *Chem. Lett.* **1991**, 1933. (d) Naruta, Y.; Tani, F.; Maruyama, K. *Tetrahedron Lett.* **1992**, *33*, 6323. (e) Naruta, Y.; Ishihara, N.; Tani, F.; Maruyama, K. *Bull. Chem. Soc. Jpn.* **1993**, *66*, 158.

99. Collman, J. P.; Wang, Z.; Straumanis, A.; Quelquejeu, M. *J. Am. Chem. Soc.* **1999**, *121*, 460.

100. Tu, Y.; Wang, Z.; Shi, Y. *J. Am. Chem. Soc.* **1996**, *118*, 9806.

101. Frohn, M.; Dalkiewicz, M.; Tu, Y.; Wang, Z.; Shi, Y. *J. Org. Chem.* **1998**, *63*, 2948.

102. Cao, G.; Wang, Z.; Tu, Y.; Shi, Y. *Tetrahedron Lett.* **1998**, *38*, 4425.

103. Wang, Z.; Shi, Y. *J. Org. Chem.* **1998**, *63*, 3099.

104. Miaskiewicz, K.; Smith, D. A. *J. Am. Chem. Soc.* **1998**, *120*, 1872.

105. (a) Yang, D.; Wang, X.; Wong, M.; Yip, Y.; Tang, M. *J. Am. Chem. Soc.* **1996**, *118*, 11311. (b) Yang, D.; Yip, Y.; Tang, M.; Wong, M.; Zheng, J.; Cheung, K. *J. Am. Chem. Soc.* **1996**, *118*, 491.

106. Yang, D.; Wong, M.; Yip, Y.; Wang, X.; Tang, M.; Zheng, J.; Cheung, K. *J. Am. Chem. Soc.* **1998**, *120*, 5943.

107. For a review, see Li, A.; Dai, L.; Aggarwal, V. K. *Chem. Rev.* **1997**, *97*, 2341.

108. Aggarwal, V. K.; Ford, J. G.; Fonquerna, S.; Adams, H.; Jones, R. V. H.; Fieldhouse, R. *J. Am. Chem. Soc.* **1998**, *120*, 8328.

109. (a) Evans, D. A.; Takacs, J. M. *Tetrahedron Lett.* **1980**, *21*, 4233. (b) Davis, F. A.; Sheppard, A. C. *Tetrahedron* **1989**, *45*, 5703.

110. Evans, D. A.; Morrissey, M. M.; Dorow, R. L. *J. Am. Chem. Soc.* **1985**, *107*, 4346.

111. Davis, F. A.; Haque, M. S.; Ulatowski, T. G.; Towson, J. C. *J. Org. Chem.* **1986**, *51*, 2402.

112. (a) Davis, F. A.; Reddy, R. T.; Han, W.; Reddy, R. E. *Pure Appl. Chem.* **1993**, *65*, 633. (b) Davis, F. A.; Weismiller, M. C. *J. Org. Chem.* **1990**, *55*, 3715. (c) Davis, F. A.; Haque, M. S. *J. Org. Chem.* **1986**, *51*, 4083.

113. Zhu, Y.; Tu, Y.; Yu, H. W.; Shi, Y. *Tetrahedron Lett.* **1998**, *39*, 7819.

114. Tanner, D. *Angew. Chem. Int. Ed. Engl.* **1994**, *33*, 599.

115. Evans, D. A.; Woerpel, K. A.; Hinman, M. M.; Faul, M. M. *J. Am. Chem. Soc.* **1991**, *113*, 726.

116. Evans, D. A.; Faul, M. M.; Bilodeau, M. T.; Anderson, B. A.; Barnes, D. M. *J. Am. Chem. Soc.* **1993**, *115*, 5328.

117. Li, Z.; Conser, K. R.; Jacobsen, E. N. *J. Am. Chem. Soc.* **1993**, *115*, 5326.

118. Minakata, S.; Ando, T.; Nishimura, M.; Ryu I.; Komatsu, M. *Angew. Chem. Int. Ed. Engl.* **1998**, *37*, 3392.

119. Jeong, J. U.; Tao, B.; Sagasser, I.; Henniges, H.; Sharpless, K. B. *J. Am. Chem. Soc.* **1998**, *120*, 6844.

120. (a) Groves, J. T.; Takahashi, T. *J. Am. Chem. Soc.* **1983**, *105*, 2073. (b) Mansuy, D.; Battioni, P.; Mahy, J. *J. Am. Chem. Soc.* **1982**, *104*, 4487. (c) Mahy, J.; Battioni, P.; Mansuy, D. *J. Am. Chem. Soc.* **1986**, *108*, 1079.

121. (a) Noda, K.; Hosoya, N.; Irie, R.; Ito, Y.; Katsuki, T. *Synlett* **1993**, 469. (b) Martres, M.; Gili, G.; Meou, A. *Tetrahedron Lett.* **1994**, *35*, 8787. (c) Tanner, D.; Andersson, P. G.; Harden, A.; Somfai, P. *Tetrahedron Lett.* **1994**, *35*, 4631. (d) Zhang, W.; Lee, N. H.; Jacobsen, E. N. *J. Am. Chem. Soc.* **1994**, *116*, 425. (e) Li, Z.; Quan, R. W.; Jacobsen, E. N. *J. Am. Chem. Soc.* **1995**, *117*, 5889. (f) Nishikori, H.; Katsuki, T. *Tetrahedron Lett.* **1996**, *37*, 9245. (g) Lowenthal, R. E.; Masamune, S. *Tetrahedron Lett.* **1991**, *32*, 7373. (h) Harm, A. M.; Knight, J. G.; Stemp, G. *Synlett* **1996**, *677*. (i) Lai, T.; Kwong, H.; Che, C.; Peng, S. *J. Chem. Soc. Chem. Commun.* **1997**, 2373. (j) Södergren, M. J.; Alonso, D. A.; Andersson, P. G. *Tetrahedron Asymmetry* **1997**, *8*, 3563. (k) Atkinson, R. S.; Gattrell, W. T.; Ayscough, A. P.; Raynham, T. M. *J. Chem. Soc. Chem. Commun.* **1996**, 1935.

122. Lim, Y.; Lee, W. K. *Tetrahedron Lett.* **1995**, *36*, 8431.

123. Nakajima, K.; Tanaka, T.; Morita, K.; Okawa, K. *Bull. Chem. Soc. Jpn.* **1980**, *53*, 283.

124. Wipf, P.; Venkatraman, S.; Miller, C. P. *Tetrahedron Lett.* **1995**, *36*, 3639.

125. (a) Ploux, O.; Caruso, M.; Chassaing, G.; Marquet, A. *J. Org. Chem.* **1988**, *53*, 3154. (b) Nakajima, K.; Oda, H.; Okawa, K. *Bull. Chem. Soc. Jpn.* **1983**, *56*, 520.

126. Osborn, H.; Sweeney, J. B.; Howson, W. *Tetrahedron Lett.* **1994**, *35*, 2739.

127. Kim, D. Y.; Rhie, D. Y. *Tetrahedron* **1997**, *53*, 13603.

CHAPTER 5

Asymmetric Diels-Alder and Other Cyclization Reactions

The Diels-Alder reaction is a powerful synthetic process for constructing complex molecules. The reaction has been extensively studied and refined since its discovery in 1928.[1] The most attractive feature of the Diels-Alder reaction is its simultaneous, regioselective construction of two bonds, resulting in the creation of up to four chiral centers with largely predictable relative stereochemistry at the bond formation sites. Theoretically, there are a total of $2^4 = 16$ stereoisomers when atoms marked with an asterisk are all chiral centers (Scheme 5–1); therefore, the complete control of the reaction process to obtain enantiomerically pure products has been the object of active research in many laboratories.

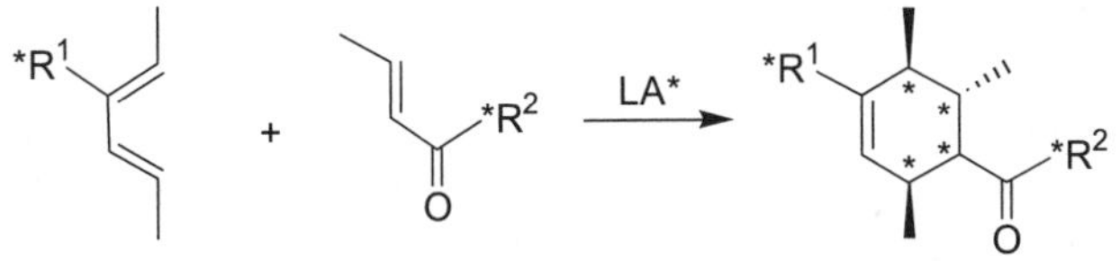

Scheme 5–1

In addition to the *syn*-facial addition of the reaction, considerable advances have been made in achieving asymmetric induction through the following three methods: (1) attaching chiral auxiliaries to dienophiles, such as R^{*2} in Scheme 5–1; (2) attaching a chiral auxiliary to the diene, such as R^{*1} in Scheme 5–1; and (3) employing a chiral catalyst, usually a Lewis acid, such as LA* in Scheme 5–1. The first and the second approaches have been the most commonly employed method for achieving asymmetric induction in the Diels-Alder reaction during the past decade. However, applying chiral catalytic Lewis acids has shown widespread utility, with several excellent catalysts readily available. Indeed, the search for efficient chiral Lewis acids has been the prevailing issue in the study of asymmetric Diels-Alder reactions.

This chapter focuses on some typical examples, starting with the usual cycloaddition reactions and then the catalytic asymmetric Diels-Alder reactions, hetero Diels-Alder reactions, retro Diels-Alder reactions, and intramolecular

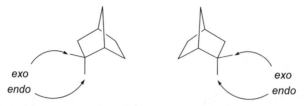

Figure 5–1. Definition of *endo*- and *exo*-substituents in a bicyclic system.

Diels-Alder reactions. In the last two sections, the asymmetric 1,3-dipolar reaction [2+3] and cyclopropanation reactions [1+2] are discussed.

Before discussing asymmetric cycloaddition reactions, it is necessary to introduce the concepts of *exo* and *endo*. These are the stereochemical prefixes that describe the relative configurations of a substituent on a bridged cyclic compound. (Note that they are applied only to the configuration of substituents not on the bridge head.) Given a molecule where the two bridges that do not contain the substituent are of unequal length, the prefix *endo* refers the substituent that is closer to the longer of the two unsubstituted bridges; and the prefix *exo* refers to the substituent that is closer to the shorter bridge, as depicted in Figure 5–1.

5.1 CHIRAL DIENOPHILES

Chiral dienophiles comprise the majority of asymmetric Diels-Alder reactions. As the most common chiral dienophiles, acrylates have been classified into three categories, types I, II, and III (see Fig. 5–2). Type I reagents are chiral acrylates that incorporate the chiral group in a simple and straightforward manner. Type II reagents are those in which the chiral group is, in comparison with type I, one atom closer to the double bond. This type of compound typically requires more complex synthesis and the subsequent removal of the stereogenic center present in the compound. Furthermore, the recycling of the chiral group may be cumbersome. Type III reagents are acrylamide compounds bearing a chiral auxiliary connected via an amide linkage. This type of reagent exhibits high activity due to the positive electronic effect at the nitrogen atom of the corresponding iminium salt.

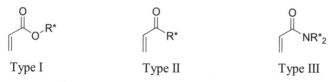

Figure 5–2. Three types of chiral dienophiles.

5.1.1 Acrylate

In general, compared with uncatalyzed cycloaddition reactions that may require a high temperature or high pressure, Lewis acid–catalyzed reactions can proceed at significantly lower temperatures with high selectivity. Factors that improve the selectivity of catalyzed reactions generally include low temperature and a more organized coordination transition state between the Lewis acid and the substrate. Scheme 5–2 shows that, in the presence of a Lewis acid catalyst, chiral dienophiles **1–4** undergo a Diels-Alder reaction with a stereoselectivity as high as 200:1 or even better.[2]

For acrylates, or type I reagents, applied in asymmetric Diels-Alder reactions, several chiral auxiliaries such as menthol derivatives, camphor derivatives,[1e,3] and oxazolidinones[4] are available. Carbohydrate compounds have also been reported as chiral auxiliaries in a recent publication, although the stereoselectivity was not good.[5] Here are examples in which asymmetric Diels-

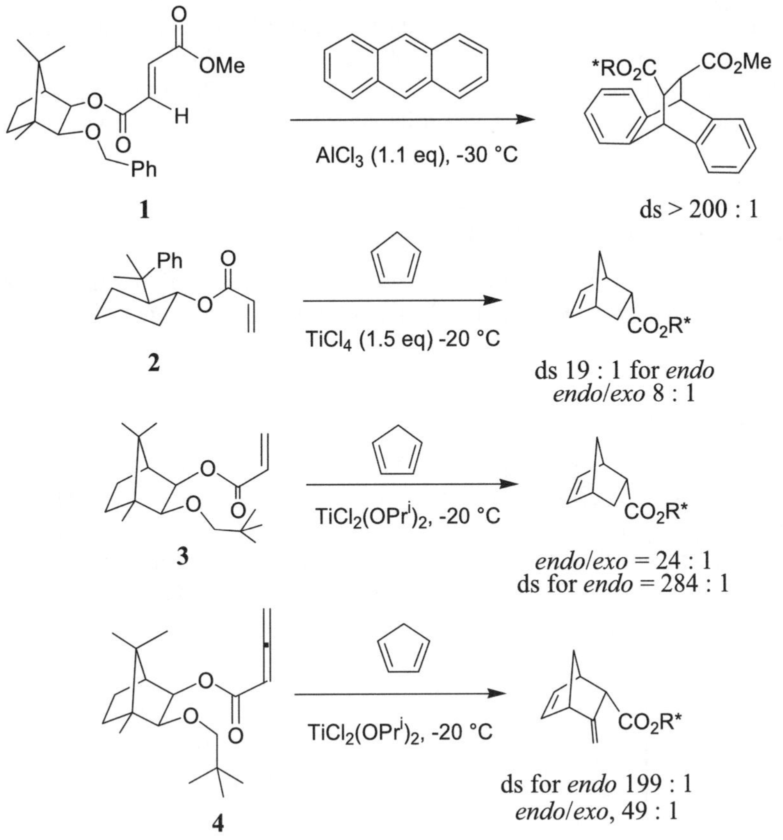

Scheme 5–2

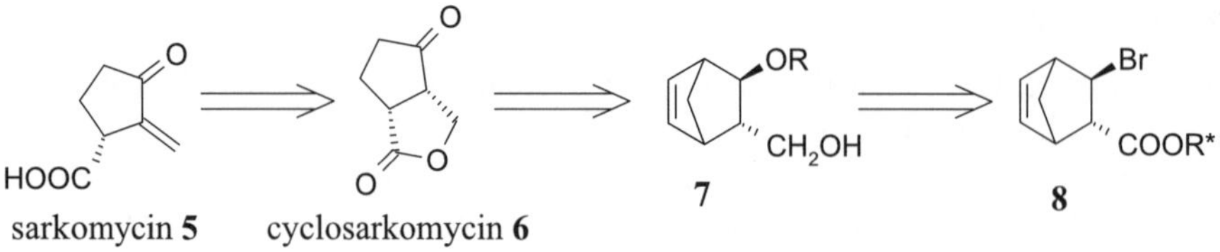

sarkomycin **5**   cyclosarkomycin **6**

Scheme 5–3 Retro synthetic analysis of the synthesis of sarkomycin.

Alder reactions involving type I reagents serve as a key step in the syntheses of important bioactive compounds.

The antibiotic sarkomycin (**5**), an antitumor agent, is a good target for synthetic efforts based on a Diels-Alder reaction. The retro synthesis of this compound is depicted in Scheme 5–3.

Linz et al.[6] report the synthesis of enantiomerically pure cyclosarkomycin **6**, a stable crystalline precursor of sarkomycin **5**. As described in Scheme 5–3, **6** can be obtained from **8**, an asymmetric Diels-Alder adduct of (*E*)-bromoacrylate. (*E*)-3-bromoacrylate **9a** [the acrylate of (*R*)-pentolactone **11**] and **9b** [the acrylate of (*S*)-*N*-methyl hydroxyl succinimide **12**] undergo TiCl₄-mediated Diels-Alder reactions giving **10a** or **10b′**, the *endo*-product, with high diastereoselectivity (Scheme 5–4). With the key intermediate **10a** in hand, synthesis of compound **6** is accomplished by following the reaction sequence shown in Scheme 5–5.

5.1.2 α,β-Unsaturated Ketone

The design of type II substrates is based on the assumption that the chiral auxiliary R* may act as a highly effective chiral promoter due to its proximity

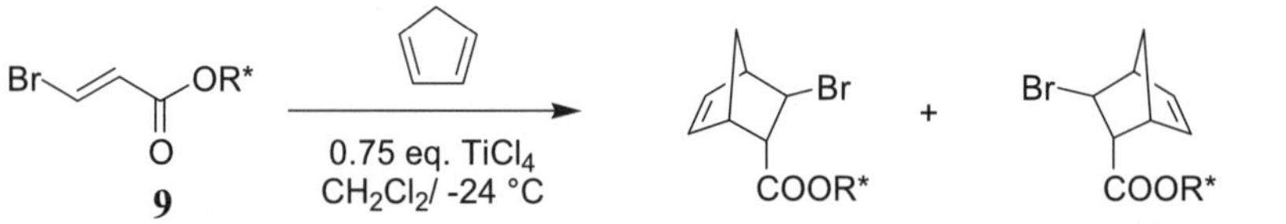

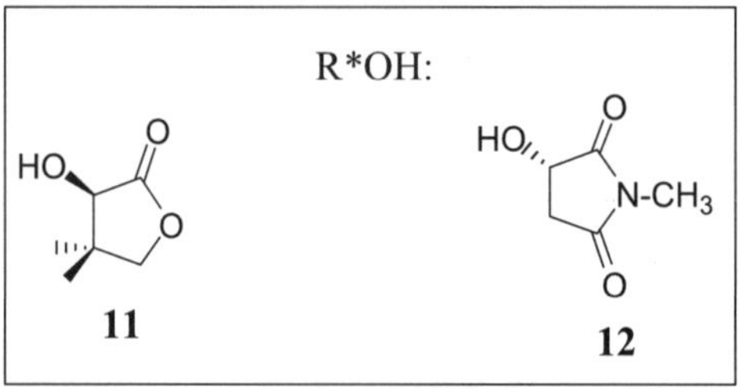

Scheme 5–4

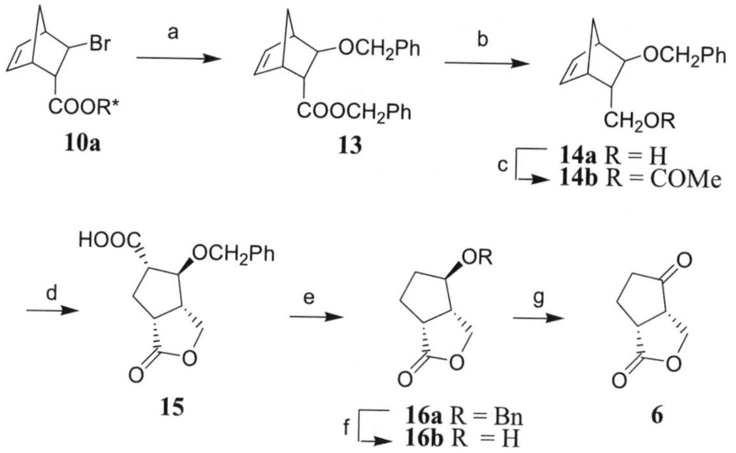

Scheme 5–5. Reagents and conditions: a: NaOCH$_2$Ph, THF/DMF 5:1, $-10°$C to room temperature. b: LiAlH$_4$, Et$_2$O, reflux. c: Ac$_2$O, Py. d: (1) Cat. RuCl$_3$/NaIO$_4$, CCl$_4$/ MeCN/H$_2$O 1:1:1.5, room temperature. (2) LiOH, THF/H$_2$O 5:4, 0°C. (3) 2N HCl, Et$_2$O, 0°C to room temperature. e: (1) N-methylmorpholine, isobutylchloroformate, N-hydroxy-2-thiopyridone, THF, NEt$_3$, THF, $-15°$C, (2) t-butyl mercaptan, irradiation, THF, room temperature. f: H$_2$, 10% Pd/C, EtOH. g: Cat. (n-Pr)$_4$NRuO$_4$, NMO, 4 Å MS, CH$_2$Cl$_2$, room temperature.

to the reacting site. Thus, compound **17** reacts with cyclopentadiene in the presence of ZnCl$_2$ at $-43°$C, affording exclusively *endo*-adduct **18**. The reaction proceeds quickly and completes within an hour. Similarly, compound **19**, a homologue of **17**, undergoes a Diels-Alder reaction giving compound **20** with similarly high stereoselectivity (Scheme 5–6).[7]

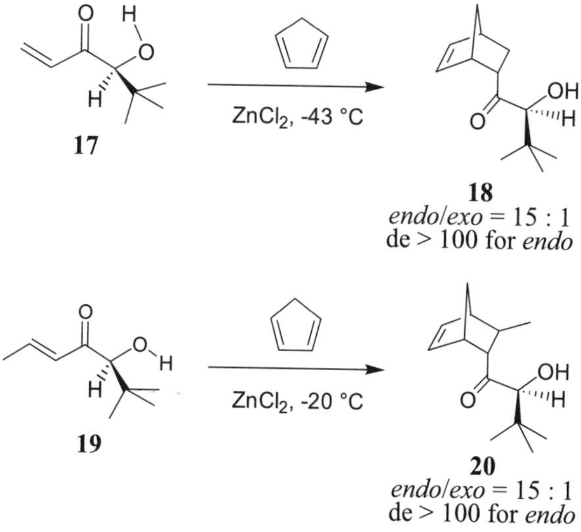

Scheme 5–6

The high enantioselectivity shown in the above reactions can be attributed to two important factors. First, coordination of the Lewis acid with the α-hydroxy ketone moiety of dienophile **17** or **19** leads to the formation of a rigid five-membered chelate **21**. This chelate causes the differentiation of the two diastereotopic faces of the enone system. Second, arising from the established absolute configuration of **17** and **19**, within **21**, the Diels-Alder reaction proceeds with the enone fragment at its cisoid position (*syn*-planar).

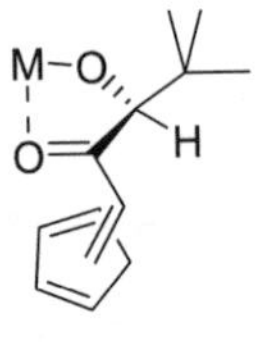

21

A variety of dienes have been subjected to Diels-Alder reactions with (*S*)-**17** and (*R*)-**28**, giving almost pure single adducts in each case. Oxidative removal of the chiral auxiliary from the adduct provides the desired optically pure building blocks **23**, **26**, and **30**. Subsequent conversions complete the synthesis of the desired natural products **5**, **27**, and **32** (Scheme 5–7).[7]

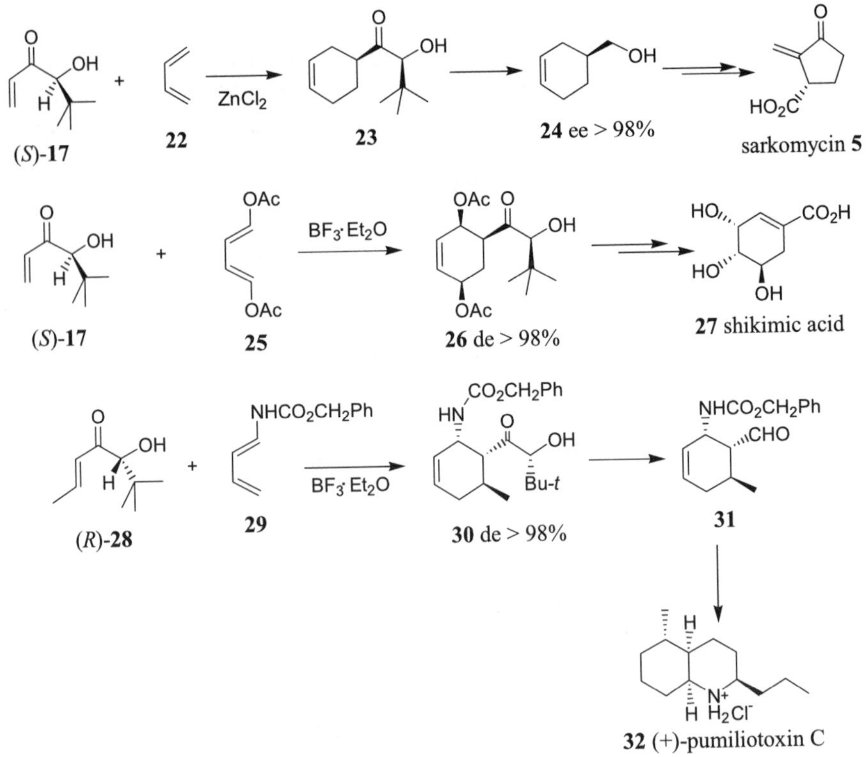

Scheme 5–7

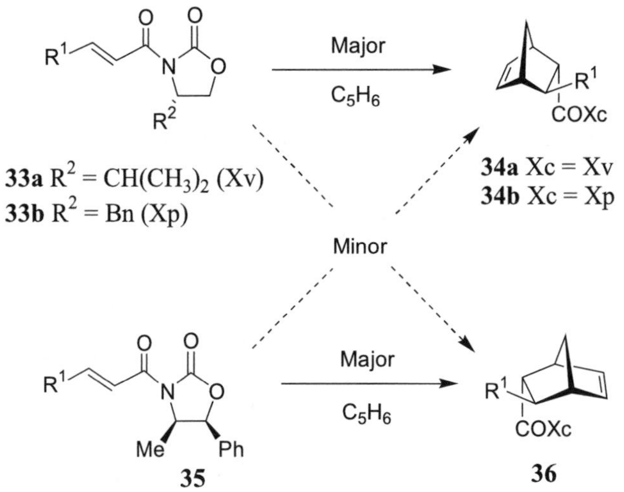

33a $R^2 = CH(CH_3)_2$ (Xv)
33b $R^2 = Bn$ (Xp)

34a Xc = Xv
34b Xc = Xp

35

36

Scheme 5–8

5.1.3 Chiral α,β-Unsubstituted *N*-Acyloxazolidinones

Chiral α,β-unsaturated *N*-acyloxazolidinones have been regarded as a complement for type I dienophile reagents. Evans et al.[4] reported a Diels-Alder reaction promoted by dialkyl aluminum chloride. In this reaction, chiral α,β-unsaturated *N*-acyloxazolidinones were used as highly reactive and diastereoselective dienophiles. The stereogenic outcome of the Diels-Alder adducts **34** and **36** depends on the chirality at C-4 of the oxazolidinone auxiliaries. As shown in Scheme 5–8, all the diastereomeric adducts of **34a**, **34b**, and **36** can be obtained from the dienophiles **33a**, **33b**, or **35**. The high stereoselectivity is a consequence of chloride ionization from 1:1 complex **38** (taking **33a** as an example, where $R^1 = CH_3$). As illustrated by the model in Figure 5–3, the resulting ionic dienophile **38** favors the C_α–*Si*-face of the cyclopentadiene molecule leading to **34a** as the major product. In accordance with this model, cycloaddition should occur selectively from the C_α–*Re*-face of the crotonyl imide **35** to afford the enantiomer of **34a** (Xc = OH) via adduct **36**. The disadvantage of this Diels-Alder reaction is that it is limited to monosubstituted and (*E*)-disubstituted imide dienophiles.

Chiral oxazolidine compounds have also been used as chiral auxiliaries for asymmetric Diels-Alder reactions. Adam et al.[8] demonstrated the cycloaddition of optically active 2,3-dimethyloxazolidine derivatives with singlet oxygen. As shown in Scheme 5–9, the reaction of chiral substrate **39** with singlet oxygen provides product **40** in high diastereomeric ratio.

5.1.4 Chiral Alkoxy Iminium Salt

As depicted in Scheme 5–10, iminium salts such as **42** have been found to exhibit substantially increased reactivity in Diels-Alder reactions.[9] This means

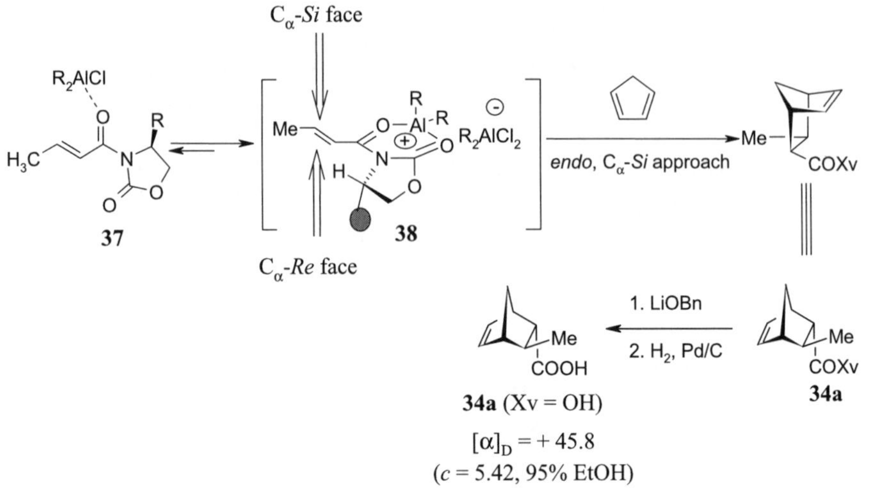

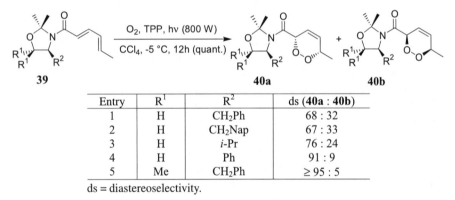

Figure 5–3. Diastereoface selection in the cycloaddition process. Reprinted with permission by Am. Chem. Soc., Ref. 4.

Entry	R¹	R²	ds (**40a** : **40b**)
1	H	CH₂Ph	68 : 32
2	H	CH₂Nap	67 : 33
3	H	i-Pr	76 : 24
4	H	Ph	91 : 9
5	Me	CH₂Ph	≥ 95 : 5

ds = diastereoselectivity.

Scheme 5–9. Reprinted with permission by Am. Chem. Soc., Ref. 8.

that a chiral amide such as **41** might be used for asymmetric Diels-Alder reactions. The asymmetric Diels-Alder reaction of **41a–c** in the presence of TMS-triflate yields amide compounds **43a–c** in high yield and with the *endo*-compound as the major product.

The high stereoselectivity of the products obtained can be explained by postulating transition states A and B (Fig. 5–4). In transition state A, the diene approaches in an *endo* fashion, from the more sterically hindered side of the dienophile; while in transition state B, although the diene again approaches in an *endo* fashion, it approaches from a less hindered side. Thus, transition state B has a lower energy and leads to product **43a** with (*S*)-configuration α to the

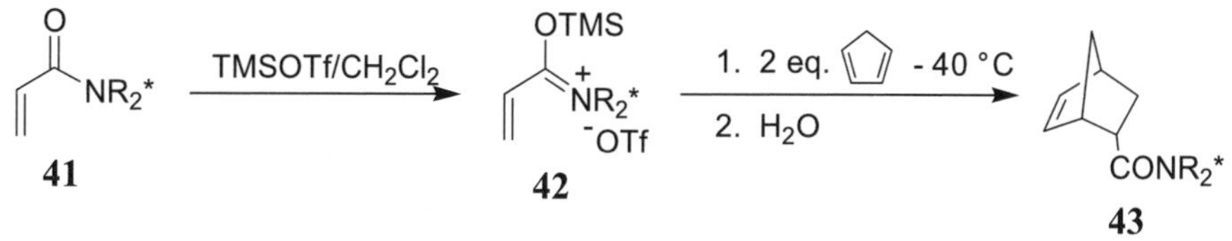

Scheme 5–10

	NR$_2$* in **41a-c** and **42a-c**	adduct **43**		
		endo/*exo*	yield (%)	de (%)
a	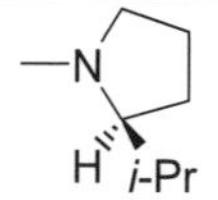 (pyrrolidine with H *i*-Pr)	> 10	90 - 95	>92
b	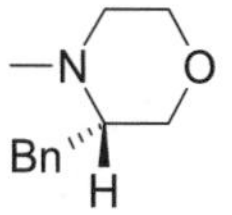 (morpholine with Bn H)	13	91 - 97	90 - 96
c	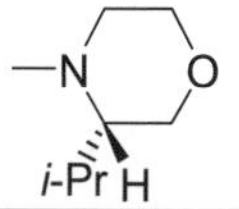 (morpholine with *i*-Pr H)	18	94 - 98	74 - 76

carbonyl group. The same rationale is applicable to **41b** and **41c**, which result in Diels-Alder products **43b** and **43c** with predominantly (*R*)-configuration (Scheme 5–10). Because the morpholine and pyrrolidine chiral auxiliaries can readily be prepared in enantiomerically pure form, this method provides a route to both the (*R*)- and (*S*)-acids, which can be obtained from compound **43** upon removal of the chiral auxiliary.

Recently, two new axially chiral compounds **44** and **45** have been prepared [**44**, *N*-acryl-*N*-allyl-*o*-*t*-butylanilide; **45**, *N*-(*o*-*t*-butylphenyl)-2-methyl-maleimide]. This represents the first instance of using nonbiaryl axially chiral ligands in asymmetric Dield-Alder reactions. In the presence of iodine, high *endo*-facial and diastereofacial selectivities have been obtained in **44**/**45**-mediated reactions.[10]

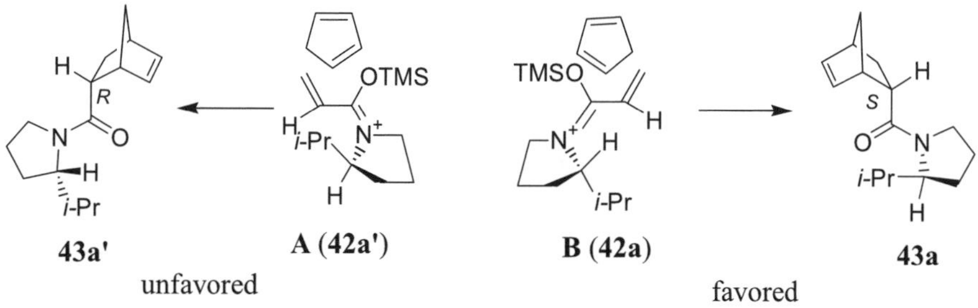

43a'   **A (42a')**   **B (42a)**   **43a**

unfavored   favored

Figure 5–4. The proposed transition states of **42a** and **42a'** with cyclopentadiene.

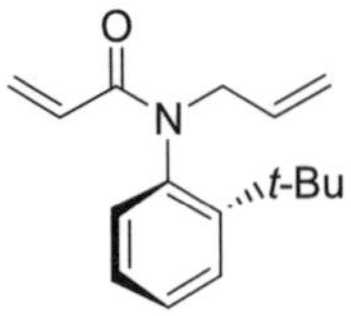

44 96% ee

$[\alpha]_D^{28} = +182 \ (CHCl_3)$

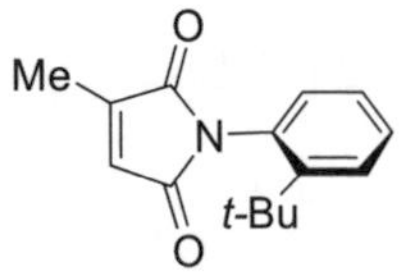

45 96% ee

$[\alpha]_D^{28} = +1.3 \ (CHCl_3)$

Asymmetric Diels-Alder reactions have been carried out to investigate the efficacy of anilide **44** and imide **45** as dienophiles. In the presence of iodine, the asymmetric Diels-Alder reaction of anilide (+)-**44** shows a remarkable improvement in both reactivity and stereoselectivity (Table 5–1). Therefore, it is believed that, in the presence of I_2, Diels-Alder reactions of N-allylic enamides take place through an activating process involving the formation of a cationic iodocyclization intermediate.

The Diels-Alder reaction of maleimide (+)-**45** with cyclopentadiene requires a prolonged reaction time (3 days) at room temperature in the absence of activating agents. The yield is around 80%. However, in the presence of Et_2AlCl, the reaction can be completed with a quantitative yield within 20 minutes at room temperature (Scheme 5–11).

TABLE 5–1. Iodine-Mediated Asymmetric Diels-Alder Reaction of (+)-44

Entry	Diene	Major Product	Yield (%)	endo/exo	ds (%)
1		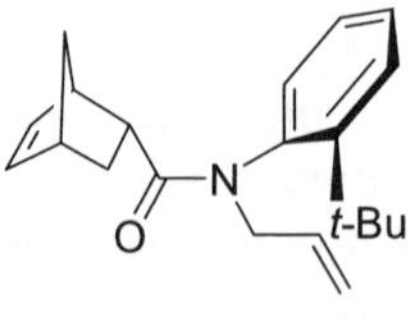	92	30	29/1
2		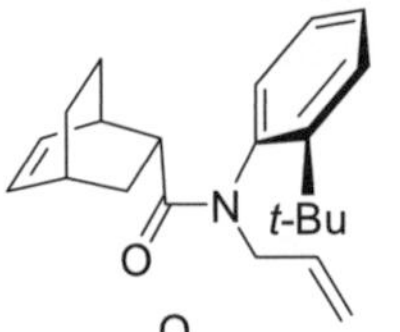	84	>50	14/1
3	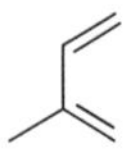	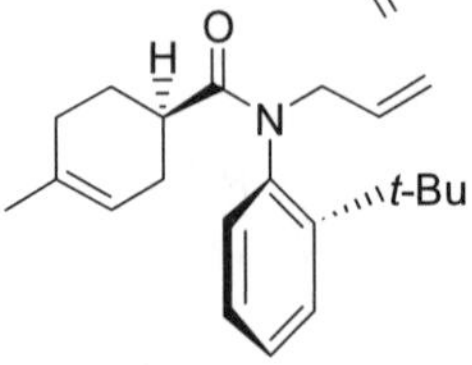	87	20	

Conditions: **44**(1 mmol), I_2 and diene (2 mmol), AcOEt (5 ml), −78°C to room temperature. ds = diastereoselectivity.

Reprinted with permission by Am. Chem. Soc., Ref. 10(b).

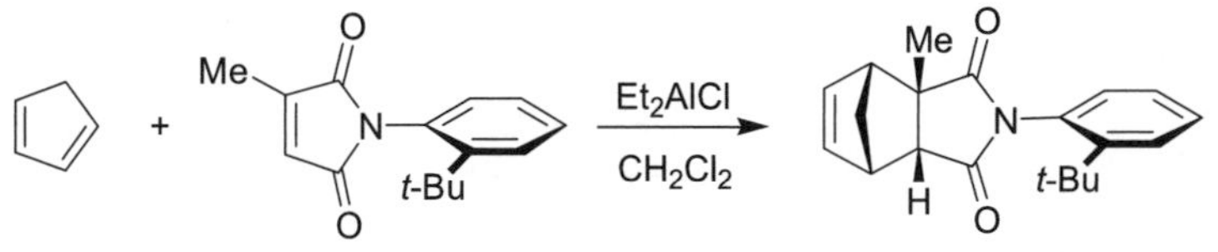

(+)-**45**   endo/exo > 97 : 3, 96% ee

Scheme 5–11

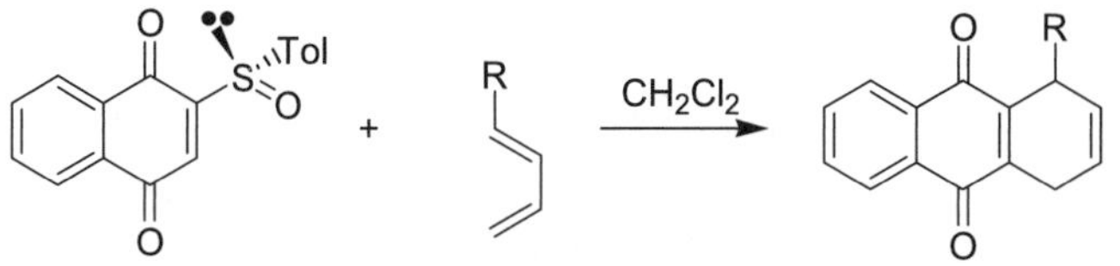

46   **47**   **48**

Scheme 5–12

5.1.5 Chiral Sulfinyl-Substituted Compounds as Dienophiles

Carreño et al.[11] have reported an enantioselective Diels-Alder reaction using chiral *p*-toluenesulfinyl-naphthoquinone as a dienophile. The reaction proceeded readily, yielding the product in about 50% de and 68–82% ee. The advantage of using sulfinyl compounds as dienophiles is the ease of subsequent removal of the sulfinyl group from the resulting compound by pyrolytic elimination. As shown in Scheme 5–12, an asymmetric Diels-Alder reaction using chiral sulfinyl compound **46** as the substrate results in product **48** in moderate to good enantiomeric excess after the subsequent removal of the sulfinyl group.

5.2 CHIRAL DIENES

There are few reports about chiral auxiliary–attached diene components due, in part, to the difficulty of the preparation of the modified dienes. Schemes 5–13,[12] 5–14,[12b] 5–15,[13] and 5–16[13] show some selected examples.

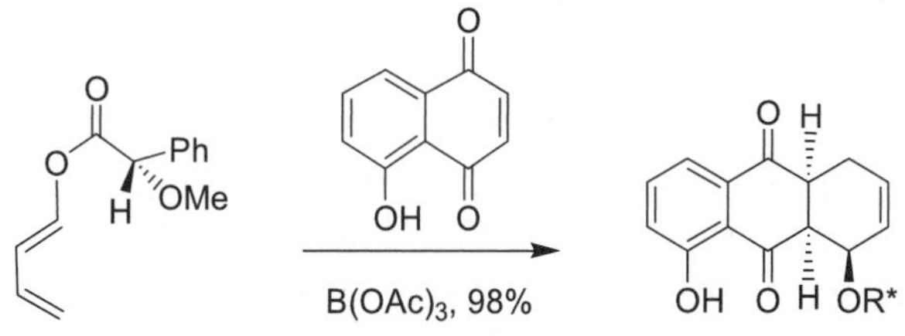

B(OAc)$_3$, 98%

> 97% de

Scheme 5–13

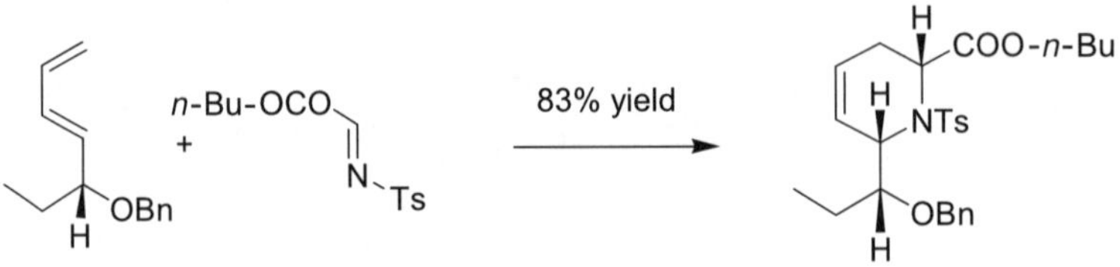

Scheme 5–14

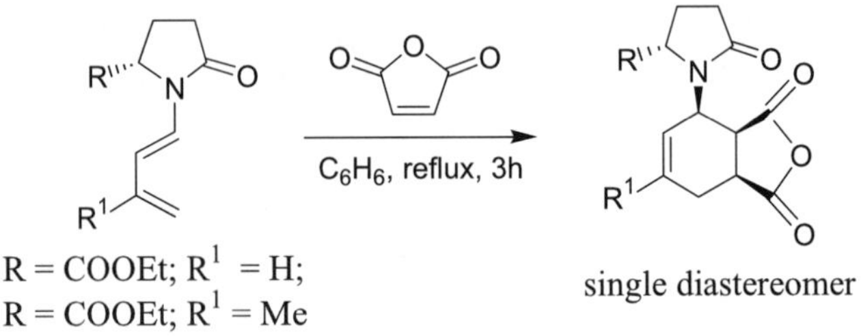

R = COOEt; R^1 = H;
R = COOEt; R^1 = Me

single diastereomer

Scheme 5–15

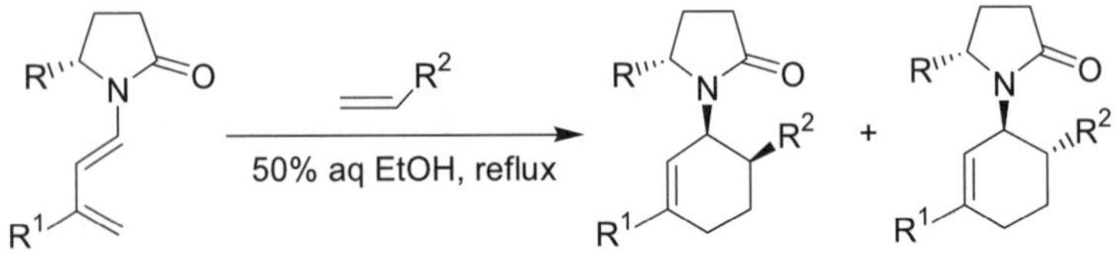

	cis (ds)		*trans* (ds)
(a) R = CO$_2$Et, R^1 = H, R^2 = CO$_2$Et	93 (91.5/1)	:	7 (65 : 1)
(b) R = CO$_2$Et, R^1 = Me, R^2 = CO$_2$Et	92 (18/1)	:	8 (84 : 1)
(c) R = CO$_2$Et, R^1 = H, R^2 = COMe	82 (40/1)	:	18 (180 : 1)

Scheme 5–16

5.3 DOUBLE ASYMMETRIC CYCLOADDITION

As mentioned in the previous section, chiral dienophile **17** is highly enantiose-lective and meets the criteria for double asymmetric induction. A set of asymmetric reactions have been performed by Masamune's group using **49** as an achiral diene and (S)/(R)-methyl mendelates (S)/(R)-**51** as chiral dienes.[2a] The Diels-Alder reaction between (S)-**17** and **49** exhibits the high diastereoselective potential of **17**. As shown in Scheme 5–17, in the presence of a catalytic amount of BF$_3$ · Et$_2$O, the reaction of (S)-**17** and **49** gives compound **50** as the major product with a diastereoselectivity of over 100:1. The reaction of (S)-**17** with

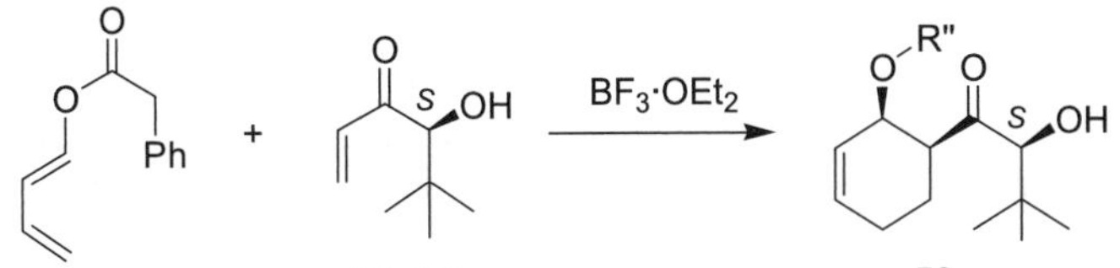

49 (*S*)-**17** **50**
R" = $PhCH_2CO$
selectivity > 100 : 1

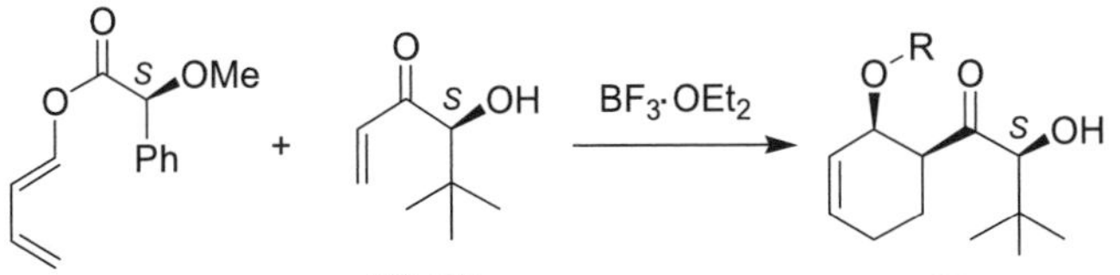

(*S*)-**51** (*S*)-**17** **52**
matched pair
R = (*S*)-PhCH(OMe)CO
selectivity > 130 : 1

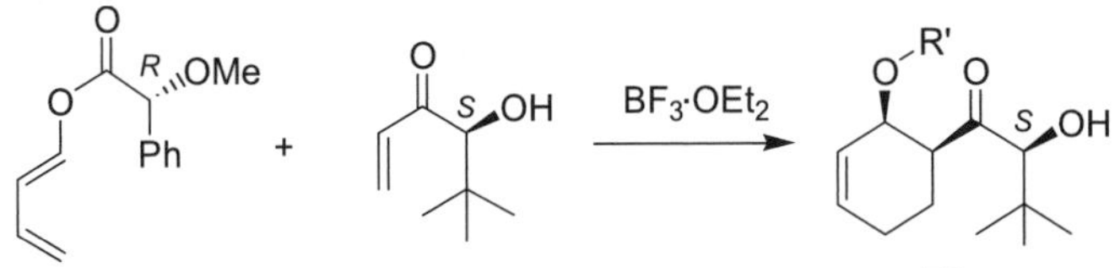

(*R*)-**51** (*S*)-**17** **52**
mismatched pair
R' = (*R*)-PhCH(OMe)CO
selectivity, 35 : 1

Scheme 5–17

(*S*)-**51** or (*R*)-**51** in the presence of $BF_3 \cdot Et_2O$ gives product **52** with a diastereoselectivity of 130:1 or 35:1, referring to the matched and mismatched pairs, respectively (Scheme 5–17).

5.4 CHIRAL LEWIS ACID CATALYSTS

Perhaps the most attractive method of introducing enantioselectivity into the Diels-Alder reaction is to use a chiral catalyst in the form of a Lewis acidic metal complex. In recent years, this area has shown the greatest progress, with the introduction of many excellent catalytic processes. Quite a number of ligand–metal combinations have been evaluated for their potential as chiral catalysts in Diels-Alder reactions. The most commonly used metals are boron, titanium, and aluminum. Copper, magnesium, and lanthanides have also been used in asymmetric catalytic Diels-Alder reactions.

The most common sources of the chiral ligands employed for making a chiral Lewis acid complex are chiral diols with a C_2-symmetric axis. This C_2-symmetric feature reduces the number of competing transition states, which is especially important when the sphere coordination number is greater than 4. Earlier examples[14] of highly selective asymmetric induction in chiral Lewis acid–mediated Diels-Alder reactions were of this type. The main drawback of this type of promoter used to be its noncatalytic property. In other words, a stoichiometric amount of Lewis acid had to be used in the reaction. Nowadays, with modifications to the catalyst structure and the reaction procedure, the second generation of Lewis acids has been used in a catalytic manner, providing high yields and high enantioselectivities.[15]

5.4.1 Narasaka's Catalyst

Narasaka et al.[16] reported that **53** catalyzes Diels-Alder reactions of **54**-type substrates with diene in the presence of 4 Å molecular sieves (Scheme 5–18). A remarkable solvent effect on the enantioselectivity is observed. High enantio-selectivity is attained using mesitylene as the solvent. As shown in Scheme 5–18, the reaction of **54a** with isoprene proceeds smoothly in this solvent, affording product **55a** with 92% ee. Other 3-(3-substituted acryloyl)-1,3-oxazolidin-2-ones **54b–d** also give good results (75–91% ee) when reacted with cyclopentadiene.

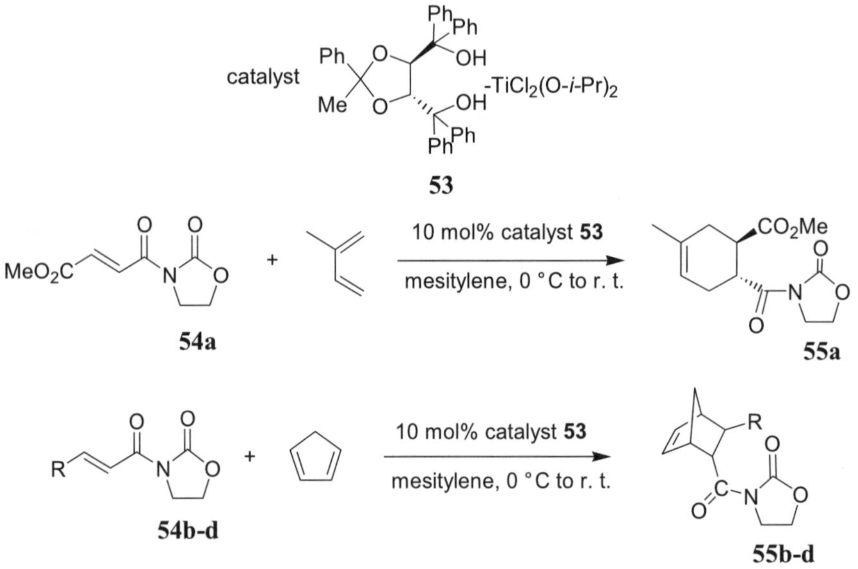

Dienophile	Yield(%) (*endo* : *exo*)	Product	ee (%)
54b R = Me	90 (91:9)	**55b**	91
54c Ph	97 (92:8)	**55c**	82
54d *n*-Pr	75 (91:9)	**55d**	75

Scheme 5–18. Diels-Alder reaction of **54** with cyclopentadiene.

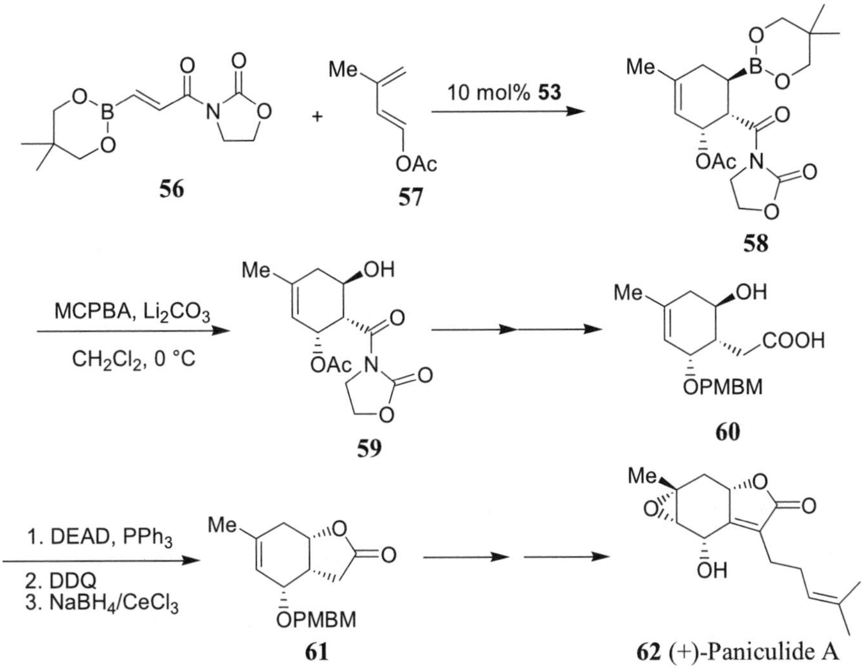

Scheme 5–19. Asymmetric synthesis of (+)-paniculide A.

This type of chiral diol has received considerable attention as a chiral ligand not only in Diels-Alder reactions but also in other reactions such as alkylation and allylation of carbonyl groups, mainly because of its ease of preparation from tartrate. Yamamoto and Narasaka[17] described the application of this chiral titanium catalyst in the synthesis of (+)-paniculide A (**62**), a highly oxidized sesquiterpene. Scheme 5–19 outlines the synthesis route to this compound. The enantioselective cycloaddition of [(*E*)-3-(3-(5,5-dimethyl-1,3-dioxaborinan-2-yl)propenoyl)]-1,3-oxazolidin-2-one (**56**) to 1-acetoxy-3-methyl-1,3-butadiene (**57**) is a key step. The 3-hydroxyl group in **56** is masked as its borate and can be regenerated by *m*-CPBA oxidation of the Diels-Alder reaction product **58**. Highly enantiomerically pure product **59** can be obtained with the desired stereochemistry of (+)-paniculide A.

The Lewis acid catalyst **53** is now referred to as the *Narasaka catalyst*. This catalyst can be generated in situ from the reaction of dichlorodiisopropoxytitanium and a diol chiral ligand derived from tartaric acid. This compound can also catalyze [2+2] cycloaddition reactions with high enantioselectivity. For example, as depicted in Scheme 5–20, in the reaction of alkenes bearing alkylthio groups (ketene dithioacetals, alkenyl sulfides, and alkynyl sulfides) with electron-deficient olefins, the corresponding cyclobutane or methylenecyclobutene derivatives can be obtained in high enantiomeric excess.[18]

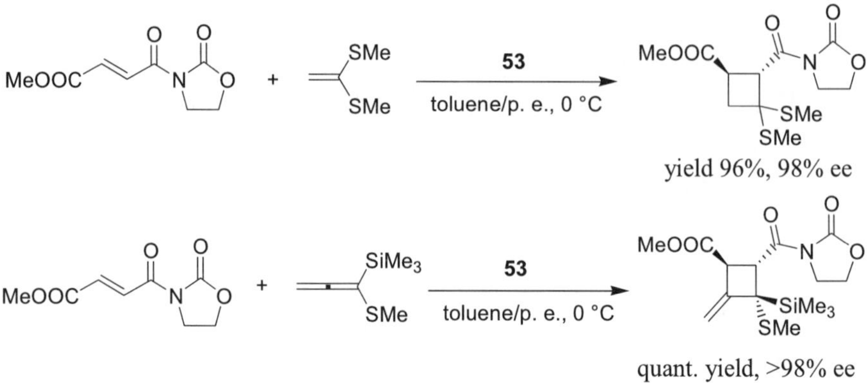

yield 96%, 98% ee

quant. yield, >98% ee

Scheme 5–20

5.4.2 Chiral Lanthanide Catalyst

In the presence of a catalytic amount of chiral lanthanide triflate **63**, the reaction of 3-acyl-1,3-oxazolidin-2-ones with cyclopentadiene produces Diels-Alder adducts in high yields and high ee. The chiral lanthanide triflate **63** can be prepared from ytterbium triflate, (R)-$(+)$-binaphthol, and a tertiary amine. Both enantiomers of the cycloaddition product can be prepared via this chiral lanthanide(III) complex–catalyzed reaction using the same chiral source [(R)-$(+)$-binaphthol] and an appropriately selected achiral ligand. This achiral ligand serves as an additive to stabilize the catalyst in the sense of preventing the catalyst from aging. Asymmetric catalytic aza Diels-Alder reactions can also be carried out successfully under these conditions (Scheme 5–21).[19]

5.4.3 Bissulfonamides (Corey's Catalyst)

Sulfonamide derivatives of α-amino acids and the similar bissulfonamide derivatives of diamines can be used to prepare reactive Lewis acid complexes. Corey[20] reported the Lewis acid (R,R)- or (S,S)-complex **69**, which can be employed at 10 mol% level to catalyze the Diels-Alder reaction of cyclopentadiene and imide. Reactions catalyzed by this complex give an *endo:exo* ratio of over 50:1, as well as a high ee (91%) at −78°C, and this can be further improved to 95% by carrying out the reaction at −90°C.[20] The related aluminum complex **69b** shows very similar reactivity at −78°C, with generally higher ee values, typically over 95%, for the reaction of cyclopentadiene derivatives with imide.[20,21]

Thus, as shown in Scheme 5–22,[20] the reaction of **71** with cyclopentadiene in the presence of 10 mol% of (S,S)-**69b** at −90°C generates **72a** at 92% yield with 95% ee (*endo:exo* > 50:1). The corresponding reaction of **64b** with cyclopentadiene gives **72b** at 88% yield with 94% ee (*endo:exo* = 96:4). The reaction of **71** with **70b** in the presence of 0.1 mol% of (S,S)-**69b** at −78°C gives the adduct **72c** at 94% yield with 95% ee.

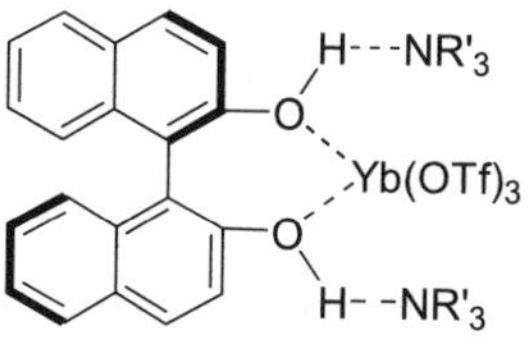

63
chiral ytterbium triflate

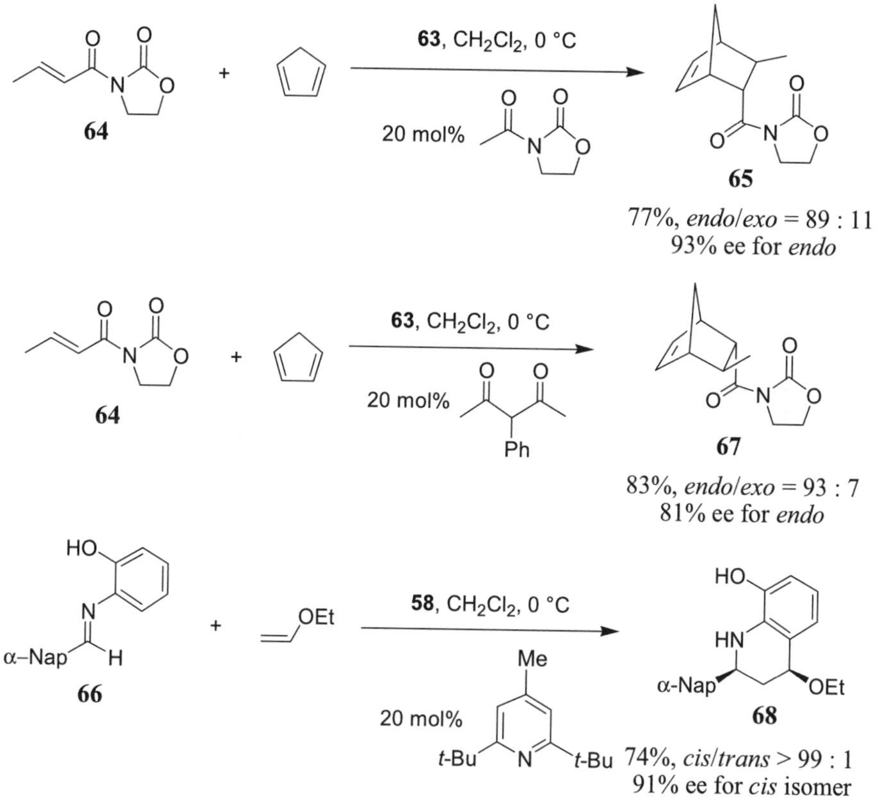

64

63, CH₂Cl₂, 0 °C

20 mol%

65

77%, *endo/exo* = 89 : 11
93% ee for *endo*

64

63, CH₂Cl₂, 0 °C

20 mol%

Ph

67

83%, *endo/exo* = 93 : 7
81% ee for *endo*

α–Nap

66

+ ⟋OEt

58, CH₂Cl₂, 0 °C

Me

20 mol%

t-Bu N t-Bu

HO

HN

α-Nap OEt
68

74%, *cis/trans* > 99 : 1
91% ee for *cis* isomer

Scheme 5–21

5.4.4 Chiral Acyloxy Borane Catalysts

Stable aryl boronates derived from tartaric acid catalyze the reaction of cyclo-pentadiene with vinyl aldehyde with high selectivity. Chiral acyloxy borane (CAB), derived from tartaric acid, has proved to be a very powerful catalyst for the enantioselective Diels-Alder reaction and hetero Diels-Alder reaction. Scheme 5–23 presents an example of a CAB **73** (R = H) catalyzed Diels-Alder reaction of α-bromo-α,β-enal **74** with cyclopentadiene. The reaction product is another important intermediate for prostaglandin synthesis. In the presence of

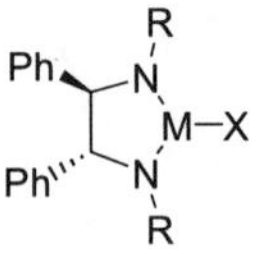

(*R, R*) or (*S, S*)-**69**

M = B, X = Br, R = SO$_2$CF$_3$, *p*-MeC$_6$H$_4$SO$_2$ or *p*-NO$_2$C$_6$H$_4$SO$_2$,
M = Al, X = CH$_3$/*i*-Bu, R = SO$_2$CF$_3$, *p*-MeC$_6$H$_4$SO$_2$ or *p*-NO$_2$C$_6$H$_4$SO$_2$

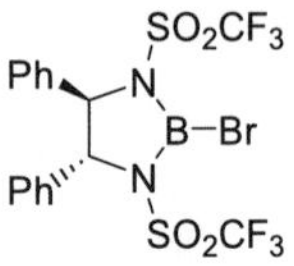

69a 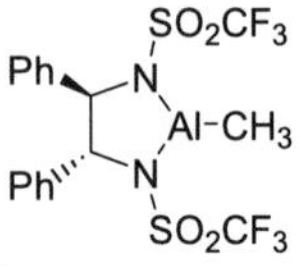   **69b**

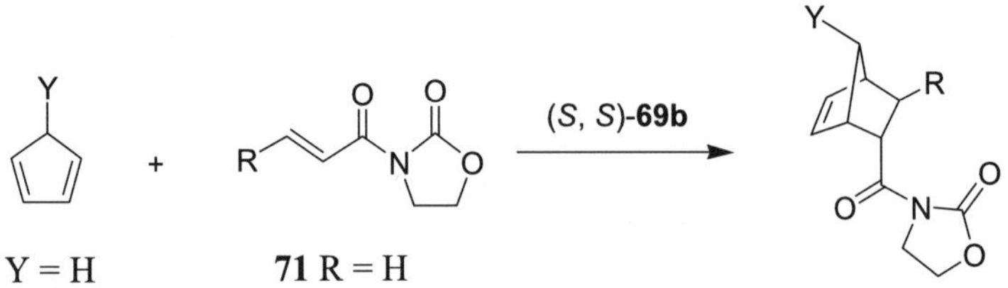

70a Y = H   **71** R = H
70b Y = OBn   **64b** R = Me

72a R = H, Y = H
72b R = Me, Y = H
72c R = H, Y = OBn

Product	Temp. (°C)	Yield (%)	*endo/exo*	ee (%)
72a	-90	92	> 50 : 1	95
72b	-78	88	96 : 4	94
72c	-78	94		95%

Scheme 5–22

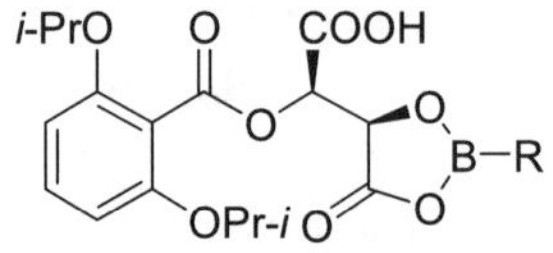

CAB **73**

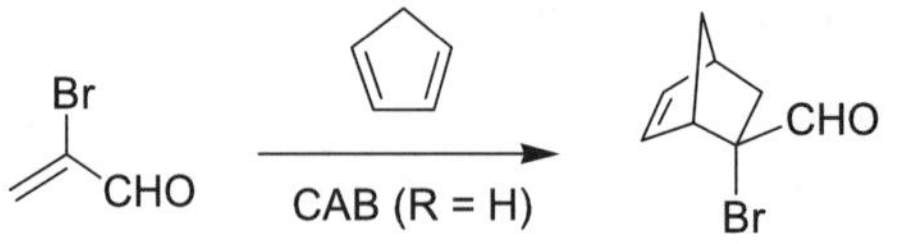

74

(*S*)-**75**
exo/endo = 94 : 6
95% ee

Scheme 5–23

10 mol% of **73** (R = H), the reaction of **74** with cyclopentadiene proceeds smoothly, giving (*S*)-**75** at a quantitative yield and with 95% ee.[22]

5.4.5 Brønsted Acid–Assisted Chiral Lewis Acid Catalysts

Lewis acids of chiral metal aryloxides prepared from metal reagents and optically active binaphthol derivatives have played a significant role in asymmetric synthesis and have been extensively studied.[23] However, in Diels-Alder reactions, the asymmetric induction with chiral metal aryloxides is, in most cases, controlled by steric interaction between a dienophile and a chiral ligand. This kind of interaction is sometimes insufficient to provide a high level of enantioselectivity.

In recent years, a number of groups (e.g., Hawkins and Loren,[24] Corey and

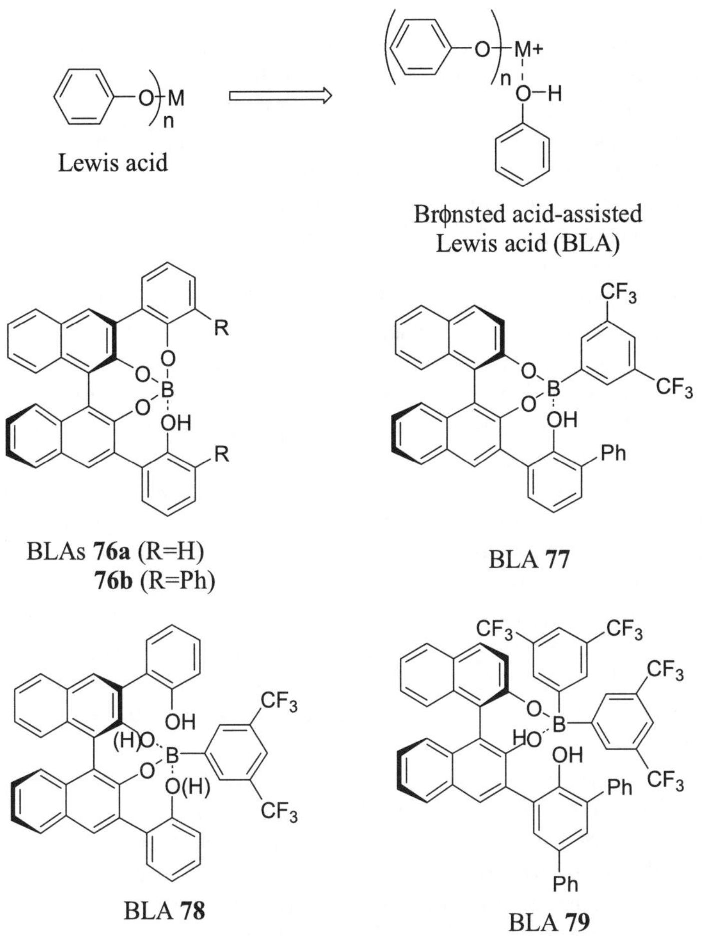

Figure 5–5. Brønsted acid-assisted lewis acids.

TABLE 5–2. Enantioselective Diels-Alder Reaction of Various α-Substituted α,β-Enals with Cyclopentadiene Catalyzed by (R)-BLA 76a

Entry	Dienophile	Yield (%) (exo:endo)	ee (%) (config.)
1	α-Bromoacrolein	>99 (>99:1)	99 (S)
2	CH$_2$=CMeCHO	>99 (>99:1)	99 (R)
3	CH$_2$=CEtCHO	>99 (97:3)	92
4	(E)-MeCH=CHMeCHO	>99 (>99:1)	98
5	1-Formylcyclopentene	>99 (98:2)	93
6	CH$_2$=CHCHO	91 (9:91)	40 (R)
7	CH$_2$=CHCHO	85 (14:86)	92 (R)[a]
8	(E)-MeCH=CHCHO	12 (11:89)	36 (R)

ee = Enantiomeric excess. [a] (R)-**76b** was used instead of (R)-**76a**.

Reprinted with permission by Am. Chem. Soc., Ref. 27.

Lo,[25] and Ishihara et al.[26]) have reported that the attractive π–π donor–acceptor interaction between a dienophile and a chiral ligand is highly effective for inducing asymmetry. Encouraged by this finding, Ishihara et al.[27] developed Brønsted acid–assisted chiral Lewis acid (BLA) catalysts **76–79** (Fig. 5–5). This new class of catalysts can promote highly enantioselective Diels-Alder reactions through a combination of intramolecular hydrogen bonding and the attractive π–π donor–acceptor interaction of the hydroxyl aromatic groups in the chiral Lewis acids in the transition state assembly.

Yamamoto and co-workers found that BLA **76a** is one of the best catalysts for the enantioselective and *exo*-selective cycloaddition of α-substituted α,β-enals to highly reactive dienes such as cyclopentadiene. The results in the presence of (R)-**76a** are summarized in Table 5–2. The major enantiomer has, in several cases, been demonstrated to have the predicted absolute configuration.

Entries 6–8 show that with BLA **76**, the corresponding reaction with α-substituted α,β-enals such as acrolein and crotonaldehyde exhibits low enantioselectivity and/or reactivity. However, the enantioselectivity can be raised from 40% to 90% ee by using (R)-**76b** instead of (R)-**76a** (Entries 6 and 7).

A more practical BLA has been designed to improve the scope of dienophiles in reactions with less reactive dienes. 3,5-Bis(trifluoromethyl)-phenylboronic acid (**80**) was chosen as the Lewis acidic metal component in the new BLAs **77–79**. These new BLAs show higher catalytic activity in enantioselective cycloaddition of both α-substituted and α-unsubstituted α,β-enal with various dienes.

The results suggest that BLA **77** is the best catalyst. For example, the reaction of (E)-methyl acrolein with cyclopentadiene catalyzed by **77** (5 mol%) gives the adduct at 96% yield with 99% ee [(S)-configuration].

Studies also show that by using catalyst **76a** or **78** derived from one single chiral tetraol with the same absolute configuration, both enantiomers of the Diels-Alder reaction product can be obtained. For example, reaction of 2-methyl-

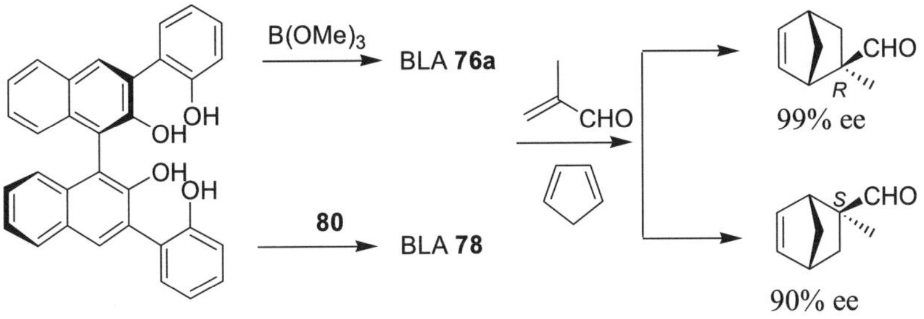

Scheme 5–24

acrolein with cyclopentadiene catalyzed by **76a** yields the product with an (R)-configuration, while reaction of 2-methylacrolein with cyclopentadiene catalyzed by **78** (5 mol%) yields the adduct with an (S)-configuration (Scheme 5–24).

5.4.6 Bis(Oxazoline) Catalysts

C_2-symmetric bis(oxazoline) and salen–metal complexes have recently been shown to be very effective ligands for iron(III),[28] magnesium(II),[29] copper(II),[30] and chromium(III)[31] complex-catalyzed reactions.

Corey and Ishihara[29] report the synthesis of a new bis(oxazoline). This catalyst effects Diels-Alder reaction via a tetracoordinated metal complex. Ligand (S)-**81** is synthesized from (S)-phenylglycine, as depicted in Scheme 5–25. Treatment of **81** with $MgI_2 \cdot I_2$ gives a dark solution of complex **82**, which can be utilized as a Diels-Alder reaction catalyst. Thus, reaction of cyclopentadiene with **71** in the presence of **82** yields product **72a** with an enantiomeric ratio of over 20:1 (Scheme 5–26).

When a copper complex of **83** is used as the catalyst, the reaction can also proceed with excellent selectivity (98% de and >98% ee for the endo-isomer).[30] Note that **83** is similar to **81**, but in this compound the two phenyl groups are replaced by two t-butyl groups.

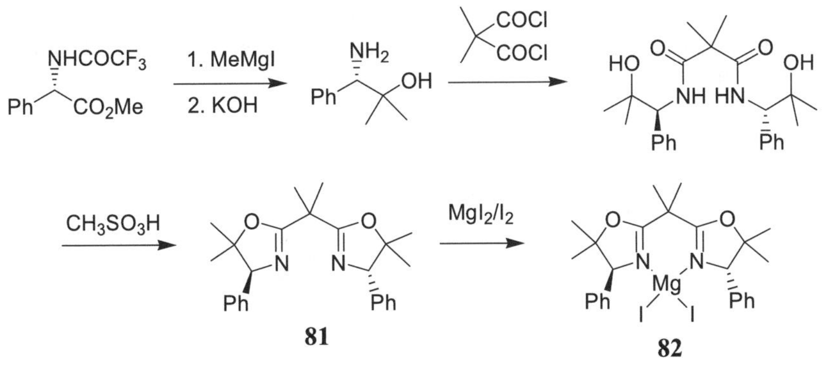

Scheme 5–25. Preparation of bis(oxazoline) magnesium(II) diiodide.

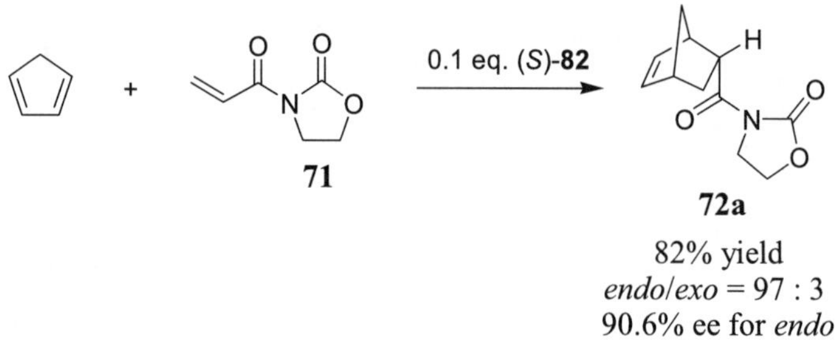

71

72a

82% yield
endo/exo = 97 : 3
90.6% ee for *endo*

Scheme 5–26

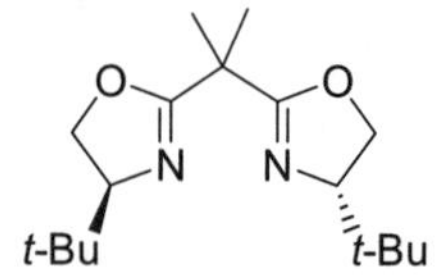

83

The explanation for the topicity of the two complexes that lead to opposite stereo-outcomes is presented in Figure 5–6. Even though they are similar C_2 symmetric ligands, they coordinate with metals in different tetravalent geometry, which then follow different pathways to the products. The tetrahedral magnesium complex **84** facilitates cycloaddition to the C_2 *Si*-face because of rear phenyl blocking of the *Re*-face. In contrast, the square–planar copper complex **85** favors *Re*-addition due to its *Si*-face being blocked by a *t*-butyl group.

Ghosh et al.[32] have demonstrated another bis(oxazoline) derivative chiral ligand **86** for asymmetric Diels-Alder reaction and obtained excellent results. Reaction of an equimolar mixture of chiral ligand **86** and $Cu(ClO_4)_2 \cdot 6H_2O$ produces the aqua complex **87** (w being water molecule), which shows excellent catalytic power in asymmetric Diels-Alder reactions. As depicted in Scheme 5–27, the reaction of **88** with cyclopentadiene gives product **89** with more than 80% yield, over 99:1 diastereoselectivity and up to 99% ee.

For a recent review of chiral bis(oxazoline)-mediated asymmetric syntheses, see Ghosh et al.[33]

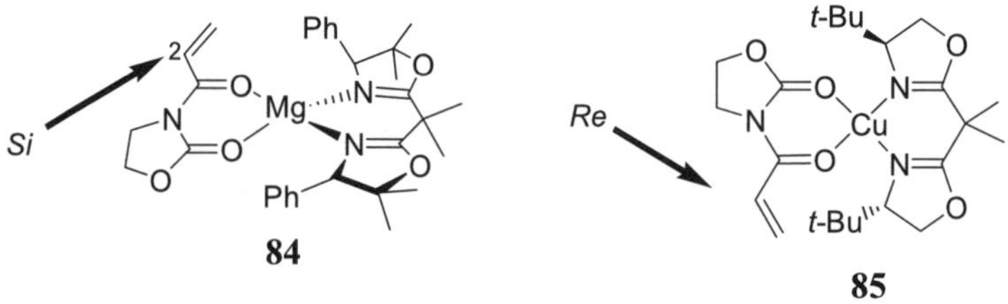

84

85

Figure 5–6. Rationale for the different topicities of bis(oxazoline) complexes.

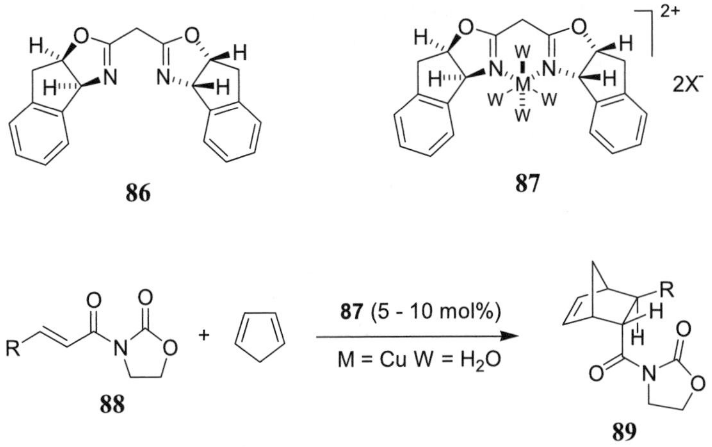

Scheme 5–27

5.4.7 Amino Acid Salts as Lewis Acids for Asymmetric Diels-Alder Reactions

Besides the Lewis acid catalysts discussed above, chiral amino acid copper salts have also been used for inducing asymmetric Diels-Alder reactions. In most of the above reactions, aprotic organic solvents were used as the reaction media, and donor solvents such as water or alcohol were generally avoided. This is partially explained by the fact that these solvents exert detrimental effects, such as coordinating with the catalytic Lewis acid and thus deactivating the catalyst or diminishing the asymmetric catalytic activity of the catalyst. Otto et al.[34] report that asymmetric Diels-Alder reactions catalyzed by amino acid copper salts can proceed with higher enantioselectivity in the presence of water. They studied the asymmetric Diels-Alder reaction of 3-phenyl-1-(2-pyridyl)-2-propen-1-one with cyclopentadiene in the presence of a series of amino acid salts. The Diels-Alder adducts can normally be obtained at over 90% yield after 48 hours of reaction. Significant enantioselectivity has been observed with α-amino acids bearing aromatic side groups. This may be due to the possible π-stacking effect in the transition state, as shown in Figure 5–7.

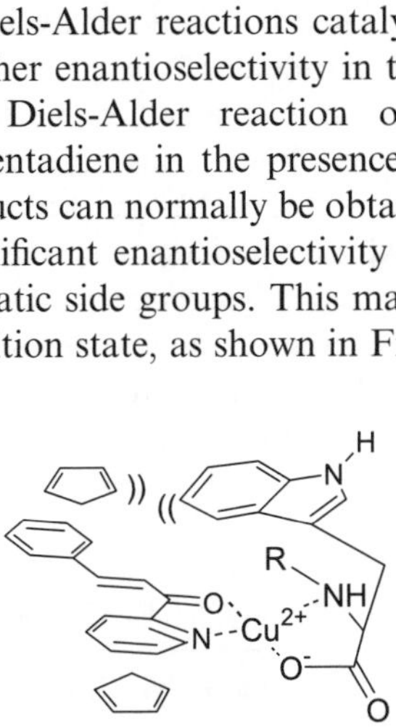

Figure 5–7. Transition state of amino acid salt catalyzed Diels-Alder reaction. Reprinted with permission by Am. Chem. Soc., Ref. 34.

TABLE 5–3. Solvent Effect of Amino Acid Salt-Catalyzed Diels-Alder Reaction

Entry	Solvent	ee (%)
1	Acetonitrile	17
2	THF	24
3	Ethanol	39
4	Chloroform	44
5	Water	74

ee = Enantiomeric excess; THF = tetrahydrofuran.

Reprinted with permission by Am. Chem. Soc., Ref. 34.

Otto et al. studied asymmetric Diels-Alder reactions in the presence of the copper salts of glycine, L-valine, L-leucine, L-phenylalanine, L-tyrosine, L-tryptophan, and N-α-L-tryptophan (L-abrine). The copper salt of L-abrine gave the highest enantioselectivity. Table 5–3 compares the solvent effect in this reaction, and clearly water is the best solvent among the solvent systems studied.

5.5 HETERO DIELS-ALDER REACTIONS

5.5.1 Oxo Diels-Alder Reactions

When aldehyde containing an electron-withdrawing group is employed, or when a Lewis acid promoter is present, C=O double bonds can readily undergo Diels-Alder–type reactions. This process is referred to as the *oxo Diels-Alder reaction*, and it has been explored by Danishefsky and DeNinno[35] for the synthesis of a wide range of saccharide derivatives.

The oxo Diels-Alder reaction, or the hetero Diels-Alder reaction involving a carbonyl compound as the dienophile, constitutes a useful method for carbon skeleton construction. Much attention has focused on developing effective new catalysts for hetero Diels-Alder reactions over the last few years.[36] Mikami et al.[37] have reported that by using isoprene as a diene component, Diels-Alder product can be obtained in extremely high enantiomeric excess. For example, the reaction of 1-methoxy-1,3-butadiene proceeds smoothly in the presence of BINOL–TiCl$_2$, giving the *cis*-product in high ee (Scheme 5–28).

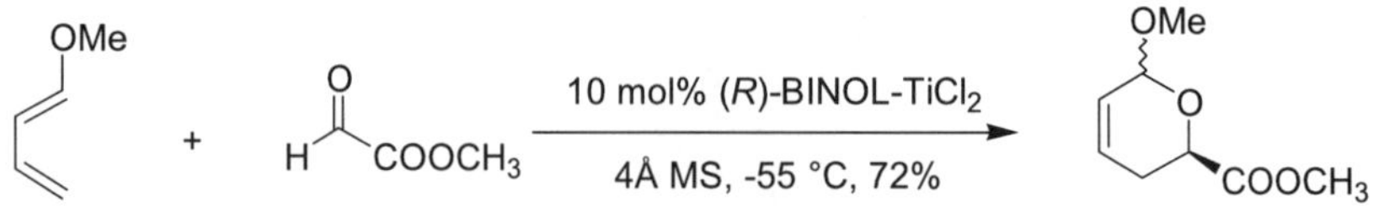

Scheme 5–28

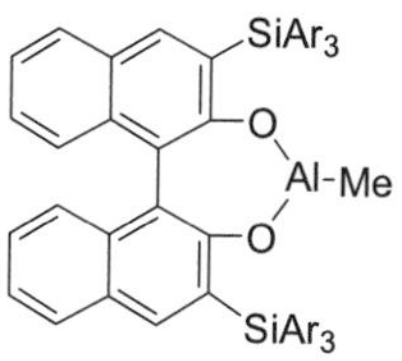

(*R*)- or (*S*)-**90**

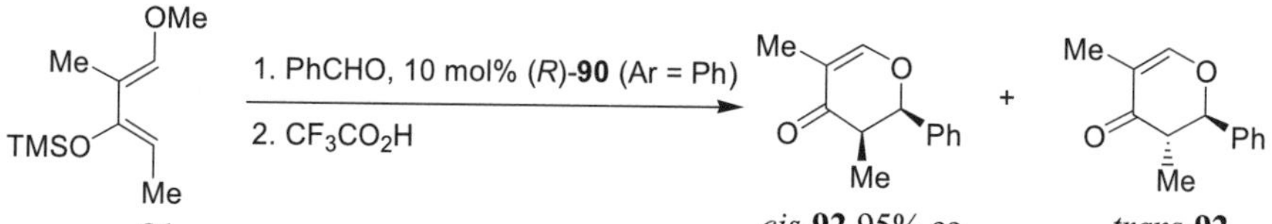

91 *cis*-**92** 95% ee *trans*-**92**

Scheme 5–29

In the course of asymmetric hetero Diels-Alder reaction studies, Maruoka et al.[38] developed a new chiral organometallic catalyst of type **90**. These catalysts are applicable to various reactions of siloxydienes and aldehydes, promoting high enantioselectivity. For example, treating benzaldehyde with siloxydiene in the presence of (*R*)-**90** (Ar = Ph, 10 mol%) at −20°C for 2 hours yields *cis*-dihydropyrane **92** (77%) and its *trans*-isomer (7%) after exposing the adduct to trifluoroacetic acid (Scheme 5–29). A variety of siloxydienes have been treated with aldehydes in the same manner, yielding adducts in moderate to high enantiomeric excess (Table 5–4).

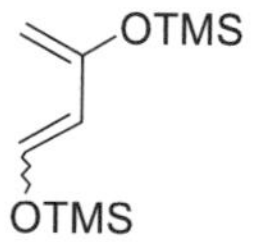

93a

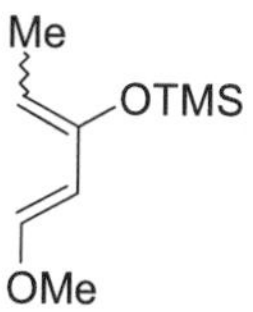

93b

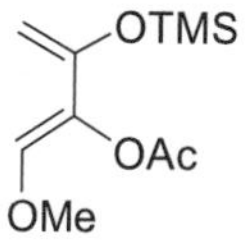

93c

TABLE 5–4. (*R*)-90 Catalyzed Diels-Alder Reaction

Aldehyde	Diene	(*R*)-**92**	Product	Yield (%)	ee (%, config.)
c-C_6H_{11}CHO	**91**	Ar = Ph	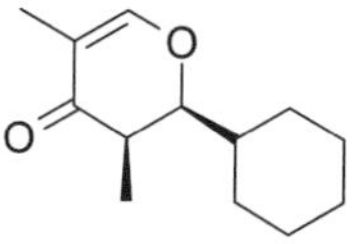	65	91 (2*S*,3*R*)
PhCHO	**93a**	Ar = Ph	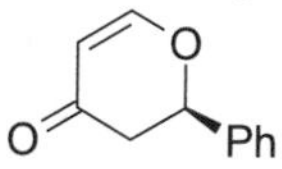	71	67 (*R*)

ee = Enantiomeric excess.

Reprinted with permission by Am. Chem. Soc., Ref. 38.

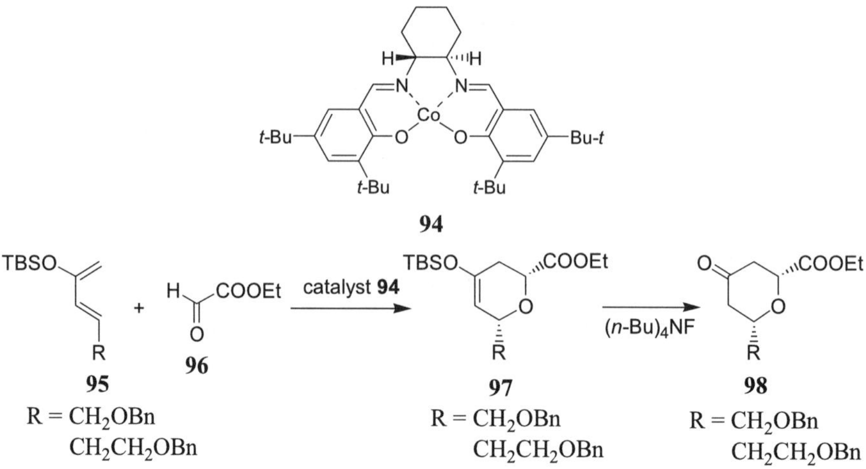

94

R = CH$_2$OBn
CH$_2$CH$_2$OBn

R = CH$_2$OBn
CH$_2$CH$_2$OBn

R = CH$_2$OBn
CH$_2$CH$_2$OBn

Scheme 5–30

Li et al.[39] reported the hetero Diels-Alder reaction of alkyl-3-(*t*-butyldimethylsilyl) oxy-1,3-butadiene **95** with ethyl glyoxylate **96** in the presence of a salen–Co(II) catalyst **94** (2 mol%). Product **97** was obtained in 75% isolated yield with an *endo:exo* ratio >99:1. The enantiomeric excess of the *endo*-form was up to 52% (Scheme 5–30).

Upon treatment with *n*-Bu$_4$NF, product **97** is converted to **98**, which is a key intermediate for the synthesis of the monosaccharide 3-deoxy-D-manno-2-octulosonic acid (KDO)—an essential constituent of the outer membrane lipopolysaccharide (LPS) of gram-negative bacteria.[40]

Schaus et al.[41] have also reported an asymmetric hetero Diels-Alder reaction of Danishefsky's diene **100**[42] with aldehyde **101** catalyzed by chromium(III) complex **99** bearing a similar chiral salen ligand. Product **102** is obtained in moderate to good yield and stereoselectivity (Scheme 5–31 and Table 5–5).

The C$_2$-symmetric bis(oxazoline)–Cu(II) complexes have proved to be very effective in asymmetric aldol reactions (see Section 3.4.3), as well as Diels-Alder reactions (see Section 5.4.6). These compounds are also powerful catalysts in hetero Diels-Alder reactions. Figure 5–8 shows some of the bis(oxazoline) ligands applied in asymmetric hetero Diels-Alder reactions.

Evans et al.[43] and Thorhauge et al.[44] report that hetero Diels-Alder reactions in the presence of **81** or **83** coordinated copper complex proceed smoothly, resulting in excellent yields and enantiomeric excess.

Jørgensen's group[44a] carried out the reaction using the anhydrous form of chiral bis(oxazoline) coordinated copper complex. Complex **106** containing **83** as the chiral ligand was found to be the most effective. As shown in Scheme 5–32, the asymmetric hetero Diels-Alder reaction of β,γ-unsaturated α-keto esters with acyclic enol ethers results in products with excellent yield and enantioselectivity. This catalyst is also effective with cyclic enol ethers.[44a] As shown in Scheme

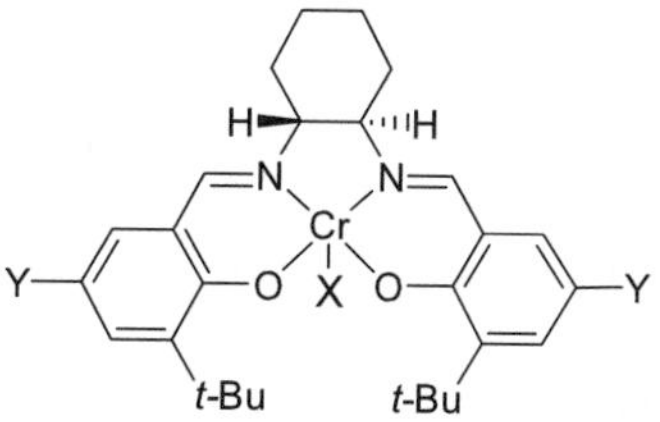

(R, R)-**99a** X = Cl, Y = t-Bu
(R, R)-**99b** X = N$_3$, Y = t-Bu
(R, R)-**99c** X = F, Y = t-Bu
(R, R)-**99d** X = BF$_4$, Y = t-Bu
(R, R)-**99e** X = BF$_4$, Y = OMe

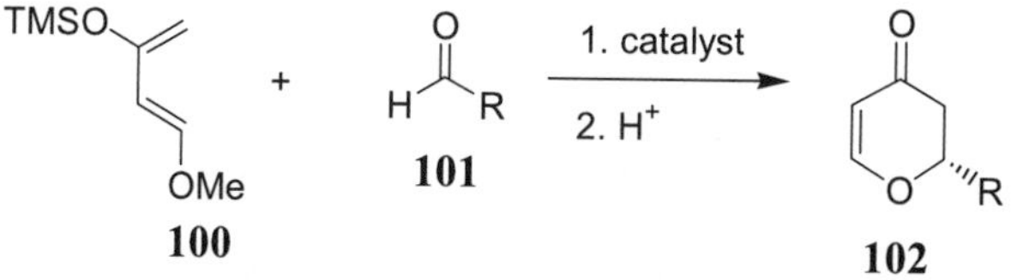

Scheme 5–31

TABLE 5–5. Hetero Diels-Alder Reaction Catalyzed by 99

Entry	R in Aldehyde	ee (%)	
		Catalyst **99d**	Catalyst **99e**
1	Ph	87	65
2	C$_6$H$_{11}$	93	85
3	n-C$_5$H$_{11}$	83	62
4	2-Furyl	76(99)	68
5	(E)-PhCH=CH	70	73(99)
6	p-BrC$_6$H$_4$CH$_2$OCH$_2$	79	84(99)
7	o-ClC$_6$H$_4$COOCH$_2$	83(99)	72

ee = Enantiomeric excess.

Reprinted with permission by Am. Chem. Soc., Ref. 41.

5–33, the hetero Diels-Alder reaction of β,γ-unsaturated α-keto ester with cyclic enol ether dihydrofuran also gives high stereoselectivity.

The procedure reported by Jørgensen's group used anhydrous copper complex. The complex is hygroscopic, and special attention must be paid to handling the catalyst, as well as when carrying out the reaction. Evans' group[43a] found that the hetero Diels-Alder reaction can proceed as well using the hydrate of complex **108** as the catalyst in the presence of molecular sieves. This greatly simplifies the reaction. As shown in Scheme 5–34, the asymmetric hetero Diels-Alder reaction of acyclic or cyclic enol ethers with β,γ-unsaturated

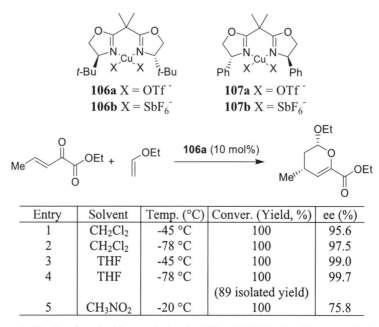

Figure 5–8. Bis(oxazoline) ligands used in asymmetric hetero Diels-Alder reactions.

Entry	Solvent	Temp. (°C)	Conver. (Yield, %)	ee (%)
1	CH₂Cl₂	-45 °C	100	95.6
2	CH₂Cl₂	-78 °C	100	97.5
3	THF	-45 °C	100	99.0
4	THF	-78 °C	100 (89 isolated yield)	99.7
5	CH₃NO₂	-20 °C	100	75.8

Scheme 5–32. Reprinted with permission by Wiley-VCH Verlag Germany, Ref. 44(a).

keto esters generates product **109** or **110** in excellent yield and with excellent diastereoselectivity as well as enantioselectivity.

The hetero Diels-Alder reactions discussed thus far use 2–10 mol% of catalyst. Jørgensen's group[44b] found that the reaction could be carried out even at very low catalyst loading. The catalyst can conveniently be prepared in situ by mixing the chiral ligand **83** and copper triflate in the reaction system. Scheme 5–35 shows that product **112** can be obtained with good yield and high enan-

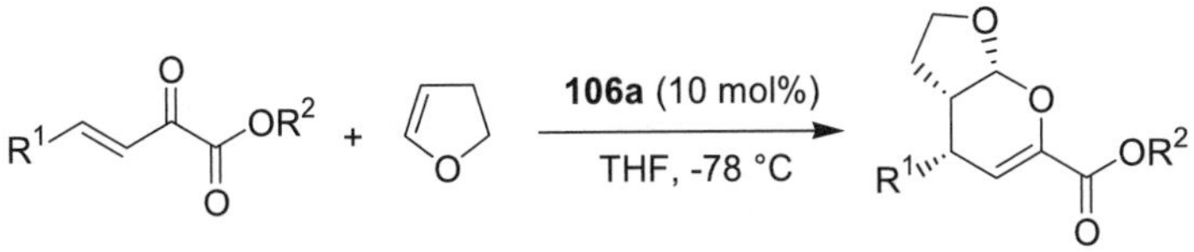

Entry	R^1	R^2	Yield (%)	ee (%)
1	Me	Et	51	> 99.5
2	Ph	Me	96	99.5
3	OEt	Et	84	97.5

Scheme 5–33. Reprinted with permission by Wiley-VCH Verlag Germany, Ref. 44(a).

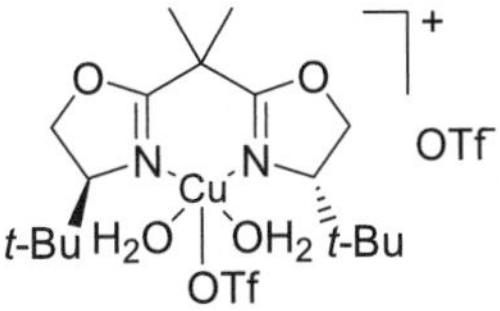

108

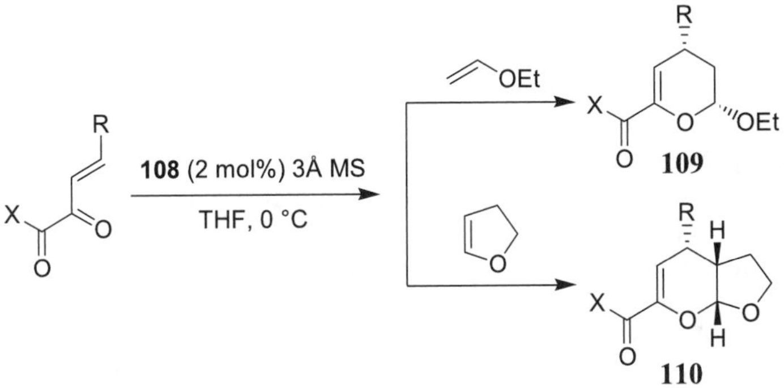

Entry	X	R	Enol Ether	*endo/exo*	Yield (%)	ee (%)
1	OEt	Ph	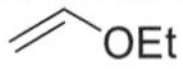 OEt	> 20 : 1	93	97
2	OEt	*i*-Pr	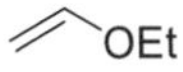 OEt	22 : 1	95	96
3	OEt	Me	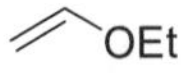 OEt	24 : 1	87	97
4	OEt	OMe	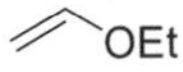 OEt	59 : 1	90	98
5	OEt	OEt	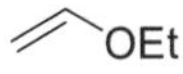 OEt	55 : 1	98	98
6	OEt	SBn	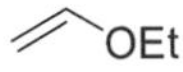 OEt	> 20 : 1	97	99
7	OEt	Ph	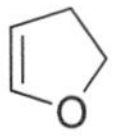 (furan)	16 : 1	96	97
8	OEt	*i*-Pr	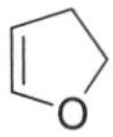 (furan)	16 : 1	94	95
9	N(OMe)Me	Me	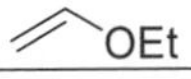 OEt	64 : 1	99	99

Scheme 5–34. Reprinted with permission by Wiley-VCH Verlag Germany, Ref. 43a.

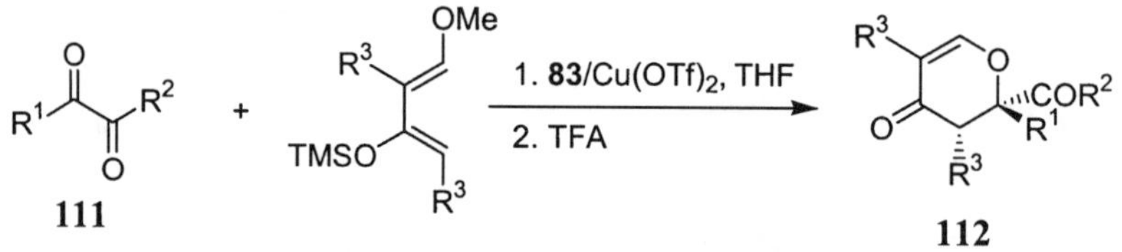

Entry	R^1	R^2	Diene (R^3)	Catalyst (%)	Temp. (°C)	Time (h)	Yield (%)
1	Me	OMe	100 (H)	0.05	-78 to -40	20	90
2	Me	Me	100 (H)	0.05	-78	18	88
3	Me	Et	100 (H)	0.05	-78	20	76
4	Me	Ph	100 (H)	0.05	-78	20	25
5	Et	OMe	100 (H)	0.5	-78	30	70
6	Me	OMe	91 (Me)	2.5	-40	12	85
7	Ph	OEt	91 (Me)	2.5	-40	12	65
8	Me	Me	91 (Me)	2.5	-40	12	81

Scheme 5–35. Reprinted with permission by Am. Chem. Soc., Ref. 44b.

tiomeric excess after treating the hetero Diels-Alder reaction product with TFA.

The copper complex of these bis(oxazoline) compounds can also be used for hetero Diels-Alder reactions of acyl phosphonates with enol ethers.[43b] A favorable acyl phosphonate–catalyst association is achieved via complexation between the vicinal C=O and P=O functional groups. The acyl phosphonates are activated, leading to facile cycloaddition with electron-rich alkenes such as enol ethers. The product cyclic enol phosphonates can be used as building blocks in the asymmetric synthesis of complicated molecules. Scheme 5–36 shows the results of such reactions.

5.5.2 Aza Diels-Alder Reactions

In Section 5.5.1, we discussed the oxo hetero Diels-Alder reaction, or the hetero Diels-Alder reaction involving oxygen as the hetero atom. There is another type of hetero Diels-Alder reaction in which nitrogen-containing compounds are involved, and these are referred to as *aza Diels-Alder reactions*.

Asymmetric aza Diels-Alder reactions provide a useful route to optically active heterocyclics such as piperidines and tetrahydroquinolines.[45] Although successful examples of diastereoselective approaches had been reported as early as 10 years ago,[46] only recently have enantioselective reactions been accomplished.[47] For example, the reaction of chiral amine-derived aromatic imine **115** with Brassard's diene **116** gives adduct **117** with up to 95% diastereoselectivity (Scheme 5–37).[48]

A stoichiometric amount of BINOL–boron chiral Lewis acid **118** activates

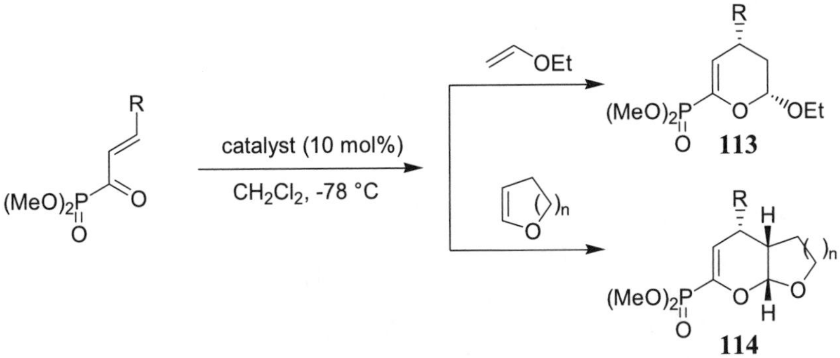

Entry	R	Enol Ether	Catalyst	*endo/exo*	Yield (%)	ee (%)
1	Me	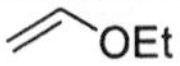 OEt	**106a**	99 : 1	89	99
2	Me	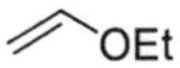 OEt	**106b**	69 : 1	84	93
3	Me	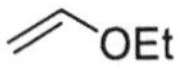 OEt	**107a**	> 99 : 1	85	94
4	Me	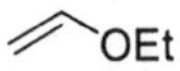 OEt	**107b**	> 99 : 1	100	93
5	Me	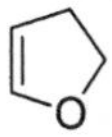 (furan)	**106a**	> 99 : 1	91	95
6	Me	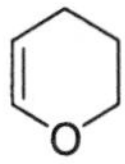 (pyran)	**106b**	98 : 2	55	92
7	Ph	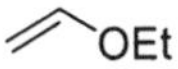 OEt	**107b**	167 : 1	98	98
8	Ph	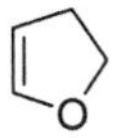 (furan)	**107a**	171 : 1	100	93
9	*i*-Pr	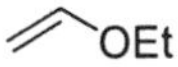 OEt	**107b**	146 : 1	99	96
10	*i*-Pr	 (furan)	**106b**	98 : 2	79	90
11	OEt	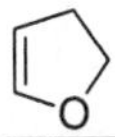 (furan)	**106b**	> 99 : 1	98	97

Scheme 5–36. Reprinted with permission by Am. Chem. Soc., Ref. 43b.

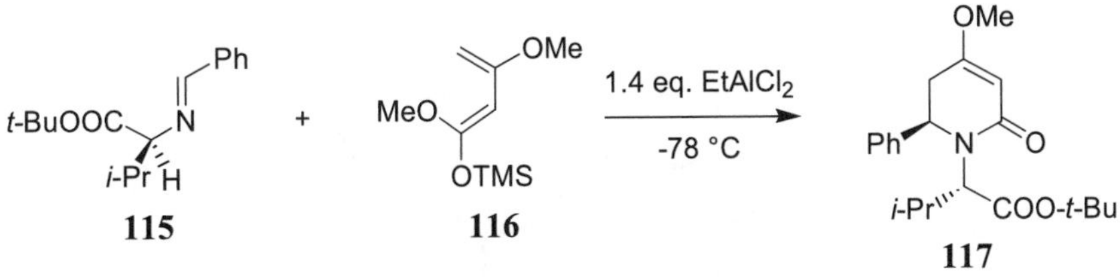

Scheme 5–37

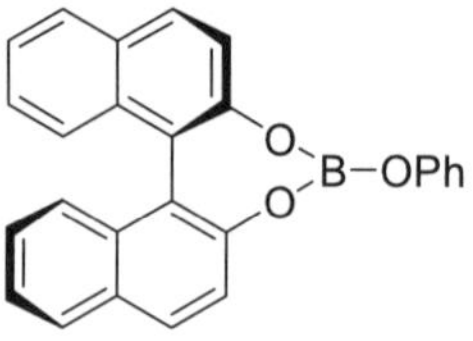

118

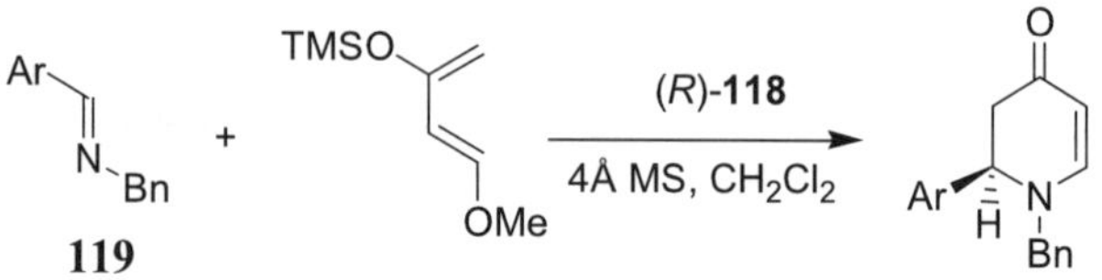

Scheme 5–38

the reaction of aromatic imine **119** with Danishefsky's diene **100**, affording adduct **120** in up to 90% ee (Scheme 5–38).[49]

In the presence of a catalytic amount of Lewis acid, chiral esters **121** and **122** show improved diastereoselectivity. When they are allowed to react with cyclopentadiene, adducts of type **123** are obtained with good stereoselectivity.[50]

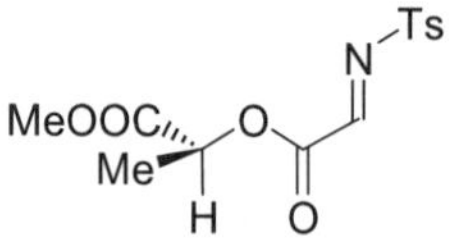

121 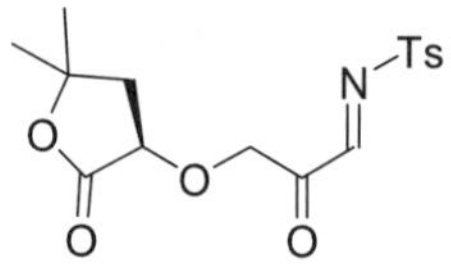 **122** 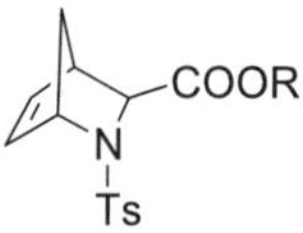 **123**

Kobayashi et al.[51] have reported an asymmetric Mannich-type reaction using chiral zirconium catalysts of type **124** (see Section 3.7). This catalyst is also effective for asymmetric aza Diels-Alder reactions. Kobayashi's study showed that the ligand had a profound influence on the yields and enantiose-lectivities of the reaction, and NMI (1-methylimidazole) proved to be the best ligand.[51] With an increase in the amount of catalyst, both the chemical yields and enantioselectivities of the product can be enhanced. Scheme 5–39 depicts such aza Diels-Alder reactions, and its table shows that good to excellent enantioselectivity can be obtained for most reactions.

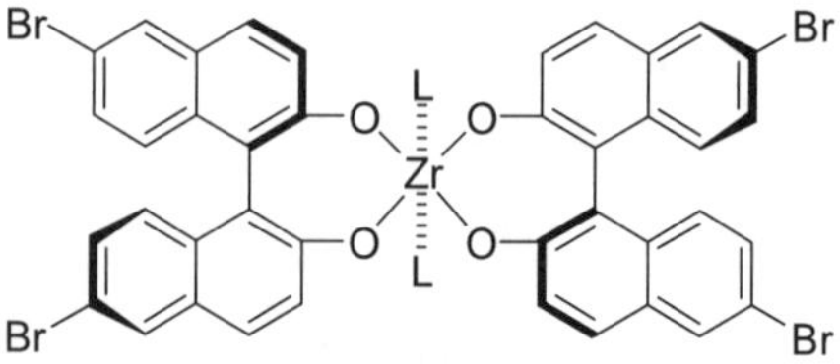

124 L = ligands

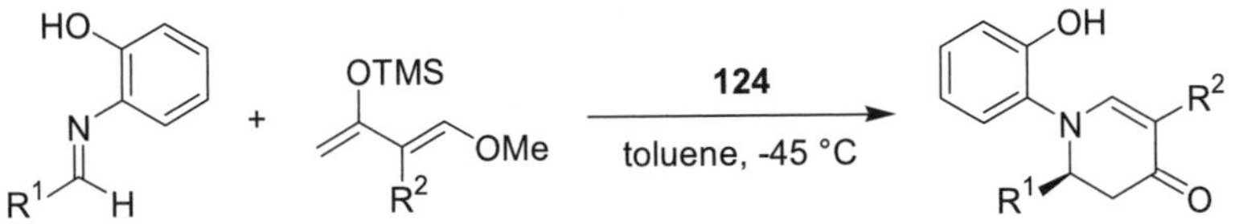

Entry	R^1	R^2	Catalyst (mol%)	Yield (%)	ee (%)
1	α-Nap	H	5	72	67
2	α-Nap	H	10	86	82
3	α-Nap	H	20	96	88
4	α-Nap	H	30	98	89
5	α-Nap	H	50	88	90
6	α-Nap	Me	10	79	89
7	α-Nap	Me	20	93	93
8	o-MePh	H	20	83	82

Scheme 5–39. Reprinted with permission by Wiley-VCH Verlag Germany, Ref. 51.

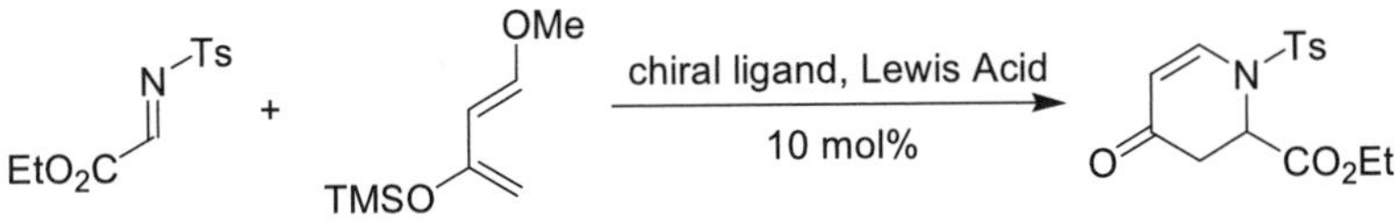

Entry	Ligand	Lewis acid	Yield (%)	ee (%)
1	**81**	$Zn(OTf)_2$	74	17
2	**83**	$2CuOTf \cdot C_6H_6$	74	12
3	**83**	$Cu(OTf)_2$	60	10
4	**103**	$Zn(OTf)_2$	70	8

Scheme 5–40. Reprinted with permission by Wiley-VCH Verlag GmbH, Ref. 52.

Jørgensen's group reported the aza Diels-Alder reactions in the presence of several chiral catalysts.[52] They found that chiral bis(oxazoline) ligands **81**, **83**, **103**, **104**, and **105**, which were effective in asymmetric oxo hetero Diels-Alder reactions, induced the aza Diels-Alder reaction of α-imino ester with Danishefsky's diene with only poor to moderate enantioselectivity. Selected results are listed in Scheme 5–40.

On the other hand, the combination of Tol–BINAP with CuClO$_4$ has been shown to be very effective for aza Diels-Alder reactions. As shown in Scheme 5–41, moderate yield and good to excellent enantioselectivity can be obtained in the reaction of α-imino ester with diene **91** or **100**.

Another successful aza Diels-Alder reaction involves 2-azadienes of type **125** with dienophile **126** in the presence of bis(oxazoline) catalyst **106a**.[53] Product **127** is obtained in a high *exo:endo* ratio, as well as high enantiomeric excess for the *exo*-isomer of **127**. The high enantioselectivity and high yield rely on the chiral catalyst **106a**, which activates the dienophile by complexation with an appropriate functional group and does not irreversibly coordinate with the

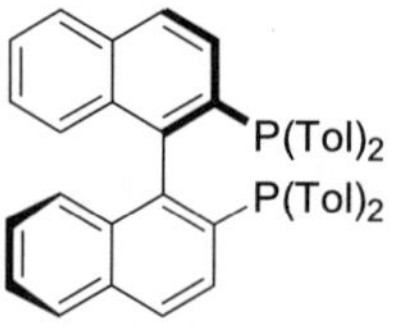

(R)-Tol-BINAP

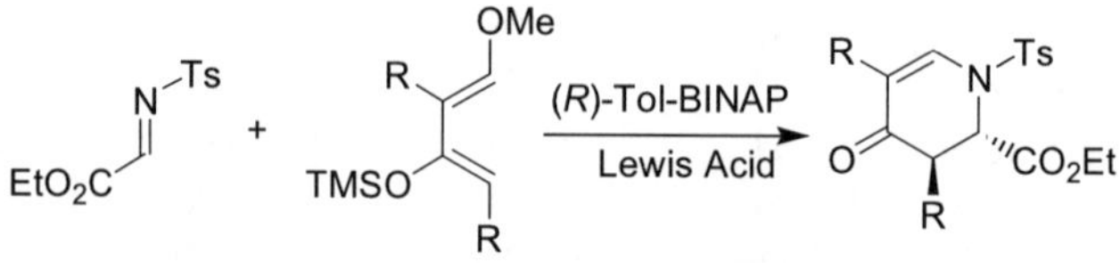

Entry	Lewis acid	Catalyst (mol%)	Diene (R)	Yield (%)	ee (%)
1	CuClO$_4$.4MeCN	10	**100** (H)	68	80
2	CuClO$_4$.4MeCN	10	**91** (Me)	67	94
3	CuClO$_4$.4MeCN	5	**91** (Me)	70	94
4	CuClO$_4$.4MeCN	1	**91** (Me)	70	96
5	CuClO$_4$.4MeCN	10	**91** (Me)	70	81

Scheme 5–41. Reprinted with permission by Wiley-VCH Verlag GmbH, Ref. 52.

nucleophilic nitrogen atom of the azadiene. The experimental results are shown in Scheme 5–42.

The high stereoselectivity of the reaction can be explained by the transition state shown in Figure 5–9. An *exo*-approach of the diene to the less hindered face of the square–planar complex causes the high enantioselectivity.

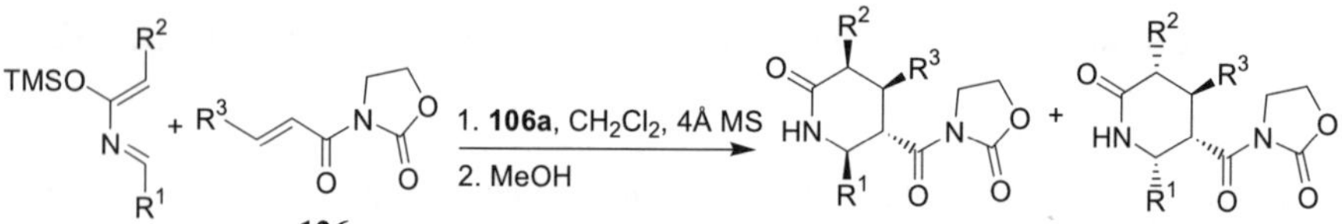

Entry	R^1	R^2	R^3	Catalyst (mol%)	Temp. (°C)	*exo/endo*	Yield (%)	ee (%)
1	Ph	Me	Me	8	-78	-	0	-
2	Ph	Me	Me	8	-45	> 99 : 1	80	95.1
3	Ph	Me	Me	8	r. t.	> 99 : 1	96	94
4	Ph	Me	Me	5	r. t.	> 99 : 1	85	93.4
5	Ph	H	H	8	-45	6.1 : 1	83	98.3
6	Ph	Me	H	8	-45	> 99 : 1	96	98.3
7	Ph	H	Me	8	r. t.	> 99 : 1	80	93
8	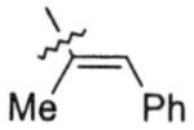 (Me, Ph group)	Me	Me	8	r. t.	> 99 : 1	98	90
9	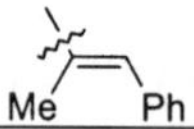 (Me, Ph group)	Me	Me	8	-45	> 99 : 1	62	95.4

Scheme 5–42. Reprinted with permission by Am. Chem. Soc., Ref. 53.

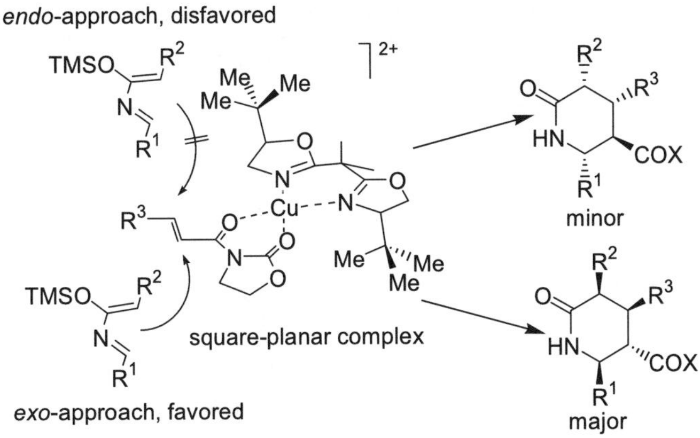

endo-approach, disfavored

exo-approach, favored

Figure 5–9. Transition state for bis(oxazoline) catalyzed aza Diels-Alder reaction. Reprinted with permission by Am. Chem. Soc., Ref. 53.

5.6 FORMATION OF QUATERNARY STEREOCENTERS THROUGH DIELS-ALDER REACTIONS

Besides the methods discussed in Chapter 2, some quaternary stereocenters can also be conveniently constructed through the enantioselective Diels-Alder reaction of the 2-substituted acroleins **75** and **128–130**.

In particular, Diels-Alder adducts from the enantioselective reaction of 2-bromo-acrolein and 2-chloroacrolein with a variety of dienes are of exceptional synthetic versatility. Readers are advised to consult the review article by Corey and Guzman-Perez.[54] For the purpose of quick reference, chiral ligands commonly used in Diels-Alder reactions are listed in Table 5–6.

5.7 INTRAMOLECULAR DIELS-ALDER REACTIONS

Intermolecular cycloadditions or Diels-Alder reactions have proved to be a successful route to several valuable intermediates for natural product syntheses. In creating new chiral centers, most of these reactions apply single asymmetric induction. As mentioned in Chapter 3, in the asymmetric synthesis of the octahydronaphthalene fragment, the Roush reaction is used twice. Subsequent intramolecular cyclization leads to the key intermediate, the aglycones, of several natural antitumor antibiotics. On the other hand, the Diels-Alder reaction of a dienophile-bearing chiral auxiliary can also be used intramolecularly to build

TABLE 5–6. List of Chiral Catalysts for the Diel-Alder Reaction

Chiral Catalyst	References	Chiral Catalyst	References
	55		20, 21b
			56
	57		58
M = Al, B, Ti	38, 59	M = Al, B, Ti	60
M = Al, B, Ti	61		22, 61c, 62
M = Cu, Fe	28, 29, 30	M = Cu	63
	64		60b
	65	M = Al, B, Sn, Ti	59a, 60b, 66

TABLE 5–6 *(Continued)*

Chiral Catalyst	References	Chiral Catalyst	References
	60b		39, 41
 131	67	 **132a** $R^1 = n$-Bu, $R^2 = $ H **132b** $R^1 = n$-Bu, $R^2 = CH_3$ **132c** $R^1 = $ H, $R_2 = $ H	25a
 133	59d	 **76a**	68
 134	69	 **135a** X = Cl **135b** X = Br	37a
 136	30c	 **77a**	70
 137a X = Br **137b** X = B[3, 5-$(CF_3)_2C_6H_4]_4$	71		

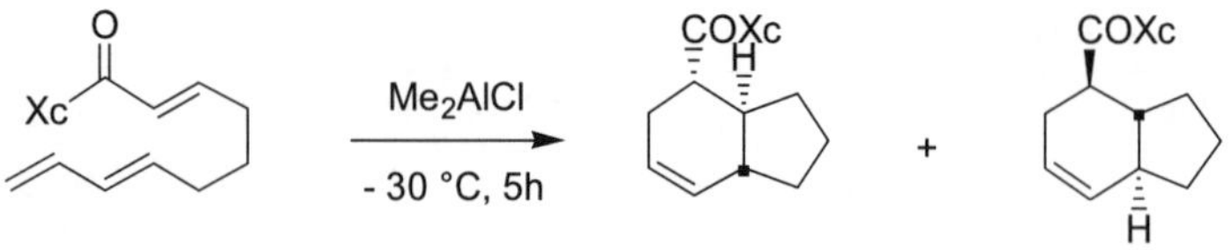

138a Xc = X$_V$
138b Xc = X$_P$
138c Xc = X$_N$
138d Xc = X$_{PPG}$

endo I
139a Xc = X$_V$
139b Xc = X$_P$

endo II
140a Xc = X$_N$
140b Xc = X$_{PPG}$

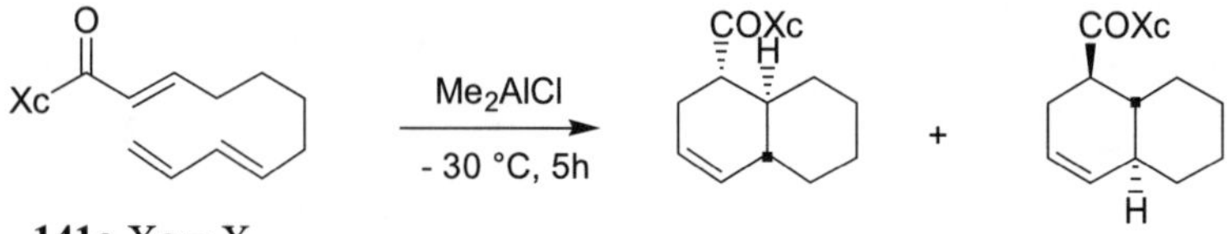

141a Xc = X$_V$
141b Xc = X$_V$
141c Xc = X$_N$
141d Xc = X$_{PPG}$

142a Xc = X$_V$
142b Xc = X$_P$

143a Xc = X$_N$
143b Xc = X$_{PPG}$

chiral auxiliary Xc

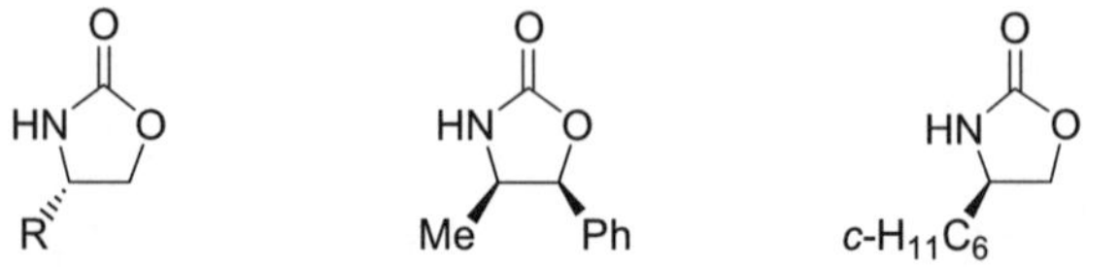

144a R = (CH$_3$)$_2$CH- (HX$_V$)
144b R = PhCH$_2$- (HX$_P$)

145 HX$_N$

146 HX$_{PPG}$

Scheme 5–43

up important natural product subunits. In Scheme 5–43, Xc, which represents various chiral auxiliaries such as chiral amines, alcohols, or imides, can be removed afterward to complete the synthesis. In the given example, the bicyclic ring system is *trans*-fused. Table 5–7 gives some results for this type of intramolecular cyclization.[4]

Camphor sultam derivatives have proved to be effective chiral auxiliaries in many different types of asymmetric reactions. As shown in Scheme 5–44, chiral camphor sulfam can be applied in the synthesis of (−)-pulo'upone precursor **151** using an intramolecular Diels-Alder reaction. A Wittig reaction of **148** with **147** connects the chiral auxiliary to the substrate, and subsequent intramolecular Diels-Alder reaction via transition state **150** affords product **151**. Compound **151** already has the stereochemistry of (−)-pulo'upone **153**.[72]

Apart from its application in intermolecular Diels-Alder reactions, chiral acyloxy boron (CAB) can also be used to effect intramolecular Diels-Alder reactions with excellent stereoselectivity (Scheme 5–45).[73]

TABLE 5–7. Me$_2$AlCl-Promoted Intramolecular Diels-Alder Reactions of Trienimides 138 and 141

Imide (Xc)	*endo* I:*endo* II	*endo*:*exo*	Purified Ratio	Yield (%)
138a (X$_V$)	83:17	>99:1	>99:1	60 (**139a**)
138b (X$_P$)	95:5	>99:1	>99:1	73 (**139b**)
138c (X$_N$)	15:85	>99:1	>99:1	70 (**140a**)
138d (X$_{PPG}$)	3:97	>99:1	>99:1	65 (**140a**)
141a (X$_V$)	92:8	>30:1	>99:1	65 (**142a**)
141b (X$_P$)	97:3	>50:1	>99:1	88 (**142b**)
141c (X$_N$)	9:91	>50:1	>99:1	70 (**143a**)
141d (X$_{PPG}$)	6:94	>30:1	>99:1	70 (**143b**)

Reprinted with permission by Am. Chem. Soc., Ref. 4.

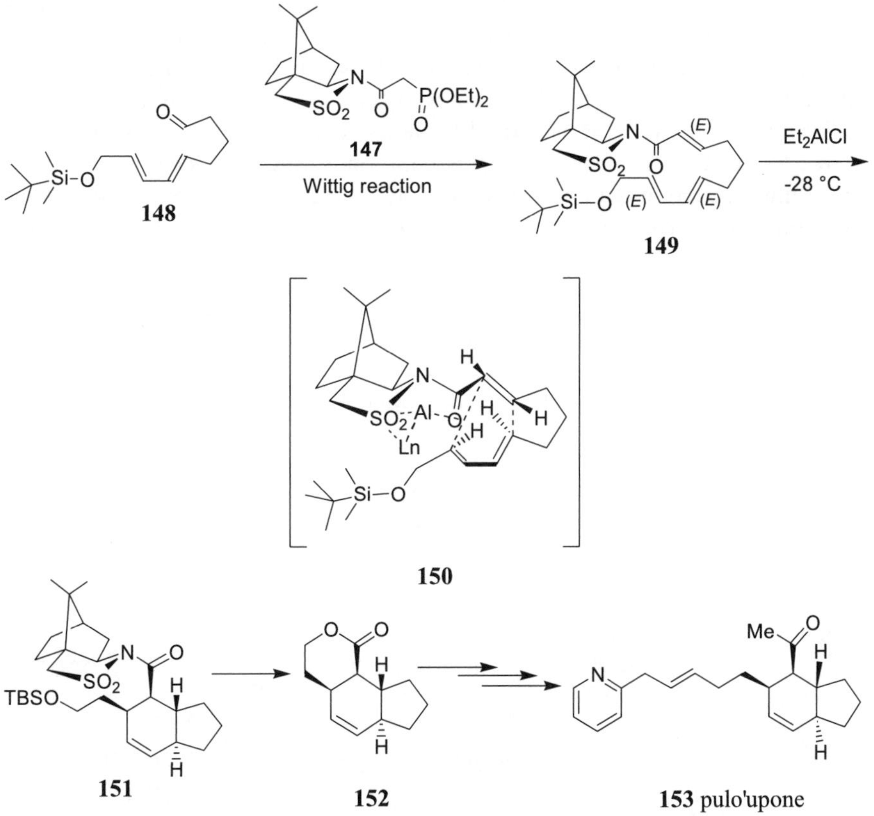

Scheme 5–44

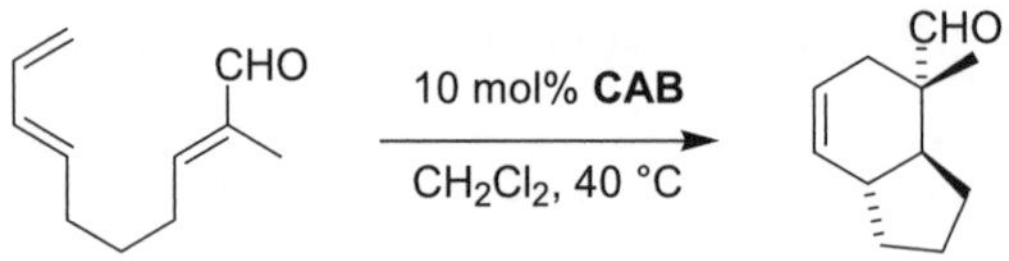

Scheme 5–45

5.8 RETRO DIELS-ALDER REACTIONS

The term *Diels-Alder reaction* in a general sense refers to the reaction between a diene and a dienophile. Retro Diels-Alder reaction is a process that, under certain conditions, produces diene and olefin or a compound containing a C=C bond. The application of flash vacuum pyrolysis to effect the retro Diels-Alder reaction, as shown in Schemes 5–46 and 5–47, has become the standard procedure since the introduction of the method by Stork et al.[74] in the 1970s. Therefore, alkenes that are difficult to access by conventional methods may be obtained via retro Diels-Alder reactions.[75] In particular, this reaction allows the preparation of thermodynamically less stable compounds such as 4,5-dialkyl cyclopenta-2-en-one. In this case, the alkene functional group can be regarded as being protected by cyclopentadiene (as shown in **154** or **157**), which, after subsequent reaction, can easily be removed through quick pyrolysis.

Because compound **156** is a rigid framework that can be readily generated from a dimer of cyclopentadiene, this compound offers an ideal structure on which many reactions can be carried out stereoselectively. Scheme 5–47 shows the synthesis of prostaglandin A_1 or A_2 (**158**). After the two side chains have

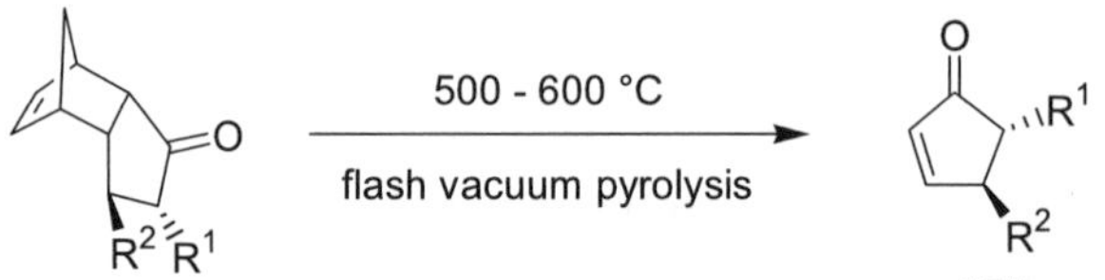

Scheme 5–46

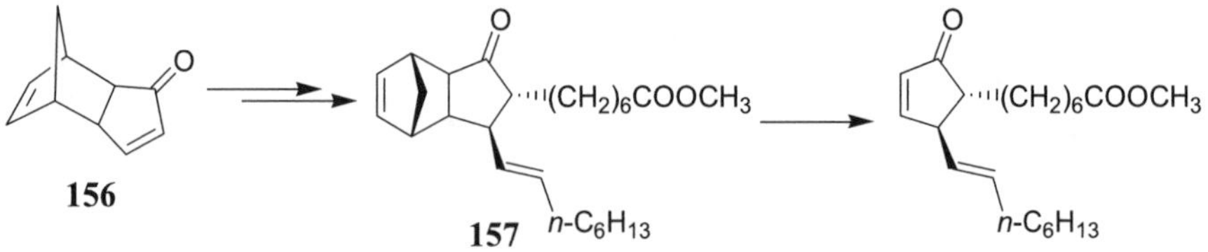

Scheme 5–47

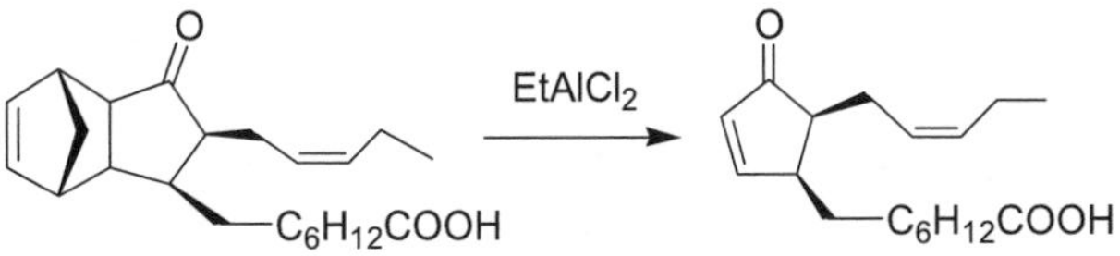

Scheme 5–48

been attached to **156** to give compound **157** via a three-component coupling process,[76] the desired compound **158** can be obtained after a quick pyrolysis.[77]

In the presence of a more reactive dienophile, a retro Diels-Alder reaction can be carried out at or below room temperature when catalyzed by a Lewis acid.[78] In fact, this process can be regarded as a trans-Diels-Alder reaction in which the C=C bond is replaced by another more reactive functionality. Thus, when treated with fumaronitrile in the presence of $EtAlCl_2$ at ambient temperature for 2 hours, compound **159** can easily be converted to compound **160** with the removal of cyclopentadiene (Scheme 5–48).

Sometimes [4+2] reversion of highly functionalized tricyclo[5.2.1.0.2.6]-decanones is not compatible with high temperature or Lewis acid treatment due to double bond isomerization, rearrangement, and/or extensive decomposition. Grieco and Abood[79] applied pentamethylcyclopentadiene derivative **161** in replacement of **156** and found that [4+2] reversion could be accelerated. Scheme 5–49 shows the application of this reagent in the synthesis of (+)-(15S)-prostaglandin A_2. Stirring a 0.06 M solution of **162** in $ClCH_2CH_2Cl$ containing 3.5 equivalents of Me_2AlCl and 5.0 equivalents of fumaronitrile at 10°C for over 48 hours gave **163** with 70% yield (Scheme 5–49).

Compound **168** is a key intermediate for the synthesis of prostaglandin or prostacyclin compounds. Scheme 5–50 shows its preparation via a retro Diels-Alder reaction and subsequent treatment. Using enzyme-catalyzed acetylation, Liu et al.[80] succeeded in the asymmetric synthesis of enantiomerically pure (+)/(−)-**156** and (−)-**168** from the meso-diol **164**. When treated with vinyl acetate, meso-diol **164** can be selectively acetylated to give (+)-**165** in the presence of Candida cyclindracea lipase (CCL). The yield for the reaction is 81%, and the enantiomeric excess of the product is 98.3%.

As previously mentioned, the optically pure synthons (+)- and (−)-**156** can be used in the asymmetric synthesis of many important intermediates via

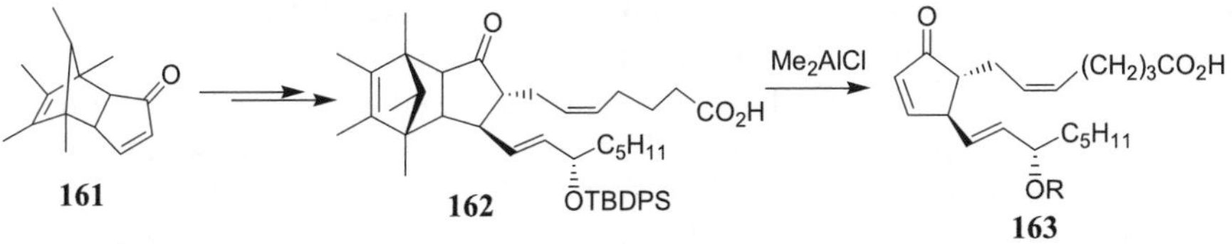

Scheme 5–49

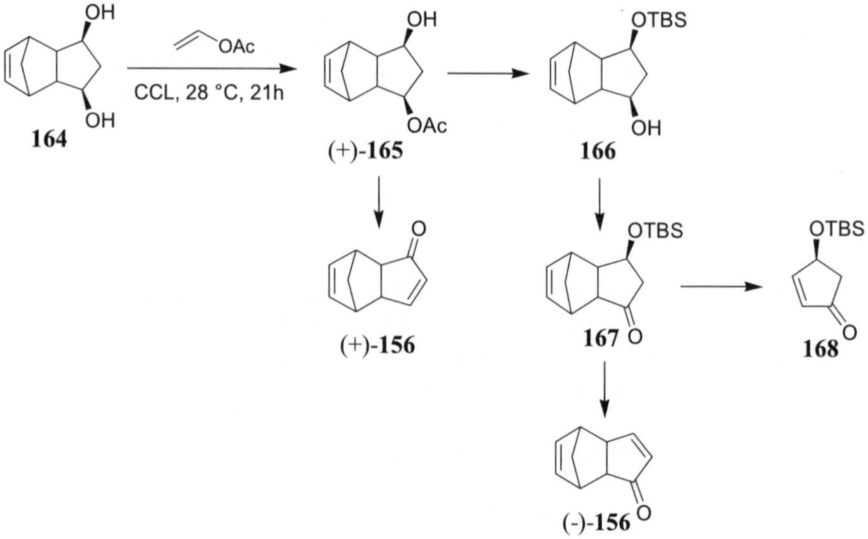

Scheme 5–50

retro Diels-Alder reactions. Indeed, ($\pm$)-chromomoric acid methyl ester, ($\pm$)-invictolide, and ($\pm$)-epiinvictolide have all been synthesized using ($\pm$)-**156** as the starting material.[81]

5.9 ASYMMETRIC DIPOLAR CYCLOADDITION

1,3-Dipolar addition is closely related to the Diels-Alder reaction, but allows the formation of five-membered adducts, including cyclopentane derivatives. Like Diels-Alder reactions, 1,3-dipolar cycloaddition involves [4+2] concerted reaction of a 1,3-dipolar species (the 4π component and a dipolar 2π component). Very often, condensation of chiral acrylates with nitrile oxides or nitrones gives only modest diastereoselectivity.[82] 1,3-Dipolar cycloaddition between nitrones and alkenes is most useful and convenient for the preparation of isoxazolidine derivatives, which can then be readily converted to 1,3-amino alcohol equivalents under mild conditions.[83] The low selectivity of the 1,3-dipolar reaction can be overcome to some extent by introducing a chiral auxiliary to the substrate. As shown in Scheme 5–51, the reaction of **169** with acryloyl chloride connects the chiral sultam to the acrylic acid substrate, and subsequent cycloaddition yields product **170** with a diastereoselectivity of 90:10.[84]

Mukai et al.[85] reported an asymmetric 1,3-dipolar cycloaddition of chromium(0)-complexed benzaldehyde derivatives. As shown in Scheme 5–52, heating chiral nitrone **171a**, derived from Cr(CO)$_3$-complexed benzaldehyde, with electron-rich olefins such as styrene (**173a**) or ethyl vinyl ether (**173b**) generates the corresponding chiral *cis*-3,5-disubstituted isoxazolidine adduct **174** or

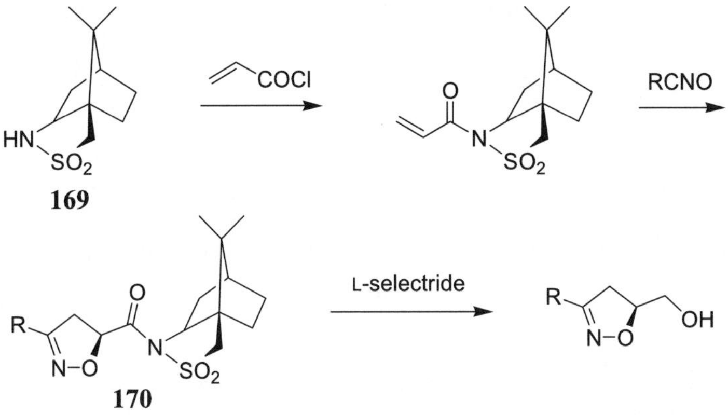

Scheme 5–51

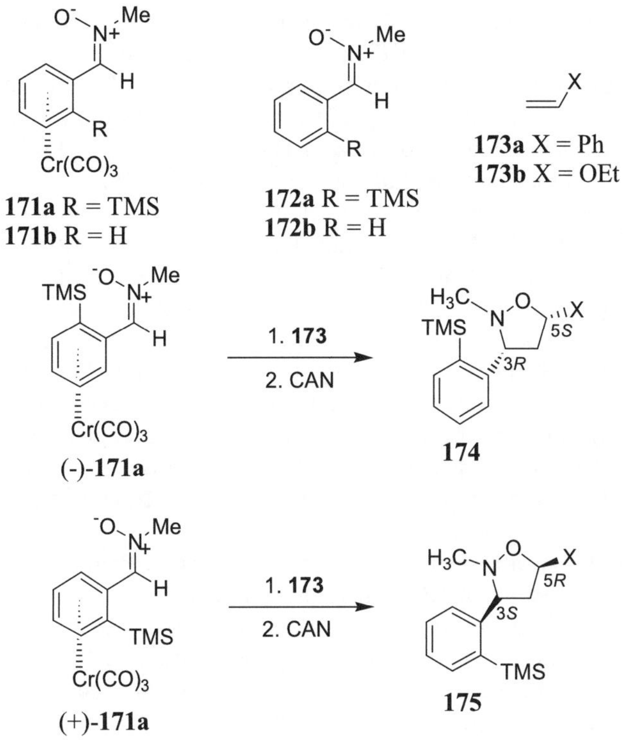

Scheme 5–52

175 with ee in the range of 96–98%. These isoxazolidines can be further converted into chiral amino alcohols.

As compound **171** is a special planar chiral molecule, the dipolar reaction of compound **171** can be regarded as a substrate-controlled reaction. In 1,3-dipo-

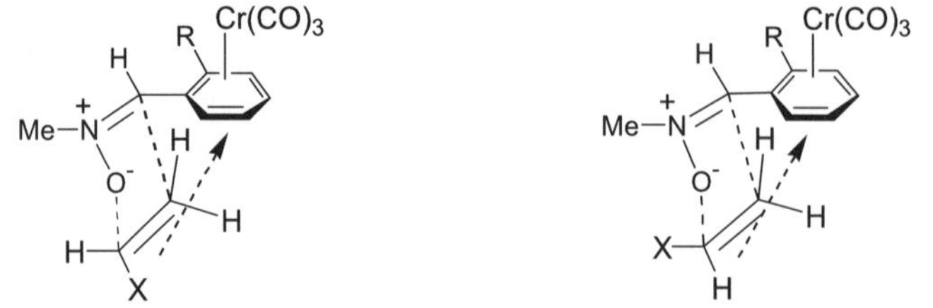

endo transition state (A) ⟶ *cis*   *exo*-transition state (B) ⟶ *trans*

Figure 5–10

lar cycloaddition of the chromium-complexed nitrone [e.g., (−)-**171a** or (+)-**171a** with electron-rich olefins **173a** or **173b**], the *ortho*-TMS group in the Cr complex governs the geometry of the aldehyde group, whereas the chromium complexation controls the facial selectivity. These two factors in the Cr complex seem to contribute to the observed high *cis*-stereoselectivity (Fig. 5–10). For example, when nitrone (+)-**171a** is treated with styrene (**173a**), followed by decomplexation with cerium ammonium nitrate (CAN), the optically active *cis*-3,5-disubstituted isoxazolidine (−)-**174** is produced with 65% yield. An asymmetric 1,3-dipolar cycloaddition of (+)-**171a** with ethyl vinyl ether **173b** also results in formation of an optically active isoxazolidine ring.

The high stereoselectivity can be explained by the proposed transition states (Fig. 5–10). The *endo*-transition state (A) offers the advantage of releasing electrons through space from the electron-rich substituent on the dipolarophile to the electron-deficient aromatic ring, thus resulting in a more stable (lower energy) transition state. The corresponding *exo*-transition state (B) does not have such a stabilizing factor. The high ee observed for **174** or **175** can best be explained by the approach of dipolarophiles to the chromium complexed nitrone, which must be from the face opposite that occupied by the Cr complexation. When uncomplexed nitrone **172a** or **172b** is considered, rather than (*S*)-**171a** or **171b**, repulsion between the *p*-orbital electrons of the benzene ring and the electron-rich substituent on the dipolarophile in the *endo*-transition state (A) would become a strong destabilizing element. Preference for the *endo*-type (A) over the *exo*-type (B) is therefore expected.

Using a stoichiometric amount of (*R*,*R*)-DIPT as the chiral auxiliary, optically active 2-isoxazolines can be obtained via asymmetric 1,3-dipolar addition of achiral allylic alcohols with nitrile oxides or nitrones bearing an electron-withdrawing group (Scheme 5–53).[86a] Furthermore, the catalytic 1,3-dipolar cycloaddition of nitrile oxide has been achieved by adding a small amount of 1,4-dioxane (Scheme 5–53, Eq. 3).[86b] The presence of ethereal compounds such as 1,4-dioxane is crucial for the reproducibly higher stereoselectivity.

Kobayashi's chiral lanthanide complex **63** has been used for asymmetric Diels-Alder reactions, and very good results have been obtained (see Section 5.4.2). This kind of complex is also effective in asymmetric 1,3-dipolar reactions.[87] The chiral ligand is prepared in situ by mixing Yb(OTf)$_3$,

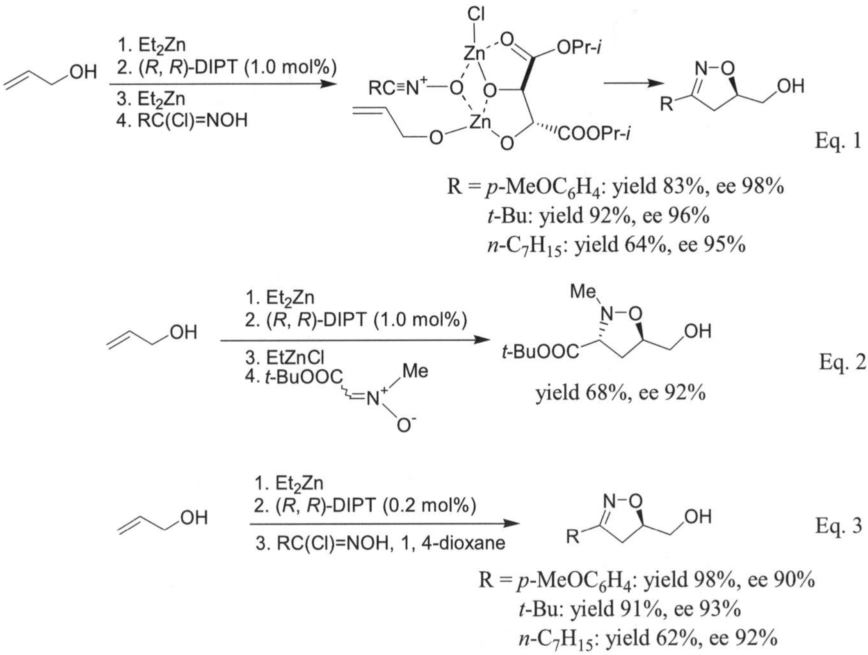

Eq. 1

R = p-MeOC$_6$H$_4$: yield 83%, ee 98%
t-Bu: yield 92%, ee 96%
n-C$_7$H$_{15}$: yield 64%, ee 95%

Eq. 2

yield 68%, ee 92%

Eq. 3

R = p-MeOC$_6$H$_4$: yield 98%, ee 90%
t-Bu: yield 91%, ee 93%
n-C$_7$H$_{15}$: yield 62%, ee 92%

Scheme 5–53

(S)-BINOL and a tertiary amine together. The reaction proceeds smoothly at room temperature to afford the corresponding isoxazolidine derivatives with good yield and diastereoselectivity, as well as enantioselectivity. Kobayashi's study shows that combining the chirality of BINOL and the chiral amine is crucial for getting better enantioselectivity. Use of N,N-di-[(1R)-(α-naphthyl)ethyl]-N-methylamine [**176**, (R)-MNEA] resulted in the highest enantioselectivity. As shown in Scheme 5–54, in the presence of chiral ytterbium triflate complex **63** containing (R)-MNEA **176** as the chiral tertiary amine, the asymmetric 1,3-dipolar reaction of nitrone **177** with dienophile **178** proceeds smoothly, providing *endo*-product **179** in excellent yield and diastereoselectivity as well as enantiomeric excess.

Bis(oxazoline)-type complexes, which have been found useful for asymmetric aldol reactions, Diels-Alder, and hetero Diels-Alder reactions can also be used for inducing 1,3-dipolar reactions. Chiral nickel complex **180**, which can be prepared by reacting equimolar amounts of Ni(ClO)$_4$ · 6H$_2$O and the corresponding (R,R)-4,6-dibenzofurandiyl-2,2′-bis(4-phenyloxazoline) (DBFOX/Ph) in dichloromethane, can be used for highly *endo*-selective and enantioselective asymmetric nitrone cycloaddition. The presence of 4 Å molecular sieves is essential to attain high selectivities.[88] In the absence of molecular sieves, both the diastereoselectivity and enantioselectivity will be lower. Representative results are shown in Scheme 5–55.

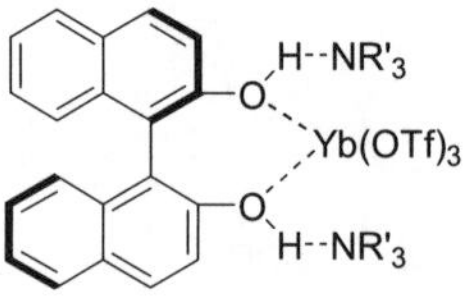

63 chiral ytterbium triflate 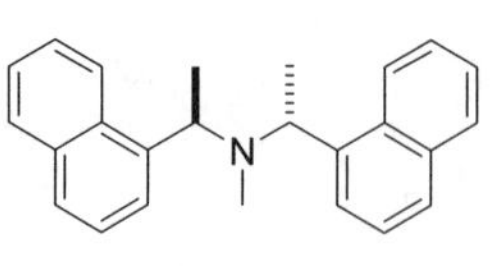   **176** (R)-MNEA

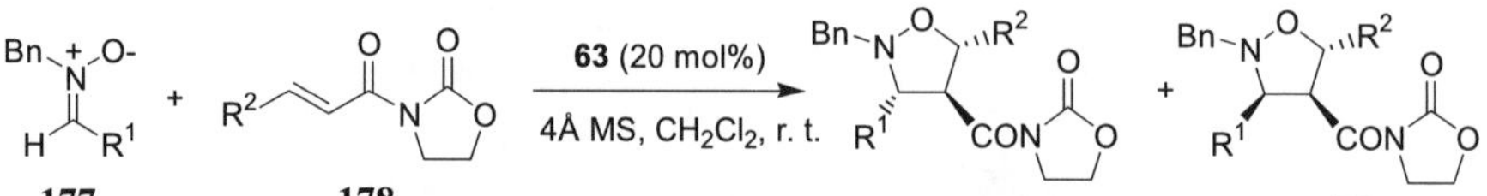

Scheme 5–54. Reprinted with permission by Am. Chem. Soc., Ref. 87.

Entry	R¹	R²	Yield (%)	endo/exo	ee (%)
1	Ph	CH₃	92	99 : 1	96
2	p-ClPh	CH₃	93	99 : 1	92
3	p-MePh	CH₃	82	95 : 5	90
4	2-Furyl	CH₃	89	95 : 5	89
5	1-Naphthyl	CH₃	88	98 : 2	85
6	Ph	H	91	> 99 : 1	79
7	Ph	C₃H₇	89	98 : 2	93
8	C₂H₅	CH₃	88	54 : 47	96

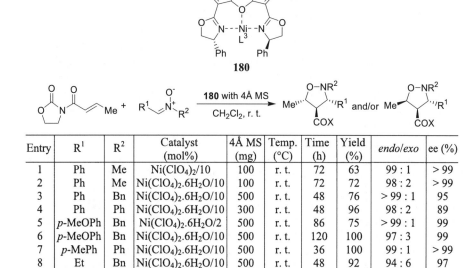

180

Entry	R¹	R²	Catalyst (mol%)	4Å MS (mg)	Temp. (°C)	Time (h)	Yield (%)	endo/exo	ee (%)
1	Ph	Me	Ni(ClO₄)₂/10	100	r. t.	72	63	99 : 1	> 99
2	Ph	Me	Ni(ClO₄)₂.6H₂O/10	100	r. t.	72	72	98 : 2	> 99
3	Ph	Bn	Ni(ClO₄)₂.6H₂O/10	500	r. t.	48	76	> 99 : 1	95
4	Ph	Ph	Ni(ClO₄)₂.6H₂O/10	300	r. t.	48	96	98 : 2	89
5	p-MeOPh	Bn	Ni(ClO₄)₂.6H₂O/2	500	r. t.	86	75	> 99 : 1	99
6	p-MeOPh	Bn	Ni(ClO₄)₂.6H₂O/10	500	r. t.	120	100	97 : 3	99
7	p-MePh	Ph	Ni(ClO₄)₂.6H₂O/10	500	r. t.	36	100	99 : 1	> 99
8	Et	Bn	Ni(ClO₄)₂.6H₂O/10	500	r. t.	48	92	94 : 6	97

Scheme 5–55. Reprinted with permission by Am. Chem. Soc., Ref. 88.

5.10 ASYMMETRIC CYCLOPROPANATION

Cyclopropane subunit has been found in many natural and synthetic compounds possessing important biological properties. For example, the two compounds FR-900848[89] and U-106305[90] show strong bioactivity.

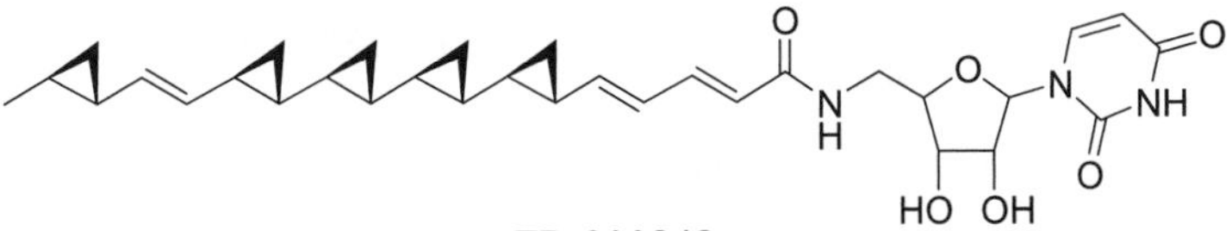

FR-900848

U-106305, cholesteryl ester transfer protein inhibitor

Furthermore, cyclopropane structures have often served as intermediates in organic synthesis. For these reasons, olefin cyclopropanation has proved to be a useful tool for synthetic organic chemists. This has led to the development of several methods for cyclopropanation reactions,[91] including the metal-catalyzed reactions of diazo compounds with olefins, as well as the Simmons-Smith reaction.

Taking 1,2-disubstituted cyclopropane as an example, retro synthesis analysis shows that there are three possible ways to disconnect the three-membered ring—*a*, *b*, and *c* as shown in Figure 5–11. Route *a* involves the addition of methylene across a double bond, and this is often a stereospecific conversion or Simmons-Smith reaction.[92] One can clearly see that route *b* or *c* will encounter the issue of *cis/trans*-product formation.

Two strategies have been adopted for asymmetric cyclopropanation. First, there are auxiliary-based methods, involving a covalently attached adjacent chiral moiety on either the olefin or the cyclopropylating agent. The second process, on the other hand, employs a chiral ligand on a metal catalyst. This method is more applicable to route *b* or *c*, and this is an issue that warrants further discussion.

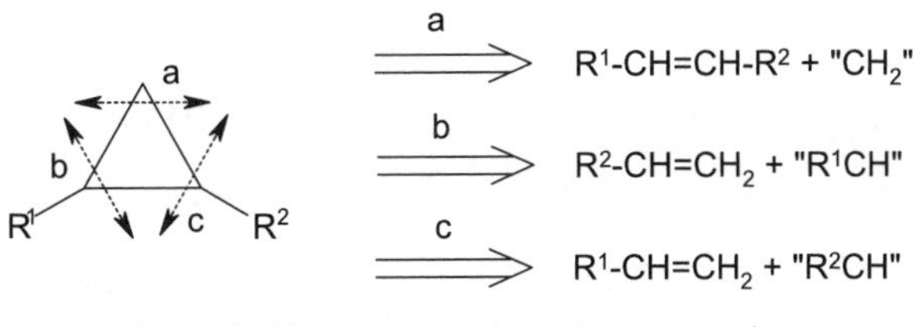

a $R^1\text{-CH=CH-}R^2 + \text{"CH}_2\text{"}$

b $R^2\text{-CH=CH}_2 + \text{"}R^1\text{CH"}$

c $R^1\text{-CH=CH}_2 + \text{"}R^2\text{CH"}$

Figure 5–11. Asymmetric cyclopropanation.

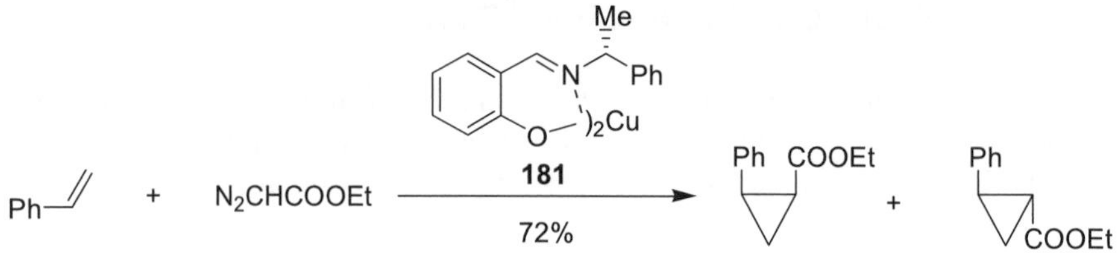

cis-**182**, minor *trans*-**182** major, 6% ee

Scheme 5–56

5.10.1 Transition Metal Complex–Catalyzed Cyclopropanations

Certain transition metal complexes catalyze the decomposition of diazo compounds. The metal-bonded carbene intermediates behave differently from the free species generated via photolysis or thermolysis of the corresponding carbene precursor. The first catalytic asymmetric cyclopropanation reaction was reported in 1966 when Nozaki et al.[93] showed that the cyclopropane compound *trans*-**182** was obtained as the major product from the cyclopropanation of styrene with diazoacetate with an ee value of 6% (Scheme 5–56). This reaction was effected by a copper(II) complex **181** that bears a salicyladimine ligand.

This report initiated activities that resulted in the discovery of **183**, the Aratani catalyst. It is now applied in the enantioselective synthesis of ethyl-2,3-dimethoxycyclopropanecarboxylate, the key intermediate in the preparation of cilastatin (Scheme 5–57).[94]

The next major contribution in asymmetric cyclopropanation was the introduction of chiral semicorrin ligands **184** by Fritschi et al.[95] This ligand has been used for coordinating with copper and has been found to provide improved enantiocontrol in the cyclopropanation of monosubstituted olefins. Copper(I), coordinated by only one semicorrin ligand, is believed[96] to be the catalytically active oxidation state. The copper(I) oxidation state can be reached directly

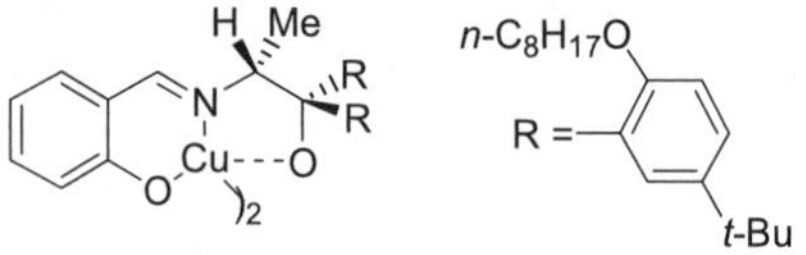

183

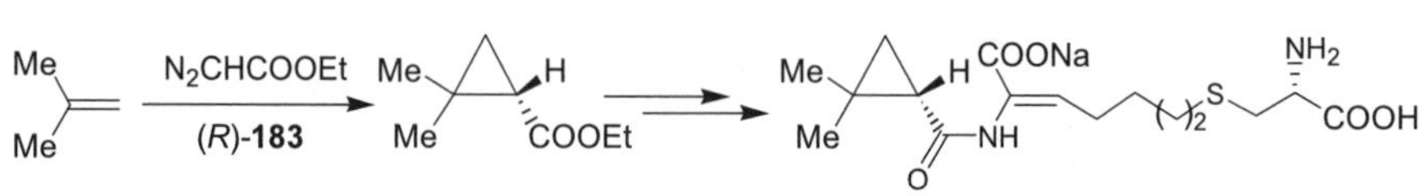

cilastatin

Scheme 5–57

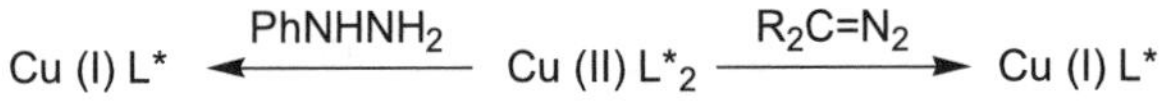

Scheme 5–58

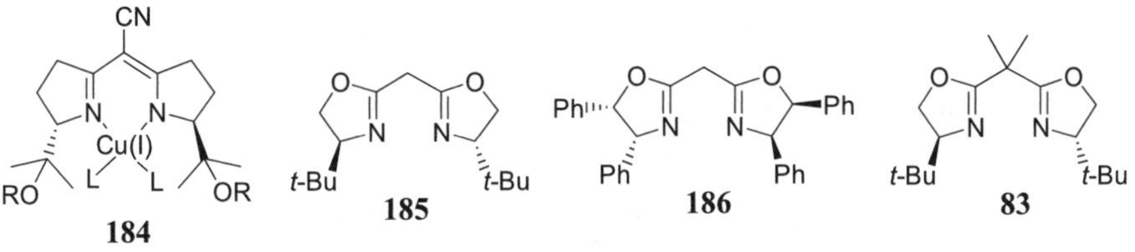

Figure 5–12. Chiral bis(oxazoline) ligands for asymmetric cyclopropanation.

through the reduction of Cu(II)L₂ by the diazo compound or by the treatment of copper(II)–semicorrin with phenylhydrazine (Scheme 5–58).

Subsequently, Lowenthal and co-workers,[31a,97] Evans et al.,[31b] and Müller et al.[98] reported chiral bis(oxazoline) ligands **185**, **186**, and **83** as shown in Figure 5–12. The *gem*-dimethyl [(bis)oxazoline] **83**–coordinated copper catalyst is the most widely used ligand. The catalyst is prepared in situ by mixing ligand **83** with an equal molar amount of CuOTf. Asymmetric cyclopropanation of isobutylene with ethyl diazoacetate (EDA) gives ethyl 2,3-dimethylcyclopropane carboxylate with >99% ee.

Unlike the catalytic epoxidation or aziridination reactions of simple alkenes, where enantiocontrol is the only stereochemical differentiation, synthetically effective intermolecular cyclopropanation requires both diastereocontrol and enantiocontrol. High diastereoselectivity for the *trans*-isomer can be achieved with the use of bulky diazoacetates such as BDA[99] **187** or DCM[97] **188**.

BDA **187**

DCM **188**

For example, the asymmetric cyclopropanation of styrene with BDA catalyzed by CuOTf–**83** yields cyclopropane with a *trans:cis* ratio of 94:6 and 99% ee for the *trans*-isomer.[31]

Ito and Katsuki[100] report that compounds **189a–191** can also be used as chiral ligands in asymmetric cyclopropanation reactions and that the bipyridine ligand has a potential for catalyzing the reaction (see Fig. 5–13 for the structure of these chiral ligands). Scheme 5–59 shows one example of using bipyridine compound **189b** as a chiral ligand in the cyclopropanation reaction of styrene with *t*-butyl diazoacetate; 92% ee for the *trans*-product can be obtained.

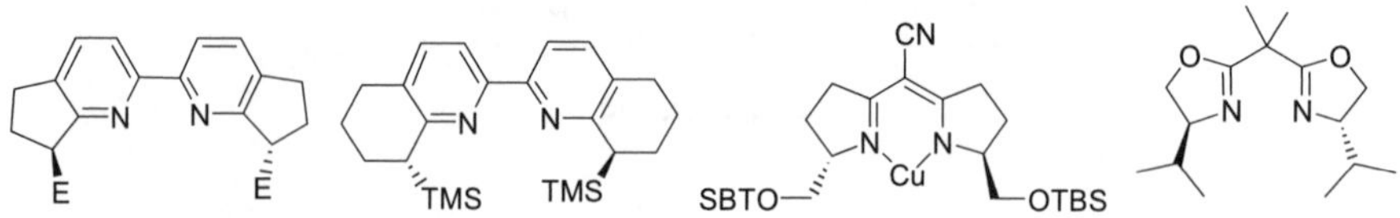

189a

E = TMS, bipyridine

189b

190 semicorrine

191 bis(oxazoline)

Figure 5–13. Chiral ligands for asymmetric cyclopropanation.

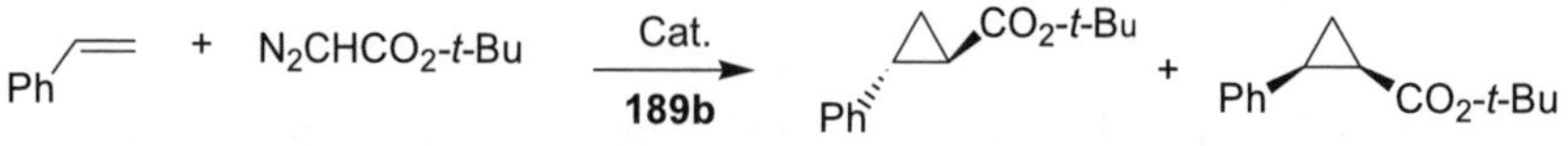

86 (92% ee) : 14

Scheme 5–59

In addition to the above ligands **189–191**, bis(oxazolinyl)pyridine compounds **192–194** have also been applied in asymmetric cyclopropanation reactions, but only moderate enantioselectivity has been achieved.[101]

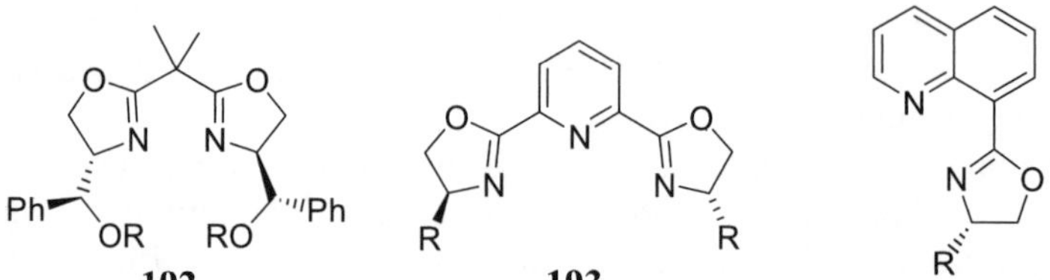

192

193

194

Use of chiral semicorrins and related nitrogen heterocyclic compounds as chiral ligands in asymmetric catalysis has been reviewed periodically. Interested readers are referred to the review by Pfaltz.[102]

Chiral dirhodium(II) catalysts with carboxylate or carboxamidate ligands have recently been developed to take advantage of their versatility in metal carbene transformation, and these have now become the catalysts of choice for cyclopropanation. Chiral carboxylate ligands **195**,[103] **196**,[104] and **197**[105] have been used for tetrasubstitution around a dirhodium(II) core. However, the enantioselectivity in intermolecular reactions with simple ketenes is marginal.

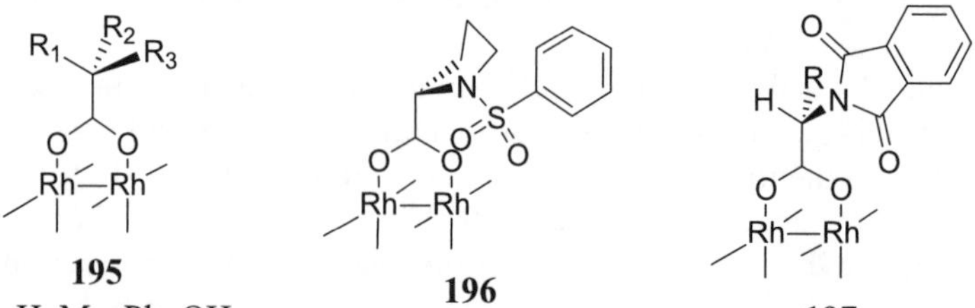

195

R = H, Me, Ph, OH
NHAc, CF$_3$

196

197

R = PhCH$_2$, t-Bu

Chiral carboxamidate-ligated dirhohdium(II) compounds **198**,[106] **199**,[107] **200**,[108] and **201**[109] are not oxygen sensitive and possess a long shelf life. Although their initial application to intermolecular cyclopropanation reactions between styrene and *l*- or *d*-menthyl diazoacetate provided lower stereocontrol than that of the Aratani catalyst, they have proved to be the most effective catalysts for intramolecular cyclopropanation reactions of diazoacetate and diazoacetamide.[110] These compounds can also be used to induce asymmetric cyclopropenation reactions.[111]

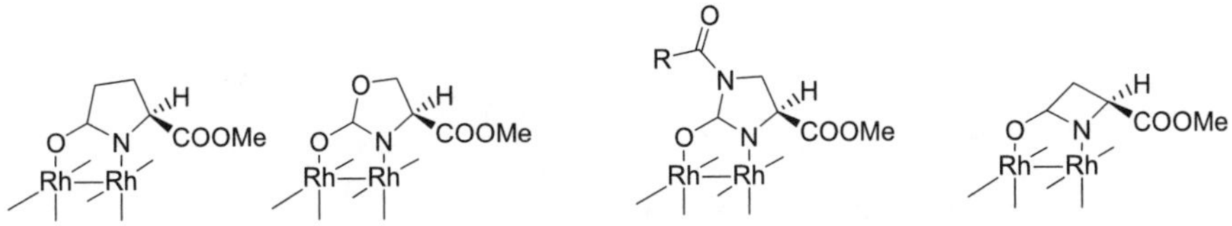

Rh₂(5S-MEPY)₄
198

Rh₂(4S-MEOX)₄
199

200a R = Me, Rh₂(4S-MACIM)₄
200b R = BnCH₂, Rh₂(4S-MPPIM)₄

Rh₂(4S-IBAZ)₄
201

Intramolecular cyclopropanation has a noteworthy advantage. Unlike intermolecular asymmetric cyclopropanation, the intramolecular reaction produces only one diastereomer due to geometric constrains on the fused bicyclic products. Doyle has extensively studied the intramolecular enantioselective reactions of a variety of alkenyl diazoacetates catalyzed by chiral rhodium carboxamides **198** and **200** and has achieved excellent results.

In the simplest case, the reaction of allyl diazoacetate, the catalyst (*S*)-**198** or (*R*)-**198** in a concentration as low as 0.1 mol% can still catalyze the formation of enantiomeric-3-oxabicyclo[3.1.0]hexan-2-ones with 95% ee (Scheme 5–60). Substituted alkyl diazoacetates undergo intramolecular cyclopropanation, with similarly high enantiomeric excess (Scheme 5–61).[110]

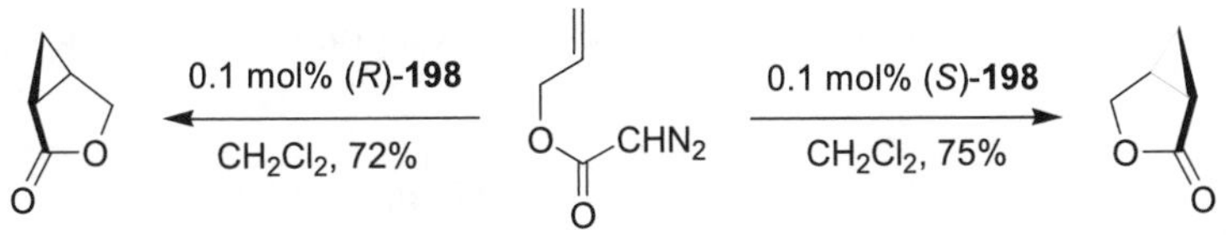

(1S, 5R)-**202**, 95% ee

Scheme 5–60

(1R, 5S)-**202**, 95% ee

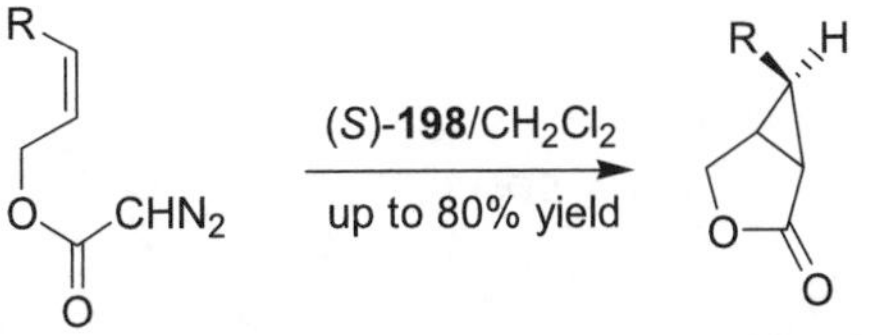

R = Et, *i*-Pr, *i*-Bu
Ph, Bn, (*n*-Bu)₃Sn, I

up to 90 ee%

Scheme 5–61

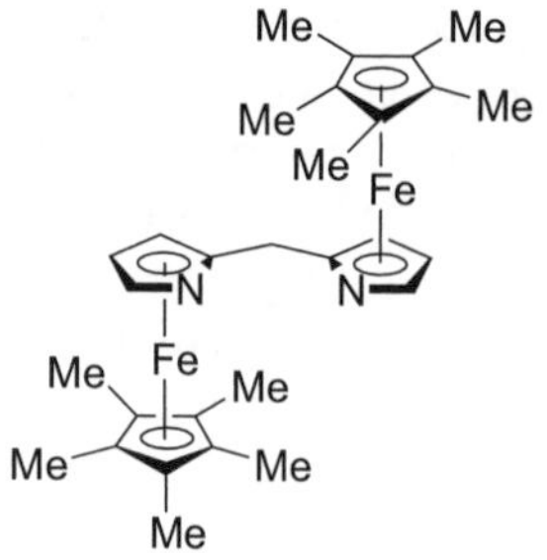

203

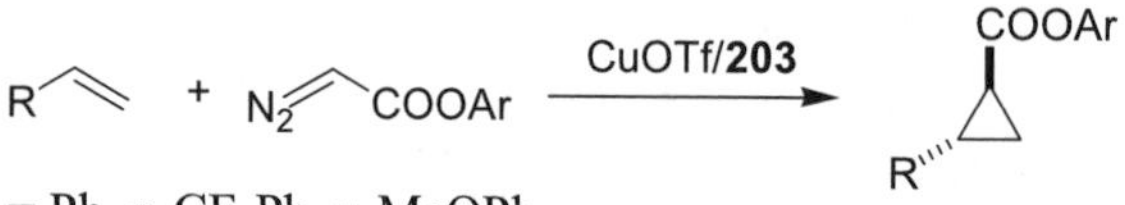

$$R = Ph, \; p\text{-}CF_3Ph, \; p\text{-}MeOPh,$$
$$PhCH_2, \; n\text{-}C_6H_{13}, \; TMS$$

87 - 96 % ee

Scheme 5-62

Lo and Fu[112] have reported a new type of planar–chiral ligand **203** for the enantioselective cyclopropanation of olefins. As shown in Scheme 5–62, asymmetric cyclopropanation in the presence of chiral ligand **203** proceeds smoothly, giving the cyclopropanation product with high diastereoselectivity and enantioselectivity.

The Cope rearrangement of divinyl cyclopropanes is an attractive method for the stereoselective synthesis of seven-membered rings.[113] Due to the requirement of a boat transition state for this Cope rearrangement, multiple stereogenic centers can be formed in a well-defined manner. As shown in Scheme 5–63, compound rhodium(II) N-dodecylbenzenesulfonyl prolinate (**204**) catalyzes the asymmetric cyclopropanation. Subsequent Cope rearrangement gives the corresponding seven-membered ring compound 1,4-cycloheptadiene

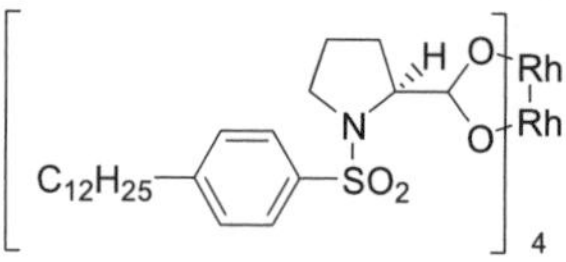

204 Rh$_2$(S-DOSP)$_4$

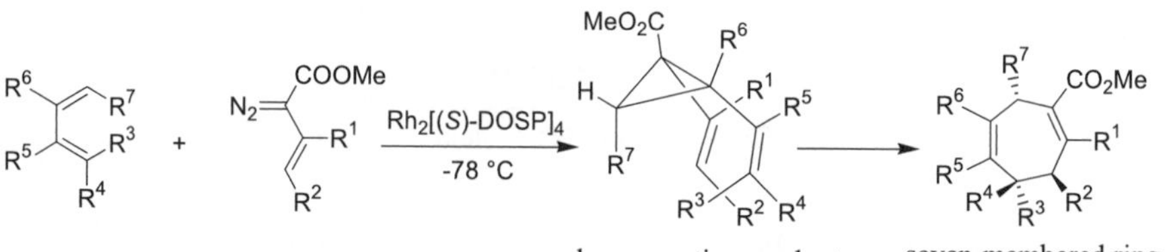

cyclopropanation product

seven-membered ring
product up to 98% ee

Scheme 5-63

with high yield and stereoselectivity.[114] This example illustrates the beauty of tandem reactions.

5.10.2 The Catalytic Asymmetric Simmons-Smith Reaction

Among methods of preparing optically active cyclopropane compounds, the Simmons-Smith reaction, first reported in 1958, is of significance. This reaction refers to the cyclopropanation of alkene with a reagent prepared in situ from a zinc–copper alloy and diiodomethane. The reaction is stereospecific with respect to the geometry of the alkene and is generally free from side reactions in contrast to reactions involving free carbenes.

In the Simmons-Smith reaction, the purpose of copper is simply to activate the Zn surface and has no other role. Iodomethyl zinc behaves as a weak nucleophile. As generally expected, the presence of an allylic oxygen gives a large rate enhancement, and more remote neighboring oxygen atoms also influence the stereochemical course of the reaction.

The reaction was first carried out with the substrate bearing the chiral auxiliary. Scheme 5–64 shows the asymmetric cyclopropanation reaction using 2,4-pentandiol as a chiral auxiliary.[115] Scheme 5–65 illustrates the use of optically pure 1,2-*trans*-cyclohexanediol as a chiral auxiliary in asymmetric Simmons-Smith cyclopropanation.[116] Excellent yield and diastereoselectivity are obtained in most cases.

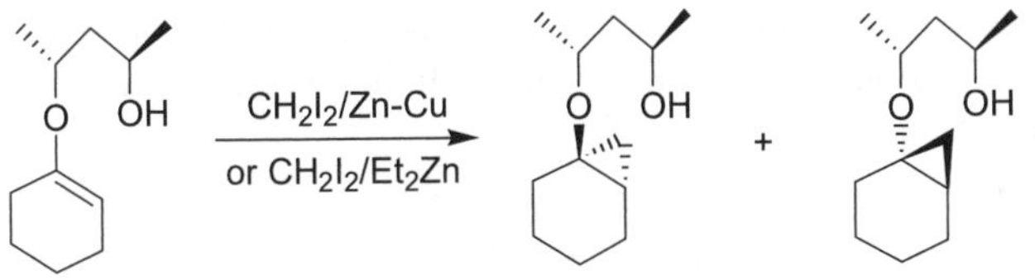

97.6 : 2.4 (when carried out in DME)

Scheme 5–64

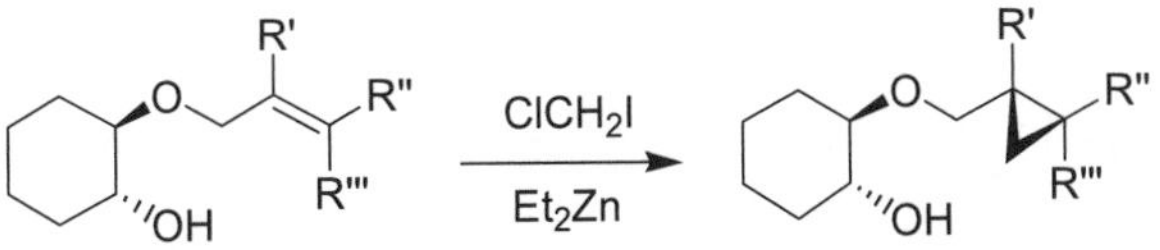

R'	R"	R'''	Temp. (°C)	Yield (%)	ds
H	Pr	H	-30	98	21 : 1
H	H	Pr	-20	97	24 : 1
H	Me	Me	-30	98	23 : 1
H	Ph	H	-20	97	24 : 1
Me	Ph	H	-10 → 0	90	15 : 1
H	H	CH$_2$OTIPS	-20	95	>20 : 1

Scheme 5–65. Reprinted with permission by Pergamon Press Ltd., Ref. 116.

Ukaji et al.[117] reported an enantioselective cyclopropanation reaction in which moderate enantiomeric excess was obtained when a stoichiometric amount of diethyl tartrate was used as a chiral modifier. Takahashi et al.[118] achieved better results using the C_2-symmetric chiral disulfonamide **205** as the chiral ligand.

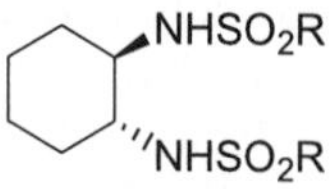

205

Figure 5–14 illustrates the transition state in the reaction. The free hydroxyl group is necessary for producing an effective chiral environment, probably through complexation as a zinc alkoxide.[118]

Scheme 5–66 shows another example of chiral bis(sulfonamide) **205**–catalyzed asymmetric cyclopropanation of allylic alcohol.[119]

Another chiral ligand that can be used in asymmetric Simmons-Smith reactions to build up a cyclopropane moiety is the dioxaborolane compound **206**, containing tartaric acid diamide as a chiral ligand.[120] This compound is an efficient chiral controller for the enantioselective conversion of allylic alcohols to substituted cyclopropylmethanols with both high yields and high enantiomeric excesses.[121] The design of this dioxaborolane compound relies on the presence of an acidic (boron) and a basic (amide) site that allow the simultaneous complexation of the acidic halomethylzinc reagent and the basic allylic metal alkoxide. A wide range of substrates such as allylic alcohols, polyenes, 2,4-dien-1-ols, homoallylic alcohols, and allylic amines can then undergo cyclopropanation reactions.[120]

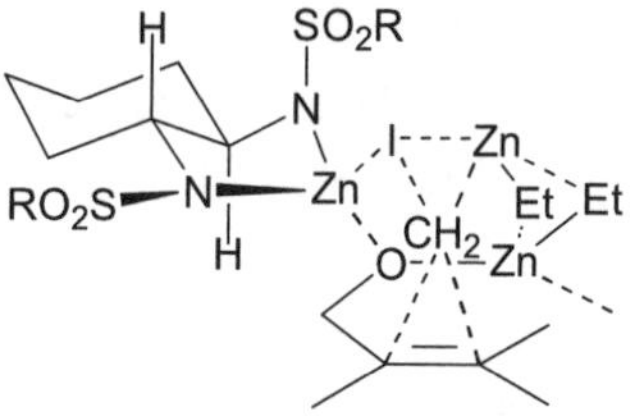

Figure 5–14. Transition State for **205** Catalyzed Cyclopropanation.

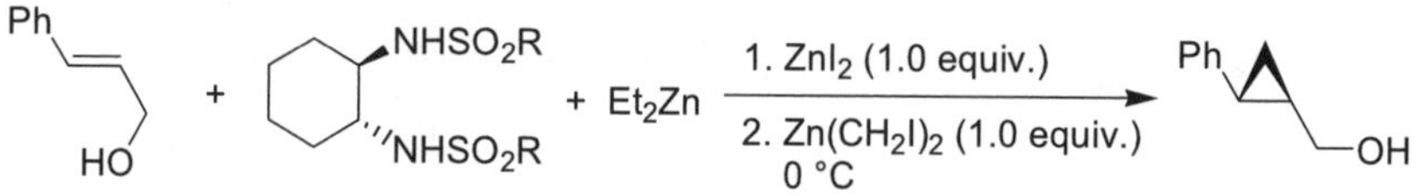

0.1 equiv. 1.1 equiv.

R = Me 99% yield, 89% ee
R = *n*-Bu 98% yield, 84% ee

Scheme 5–66

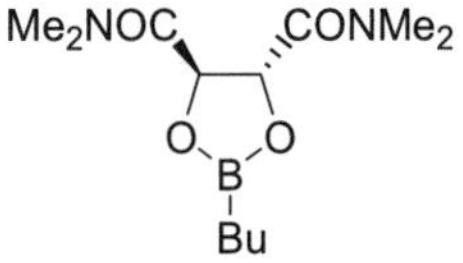

206

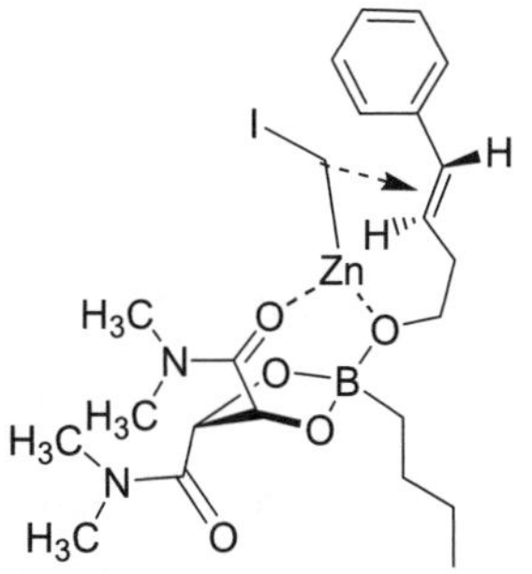

Figure 5–15. Transition state for **206** catalyzed cyclopropanation.

Figure 5–15 shows a possible transition state for the enantioselective cyclopropanation of cinnamyl alcohol in the presence of dioxaborolane **206**. This model predictes the absolute configuration of the products.

Finally, let us look at some important examples in which asymmetric cyclopropanation reactions are used in the asymmetric construction of subunits of biologically active organic molecules.

The compound curacin A **207** is a novel antimitotic agent isolated from the Caribbean cyanobacterium *Lyngbya majuscula*. The compound consists of a disubstituted thiazoline bearing a chiral cyclopropane ring and an aliphatic side chain. Scheme 5–67 depicts the construction of the cyclopropane ring using an asymmetric cyclopropanation reaction.[122]

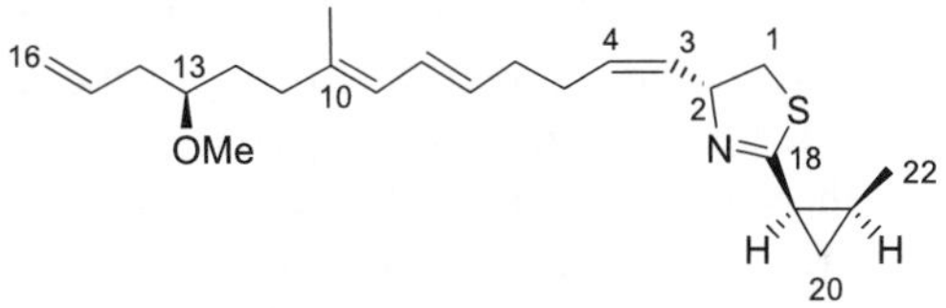

207 curacin A, a novel antimitotic agent

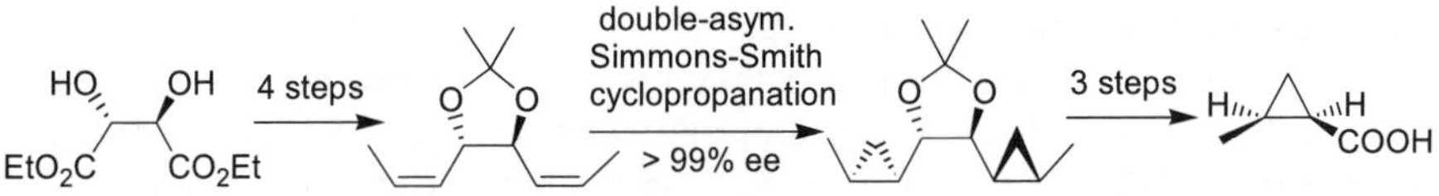

Scheme 5–67

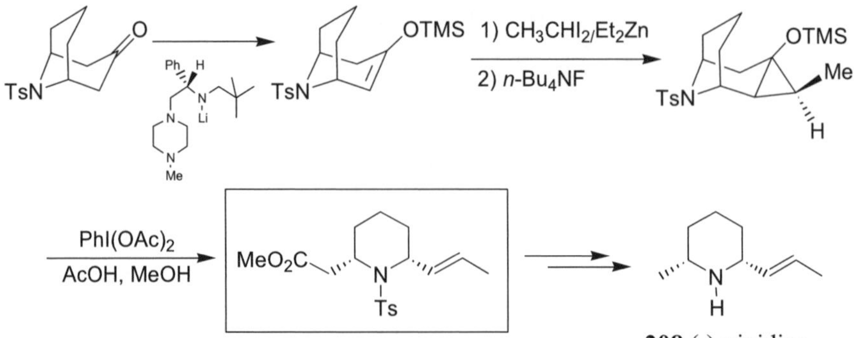

Scheme 5–68

Another example is the asymmetric synthesis of $(-)$-pinidine **208** and its isomers. These syntheses are achieved via asymmetric enolization, stereoselective cyclopropanation, and oxidative ring cleavage of the resulting cyclopropanol system (Scheme 5–68).[123]

5.11 SUMMARY

In summary, asymmetric cycloadditions are powerful methods for the synthesis of complex chiral molecules because multiple asymmetric centers can be constructed in one-step transformations. Among them, reactions using chiral catalysts are the most effective and promising, and fruitful results have been reported in asymmetric Diels-Alder reactions.

Hetero Diels-Alder reactions are very useful for constructing heterocyclic compounds, and many important chiral molecules have thus been synthesized. Although the retro Diels-Alder reaction does not itself involve the asymmetric formation of chiral centers, this reaction can still be used as an important tool in organic synthesis, especially in the synthesis of some thermodynamically less stable compounds. The temporarily formed Diels-Alder adduct can be considered as a protected active olefin moiety. Cyclopentadiene dimer was initially used, but it proved difficult to carry out the pyrolytic process. Pentamethyl cyclopentadiene was then used, and it was found that a retro Diels-Alder reaction could easily be carried out under mild conditions.

As a kind of special case, the asymmetric 1,3-dipolar reaction of nitrile oxides or nitrones constitutes one of the most useful and convenient methods for preparing isoxazolidine derivatives.

Because the chiral cyclopropane subunit is present in a wide range of natural and synthetic products showing important biological properties, asymmetric construction of the cyclopropane moiety via asymmetric cyclopropanation is of commercial interest. Many chiral catalysts and chiral ligands have been prepared and utilized in asymmetric synthesis of these cyclopropane compounds, and very good results have been obtained. An excellent monograph of this subject has recently been published by Doyle et al.[124]

5.12 REFERENCES

1. For references on simple Diels-Alder reactions, see (a) Furuta, K.; Miwa, Y.; Iwanaga, K.; Yamamoto, H. *J. Am. Chem. Soc.* **1988**, *110*, 6254. (b) Chapuis, C.; Jurczak, J. *Helv. Chim. Acta* **1987**, *70*, 436. (c) Narasaka, K.; Inoue, M.; Okada, N. *Chem. Lett.* **1986**, 1109. (d) Oppolzer, W. *Angew. Chem. Int. Ed. Engl.* **1984**, *23*, 876.

2. For detailed discussion: (a) Masamune, S.; Choy, W.; Paterson, J. S.; Sita, L. R. *Angew. Chem. Int. Ed. Engl.* **1985**, *24*, 1. (b) Helmchen, G.; Schmierer, R. *Angew. Chem. Int. Ed. Engl.* **1981**, *20*, 205. (c) Corey, E. J.; Ensley, H. E. *J. Am. Chem. Soc.* **1975**, *97*, 6908. (d) Oppolzer, W.; Chapuis, C.; Dao, G. M.; Reichlin, D.; Godel, T. *Tetrahedron Lett.* **1982**, *23*, 4781. (e) Oppolzer, W.; Chapuis, C. *Tetrahedron Lett.* **1983**, *24*, 4665. (f) Oppolzer, W.; Kurth, M.; Reichlin, D.; Chapuis, C.; Mohnhaupt, M.; Moffatt, F. *Helv. Chim. Acta* **1981**, 2802. (g) Oppolzer, W.; Kurth, M.; Reichlin, D.; Moffatt, F. *Tetrahedron Lett.* **1981**, *22*, 2545.

3. Tanaka, K.; Uno, H.; Osuga, H.; Suzuki, H. *Tetrahedron Asymmetry* **1993**, *4*, 629.

4. Evans, D. A.; Chapman, K. T.; Bisaha, J. *J. Am. Chem. Soc.* **1988**, *110*, 1238.

5. Ferreira, M. L. G.; Pinheiro, S.; Perrone, C. C.; Costa, P. R. R.; Ferreira, V. F. *Tetrahedron Asymmetry* **1998**, *9*, 2671.

6. Linz, G.; Weetman, J.; Hady, A. F. A.; Helmchen, G. *Tetrahedron Lett.* **1989**, *30*, 5599.

7. (a) Masamune, S.; Reed, L. A.; Davis, J. T.; Choy, W. *J. Org. Chem.* **1983**, *48*, 4441. (b) Choy, W.; Reed, L. A.; Masamune, S. *J. Org. Chem.* **1983**, *48*, 1137.

8. Adam, W.; Güthlein, M.; Peters, E.; Peters, K.; Wirth, T. *J. Am. Chem. Soc.* **1998**, *120*, 4091.

9. Jung, M. E.; Vaccaro, W.; Buszek, K. R. *Tetrahedron Lett.* **1989**, *30*, 1893.

10. (a) Kitagawa, O; Aoki, K.; Inoue, T.; Taguchi, T. *Tetrahedron Lett.* **1995**, *36*, 593. (b) Kitagawa, O.; Izawa, H.; Sato, K.; Dobashi, A.; Taguchi, T. *J. Org. Chem.* **1998**, *63*, 2634.

11. Carreño, M. C.; García-Cerrada, S.; Urbano, A.; Di Vitta, C. *Tetrahydron Asymmetry* **1998**, *9*, 2965.

12. (a) Trost, B. M.; O'Krongly, D.; Belletire, J. L. *J. Am. Chem. Soc.* **1980**, *102*, 7595. (b) Hamada, T.; Sato, H.; Hikota, M.; Yonemitsu, O. *Tetrahedron Lett.* **1989**, *30*, 6405.

13. Menezes, R. F.; Zezza, C. A.; Sheu, J.; Smith, M. B. *Tetrahedron Lett.* **1989**, *30*, 3295.

14. Kelly, T. R.; Whiting, A.; Chandrakumar, N. S. *J. Am. Chem. Soc.* **1986**, *108*, 3510.

15. For reviews about catalytic enantioselective Diels-Alder reactions, see (a) Kagan, H. B.; Riant, O. *Chem. Rev.* **1992**, *92*, 1007. (b) Togni, A.; Venanzi, L. M. *Angew. Chem. Int. Ed. Engl.* **1994**, *33*, 497. (c) Deloux, L.; Srebnik, M. *Chem. Rev.* **1993**, *93*, 763. (d) Pindur, U.; Lutz, G.; Otto, C. *Chem. Rev.* **1993**, *93*, 741.

16. Narasaka, K.; Inoue, M.; Yamada, T.; Sugimori, J.; Iwasawa, N. *Chem. Lett.* **1987**, 2409.

17. Yamamoto, I.; Narasaka, K. *Bull. Chem. Soc. Jpn.* **1994**, *67*, 3327.

18. Narasaka, K.; Hayashi, Y.; Shimadzu, H.; Niihata, S. *J. Am. Chem. Soc.* **1992**, *114*, 8869.

19. Kobayashi, S.; Ishitani, H. *J. Am. Chem. Soc.* **1994**, *116*, 4083.

20. Corey, E. J.; Imwinkelried, R.; Pikul, S.; Xiang, Y. *J. Am. Chem. Soc.* **1989**, *111*, 5493.

21. (a) Corey, E. J.; Imai, N.; Pikul, S. *Tetrahedron Lett.* **1991**, *32*, 7517. (b) Corey, E. J.; Sarshar, S.; Bordner, J. *J. Am. Chem. Soc.* **1992**, *114*, 7938.

22. Ishihara, K.; Gao, Q.; Yamamoto, H. *J. Org. Chem.* **1993**, *58*, 6917.

23. For enantioselective Diels-Alder reactions catalyzed by chiral metal aryloxide, see B(OAr)$_3$: (a) Kaufmann, D.; Boese, R. *Angew. Chem. Int. Ed. Engl.* **1990**, *29*, 545. MeAl(OAr)$_2$: (b) Maruoka, K.; Concepcion, A. B.; Yamamoto, H. *Bull. Chem. Soc. Jpn.* **1992**, *65*, 3501. ClAl(OAr)$_2$: (c) Bao, J.; Wulff, W. D.; Rheingold, A. L. *J. Am. Chem. Soc.* **1993**, *115*, 3814. (d) Heller, D. P.; Goldberg, D. R.; Wulff, W. D. *J. Am. Chem. Soc.* **1997**, *119*, 10551. Ti(OAr)$_4$: (e) Maruoka, K.; Murase, N.; Yamamoto, H. *J. Org. Chem.* **1993**, *58*, 2938. Cl$_2$Ti(OAr)$_2$: (f) Harada, T.; Takeuchi, M.; Hatsuda, M.; Ueda, S.; Oku, A. *Tetrahedron Asymmetry* **1996**, *7*, 2479.

24. (a) Hawkins, J. M.; Loren, S. *J. Am. Chem. Soc.* **1991**, *113*, 7794. (b) Hawkins, J. M.; Loren, S.; Nambu, M. *J. Am. Chem. Soc.* **1994**, *116*, 1657.

25. (a) Corey, E. J.; Loh, T. *J. Am. Chem. Soc.* **1991**, *113*, 8966. (b) Corey, E. J.; Loh, T.; Roper, T. D.; Azimioara, M. D.; Noe, M. C. *J. Am. Chem. Soc.* **1992**, *114*, 8290.

26. Ishihara, K.; Gao, Q.; Yamamoto, H. *J. Am. Chem. Soc.* **1993**, *115*, 10412.

27. Ishihara, K.; Kurihara, H.; Matsumoto, M.; Yamamoto, H. *J. Am. Chem. Soc.* **1998**, *120*, 6920.

28. Corey, E. J.; Imai, N.; Zhang, H. *J. Am. Chem. Soc.* **1991**, *113*, 728.

29. Corey, E. J.; Ishihara, K. *Tetrahedron Lett.* **1992**, *33*, 6807.

30. (a) Evans, D. A.; Miller, S. J.; Lectka, T. *J. Am. Chem. Soc.* **1993**, *115*, 6460. (b) Evans, D. A.; Lectka, T.; Miller, S. J. *Tetrahedron Lett.* **1993**, *34*, 7027. (c) Evans, D. A.; Murry, J. A.; von Matt, P.; Norcross, R. D.; Miller, S. J. *Angew. Chem. Int. Ed. Engl.* **1995**, *34*, 798. (d) Evans, D. A.; Johnson, J. S. *J. Org. Chem.* **1997**, *62*, 786.

31. For the other studies of chiral bis(oxazolines) as enantioselective catalysts, see (a) Lowenthal, R. E.; Abiko, A.; Masamune, S. *Tetrahedron Lett.* **1990**, *31*, 6005. (b) Evans, D. A.; Woerpel, K. A.; Hinman, M. M.; Faul, M. M. *J. Am. Chem. Soc.* **1991**, *113*, 726. (c) Nishiyama, H.; Kondo, M.; Nakamura, T.; Itoh, K. *Organometallics* **1991**, *10*, 500. (d) Lowenthal, R. E.; Masamune, S. *Tetrahedron Lett.* **1991**, *32*, 7373.

32. Ghosh, A, K.; Cho, H.; Cappiello, J. *Tetrahydron Asymmetry* **1998**, *9*, 3687.

33. Ghosh, A. K.; Mathivanan, P.; Cappiello, J. *Tetrahedron Asymmetry* **1998**, *9*, 1.

34. Otto, S.; Boccaletti, G.; Engberts, J. B. F. N. *J. Am. Chem. Soc.* **1998**, *120*, 4238.

35. Danishefsky, S. J.; DeNinno, M. P. *Angew. Chem. Int. Ed. Engl.* **1987**, *26*, 15.

36. (a) Keck, G. E.; Li, X. Y.; Krishnamurthy, D. *J. Org. Chem.* **1995**, *69*, 5998. (b) Ghosh, A. K.; Mathivanan, P.; Cappiello, J. *Tetrahedron Lett.* **1997**, *38*, 2427. (c) Matsukawa, S.; Mikami, K. *Tetrahedron Asymmetry* **1997**, *8*, 815. (d) Jahannsen, M.; Yao, S.; Jorgensen, K. A. *Chem. Commun.* **1997**, 2169. (e) Saito, T.; Takekawa, K.; Nishimura, J.; Kawamura, M. *J. Chem. Soc. Perkin Trans. 1* **1997**, 2957.

37. (a) Mikami, K.; Motoyama, Y.; Terada, M. *J. Am. Chem. Soc.* **1994**, *116*, 2812. (b) Terada, M.; Mikami, K.; Nakai, T. *Tetrahedron Lett.* **1991**, *32*, 935.

38. Maruoka, K.; Itoh, T.; Shirasaka, T.; Yamamoto, H. *J. Am. Chem. Soc.* **1988**, *110*, 310.

39. Li, L.; Wu, Y.; Hu, Y.; Xia, L.; Wu, Y. *Tetrahedron Asymmetry* **1998**, *9*, 2271.

40. Goldman, R.; Kohlbrenner, W.; Lartey, P.; Parnet, A. *Nature* **1987**, *329*, 162.

41. Schaus, S. E.; Brånalt, J.; Jacobsen, E. N. *J. Org. Chem.* **1998**, *63*, 403.

42. (a) Danishefsky, S.; Kitahara, T. *J. Am. Chem. Soc.* **1974**, *96*, 7807. (b) Kerwin, J. F.; Danishefsky, S. *Tetrahedron Lett.* **1982**, *23*, 3739.

43. (a) Evans, D. A.; Olhava, E. J.; Johnson, J. S.; Janey, J. M. *Angew. Chem. Int. Ed. Engl.* **1998**, *37*, 3372. (b) Evans, D. A.; Johnson, J. S. *J. Am. Chem. Soc.* **1998**, *120*, 4895.

44. (a) Thorhauge, J.; Johannsen, M.; Jørgensen, K. A. *Angew. Chem. Int. Ed. Engl.* **1998**, *37*, 2404. (b) Yao, S.; Johannsen, M.; Audrain, H.; Hazell, R. G.; Jørgensen, K. A. *J. Am. Chem. Soc.* **1998**, *120*, 8599.

45. For principles and applications of aza hetero Diels-Alder reactions, see (a) Waldmann, H. *Synthesis* **1994**, 535. (b) Waldmann, H. ed. *Organic Synthesis Highlights* **1995**, 37. (c) Tietze, L. F.; Kettschau, G. *Top. Curr. Chem.* **1997**, *189*, 1. (d) Weinreb S. M., in Trost, B. M.; Fleming, I.; Semmelhock. M. F. eds. *Comprehensive Organic Synthesis*, Pergamon, Oxford, **1991**, vol. 5, p 401. (e) Boger, D. L.; Weinreb, S. M. *Hetero Diels-Alder Methodology in Organic Synthesis*, Academic Press, San Diego, **1987**.

46. For example, (a) Borrione, E.; Prato, M.; Scorrano, G.; Stivanello, M. *J. Chem. Soc. Perkin Trans. 1* **1989**, 2245. (b) Waldmann, H.; Braun, M.; Dräger, M. *Angew. Chem. Int. Ed. Engl.* **1990**, *29*, 1468. (c) Bailey, P. D.; Londesbrough, D. J.; Hancox, T. C.; Heffernan, J. D.; Holmes, A. B. *J. Chem. Soc. Chem. Commun.* **1994**, 2543. (d) McFarlane, A. K.; Thomas, G.; Whiting, A. *J. Chem. Soc. Perkin Trans. 1* **1995**, 2803. (e) Kündig, E. P.; Xu, L. H.; Romanens, P.; Bernardinelli, G. *Synlett* **1996**, 270.

47. Ishitani, H.; Kobayashi, S. *Tetrahedron Lett.* **1996**, *37*, 7357.

48. Waldmann, H.; Braun, M.; Dräger, M. *Tetrahedron Asymmetry* **1991**, *2*, 1231.

49. (a) Hattori, K.; Yamamoto, H. *Synlett* **1993**, 129. (b) Hattori, K.; Yamamoto, H. *Tetrahedron* **1993**, *49*, 1749.

50. Hamley, P.; Helmchen, G.; Holmes, A. B.; Marshall, D. R.; MacKinnon, J. W. M.; Smith, D. F.; Ziller, J. W. *J. Chem. Soc. Chem. Commun.* **1992**, 786.

51. Kobayashi, S.; Komiyama, S.; Ishitani, H. *Angew. Chem. Int. Ed. Engl.* **1998**, *37*, 979.

52. Yao, S.; Johannsen, M.; Hazell, R. G.; Jørgensen, K. A. *Angew. Chem. Int. Ed. Engl.* **1998**, *37*, 3121.

53. Jnoff, E.; Ghosez, L. *J. Am. Chem. Soc.* **1999**, *121*, 2617.

54. Corey, E. J.; Guzman-Perez, A. *Angew. Chem. Int. Ed. Engl.* **1998**, *37*, 388.

55. Maruoka, K.; Murase, N.; Yamamoto, H. *J. Org. Chem.* **1993**, *58*, 2938.

56. Yao, S.; Johannsen, M.; Audrain, H.; Hazell, R. G.; Jørgensen, K. A. *J. Am. Chem. Soc.* **1998**, *120*, 8599.

57. Rebiere, F.; Riant, O.; Kagan, H. B. *Tetrahedron Asymmetry* **1990**, *1*, 199.

58. (a) Devine, P. N.; Oh, T. *Tetrahedron Lett.* **1991**, *32*, 883. (b) Devine, P. N.; Oh, T. *J. Org. Chem.* **1992**, *57*, 396.

59. (a) Yamamoto, H.; Maruoka, K.; Furuta, K.; Naruse, Y. *Pure Appl. Chem.* **1989**, *61*, 419. (b) Terada, M.; Mikami, K.; Nakai, T. *Tetrahedron Lett.* **1991**, *32*, 935. (c) Kaufmann, D.; Boese, R. *Angew. Chem. Int. Ed. Engl.* **1990**, *29*, 545. (d) Bao, J.; Wulff, W. D.; Rheingold, A. L. *J. Am. Chem. Soc.* **1993**, *115*, 3814. (e) Hattori, K.; Yamamoto, H. *J. Org. Chem.* **1992**, *57*, 3264.

60. (a) Narasaka, K.; Iwasawa, N.; Inoue, M.; Yamada, T.; Nakashima, M.; Sugimori, J. *J. Am. Chem. Soc.* **1989**, *111*, 5340. (b) Ketter, A.; Glahsl, G.; Herrmann, R. *J. Chem. Res. (S)* **1990**, 278.

61. (a) Takasu, M.; Yamamoto, H. *Synlett* **1990**, 194. (b) Corey, E. J.; Loh, T. *Tetrahedron Lett.* **1993**, *34*, 3979. (c) Corey, E. J.; Loh, T.; Roper, T. D.; Azimioara, M. D.; Noe, M. C. *J. Am. Chem. Soc.* **1992**, *114*, 8290. (d) Sartor, D.; Saffrich, J.; Helmchen, G. *Synlett* **1990**, 197. (e) Seerden, J. G.; Scheeren, H. W. *Tetrahedron Lett.* **1993**, *34*, 2669.

62. Gao, Q.; Maruyama, T.; Mouri, M.; Yamamoto, H. *J. Org. Chem.* **1992**, *57*, 1951.

63. Evans, D. A.; Lectka, T.; Miller, S. J. *Tetrahedron Lett.* **1993**, *34*, 7027.

64. Uemura, M.; Hayashi, Y.; Hayashi, Y. *Tetrahedron Asymmetry* **1993**, *4*, 2291.

65. Corey, E. J.; Roper, T. D.; Ishihara, K.; Sarakinos, G. *Tetrahedron Lett.* **1993**, *34*, 8399.

66. Furuta, K.; Miwa, Y.; Iwanaga, K.; Yamamoto, H. *J. Am. Chem. Soc.* **1988**, *110*, 6254.

67. Furuta, K.; Shimizu, S.; Miwa, Y.; Yamamoto, H. *J. Org. Chem.* **1989**, *54*, 1481.

68. Ishihara, K.; Yamamoto, H. *J. Am. Chem. Soc.* **1994**, *116*, 1561.

69. Kündig, E. P.; Bourdin, B.; Bernardinelli, G. *Angew. Chem. Int. Ed. Engl.* **1994**, *33*, 1856.

70. Ishihara, K.; Kurihara, H.; Yamamoto, H. *J. Am. Chem. Soc.* **1996**, *118*, 3049.

71. Hayashi, Y.; Rohde, J. J.; Corey, E. J. *J. Am. Chem. Soc.* **1996**, *118*, 5502.

72. (a) Oppolzer, W.; Dupuis, D.; Poli, G.; Raynham, T. M.; Bernardinelli, G. *Tetrahedron Lett.* **1988**, *29*, 5885. (b) Oppolzer, W.; Dupuis, D. *Tetrahedron Lett.* **1985**, *26*, 5437.

73. Furuta, K.; Kanematsu, A.; Yamamoto, H.; Takaoka, S. *Tetrahedron Lett.* **1989**, *30*, 7231.

74. Stork, G.; Nelson, G. L.; Rouessac, F.; Gringore, O. *J. Am. Chem. Soc.* **1971**, *93*, 3091.

75. For reviews about retro Diels-Alder reactions, see (a) Lasne, M.; Ripoll, J. *Synthesis* **1985**, 121. (b) Ichihara, A. *Synthesis* **1987**, 207. (c) Ripoll, J.; Rouessac, A.; Rouessac, F. *Tetrahedron* **1978**, *34*, 19.

76. Suzuki, M.; Yanagisawa, A.; Noyori, R. *J. Am. Chem. Soc.* **1985**, *107*, 3348.

77. Weil, J. B.; Rouessac, F. *Bull. Soc. Chim. Fr.* **1979**, II-273.

78. Grieco, P. A.; Abood, N. *J. Org. Chem.* **1989**, *54*, 6008.

79. Grieco, P. A.; Abood, N. *J. Chem. Soc. Chem. Commun.* **1990**, 410.

80. Liu, Z.; He, L.; Zheng, H. *Tetrahedron Asymmetry* **1993**, *4*, 2277.

81. (a) Liu, Z., Chu, X. *Tetrahedron Lett.* **1993**, *34*, 349. (b) Liu, Z., He, L.; Zheng, H. *Synlett* **1993**, 191.

82. Olsson, T.; Stern, K.; Westmann, G.; Sundell, S. *Tetrahedron* **1990**, *46*, 2473.

83. (a) Tufariello, J. J. in Padwa, A. ed. *1,3-Dipolar Cycloaddition Chemistry*, John Wiley & Sons, Chichester, **1984**, vol. 2, p 83. (b) Torssell, K. B. G. *Nitrile Oxides, Nitrones and Nitronates in Organic Synthesis*, VCH, Weinheim, **1988**.

84. (a) Kim, B. H.; Curran, D. P. *Tetrahedron* **1993**, *49*, 293. (b) Curran, D. P.; Heffner, T. A. *J. Org. Chem.* **1990**, *55*, 4585.

85. Mukai, C.; Kim, I. J.; Cho, W. J.; Kido, M.; Hanaoka, M. *J. Chem. Soc. Perkin Trans. 1* **1993**, 2495.

86. (a) Ukaji, Y.; Sada, K.; Inomata, K. *Chem. Lett.* **1993**, 1847. (b) Shimizu, M.; Ukaji, Y.; Inomata, K. *Chem. Lett.* **1996**, 455.

87. Kobayashi, S.; Kawamura, M. *J. Am. Chem. Soc.* **1998**, *120*, 5840.

88. Kanemasa, S.; Oderaotoshi, Y.; Tanaka, J.; Wada, E. *J. Am. Chem. Soc.* **1998**, *120*, 12355.

89. (a) Barrett, A. G. M.; Doubleday, W. W.; Kasdorf, K.; Tustin, G. J. *J. Org. Chem.* **1996**, *61*, 3280. (b) Barrett, A. G. M.; Kasdorf, K.; Tustin, G. J.; Williams, D. J. *J. Chem. Soc. Chem. Commun.* **1995**, 1143.

90. Barrett, A. G. M.; Hamprecht, D.; White, A. J. P.; Williams, D. J. *J. Am. Chem. Soc.* **1996**, *118*, 7863.

91. (a) Kanemasa, S.; Hamura, S.; Harada, E.; Yamamoto, H. *Tetrahedron Lett.* **1994**, *35*, 7985. (b) Nale, D. G.; Geralds, R. S.; Yoo, H.; Gerwick, W. H.; Kim, T. S.; Nambu, M.; White, J. D. *Tetrahedron Lett.* **1995**, *36*, 1189. (c) Mitome, H.; Miyaoka, H.; Nakano, M.; Yamada, Y. *Tetrahedron Lett.* **1995**, *36*, 8231. (d) Armstrong, R. W.; Maurer, K. W. *Tetrahedron Lett.* **1995**, *36*, 357. (e) Harm, A. M.; Knight, J. G.; Stemp, G. *Tetrahedron Lett.* **1996**, *37*, 6189. (f) Charette, A. B.; Juteau, H. *J. Am. Chem. Soc.* **1994**, *116*, 2651. (g) Murali, R.; Ramana, C. V.; Nagarajan, M. *J. Chem. Soc. Chem. Commun.* **1995**, 217. (h) Barrett, A. G. M.; Kasdorf, K.; White, A. J. P.; Williams, D. J. *J. Chem. Soc. Chem. Commun.* **1995**, 649. For reviews, see (i) Charette, A. B.; Marcoux, J. *Synlett* **1995**, 1197. (j) Doyle, M. P.; Protopopova, M. N. *Tetrahedron* **1998**, *54*, 7919. (k) Padwa, A.; Austin, D. J. *Angew. Chem. Int. Ed. Engl.* **1994**, *33*, 1797.

92. Simmons, H. E.; Cairns, T. L.; Vladuchick, S. A.; Hoiness, C. M. *Org. React.* **1973**, *20*, 1.

93. Nozaki, H.; Moriuti, S.; Takaya, H.; Noyori, R. *Tetrahedron Lett.* **1966**, 5239.

94. Aratani, T. *Pure Appl. Chem.* **1985**, *57*, 1839.

95. Fritschi, H.; Leutenegger, U.; Pfaltz, A. *Angew. Chem. Int. Ed. Engl.* **1986**, *25*, 1005.

96. Fritschi, H.; Leutenegger, U.; Pfaltz, A. *Helv. Chim. Acta* **1988**, *71*, 1553.

97. Lowenthal, R. E.; Masamune, S. *Tetrahedron Lett.* **1991**, *32*, 7373.

98. Müller, D.; Umbricht, G.; Weber, B.; Pfaltz, A. *Helv. Chim. Acta* **1991**, *74*, 232.

99. Doyle, M. P.; Bagheri, V.; Wandless, T. J.; Harn, N. K.; Brinker, D. A.; Eagle, C. T.; Loh, K. *J. Am. Chem. Soc.* **1990**, *112*, 1906.

100. Ito, K.; Katsuki, T. *Tetrahedron Lett.* **1993**, *34*, 2661.

101. (a) Hoarau, O.; Aït-Haddou, H.; Castro, M.; Balavoine, G. G. A. *Tetrahedron Asymmetry* **1997**, *8*, 3755. (b) Nishiyama, H.; Soeda, N.; Naito, T.; Motoyama, Y. *Tetrahedron Asymmetry* **1998**, *9*, 2865. (c) Wu, X.; Li, X.; Zhou, Q. *Tetrahedron Asymmetry* **1998**, *9*, 4143.

102. Pfaltz, A. *Acc. Chem. Res.* **1993**, *26*, 339.

103. Brunner, H.; Kluschanzoff, H.; Wutz, K. *Bull. Chem. Soc. Belg.* **1989**, *98*, 63.

104. Kennedy, M.; McKervey, M. A.; Maguire, A. R.; Roos, G. H. P. *J. Chem. Soc. Chem. Commun.* **1990**, 361.

105. Hashimoto, S.; Watanabe, N.; Ikegami, S. *Tetrahedron Lett.* **1990**, *31*, 5173.

106. Doyle, M. P.; Winchester, W. R.; Hoorn, J. A. A.; Lynch, V.; Simonsen, S. H.; Ghosh, R. *J. Am. Chem. Soc.* **1993**, *115*, 9968.

107. Doyle, M. P.; Dyatkin, A. B.; Protopopova, M. N.; Yang, C. I.; Miertschin, C. S.; Winchester, W. R.; Simonsen, S. H.; Lynch, V.; Ghosh, R. *Recl. Trav. Chim. Pays-Bas* **1995**, *114*, 163.

108. Doyle, M. P.; Zhou, Q. L.; Raab, C. E.; Roos, G. H. P.; Simonsen, S. H.; Lynch, V. *Inorg. Chem.* **1996**, *35*, 6064.

109. Doyle, M. P.; Zhou, Q. L.; Simonsen, S. H.; Lynch, V. *Synlett* **1996**, 697.

110. Doyle, M. P.; Austin R. E.; Bailey, A. S.; Dwyer, M. P.; Dyatkin, A. B.; Kalinin, A. V.; Kwan, M. M. Y.; Liras, S.; Oalmann, C. J.; Pieters, R. J.; Protopopova, M. N.; Raab, C. E.; Roos, G. H. P., Zhou, Q.; Martin, S. F. *J. Am. Chem. Soc.* **1995**, *117*, 5763.

111. Müller, P.; Imogaï, H. *Tetrahedron Asymmetry* **1998**, *9*, 4419.

112. Lo, M. M.; Fu, G. C. *J. Am. Chem. Soc.* **1998**, *120*, 10270.

113. For general reviews on the Cope rearrangement of divinylcyclopropanes, see (a) Wong, H. N. C.; Hon, M.; Tse, C.; Yip, Y. C.; Tanko, J.; Hudlicky, T. *Chem. Rev.* **1989**, *89*, 165. (b) Rhoads, S. J.; Raulinus, N. R. *Org. React.* **1975**, *21*, 1. (c) Hudlicky, T.; Fan, R.; Reed, J. W.; Gadamasetti, K. G. *Org. React.* **1992**, *41*, 1. (d) Piers, E. in Trost, B. M., Fleming I. ed. *Comprehensive Organic Synthesis*, Pergamon Press, New York, **1991**, vol. 5, p 971.

114. Davies, H. M. L.; Stafford, D. G.; Doan, B. D.; Houser, J. H. *J. Am. Chem. Soc.* **1998**, *120*, 3326.

115. Sugimura, T.; Futagawa, T.; Tai, A. *Tetrahedron Lett.* **1988**, *29*, 5775.

116. Charette, A. B.; Marcoux, J. *Tetrahedron Lett.* **1993**, *34*, 7157.

117. Ukaji, Y.; Nishimura, M.; Fujisawa, T. *Chem. Lett.* **1992**, 61.

118. Takahashi, H.; Yoshioka, M.; Ohno, M.; Kobayashi, S. *Tetrahedron Lett.* **1992**, *33*, 2575.

119. For more examples of chiral bis(sulfonamide) catalyzed asymmetric cyclopropanation, see Denmark, S. E.; O'Connor, S. P.; Wilson, S. R. *Angew. Chem. Int. Ed. Engl.* **1998**, *37*, 1149, and the references cited therein.

120. Charette, A. B.; Juteau, H.; Lebel, H.; Molinaro, C. *J. Am. Chem. Soc.* **1998**, *120*, 11943.

121. (a) Charette, A. B.; Juteau, H. *J. Am. Chem. Soc.* **1994**, *116*, 2651. (b) Kitajima,

H.; Ito, K.; Aoki, Y.; Katsuki, T. *Bull. Chem. Soc. Jpn.* **1997**, *70*, 207. (c) Kitajima, H.; Aoki, Y.; Ito, K.; Katsuki, T. *Chem. Lett.* **1995**, 1113.

122. Onoda, T.; Shirai, R.; Koiso, Y.; Iwasaki, S. *Tetrahedron* **1996**, *52*, 14543.

123. Momose, T.; Nishio, T.; Kirihara, M. *Tetrahedron Lett.* **1996**, *37*, 4987.

124. Doyle, M. P.; McKervey, M. A.; Ye, T. *Modern Catalytic Methods for Organic Synthesis with Diazo Compounds*, John Wiley & Sons, New York, **1998**.

CHAPTER 6

Asymmetric Catalytic Hydrogenation and Other Reduction Reactions

Previous chapters have discussed a variety of reactions on the carbonyl group, focusing on C–C bond formation via alkylation at the α-position of carbonyl groups, via asymmetric aldol reaction, and via Diels-Alder reactions. We also introduced asymmetric oxidations, such as asymmetric epoxidation, dihydroxylation, aminohydroxylation, and aziridination of C=C bonds. Asymmetric addition of hydrogen to sp² carbon is the main theme of this chapter. Unsaturated bonds like C=C, C=O, or C=N are converted to the corresponding saturated CH–CH, CH–OH, and CH–NH bonds. Chiral metal–hydride reduction and the catalytic transfer hydrogenation of ketones are also discussed. These reactions are of great interest industrially. Indeed, many of them have been used for the production of highly enantiomerically pure amino acids, flavor and fragrance materials, and important pharmaceuticals and agrochemicals, which are all highly valued.

As there are already several books covering these topics, the present coverage is necessarily an informed selection to highlight particularly important and efficient reagents, as well as the most recent developments in these areas.

6.1 INTRODUCTION

Before the 1960s, heterogeneous catalysis was a topic of indisputable importance in chemical research. The first asymmetric reaction was the application of chiral supports in the catalytic dehydrogenation of racemic 2-butanol by Schwab in 1932. Attempts to hydrogenate olefins with the aid of heterogeneous catalysts produced chiral products with only 10–15% ee in the 1950s. By the mid-1960s, it was becoming apparent that heterogeneous catalysts in general were not capable of providing satisfactory results in the hydrogenation of prochiral olefins.

A new approach to asymmetric hydrogenation emerged in the late 1960s. In 1965, Wilkinson discovered a practical homogenous catalyst, $Rh(PPh_3)_3Cl$, which showed very high activity in the hydrogenation of alkenes under mild conditions, and more attention has since been focused on modifying this catalyst by replacing the common triphenyl phosphine with chiral phosphine ligands.

The first successful examples of homogeneous asymmetric hydrogenation were reported independently by Hörner et al.[1] and Knowles and Sabacky[2] in 1968. The Wilkinson compound and related complexes modified by the incorporation of a chiral tertiary phosphine, such as $P(C_6H_5)(n-C_3H_7)(CH_3)$, catalyzed the hydrogenation of certain hydrocarbon olefins with an optical yield of 3–15%. Although the ee values of the hydrogenated products were not very high, the results established a solid foundation for developing the concept of homogeneous asymmetric hydrogenation.

A breakthrough in this area came when Dang and Kagan[3] synthesized DIOP, a C_2 chiral diphosphine obtained from tartaric acid (Fig. 6–1). DIOP–Rh(I) complex catalyzed the enantioselective hydrogenation of α-(acylamino)acrylic acids and esters to produce the corresponding amino acid derivatives with up to 80% ee. These achievements stimulated research on a variety of bidentate chiral diphosphines, and numerous chiral ligands bearing C_2 symmetry have been developed as a result (see Fig. 6–1 for examples).

Thirty years have passed, and bisphosphines containing chiral substituents have proved to be the most useful and versatile ligands in organotransition metal–catalyzed reactions. The mechanism of Rh complex–catalyzed asymmetric hydrogenation was elucidated in detail by Chan and Halpern[8] and Brown et al.[9] The successful commercial application of Rh-catalyzed hydrogenation of prochiral enamides in the synthesis of L-DOPA[10] made asymmetric catalytic hydrogenation a popular subject of research.

6.1.1 Chiral Phosphine Ligands for Homogeneous Asymmetric Catalytic Hydrogenation

Asymmetric catalytic hydrogenation is one of the most efficient and convenient methods for preparing a wide range of enantiomerically pure compounds, and Ru-BINAP–catalyzed asymmetric hydrogenation of 2-arylacrylic acids has attracted a great deal of attention,[11] as the chiral 2-arylpropionic acid products constitute an important class of nonsteroidal antiinflammatory drugs.

Historically, the desire for practical routes to α-amino acids ultimately led to the development of effective chiral diphosphine rhodium catalysts for the enantioselective hydrogenation of α-amidoacrylates (α-enamides) (see Scheme 6–1).[12] It had been found that many ligands with C_2 symmetry were effective in asymmetric hydrogenation reactions. DIOP is a ligand with two sp^3 asymmetric carbons and a C_2-symmetric axis, while DIPAMP possesses two asymmetric phosphorous atoms. The aromatic BINAP ligand possesses C_2 axial chirality and has shown great asymmetric induction potential. It has been suggested that the highly skewed position of the naphthyl rings in BINAP is the determining factor in its effectiveness in asymmetric catalytic reactions.[13] Indeed, in most of the phosphine ligands, the linkage between the chiral part and the phosphine part is a C–P bond, while in some of the promising ligands C–O–P or C–N–P linkages are also involved.

Design of such diphosphine ligands remains an active area of research, and

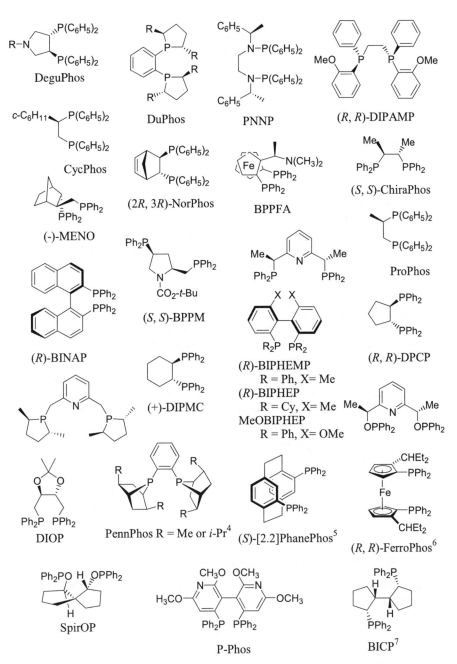

Figure 6–1. Selected ligands for catalytic asymmetric hydrogenation.

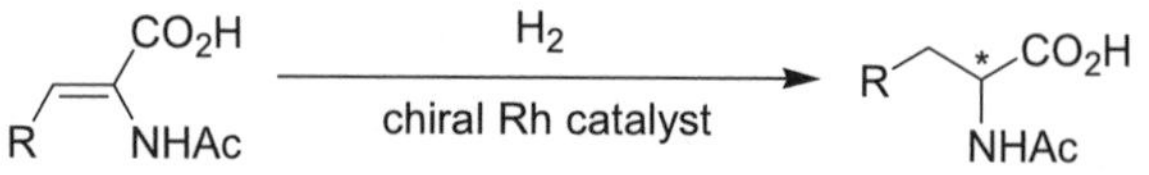

Scheme 6–1

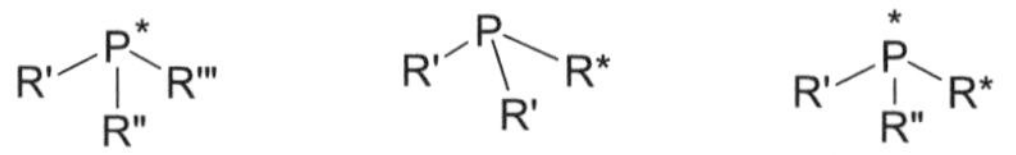

Figure 6–2. Possible types of chiral phosphine ligands.

the homogeneous asymmetric catalytic hydrogenation of prochiral C=X (X = C, N, O, and so forth) double bonds is one of the most important applications of these enantioselective catalytic technologies. There are three ways to design a chiral phosphine. The chirality can be located on the P atom, on a side chain, or on both. Figure 6–2 depicts these three possible chiral phosphine ligands.

Much effort has been devoted to investigating chiral phosphine ligands for their synthesis and asymmetric catalytic hydrogenation potential, and such chiral phosphine ligands have been extensively used for catalytic asymmetric hydrogenation, both academically and industrially.[14]

Numerous practical advantages are associated with asymmetric catalytic processes that allow the conversion of prochiral substrates to valuable enantiopure products, and homogeneous asymmetric hydrogenation constitutes one of the most efficient and versatile methods for making such conversions. A variety of chiral compounds can be obtained via the catalytic asymmetric hydrogenation of C=C, C=N, and C=O bonds with outstanding levels of efficiency and enantioselectivity. The yield and the enantiomeric excess of the product are influenced not only by chiral ligands but also by counterions, by substrate-to-catalyst ratio and solvent, as well as by reaction time and reaction temperature. The following sections present some examples to show the influence of these factors.

6.1.2   Asymmetric Catalytic Hydrogenation of C=C Bonds

6.1.2.1   *Reaction Mechanisms for the Asymmetric Hydrogenation of Enamides.*   Over the past three decades, use of chiral catalysts to synthesize highly enantiomerically pure compounds has been one of the most impressive achievements in the asymmetric hydrogenation of prochiral olefinic substrates. Homogeneous asymmetric hydrogenation of enamides catalyzed by transition metal complexes has become one of the most powerful methods for synthesizing optically active organic compounds. High enantioselectivities have been obtained in the synthesis of amino acids and related compounds through Rh complex–catalyzed asymmetric hydrogenation. Representative asymmetric catalytic hydrogenation reactions are shown in the table of Scheme 6–2[15] in

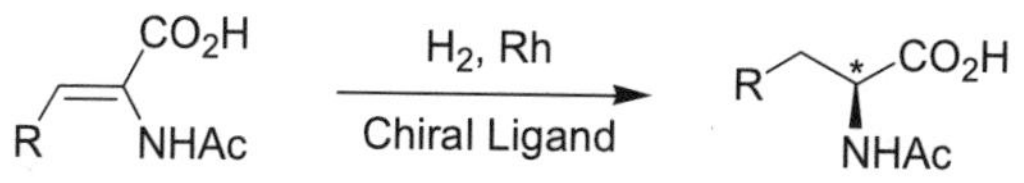

Phosphine	Product ee (%)	
ligand	$R = C_6H_5$	$R = H$
(R, R)-DIPAMP	96 (S)	94 (S)
(S, S)-ChiraPhos	99 (R)	91 (R)
(S, S)-NorPhos	95 (S)	90 (R)
(R, R)-DIOP	85 (R)	73 (R)
(S, S)-BPPM	91 (R)	98.5 (R)[a]
(S)-BINAP	100 (R)[a]	98 (R)
(S)-(R)-BPPFA	93 (S)	
(S, S)-SkewPhos	92 (R)	
(S, S)-CycPhos	88 (R)	
(S, S)-Et-DuPhos	99 (S)	99.4(S)

a. hydrogenation of the N-benzoyl derivative

Scheme 6–2. Catalytic hydrogenation of enamindes.

which the results for various Rh complex–catalyzed enantioselective hydrogenations of α-acylaminoacrylic acids to the corresponding amino acids are summarized.

The reaction mechanism in phosphine–Rh complex catalyzed hydrogenation was elucidated by Halpern[16] and Brown and Maddox[17] on the basis of NMR and X-ray crystallographic studies of the reaction intermediates, as well as detailed kinetic analyses. The well-recognized mechanism proposed by Halpern is presented in Figure 6–3.

First, solvent molecules, referred to as S in the catalyst precursor, are displaced by the olefinic substrate to form a chelated Rh complex in which the olefinic bond and the amide carbonyl oxygen interact with the Rh(I) center (rate constant k_1). Hydrogen then oxidatively adds to the metal, forming the Rh(III) dihydride intermediate (rate constant k_2). This is the rate-limiting step under normal conditions. One hydride on the metal is then transferred to the coordinated olefinic bond to form a five-membered chelated alkyl–Rh(III) intermediate (rate constant k_3). Finally, reductive elimination of the product from the complex (rate constant k_4) completes the catalytic cycle.

When an appropriate chiral phosphine ligand and proper reaction conditions are chosen, high enantioselectivity is achievable. If a diphosphine ligand with C_2 symmetry is used, two diastereomers for the enamide-coordinated complex can be formed because the olefin can interact with the metal from either the Re- or Si-face. Therefore, enantioselectivity is determined by the relative concentrations and reactivities of the diastereomeric substrate–Rh complexes. It should be mentioned that in most cases it is not the preferred mode of initial binding of the prochiral olefinic substrate to the catalyst that dictates the final stereoselectivity of these catalyst systems. The determining factor is the differ-

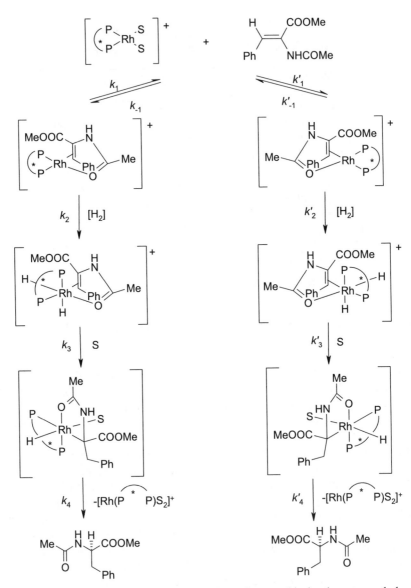

Figure 6–3. Mechanism for the hydrogenation of a prochiral substrate methyl acetamidocinnamate (MAC) with a catalyst containing a chiral chelating diphosphine ligand (P * P). S = Methanol.

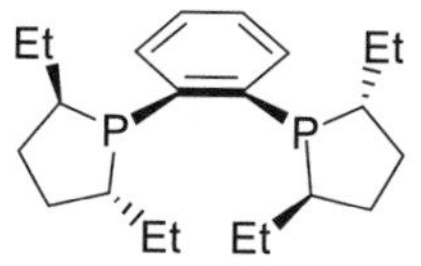

(R, R)-Et-DuPhos

1

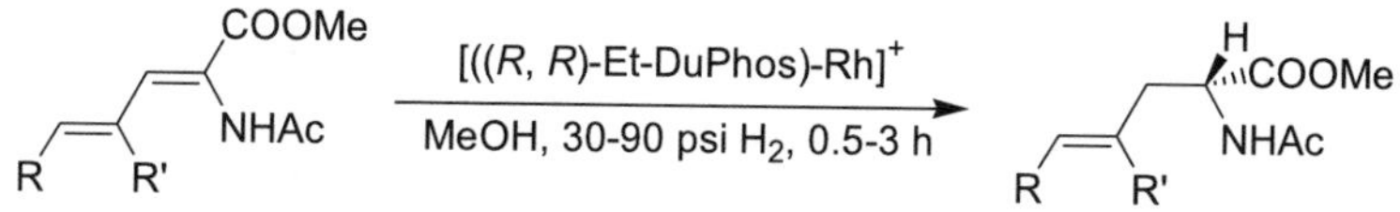

Scheme 6–3

ence in the rate of the subsequent reactions of the diastereomeric catalyst–substrate adducts with hydrogen. Interestingly it is the minor diastereomer rather than the major one that determines the predominate chirality of the product because the minor diastereomer is much more reactive than the major one. The following are some examples of using catalytic asymmetric hydrogenation to synthesize optically active compounds.

Unsaturated amino acids are an important class of natural products that have shown interesting biological properties.[18] Among them, γ,δ-unsaturated amino acids have been isolated from a variety of natural sources[19] and served as intermediates in the synthesis of complex amino acids and peptides.[20] This type of compound can be synthesized via the catalytic hydrogenation of the corresponding substrate. As shown in Scheme 6–3, the asymmetric hydrogenation of α,γ-dienamide ester catalyzed by [Rh-(R,R)-Et-DuPhos]$^+$ affords the corresponding γ,δ-unsaturated amino acid with excellent enantioselectivity.[21]

In this reaction, the Et-DuPhos-Rh catalyst is applied at a substrate-to-catalyst ratio S/C = 500 under an initial hydrogen pressure ranging from 30 to 90 psi. Full conversion of the substrate can be obtained in 0.5 to 3 hours. In all cases, less than 2% over-reduction was detected, and the products were isolated with better than 95% yield. The amount of over-reduction product varied but could be minimized by careful control of the S/C ratio, reaction time, and, to a lesser degree, initial hydrogen pressure. In all cases, the hydrogenation of the enamide double bond preceded the reduction of the distal double bond. When (R,R)-Et-DuPhos was used as the chiral ligand, the product in (R)-configuration was obtained, while product in (S)-configuration was obtained when (S,S)-DuPhos was used.

α-Aminophosphinic acids, structural analogs of α-aminocarboxylic acids, have interesting biological properties and serve as active herbicides, bactericides, and antibiotics. The optically active α-aminophosphinic acids and their derivatives can also be synthesized via catalytic asymmetric hydrogenation of the corresponding phosphinic substrates. As shown in Scheme 6–4, optically active α-aminophosphinic acids are synthesized via asymmetric hydrogenation of the

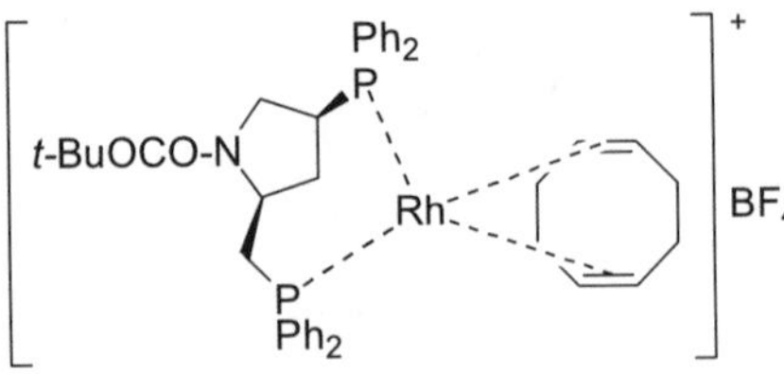

2

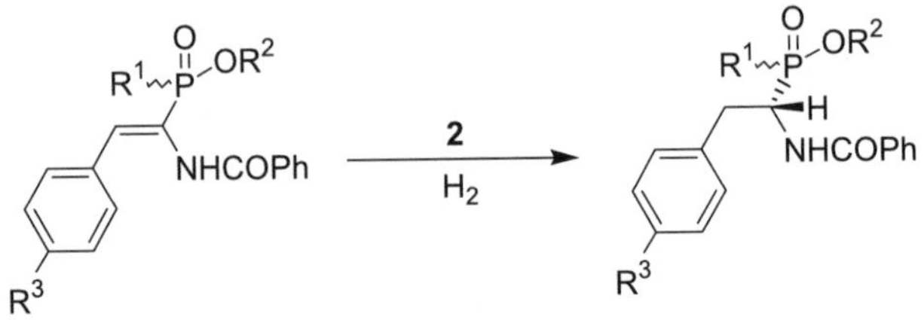

Scheme 6–4

corresponding unsaturated precursor catalyzed by Rh(I) complex **2**. Enantiomeric excess of 98% can be obtained when (S,S)-BPPM is used as the chiral ligand.[22]

Optically active phosphines can play an important role as chiral ligands in various metal-catalyzed asymmetric reactions, and numerous chiral phosphines have been designed and synthesized. Some P-chiral phosphines were landmark discoveries at an early stage in the history of asymmetric hydrogenation reactions. However, less attention has been paid to P-chiral phosphine ligands in asymmetric catalysis for two reasons. First, there is synthetic difficulty in getting highly enantiomerically enriched P-chiral phosphines; second, this class of chiral phosphines, especially the diarylphosphines and triarylphosphines, are configurationally unstable and gradually racemize at high temperatures.

Optically active trialkylphosphines are known to be configurationally stable toward racemization, and ligands of type **3** have been synthesized and applied in the asymmetric catalytic hydrogenation of amidoacrylic acids (Scheme 6–5).

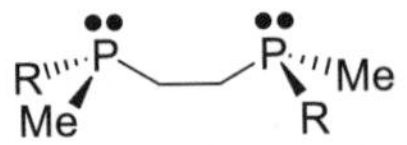

$R = t\text{-Bu, Et}_3C,$
1-adamantyl
$c\text{-C}_5H_9, c\text{-C}_6H_{11}$

3

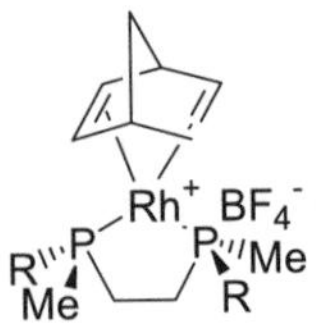

4

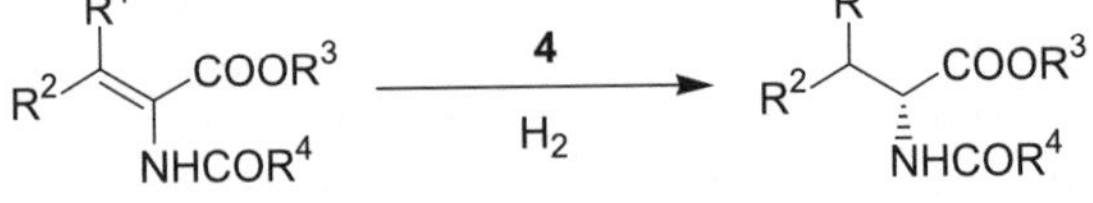

Scheme 6–5

This type of ligand forms a five-membered chelate with transition metals, and this feature may lead to high enantioselectivity in asymmetric synthesis. Furthermore, as trialkylphosphines are electron-rich ligands, reactions catalyzed by complexes containing these chiral *P*-coordinated ligands are anticipated to show high efficiency. A variety of substrates have been hydrogenated in the presence of chiral catalyst **4** (Scheme 6–5). Higher than 99.9% ee has been obtained in some specific reactions.[23]

6.1.2.2 Asymmetric Catalytic Hydrogenation of Acrylic Acids and Derivatives.

Enantiomerically pure 2-substituted succinic acid derivatives have attracted great interest because of their utility as chiral building blocks for peptidomimetics in the design of pharmaceuticals, flavors, fragrances, and agrochemicals. Asymmetric catalytic hydrogenation of β-substituted itaconic acid derivatives offers one of the most practical and convenient routes to these compounds.

Burk et al. reported an asymmetric hydrogenation catalyzed by [(Et-DuPhos)Rh]$^+$ catalyst. Very high enantioselectivity was obtained. When R = *i*-Pr, the minor enantiomer could not be detected by chiral GC methods. The results are shown in Scheme 6–6.[24]

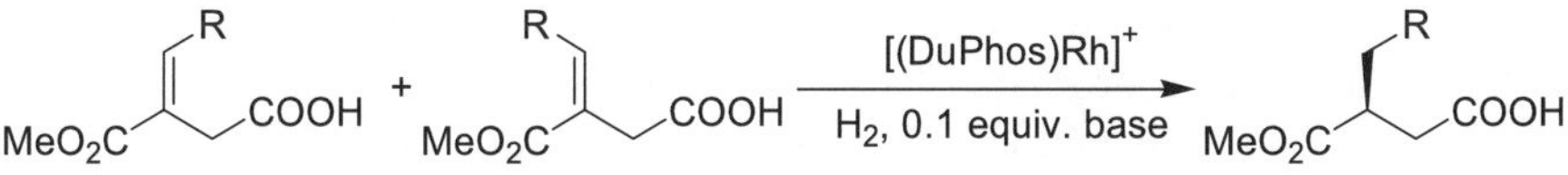

Entry	R	S/C	Time (h)	Ee (%)
1	H	1000	1	97
2	Et	1000	1	99
3	*n*-Bu	1500	2	97
4	CH$_2$CH$_2$Ph	2000	2	99
5	(*E*)-CH=CHPh	1000	20	99
6	*i*-Pr	3000	2	99
7	Cyclopropyl	2000	2	99
8	Cyclohexyl	1500	3	98
9	*t*-Bu	5000	4	99
10	Ph	3000	12	97
11	1-Naphthyl	3000	12	98
12	2-Naphthyl	3000	12	97
13	3-Thienyl	1000	15	99

Scheme 6–6. Reprinted with permission by Wiley-VCH Verlag GmbH, Ref. 24.

Generally, stereoselectivity can be enhanced by increasing the interactions between functional groups in a substrate and the chiral ligands. In light of this concept, Hayashi et al.[25] designed chiral aminoalkylferrocenyl phosphine ligands in which an amino group was introduced. This type of ligand offers high stereoselectivity as well as reasonable catalytic activity in the hydrogenation of trisubstituted acrylic acids (tetrasubstituted olefins) (Scheme 6–7).

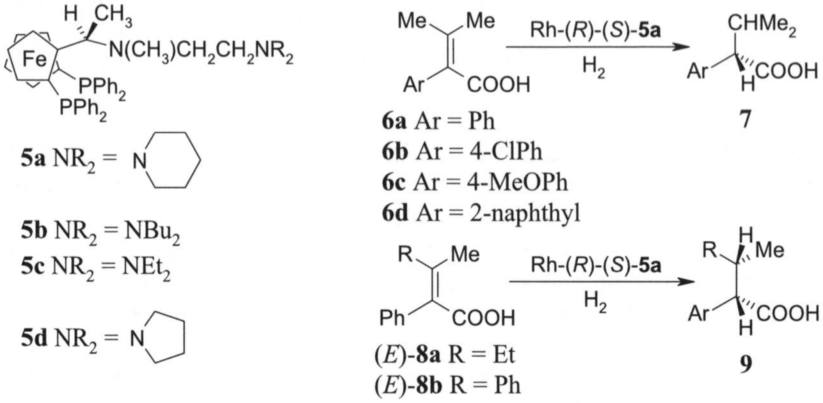

Scheme 6–7

Rh complexes with ChiraPhos, PyrPhos, or ferrocenyl phosphines lacking amino alkyl side chains (such as BPPFA) are much less active toward tetrasubstituted olefins. Table 6–1 shows that in asymmetric hydrogenations catalyzed by **5a–d**, the coordinated Rh complex exerts high selectivity on various substrates. It is postulated that the terminal amino group in the ligand forms an ammonium carboxylate with the olefinic substrates and attracts the substrate to the coordination site of the catalyst to facilitate the hydrogenation.

TABLE 6–1. Asymmetric Hydrogenation of Trisubstituted Acrylic Acids Catalyzed by Chiral Ferrocenylphosphine–Rhodium Complexes

Entry	Olefin	Ligand	Solvent	Time (h)	Product	ee (%, config.)
1	**6a**	**5a**	THF/MeOH (90/10)	30	**7a**	98.4(S)
2	**6a**	**5a**	THF/MeOH (80/20)	20	**7a**	97.6(S)
3	**6a**	**5a**	i-PrOH	20	**7a**	97.0(S)
4	**6a**	**5a**	MeOH	5	**7a**	95.8(S)
5	**6a**	**5b**	THF/MeOH (80:20)	20	**7a**	97.9(S)
6	**6a**	**5c**	THF/MeOH (80:20)	30	**7a**	98.1(S)
7	**6a**	**5d**	THF/MeOH (80:20)	30	**7a**	98.2(S)
8	**6b**	**5a**	THF/MeOH (80:20)	40	**7b**	97.4(S)
9	**6c**	**5a**	THF/MeOH (80:20)	40	**7c**	96.7(S)
10	**6d**	**5a**	THF/MeOH (80:20)	65	**7d**	97.3(S)
11	(*E*)-**8a**	**5a**	i-PrOH	100	**9a**	97.3(2S,3S)
12	(*E*)-**8b**	**5a**	THF/MeOH (80:20)	100	**9b**	92.1(2S,3R)

ee = Enantiomeric excess.

Reprinted with permission by Am. Chem. Soc., Ref. 25.

Kang et al.[6] reported a practical synthesis of an air-stable ferrocenyl bis-(phosphine) (pS,pS)-1,1'-bis-(diphenylphosphino)-2,2'-di-3-pentyl ferrocene ([S,S]-FerroPhos, **10a**) and its application in the rhodium(I)-catalyzed enantioselective hydrogenation of dehydroamino acid derivatives.

In the presence of a catalyst prepared in situ from [Rh(COD)$_2$]BF$_4$ and **10a**, the hydrogenation of various dehydroamino acid derivatives proceeds smoothly under mild conditions (2 atm, 20°–30°C) with high enantioselectivity (over 99.9% ee can be obtained). The enantioselectivity in the hydrogenation of α-acetamidocinnamic acid compares favorably with the reported ee from the reactions catalyzed by other air-stable triarylsubstituted ligands such as BINAP (84%)[26] and 2,2-PhanePhos (98%).[27] Due to the high enantioselectivity of this new ligand and its high stability toward air, it may have good potential for industrial application.

Several ferrocenyl phosphines have found industrial applications in the synthesis of chiral pharmaceuticals and agrochemicals. A process developed by Ciba-Geigy (now Novartis) for the production of a herbicide (S)-Metolachlor involves the asymmetric hydrogenation of an imine using an Ir complex of **10b** as catalyst.[28a] Similar technologies involving an Rh catalyst–containing ligand **10c** and an Ir catalyst–containing ligand **10d** were developed by Lonza Fine Chemicals in partnership with Ciba-Geigy for the production of the vitamin (+)-biotin and the cough medicine dextromethorphan, respectively.[28b,c] Firmenich developed a process for the production of (+)-cis-Hedione®, a perfume ingredient, by using an Ru complex of **10e**.[28d] The key steps of these processes are illustrated in Scheme 6–8.

Many attempts have been made to develop novel nonracemic ferrocenyl phosphine derivatives as asymmetric hydrogenation catalysts. Interested readers will find the design and synthesis of these chiral ferrocenyl phosphine ligands in a recent review by Richards and Locke.[28e]

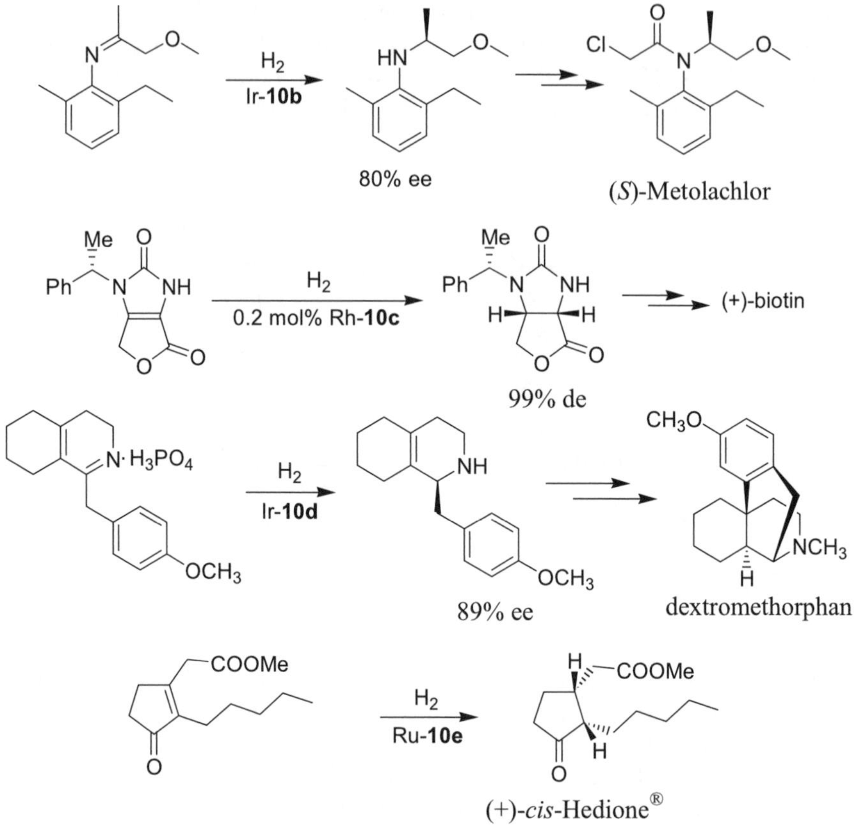

80% ee

(S)-Metolachlor

99% de

(+)-biotin

89% ee

dextromethorphan

(+)-cis-Hedione®

Scheme 6–8

6.1.2.3 Sodium Borohydride Reduction of α,β-Unsaturated Esters.

In 1989, Leutenegger et al.[29] reported a semicorrin-type chiral C_2-symmetric ligand bearing the structural features of **11a–c**. Such chiral ligands are readily accessible in enantiomerically pure form from pyroglutamic acid.[30] By varying substituent R at the two chiral centers in the compound, a series of chiral ligands can be obtained.

A cobalt complex containing this type of ligand is effective in the sodium borohydride–mediated enantioselective reduction of a variety of α,β-unsaturated carboxylates. As can be seen from Scheme 6–8, in the presence of a catalytic amount of a complex formed in situ from $CoCl_2$ and chiral ligand **11**, reduction proceeds smoothly, giving product with up to 96% ee. The chiral ligand can easily be recovered by treating the reaction mixture with acetic acid.

The reaction can be used on a laboratory scale in vitamin synthesis. The enantioselectivity of this method lies in the same range as that observed in the catalytic hydrogenation of structurally related substrates. In contrast with chiral Rh or Ru complex–mediated catalytic hydrogenation, reduction of α,β-unsaturated carboxylate can proceed with excellent ee even in the absence of the α-acylamido group in the substrate (Scheme 6–9).

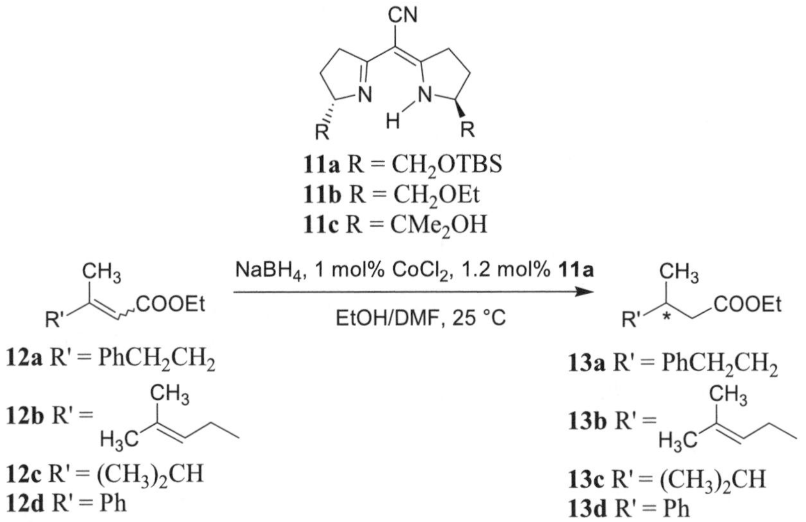
CN
N
N
H
R
R
11a R = CH_2OTBS
11b R = CH_2OEt
11c R = CMe_2OH
CH_3
R'
COOEt
$NaBH_4$, 1 mol% $CoCl_2$, 1.2 mol% **11a**
EtOH/DMF, 25 °C
CH_3
R'
*
COOEt
12a R' = $PhCH_2CH_2$
12b R' =
CH_3
H_3C
12c R' = $(CH_3)_2CH$
12d R' = Ph
13a R' = $PhCH_2CH_2$
13b R' =
CH_3
H_3C
13c R' = $(CH_3)_2CH$
13d R' = Ph

Enantioselective reduction of the α,β-unsaturated carboxylates (*Z*)-/(*E*)-**12a-d**

Substrate	Yield (%)	ee (%)	Config. of **13**
(*E*)-**12a**	97	94	(*R*)-(+)
(*Z*)- **12a**	95	94	(*S*)-(-)
(*E*)- **12b**	95	94	(*R*)-(+)
(*Z*)- **12b**	94	94	(*S*)-(-)
(*E*)- **12c**	84	96	(*S*)-(-)
(*Z*)- **12c**	86	90	(*R*)-(+)
(*E*)- **12d**	95	81	(*S*)-(+)
(*Z*)- **12d**	97	73	(*R*)-(-)

Scheme 6–9. Reprinted with permission by VCH, Ref. 29.

6.1.2.4 Asymmetric Hydrogenation of Enol Esters. Prochiral ketones represent an important class of substrates. A broadly effective and highly enantioselective method for the asymmetric hydrogenation of ketones can produce many useful chiral alcohols. Alternatively, the asymmetric hydrogenation of enol esters to yield α-hydroxyl compounds provides another route to these important compounds.

The asymmetric hydrogenation of enol esters generally proceeds with moderate to high enantioselectivities. In the presence of Rh-DuPhos catalyst,[31] highly enantioselective hydrogenation of enol esters bearing various substituents can be achieved, giving chiral alcohol derivatives at high optical purity. As shown in Scheme 6–10, a series of α-enol esters is hydrogenated in the presence

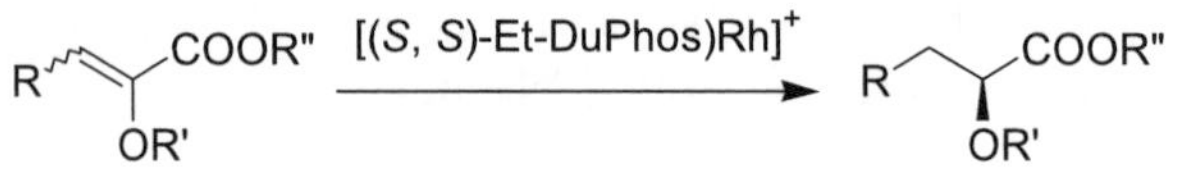

Entry	R	R'	R"	$(Z)/(E)$	ee (%)
1	H	Ac	Et		> 99 (S)
2	Me	Bz	Et	3	96.0 (S)
3	n-Pr	Bz	Me	3	98 (S)
4	i-Pr	Bz	Et	6	96.9 (S)
5	c-Pr	Bz	Me	9	97.5 (S)
6	CH_2-i-Pr	Bz	Et	2.5	> 99 (S)
7	n-C_5H_{11}	Ac	Et	3.5	> 99 (R)
8	c-C_6H_{11}	Bz	Me	3	95 (S)
9	Ph	Bz	Me	10	98 (S)
10	1-Naphthyl	Bz	Et	3	93.2 (S)
11	2-Thienyl	Bz	Me	4	97.5 (S)

Scheme 6–10. Reprinted with permission by Am. Chem. Soc., Ref. 32.

of $[(S,S)\text{-}(Et\text{-}DuPhos)Rh]^+$, providing the corresponding product with high enantioselectivity irrespective of the $(Z)/(E)$ ratio of the starting material.[32]

Allylic alcohol derivatives are quite useful in organic synthesis, so the asymmetric synthesis of such compounds via asymmetric hydrogenation of dienyl (especially enynyl) esters is desirable. The olefin functionality preserves diverse synthetic potential by either direct or remote functionalization. Boaz[33] reported that enynyl ester and dienyl ester were preferred substrates for asymmetric hydrogenation using Rh-(Me-DuPhos) catalyst [Rh(I)-(R,R)-**14**], and products with extremely high enantioselectivity (>97%) were obtained (Schemes 6–11 and 6–12).

While Rh-DuPhos–mediated asymmetric hydrogenation of acyclic enol esters shows high levels of enantioselectivity, it does not provide the same high

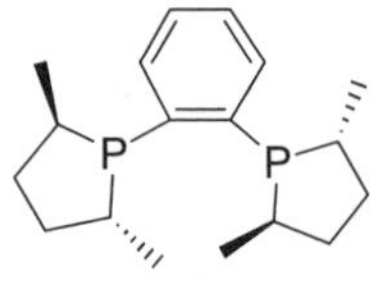

14

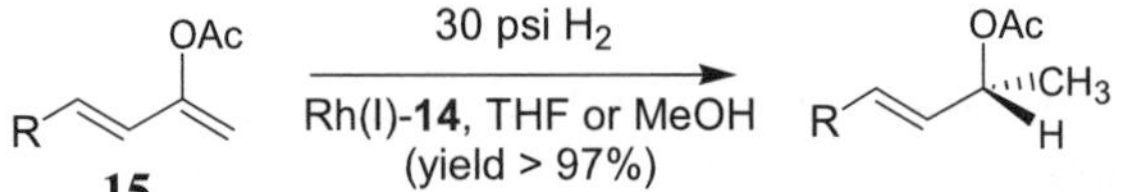

15

30 psi H_2

Rh(I)-**14**, THF or MeOH
(yield > 97%)

16a R = n-C_5H_{11}, 94% ee
16b R = Ph, 94% ee

Scheme 6–11. Asymmetric hydrogenation of dienyl esters.

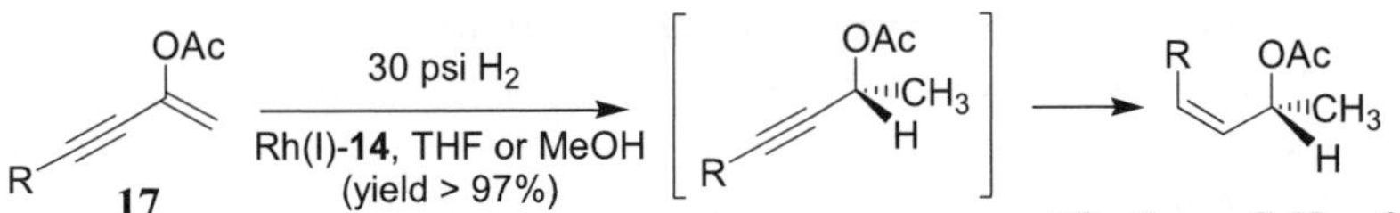

17

18a R = n-C_5H_{11}, 98.5% ee
18b R = Ph, 97.8% ee
18c R = $CH_2CH_2OCH_2Ph$, > 98% ee

Scheme 6–12. Asymmetric hydrogenation of enynyl esters.

selectivity with cyclic enol ester substrates. Changes in the steric and electronic properties of the substrates sometimes lead to unexpected results. Jiang et al.[34] achieved excellent enantioselectivity in the asymmetric hydrogenation of cyclic enol ethers using a PennPhos series developed by their group as the chiral ligand. In the model reaction shown in Scheme 6–13, asymmetric hydrogenation of 3,4-dihydronaphth-1-yl acetate catalyzed by Me-PennPhos (**19**, R = Me) gives the corresponding acetate with 100% conversion and over 99% ee. Other chiral ligands such as BINAP and DuPhos give only poor results.

The asymmetric hydrogenation of enol esters can also be catalyzed by chiral amidophosphine phosphinite catalysts derived from chiral amino acids, but the enantioselectivity of these reactions has thus far been only moderate.[35]

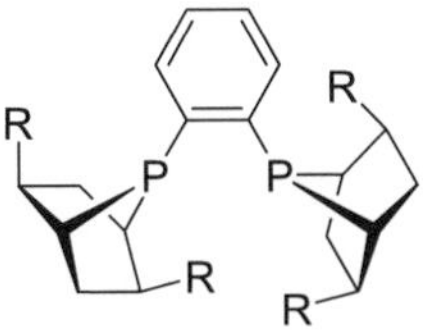

19

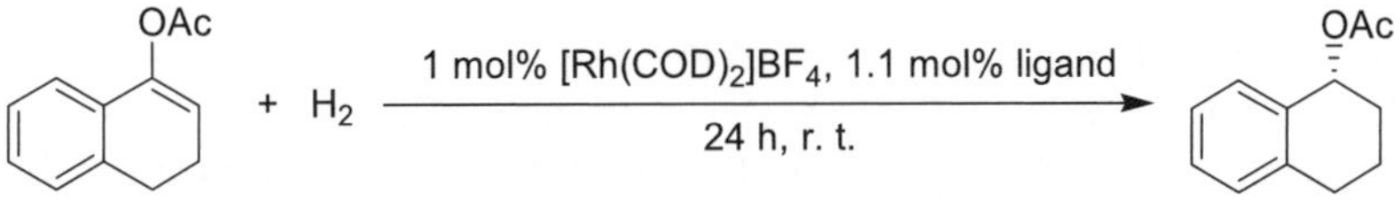

Entry	Ligand	Solvent	Conver. (%)	ee (%)	Config.
1	Me-PennPhos	toluene	64	98.3	*R*
2	Me-PennPhos	CH_2Cl_2	100	86.8	*R*
3	Me-PennPhos	THF	100	98.7	*R*
4	Me-PennPhos	MeOH	100	99.1	*R*
5	(*R*)-BINAP	THF	2.4	18.0	*R*
6	(*R, R*)-Me-DuPhos	THF	1.3	12.3	*S*
7	(*R, R*)-Me-DuPhos	MeOH	No reaction	No reaction	

Scheme 6–13. Reprinted with permission by Wiley-VCH Verlag GmbH, Ref. 34.

6.1.2.5 Asymmetric Hydrogenation of Unfunctionalized Olefins.

Enantioselective hydrogenation with catalysts bearing rhodium or ruthenium as the metal and chiral diphosphane as the ligand is one of the most powerful methods in asymmetric catalysis. However, the range of substrates is still limited to certain classes of olefins bearing polar groups that can coordinate with the catalyst. As has been discussed thus far, most of the substrates have polar functional groups. Examples of the asymmetric hydrogenation of unfunctionalized olefins with high enantioselectivity are rare.

Lightfoot et al.[36] have reported a series iridium complexes **21** containing phosphanodihydroxazole ligands **20**. These complexes give high enantioselectivity in the asymmetric hydrogenation of unfunctionalized olefins. Selected results are given in Table 6–2.

Broene and Buchwald[37] synthesized chiral titanocene compound **22** for the asymmetric hydrogenation of trisubstituted olefins.

The reaction was carried out by addition of 1.95 equivalents of *n*-BuLi to a THF solution of **22** at 0°C to generate the active catalyst, which was then combined with substrate (S/C about 20:1) under an inert atmosphere using phenylsilane as the stabilizing agent. Trisubstituted unfunctionalized olefins can be hydrogenated in good yield with high ee. Representative results are listed in Table 6–3.

6.1.2.6 New Developments in the Asymmetric Hydrogenation of Enamides.

Besides the chiral phosphine ligands mentioned above, chiral phosphinites and chiral phosphinamidites also have emerged as powerful ligands for catalytic asymmetric hydrogenation of a variety of substrates. Selke et al.[38a] and RajanBabu et al.[38c] developed a series of phosphinites derived from D-(+)-glucose and found them to be effective ligands for the rhodium-catalyzed

TABLE 6–2. Asymmetric Hydrogenation of Unfunctionalized Olefins

Entry	Substrate	Catalyst (mol%)	Yield (%)	ee (%)
1		**21d** (4)	96	84
2		**21d** (1)	95	96
3		**21f** (0.1)	>99	97
4		**21f** (0.5)	98	95
5		**21f** (0.3)	97	95
6		**21f** (0.3)	>99	61
7		**21f** (1)	97	42
8		**21f** (0.5)	>99	91
9		**21g** (2)	>99	81

ee = Enantiomeric excess.

asymmetric hydrogenation of dehydroamino acid derivatives. The most important advantage of chiral phosphinite ligands over their corresponding phosphine counterparts is their ease of preparation, which can be achieved by reacting the corresponding chiral alcohol with chlorophosphines. Chan et al.[39] have developed spiro ligand **23** (SpirOP), which has shown great efficiency in catalytic hydrogenation reactions. Chiral ligand **24** is also a chiral phosphinite ligand that has shown catalytic results as good as chiral phosphine ligands.[40]

TABLE 6–3. Chiral Titanocene-Catalyzed Asymmetric Hydrogenation of Unfunctionalized Trisubstituted Olefins

Entry	Substrate	Time (h)	Yield (%)	ee (%)
1		48	91	>99
2		48	79	95
3		44	77	92
4		132	70	93
5		184	70	83
6		169	87	83

Reprinted with permission by Am. Chem. Soc., Ref. 37.

(1R,5R,6R)-**23**   (1S,5S,6S)-**23**   (1R,1'R,2S,2'S)-**24**

The most difficult aspect of research in asymmetric catalysis is finding effective ligands. As suggested by Noyori, the highly skewed naphthyl rings in BINAP are the determining features that made the ligand so effective in asymmetric catalytic reactions. A rigid ligand–metal complex structure is thus essential for obtaining effective chiral recognition. In designing novel ligands for catalytic asymmetric hydrogenation, one should always keep this in mind. In the above phosphinite ligand **23**, the spiro backbone, which mimics the skewed positions of the binaphthyl rings in BINAP, possesses a highly rigid structure, and this makes it possible to compensate for the conformational flexibility caused by the introduction of the C–O–P bond. The asymmetric hydrogenation of prochiral enamides shows that the spiro phosphinite ligand **23** is superior to the less rigid chiral phosphinite and phosphines. Representative results are shown in Scheme 6–14.

The Rh-**23**–catalyzed asymmetric hydrogenation of (Z)-2-acetamidoacrylic

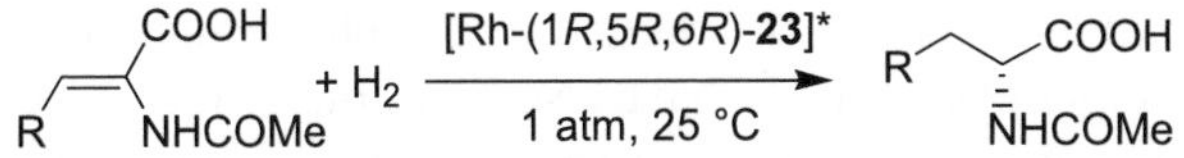

$$R\text{—}\overset{\text{COOH}}{\underset{\text{NHCOMe}}{=}} + H_2 \xrightarrow[\text{1 atm, 25 °C}]{[\text{Rh-}(1R,5R,6R)\text{-23}]^*} R\text{—}\overset{\text{COOH}}{\underset{\text{NHCOMe}}{\diagdown}}$$

Entry	Substrate	ee (%)
1	H	> 99.9
2	Ph	97.9
3	4-ClPh	97.3
4	2-ClPh	97.3
5	3-ClPh	97.4
6	4-O₂NPh	97.0

Scheme 6–14. Spiro phosphinite–catalyzed reactions. Reprinted with permission by Am. Chem. Soc., Ref. 39b.

acid gives very high enantioselectivity (>99.9% ee) in the product. When (1R,5R,6R)-**23** is used as the ligand, the (R)-configuration is obtained in all products. It is not only the unsaturated acid that can be reduced with high enantioselectivity; excellent results can also be obtained for the corresponding esters as well.

In chiral ligand **24**, the two cyclopentane rings restrict the conformational flexibility of the nine-membered ring, and the four stereogenic centers in the backbone dictate the orientation of the four P-phenyl groups. Scheme 6–15 shows the application of Rh-**24** in the asymmetric hydrogenation of dehydroacylamino acids.

The somewhat higher enantioselectivity of Rh-**23** as compared with Rh-**24** may be due to the different rigidities of these two chiral ligands: The carbon backbone of chiral compound **23** is an absolutely rigid structure, while that of compound **24** is slightly more flexible because of the possibility of free rotation along the C–C bond linking the two five-membered rings.

In contrast to the success in the synthesis of optically active amino acids and related compounds, only limited success has been achieved in the asymmetric synthesis of chiral amines or related compounds. One breakthrough is the asymmetric hydrogenation of arylenamides with Rh catalysts containing

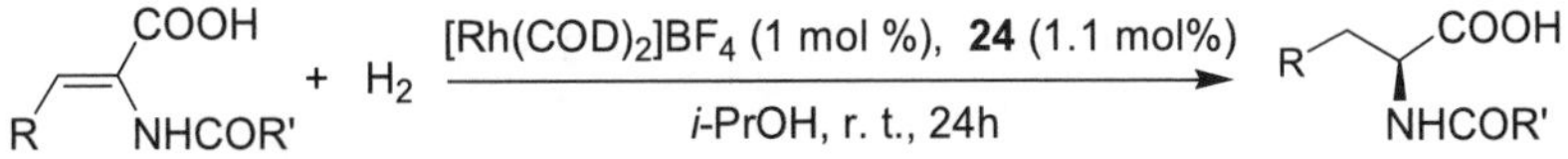

$$R\text{—}\overset{\text{COOH}}{\underset{\text{NHCOR'}}{=}} + H_2 \xrightarrow[\text{i-PrOH, r. t., 24h}]{[\text{Rh(COD)}_2]\text{BF}_4 \ (1 \ \text{mol \%}), \ \textbf{24} \ (1.1 \ \text{mol\%})} R\text{—}\overset{\text{COOH}}{\underset{\text{NHCOR'}}{\diagdown}}$$

Entry	Substrate	ee (%)
1	R = H, R' = CH₃	94.8
2	R = Ph, R' = CH₃	94.7
3	R = Ph, R' = Ph	89.2
4	R = 3-BrPh, R' = CH₃	93.5
5	R = 2-ClPh, R' = CH₃	92.9

Scheme 6–15. Reprinted with permission by Am. Chem. Soc., Ref. 40.

DuPhos and BPE ligands.[41] In addition to the phosphinite ligands, the easily prepared bis-aminophosphines[42] have also been found to be effective ligands in catalytic hydrogenation reactions. From a practical point of view, it is always desirable to develop effective ligands that can be easily prepared and manipulated. Zhang et al.[43] have used chiral bis-aminophosphine ligands **25** and **26** for asymmetric synthesis of chiral amine derivatives.

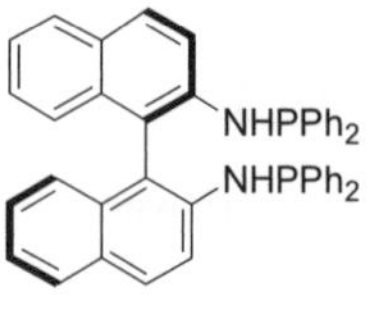

| (R)-BDPAB | (S)-BDPAB | (R)-H-8-BDPAB | (S)-H-8-BDPAB |
| (R)-**25** | (S)-**25** | (R)-**26** | (S)-**26** |

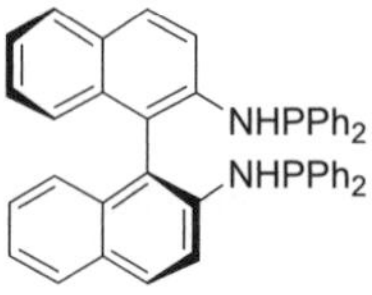

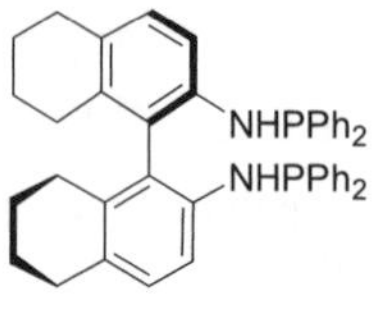

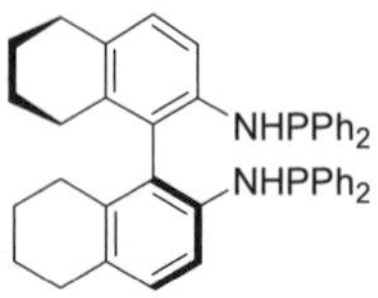

With these ligands, the catalytic hydrogenation of α-arylenamides gives a fast rate of reaction and up to 99% product ee when the reaction is carried out at 5°C under 1 atm of H_2 (Scheme 6–16).

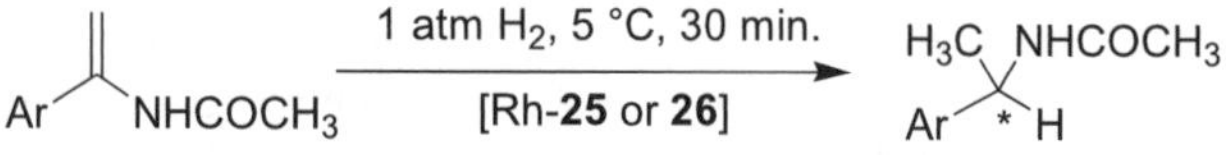

Scheme 6–16. Asymmetric hydrogenation of enamide using **25** or **26** as the chiral ligand.

Chirality transfer in catalytic asymmetric hydrogenation can be achieved not only by using powerful chiral ligands such as BINAP or DuPhos but also by the formation of a dynamic conformational isomer. The availability of many enantiomerically pure diols allows the production of electron-deficient, bidentate phosphate in the form of **27**. The backbone O–R^1–O can define the chirality of the O–R^2–O in complex **28**, hence realizing the chirality transfer.[44]

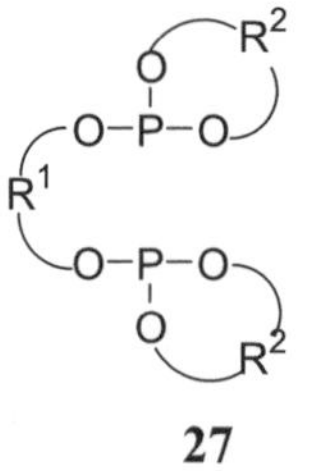

| | |
| 27 | 28 |

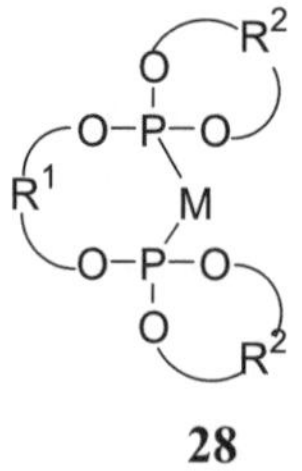

When a commercially available C_2-symmetric 1,4:3,6-dianhydro-D-mannite **29** is chosen as the backbone, reaction of this diol compound with chlorophosphoric acid diaryl ester gives a series of phosphorate ligands **30**. These were tested using the asymmetric hydrogenation of dimethyl itaconate as a model reaction. When **30** bearing two achiral β-naphthoxy residues at each phosphorus center was used, the stereoselectivity of the reaction was only 21%. This means that the backbone diol is only a poor ligand for chirality transfer. When **30** containing (S)- or (R)-binaphthol was used as the chiral ligand, ee values of 88% and 95% were obtained, consistent with the mismatched and matched pairs, respectively.

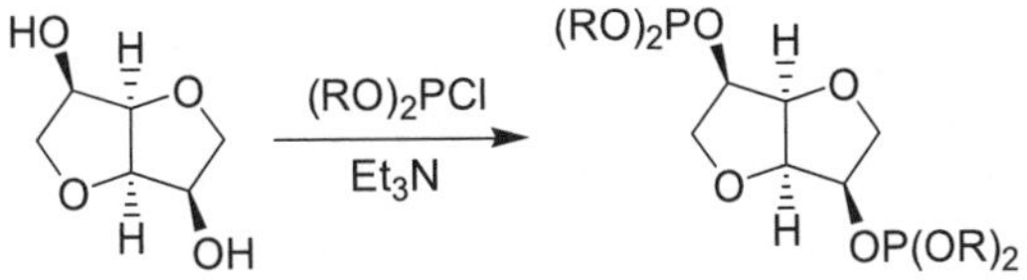

When atropisomeric biphenol units are present in the P/O hetereocycle, the enantioselectivity of the reaction is not expected to be very high because these biphenol moieties themselves are not chiral. However, enhancement of enantioselectivity has been observed, especially when 2,2'-dihydroxy-3,3'-dimethyl-1,1'-biphenyl is used as an achiral diol. A very high enantioselectivity of 98.2% has been found, even though 2,2'-dihydroxy-3,3'-dimethyl-1,1'-biphenyl itself is achiral. The explanation for this high enantioselectivity is that a dynamic chirality is induced into 2,2'-dihydroxy-3,3'-dimethyl-1,1'-biphenyl by the mannite backbone, and this dynamic chirality is the cause of the high enantioselectivity of the hydrogenation reaction. There are three defined diastereomeric metal complexes possible $(R/R, S/S, \text{ and } R/S$ combinations in the biphenol moieties), and these three atropisomers are interconvertible due to the low energy barrier for free rotation along the biaryl axis. One of the atropisomers is more stable than the others under the preexisting chirality in the mannite, and it is this atropisomer that induces the high enantioselectivity in asymmetric hydrogenation. The results are shown in Scheme 6–17.

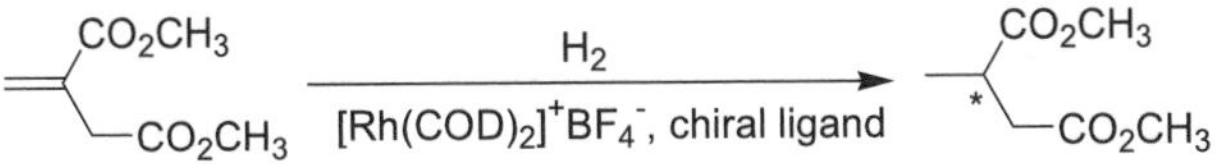

Entry	Diol in ligand	Temp. (°C)	Conver. (%)	ee (%)	Config.
1	2-naphthol	20	65	21.0	(S)
2	(S)-binol	20	> 99	87.8	(S)
3	(R)-binol	20	> 99	94.5	(R)
4	(R)-binol	-10	> 99	96.2	(R)
5	2,2'-dihydroxy-biphenyl	20	74	38.9	(S)
6	2,2'-dihydroxy-3,3'-dimethyl biphenyl	20	> 99	96.8	(R)
7	2,2'-dihydroxy-3,3'-dimethyl biphenyl	-10	> 99	98.2	(R)

Scheme 6–17. Reprinted with permission by Wiley-VCH Verlag GmbH, Ref. 44.

6.1.2.7 Examples of Potential Industrial Applications. Ru–BINAP dicarboxylate complexes catalyze the hydrogenation of a variety of functionalized prochiral olefins with high enantioselectivity. The pure diacetate complex can be prepared in good yield[11] by treating $[RuCl_2(COD)]_n$ first with (*R*)- or (*S*)-BINAP and triethylamine in toluene at 110°C, followed by sodium acetate treatment in *t*-butyl alcohol at 80°C or, more conveniently, by treatment of $RuCl_2(benzene)_2$ with BINAP in DMF at 100°C and then with excess sodium acetate.[45]

Takaya and co-workers[46] found that BINAP-based Ru(II) dicarboxylate complexes **31** can serve as efficient catalyst precursors for enantioselective hydrogenation of geraniol (2*E*)-**32** and nerol (2*Z*)-**32**. (*R*)- or (*S*)-citronellal **33** is obtained in nearly quantitative yield with 96–99% ee. The nonallylic double bonds in geraniol and nerol were intact. Neither double bond migration nor (*E*)-/(*Z*)-isomerization occurred during the catalytic process. Furthermore, the S/C ratio was extremely high, and the catalyst could easily be recovered (Scheme 6–18). This process can be applied to the asymmetric synthesis of a key intermediate for vitamin E.

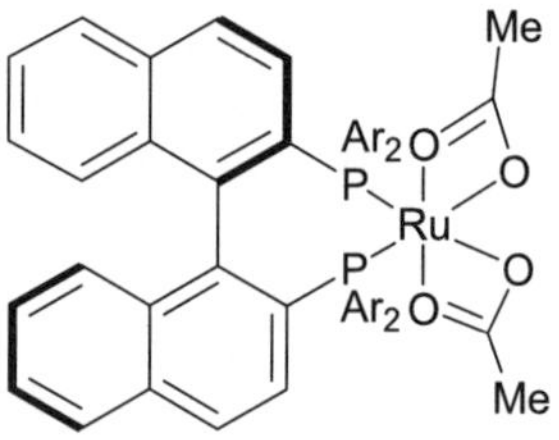
Me
Ar2
O
P
O
Ru
P
O
Ar2
O
Me

31 [Ru-(*R*)-BINAP(OAc)$_2$]

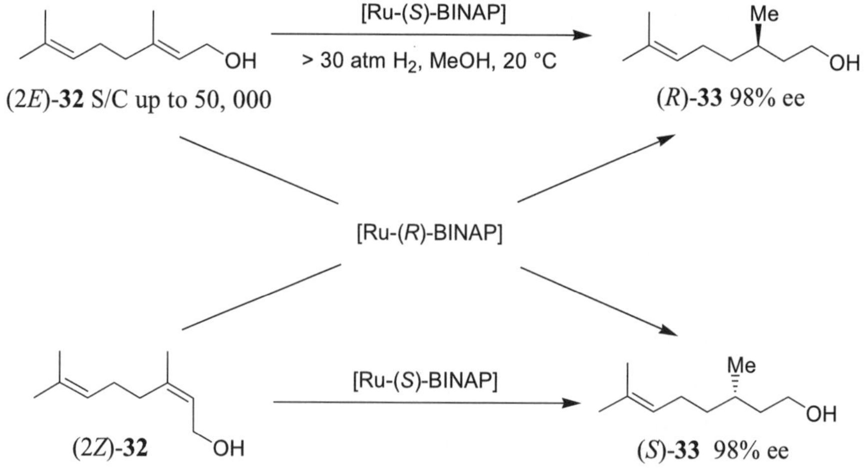
OH
[Ru-(S)-BINAP]
> 30 atm H2, MeOH, 20 °C
Me
OH
(2E)-32 S/C up to 50, 000
(R)-33 98% ee
[Ru-(R)-BINAP]
[Ru-(S)-BINAP]
Me
OH
(2Z)-32
OH
(S)-33 98% ee

Scheme 6–18

The catalytic hydrogenation of enamide **34** in the presence of 0.5–1 mol% of (*R*)-BINAP complex in a 5:1 mixture of ethanol and dichloromethane under 1–4 atm of hydrogen affords **35** with quantitative yield and higher than 99.5% ee. The (*E*)-isomer is not reduced under similar reduction conditions. This approach provides a route to a number of alkaloid compounds (Scheme 6–19).[47]

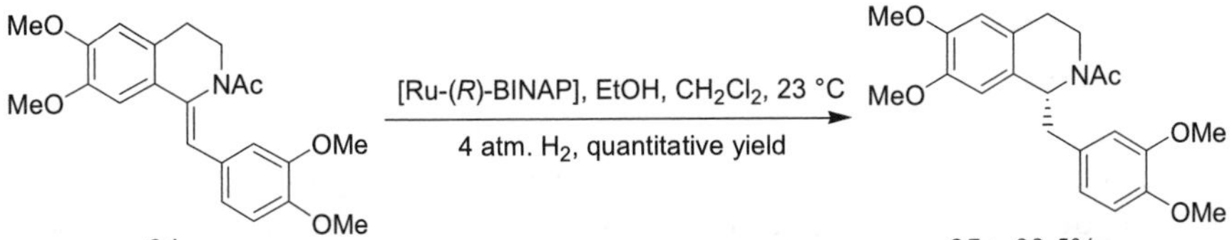

Scheme 6–19

Naproxen is a nonsteroidal antiinflammatory drug, and its (*S*)-form is about 30 times more active than its (*R*)-form. The Ru–BINAP-catalyzed asymmetric hydrogenation of substrate **36** offers an entry to an enantiomerically pure (*S*)-form of the drug **37** (Scheme 6–20).[11a]

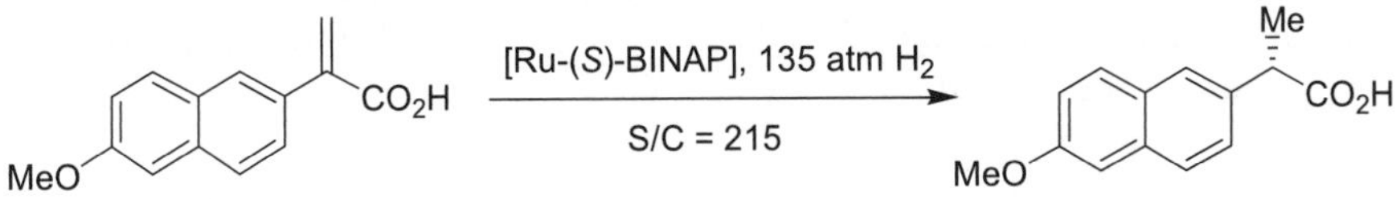

Scheme 6–20

Because naproxen is an extremely attractive target product for catalytic asymmetric synthesis due to its large volume and high value, the development of this chemistry is expected to be of high commercial interest. Extensive effort in repeating this experiment by Chan et al.[11b–d] showed that, under the above conditions, generally 93–94% ee could be obtained. The small difference in ee might be due to analytical discrepancy. The ee values increased somewhat when the reaction was carried out at lower temperature. It should be pointed out that, while there are many good catalysts for the asymmetric hydrogenation of enamides, only Ru–BINAP-type catalysts are effective for the hydrogenation of 2-arylacrylic acids (with ee >90%). Further development of the Noyori chemistry is clearly of high scientific and practical interest. To expand the scope of the BINAP (or chiral biaryl) chemistry to include pyridyl species, Chan et al. developed a new class of chiral pyridylphosphine ligand, 2,2′,6,6′-tetramethoxy-4,4′-bis(diphenylphosphino)-3,3′-bipyridine (P-Phos, **38a**).[48a]

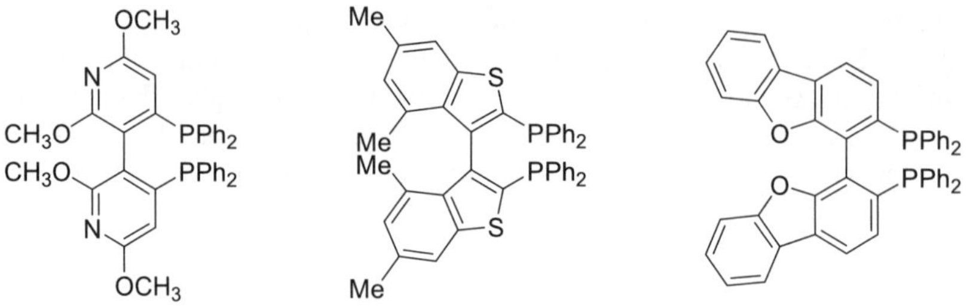

38a   **38b**   **38c**

The Ru(P-Phos) catalyst was found to be highly enantioselective in asymmetric hydrogenation of **36** leading to **37**. The best ee obtained at 0°C under 1000 psig H_2 was 96.2%. (A side-by-side comparison study using Ru–BINAP catalyst gave 94.8% ee.) The asymmetric hydrogenation of β-keto esters using Ru(P-Phos) catalyst also gave up to 99% ee. Other chiral biaryl ligands with heterocyclic moieties may also be considered as offshoots of the BINAP family. Bennincori et al.[48b] synthesized (+)- and (−)-2,2′-bis(diphenylphosphino)-4,4′,6,6′-tetramethyl-3,3′-bis(benzothiophene) **38b** and found its Ru complex catalyzed the asymmetric hydrogenation of β-keto esters to give β-hydroxy esters in up to >99% ee. Another ligand, 2,2′-bis(diphenylphosphino)-1,1′-bis-(dibenzofuranyl) **38c**, was developed by Bayer for the asymmetric hydrogenation leading to (S)-ketoprofen (Scheme 6–21).[48c] It is expected that Noyori's innovative concept is a good foundation based on which more effective new catalysts can be developed.

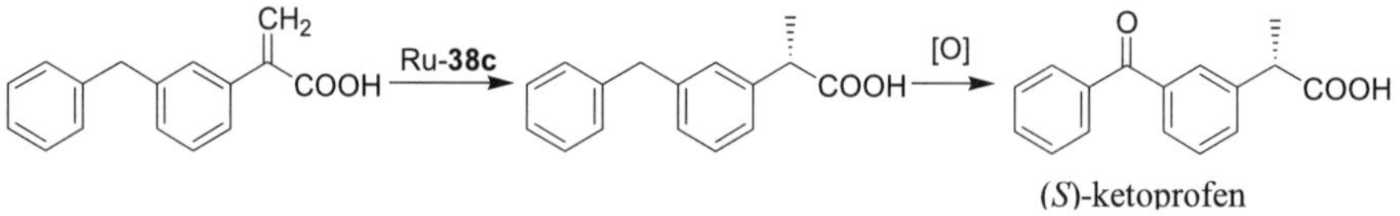

(S)-ketoprofen

Scheme 6–21

Isomerization of allylic amines is another example of the application of the BINAP complex. Rh–BINAP complex catalyzes the isomerization of N,N-diethylnerylamine **40** generated from myrcene **39** with 76–96% optical yield. Compound (R)-citronellal [(R)-**42**], prepared through hydrolysis of (R)-**41**, is then cyclized by zinc bromide treatment.[49] Catalytic hydrogenation then completes the synthesis of (−)-menthol. This enantioselective catalysis allows the annual production of about 1500 tons of menthol and other terpenic substances by Takasago International Corporation.[50]

(S)-citronellal **42** can also be prepared similarly from **40**. Asymmetric hydrogenation of (R)-**43** provides **44**, which can be used to make the side chain of vitamins E and K (Scheme 6–22).

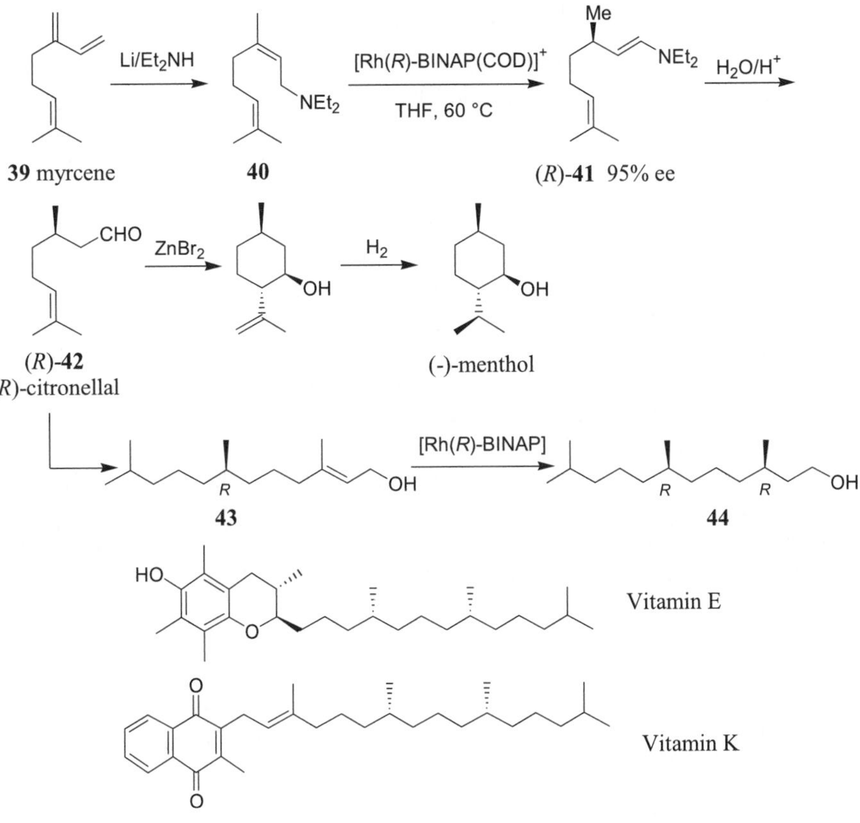
Li/Et2NH
[Rh(R)-BINAP(COD)]+
THF, 60 °C
Me
NEt2
NEt2
H2O/H+
39 myrcene
40
(R)-41 95% ee
CHO
ZnBr2
OH
H2
OH
(R)-42
(R)-citronellal
(-)-menthol
R
OH
[Rh(R)-BINAP]
R
R
OH
43
44
HO
O
Vitamin E
O
O
Vitamin K

Scheme 6–22

6.2 ASYMMETRIC REDUCTION OF CARBONYL COMPOUNDS

Asymmetric reduction of carbonyl compounds can usually be achieved either through direct catalytic hydrogenation or by metal hydride reduction. It should be mentioned here that reduction of carbonyl compounds by catalytic hydrogenation may not be chemoselective. Other co-existing functional groups such as the C=C bond may also undergo hydrogenation.

$LiAlH_4$ is a very powerful reducing agent and hence does not show much chemoselectivity compared with other metal hydrides. Replacing some of the hydrogen atoms in $LiAlH_4$ with alkoxyl groups makes it less reactive and more selective. Similarly, to achieve the stereoselective delivery of hydride to one prochiral face of prochiral ketones, $LiAlH_4$, $NaBH_4$, and borane-tetrahydrofuran ($BH_3 \cdot THF$) have been modified with chiral ligands. This line of work has been extensively reviewed. Interested readers may refer to Midland[51a] and Srebnik.[51b]

6.2.1   Reduction by BINAL–H

One approach to enantioselective reduction of prochiral carbonyl compounds is to utilize chiral ligand-modified metal hydride reagents. In these reagents, the number of reactive hydride species is minimized in order to get high chemoselectivity. Enantiofacial differentiation is due to the introduced chiral ligand.

The first effort to modify LiAlH$_4$ with a chiral ligand was reported in 1951,[52] but significant results had not been achieved until 1979 when Noyori[53] developed a binaphthol-modified aluminum hydride reagent (abbreviated to BINAL–H) of type **45**. The compound can be generated in situ by mixing LiAlH$_4$ with equimolar amounts of optically pure (S)-$(-)$-/(R)-$(+)$-binaphthol and another hydroxylic component R"OH. As would be expected from the excellent stereoselectivity of other reactions induced by BINAP complexes or catalysts derived from binaphthol, compound **45** also gives excellent results in reducing ketones. Binaphthoxy group binding in a bidentate fashion in a metal complex can provide excellent differentiation between the prochiral faces of a substrate. A reagent with a simple alkoxyl group in **45**, such as CH$_3$O or C$_2$H$_5$O, exhibits high enantioselectivity, and the optical yields can be further enhanced by lowering the reaction temperature to $-78°$C or lower.

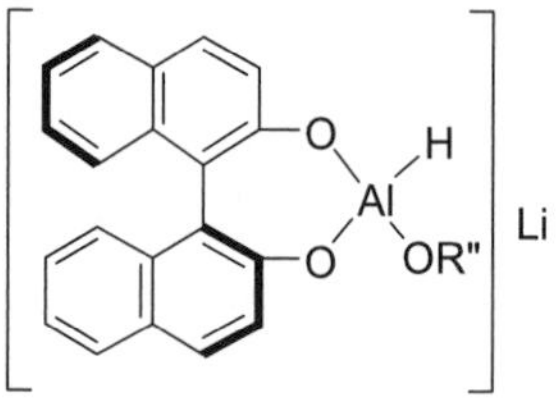

(S)-BINAL-H, (S)-**45** 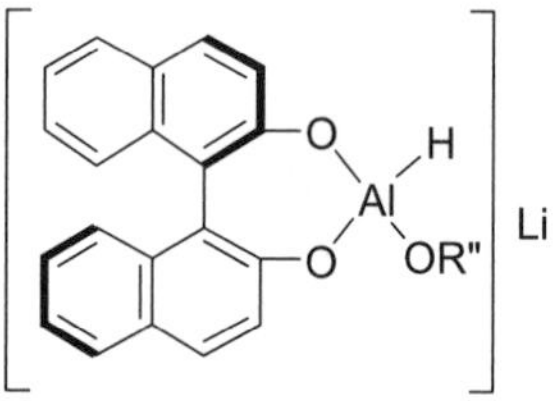 (R)-BINAL-H, (R)-**45**

In general, (R)-**45** reduction gives (R)-carbinol preferentially, and (S)-**45** provides the (S)-enantiomer predominantly (Scheme 6–23). This can be explained by the reaction transition state shown in Figure 6–4. The oxygen atom of the R"O group, which has the highest basicity, acts as the binding atom in the quasi-aromatic, six-membered ring transition state.

The two chair-like transition states **48** and **49** have been suggested to explain the stereochemistry in the reaction. Here structure **48**, leading to (S)-carbinol, is favored over the diastereomeric **49**, which gives the (R)-enantiomer, because the latter structure with axial-R' and equatorial-R group is destablized by the n-π

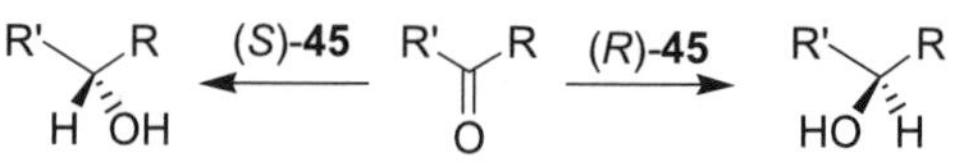

Scheme 6–23.   R' is an unsaturated group.

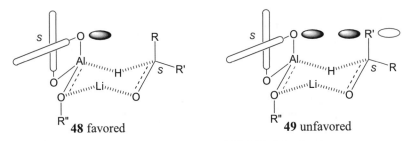

Figure 6–4. Transition state of BINAL–H reduction.

electron repulsion between the axially orientated binaphthoxy oxygen and the unsaturated moiety. It should be noted that the overwhelming kinetic preference is primarily determined by the difference in electronic properties of the R' and R attached to the carbonyl group. Steric factors are also of some significance, but do not overbalance the electronic one.

Reduction of aromatic ketones by **45** normally gives satisfactory results. Scheme 6–24 and Table 6–4 show the results of some such reactions.

Similarly, high stereoselectivity has also been observed in acetylenic ketone or olefinic ketone reductions (Scheme 6–25).

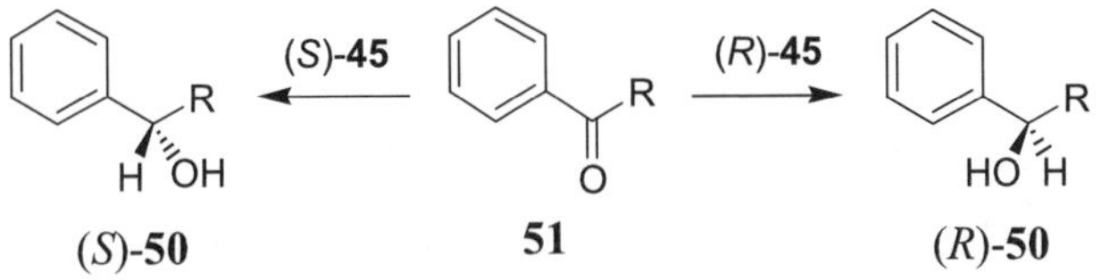

$$(S)\text{-}\mathbf{50} \quad \xleftarrow{(S)\text{-}\mathbf{45}} \quad \mathbf{51} \quad \xrightarrow{(R)\text{-}\mathbf{45}} \quad (R)\text{-}\mathbf{50}$$

Scheme 6–24

TABLE 6–4. Enantioselective Reduction of Aromatic Ketones with BINAL–H ($R''O=C_2H_5O$)

Ketone	Config. of BINAL–H	Carbinol Product		
		Yield (%)	ee (%)	Config.
$C_6H_5COCH_3$	R	61	95	R
$C_6H_5COC_2H_5$	S	62	98	S
$C_6H_5CO\text{-}n\text{-}C_3H_7$	S	78	100	S
$C_6H_5CO\text{-}n\text{-}C_4H_9$	S	64	100	S
$C_6H_5COCH(CH_3)_2$	S	68	71	S
$C_6H_5COC(CH_3)_3$	R	80	44	R
α-Tetralone	R	91	62	R

ee = Enantiomeric excess.

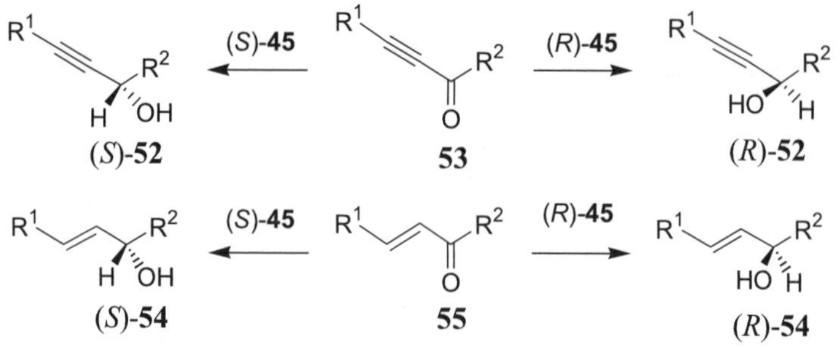

Scheme 6–25

The optically active propargylic and allylic alcohols thus obtained are important synthetic intermediates in the enantioselective synthesis of insect pheromones, prostaglandins, prostacyclins, and many other bioactive compounds (Scheme 6–26).[53]

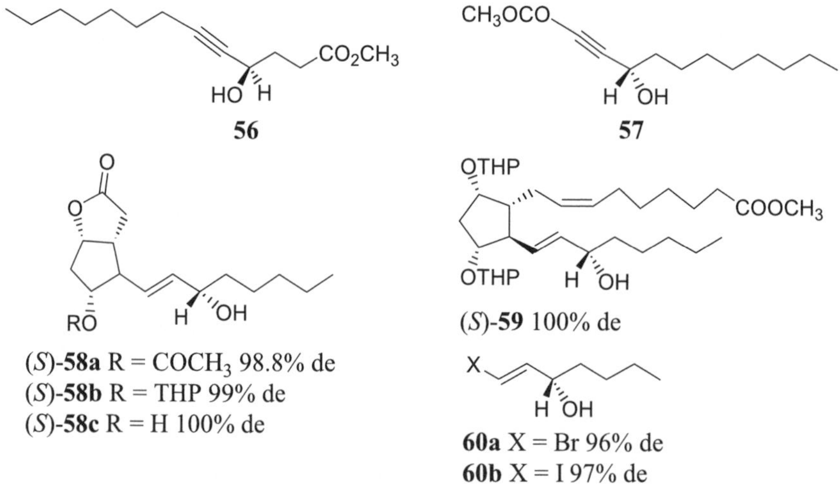

Scheme 6–26. Some important compounds prepared via BINAL–H reduction.

BINAL–H reagents **45** are not effective in the enantioselective reduction of dialkyl ketones.[53] For example, reaction of benzyl methyl ketone with (S)-**45** gives (S)-1-phenyl-2-propanol in only 13% ee (71% yield). Reaction of 2-octanone with (R)-**45** produces (S)-2-octanol in 24% ee (67% yield).[53] This drop of ee values in the reaction may be explained by the lower energy difference between the favored transition state **48** and unfavored transition state **49** caused by the lack of the above-mentioned n-π repulsion between the reductant and the substrate dialkyl ketone.

If we compare the above result with the following example, we can easily see how important this n-π repulsion is for getting high enantioselectivity. In the reduction of 1-deuterium aldehydes, good results can still be obtained, even though there is only a small steric difference between H and D. The enantio-selectivity, of course, is not as high as that for acetylenic or olefinic ketones, but it is still far higher than that for dialkyl ketone reactions (see Scheme 6–27). All these results support the postulated mechanism in Figure 6–4 where transition state **48** is favored over **49**.

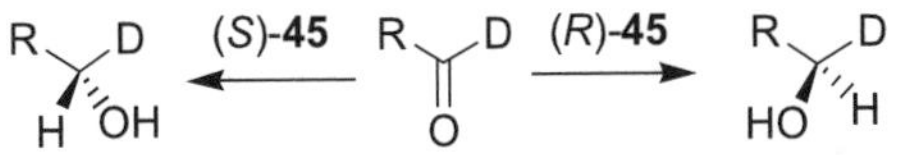

Scheme 6–27. Asymmetric reduction of 1-deutero aldehyde.

TABLE 6–5. Enantioselective Reduction of Deterium-Labeled Aldehydes with BINAL–H (R″O=C₂H₅O)

		Carbinol Product		
Aldehydes	Config. of BINAL–H	Yield (%)	ee (%)	Config.
Geranial-1-d	S	91	91 or 84	S
Neral-1-d	S	90	72	S
(E,E)-Farnesal-1-d	R	91	88	R
(Z,E)-Farnesal-1-d	R	93	82	R
Benzaldehyde-α-d	R	75	82	R

ee = Enantiomeric excess.

6.2.2 Transition Metal–Complex Catalyzed Hydrogenation of Carbonyl Compounds

Asymmetric hydrogenation of ketones is one of the most efficient methods for making chiral alcohols. Ru–BINAP catalysts are highly effective in the asym-metric hydrogenation of functionalized ketones,[54,55] and this may be used in the industrial production of synthetic intermediates for some important anti-biotics. The preparation of statine **65** (from **63b**: R = i-Bu) and its analog is one example (Scheme 6–28).[56] Table 6–6 shows the results when asymmetric hy-drogenation of **63** catalyzed by RuBr₂[(R)-BINAP] yields *threo*-**64** as the major product.

Chiral diols are highly useful ligands for preparing chiral reagents, catalysts, and other chiral ligands. For example, enantiomerically pure 1,2-, 1,3-, and 1,4-diols are important intermediates for preparing useful chiral diphosphine ligands such as Chiraphos,[57] Skewphos,[58] and DuPhos.[59] These diols can be prepared through asymmetric reduction of the corresponding diones via hydrogena-tion,[55,60] borane reduction,[61] hydrosilylation,[62] or enzymatic reduction.[63]

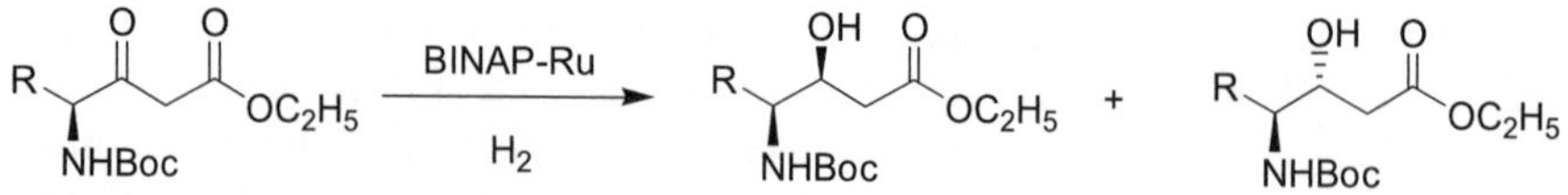

63a: R = PhCH$_2$
63b: R = (CH$_3$)$_2$CHCH$_2$
63c: R = cyclohexylmethyl

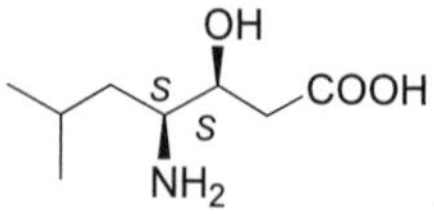

threo-**64**   *erythro*-**64**

Scheme 6–28. Statine **65**, part of aspartic proteinase inhibitor.

TABLE 6–6. Asymmetric Hydrogenation of 64

Substrate	Catalyst	Product **64**		*threo*-**64** ee (%)
		Yield (%)	*threo:erythro*	
63a	RuBr$_2$[(R)-BINAP]	97	>99:1	99
63a	RuBr$_2$[(S)-BINAP]	96	9:91	>99
63b	RuBr$_2$[(R)-BINAP]	99	>99:1	97
63c	RuBr$_2$[(R)-BINAP]	92	>99:1	100

From a practical point of view, the catalytic asymmetric hydrogenation of the corresponding diones will be the preferred method if high yields and high enantioselectivity can be ensured. Recently, over 98% yield with more than 99% ee has been achieved by optimizing the reaction conditions.[64] For example, asymmetric hydrogenation of 2,4-pentanedione catalyzed by Ru–BINAP complex in the presence of hydrochloric acid gave 2,4-pentanediol in more than 95% yield and over 99% ee (Scheme 6–29).[64]

(R, R)- or (S, S)-**67**
> 95% yield, > 99% ee

Scheme 6–29

The Ru–BINAP diacetate complex, which gives good results in the enantioselective hydrogenation of various ketones, is ineffective in the hydrogenation of methyl 3-oxobutanoate. Reactivity in methanol is low, and the enantioselectivity is discouragingly poor. As the carboxylate ligands in Ru complexes may also be replaced by other anions, it is possible to introduce a strong acid anion by

TABLE 6–7. Ru–BINAP–Catalyzed Enantioselective Hydrogenation Reactions of Methyl 3-Oxobutanoate

Catalyst System	S/C	Time (h)	Yield (%)	ee (%)
Ru(OCOCH$_3$)$_2$(BINAP)	1400	60	1	—
Ru(OCOCH$_3$)$_2$(BINAP) + 2CF$_3$COOH	1620	32	99	15
Ru(OCOCH$_3$)$_2$(BINAP) + 2HClO$_4$	1620	32	99	51
Ru(OCOCH$_3$)$_2$(BINAP) + 2HCl	1800	32	99	51
Ru(OCOCH$_3$)$_2$(BINAP) + 2HCl	10,000	64*	98	96
RuCl$_2$(BINAP)	2000	36	99	99
RuBr$_2$(BINAP)	2100	43	99	99
RuI$_2$(BINAP)	1400	40	99	99

* Reaction carried out at 100°C.

ee = Enantiomeric excess.

adding strong acids under acid–base thermodynamic equilibration.[65] Indeed, the addition of 2 equivalents of trifluoroacetic acid or aqueous perchloric acid to remove the acetate ligands greatly increases the catalytic activity, but the enantioselectivity remains moderate. Addition of hydrochloric acid results in a remarkable enhancement of catalytic efficiency.[55,66]

Scheme 6–30 shows that the halogen-containing complexes RuX$_2$(BINAP) are excellent catalysts: With an S/C of over 10^3 or even 10^4, the enantioselective hydrogenation of methyl 3-oxobutanoate can still proceed well in methanol. The yield of the enantioselective reaction is almost 100%.

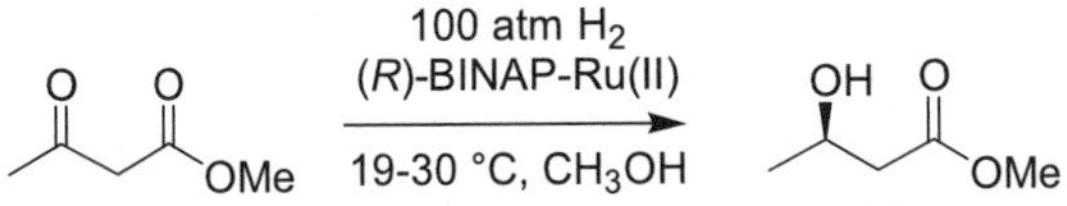

Scheme 6–30

In the case of hydrogenation using [Ru(BINAP)Cl$_2$]$_n$ as the catalyst precursor, the reaction seems to occur by a monohydride mechanism as shown in Scheme 6–31. On exposure to hydrogen, RuCl$_2$ loses chloride to form RuHCl species A, which in turn reversibly forms the keto ester complex B. Hydride transfer occurs in B from the Ru center to the coordinated ketone to form C. The reaction of D with hydrogen completes the catalytic cycle.[67]

This asymmetric catalytic reaction has found wide application in converting functionalized ketones to the corresponding secondary alcohols with high ee. A general illustration is given in Scheme 6–32. Five- to seven-membered chelate complexes, formed by the interaction of the Ru atom with carbonyl oxygen and a heteroatom X, Y, or Z may be the key intermediates that cause the high enantioselectivity in the reaction.[67]

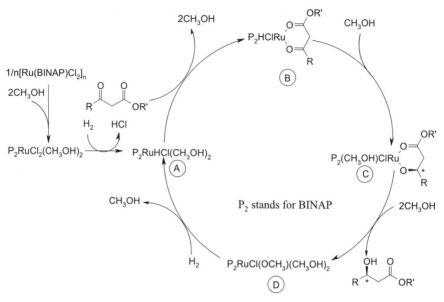

Scheme 6–31. Catalytic cycle of BINAP–Ru-catalyzed hydrogenation of β-keto esters.

Scheme 6–32. X, Y, Z = hetereoatom; C = sp^2 or nonstereogenic sp^3 carbon.

In contrast to their success in the asymmetric hydrogenation of functionalized ketones, BINAP–Ru catalysts fail to give good results with simple ketone because such substrates lack heteroatoms that enable the substrate to anchor strongly to the Ru metal.

It is well accepted that the asymmetric reduction of simple dialkyl ketones generally proceeds with low enantioselectivity.[68] Ohkuma et al.[69] reported that hydrogenation of simple ketones can be achieved using Ru(II) catalysts in the presence of diamine and alcoholic KOH in 2-propanol. Promising results have been achieved in the asymmetric hydrogenation of alkyl aryl ketones with a mixture of an Ru–BINAP complex, chiral diamine, and KOH (Scheme 6–33).

For example, the hydrogenation of 1-acetonaphthone with a catalyst system consisting of RuCl$_2$[(S)-BINAP](DMF)$_n$, (S,S)-1,2-diphenylethylenediamine

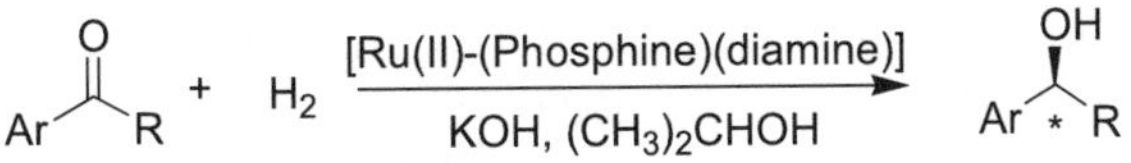

Scheme 6–33

and KOH (1:1.2 mol ratio) in 2-propanol (S/C $= 500$, 4 atm of H_2, 28°C, 6 hours) afforded (R)-1-(1-naphtyl)ethanol in 97% ee and >99% yield.[70]

Following their success in the catalytic hydrogenation of simple ketones by a combination of Ru complex, 1,2-diamine, and an inorganic base, Doucet et al.[71] further designed a type of shelf-stable catalyst for rapid and stereoselective hydrogenation of ketones. Complex **70** can be obtained by treating oligomeric Ru complex $[RuCl_2(diphosphane)(DMF)_n]$ with 1.1 equivalent of diamine in DMF at 25°C for 3 hours. These isolated Ru complexes are among the most reactive catalysts for homogeneous hydrogenation of ketones thus far reported.[72]

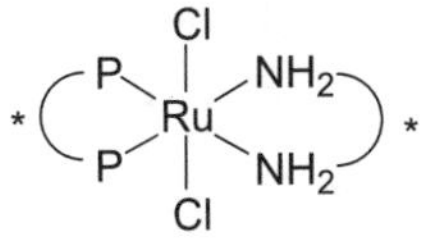

70

Rapid, highly enantioselective asymmetric hydrogenation of simple ketones has been achieved using compound **70** as a catalyst precursor. For example, the hydrogenation of **71** takes place on the carbonyl group only, leaving the C=C double bond intact. Thus 94% ee and 100% yield can be obtained (Scheme 6–34). This process is of industrial significance because of the usefulness of the resulting compound **72** as an intermediate in the synthesis of various carotenoid-derived bioactive terpene and fragrance materials.

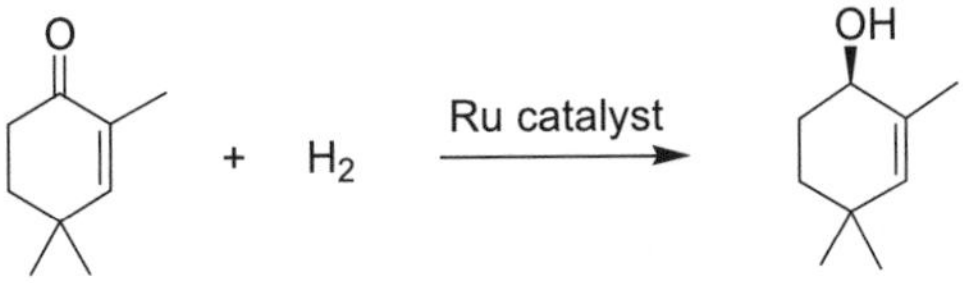

Scheme 6–34

This new hydrogenation procedure is clean, mild, and effective. It offers a very practical method for chiral alcohol synthesis. Isolated Ru complexes are fairly air and moisture stable and can be stored in an ordinary vial for quite a long time. Compared with the catalysts prepared in situ, the reaction rates in the asymmetric hydrogenations catalyzed by **70** are higher by two orders of magnitude.

It is of interest to note that even with the co-existing C=C bond in the ketone substrate, the C=O bond can still be hydrogenated with high stereo-selectivity, leaving the C=C bond intact. Ohkuma et al.[73] further found that using complex **73** as the catalyst in which BINAP was replaced by XylBINAP as the chiral diphosphine, many substrates could be hydrogenated with increased stereoselectivity.

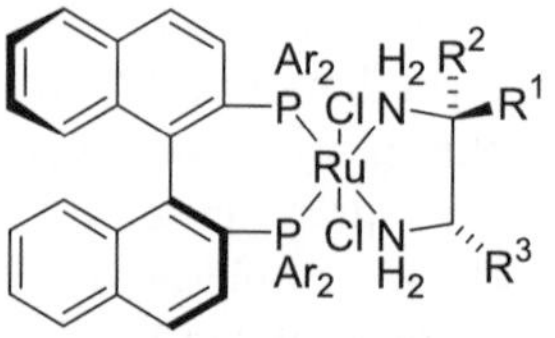

73a Ar = 3,5-(Me)$_2$C$_6$H$_3$
   R^1 = R^2 = 4-MeOC$_6$H$_4$, R^3 = (CH$_3$)$_2$CH
73b Ar = 3,5-(Me)$_2$C$_6$H$_3$
   R^1 = R^3 = -(CH$_2$)$_4$-, R^2 = H
73c Ar = 3,5-(Me)$_2$C$_6$H$_3$
   R^1 = R^3 = Ph, R^2 = H

Using **73a–c** as the catalyst, simple α,β-unsaturated ketones can be hydrogenated to give allylic alcohols in high enantiomeric excess with the C=C double bond intact. Cyclopropyl methyl ketone can also be hydrogenated without opening of the cyclopropane ring. Furthermore, a great improvement in enantioselectivity has been observed in the asymmetric hydrogenation of aromatic ketones, and up to 100% ee has been obtained.[73]

Jiang et al.[4] have recently succeeded in hydrogenating both aryl alkyl and dialkyl ketones. High enantioselectivity was obtained using PennPhos (**19**)–coordinated Rh–complex as the catalyst. This success is based on the finding that a weak base (such as 2,6-lutidine) can facilitate the Rh-catalyzed hydrogenation of simple ketones (Scheme 6–35).

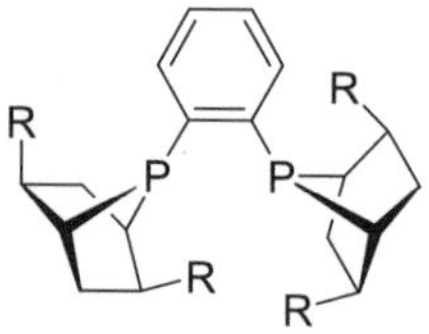

PennPhos **19** (R = Me, i-Pr)

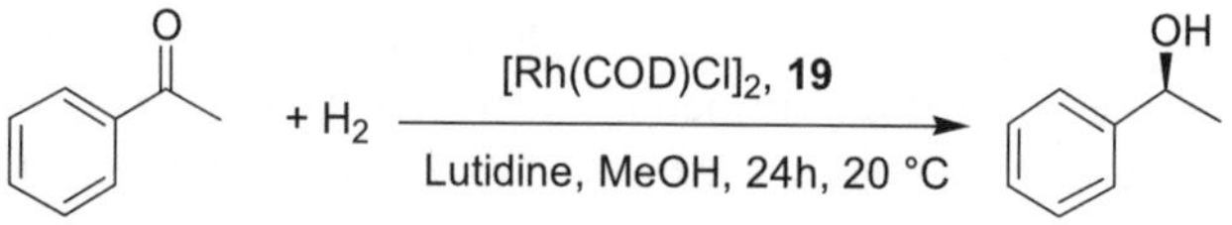

The hydrogenation was carried out at room temperature, 30 atm of H$_2$, 1.0 mmol scale; substrate:[Rh(COD)Cl$_2$]$_2$:**19** = 1.0:0.005:0.01.

Scheme 6–35. PennPhos catalyzed ketone reduction.

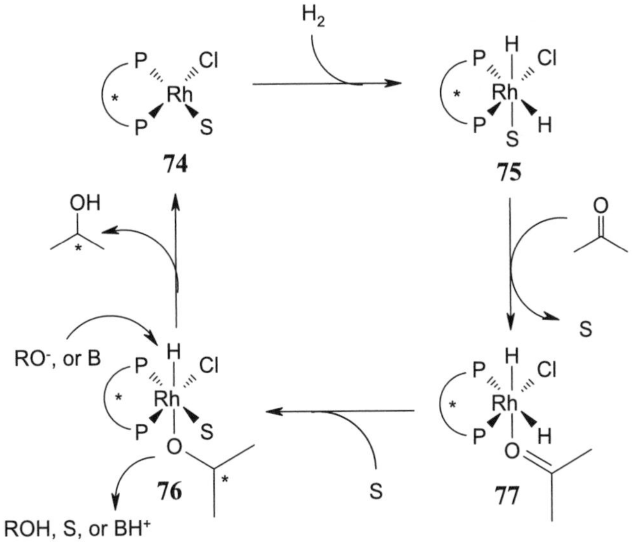

Figure 6–5. Proposed mechanism for weak-base–promoted Rh-catalyzed asymmetric hydrogenation. Reprinted with permission by Wiley-VCH Verlag GmbH, Ref. 4.

Chiral ligand 19 was designed based on the DuPhos structure. It shares some general features with DuPhos, such as its electron-donating properties and a modular structure, but it has its own individual characteristics as well. Compared with DuPhos, PennPhos is bulkier and more rigid. The high reactivity of PennPhos may come from its high electron-donating ability in comparison with the corresponding triarylphosphines, and the stereoselectivity may come from its relatively rigid structure. The chiral ligand is air stable and can be synthesized in large quantities from inexpensive starting materials.

The proposed mechanism for weak-base–promoted Rh-catalyzed asymmetric hydrogenation is shown in Figure 6–5.[4]

For most aryl methyl ketones, high enantioselectivity (93–96%) can be attained, as shown in Table 6–8. Increasing the bulk of the alkyl group by varying the alkyl group from methyl to ethyl or isopropyl in the alkyl aryl ketone dramatically decreases the reactivity and enantioselectivity. This indicates that the chiral environment around the Rh complex can effectively discriminate the small methyl group from other larger alkyl groups. Enantiomeric excess of up to 94% for t-butyl methyl ketone (Entry 9) and 92% for cyclohexyl methyl ketone have been observed. With i-propyl methyl ketone and i-butyl methyl ketone (Entries 6 and 7), 84% and 85% ee have been achieved, respectively. Even with unbranched alkyl groups, ee values of 73% (β-phenylethyl methyl ketone, Entry 4) and 75% (n-butyl methyl ketone, Entry 5) can still be achieved.

Chiral ligand 78, bearing structural features similar to those of DuPhos, has also been synthesized and gives moderate to high enantioselectivity in the catalytic asymmetric hydrogenation of functionalized carbonyl groups. High levels

TABLE 6–8. Asymmetric Hydrogenation of Simple Ketones Catalyzed by Rh–PennPhos

Entry	Ketone	Equiv. of Lutidine	Equiv. of KBr	Time (h)	Yield (%)	ee (%)
1		0.4	—	24	97	95
2		0.4	—	53	94	95
3		0.8	1.0	88	95	93
4		0.8	1.0	56	99	73
5		0.8	1.0	48	96	75
6		0.8	1.0	75	66	85
7		0.8	1.0	94	99	84
8		0.8	1.0	106	90	92
9		0.8	1.0	96	51	94

ee = Enantiomeric excess.

of diastereoselectivity and enantioselectivity have been achieved in the hydrogenation of β-diketones to the corresponding *anti*-1,3-diols.[74]

78

6.2.3 The Oxazaborolidine Catalyst System

Boranes have opened the door to asymmetric reduction of carbonyl compounds. The first attempt at modifying borane with a chiral ligand was reported by Fiaud and Kagan,[75] who used amphetamine–borane and desoxyephedrine–borane to reduce acetophenone. The ee of the 1-phenyl ethanol obtained was quite low ($\leq 5\%$). A more successful borane-derived reagent, oxazaborolidine, was introduced by Hirao et al.[76] in 1981 and was further improved by Itsuno and Corey.[77] Today, this system can provide high stereoselectivity in the asymmetric reduction of carbonyl compounds, including alkyl ketones.

Oxazaborolidine catalysts behave like an enzyme in the sense of binding with both ketone and borane, bringing them close enough to undergo reaction and releasing the product after the reaction. Thus these compounds are referred to as *chemzymes* by Corey.[78] The oxazaborolidines listed in Figure 6–6 are representative catalysts for the asymmetric reduction of ketones to secondary alcohols.

Addition of triethylamine to the oxazaborolidine reaction system can significantly increase the enantioselectivity, especially in dialkyl ketone reductions.[79] In 1987, Corey et al.[80] reported that the diphenyl derivatives of **79a** afford excellent enantioselectivity ($>95\%$) in the asymmetric catalytic reduction of various ketones. This oxazaborolidine-type catalyst was named the CBS system based on the authors' names (Corey, Bakshi, and Shibata). Soon after, Corey's group[81] reported that another *B*-methyl oxazaborolidine **79b** (Fig. 6–6) was easier to prepare and to handle. The enantioselectivity of the **79b**-catalyzed reaction is comparable with that of the reaction mediated by **79a** (Scheme 6–36).[81] The *β*-naphthyl derivative **82** also affords high enantioselectivity.[78] As a general procedure, oxazaborolidine catalysts may be used in 5–10 mol%

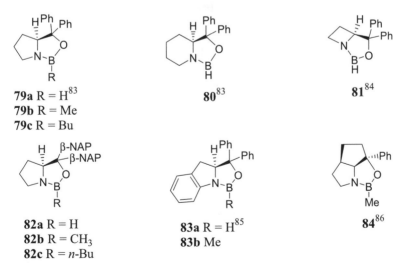

Figure 6–6. Representative oxazaborolidines for the asymmetric reduction of prochiral ketones.

$$2 R^1R^2CO + BH_3 \xrightarrow[\text{1 min, 25 °C}]{\text{(S)-79a, THF}} R^1R^2CHOH$$

Ketone	Equiv. of BH$_3$	Equiv. of catalyst	Config., ee (%)
C$_6$H$_5$COCH$_3$	2.0	1	R (97)
C$_6$H$_5$COCH$_3$	1.0	0.1	R (97)
C$_6$H$_5$COCH$_3$	1.2	0.025	R (95)
C$_6$H$_5$COC$_2$H$_5$	1.2	0.05	R (86)
C$_6$H$_5$COC$_2$H$_5$	1.0	0.05	R (88)
C$_6$H$_5$COC$_2$H$_5$	0.6	0.05	R (90)
t-BuCOCH$_3$	0.6	0.05	R (88)
t-BuCOCH$_3$	0.6	0.1	R (92)
α-tetralone	0.6	0.05	R (89)
C$_6$H$_5$COCH$_2$Cl	0.6	0.05	S (97)

Scheme 6–36. Reduction catalyzed by (S)-**79a**. Reprinted with permission by Am. Chem. Soc., Ref. 80.

with either borane or catechol–borane[82] as the stoichiomeric reductant under moisture-free conditions. Scheme 6–34 shows representative results of using (S)-**79b** as a catalyst for the asymmetric reduction of various ketones.

The proposed mechanism for B-methyl **79a** or **79b** catalyzed reduction is illustrated in Scheme 6–37.[80,87]

In the first step, borane coordinates to the nitrogen atom in oxazaborolidine **79b** from the less hindered side of the fused bicyclic system, resulting in the

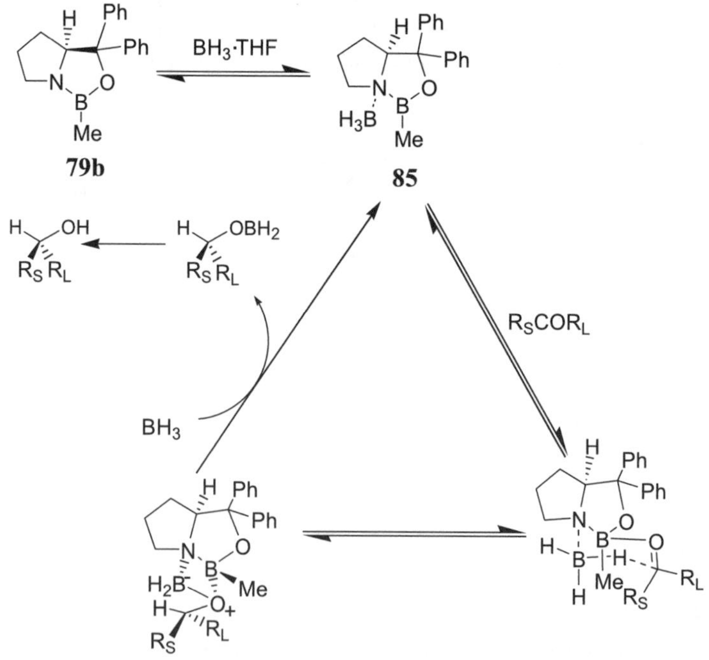

Scheme 6–37

formation of **85**. In the Lewis acid–base adduct **85**, the boron atom of the oxazaborolidine moiety coordinates with the prochiral ketone *cis* to the BH_3 molecule. The Lewis acid (boron atom) locates *trans* to the larger substituent of the ketone. After a generally fast and reversible formation of the ketone–borane–oxazaborolidine complex, intramolecular hydride transfer takes place via a six-membered chair-like transition state. [The (*S*)-enantiomer of the catalyst favors the *Re*-face of the carbonyl group.] This is an irreversible, rate-limiting step. The labels R_S and R_L refer to the effective steric sizes of these groups with respect to their effect on the equilibrium rate of coordination of the *syn*-carbonyl lone pair with the catalytic oxazaborolidine–borane complex.

The power of CBS-catalyzed enantioselective reduction in organic synthesis is well illustrated by the following examples. In prostaglandin synthesis, chiral ester keto-lactone **86**, a standard intermediate, can undergo selective reduction at the keto group catalyzed by **79b**, giving (15*R*)-**87** and its (*S*)-counterpart in a ratio of 91:9 (Scheme 6–38).[81b] This catalytic reduction presents a very practical solution to the problem of controlling C-15 stereochemistry in prostaglandin synthesis. This method can also be used for preparing (*R*)-fluoxetine **90**,[88] the therapeutic serotonin-uptake inhibitor. The key step involves the enantioselective reduction of β-chloropropiophenone **88** with CBS catalyst to yield compound **89** with over 99% yield and 94% ee (Scheme 6–39). The CBS

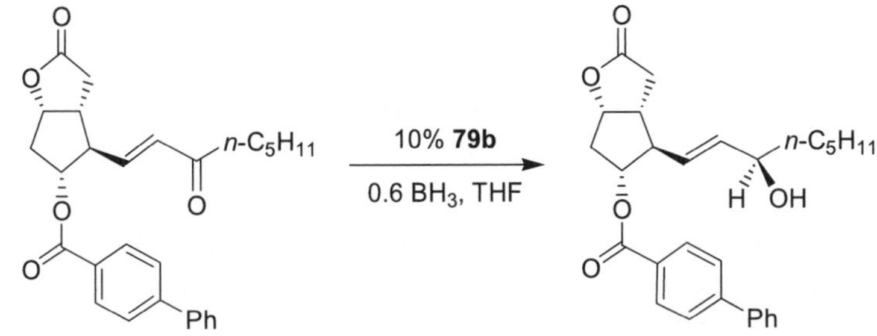

Scheme 6–38. Synthesis of prostaglandin intermediate.

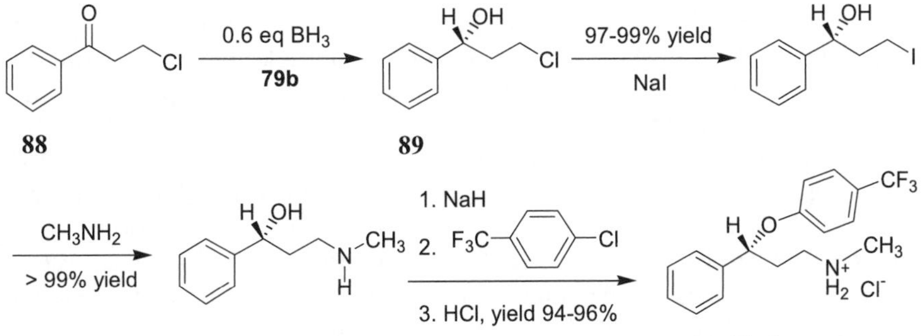

Scheme 6–39. Synthesis of (*R*)-fluoxetine.

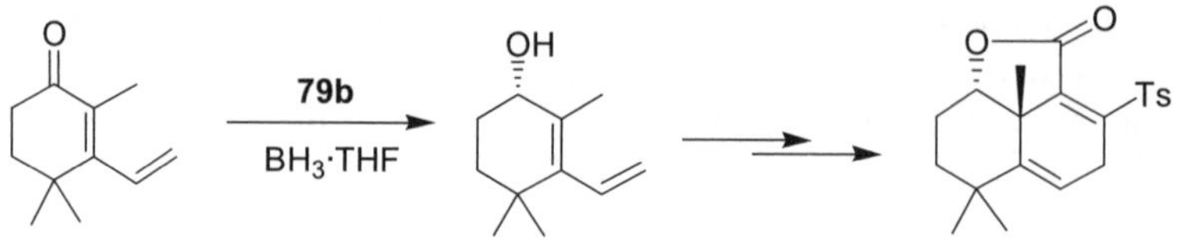

Scheme 6–40. Synthesis of forskolin intermediate **93**.

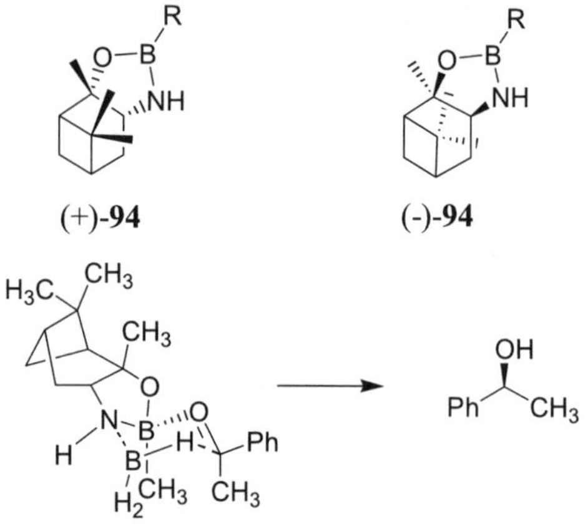

Scheme 6–41

reagent can also be used in the enantioselective synthesis of ginkgolide A and B,[89] forskolin,[90] and *anti*-PAF *trans*-2,5-diarylfurans.[91] Scheme 6–40 depicts the synthesis of forskolin intermediate.

New chiral oxazaborolidines that have been prepared from both enantiomers of optically active inexpensive α-pinene have also given quite good results in the asymmetric borane reduction of prochiral ketones.[92] Borane and aromatic ketone coordinate to this structurally rigid oxazaborolidine (+)- or (−)-**94**, forming a six-membered cyclic chair-like transition state (Scheme 6–41). Following the mechanism shown in Scheme 6–37, intramolecular hydride transfer occurs to yield the product with high enantioselectivity. With aliphatic ketones, poor ee is normally obtained (see Table 6–9).

Since the discovery of the CBS catalyst system, many chiral β-amino alcohols have been prepared for the synthesis of new oxazoborolidine catalysts. Compounds **95** and **96** have been prepared[93] from L-cysteine. Aziridine carbinols **97a** and **97b** have been prepared[94] from L-serine and L-threonine, respectively. When applied in the catalytic borane reduction of prochiral ketones, good to excellent enantioselectivity can be attained (Schemes 6–42 and 6–43).

Chiral boranes, such as isocamphenyl derivatives (Ipc)₂BH, IpcBH₂, and

TABLE 6–9. Asymmetric Reduction of Prochiral Ketones Using 10 mol% of (+)-94

Substrate	Product	Temp. (°C)	Time (h)	Yield (%)	ee (%)
Ph Et (O)	Ph Et (OH)	0–5	4	95	92
		25–30	4	93	81
Ph CH₂Cl (O)	Ph CH₂Cl (OH)	0–5	1	93	76
		25–30	1	96	90
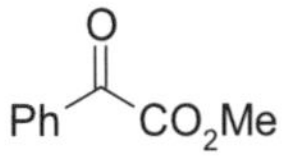 Ph CO₂Me (O)	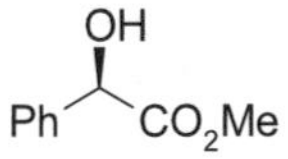 Ph CO₂Me (OH)	0–5	6	65	59
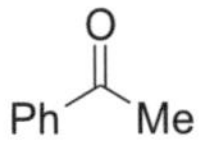 Ph Me (O)	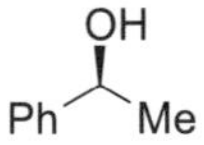 Ph Me (OH)	0–5	2	>90	93
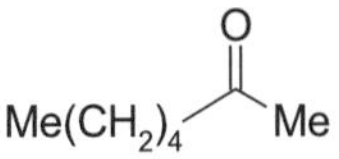 Me(CH₂)₄ Me (O)	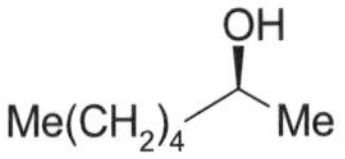 Me(CH₂)₄ Me (OH)	0–5	2	>90	37

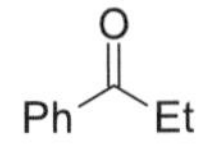

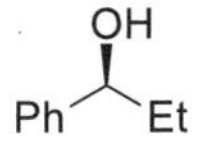

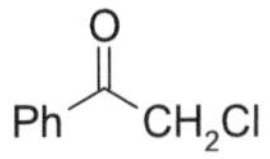

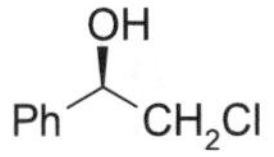

ee = Enantiomeric excess.

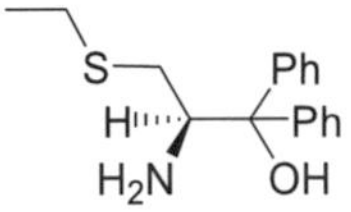

95

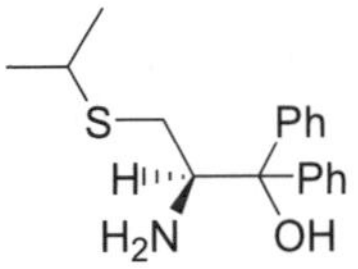

96

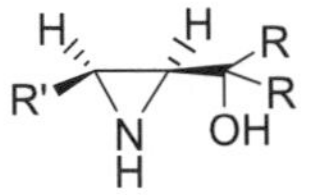

97a R' = H, from L-serine
97b R' = Me, from L-threonine

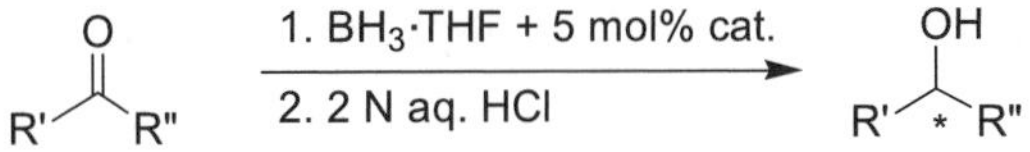

$$\text{R'} \overset{O}{\underset{}{\big\|}} \text{R''} \xrightarrow[\text{2. 2 N aq. HCl}]{\text{1. BH}_3\text{·THF + 5 mol% cat.}} \text{R'} \overset{OH}{\underset{*}{|}} \text{R''}$$

Aromatic ketone	Catalyst	ee (%)	Config.
Acetophenone	95	83	R
ω-Cl-acetophenone	95	97	S
ω-Br-acetophenone	95	100	S
Methyl-2-naphthylketone	95	88	R
Acetophenone	96	70	R
ω–Cl-acetophenone	96	100	S
ω–Br-acetophenone	96	100	S
Methyl-2-naphthylketone	96	83	R

Scheme 6–42. Reprinted with permission by Pergamon Press Ltd., Ref. 93.

trialkylboranes, are also good chiral reductants. In the presence of triethyl amine, (+)-Ipc₂BCl can also be used to reduce α-oxocarboxylic acids to the corresponding α-hydroxy carboxylic acids with high enantioselectivity (Scheme 6–44). It is presumed that the reaction proceeds in an intramolecular fashion through a rigid bicyclic transition state.[95]

$$Ph\overset{O}{\underset{}{\Vert}}\xrightarrow[\text{Cat. 10 mol\%}]{\text{BH}_3\cdot\text{SMe}_2}Ph\overset{OH}{\underset{*}{\cdot}}$$

Entry	Catalyst	Solvent	ee (%)	Config.
1	(S)-**97a**	THF	94	R
2	(R)-**97a**	THF	92	S
3	(S)-**97b**	Toluene	94	R
4	(R)-**97b**	Toluene	93	S

Scheme 6–43. Reprinted with permission by Elsevier Science Ltd., Ref. 94.

$$R\overset{O}{\underset{\Vert O}{\Vert}}OH \xrightarrow{\text{(+)-Ipc}_2\text{BCl, Et}_3\text{N/THF}} R\overset{OH}{\underset{\Vert O}{\cdot}}OH$$

ee 85-99%

Scheme 6–44. R = Ph, Bn, substituted Ph, i-Pr, i-Bu, and c-Hexyl.

In summary, many attempts have been made at achieving enantioselective reduction of ketones. Modified lithium aluminum hydride as well as the oxazaborolidine approach have proved to be very successful. Asymmetric hydrogenation catalyzed by a chiral ligand-coordinated transition metal complex also gives good results. Figure 6–7 lists some of the most useful chiral compounds relevant to the enantioselective reduction of prochiral ketones, and interested readers may find the corresponding applications in a number of review articles.[77,96,97]

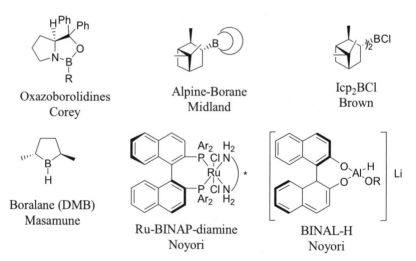

Oxazoborolidines
Corey

Alpine-Borane
Midland

Icp$_2$BCl
Brown

Boralane (DMB)
Masamune

Ru-BINAP-diamine
Noyori

BINAL-H
Noyori

Figure 6–7. Useful compounds for ketone reduction.

6.3 ASYMMETRIC REDUCTION OF IMINES

Preparation of enantiomerically pure secondary amines by catalytic asymmetric hydrogenation or hydrosilylation of imines is as important as the preparation of alcohols from ketones. However, asymmetric hydrogenation of prochiral C=N double bonds has received relatively less attention despite the obvious preparative potential of this process.[98]

It has been reported that the hydrogenation of imine ArC(Me)=NCH$_2$Ph proceeds with enantioselectivity of up to 96% when Rh(I)-sulfonated BDPP is used in a two-phase system. However, the asymmetric reaction of C=N bonds with ruthenium(II) catalyst is rather rare.[99] Willoughby and Buchwald[100] demonstrated a titanocene catalyst that shows good to excellent enantioselectivity in the hydrogenation of imine.

Several successful results have been obtained in the asymmetric hydrogenation and asymmetric hydrosilylation of imines.[101] An efficient enantioselective hydrogenation of the C=N double bond was developed by Burk and Feaster,[101a] who used $[Rh(COD)(DuPhos)]^+CF_3SO_3^-$ in the hydrogenation of N-aroylhydrazone **98**.

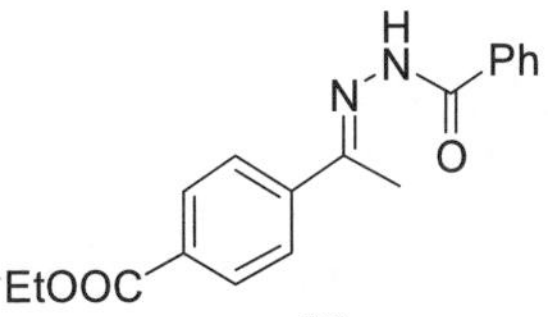

98

One problem for the asymmetric hydrogenation of imine is $(E)/(Z)$ isomerism of the substrate, which may have a significant effect on the enantioselectivity of the reaction. This problem was difficult to address because of the rapid interconversion of the E and Z isomers of the imines under reaction conditions (Fig. 6–8).

Figure 6–8. Interconversion of $(Z)/(E)$ isomers of an imine.

Becalski et al.[98b] found that the $(E)/(Z)$ ratio of imines under rhodium-catalyzed hydrogenation conditions stayed essentially constant during the course of the reaction. This observation was explained in terms of two possibilities: (1) the rates of hydrogenation of the two isomers were identical; or, more likely, (2) the rate of interconversion of the two isomers was faster than the rate of hydrogenation, leaving the $(E)/(Z)$ ratio of the substrate constant.

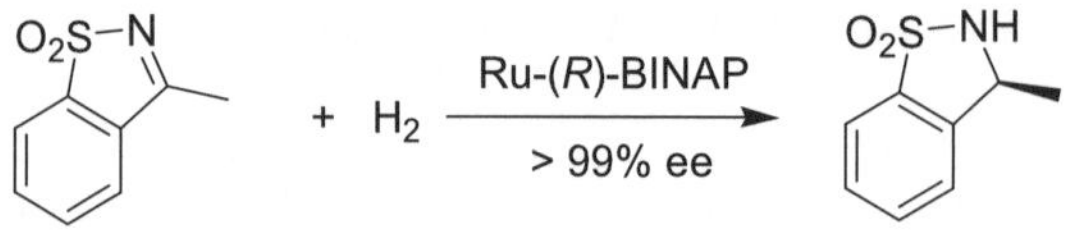

Scheme 6–45

When the C=N bond is fixed in a ring system in which no $(E)/(Z)$ isomerization can take place, the asymmetric hydrogenation of the C=N bond can be highly enantioselective. Oppolzer et al.[99] found that cyclic sulfonimide was hydrogenated with an Ru(BINAP) catalyst to give a product with essentially quantitative optical yield (Scheme 6–45).

Similar success was also achieved by Willoughby and Buchwald[100a] with a chiral titanocene catalyst. The high ee obtained by Burk and Feaster[101a] in the asymmetric hydrogenation of **98** was also consistent with the preferred coordination of one isomer forced by the bidentate chelation of the hydrazones.

To investigate the effect of $(E)/(Z)$ isomerization of C=N bonds on asymmetric hydrogenation, Chan et al.[98g] chose to study the asymmetric hydrogenation of (E) and (Z) isomers of oximes. The advantage of oximes is that the (E) and (Z) isomers may be isolated, separated, and unambiguously characterized. By choosing suitable substrates, the effect of $(E)/(Z)$ isomerism on enantioselectivity can be clearly established. A comparison of the asymmetric hydrogenation of (E)- and (Z)-1-acetophenone oxime showed a significant $(E)/(Z)$ isomeric effect. For example, with {Rh(NBD)[(S)-BINAP]}BF$_4$ used as a catalyst precursor, the product ee was 66% for the (Z) isomer and only 30% for the (E) isomer. Both the (Z) and (E) isomers gave predominantly the product with (S)-configuration. When {Rh(NBD)[(R,R)-SkewPhos]}BF$_4$ was used, the (Z) isomer gave a product in 35% ee in (R)-configuration, while the (E) isomer gave 12% ee in the opposite configuration. These results reveal the importance of $(E)/(Z)$ isomerism in the asymmetric hydrogenation of C=N bonds.

Hydrosilylation of imine compounds was also an efficient method to prepare amines. The hydrosilylation product N-silylamines can readily be desilylated upon methanol or water treatment, yielding the corresponding amines. The amines can be converted to their corresponding amides by subsequent acyl anhydride treatment. The first attempt to hydrogenate prochiral imines with Rh(I) chiral phosphine catalysts was made by Kagan[102] and others. These catalysts exhibited low catalytic activity, and only moderate ee was obtained.

Particularly noteworthy is the discovery of a new type of the active catalyst **99**,[103,104] a crystalline, air-stable yellow-orange solid, which can serve as a highly enantioselective tool in the titanium-catalyzed hydrosilylation of imines. The reaction can be highly stereoselective for both acyclic and cyclic imines under a wide range of hydrogen pressures (Scheme 6–46).

Treatment of this precatalyst with phenylsilane yields a very active catalytic system that can be used for the hydrosilylation of imines. This discovery was initiated by the observed hydrosilylation of imines when phenylsilane was added to Cp$_2$TiF$_2$. It was presumed that it might be possible to break the strong

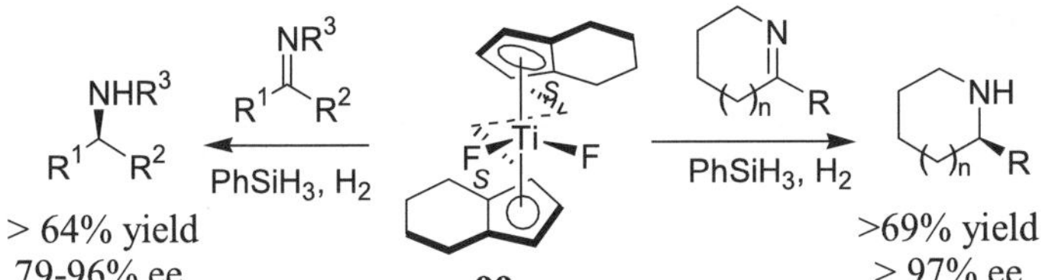

Scheme 6–46. Titanocene-catalyzed asymmetric hydrosilylation of imines.

Ti–F bond and generate a Ti–H species when **99** was treated with phenylsilane. The chirality transfer may take place through imine insertion into the Ti–H bond, similar to that in the catalytic hydrogenation process.[100c] The reaction can be carried out by the subsequent addition of imines. The corresponding silylated amines can be obtained and further converted to enantiomerically enriched amines upon acid treatment. For example, in the presence of **99**, N-methylimine **100** undergoes complete hydrosilylation within 12 hours at room temperature, with 97% ee and up to 5000 turnovers.[103]

The important feature of this reaction system is its experimental simplicity.[105] Although hydrogenation requires elevated pressure and high temperature, this reaction can be carried out at room temperature in an inert atmosphere. The catalyst can be activated before the addition of the substrate or, even more conveniently, in the presence of the substrate. After completion of the reaction and the subsequent acidic workup, the secondary amine product can be obtained in high yield and purity as well as very high enantioselectivity.

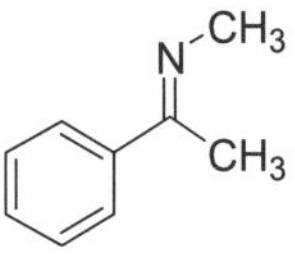

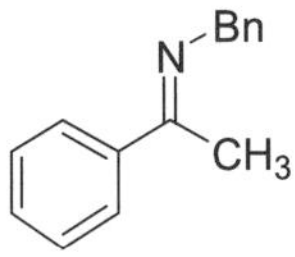

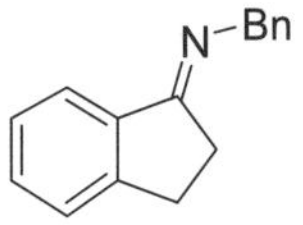

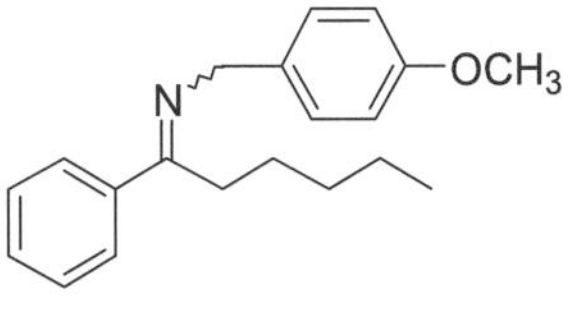

The drawback of this reaction is its extremely high sensitivity to the steric bulk of the nitrogen substitutent. When the analogous N-benzylimine **101** was subjected to the standard hydrosilylation conditions, the reaction gave 55% conversion and 47% ee with prolonged reaction time (96 hours). A practical way to overcome this problem is to employ a nucleophilic promoter to convert the intermediate into a more reactive species. In this case, it has been found that primary amines have the most pronounced effect on the reaction. For example,

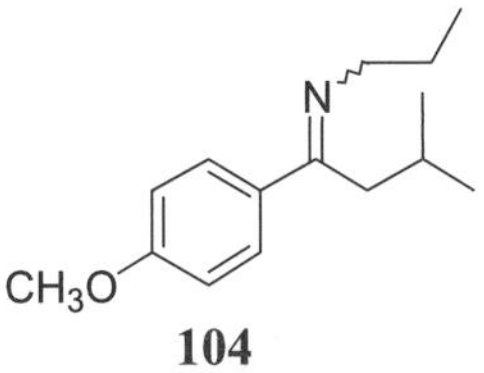

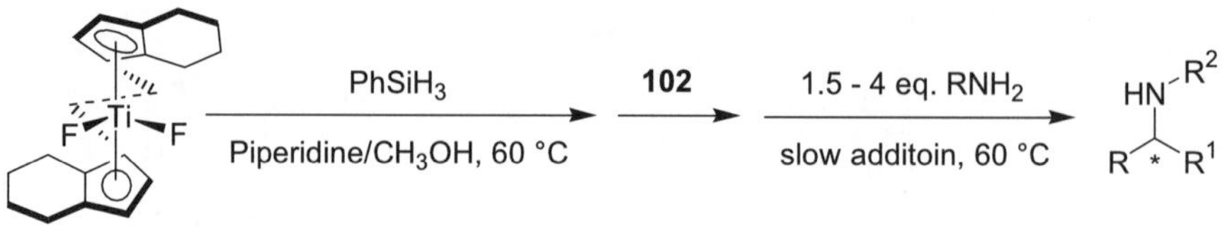

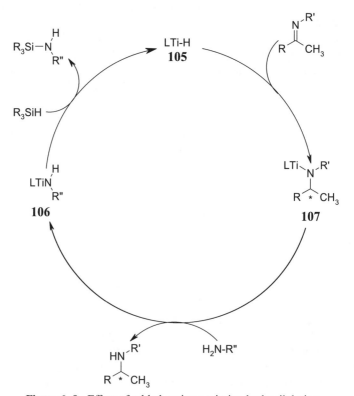

Scheme 6–47

slow addition of four equivalents of *n*-hexylamine to the hydrosilylation reaction mixture at 60°C results in the complete reduction of **102** within 2 hours (85% ee, Scheme 6–47).

Generally, amine products can be obtained with 91–99% ee. It is noteworthy that the ee of the product does not correspond to the *E:Z* ratio of the starting imines. Imines **103** and **104** exist as 2.5:1 and 1.8:1 $(E)/(Z)$ isomers, but titanocene-catalyzed reduction produces amines with 93% and 97% ee, respectively.[105]

A possible explanation for this effect of added amine is shown in Figure 6–9.[105]

Figure 6–9. Effect of added amine on imine hydrosilylation.

The added primary amine may facilitate the cleavage of the Ti–N bond of the key intermediate **107** through the coordination to the titanium center followed by σ-bond metathesis. Such an intramolecular exchange process is expected to be facile. The amine exchange product is **106**, which can then be rapidly converted to **105** and the corresponding silylated amine to complete the catalytic cycle.

6.4 ASYMMETRIC TRANSFER HYDROGENATION

The Meerwein-Ponndorf-Verley reaction is a classic method for ketone/ aldehyde carbonyl group reduction, which involves at least 1 equivalent of aluminum alkoxide as a promoter. In this reaction, the hydrogen is transferred from isopropanol to the ketone/aldehyde substrate, so the reaction can also be referred to as a *transfer hydrogenation reaction*.

Evans et al.[106] report an asymmetric transfer hydrogenation of ketones using samarium(III) complex (**108**) as the catalyst at ambient temperature in 2-propanol. The products showed ee comparable with those obtained through enantioselective borane reduction (Scheme 6–48).

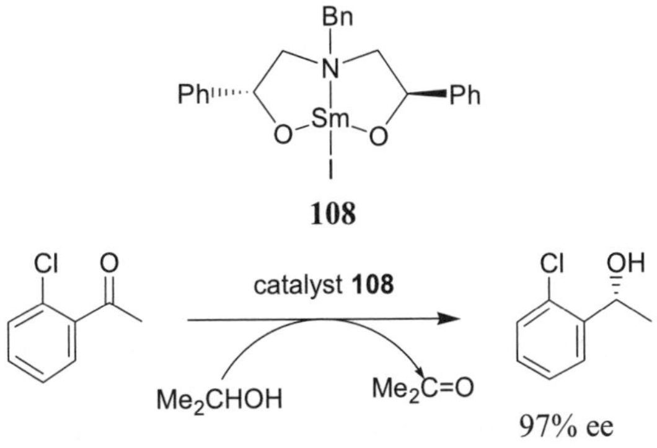

Scheme 6–48. Meerwein-Ponndorf-Verley reaction.

Increasing effort has been applied to develope asymmetric transfer hydrogenations for reducing ketones to alcohols because the reaction is simple to perform and does not require the use of reactive metal hydrides or hydrogen. Ruthenium-catalyzed hydrogen transfer from 2-propanol to ketones is an efficient method for the preparation of secondary alcohols.

Chowdhury and Bäckvall[107] showed that $RuCl_2(PPh_3)_3$ was an effective catalyst for transfer hydrogenation in the presence of 2% NaOH. This finding initiated the development of asymmetric processes that utilized chiral ligand-modified transition metal complexes as catalysts to achieve high stereoselectivity.[108]

Chiral propargylic alcohols are useful building blocks for the synthesis of various biologically active and structurally interesting compounds. These alcohols can be prepared by stoichiometric asymmetric reduction of the corresponding acetylenic ketones with chiral metal hydrides (Section 6.2) or through other asymmetric synthesis methods.[109] Noyori and co-workers report that this conversion can also be realized via asymmetric transfer hydrogenation using chiral Ru(II) complex as the catalyst and 2-propanol as the hydrogen donor.[110] Compound **109** bearing a chiral diamine derivative as ligand has been demonstrated to be an efficient and highly enantioselective catalyst (Scheme 6–49).

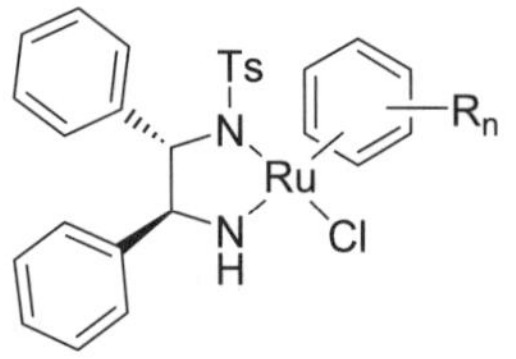

109a η^6-arene = mesitylene
109b η^6-arene = cymene

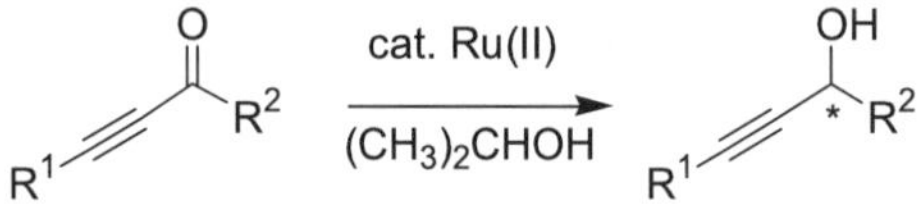

Scheme 6–49

With **109a** or **109b**, the reaction proceeds smoothly to provide the corresponding alkynyl alcohol with consistently high ee (up to 99%). In the presence of bulky R^2 substituents, the substrate can undergo transfer hydrogenation under neutral conditions at room temperature, with an S/C molar ratio of 100 to 200, or even higher. The ee values are very high regardless of the bulkiness of the R^2 substituent, varying from a methyl to a t-butyl group.

Furthermore, asymmetric transfer hydrogenation proceeds equally well when there is a preexisting stereogenic center in the acetylenic ketone. The stereoselectivity is not influenced by this preexisting stereogenic center. For example, transfer hydrogenation of compound (S)-**112** catalyzed by (R,R)-**109a** gives (3S,4S)-**113** with 99% ee and over 97% yield. Similarly, the (S,S)-**109a**–catalyzed reduction of (S)-**112** gives over 97% yield of (3R,4S)-**113** in more than 99% ee (Scheme 6–50). The adjacent nitrogen-substituted stereogenic center does not interfere with the stereoselectivity of the reaction.

Complex **109** can also be used for the asymmetric transfer hydrogenation

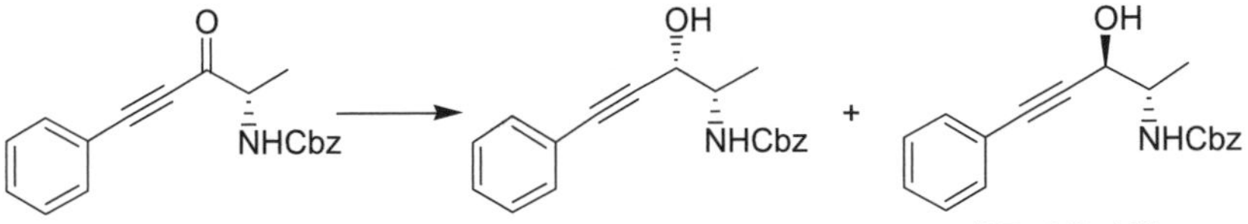

112   (3*S*, 4*S*)-**113**   (3*R*, 4*S*)-**113**

Scheme 6–50

of imines, providing the corresponding chiral amines in high enantiomeric excess.[111] The reaction can be carried out in various aprotic solvents such as DMF, DMSO, or CH_2Cl_2, but not in ethereal or alcoholic solvents. Reaction in neat formic acid/triethyl amine is very slow. Table 6–10 presents the results of the transfer hydrogenation of a variety of imines in the presence of an HCO_2H–Et_3N mixture.

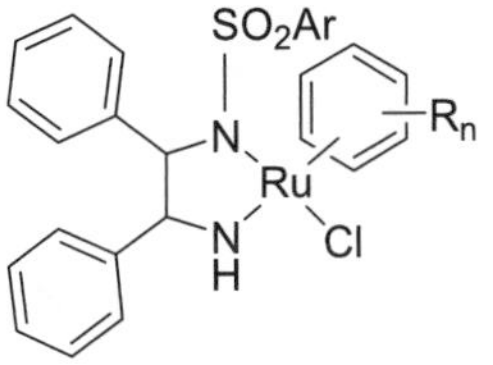

109b η^6-arene = *p*-cymene, Ar = 4-$CH_3C_6H_4$
109c η^6-arene = *p*-cymene, Ar = 2,4,6-$(CH_3)_3C_6H_2$
109d η^6-arene = benzene, Ar = 2,4,6-$(CH_3)_3C_6H_2$
109e η^6-arene = benzene, Ar = 1-naphthyl

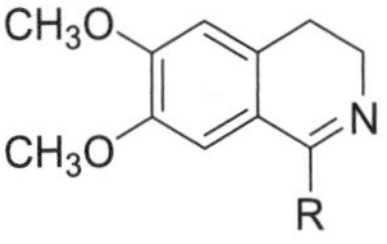

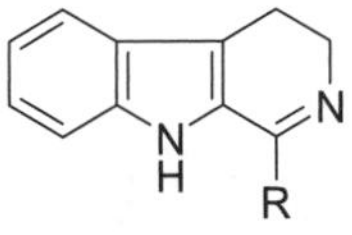

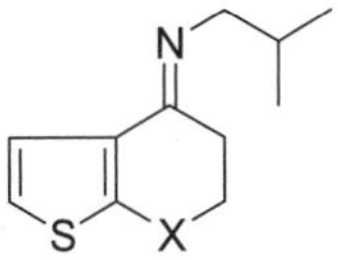

114a R = CH_3 **115a** R = CH_3 **116a** X = S
114b R = 3,4-$(CH_3O)_2C_6H_3CH_2$ **115b** R = C_6H_5 **116b** X = SO_2
114c R = 3,4-$(CH_3O)_2C_6H_3(CH_2)_2$
114d R = C_6H_5
114e R = 3,4-$(CH_3O)_2C_6H_3$

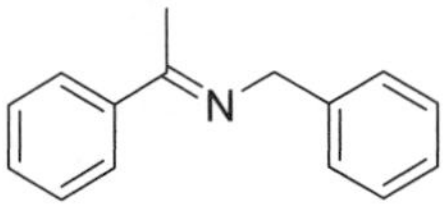

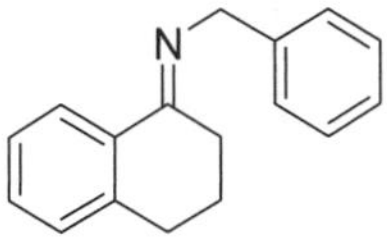

117   **118**

TABLE 6–10. Asymmetric Transfer Hydrogenation of a Variety of Imines

Entry	Imine	Catalyst	S/C	Solvent	Time (h)	Yield (%)	ee (%)	Config.
1	**114a**	(S,S)-**109b**	200	CH₃CN	3	>99	95	R
2	**114b**	(R,R)-**109c**	200	DMF	7	90	95	S
3	**114c**	(R,R)-**109c**	200	CH₂Cl₂	12	99	92	S
4	**114d**	(S,S)-**109e**	200	CH₂Cl₂	8	99	84	R
5	**114e**	(R,R)-**109e**	100	CH₂Cl₂	12	>99	84	S
6	**115a**	(S,S)-**109b**	200	DMF	5	86	97	R
7	**115b**	(S,S)-**109b**	200	DMF	5	83	96	R
8	**116a**	(S,S)-**109e**	100	CH₃CN	12	82	85	S
9	**116b**	(S,S)-**109e**	100	CH₃CN	5	84	88	S
10	**117**	(S,S)-**109d**	200	CH₂Cl₂	36	72	77	S
11	**118**	(S,S)-**109e**	100	CH₂Cl₂	6	90	89	S

ee = Enantiomeric excess.

Reprinted with permission by Am. Chem. Soc., Ref. 111.

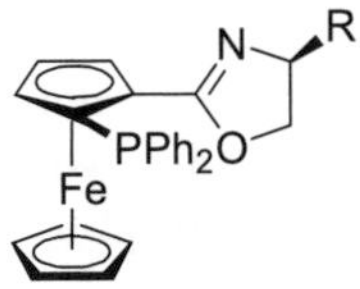

119

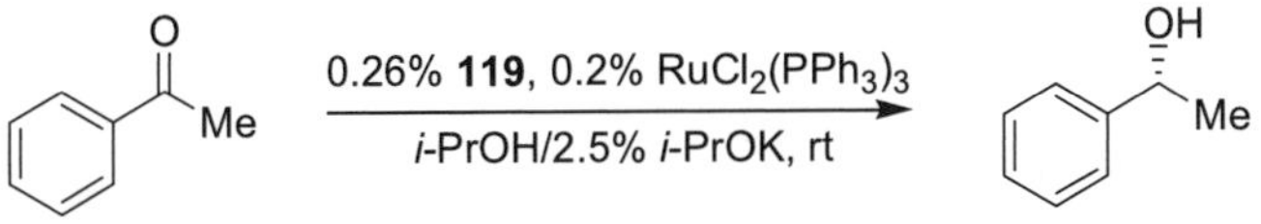

Entry	R in **119**	Time (h)	Conver. (%)	ee (%)
1	Me	8	92	92
2	Bn	6	93	90
3	*i*-Pr	3	93	92
4	Ph	6	93	94
5	*t*-Bu	6	51	94

Scheme 6–51. Asymmetric transfer hydrogenation of acetophenone in the presence of **119**. Reprinted with permission by Am. Chem. Soc., Ref. 112.

As transfer hydrogenation uses inexpensive reagents and is usually easy to perform, the asymmetric transfer hydrogenation using 2-propanol as the hydrogen source offers an attractive method for reducing unfunctionalized ketones to chiral alcohols. The chiral ligand used by Noyori et al. contains an NH moiety that may promote a cyclic transition state through hydrogen bonding to a ketone substrate. Therefore, it is desirable to design a chiral ligand that can form a metal–ligand bifunctional catalyst to increase the affinity of the substrate to the active site of the catalyst and thus to increase the stereoselectivity.

Sammakia and Stangeland[112] reported another type of chiral ligand **119** bearing a chiral oxazoline moiety and tested its asymmetric induction capability in transfer hydrogenation. As shown in Scheme 6–51, 90% or higher ee values were observed in all cases, along with high conversion of the substrates.

It is well recognized that chiral tridentate ligands generally form a deeper chiral cavity around the metal center than a bidentate ligand. For example, as mentioned in previous chapters, the chiral tridentate ligand Pybox **120** has been used in asymmetric aldol reactions (see Section 3.4.3) and asymmetric Diels-Alder reactions (see Section 5.7). The two substituents on the oxazoline rings of **120** form a highly enantioselective chiral environment that can effectively differentiate the prochiral faces of many substrates.

Jiang et al.[113] synthesized another tridentate ligand **121** for asymmetric transfer hydrogenation. Ru–**121**-catalyzed asymmetric transfer hydrogenation gives comparable enantioselectivity to Noyori's catalyst **109** but shows more temperature tolerance. Using Ru–**121** as the catalyst, asymmetric transfer hydrogenation proceeds much faster at higher temperature without loss of enantioselectivity. Scheme 6–52 summarizes the asymmetric transfer hydrogenation of a series of ketones. Clearly, the reactions proceed at high rates. Some are completed within 5 minutes.

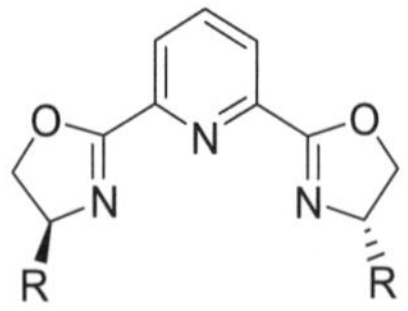

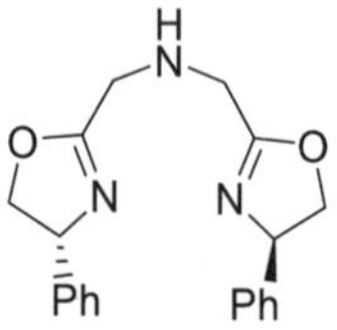

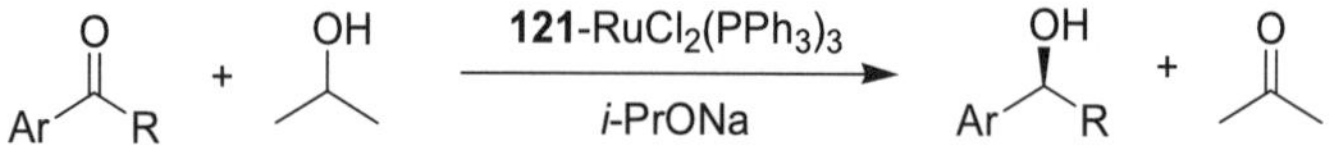

Entry	Ketone		Time (min)	Conver. (%)	ee (%)
1	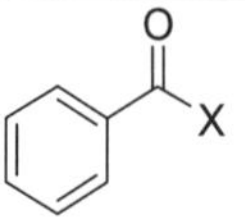	X = CH₃	5 (10)	80 (91)	98 (97)
2		X = Et	10 (20)	77 (92)	95 (92)
3		X = i-Pr	10	15	78
4	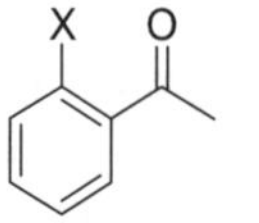	X = CH₃	40	96	98
5		X = Cl	5	>99	97
6		X = CH₃O	240	3	19
7	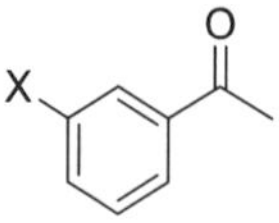	X = CH₃	5 (7)	75 (90)	96 (94)
8		X = Cl	10	5	92
9		X = CH₃O	7 (10)	91 (94)	95 (93)
10	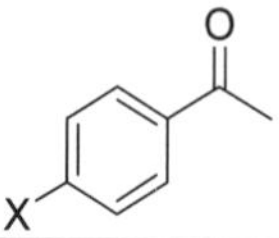	X = CH₃	4	68	95
11		X = Cl	10	97	90
12		X = CH₃O	10	41	98
13	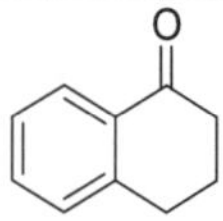		10	42	95
14	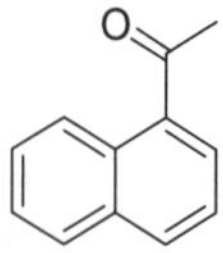		2 (5)	72 (98)	96 (94)
15	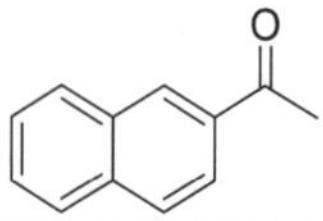		2 (7)	55 (91)	96 (92)

Scheme 6–52. Reprinted with permission by Am. Chem. Soc., Ref. 113.

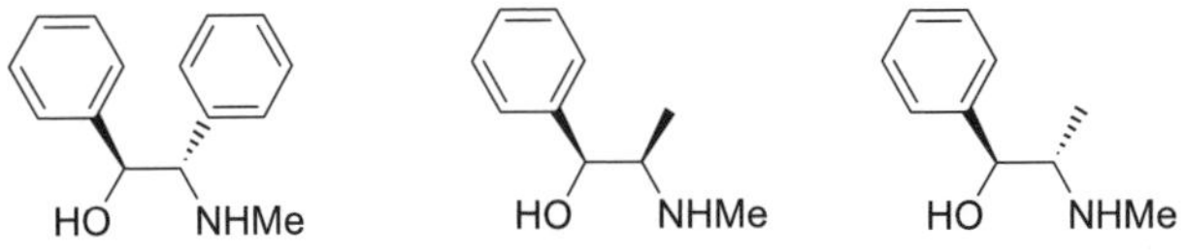

Scheme 6–53. Reprinted with permission by Royal Chem. Soc., Ref. 115.

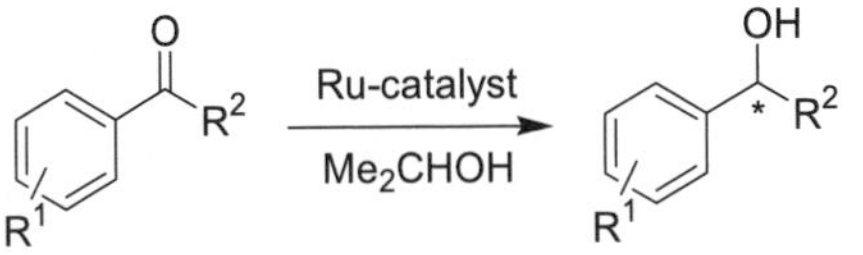

Entry	R^1	R^2	Chiral Ligand	Time (h)	Yield (%)	ee (%)	Config.
1	H	Me	(1S, 2S)-**122**	1	94	92	S
2	H	Me	(1S, 2R)-**123**	1	95	91	S
3	H	Me	(1S, 2S)-**122**	8	92	92	S
4	H	Et	(1S, 2S)-**122**	2	95	82	S
5	H	i-Pr	(1S, 2S)-**122**	15	93	5	S
6	H	t-Bu	(1S, 2S)-**122**	20	22	40	R
7	o-Me	Me	(1S, 2S)-**122**	6	96	83	S
8	o-Cl	Me	(1S, 2S)-**123**	1	99	89	S
9	p-OMe	Me	(1S, 2S)-**122**	4	73	79	S

Transition metal complexes with chiral phosphorous and nitrogen ligands have also been used for promoting asymmetric transfer hydrogenation. Moderate to good results have been obtained.[114]

Chiral β-amino alcohols, which can be used for enantioselective alkylation of carbonyl groups (see Chapter 2 for β-amino alcohol–mediated asymmetric alkylation), are also effective chiral ligands in asymmetric transfer hydrogenation. Takehara et al.[115] report Ru(II)-catalyzed asymmetric transfer hydrogenation in the presence of chiral amino alcohol. A particularly high ligand-acceleration effect in the reduction of aromatic ketones was observed. Highly enantioselective results were achieved when using a chiral ligand with suitable configuration and functionality.[115] As shown in Scheme 6–53, excellent enantioselectivities are observed in the asymmetric transfer hydrogenations of aromatic ketones catalyzed by a combination of ruthenium complex and chiral β-amino alcohol **122** or **123**.

Other amino alcohols have also been used as chiral ligands in asymmetric catalytic hydrogen transfer. Scheme 6–54 depicts another example. Ruthenium complex bearing 2-azanorbornyl methanol was used as the chiral ligand, and the corresponding secondary alcohols were obtained in excellent ee.[116]

The asymmetric transfer hydrogenation of ketones is further described elsewhere.[117]

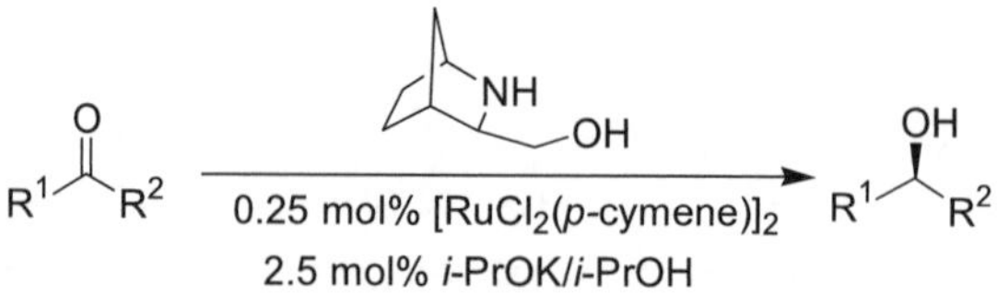

Entry	R^1	R^2	Time (h)	ee (%)
1	Ph	Me	1.5	94
2	1-Nap	Me	1.5	97
3	Ph	Et	1.5	93
4	Ph	n-Pr	1.5	92
5	Ph	n-Bu	1.5	95
6	Ph	n-C$_6$H$_{13}$	1.5	95

Scheme 6–54. Reprinted with permission by Am. Chem. Soc., Ref. 116.

6.5 ASYMMETRIC HYDROFORMYLATION

Thus far, we have discussed the transition metal complex–catalyzed hydrogenation of C=C, C=O, and C=N bonds. In this section, another type of transition metal complex–mediated reaction, namely, the hydroformylation of olefins, is presented.

Optically active aldehydes are important precursors for biologically active compounds, and much effort has been applied to their asymmetric synthesis. Asymmetric hydroformylation has attracted much attention as a potential route to enantiomerically pure aldehyde because this method starts from inexpensive olefins and synthesis gas (CO/H_2). Although rhodium-catalyzed hydrogenation has been one of the most important applications of homogeneous catalysis in industry, rhodium-mediated hydroformylation has also been extensively studied as a route to aldehydes.

Cobalt and rhodium complexes are the most important catalysts for hydroformylation. The latter is more favored because the cobalt complex–catalyzed reactions usually involve high temperature and high pressure (e.g., up to 190°C, 250 bar), which may not be desirable for industrial processes. Another advantage of using a rhodium complex may be the possibility of using a variety of phosphorous ligands to optimize the regioselectivity as well as the stereoselectivity of the reaction. Thus, the rhodium–phosphine-catalyzed hydroformylation reaction, which was first reported by Evans et al.[118] in the late 1960s, remains one of the most attractive homogeneous catalytic processes. The mechanism of rhodium–phosphine-catalyzed hydroformylation is shown in Scheme 6–55.

The mechanism of the hydroformylation reaction suggests that aldehyde regioselectivity is determined in the hydride addition step, which converts the five-coordinated H(alkene)-Rh(CO)L$_2$ into either a primary or a secondary four-coordinated (alkyl)Rh(CO)L$_2$. For the linear rhodium alkyl species, this

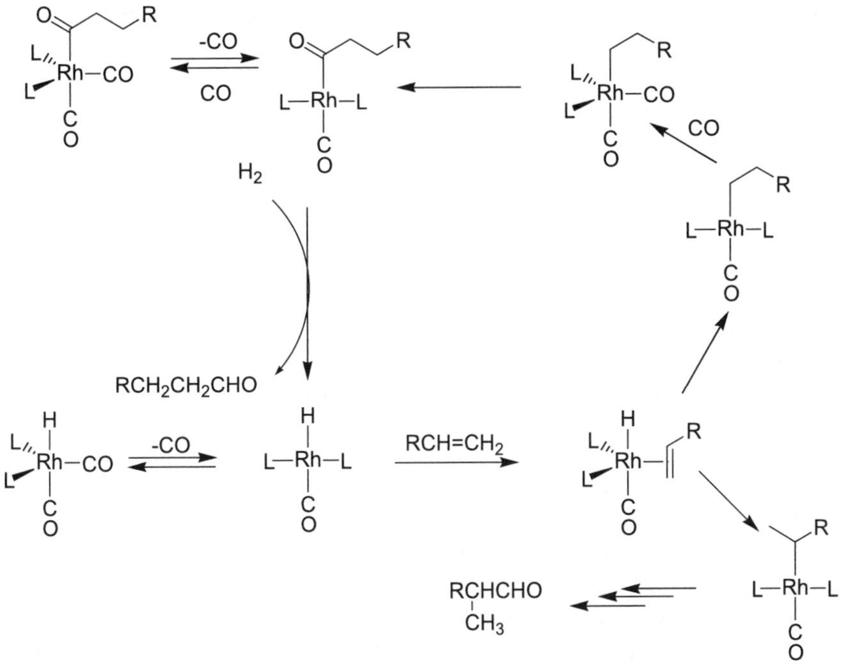

Scheme 6–55

step is irreversible at moderate temperatures and sufficiently high pressures of CO. The structure of the alkene complex is therefore thought to play a crucial role in controlling the regioselectivity.[119]

The first highly enantioselective asymmetric hydroformylation was the asymmetric hydroformylation of styrene.[120] In 1991, Stille et al.[121] reported the achievement of up to 96% ee using a chiral bisphosphine complex of PtCl$_2$ as the catalyst in combination with SnCl$_2$. However, the Pt(II)-catalyzed hydroformylation of arylethenes and some functionalized olefins has several disadvantages, such as low reaction rates, a tendency for the substrates to undergo hydrogenation, and poor branched-to-linear ratio.

A big problem in asymmetric hydroformylation is that the chiral aldehyde products may be unstable and may undergo racemization during the reaction. This problem is even more serious for the Pt catalyst systems, which are usually plagued by slow reaction rates. Stille et al.[121] tackled this problem by using triethyl orthoformate to trap the aldehyde products as their diethyl acetals and consequently increased the product ee values significantly.

Using a chiral diphosphine–Rh(I) complex as the catalyst, high catalytic activity and a good branched/linear ratio can be achieved, but in most cases the ee values remain moderate.[122] The enantioselectivity of Rh(I)-catalyzed reactions depends on the amount of added chiral ligand because of the much higher catalytic activity of ligand-free rhodium species. Usually 4–6 equivalents of the ligand relative to the Rh(I) species has to be applied to maintain reasonable stereoselectivity.[122a]

Phosphite ligands have attracted attention for Rh(I)-catalyzed hydroformylation because their complexes often show higher catalytic activity than do phosphine complexes.[123] Bisphosphite ligands with bulky substituents are often used to produce linear aldehydes from 1-alkenes, while the hydroformylation of styrene using these ligands results in branched aldehydes as major products. A bis(triarylphosphite) Rh(I) catalyst system shows comparable enantioselectivity but superior catalytic activity when compared with the corresponding triaryl phosphine–Rh catalyst. One feature of the phosphite system that differs from the corresponding bisphosphine ligand is that smaller amounts of chiral ligand are needed to attain the maximum ee.[124]

The stereoelectronic properties of phosphorous ligands have a dramatic influence on the reactivity of hydroformylation catalysts. By using a series of p-substituted triphenylphosphines, Moser et al.[125] investigated the effect of phosphine basicity on the hydroformylation of olefins. Their study showed that less basic phosphines afforded higher reaction rates and a higher ratio of linear to branched product.[125] Electron-withdrawing substituents on phosphines in the equatorial position give high linear/branched ratios, whereas electron-withdrawing substituents on phosphines in the apical position favor the production of branched aldehydes. Rossi and Hoffmann's molecular orbital analysis[126] of ligand electronic site preferences in five coordinated d^8 ML$_5$ complexes suggests that weaker σ-donor ligands prefer to stay at the equatorial position while strong σ-donor ligands prefer to occupy the apical position.

In the mechanism proposed by Evans et al.,[118] a five-coordinated trigonal bipyramidal bis(phosphine)rhodium species is the most important intermediate. In this species, two phosphine ligands occupy either two equatorial sites or one equatorial and one apical site. Casey et al.[119a,127] found that the intermediate in which both phosphorous atoms of the ligand occupy an equatorial position is essential for achieving high normal/iso selectivity in the hydroformylation of 1-hexene, whereas intermediate RhH(CO)$_2$(bisphosphite), bearing both of the phosphorous atoms at equatorial positions, is the active species for achieving high branched/linear selectivity and high enantioselectivity for the hydroformylation of styrene.[124f,g]

van der Veen et al.[128] studied the electronic effect on rhodium–diphosphine-catalyzed hydroformylation by using a series of thixantphos **124** as ligands. They studied the catalytic performance and coordination chemistry of the diphosphine ligands in rhodium complexes. In this series of ligands, steric differences are mininal, so purely electronic effects can be investigated. In the hydroformylation of 1-octene and styrene, decreasing the phosphine basicity led to an increase in both linear/branched ratio and reactivity, while increasing the phosphine basicity gave the opposite result. Decreasing the phosphine basicity facilitates CO dissociation from the (diphosphine)Rh(CO)$_2$H complex, hence enhancing the alkene coordination to form the (diphosphine)Rh(CO)H(alkene)

complex and therefore increasing the reaction rate. The weakening of the Rh–
CO bond in the rhodium–carbonyl complex by electron-withdrawing substituents on the ligand can be visualized as an increase in the carbonyl stretch-
ing frequency. They argue that the natural bite angle of the ligand might not be
as effective as previously claimed.[128]

124

The asymmetric hydroformylation of aryl ethenes such as substituted styrene
or naphthylethene is of industrial interest because the hydroformylation products of these substrates are precursors to important nonsteroidal antiinflammatory drugs such as (S)-ibuprofen and (S)-naproxen. Strong efforts have been
made to improve the branched/linear ratio, as well as the enantioselectivity of
the product.

(S)-ibuprofen

(S)-naproxen

Nozaki et al.[129] report the asymmetric hydroformylation of aryl ethene
using **125** as a ligand. The ligand has a new feature in that it contains both a
phosphine and a phosphite moiety. High branched/linear ratio as well as
enantioselectivity results when its Rh(I) complex is applied to the hydroformylation of styrene substrates. This study shows that **126** is the single species
in the reaction. In this structure, phosphine occupies the equatorial position and
the phosphite occupies the apical position, and the conversion of the equatorial–
apical configuration to the equatorial–equatorial configuration is not observed.
Indeed, this may be the cause of the high regioselectivity as well as enantioselectivity of the reaction.

The asymmetric hydroformylation of styrene and its derivatives is shown in
Scheme 6–56. The results of the reaction are governed by several factors, such
as solvent, temperature, reaction time, and configuration of the ligand. The
configuration of the product, in turn, is governed mainly by the configuration
of the phosphine moiety (compare Entries 1 and 5 in Scheme 6–56). Low temperature gives a better branched/linear ratio, as well as better enantioselectivity,
but normally it leads to a drop in reaction rate. The reaction is sensitive to the

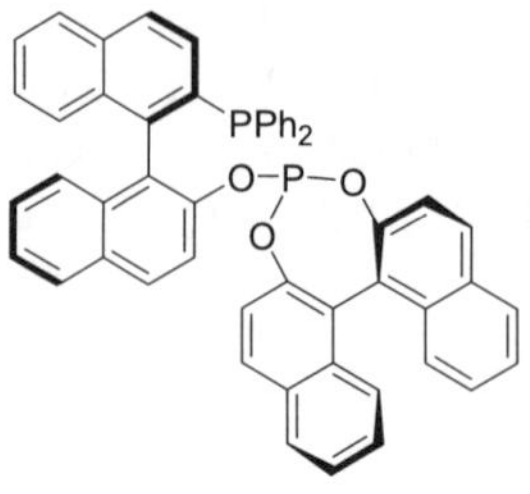

Scheme 6–56. Reprinted with permission by Am. Chem. Soc., Ref. 129.

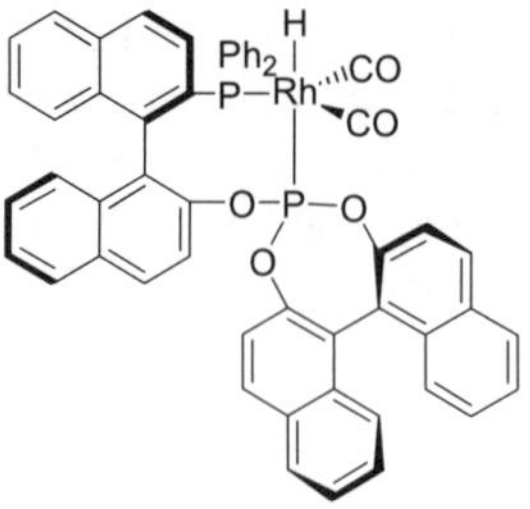

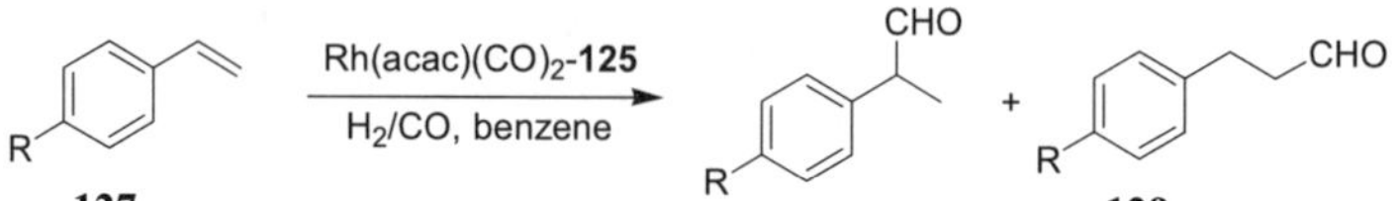

Entry	R	CO/H$_2$ atm/atm	Solvent	Ligand	Temp. (°C)	Time (h)	Conver. (%)	128/129	ee (%)
1	H	50/50	C$_6$H$_6$	(S, R)-**125**	60	43	> 99	88/12	94 (S)
2	H	50/50	C$_6$H$_6$	(R, S)-**125**	80	16	> 99	86/14	89 (R)
3	H	50/50	C$_6$H$_6$	(R, S)-**125**	rt	39	24	92/8	95 (R)
4	H	50/50	C$_6$H$_6$	(R, S)-**125**	60	59	> 99	90/10	45 (R)
5	H	50/50	C$_6$H$_6$	(R, R)-**125**	60	38	>99	86/14	25 (R)
6	H	63/8	C$_6$H$_6$	(R, S)-**125**	60	40	93	88/12	92 (R)
7	H	10/90	C$_6$H$_6$	(R, S)-**125**	60	40	> 99	88/12	92 (R)
8	H	50/50	c-C$_6$H$_{12}$	(R, S)-**125**	60	40	> 98	87/13	89 (R)
9	H	50/50	CH$_2$Cl$_2$	(R, S)-**125**	60	41	> 99	89/11	84 (R)
10	H	50/50	THF	(R, S)-**125**	60	41	> 99	92/8	41 (R)
11	H	50/50	MeOH	(R, S)-**125**	60	98	98	92/8	25 (R)
12	CH$_3$	50/50	C$_6$H$_6$	(S, R)-**125**	60	20	97	86/14	95 (+)
13	CH$_3$O	50/50	C$_6$H$_6$	(S, R)-**125**	60	34	> 99	87/13	88 (+)
14	Cl	50/50	C$_6$H$_6$	(S, R)-**125**	60	34	> 99	87/13	93 (+)
15	i-PrCH$_2$	50/50	C$_6$H$_6$	(S, R)-**125**	60	66	> 99	88/12	92 (S)

solvent used. Reactions carried out in an electron-donating solvent such as THF or methanol normally give poor results. The Rh catalyst **126** have given the best overall results thus far for the asymmetric hydroformylation of aryl olefins.

6.6 SUMMARY

Homogeneous enantioselective hydrogenation constitutes one of the most versatile and effective methods to convert prochiral substrates to valuable optically active products. Recent progress makes it possible to synthesize a variety of chiral compounds with outstanding levels of efficiency and enantioselectivity through the reduction of the C=C, C=N, and C=O bonds. The asymmetric hydrogenation of functionalized C=C bonds, such as enamide substrates, provides access to various valuable products such as amino acids, pharmaceuticals, and important building blocks for organic synthesis. The asymmetric hydrogenation or reduction of ketones allows the production of a large number of useful chiral alcohols that are ubiquitous in nature and are in great demand for flavors and fragrances, for pharmaceuticals, and for agrochemicals.

In asymmetric hydrogenation, the pressure of hydrogen may have a substantial impact on both the rates and the stereoselectivities of the reaction. These effects may be attributed either to the formation of different catalytically competing species in solution or to the operation of kinetically distinct catalytic cycles at different pressures.

The solvent employed in asymmetric catalytic reactions may also have a dramatic influence on the reaction rate as well as the enantioselectivity, possibly because the solvent molecule is also involved in the catalytic cycle. Furthermore, the reaction temperature also has a profound influence on stereoselectivity. The goal of asymmetric hydrogenation or transfer hydrogenation studies is to find an optimal condition with a combination of chiral ligand, counterion, metal, solvent, hydrogen pressure, and reaction temperature under which the reactivity and the stereoselectivity of the reaction will be jointly maximized.

The asymmetric hydroformylation of aryl ethenes such as substituted styrene and substituted β-naphthyl ethene will lead to the intermediates for important pharmaceuticals. Much concerted effort has been applied to achieve high enantioselectivity as well as high regioselectivity toward the branched aldehydes. The research work in this area is of great industrial interest, and it continues to be a dynamic field of study.

6.7 REFERENCES

1. Hörner, L.; Siegewel, H.; Büthe, H. *Angew. Chem. Int. Ed. Engl.* **1968**, *7*, 942.

2. Knowles, W. S.; Sabacky, M. J. *J. Chem. Soc. Chem. Commun.* **1968**, 1445.

3. (a) Dang, T.; Kagan, H. B. *J. Chem. Soc. Chem. Commun.* **1971**, 481. (b) Dumont, W.; Poulin, J.; Dang, T.; Kagan, H. B. *J. Am. Chem. Soc.* **1973**, *95*, 8295.

4. Jiang, Q.; Jiang, Y.; Xiao, D.; Cao, P.; Zhang, X. *Angew. Chem. Int. Ed. Engl.* **1998**, *37*, 1100.

5. Pye, P. J., Rossen, K.; Reamer, R. A.; Volante, R. P.; Reider, P. J. *Tetrahedron Lett.* **1998**, *39*, 4441.

6. Kang, J., Lee, J.; Ahn, S.; Choi, J. *Tetrahedron Lett.* **1998**, *39*, 5523.

7. Zhu, G.; Cao, P.; Jiang, Q.; Zhang, X. *J. Am. Chem. Soc.* **1997**, *119*, 1799.

8. (a) Chan, A. S. C.; Halpern, J. *J. Am. Chem. Soc.* **1980**, *102*, 838. (b) Chan, A. S. C.; Pluth, J. J.; Halpern, J. *J. Am. Chem. Soc.* **1980**, *102*, 5952. (c) Halpern, J. *Science* **1982**, *217*, 401. (d) Landis, C. R.; Halpern, J. *J. Am. Chem. Soc.* **1987**, *109*, 1746.

9. Brown, J. M.; Chaloner, P. A.; Morris, G. A. *J. Chem. Soc. Perkin Trans. 2,* **1987**, 1583.

10. Knowles, W. S. *Acc. Chem. Res.* **1983**, *16*, 106.

11. (a) Ohta, T.; Takaya, H.; Kitamura, M.; Nagai, K.; Noyori, R. *J. Org. Chem.* **1987**, *52*, 3174. (b) Chan, A. S. C. US Patent 4 994 607, **1991**. (c) Chan, A. S. C.; Laneman, S. A. US Patent 5 144 050, **1992**. (d) Chan, A. S. C. *CHEMTECH* **1993**, *March*, 46.

12. Burk, M. J.; Gross, M. F.; Martinez, J. P. *J. Am. Chem. Soc.* **1995**, *117*, 9375, and the references cited therein.

13. Ohta, T.; Takaya, H.; Noyori, R. *Inorg. Chem.* **1988**, *27*, 566.

14. For general application of these chiral ligands, see (a) Kagan, H. B. "Chiral Ligands for Asymmetric Catalysis" in Morrison, J. D. ed. *Asymmetric Synthesis*, vol. 5, Chap. 1, Academic Press, New York, **1985**. (b) Kagan, H. B., Sasaki, M. "Optically Active Phosphines: Preparation, Uses and Chiroptical Properties" in Hartley, F. R. ed. *The Chemistry of Organo Phosphorous Compounds*, John Wiley & Sons, New York, **1990**, vol. 1, Chap. 3.

15. For details, see Koenig, K. E. "Asymmetric Hydrogenation of Prochiral Olefins" in Kosak, J. R. ed. *Catalysis of Organic Reactions*, Marcel Dekker, New York, **1984**, Chap. 3.

16. For the mechanism of asymmetric hydrogenation, see Halpern, J. "Asymmetric Catalytic Hydrogenation: Mechanism and Origin of Enantioselection" in Morrision, J. D. ed. *Asymmetric Synthesis*, Academic Press, New York, **1985**, vol. 5.

17. (a) Brown J. M.; Maddox, P. J. *J. Chem. Soc. Chem. Commun.* **1987**, 1276. (b) Brown J. M.; Chaloner, P. A.; Morris, G. A. *J. Chem. Soc. Chem. Commun.* **1983**, 664. (c) Brown J. M.; Chaloner, P. A. *J. Chem. Soc. Chem. Commun.* **1980**, 344. (d) Brown J. M.; Chaloner, P. A. *J. Chem. Soc. Chem. Commun.* **1978**, 321. (e) Brown J. M. Chaloner, P. A. *Tetrahedron Lett.* **1978**, 1877. (f) Brown J. M.; Parker, D. *J. Chem. Soc. Chem. Commun.* **1980**, 342.

18. (a) Shimohigashi, Y; English, M. L.; Stammer, C. H.; Costa, T. *Biochem. Biophys. Res. Commun.* **1982**, *104*, 583. (b) Jung, G. *Angew Chem. Int. Ed. Engl.* **1991**, *30*, 1051. (c) Freud, S.; Jung, G.; Gutbrod, O.; Folkers, G.; Gibbons, W. A.; Allgaier, H.; Werner, R. *Biopolymers* **1991**, *31*, 803. (d) Jain, R.; Chauhan, V. S. *Biopolymers* **1996**, *40*, 105.

19. (a) Drinkwater, D. J.; Smith, P. W. G. *J. Chem. Soc. C* **1971**, 1305. (b) Letham, D. S.; Young, H. *Phytochemistry* **1971**, *10*, 23. (c) Davis, A. L.; Cavitt, M. B.; McCord, T. J.; Vickrey, P. E.; Shive, W. *J. Am. Chem. Soc.* **1973**, *95*, 6800. (d) Gellert, E.; Halpern, B.; Rudzats, R. *Phytochemistry* **1978**, *17*, 802. (e) Cramer, U.; Rehfeldt, A. G.; Spener, F. *Biochemistry* **1980**, *19*, 3074. (f) Baldwin, J. E.; Adlington, R. M.; Basak, A. *J. Chem. Soc. Chem. Commun.* **1984**, 1284. (g) Tsubotani, S.; Funabashi, Y.; Takamoto, M.; Hakoda, S.; Harada, S. *Tetrahedron* **1991**, *47*, 8079.

20. (a) Bartlett, P. A.; Tanzella, D. J.; Barstow, J. F. *Tetrahedron Lett.* **1982**, 619. (b) Kurokawa, N.; Ohfune, Y. *J. Am. Chem. Soc.* **1986**, *108*, 6041. (c) Ohfune, Y.; Hori, K.; Sakaitani, M. *Tetrahedron Lett.* **1986**, *27*, 6079. (d) Madau, A.; Porzi, G.; Sandri, S. *Tetrahedron Asymmetry* **1996**, *7*, 825. (e) Graziani, L.; Porzi, G.; Sandri, S. *Tetrahedron Asymmetry* **1996**, *7*, 1341. (f) Wang, Y.; Izawa, T.; Kobayashi, S.; Ohno, M. *J. Am. Chem. Soc.* **1982**, *104*, 6465. (g) Takano, S.; Iwabuchi, Y.; Ogasawara, K. *J. Chem. Soc. Chem. Commun.* **1988**, 1527. (h) Hirai, Y.; Terada, T.; Amemiya, Y.; Momose, T. *Tetrahedron Lett.* **1992**, *33*, 7893. (i) Burk, M. J.; Martinez, J. P.; Feaster, J. E.; Cosford, N. *Tetrahedron* **1994**, *50*, 4399.

21. Burk, M. J.; Allen, J. G.; Kiesman, W. F. *J. Am. Chem. Soc.* **1998**, *120*, 657.

22. Dwars, T.; Schmidt, U.; Fischer, C.; Grassert, I.; Kempe, R.; Fröhlich, R.; Drauz, K.; Oehme, G. *Angew. Chem. Int. Ed. Engl.* **1998**, *37*, 2851.

23. Imamoto, T.; Watanabe, J.; Wada, Y.; Masuda, H.; Yamada, H.; Tsuruta, H.; Matsukawa, S.; Yamaguchi, K. *J. Am. Chem. Soc.* **1998**, *120*, 1635.

24. Burk, M. J.; Bienewald, F.; Harris, M.; Zanotti-Gerosa, A. *Angew. Chem. Int. Ed. Engl.* **1998**, *37*, 1931.

25. Hayashi, T.; Kawamura, N.; Ito, Y. *J. Am. Chem. Soc.* **1987**, *109*, 7876.

26. Miyashita, A.; Yasuda, A.; Takaya, H.; Toriumi, K.; Ito, T.; Souchi, T.; Noyori, R. *J. Am. Chem. Soc.* **1980**, *102*, 7932.

27. Pye, P. J.; Rossen, K.; Reamer, R. A.; Tsou, N. N.; Volante, R. P.; Reider, P. J. *J. Am. Chem. Soc.* **1997**, *119*, 6207.

28. (a) Blaser, H. U.; Spindler, F. *Top. Catalysis* **1997**, *4(3–4)*, 275. (b) Imeinkelried, R. *Chimia* **1997**, *51*, 300. (c) Togni, A. *Angew. Chem. Int. Ed. Engl.* **1996**, *35*, 1475. (d) Dobbs, D. A.; van Hessche, K. P. M.; Rautenstrauch, V. WO 98/52687, **1998**. (e) Richards, C. J.; Locke, A. J. *Tetrahedron Asymmetry* **1998**, *9*, 2377.

29. Leutenegger, U.; Madin, A.; Pfaltz, A. *Angew. Chem. Int. Ed. Engl.* **1989**, *28*, 60.

30. (a) Fritschi, H.; Leutenegger, U.; Pfaltz, A. *Angew. Chem. Int. Ed. Engl.* **1986**, *25*, 1005. (b) Fritschi, H.; Leutenegger, U.; Siegmann, K.; Pfaltz, A.; Keller, W.; Kratky, C. *Helv. Chim. Acta* **1988**, *71*, 1541. (c) Fritschi, H.; Leutenegger, U.; Pfaltz, A. *Helv. Chim. Acta* **1988**, *71*, 1553.

31. Burk, M. J. *J. Am. Chem. Soc.* **1991**, *113*, 8518.

32. Burk, M. J.; Kalberg, C. S.; Pizzano, A. *J. Am. Chem. Soc.* **1998**, *120*, 4345.

33. Boaz, N. W. *Tetrahedron Lett.* **1998**, *39*, 5505.

34. Jiang, Q.; Xiao, D.; Zhao, Z.; Cao, P.; Zhang, X. *Angew. Chem. Int. Ed. Engl.* **1999**, *38*, 516.

35. Broger, E. A.; Burkart, W.; Hennig, M.; Scalone, M.; Schmid, R. *Tetrahedron Asymmetry* **1998**, *9*, 4043.

36. Lightfoot, A.; Schnider, P.; Pfaltz, A. *Angew. Chem. Int. Ed. Engl.* **1998**, *37*, 2897.

37. Broene, R. D.; Buchwald, S. L. *J. Am. Chem. Soc.* **1993**, *115*, 12569.

38. (a) Selke, R.; Schwarze, M.; Baudisch, H.; Grassert, I.; Michalik, M.; Oehme, G.; Stoll, N.; Costisella, B. *J. Mol. Catal.* **1993**, *84*, 223. (b) Kumar, A.; Oehme, G.; Roque, J. P.; Schwarze, M.; Selke, R. *Angew. Chem. Int. Ed. Engl.* **1994**, *33*, 2197. (c) RajanBabu, T. V.; Ayers, T. A.; Casalnuovo, A. L. *J. Am. Chem. Soc.* **1994**, *116*, 4101.

39. (a) Chan, A. S. C.; Lin, C.; Sun, J.; Hu, W.; Li, Z.; Pan, W.; Mi, A.; Jiang, Y.; Huang, T.; Yang, T.; Chen, J.; Wang, Y.; Lee, G. *Tetrahedron Asymmetry* **1995**, *6*, 2953. (b) Chan, A. S. C.; Hu, W.; Pai, C.; Lau, C.; Jiang, Y.; Mi, A.; Yan, M.; Sun, J.; Lou, R.; Deng, J. *J. Am. Chem. Soc.* **1997**, *119*, 9570.

40. Zhu, G.; Zhang, X. *J. Org. Chem.* **1998**, *63*, 3133.

41. Burk, M. J.; Wang, Y.; Lee, J. R. *J. Am. Chem. Soc.* **1996**, *118*, 5142.

42. (a) Fiorini, M.; Giongo, G. M. *J. Mol. Catal.* **1979**, *5*, 303. (b) Fiorini, M.; Giongo, G. *J. Mol. Catal.* **1980**, *7*, 411. (c) Onuma, K.; Ito, T.; Nakamura, A. *Bull. Chem. Soc. Jpn.* **1980**, *53*, 2016. (d) Miyano, S.; Nawa, M.; Hashimoto, H. *Chem.*

Lett. **1980**, 729. (e) Miyano, S.; Nawa, M.; Mori, A.; Hashimoto, H. *Bull. Chem. Soc. Jpn.* **1984**, *57*, 2171.

43. Zhang, F.; Pai, C.; Chan, A. S. C. *J. Am. Chem. Soc.* **1998**, *120*, 5808.

44. Reetz, M. T.; Neugebauer, T. *Angew. Chem. Int. Ed. Engl.* **1999**, *38*, 179.

45. Kitamura, M.; Tokunaga, M.; Noyori, R. *J. Org. Chem.* **1992**, *57*, 4053.

46. Takaya, H.; Ohta, T.; Sayo, N.; Kumobayashi, H.; Akutagawa, S.; Inoue, S.; Kasahara, I.; Noyori, R. *J. Am. Chem. Soc.* **1987**, *109*, 1596.

47. Noyori, R.; Ohta, M.; Hsiao, Y.; Kitamura, M.; Ohta, T.; Takaya, H. *J. Am. Chem. Soc.* **1986**, *108*, 7117.

48. (a) Chan, A. S. C.; Pai, C. C. US Patent 5 886 182, **1999**. (b) Benincori, T.; Brenna, E.; Sannicolo, F.; Trimarco, L.; Antognazza, P.; Cesarotti, E. *J. Chem. Soc. Chem. Commun.* **1995**, 685. (c) *Chem. Eng. News* **1995**, *73 (41)*, 72.

49. (a) Tani, K.; Yamagata, T.; Otsuka, S.; Akutagawa, S.; Kumobayashi, H.; Taketomi, T.; Takaya, H.; Miyashita, A.; Noyori, R. *J. Chem. Soc. Chem. Commun.* **1982**, 600. (b) Tani, K.; Yamagata, T.; Akutagawa, S.; Kumobayashi, H.; Taketomi, T.; Takaya, H.; Miyashita, A.; Noyori, R.; Otsuka, S. *J. Am. Chem. Soc.* **1984**, *106*, 5208. (c) Frauenrath, H.; Philipps, T. *Angew. Chem. Int. Ed. Engl.* **1986**, *25*, 274.

50. Noyori, R. *Asymmetric Catalysis in Organic Synthesis*, John Wiley & Sons, Inc., New York, **1994**, Chap. 3, p 96.

51. (a) Midland, M. M. *Chem. Rev.* **1989**, *89*, 1553. (b) Srebnik, M. *Aldrichimica Acta.* **1987**, *20*, 9.

52. Bothner-By, A. A. *J. Am. Chem. Soc.* **1951**, *73*, 846.

53. (a) Noyori, R. *Pure Appl. Chem.* **1981**, *53*, 2315. (b) Noyori, R. *Tetrahedron* **1994**, *50*, 4259.

54. (a) Noyori, R. *CHEMTECH* **1992**, *22*, 360. (b) Noyori, R.; Ikeda, T.; Ohkuma, T.; Widhalm, M.; Kitamura, M.; Takaya, H.; Akutagawa, S.; Sayo, N.; Saito, T.; Taketomi, T.; Kumobayashi, H. *J. Am. Chem. Soc.* **1989**, *111*, 9134.

55. Kitamura, M.; Ohkuma, T.; Inoue, S.; Sayo, N.; Kumobayashi, H.; Akutagawa, S.; Ohta, T.; Takaya, H.; Noyori, R. *J. Am. Chem. Soc.* **1988**, *110*, 629.

56. Nishi, T.; Kitamura, M.; Ohkuma, T.; Noyori, R. *Tetrahedron Lett.* **1988**, *29*, 6327.

57. Fryzuk, M. D.; Bosnich, B. *J. Am. Chem. Soc.* **1977**, *99*, 6262.

58. MacNeil, P. A.; Roberts, N. K.; Bosnich, B. *J. Am. Chem. Soc.* **1981**, *103*, 2273.

59. Burk, M. J.; Feaster, J. E.; Harlow, R. L. *Organometallics* **1990**, *9*, 2653.

60. Kawano, H.; Ishii, Y.; Saburi, M.; Uchida, Y. *J. Chem. Soc. Chem. Commun.* **1988**, 87.

61. Chong, J. M.; Clarke, I. S.; Koch, I.; Olbach, P. C.; Taylor, N. J. *Tetrahedron Asymmetry* **1995**, *6*, 409.

62. Kuwano, R.; Sawamura, M.; Shirai, J.; Takahashi, M.; Ito, Y. *Tetrahedron Lett.* **1995**, *36*, 5239.

63. Short, R. P.; Kennedy, R. M.; Masamune, S. *J. Org. Chem.* **1989**, *54*, 1755.

64. Fan, Q.; Yeung, C. H.; Chan, A. S. C. *Tetrahedron Asymmetry* **1997**, *8*, 4041.

65. Mashima, K.; Hino, T.; Takaya, H. *J. Chem. Soc. Dalton Trans.* **1992**, 2099.

66. Noyori, R.; Ohkuma, T.; Kitamura, M.; Takaya, H.; Sayo, N.; Kumobayashi, H.; Akutagawa, S. *J. Am. Chem. Soc.* **1987**, *109*, 5856.

67. (a) Bennet, M. A.; Matheson, T. W. "Catalysis by Ruthenium Compounds" in Wilkinson, G.; Stone, F. G. A.; Abel, E. W. ed. *Comprehensive Organometallic Chemistry*, vol. 4, Chap. 32.9, Pergamon Press, Oxford, **1982**. (b) Jardine, F. *Prog. Inorg. Chem.* **1984**, *31*, 265. (c) Noyori, R. *Asymmetric Catalysis in Organic Synthesis*, John Wiley & Sons, Inc., New York, **1994**, p 65.

68. (a) Masamune, S.; Kennedy, R. M.; Peterson, J. S. *J. Am. Chem. Soc.* **1986**, *108*, 7404. (b) Brown, H. C.; Ramachandran, R. V. *Acc. Chem. Res.* **1992**, *25*, 16.

69. Ohkuma, T.; Ooka, H.; Hashiguchi, S.; Ikariya, T.; Noyori, R. *J. Am. Chem. Soc.* **1995**, *117*, 2675.

70. Kitamura, M.; Tokunaga, M.; Ohkuma, T.; Noyori, R. *Org. Syn.* **1993**, *71*, 1.

71. Doucet, H.; Ohkuma, T.; Murata, K.; Yokozawa, T.; Kozawa, M.; Katayama, E.; England, A. F.; Ikariya, T.; Noyori, R. *Angew. Chem. Int. Ed. Engl.* **1998**, *37*, 1703.

72. For review, see Herrmann, W. A.; Cornils, B. *Angew. Chem. Int. Ed. Engl.* **1997**, *36*, 1048.

73. Ohkuma, T.; Koizumi, M.; Doucet, H.; Pham, T.; Kozawa, M.; Murata, K.; Katayama, E.; Yokozawa, T.; Ikariya, T.; Noyori, R. *J. Am. Chem. Soc.* **1998**, *120*, 13529.

74. Marinetti, A.; Genêt, J.; Jus, S.; Blanc, D.; Ratovelomanana-Vidal, V. *Chem. Eur. J.* **1999**, *5*, 1160.

75. Fiaud, J. C.; Kagan, H. B. *Bull. Soc. Chim. Fr.* **1969**, 2742.

76. (a) Hirao, A.; Itsuno, S.; Nakahama, S.; Yamazaki, N. *J. Chem. Soc. Chem. Commun.* **1981**, 315. (b) Itsuno, S.; Hirao, A.; Nakahama, S.; Yamazaki, N. *J. Chem. Soc. Perkin Trans. 1* **1983**, 1673.

77. For a review, see Wallbaum, S.; Martens, J. *Tetrahedron Asymmetry* **1992**, *3*, 1475.

78. Corey, E. J.; Link, J. O. *Tetrahedron Lett.* **1989**, *30*, 6275.

79. Cai, D.; Tschaen, D.; Shi, Y.; Verhoeven, T. R.; Reamer, R. A.; Douglas, A. W. *Tetrahedron Lett.* **1993**, *34*, 3243.

80. Corey, E. J.; Bakshi, R. K.; Shibata, S. *J. Am. Chem. Soc.* **1987**, *109*, 5551.

81. (a) Corey, E. J.; Shibata, S.; Bakshi, R. K. *J. Org. Chem.* **1988**, *53*, 2861. (b) Corey, E. J.; Bakshi, R. K.; Shibata, S.; Chen, C.; Singh, V. K. *J. Am. Chem. Soc.* **1987**, *109*, 7925.

82. Corey, E. J.; Bakshi, R. K. *Tetrahedron Lett.* **1990**, *31*, 611.

83. Rama Rao, A. V.; Gurjar, M. K.; Sharma, P. A.; Kaiwar, V. *Tetrahedron Lett.* **1990**, *31*, 2341.

84. (a) Behnan, W.; Dauelsbery, Ch.; Wallbaum, S.; Martens, J.; *Synth. Commun.* **1992**, *22*, 2143. (b) Rama Rao, A. V.; Gurjar, M. K.; Kaiwar, V. *Tetrahedron Asymmetry* **1992**, *3*, 859.

85. (a) Youn, I. K.; Lee, S.; Pak, C. S. *Tetrahedron Lett.* **1988**, *29*, 4453. (b) Martens, J.; Dauelsberg, Ch.; Behnen, W.; Wallbaum, S. *Tetrahedron Asymmetry* **1992**, *3*, 347.

86. Corey, E. J.; Chen, C.; Reichard, G. A. *Tetrahedron Lett.* **1989**, *30*, 5547.

87. For studies of these reactions, see (a) Corey, E. J.; Link, J. O. *Tetrahedron Lett.* **1992**, *33*, 4141. (b) Nevalainen, V. *Tetrahedron Asymmetry* **1991**, *2*, 1133. (c) Corey, E. J.; Link, J. O.; Sarshar, S.; Shao, Y. *Tetrahedron Lett.* **1992**, *33*, 7103. (d) Corey, E. J.; Link, J. O.; Bakshi, R. K.; Shao, Y. *Tetrahedron Lett.* **1992**, *33*, 7071. (e) Jones, D. K.; Liotta, D. C. *J. Org. Chem.* **1993**, *58*, 799.

88. Corey, E. J.; Reichard, G. A. *Tetrahedron. Lett.* **1989**, *30*, 5207.

89. (a) Corey, E. J. *Chem. Soc. Rev.* **1988**, *17*, 111. (b) Corey, E. J.; Gavai, A. V. *Tetrahedron Lett.* **1988**, *29*, 3201.

90. Corey, E. J.; Jardine, P. D. S.; Mohri, T. *Tetrahedron Lett.* **1988**, *29*, 6409.

91. Corey, E. J.; Chen, C.; Parry, M. J. *Tetrahedron Lett.* **1988**, *29*, 2899.

92. Masui, M.; Shirori, T. *Synlett* **1996**, 49.

93. Mehler, T.; Martens, J. *Tetrahedron Asymmetry* **1993**, *4*, 2299.

94. Willems, J. G. H.; Dommerholt, F. J.; Hammink, J. B.; Vaarhorst, A. M.; Thijs, L.; Zwanenburg, B. *Tetrahedron Lett.* **1995**, *36*, 603.

95. Wang, Z.; La, B.; Fortunak, J. M.; Meng, X. J.; Kabalka, G. W. *Tetrahedron Lett.* **1998**, *39*, 5501.

96. Singh, V. K. *Synthesis* **1992**, 605.

97. For a review regarding the chiral oxazoborolidine–catalyzed reduction of carbonyl compounds, see Corey, E. J.; Helal, C. J. *Angew. Chem. Int. Ed. Engl.* **1998**, *37*, 1986.

98. For literature on the asymmetric hydrogenation of imines, see (a) Bakos, J.; Tóth, I.; Heil, B.; Markó, L. *J. Organomet. Chem.* **1985**, *279*, 23. (b) Becalski, A. G.; Cullen, W. R.; Fryzuk, M. D.; James, B. R.; Kang, G.; Rettig, S. J. *Inorg. Chem.* **1991**, *30*, 5002. (c) Bakos, J.; Orosz, Á.; Heil, B.; Laghmari, M.; Lhoste, P.; Sinou, D. *J. Chem. Soc. Chem. Commun.* **1991**, 1684. (d) Chan, Y.; Ng, C.; Osborn, J. A. *J. Am. Chem. Soc.* **1990**, *112*, 9400. (e) Morimoto, T.; Achiwa, K. *Tetrahedron Asymmetry* **1995**, *6*, 2661. (f) Tani, K.; Onouchi, J.; Yamagata, T.; Kataoka, Y. *Chem. Lett.* **1995**, 955. (g) Chan, A. S. C.; Chen, C. C.; Lin, C. W.; Lin, Y. C.; Cheng, M. C.; Peng, S. M. *J. Chem. Soc. Chem. Commun.* **1995**, 1767.

99. Oppolzer, W.; Wills, M.; Starkemann, C.; Bernardinelli, G.; *Tetrahedron Lett.* **1990**, *31*, 4117.

100. (a) Willoughby, C. A.; Buchwald, S. L. *J. Am. Chem. Soc.* **1992**, *114*, 7562. (b) Willoughby, C. A.; Buchwald, S. L. *J. Am. Chem. Soc.* **1994**, *116*, 8952. (c) Willoughby, C. A.; Buchwald, S. L. *J. Am. Chem. Soc.* **1994**, *116*, 11703. (d) Bolm, C. *Angew. Chem. Int. Ed. Engl.* **1993**, *32*, 232.

101. For asymmetric hydrogenation, see (a) Burk, M. J.; Feaster, J. E. *J. Am. Chem. Soc.* **1992**, *114*, 6266. (b) Buriak, J. M.; Osborn, J. A. *Organometallics* **1996**, *15*, 3161. For asymmetric hydrosilylation, see (c) Tillack, A.; Lefeber, C.; Peulecke, N.; Thomas, D.; Rosenthal, U. *Tetrahedron Lett.* **1997**, *38*, 1533. (d) Becker, R.; Brunner, H.; Mahboobi, S.; Wiegrebe, W.; *Angew. Chem. Int. Ed. Engl.* **1985**, *24*, 995.

102. (a) Kagan, H. B. *Pure Appl. Chem.* **1975**, *43*, 401. (b) Levi, A.; Modena, G.; Scorrano, G. *J. Chem. Soc. Chem. Commun.* **1975**, 6.

103. Verdaguer, X.; Lange, U. E. W.; Reding, M. T.; Buchwald, S. L. *J. Am. Chem. Soc.* **1996**, *118*, 6784.

104. Chin, B.; Buchwald, S. L. *J. Org. Chem.* **1996**, *61*, 5650.

105. Verdaguer, X.; Lange, U. E. W.; Buchwald, S. L. *Angew. Chem. Int. Ed. Engl.* **1998**, *37*, 1103.

106. Evans, D. A., Nelson, S. G.; Gagne, M. R.; Muci, A. R. *J. Am. Chem. Soc.* **1993**, *115*, 9800.

107. Chowdhury, R. L.; Bäckvall, J. *J. Chem. Soc. Chem. Commun.* **1991**, 1063.

108. (a) Zassinovich, G.; Mestroni, G.; Gladiali, S. *Chem. Rev.* **1992**, *92*, 1051. (b) Gladiali, S.; Pinna, L.; Delogu, G.; DeMartin, S.; Zassinovich, G.; Mestroni, G. *Tetrahedron Asymmetry* **1990**, *1*, 635. (c) Müller, D.; Umbricht, G.; Weber, B.; Pfaltz, A. *Helv. Chim. Acta* **1991**, *74*, 232. (d) Krasik, P.; Alper, H. *Tetrahedron* **1994**, *50*, 4347. (e) Yang, H.; Alvarez, M.; Lugan, N.; Mathieu, R. *J. Chem. Soc. Chem. Commun.* **1995**, 1721. (f) Langer, T.; Helmchen, G. *Tetrahedron Lett.* **1996**, *37*, 1381. (g) Langer, T.; Janssen, J.; Helmchen, G. *Tetrahedron Asymmetry* **1996**, *7*, 1599. (h) Jiang, Y.; Jiang, Q.; Zhu, G.; Zhang, X. *Tetrahedron Lett.* **1997**, *38*, 215.

109. (a) Ishihara, K.; Mori, A.; Arai, I.; Yamamoto, H. *Tetrahedron Lett.* **1986**, *27*, 983. (b) Mukaiyama, T.; Suzuki, K.; Soai, K.; Sato, T. *Chem. Lett.* **1979**, 447. (c) Mukaiyama, T.; Suzuki, K. *Chem. Lett.* **1980**, 255. (d) Tombo, G. M. R.; Didier, E.; Loubinoux, B. *Synlett* **1990**, 547. (e) Niwa, S.; Soai, K. *J. Chem. Soc. Perkin Trans. 1* **1990**, 937. (f) Corey, E. J.; Cimprich, K. A. *J. Am. Chem. Soc.* **1994**, *116*, 3151.

110. (a) Matsumura, K.; Hashiguchi, S.; Ikariya, T.; Noyori, R. *J. Am. Chem. Soc.* **1997**, *119*, 8738. (b) Haack, K.; Hashiguchi, S.; Fuji, A.; Ikariya, T.; Noyori, R. *Angew. Chem. Int. Ed. Eng.* **1997**, *36*, 285.

111. Uematsu, N.; Fujii, A.; Hashiguchi, S.; Ikariya, T.; Noyori, R. *J. Am. Chem. Soc.* **1996**, *118*, 4916.

112. Sammakia, T.; Stangeland, E. L. *J. Org. Chem.* **1997**, *62*, 6104.

113. Jiang, Y.; Jiang, Q.; Zhang, X. *J. Am. Chem. Soc.* **1998**, *120*, 3817.

114. See, for example, (a) Hashiguchi, S.; Fujii, A.; Takehara, J.; Ikariya, T.; Noyori, R. *J. Am. Chem. Soc.* **1995**, *117*, 7562. (b) Jiang, Q.; Van Plew, D.; Mutuza, S.; Zhang, X. *Tetrahedron Lett.* **1996**, *37*, 797. (c) Jiang, Y.; Jiang, Q.; Zhu, G.; Zhang, X. *Tetrahedron Lett.* **1997**, *38*, 215.

115. Takehara, J.; Hashiguchi, S.; Fujii, A.; Inoue, S.; Ikariya, T.; Noyori, R. *J. Chem. Soc. Chem. Commun.* **1996**, 233.

116. Alonso, D. A.; Guijarro, D.; Pinho, P.; Temme, O.; Andersson, P. G. *J. Org. Chem.* **1998**, *63*, 2749.

117. (a) Roucoux, A.; Devocelle, M.; Carpentier, J.; Agbossou, F.; Mortreux, A. *Synlett* **1995**, 358. (b) Zhang, X.; Taketomi, T.; Yoshizumi, T.; Kumobayashi, H.; Akutagawa, S.; Mashima, K.; Takaya, H. *J. Am. Chem. Soc.* **1993**, *115*, 3318. (c) Mashima, K.; Akutagawa, T.; Zhang, X.; Takaya, H.; Taketomi, T.; Kumobayashi, H.; Akutagawa, S. *J. Organomet. Chem.* **1992**, *428*, 213. (d) Zhang, X.; Kumobayashi, H.; Takaya, H. *Tetrahedron Asymmetry* **1994**, *5*, 1179. (e) Everaere, K.; Carpentier, J.; Mortreux, A.; Bulliard, M. *Tetrahedron Asymmetry* **1998**, *9*, 2971. (f) de Bellefon, C.; Tanchoux, N. *Tetrahedron Asymmetry* **1998**, *9*, 3677. (g) Bernard, M.; Guiral, V.; Delbecq, F.; Fache, F.; Sautet, P.; Lemaire, M. *J. Am. Chem. Soc.* **1998**, *120*, 1441. (h) Murata, K.; Ikariya, T.; Noyori, R. *J. Org. Chem.* **1999**, *64*, 2186.

118. Evans, D.; Osborn, J. A.; Wilkinson, G. *J. Chem. Soc. A* **1968**, 3133.

119. (a) Casey, C. P.; Whiteker, G. T.; Melville, M. G.; Petrovich, L. M.; Gavney, J. A.; Powell, D. R. *J. Am. Chem. Soc.* **1992**, *114*, 5535. (b) Casey, C. P.; Petrovich, L. M. *J. Am. Chem. Soc.* **1995**, *117*, 6007.

120. (a) Agbossou, F.; Carpentier, J.; Mortreux, A. *Chem. Rev.* **1995**, *95*, 2485. (b) Gladiali, S.; Bayón, J. C.; Claver, C. *Tetrahedron Asymmetry* **1995**, *6*, 1453.

121. Stille, J. K.; Su, H.; Brechot, P.; Parrinello, G.; Hegedus, L. S. *Organometallics* **1991**, *10*, 1183.

122. Discussions of Rh(I)–chiral phosphine complexes: (a) Hobbs, C. F.; Knowles, W. S. *J. Org. Chem.* **1981**, *46*, 4422. (b) Gladiali, S.; Pinna, L. *Tetrahedron Asymmetry* **1990**, *1*, 693. (c) Gladiali, S.; Pinna, L. *Tetrahedron Asymmetry* **1991**, *2*, 623. (d) Becker, Y.; Eisenstadt, A.; Stille, J. K. *J. Org. Chem.* **1980**, *45*, 2145. (e) Delogu, G.; Faedda, G.; Gladiali, S. *J. Organomet. Chem.* **1984**, *268*, 167. (f) Lee, C. W.; Alper, H. *J. Org. Chem.* **1995**, *60*, 499. (g) Stanley, G. In Scaros, M. G., Prunier, M. L., eds. *Catalysis of Organic Reactions*, Marcel Dekker, Inc., New York, **1995**, p 363.

123. Discussions of Rh(I)–achiral phosphite complexes: (a) van Leeuwen, P. W. N. M.; Roobeek, C. F. (Shell). Eur. Pat. Appl. 54986, **1982**. (b) van Leeuwen, P. W. N. M.; Roobeek, C. F. *J. Organomet. Chem.* **1983**, *258*, 343. (c) Billig, E.; Abatjoglou, A. G.; Bryant, D. R. (Union Carbide). U. S. Patent 4769498, **1988**. (d) Cuny, G. D.; Buchwald, S. L. *J. Am. Chem. Soc.* **1993**, *115*, 2066. (e) van Rooy, A.; Orij, E. N.; Kamer, P. C. J.; van Leeuwen, P. W. N. M. *Organometallics* **1995**, *14*, 34.

124. Discussions of Rh(I)–chiral phosphite complexes: (a) Wink, D. J.; Kwok, T. J.; Yee, A. *Inorg. Chem.* **1990**, *29*, 5006. (b) Kwok, J.; Wink, D. J. *Organometallics* **1993**, *12*, 1954. (c) Sakai, N.; Nozaki, K.; Mashima, K.; Takaya, H. *Tetrahedron Asymmetry* **1992**, *3*, 583. (d) Buisman, G. J. H.; Kamer, P. C. J.; van Leeuwen, P. W. N. M. *Tetrahedron Asymmetry* **1993**, *4*, 1625. (e) Babin, J. E.; Whiteker, G. T. WO 93/03839, U. S. Patent 911518, **1992**; *Chem. Abstr.* **1993**, *119*, 159872h. (f) Buisman, G. J. H.; Martin, M. E.; Vos, E. J.; Klootwijk, A.; Kamer, P. C. J.; van Leeuwen, P. W. N. M. *Tetrahedron Asymmetry* **1995**, *6*, 719. (g) Buisman, G. J. H.; Vos, E. J.; Kamer, P. C. J.; van Leeuwen, P. W. N. M. *J. Chem. Soc. Dalton Trans.* **1995**, 409.

125. (a) Moser, W. R.; Papile, C. J.; Brannon, D. A.; Duwell, R. A.; Weininger, S. J. *J. Mol. Catal.* **1987**, *41*, 271. (b) Unruh, J. D.; Christenson, J. R. *J. Mol. Catal.* **1982**, *14*, 19.

126. Rossi, A. R.; Hoffmann, R. *Inorg. Chem.* **1975**, *14*, 365.

127. Casey, C. P.; Whiteker, G. T. *Isr. J. Chem.* **1990**, *30*, 299.

128. van der Veen, L. A.; Boele, M. D. K.; Bregman, F. R.; Kamer, P. C. J.; van Leeuwen, P. W. N. M.; Goubitz, K.; Fraanje, J.; Schenk, H.; Bo, C. *J. Am. Chem. Soc.* **1998**, *120*, 11616.

129. Nozaki, K.; Sakai, N.; Nanno, T.; Higashijima, T.; Mano, S.; Horiuchi, T.; Takaya, H. *J. Am. Chem. Soc.* **1997**, *119*, 4413.

CHAPTER 7

Applications of Asymmetric Reactions in the Synthesis of Natural Products

Chapter 2 to 6 have introduced a variety of reactions such as asymmetric C–C bond formations (Chapters 2, 3, and 5), asymmetric oxidation reactions (Chapter 4), and asymmetric reduction reactions (Chapter 6). Such asymmetric reactions have been applied in several industrial processes, such as the asymmetric synthesis of L-DOPA, a drug for the treatment of Parkinson's disease, via Rh(DIPAMP)-catalyzed hydrogenation (Monsanto); the asymmetric synthesis of the cyclopropane component of cilastatin using a copper complex–catalyzed asymmetric cyclopropanation reaction (Sumitomo); and the industrial synthesis of menthol and citronellal through asymmetric isomerization of enamines and asymmetric hydrogenation reactions (Takasago). Now, the side chain of taxol can also be synthesized by several asymmetric approaches.

This chapter presents some examples of the asymmetric synthesis of complicated natural products. These examples will demonstrate that building up these molecules is unlikely if we do not use the asymmetric synthesis methodology. Excellent accounts by Masamune et al.[1] and Noyori[2] give a clear picture of the strategies for stereochemical control in organic synthesis.

Finally, this chapter discusses the synthesis of two taxol components: baccatin III, the polycyclic part of taxol; and the side chain of taxol. Although several groups have completed the total synthesis of taxol, work on taxol synthesis is still far from over. Chemists working in the area of asymmetric synthesis will find challenge and opportunity in Section 7.5 of this chapter.

7.1 THE SYNTHESIS OF ERYTHRONOLIDE A

Facing the challenge of synthesizing the antibiotic erythromycin A **1a**, Woodward's group took advantage of a cyclic system to achieve diastereofacial selectivity, the so-called cyclic approach.[3] This approach was taken to deal with the common problem of low diastereoselectivity associated with acyclic substances.

The aglycone part of erythromycin A (**1b**) can be considered as pseudosymmetric (i.e., the stereochemistry of fragment C-4 to C-6 and that of C-10 to C-12 can be regarded as the same). Retro synthetic analysis suggests that the

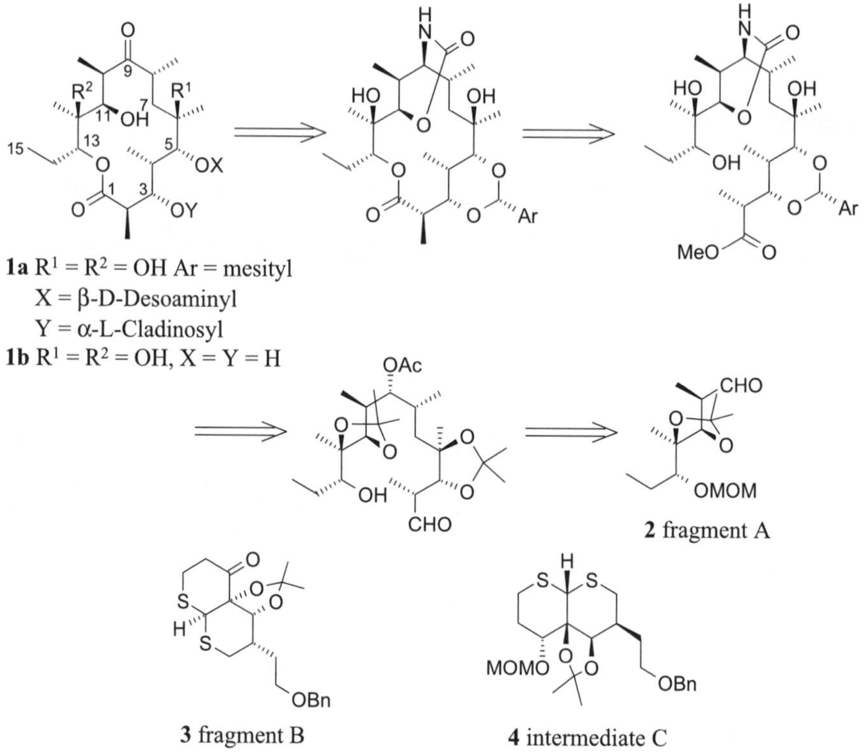

1a R^1 = R^2 = OH Ar = mesityl
X = β-D-Desoaminyl
Y = α-L-Cladinosyl
1b R^1 = R^2 = OH, X = Y = H

2 fragment A

3 fragment B   **4** intermediate C

Scheme 7–1. Retro synthetic analysis of the aglycone of erythromycin A (**1a**).

common intermediate **4** could be the ideal starting material for the construction of the two segments A (**2**) and B (**3**). With this approach, most of the stereochemistry in **1b** will then be built on a rigid dithiadecalin ring system, and the sulfur atoms are removed at the final stage (Scheme 7–1).

Woodward's achievement in constructing the 10 chiral centers of **1b** relied on the cyclic "substrate control" approach. Erythromycin A (**1a**) was finally synthesized by combining compound **1b** with a long chain residue. Although this achievement represented a historic milestone at that time, it also attested to the limitations of this popular traditional approach.

Synthesis of the common intermediate C (**4**), and its further conversion to **2** and **3** is illustrated in Scheme 7–3. Two racemic compounds, (±)-**7** and (±)-**10**, are prepared from readily available starting materials **5** and **8**, respectively (Scheme 7–2). Coupling of **7** and **10** gives a mixture of diastereomers **11**. An intramolecular aldol reaction of **11** catalyzed by D-proline yields diastereomers **12** and **13** in equal molar ratios (about 36% ee for each diastereomer). Compound **12**, the desired ketone, is converted to **14**, which is further purified by crystallization to give the compound in the desired stereochemistry in sterically pure form. Reduction of the ketone carbonyl group and subsequent methoxy

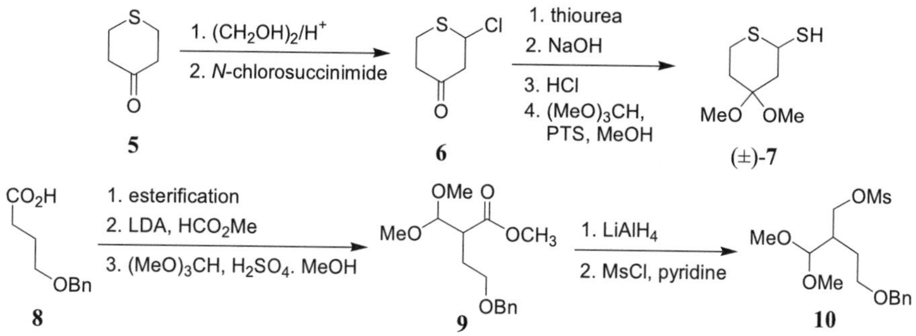

Scheme 7–2

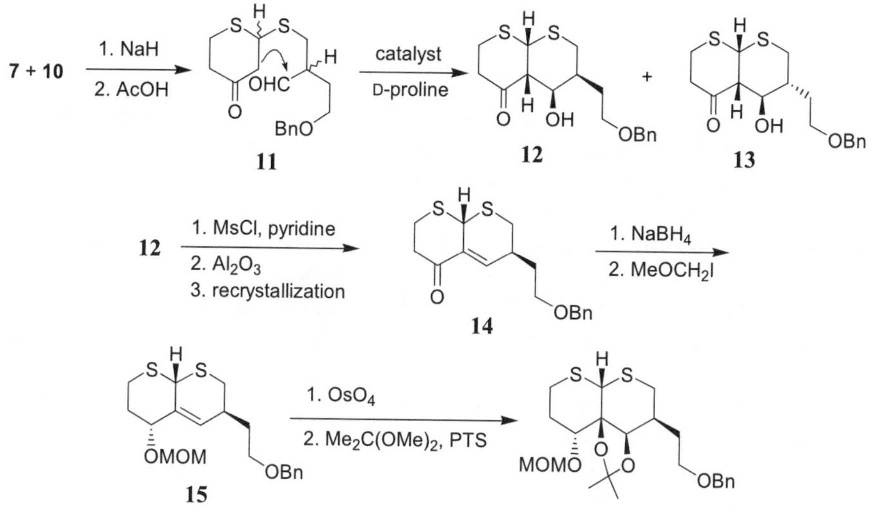

Scheme 7–3. Synthesis of intermediate C.

methyl (MOM) protection furnishes compound **15**. Osmium tetroxide oxidation and acetonide protection complete the synthesis of compound **4** (intermediate C) and thus accomplish the first round of the synthesis (Scheme 7–3).

All the new chiral centers are controlled by the rigid structure of the dithiadecalin system. Taking the NaBH$_4$ reduction of **14** as an example, the orientation of the fused rings forces the H$^-$ to attack from the *Si*-prochiral face of the carbonyl group (Fig. 7–1).

Thus, acid cleavage of the protective groups MOM and acetonide in **4** provides compound **3** (fragment B); desulfuration and oxidation of the hydroxyl group give compound **2** (fragment A) (Scheme 7–4).

The next step to erythronolide A is the coupling of fragments A and B. Asymmetric aldol reaction of aldehyde **2** with a lithium enolate generated from

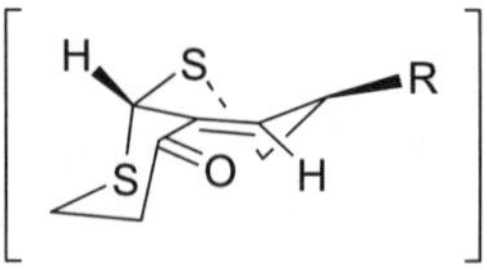

Figure 7–1. Transition state of compound **14** in NaBH$_4$ reduction.

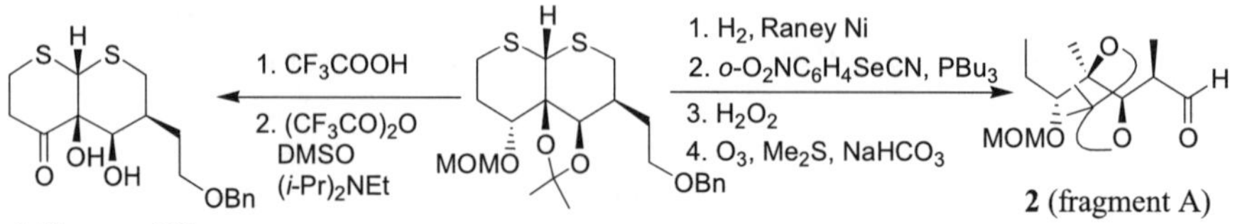

3 (fragment B) **4** (intermediate C) **2** (fragment A)

Scheme 7–4

3 followed by oxidation of the resulting hydroxy group gives the expected compound **16**. This compound can be further converted to compound **17** via C-9 enolization. The following several steps are the chemistry at C-8 and C-7. Compound **17** is subjected to NaBH$_4$ reduction and mesylation, affording compound **18**, in which the C–7 carbonyl group is converted to CH-OMs. Elimination of the mesyl group converts compound **18** to α,β-unsaturated compound **19**, and subsequent phenyl methanethiolate addition provides compound **20**. From compound **20** to aldehyde **22** is well established procedure (Scheme 7–5).

Having built up the desired stereochemistry at C-5 to C-13 as shown in compound **22**, the next step is the connection of the C-1 to C-3 fragment. An aldol reaction of **22** with lithium enolate provides **23** with the desired C-2 stereochemistry.

Finally, macrocyclization is carried out via protection and deprotection sequences and cyclization of the activated ester **26**, and this accomplishes the synthesis of **27** (Scheme 7–6), which, through deprotection and glycosidation, can be converted to erythromycin A, a compound containing a "hopelessly complex" array of chiral centers.[4]

7.2 THE SYNTHESIS OF 6-DEOXYERYTHRONOLIDE

Considering the entire synthesis illustrated in the previous section, clearly the construction of such a complicated molecule with all the desired stereogenic centers is highly tedious and demanding work. Therefore, an entirely different conceptual method based on double asymmetric induction was finally developed as a less complex synthetic strategy. A good example is the synthesis of 6-deoxyerythronolide B (**28**), which bears the same 10 chiral centers as erythronolide A (compound **1a** of the previous section).

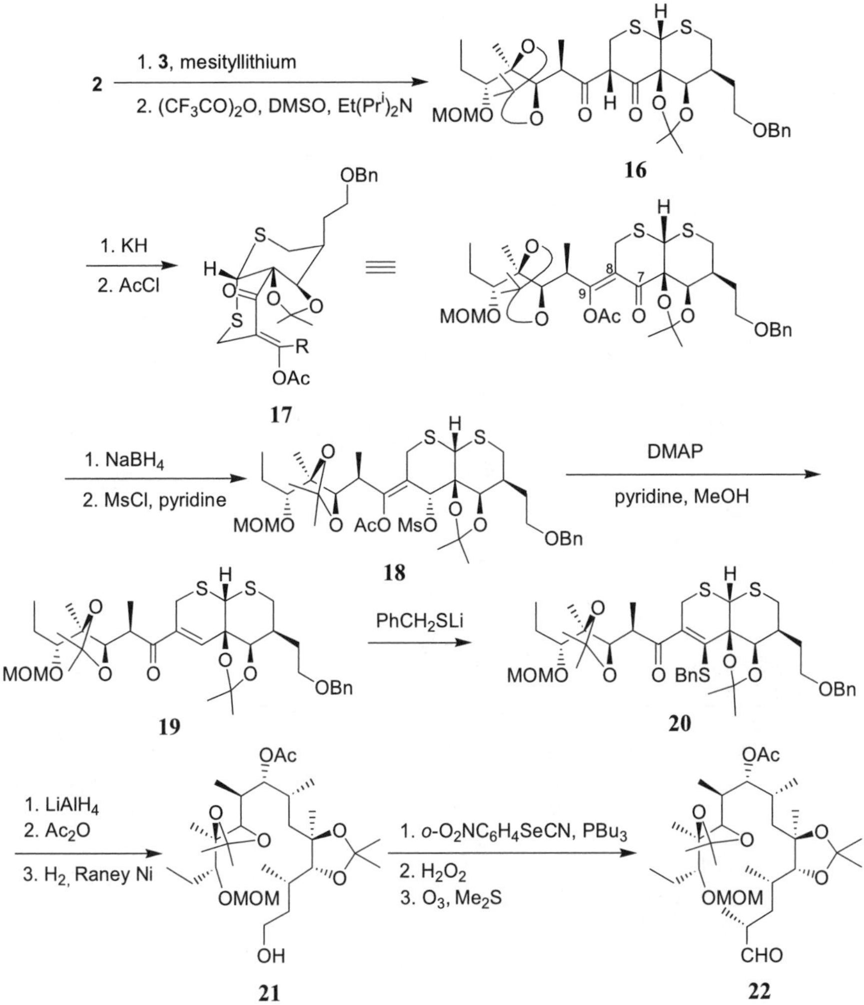

Scheme 7–5

6-Deoxyerythronolide B (**28**), produced by blocked mutants of *Streptomyces erythreus*, is a common biosynthetic precursor leading to erythromycins. A different route to this compound was developed with aldol methodology.[5] In this approach, all the crucial C–C bond formations involved in the construction of the carbon framework are exclusively aldol reactions.

Synthesis of the macrolide 6-deoxyerythronolide B **28** is one of the successful demonstrations of double asymmetric induction applied to the construction of complicated natural products.[5] Retro synthetic analysis (Scheme 7–7) shows that **28** can be obtained from thio-seco acid **29**, which consists of seven propionate building blocks. This is a typical aldol product in which a boron reagent

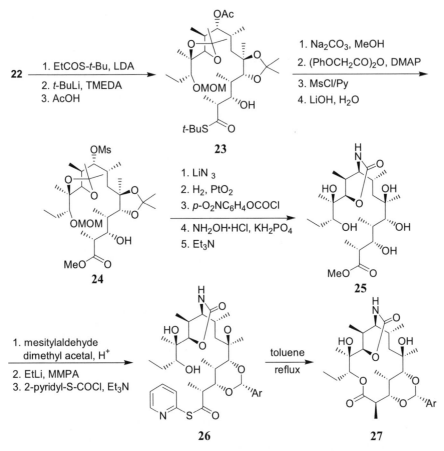

Scheme 7–6. Woodward's synthesis of erythromycin A.

can be used to fully control the stereochemistry. Splitting **29** results in fragments A (**30**) and B (**31**), which can be prepared from aldol reactions I (**30**) and II followed by III (**31**), respectively. Finally, the fragments are assembled via aldol reaction IV to give thio-seco acid **29**.

Using an appropriate enolate reagent for each reaction, the two major fragments A (**30**) and B (**31**) can be constructed through one or two steps of aldol reactions (Scheme 7–8). It should be mentioned that compound **35**, which is thoroughly discussed in Chapter 3, plays an important role in this synthesis.

Condensation of the two fragments A and B via aldol reaction, followed by macro-lactonization, completes the synthesis as shown in Scheme 7–9.

Note that in aldol reaction IV (from **31** to **42** in Scheme 7–9), the methodology differs from that used in I, II, and III (see Scheme 7–7). Aldol reaction IV is also a double asymmetric reaction involving the coupling of two structurally

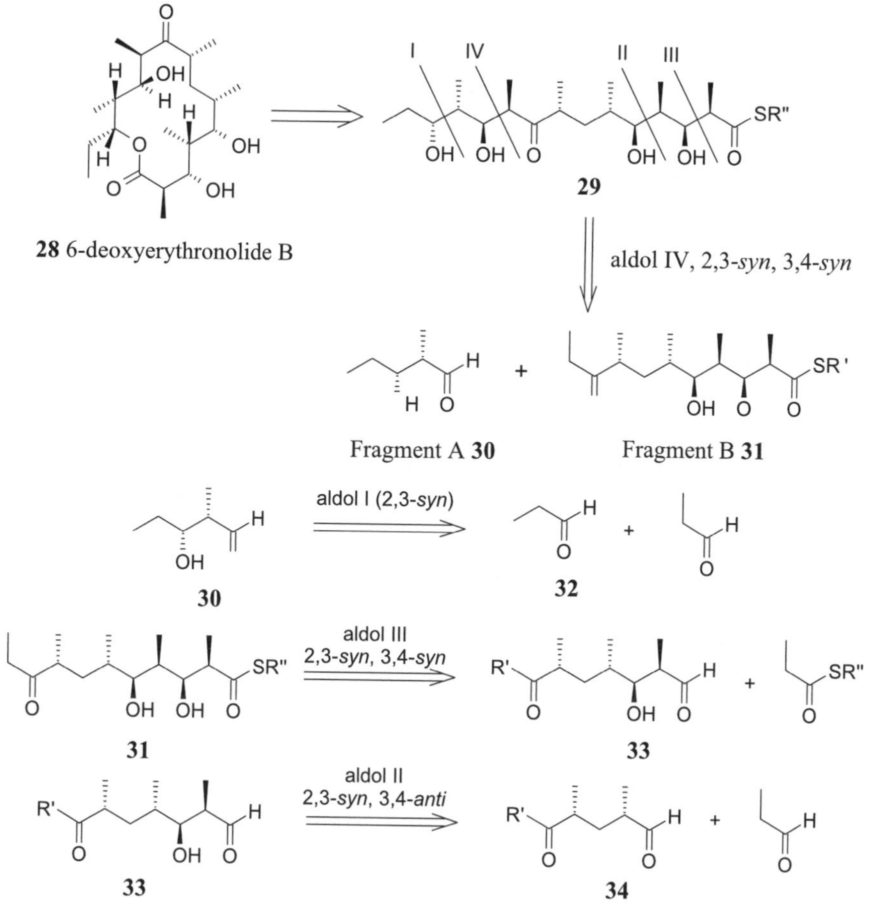

Scheme 7–7. Retro synthetic analysis of 6-deoxyerythronolide B.

prefixed components. The coordination of Li$^+$ with the β-ethereal oxygen atom in aldehyde **31** is mainly responsible for the 17:1 stereoselectivity.

Thus, eight chiral centers have been created in molecule **28** with remarkable efficiency and stereoselectivity by following the above steps.

7.3 THE SYNTHESIS OF RIFAMYCIN S

Among the syntheses of complicated natural products, the total synthesis of rifamycin S (**44**) is another example that shows how a complicated structure can be constructed by applying the concept of double asymmetric synthesis (see Section 1.5.3 for double asymmetric synthesis). Rifamycin S is one of the ansamycin antibiotics, characterized by a distinct structural feature: a macro-

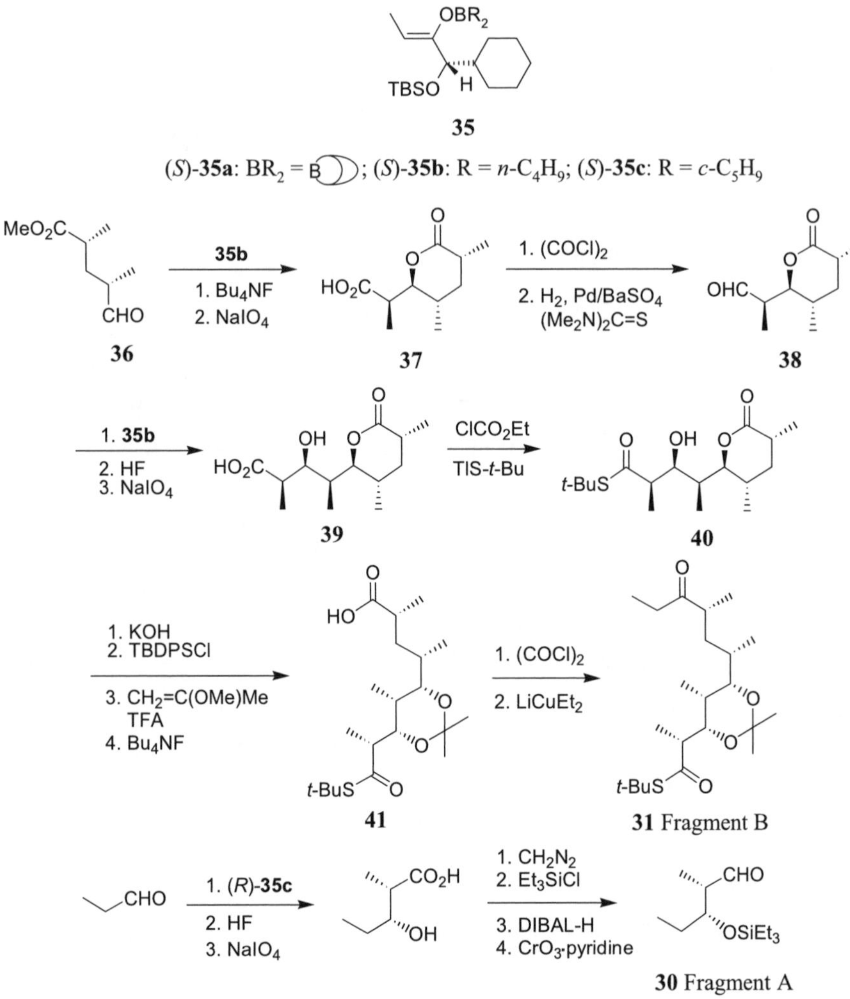

(S)-**35a**: BR$_2$ = ⬡; (S)-**35b**: R = n-C$_4$H$_9$; (S)-**35c**: R = c-C$_5$H$_9$

Scheme 7–8. Synthesis of fragment A and fragment B.

lactam with a long aliphatic ansachain (**45**) joined to an aromatic nucleus **46** at two nonadjacent positions. The intermediate **45** and its precursor **49** constitute another example of forming a subunit of the final product via typical aldol reaction (Scheme 7–10).

7.3.1 Kishi's Synthesis in 1980

The total synthesis of rifamycin S was one of Kishi's many achievements in organic synthesis.[6] Kishi recognized that a certain type of (Z)-olefin such as **51** tends to take a conformation in which C-1, C-2, C-3, and H-3 are nearly co-

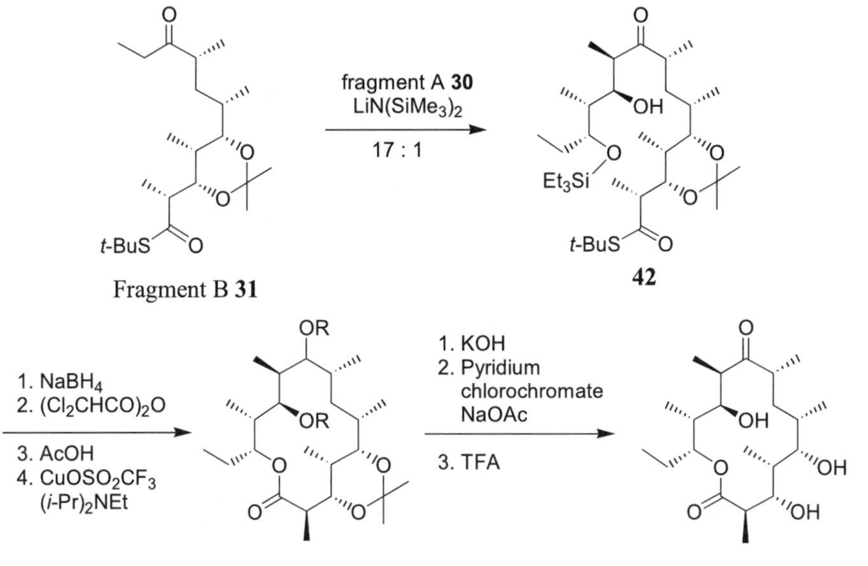

Scheme 7–9. Synthesis of a derivative of the seco acid and ring closure to 6-deoxyerythronolide B.

planar due to alkylic strains. This conformation is likely to be retained in the transition states of many reactions, and in the process the two faces of the plane are differentiated. Thus, hydroboration in Eq. 1 and epoxidation in Eq. 2 (Scheme 7–11) can be carried out with moderate diastereoselectivity.

When Kishi tried to apply these insights to rifamycin synthesis (Scheme 7–12), he at first met the difficulty that direct epoxidation of the allylic alcohol **53**, which was generated from **52** through a Wittig reaction, failed to introduce an α-epoxy group *anti* to the preexisting chiral methyl group in **52** due to interference of the OBn coordination with the oxidant. Thus, an indirect path had to be taken to prepare the rifamycin ansa chain. For his next attempts, as Scheme 7–12 indicates, a trimethylsilyl substituted allylic alcohol **55** was employed in the epoxidation reaction with *m*-CPBA, giving **56** with the introduced epoxy moiety at the α-position. Treatment of **56** with F⁻ to remove the silyl group was followed by epoxy ring opening with copper reagent, and this generated the aldol product **57**.

In the application of the above discovery, (±)-3-benzyloxy-2-methyl propionaldehyde **52** is used as the starting material in the synthesis of rifamycin A. As outlined in Scheme 7–12, compound **52** is converted to allyic alcohol **55** via a series of chemical reactions. Epoxidation of **55** proceeded stereoselectively, giving a single epoxide that affords **57** after subsequent treatment. Compound **57** may be converted to **58** upon acetonide formation.

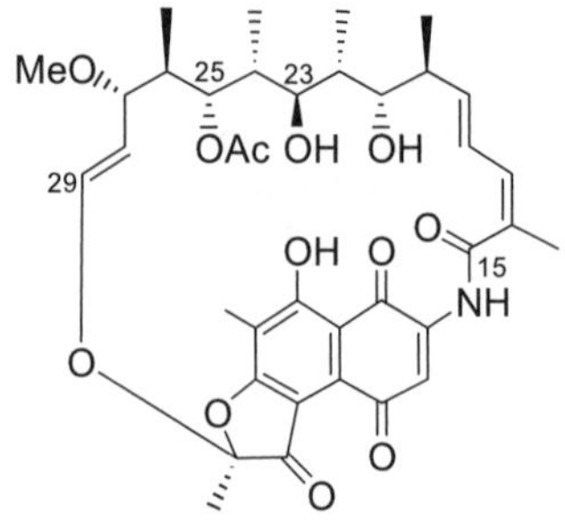

44 rifamycin S

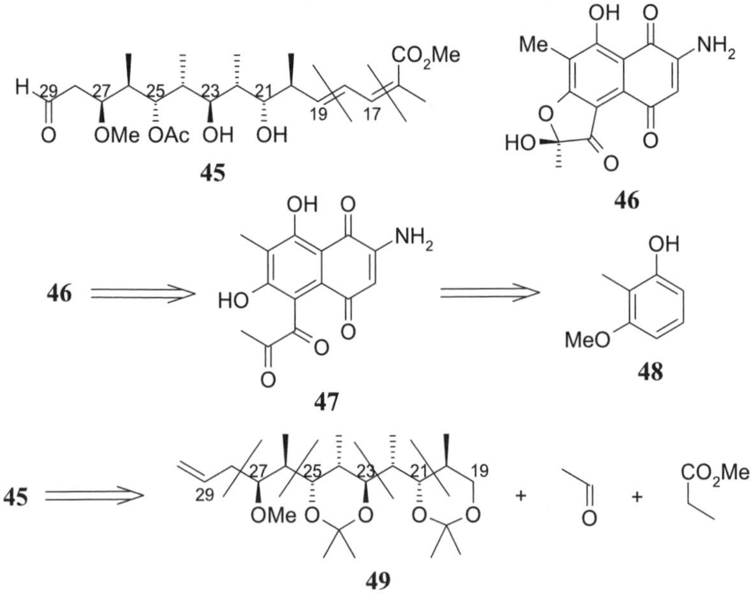

Scheme 7–10. Retro synthetic analysis of rifamycin S **44**.

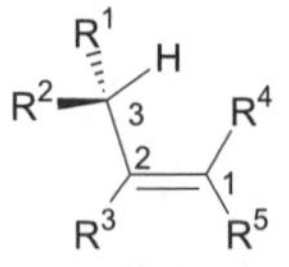

51

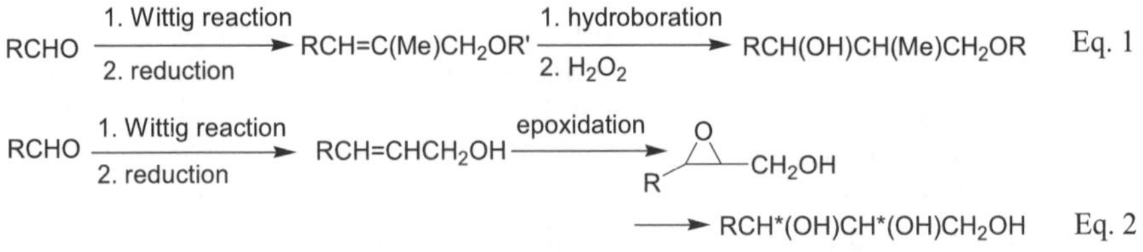

$$\text{RCHO} \xrightarrow[\text{2. reduction}]{\text{1. Wittig reaction}} \text{RCH=C(Me)CH}_2\text{OR'} \xrightarrow[\text{2. H}_2\text{O}_2]{\text{1. hydroboration}} \text{RCH(OH)CH(Me)CH}_2\text{OR} \qquad \text{Eq. 1}$$

$$\text{RCHO} \xrightarrow[\text{2. reduction}]{\text{1. Wittig reaction}} \text{RCH=CHCH}_2\text{OH} \xrightarrow{\text{epoxidation}} \underset{R}{\triangle}\text{—CH}_2\text{OH}$$

$$\longrightarrow \text{RCH*(OH)CH*(OH)CH}_2\text{OH} \qquad \text{Eq. 2}$$

Scheme 7–11

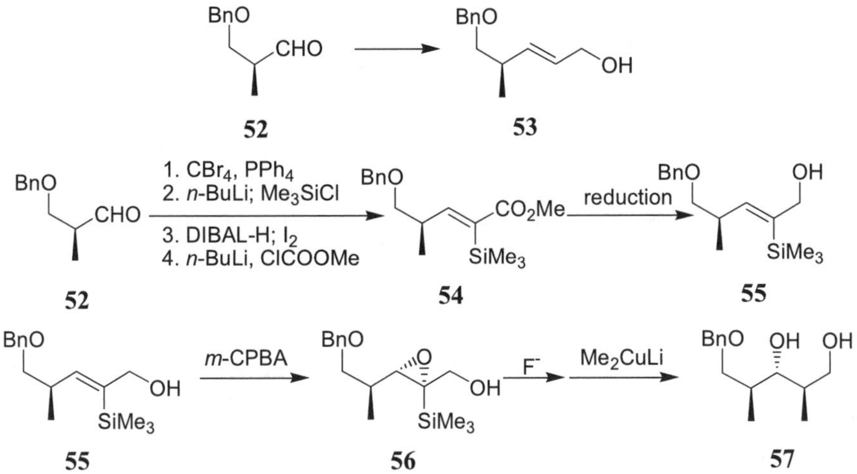

Scheme 7–12

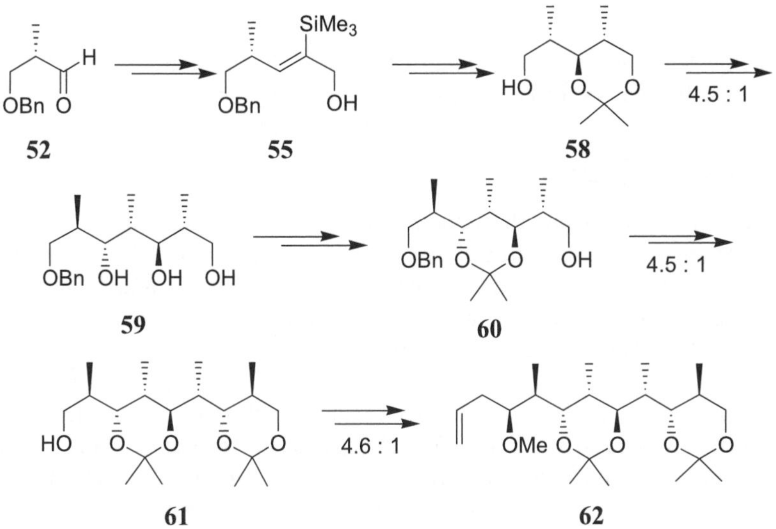

Scheme 7–13

The same strategy can be applied to complete the conversion from **58** to **61**, as described in Scheme 7–13. Nearly the same sequence is performed for the conversion from **58** to **59**, from **59** to **60**, and from **60** to **61**. Diethylzinc addition and subsequent methylation of the hydroxyl group provides **62** in a ratio of 4.6:1. In this manner, the synthesis of this key intermediate containing all eight chiral centers present in the ansa chain has been completed.

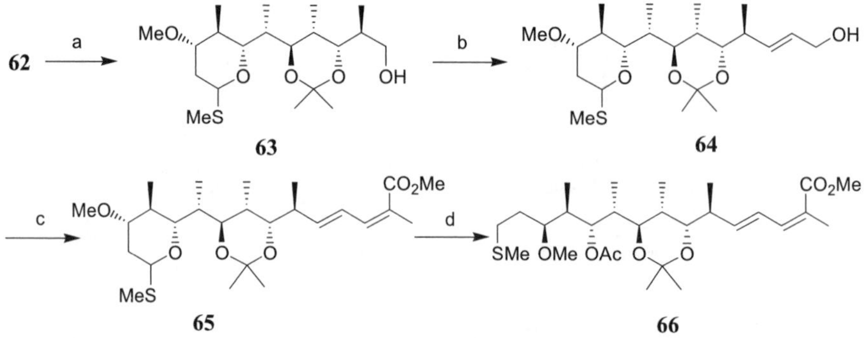

Scheme 7–14. Reagents and conditions: a (1) HCl; (2) *t*-BuCOCl, pyridine; (3) OsO$_4$, KIO$_4$; (4) MeSH, BF$_3$ · Et$_2$O; (5) Me$_2$C(OMe)$_2$, CSA; (6) LiAlH$_4$ (56% yield). b: (1) PDC; (2) Ph$_3$P=CHCO$_2$Et; (3) DIBAL–H. c: (1) PDC; (2) (MeO)$_2$P(O)CH(Me)CN, *t*-BuOK; (3) DIBAL–H; (4) NaCN, MnO$_2$, MeOH [45% yield from ($\pm$)-**63**]. d: (1) HgCl$_2$, CaCO$_3$; (2) NaBH$_4$; (3) *t*-BuPh$_2$SiCl; (4) Ac$_2$O, pyridine; (5) Bu$_4$NF; (6) methanesulfonyl chloride, Et$_3$N; (7) MeSNa (69% yield).

Further transformation of **62** to the ansa chain ($\pm$)-**66** is briefly summarized in Scheme 7–14 without a detailed discussion. Compound ($\pm$)-**66** has now been properly functionalized for coupling with the aromatic unit of rifamycin.

7.3.2 Kishi's Synthesis in 1981

One year after reporting the experiments outlined in Section 7.3.1, Kishi's group carried out an enantioselective version of the synthesis based on the experience obtained in the above synthesis of the racemate ($\pm$)-**66**.[7] This version adopted the concept of double asymmetric induction demonstrated in the synthesis of 6-deoxythronolide B (see Section 7.2).

Scheme 7–15 shows the significant improvement in overall stereoselectivity, due mainly to the adoption of the newly developed Sharpless asymmetric epoxidation. Compounds α-epoxy-**67** and β-epoxy-**67** can be readily obtained from **53** via the Sharpless reaction. Isomers of compounds **57** are then constructed via regioselective ring opening with a copper reagent.

Scheme 7–16 shows that a similar synthetic route leads to the asymmetric synthesis of optically active **62**. The synthesis that began from homochiral aldehyde (–)-**52** used this newly discovered asymmetric epoxidation three times, **52** → **58**, **58** → **68**, and **68** → **61**, finishing the conversion from **52** to **61** by following a shortened route. The last chiral center to be built is C-27, and the addition of allyltin to the aldehyde derived from **61** proceeds with high stereoselectivity to give the chiral aliphatic segment **62**.

The high stereoselectivity in the formation of **62** described in Scheme 7–16 can be explained by the *trans*-decalin type transition state **69**, which ensures the desired configuration at C-27.

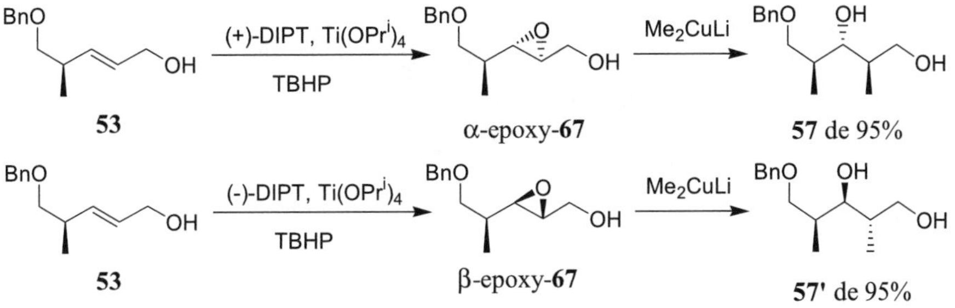

Scheme 7–15

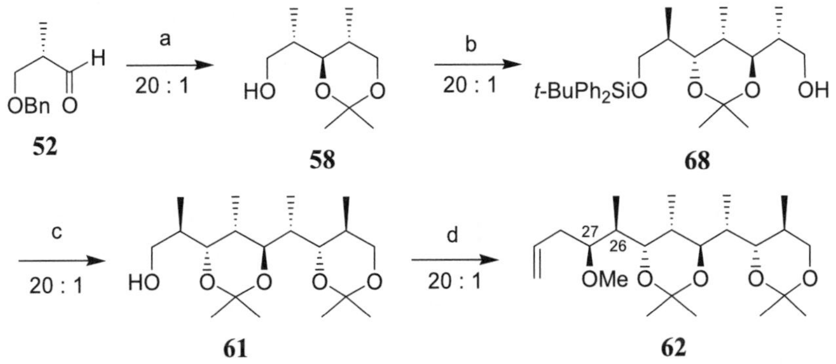

Scheme 7–16. Reagents and conditions: a: (1) $(i\text{-PrO})_2P(O)CH_2CO_2Et$, $t\text{-BuOK}$; (2) DIBAL–H; (3) (+)-DET, $Ti(OPr^i)_4$, TBHP; (4) $LiCuMe_2$; (5) Me_2CO, CSA; (6) Li, $NH_3(1)$ (57% yield). b: (1) DMSO, $(COCl)_2$, Et_3N, steps 2–5 as steps 1–4 in a; (6) $t\text{-BuPh}_2SiCl$; (7) AcOH; (8) $t\text{-BuCOCl}$, pyridine; (9) $Me_2C(OMe)_2$, CSA; (10) $LiAlH_4$ (63% yield). c: Steps 1–5 as in b but with (–)-DET; (6) $Me_2C(OMe)_2$, CSA; (7) Bu_4NF (68% yield). d: (1) DMSO, $(COCl)_2$, Et_3N; (2) $CH_2=CHCH_2I$, $SnCl_2$; (3) MeI, KH (65% yield).

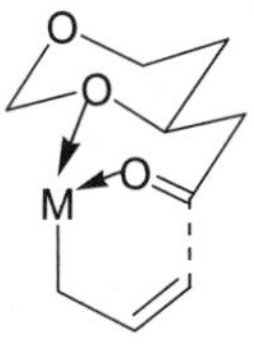

69

7.3.3 Masamune's Synthesis

The aldol reaction that establishes two chiral centers in one step has been applied to the synthesis of the ansa chain **66** by Kishi's group as discussed above. Seven chiral centers out of the eight present in the corresponding **66a** can be constructed in a different way through a convergent series of four

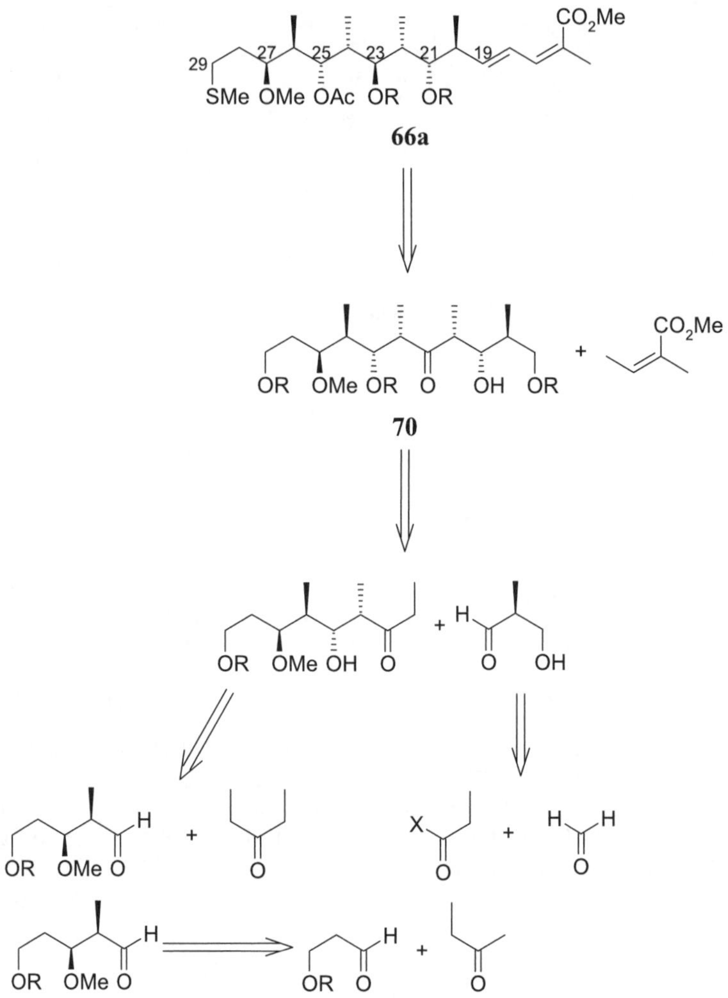

Scheme 7–17. Retro synthetic analysis of **66a**.

asymmetric cross aldol reactions. The intermediates can be prepared from a chiral boron reagent. The remaining chiral center C-23 can then be created by stereoselective reduction of the corresponding ketone **70**. Finally, the introduction of a (Z,E)-dienolate moiety (C-15 to C-19) gives product **66a**. The retro synthetic analysis is shown in Scheme 7–17.

Scheme 7–18 shows Masamune's implementation of this approach, beginning with aldehyde **71**.[8] This is reacted with boron enolate reagent (S)-**35c** (mentioned in Chapter 3; see Scheme 7–8 for its structure) and provides aldol product **72** with excellent enantioselectivity (100:1). Aldehyde **73** is obtained

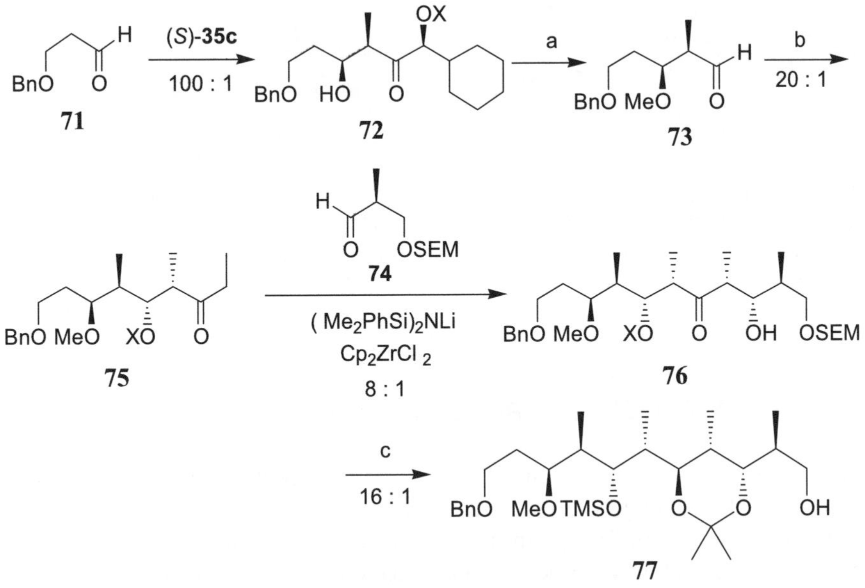

Scheme 7–18. Synthesis of precursor **77**. Reagents and conditions: a: (1) HF, CH$_3$CN; (2) NaIO$_4$; (3) CH$_2$N$_2$, HBF$_4$; (4) LiAlH$_4$; (5) PCC (75% yield). b: (1) Et$_2$CO, (Me$_2$PhSi)$_2$NLi; (2) TBSOTf (90% yield). c: (1) DIBAL–H; (2) Me$_2$C(OMe)$_2$, H$_2$SO$_4$; (3) n-Bu$_4$NF; (4) Me$_3$SiCl, aqueous workup (85% yield). X = TBS; SEM = CH$_2$OCH$_2$CH$_2$SiMe$_3$.

from **72** through cleavage of the chiral auxiliary, oxidation, and subsequent protection of the hydroxyl group. Compound **75** is then produced via an asymmetric aldol reaction between aldehyde **73** and the (Z)-enolate of 3-pentanone and subsequent silylation. Coupling of the (Z)-enolate of **75** with **74** through a zirconium-mediated aldol reaction furnishes ketone **76**, which can efficiently be converted to **77** via reduction of the carbonyl group and subsequent protection and deprotection of its hydroxyl groups (Scheme 7–18).

Based on the method used in Kishi's synthesis, a diene moiety can be connected to **77** through routine synthetic chemistry, and this finishes the C-15 to C-19 subunit of intermediate **66**. Conversion of **77** to **66** is briefly depicted in Scheme 7–19.

The interested reader is referred elsewhere for discussions of the synthesis of rifamycin.[9]

At this juncture, it is useful to look at Table 7–1, in which the syntheses of erythronolide and the ansa chain are used as examples to show that reagent-controlled syntheses are clearly more advantageous than substrate-controlled reactions in terms of three criteria: the overall yield, overall stereoselectivity, and number of steps involved in each of the syntheses. A careful examination of Table 7–1 clearly shows the advantages of this strategy.

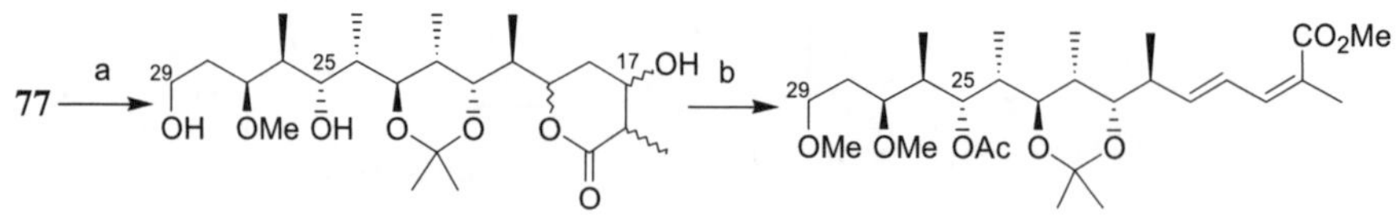

Scheme 7–19. Reagents and conditions: a: (1) $CrO_3 \cdot$ pyridine; (2) benzyl-2-methyl-acetoacetate, LDA; (3) $NaBH_4$; (4) H_2, Pd/C; (5) toluene, reflux (85% yield). b: (1) $(CF_3CO)_2O$, Et_3N; (2) hydrolysis; (3) TsCl, pyridine; (4) MeSNa; (5) CH_2N_2; (6) Ac_2O, pyridine (68% yield).

7.4 THE SYNTHESIS OF PROSTAGLANDINS

The synthesis of prostaglandins (PGs) is another good example of a preparation in which asymmetric organic reactions play an important role.

Prostaglandins are formed in human tissues, fluids, and organs and can influence the reproductive process, gastric secretion, control of blood pressure, and hypertension and respiration, as well as mediate pain and inflammation from cardiovascular problems associated with platelet aggregation. These compounds were first discovered in the 1930s but became prominent only in the 1960s. They were first found in secretions of the prostate gland, hence the name *prostaglandins*. In vivo, such compounds are synthesized from arachidonic acid, an unsaturated acid with a 20 carbon chain and four unsaturated C=C bonds or from other polyunsaturated acids.[10] Because prostaglandins are a specific type of compound showing significant bioactivity but in only limited natural supply, an efficient and flexible synthesis of PGs is essential to ensure an adequate supply of natural PGs and artificial analogs for biological, physiological, and medicinal investigations. Asymmetric synthesis has played an important role in synthesizing prostaglandin subunits.

Compound **78**, or prostanoic acid, the simplest prostaglandin compound, contains an α-side chain and also an ω-side chain. Asymmetric synthesis of prostaglandins must involve the assembly of these subunits, as well as the introduction of other functionalities.

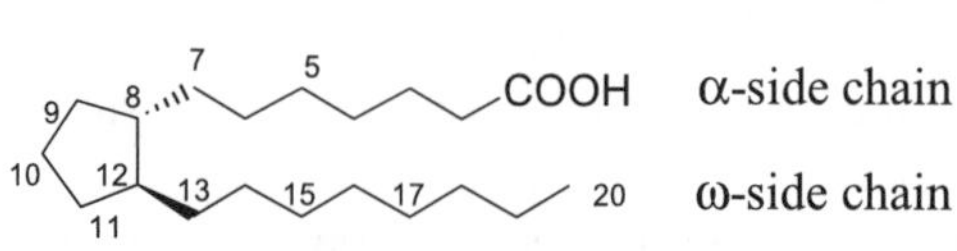

As mentioned in Chapter 2, the introduction of the α- and ω-side chains is controlled by the chirality at C-11 of **78** through 1,2-*anti*-induction. If the C-11 protected hydroxyl moiety of **78** is α-orientated (α and β are adopted from carbohydrate chemistry, indicating the relative position of substituents), the newly

TABLE 7-1. Comparison of Some Total Syntheses of Natural Products

Target Molecule	6-Deoxyerythronolide B	Erythronolide A Derivative	Ansa Chain of Rifamycin B	66	66
Approach (control)	Acyclic (reagent)	Cyclic (substrate)	Acyclic (substrate)	Acyclic (reagent)	Acyclic
Chiral centers created	8	10	7	7	8
Overall selectivity	85	46	50	75	78
Overall yield	5.7	0.3	0.2	2.1	13.8
Number of steps	23	49	49	46	24
Starting material	**36**	**5**	$(\pm)$-**52**	$(-)$-**52**	**71**

Reprinted with permission by VCH, Ref. 1.

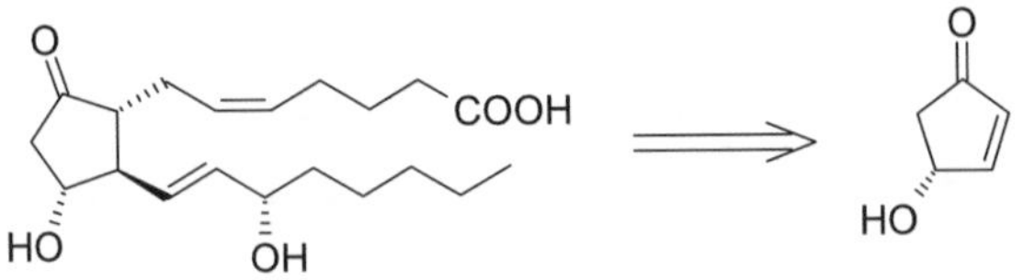

Scheme 7–20. Retro synthesis of PEG$_2$.

linked ω-side chain will be in β-induction (*anti* to the C-11 hydroxy moiety). Similarly, the α-side chain will be introduced in an α-position (*anti* to the ω-side chain).

Taking the synthesis of PEG$_2$ as an example, PGs can be assembled from an unsaturated 4-hydroxycyclopent-4-enone and some side chains (Scheme 7–20).

7.4.1 Three-Component Coupling

Among many other excellent methods, convergent three-component coupling, the consecutive linking of the α- and ω-side chains to an unsaturated 4-hydroxy-2-cyclopentenone derivative, is the most direct and effective synthesis.

To achieve the three-component coupling, O-protected 2-cyclopentanone undergoes organometallic-mediated conjugate addition of the ω-side chain unit, followed by electrophilic trapping of the enolate intermediate by an α-chain organic halide.[11] However, it is not easy to realize this simple goal because the low stereoselectivity associated with other side reactions is a major problem, and this problem remained unresolved until oranozinc chemistry was applied to this synthesis.

As shown in Scheme 7–21, an equimolar mixture of dimethylzinc and the ω-side chain vinyl lithium is treated sequentially with siloxyl-2-cyclopentenone and the propargylic iodide in the presence of HMPA. The desired product **79** is formed with 71% yield.[12] The acetylenic product is a common intermediate for the general synthesis of naturally occurring PGs.

Organocopper chemistry also provides a straightforward synthesis through a special alkylation procedure (Scheme 7–22). An organocopper reagent, generated in situ from equimolar amounts of ω-side chain vinyl lithium, CuI, and

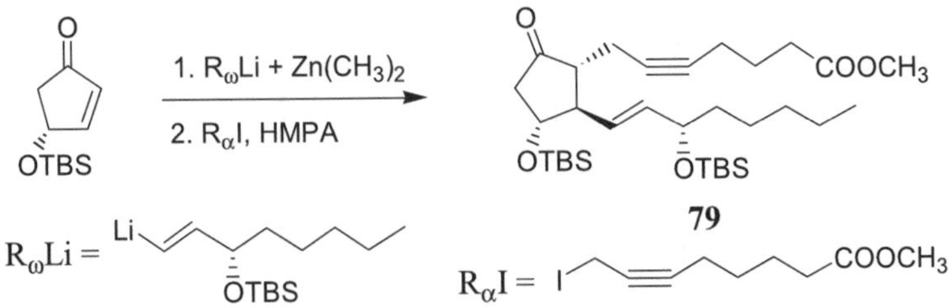

Scheme 7–21. One pot synthesis of the PG framework by an organozinc procedure.

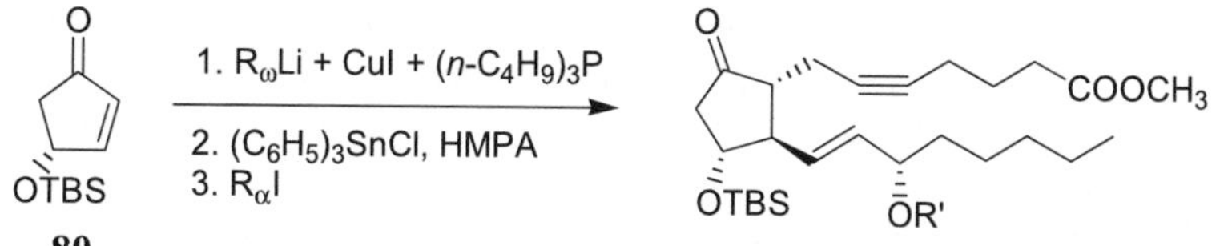

R' = TBS or THP

Scheme 7–22. One pot synthesis of PGs via an organocuprate/organotin procedure.

2.3 equivalents of $(n\text{-}C_4H_9)_3P$, undergoes conjugate addition to the chiral siloxy enone in the presence of organotin compounds. Sequential treatment of the organocopper reagent with the enone HMPA, triphenyltin chloride, and α-side chain propargylic iodide offers the desired compound **81** with more than 80% yield.[13]

As shown in Schemes 7–21 and 7–22, the desired stereochemistry at C-8 and C-12 in the PG framework can be established via three-component coupling. The remaining issue is the question of how to stereoselectively build the chiral 2-cyclopentenone and the lower ω-side chain. Sections 7.4.2 and 7.4.3 introduce some general procedures for asymmetric syntheses of these PG subunits.

7.4.2   Synthesis of the ω-Side Chain

Various catalytic or stoichiometric asymmetric syntheses and resolutions offer excellent approaches to the chiral ω-side chain. Among these methods, kinetic resolution by Sharpless epoxidation,[14] amino alcohol–catalyzed organozinc alkylation of a vinylic aldehyde,[15] lithium acetylide addition to an alkanal,[16] reduction of the corresponding prochiral ketones,[17] and BINAL–H reduction[18] are all worth mentioning.

Sharpless epoxidation reactions are thoroughly discussed in Chapter 4. This section shows how this reaction is used in the asymmetric synthesis of PG side chains. Kinetic resolution of the allylic secondary alcohol $(\pm)$-**82** allows the preparation of (R)-**82** at about 50% yield with over 99% ee (Scheme 7–23).[19]

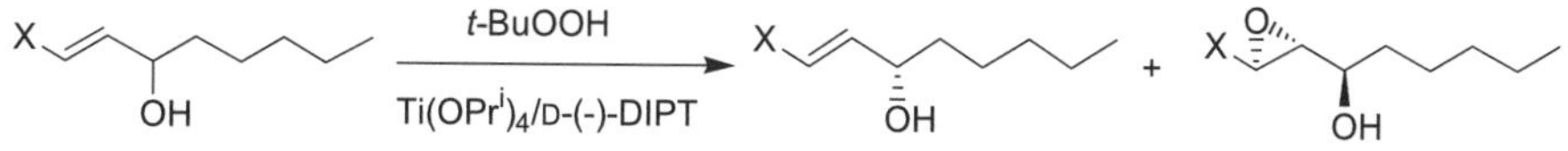

X	Yield (%)	ee (%)
$(CH_3)_3Si$	50	>99
$(n\text{-}C_4H_9)_3Sn$	38–42	>99

Scheme 7–23. Kinetic resolution of the chiral ω-side chain via Sharpless epoxidation reactions.

Alkylzinc addition to an aldehyde was introduced in Chapter 2. In this case, 2 mol% of DAIB catalyzes the addition of dialkylzinc to a tin containing α,β-unsaturated aldehyde **84**, giving the ω-side chain **85** with 84% yield and 85% ee (Scheme 7–24).

In the presence of chiral diamine **86**, addition of lithium acetylenide to an aldehyde **87** gives product **88** with 87% yield and 76% ee (Scheme 7–25).

Methods such as borane reduction and BINAL–H reduction are discussed in Chapter 6, and indeed ketone **89** can be reduced by chiral borane, providing the ω-side chain **90** at 65% yield and 97% ee (Scheme 7–26).

Another method for ketone reduction, BINAL–H asymmetric reduction, can also be used in ω-side chain synthesis. An example of applying BINAL–H asymmetric reduction in PG synthesis is illustrated in Scheme 7–27. This has been a general method for generating the alcohol with (15S)-configuration. The binaphthol chiral auxiliary can easily be recovered and reused. As shown in Scheme 7–27, when the chiral halo enone **91** is reduced by (S)-BINAL–H at −100°C, product (15S)-**92** can be obtained with high enantioselectivity.

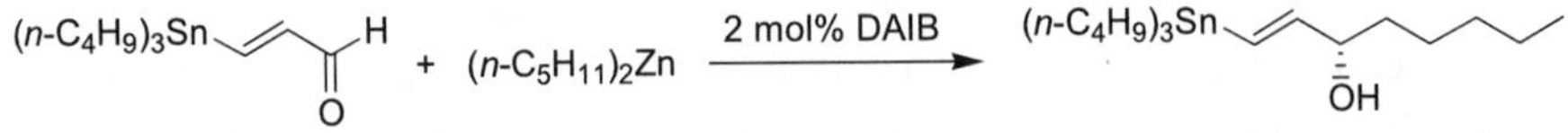

Scheme 7–24. Asymmetric synthesis of the chiral ω-side chain via dialkylzinc addition.

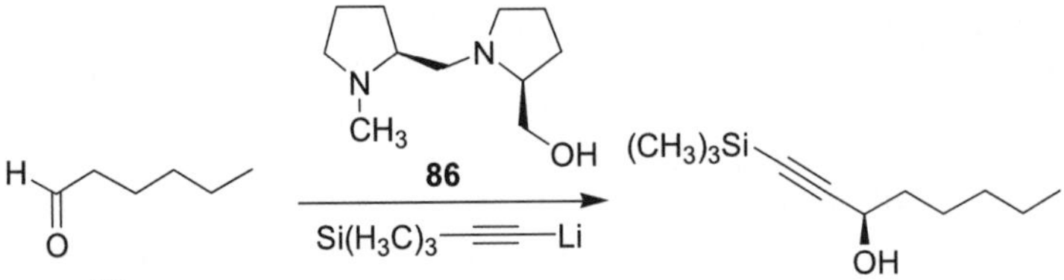

Scheme 7–25. Asymmetric synthesis of the chiral ω-side chain through chiral ligand-induced nucleophilic addition.

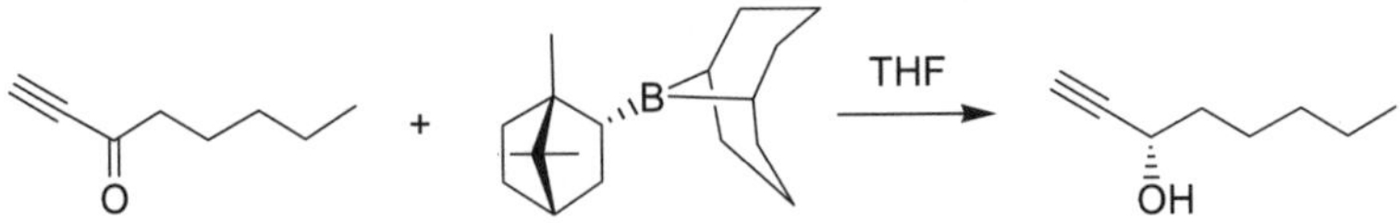

Scheme 7–26. Asymmetric synthesis of the chiral ω-side chain via chiral borane reduction.

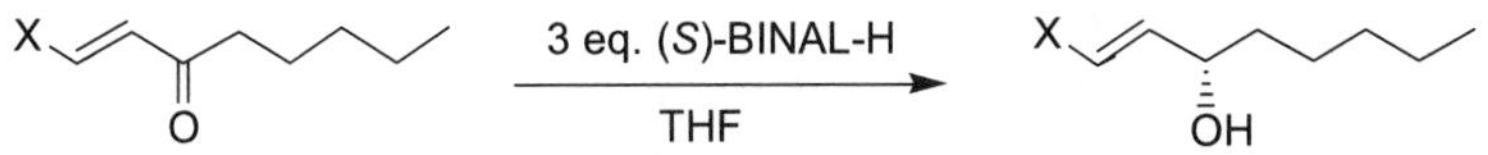

X	Yield	ee%
Br	> 95%	96%
I	> 95%	97%
$(n\text{-}C_4H_9)_3Sn$	> 90%	98%

Scheme 7–27. Asymmetric synthesis of the chiral ω-side chain via catalytic hydrogenation.

7.4.3 The Enantioselective Synthesis of (*R*)-4-Hydroxy-2-Cyclopentenone

Many methods have been reported for the enantioselective synthesis of the remaining PG building block, the (*R*)-4-hydroxy-cyclopent-2-enone. For example, the racemate can be kinetically resolved as shown in Scheme 7–28. (*S*)-BINAP–Ru(II) dicarboxylate complex **93** is an excellent catalyst for the enantioselective kinetic resolution of the racemic hydroxy enone (an allylic alcohol). By controlling the reaction conditions, the C=C double bond in one enantiomer, the (*S*)-isomer, will be prone to hydrogenation, leaving the slow reacting enantiomer intact and thus accomplishing the kinetic resolution.[20]

Asymmetric ring opening of 3,4-epoxy cyclopentanone (desymmetrization) catalyzed by 2 mol% of an (*R*)-BINOL–modified aluminum complex affords the (4*R*)-hydroxy enone in 95% ee at 98% yield (Scheme 7–29).[2]

Scheme 7–30 shows how the diketone **94** can be reduced by (*S*)-BINAL–H to give the desired (3*R*)-hydroxy-propenone **95** with 65% yield and 94% ee.

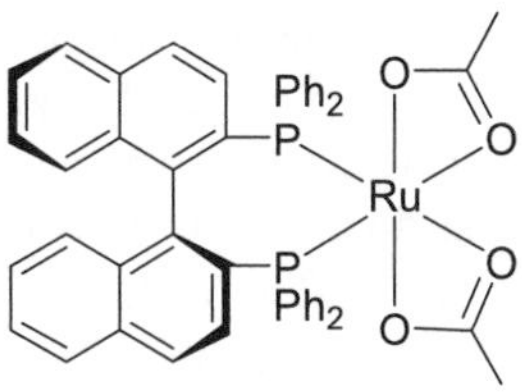

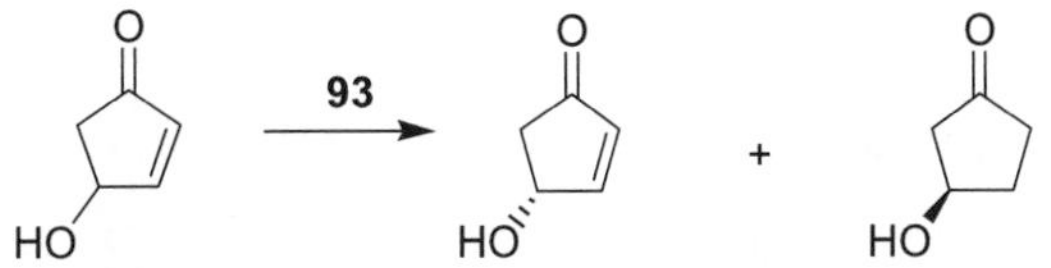

Scheme 7–28. Ru(II)–BINAP-mediated kinetic resolution of 4-hydroxy-2-cyclopentenone.

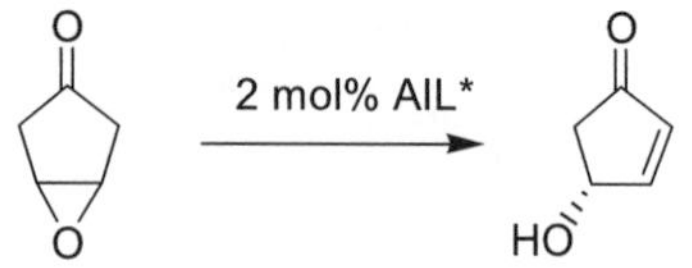

98%, 95% ee

Scheme 7–29. AIL* = Al(OPr^i)$_3$ + (R)-BINOL + 10 eq. n-BuOH.

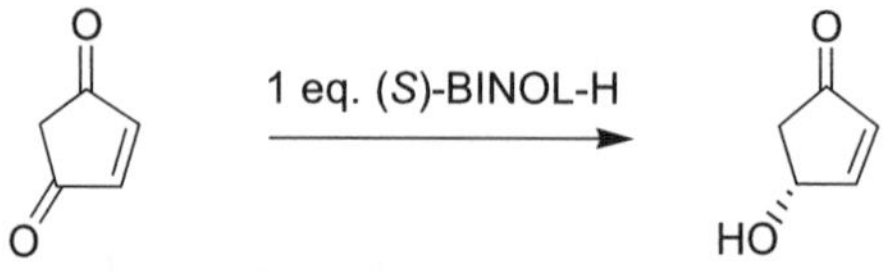

94   **95** 65%, 94% ee

Scheme 7–30. Asymmetric synthesis of a PG building block via catalytic hydrogenation.

7.5 THE TOTAL SYNTHESIS OF TAXOL—A CHALLENGE AND OPPORTUNITY FOR CHEMISTS WORKING IN THE AREA OF ASYMMETRIC SYNTHESIS

Since the discovery of the high anticancer activity of taxol, much attention has been drawn to its asymmetric synthesis. The total synthesis stood for more than 20 years as a challenge for organic chemists. The compound taxoids are diterpenoids isolated from *Taxus* species and have a highly oxidized tricyclic carbon framework consisting of a central eight-membered and two peripheral six-membered rings (see Fig. 7–2).[21]

Over the past few years, several groups have accomplished the total synthesis of taxol by way of independent and original pathways. The successful synthesis of taxol can be considered one of the landmarks of organic synthesis.[22] The total synthesis generally involves two stages: the synthesis of the side chain and the synthesis of the polycyclic ring system. Approaches to synthesizing the polycyclic skeleton of taxol can be divided into two types. One elaborates natural terpenes to generate the AB ring system of taxol by epoxy alcohol fragmenta-

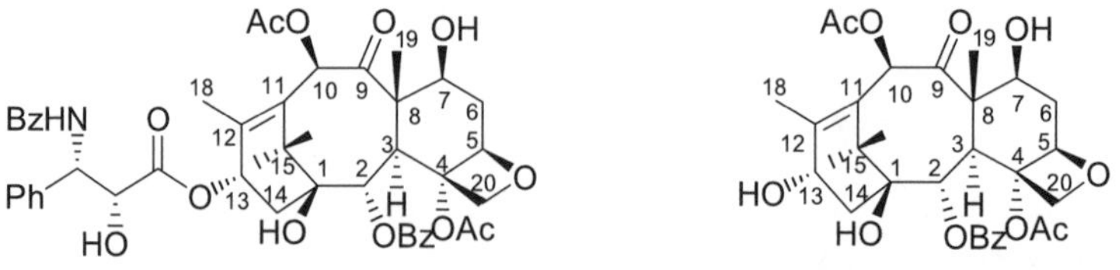

96a taxol   **96b** baccatin III

Figure 7–2. Structure and numbering of taxol and baccatin III.

tion; the other involves a convergent strategy, including a B ring-closure reaction applied to a connected AC ring system.

In Holton's and Wender's work, the total synthesis was achieved by sequentially forming the AB ring through the fragmentation of epoxy alcohols derived from (−)-camphor and α-pinene. Nicolau's, Danishefsky's, and Kuwajima's total syntheses involved B ring closure connecting the A and C rings, whereas in Mukaiyama's synthesis, the aldol reaction was extensively applied to construct the polycyclic system.

7.5.1 Synthesis of Baccatin III, the Polycyclic Part of Taxol

The structures of taxol and its polycyclic part baccatin III are shown in Figure 7–2, and the numbering of these two compounds is extensively used throughout the rest of this chapter. Because connecting the side chain to baccatin III is just routine chemistry, we introduce only the synthesis of baccatin III and the taxol side chain.

7.5.1.1 Holton's Construction of the Polycyclic Ring. In Holton's synthesis,[23] compound **97**, which can be prepared from camphor derivatives, is the starting material. Hydroxy-directed epoxidation of compound **97** gives unstable compound **98**, which fragments in situ to provide compound **99**, thus furnishing the AB skeleton of taxol (Scheme 7–31).[24]

A magnesium enolate of **99** is susceptible to aldol condensation with 4-pentenal, and the crude product can be directly protected to give its ethyl carbonate **100**. α-Hydroxylation of the carbonyl group yields the hydroxyl carbonate **101**. Reduction of the carbonyl group generates a triol, and this compound can be simultaneously converted to carbonate **102**. Swern oxidation of **102** gives ketone **103**, which can be rearranged[25] to produce lactone product **104** (Scheme 7–32).

Lactone product **104** is now susceptible to reductive C-3 hydroxyl removal, providing an enol product **105** that can be converted to the ketone **106** upon silica gel treatment. C-1 α-hydroxylation of compound **106** provides compound **107**. Compound **108** is then produced via Red-Al reduction of **107** and subsequent formation of the cyclic carbonate upon phosgene treatment (Scheme 7–33).

Ozonolysis and subsequent diazomethane treatment convert compound **108**

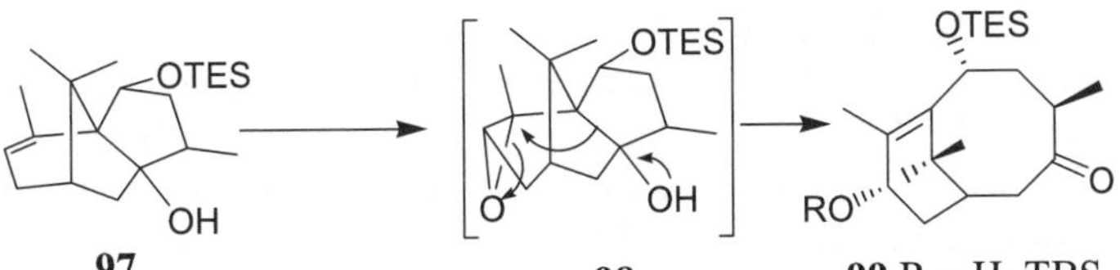

Scheme 7–31

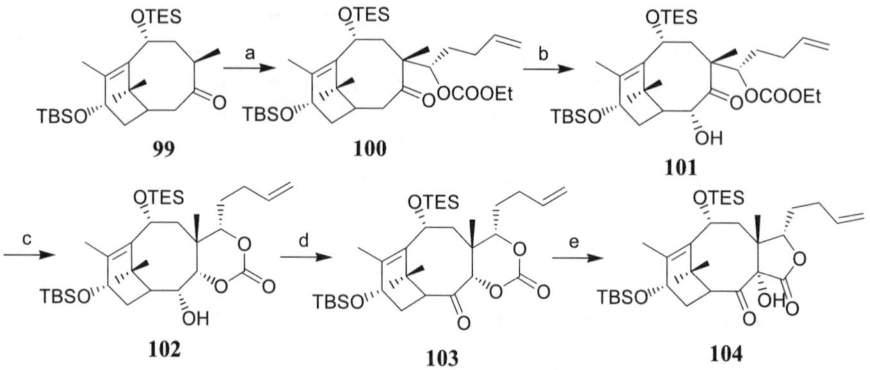

Scheme 7–32. Reagents and conditions: a: (1) Mg[N(Pr^i)$_2$]$_2$ (prepared by reaction of HN(Pr^i)$_2$ with MeMgBr in THF); (2) 4-pentenal, THF, −23°C, 1.5 hour; (3) Cl$_2$CO, CH$_2$Cl$_2$, −10°C, 0.5 hour, then ethanol 0.5 hour. b: LDA, THF, −35°C, 0.5 hour, then (+)-camphorsulphonyl oxaziridine, 0.5 hour, −78°C. c: (1) 20 eq. of Red-Al, toluene, −78°C, 6 hours, then warm to 25°C for 6 hours; (2) Cl$_2$CO, pyridine, CH$_2$Cl$_2$, −78° to 25°C, 1 hour. d: Swern oxidation. e: 1.05 eq. of LTMP, −25° to −10°C.

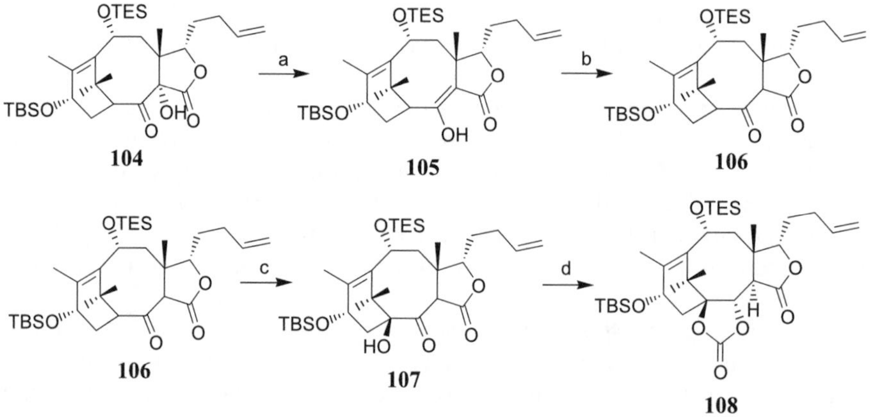

Scheme 7–33. Reagents and conditions: a: SmI$_2$ reduction. b: Silica gel treatment. c: (1) 4 eq. of LTMP, 10°C; (2) 5 eq. of (±)-camphorsulfonyl oxaziridine, −40°C. d: (1) Red-Al, −78°C, 1.5 hours; (2) 10 eq. of phosgene, pyridine, CH$_2$Cl$_2$, −23°C, 0.5 hour.

to methyl ester **109**. Dickmann cyclization of **109** gives enol ester **110**, which can be further converted to **111** via decarbomethoxylation. Compound **111** already possesses the ABC ring skeleton of a taxol compound (Scheme 7–34).

Next is the construction of the D ring. The TMS enol ether of compound **111** undergoes oxidation with *m*-CPBA, providing the C-5α trimethylsilyloxy ketone **112**. Addition of methyl Grignard reagent to the ketone group and subsequent dehydration provides compound **113**. Osmylation of the C=C double

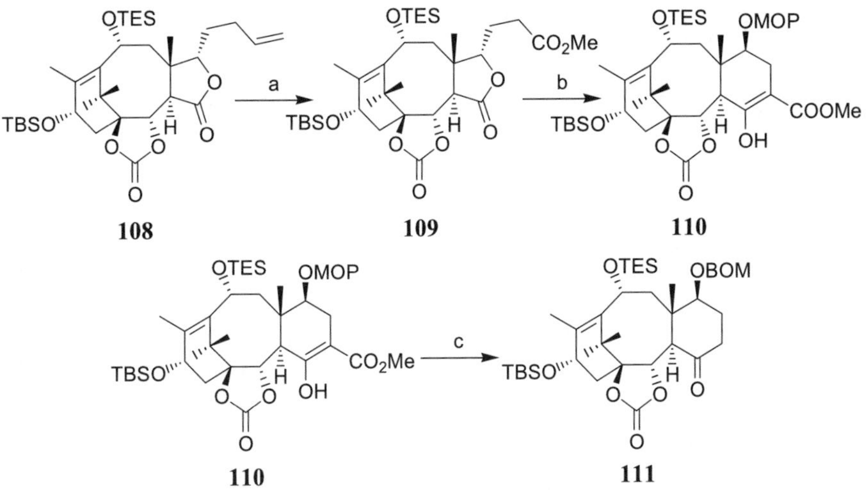

Scheme 7–34. Reagents and conditions: a: (1) Ozonolysis; (2) KMnO₄, KH₂PO₄; (3) CH₂N₂. b: (1) LDA, THF, −78°C, 0.5 hour; (2) p-TsOH, 2-methoxypropene. c: (1) PhSK, DMF, 86°C, acidic workup; (2) BOM-Cl, (i-Pr)₂NEt, CH₂Cl₂, Bu₄N⁺I⁻, reflux, 32 hours.

bond then gives the triol **114**, which can be converted to compound **115** via the mesylate or tosylate, thus furnishing the D ring of taxol. Addition of phenyl-lithium to the carbonate carbonyl group followed by TPAP oxidation provides compound **116**. Oxidation of C-9 by benzeneseleninic anhydride followed by a t-BuOK–mediated rearrangement provides the baccatin III compound **117**, which possesses all the functional groups of the polycyclic part of taxol (Scheme 7–35).

7.5.1.2 Wender's Approach.

Wender's synthesis[27] of baccatin III starts from the compound verbenone, a compound that can easily be prepared from pinene, which in turn is an abundant component of pine trees and a major constituent of the industrial solvent turpentine. The retro synthesis is outlined in Scheme 7–36.

In the synthetic procedure, verbenone **118**, the air oxidation product of pinene, is first treated with t-BuOK, followed by the addition of prenyl bromide to give a C-11 alkylated product. Selective ozonolysis of the more electron-rich double bond provides the aldehyde **119** with 85% yield. The A ring of taxane is then created through photorearrangement of the aldehyde **119**, yielding the chrysanthenone derivative **120** (85% yield, over 94% ee).

Completion of the B ring starts by introducing a two-carbon connector be-tween the C-2 and C-9 carbonyls of **120**. This is achieved in two steps. First, the lithium salt of ethyl propiolate is selectively added to the C-9 carbonyl, and the resulting alkoxide is trapped in situ with TMSCl, producing the TMS-protected product **121**. The C-9 methyl group is then introduced through conjugate ad-

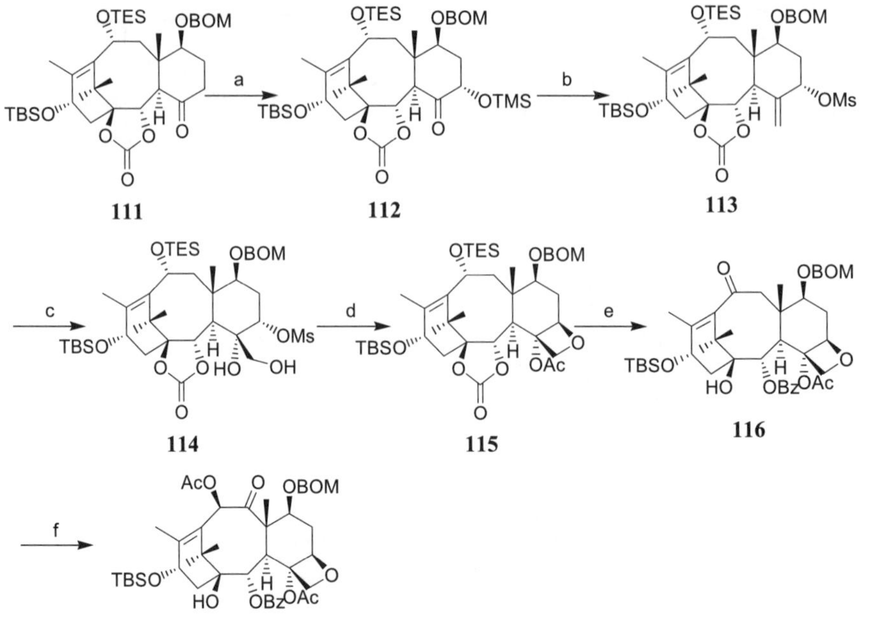

117 7-BOM-13-TBS baccatin III

Scheme 7–35. Reagents and conditions: a: (1) LDA, THF, TMSCl, −78°C; (2) *m*-CPBA. b: (1) 10 eq. of MeMgBr, CH_2Cl_2, −45°C, 15 hours; (2) Burgess's reagent[26]; (3) MsCl, pyridine. c: OsO_4, ether, pyridine. d: (1) DBU, toluene, 105°C, 2 hours; (2) Ac_2O, pyridine, DMAP, 24 hours, 25°C. e: (1) HF · pyridine, CH_2CN, 0°C, 11 hours; (2) 2.1 eq. of PhLi, THF, −78°C, 10 min; (3) TPAP, NMO, molecular sieves, CH_2Cl_2, 25°C, 15 minutes. f: (1) 4 eq. of *t*-BuOK, THF, −78 to 0°C, 10 minutes; (2) 8 eq. of benzeneseleninic anhydride, THF, 0°C, 40 minutes; (3) 4 eq. of *t*-BuOK, THF, −78°C, 10 minutes; (4) Ac_2O, pyridine, DMAP, 20 hours, 25°C.

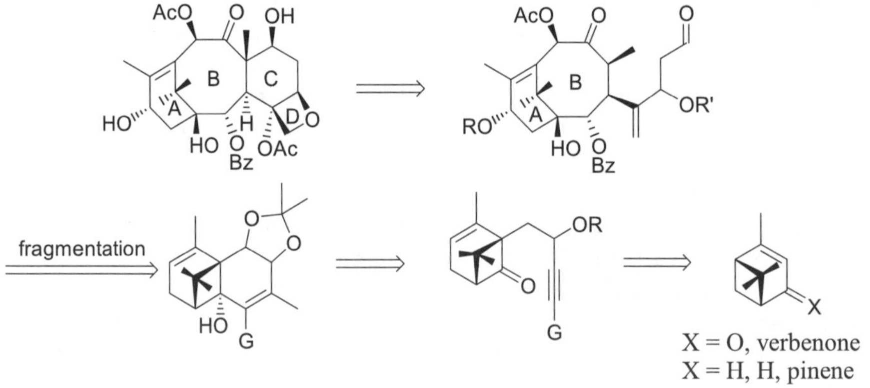

Scheme 7–36

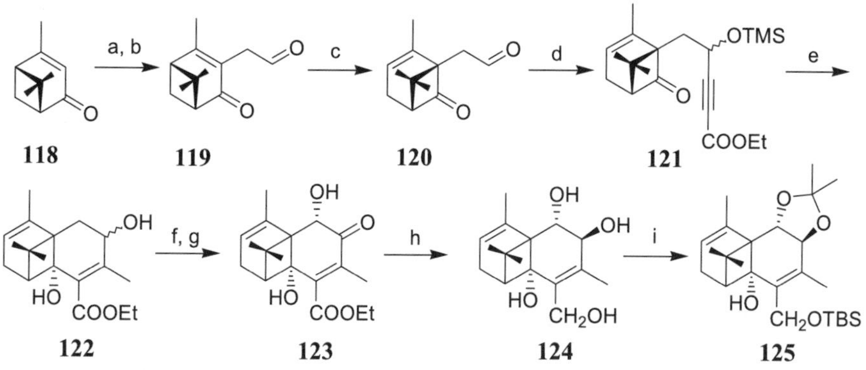

Scheme 7–37. Reagents and conditions: a: *t*-BuOK, 1-bromo-3-methyl-2-butene, DME, −78°C to room temperature (r.t.) (79% at 41% conversion). b: O₃, CH₂Cl₂, MeOH (85%). c: hν, MeOH (85%). d: LDA, ethyl propiolate, THF, −78°C; TMSCl (89%). e: Me₂CuLi, Et₂O, −78°C to r. t.; AcOH, H₂O (97%). f: RuCl₂(PPh₃)₃, NMO, acetone (97%). g: KHMDS, Davis' oxaziridine, THF, −78° to −20°C (97% at 57% conversion). h: LiAlH₄, Et₂O (74%). i: (1) TBSCl, imidazole; (2) PPTS, 2-methoxypropene, r. t. (91%).

dition of Me₂CuLi to **121**. The carbanion produced in the conjugate addition process is simultaneously added to the C-2 carbonyl group, furnishing the B ring product **122** with high yield (97%).

The hydroxyl group in alcohol **122** is then oxidized. Deprotonation of this ketone with KHMDS (1 eq.), followed by the addition of Davis' oxaziridine (see Chapter 4 for α-hydroxylation of ketones)[28] (2 eq.) allows the stereocontrolled introduction of the C-10 oxygen from the less hindered enolate face, providing only the (*R*)-hydroxyketone **123**. Subsequent reduction of **123** with excess LAH provides the tetra-ol **124**. Treatment of this compound with imidazole and TBSCl followed by PPTS and 2-methoxypropene provides in one operation the acetonide **125** with 91% yield (Scheme 7–37).

Compound **125** is subjected to epoxidation, and the epoxide compound thus formed undergoes epoxyl alcohol fragmentation, giving a compound bearing the AB ring system of baccatin III. In situ protection of the C-13 hydroxyl group of this compound provides the AB-bicyclic compound **126** with 85% yield. Treatment of ketone **126** with *t*-BuOK and P(OEt)₃ under an oxygen atmosphere, in situ removal of the TBS group with NH₄Cl/MeOH, and stereoselective reduction of the C-2 ketone by adding NaBH₄ gives the triol **127** with 91% yield. Hydrogenation of the C-3 to C-8 double bond of **127** from the desired α-face can be accomplished with Crabtree's catalyst.[29] The product is then protected in situ by treatment with TMSCl and pyridine followed by triphosgene to deliver the fully functionalized taxane AB-bicyclic system **128** with 98% yield. Compound **128** can then be further converted to aldehyde **129** for the construction of the C ring of baccatin III (Scheme 7–38).

A Wittig reaction of Ph₃P=CHOMe with **129** and subsequent one pot

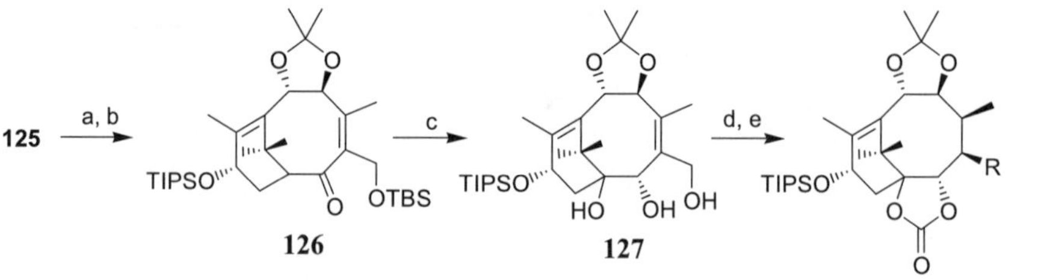

126 127

128 R = CH_2OTMS
129 R = CHO

Scheme 7–38. Reagents and conditions: a: *m*-CPBA, Na_2CO_3, CH_2Cl_2. b: (1) DABCO (cat.), CH_2Cl_2, Δ; (2) TIPSOTf, 2,6-lutidine, −78°C (85% over two steps). c: (1) *t*-BuOK, O_2, $P(OEt)_3$, THF, −40°C; (2) NH_4Cl, MeOH, room temperature (r.t.); (3) $NaBH_4$ (91%). d: (1) H_2, Crabtree's catalyst,[29] CH_2Cl_2, r. t.; (2), TMSCl, pyridine, −78°C; (3) triphosgene, 0°C, 98%. e: PCC, 4 Å molecular sieves, CH_2Cl_2 (100%).

hydrolysis converts **129** to aldehyde **130**. TESCl/pyridine treatment selectively protects the C-9 hydroxyl group, leading to compound **131**. This compound can be converted to compound **132** via Dess-Martin periodinane oxidation of the C-10 hydroxyl group, providing compound **132**. Eschenmoser's salt [(Me_2NCH_2I)/Et_3N] treatment then introduces the C-20 carbon, giving compound **133** for further functionalization (Scheme 7–39).

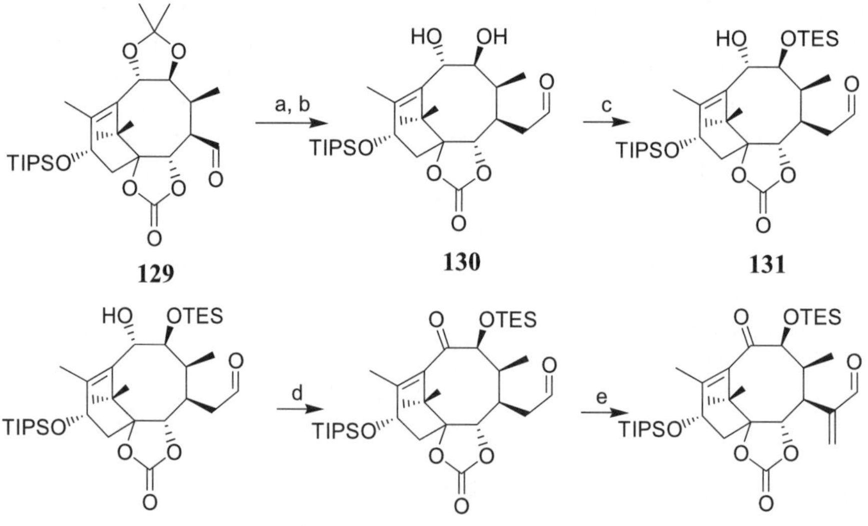

129 130 131

131 132 133

Scheme 7–39. Reagents and conditions: a: $Ph_3PCHOMe$, THF, −78°C (91%). b: 1 N HCl (aq), NaI, dioxane (94% at 90% conversion). c: TESCl, pyridine, CH_2Cl_2, −30°C (92%). d: Dess-Martin periodinane, CH_2Cl_2. e: Et_3N, Eschenmoser's salt (97%).

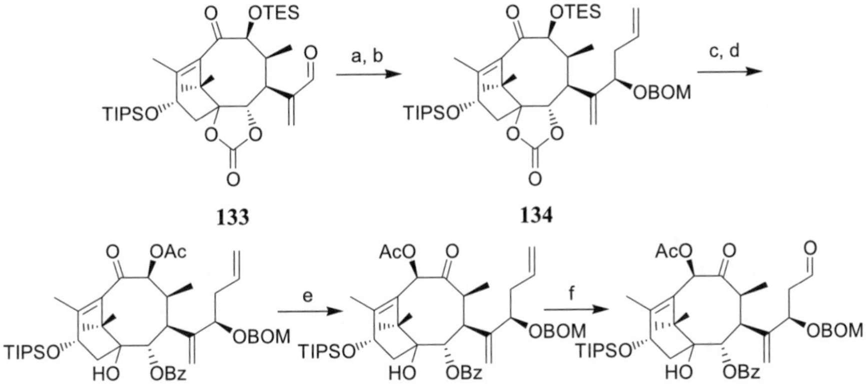

Scheme 7–40. Reagents and conditions: a: allyl-MgBr, ZnCl$_2$, THF, −78°C (89%). b: BOMCl, (*i*-Pr)$_2$NEt, 55°C. c: NH$_4$F, MeOH, room temperature (93% over two steps). d: (1) PhLi, THF, −78°C; (2) Ac$_2$O, DMAP, pyridine, 79%. e: ⟨pyrrolidinopyrimidine⟩, CH$_2$Cl$_2$, room temperature 1 hour (80% at 63% conversion). f: (1) O$_3$, CH$_2$Cl$_2$, −78°C; (2) P(OEt)$_3$ (86%).

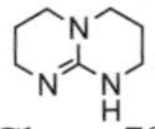

Allylation and subsequent protection of the thus formed hydroxyl group furnishes compound **134**, which bears the C-ring skeleton of baccatin III. Removal of the C-9 silyl group, PhLi treatment of the resulting hydroxyl ketone, and in situ acetylation provides compound **135**, which has the C-2 benzoate functionality. In the presence of a guanidinium base, equilibrium between the C-9 to C-10 carbonyl-acetate functional groups can be established. Thus, the desired C-9-carbonyl–C-10–acetate moiety **136** can be separated from the mixture. Compound **136** is then converted to aldehyde **137** via ozonolysis for further construction of the C-ring system (Scheme 7–40).

Intramolecular aldol reaction of **137** takes place in the presence of 4-pyrrolidinopyridine (Scheme 7–41), providing **138a** and **138b** as a mixture (11:1), from which the desired compound **138a** can be obtained as the major component. The thus formed C-7 hydroxyl group is then protected by TrocCl

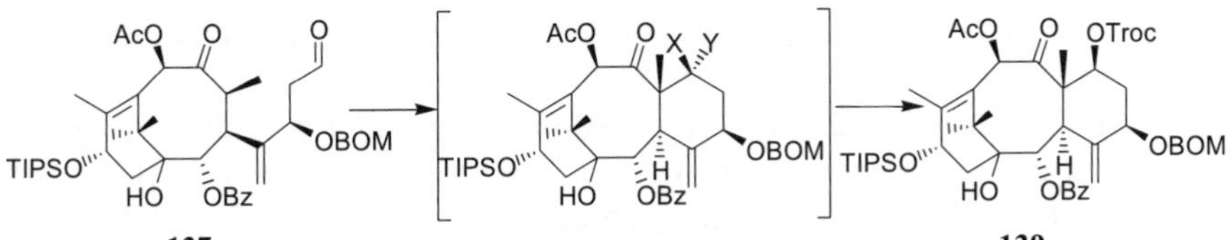

Scheme 7–41. Reagents and conditions: (1) DMAP, CH$_2$Cl$_2$; (2) TrocCl (62%).

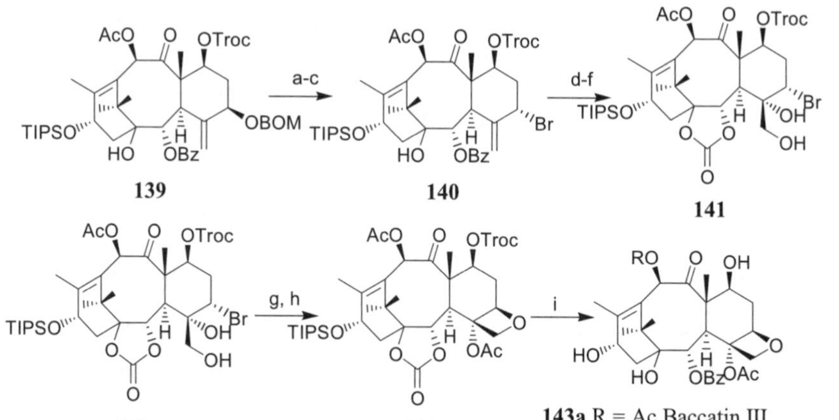

139 140 141 141 142

143a R = Ac Baccatin III
143b R = H 10-deacetylbaccatin III

Scheme 7–42. Reagents and conditions: a: NaI, HCl (aq), acetone (97% at 67% conversion). b: MsCl, pyridine, DMAP, CH_2Cl_2 (83%). c: LiBr, acetone (79% at 94% conversion). d: (1) OsO_4, pyridine, THF; (2) $NaHSO_3$; imidazole, $CHCl_3$ (76% at 94% conversion). e: Triphosgene, pyridine, CH_2Cl_2 (92%). f: KCN, EtOH, 0°C (76% at 89% conversion). g: $(i\text{-Pr})_2NEt$, toluene, 110°C (95% at 83% conversion). h: Ac_2O, DMAP (89%). i: (1) TASF, THF, 0°C; (2) PhLi, −78°C (46% 10-deacetylbaccatin III, 33% baccatin III).

(2,2,2-trichloroethyl chloroformate), providing compound **139**, which is ready for the construction of the D-ring system via dihydroxylation and intramolecular cyclization.

The β-OBOM group at C-5 is first converted to α-Br through functional group transformation reactions, furnishing compound **140** ready for osmium tetroxide–mediated dihydroxylation. As the C-2 benzoate tends to migrate to the C-20 hydroxyl group during the D-ring formation process, it is removed, and the C-1 to C-2 diol is again protected with cyclic carbonate. The thus formed compound **141** undergoes a cyclization reaction, providing the D-ring skeleton of the target baccatin III. Final PhLi treatment regenerates the C-2 benzoate, completing the full synthesis of bacctin III (Scheme 7–42).

7.5.1.3 Kuwajima's Synthesis of (±)-Taxusin. Two key transformations are involved in Kuwajima's synthesis[30]: (1) construction of the tricyclic taxane skeleton via cyclization of the eight-membered B ring between C-9 and C-10; and (2) subsequent installation of the C-19 methyl group onto the ring system.

The six-membered ring vinyl bromide **144**, corresponding to the C-ring system, is chosen as the starting material. Treatment of **144** with *t*-BuLi and CuCN produces the corresponding cyanocuprate, which reacts with 3,4-epoxy-1-hexene to give an S_N2 coupling product, compound **145**. Pyridinium dichromate (PDC) oxidation of the resulting allyl alcohol **145** affords enone **146** at 66% yield (in two steps). Conjugate addition of the lithium enolate of ethyl

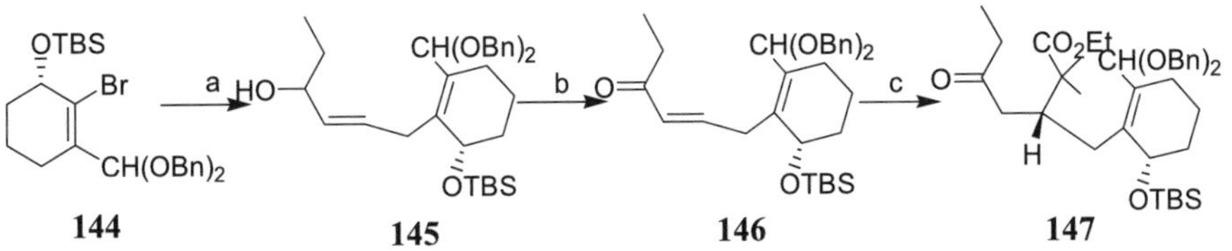

Scheme 7–43. Reagents and conditions: a: (1) *t*-BuLi, Et$_2$O, −78°C, 1.5 hours; (2) CuCN, −45°C, 1 hour; (3) 3,4-epoxy-1-hexene, −23°C, 2 hours. b: PDC, 4Å MS, CH$_2$Cl$_2$, room temperature, 1.5 hours (66% from **144**). c: LDA/Me$_2$CHCOOEt, THF, −78° to 5°C, 7 hours.

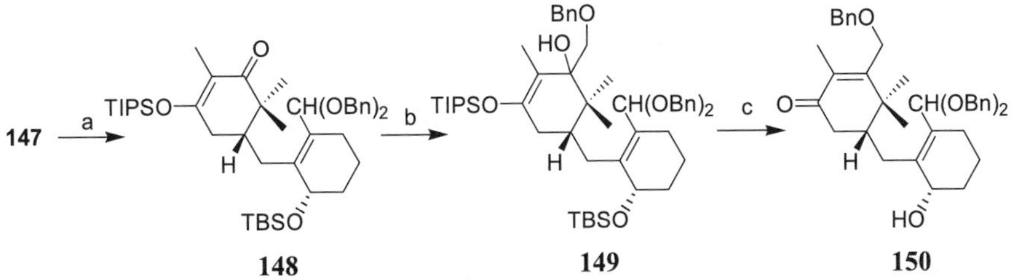

Scheme 7–44. Reagents and conditions: a: (1) *t*-BuOK, THF, 0°C, 1 hour; (2) TIPSCl, 0°C, 4 hours (60%). b: BnOCH$_2$Li, THF, −78°C, 2.5 hours (86%). c: (1) Montmorillonite K 10, 4Å MS, BnOH, CH$_2$Cl$_2$, −45°C, 1 hour (79%); (2) TBAF, THF, room temperature, overnight (83%).

isobutyrate to **146** proceeds with fairly high 1,4-asymmetric induction (4:1), yielding **147** as the major isomer (Scheme 7–43).

Upon *t*-BuOK treatment, **147** undergoes Dieckmann-type cyclization, and subsequent enolization affords compound **148** at 60% yield. Compound **148** is then converted to **149** through benyloxymethyl lithium addition. Successive deprotection and isomerization converts compound **149** to **150** for further functionalization (Scheme 7–44).

A Mitsunobu reaction then inverts the C-4 configuration to provide compound **151**, and subsequent isomerization provides compound **152**, which is ready for Lewis acid–mediated cyclization to construct the eight-membered B-ring. Using Me$_2$AlOTf as a catalyst, intramolecular cyclization occurs, giving product **153** in which A, B, and C rings have been introduced with the desired stereochemistry (Scheme 7–45).

Reduction of the C-13 keto group of **153** with Li(*t*-BuO)$_3$AlH followed by sillylation of the resulting hydroxyl group and reductive removal of the pivalate group furnishes allyl alcohol **154**. A cyclopropanation reaction then introduces a cyclopropane moiety, which can be converted to the desired C-19 methyl group upon opening of the three-membered ring. As shown in Scheme 7–46, the cyclopropane group in **156** has been converted to the desired C-19 methyl group via Li/NH$_3$(l) reduction.

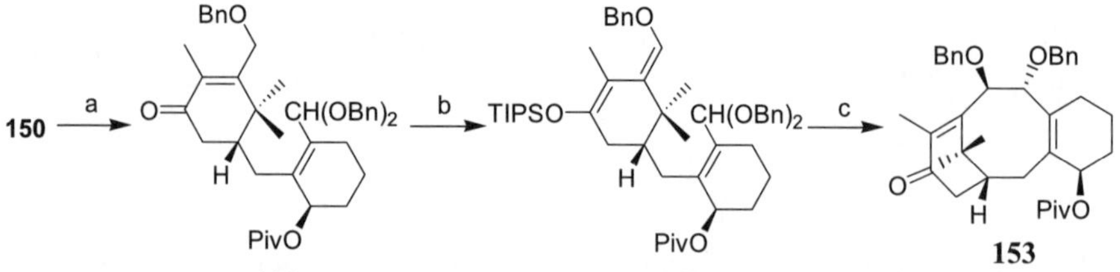

151 152 153

Scheme 7–45. Reagents and conditions: a: DEAD, Ph_3P, PivOH, THF, room temperature, 1 week (67%). b: (1) *t*-BuOK, THF, 0°C (1 hour); (2) TIPSCl, −78°C, 1 hour, quantitative. c: Me_2AlOTf (3 eq.), CH_2Cl_2, −45°C (62%).

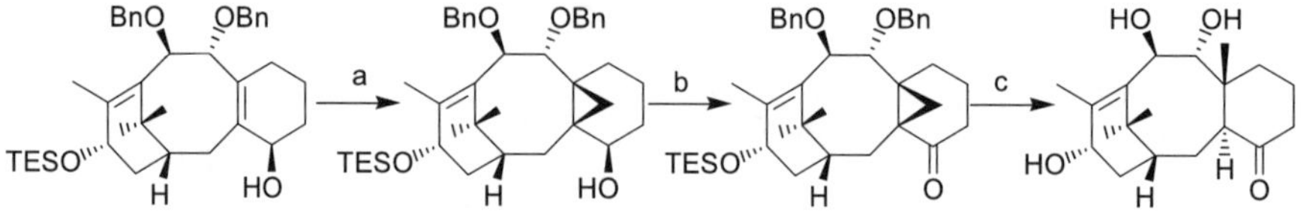

154 155 156 157

Scheme 7–46. Reagents and conditions: a: CH_2I_2, $ZnEt_2$, Et_2O, room temperature (r.t.), 6 hours (quantitative). b: PDC, 4 Å MS, CH_2Cl_2, r. t., 1.5 hours (85%). c: (1) Li, NH_3(1), *t*-BuOH, THF, −78°C, 1 hour; (2) MeOH, r. t., 1 hour (91%).

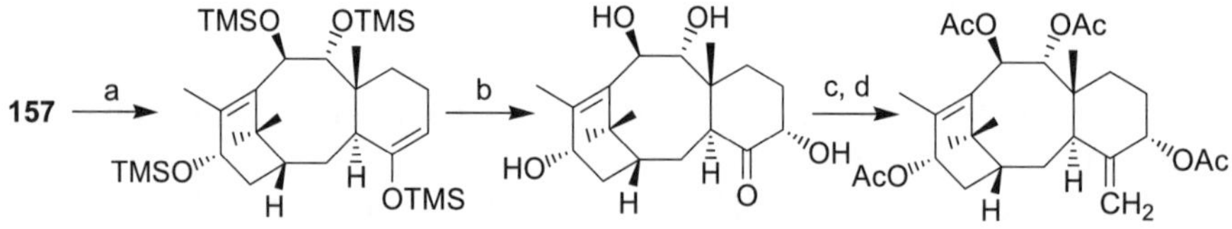

(±)-taxusin

Scheme 7–47. Reagents and conditions: a: TMSCl, LDA, THF, −78°C, 10 minutes, then 0°C, 30 minutes. b: *m*-PCBA, $KHCO_3$, CH_2Cl_2, 0°C, 10 minutes. c: Ac_2O, DMAP, Et_3N, CH_2Cl_2, room temperature, 1.5 hours. d: $Ph_3P{=}CH_2$, benzene, hexane, 0°C, 1.5 hours.

Finally, the C-20 group is introduced to the C ring, thus finishing the synthesis of (±)-taxusin in 25 steps (Scheme 7–47).

7.5.1.4 Danishefsky's Synthesis. Danishefsky's synthesis of baccatin III[31] involves a B-ring closure to connect the A and C rings. The retro synthetic route of Danishefsky's synthesis is outlined in Scheme 7–48.

Compound **158** is the starting material. This compound is first reduced with sodium borohydride to **159**, and the resulting hydroxyl group is then protected as its *tert*-butyl dimethylsilyl ether (Scheme 7–49). The resulting compound **160** appears to possess the C-ring skeleton of baccatin III.

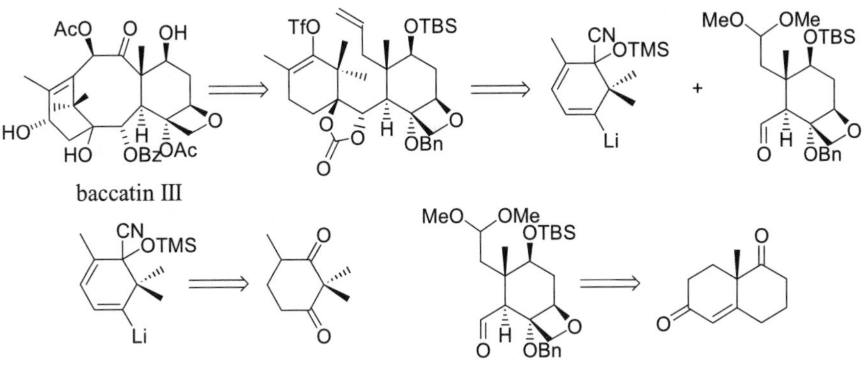

Scheme 7–48

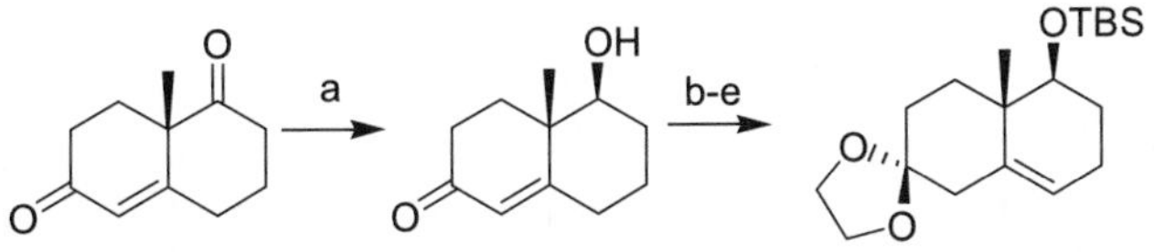

158 159 160

Scheme 7–49. Reagents and conditions: a: $NaBH_4$, EtOH, 0°C (97%). b: Ac_2O, DMAP, pyridine, CH_2Cl_2, 0°C (99%). c: $(HOCH_2)_2$, benzene, naphthalenesulfonic acid, reflux (70%). d: NaOMe, MeOH, THF (98%). e: TBSOTf, 2,6-lutidine, CH_2Cl_2 0°C (97%).

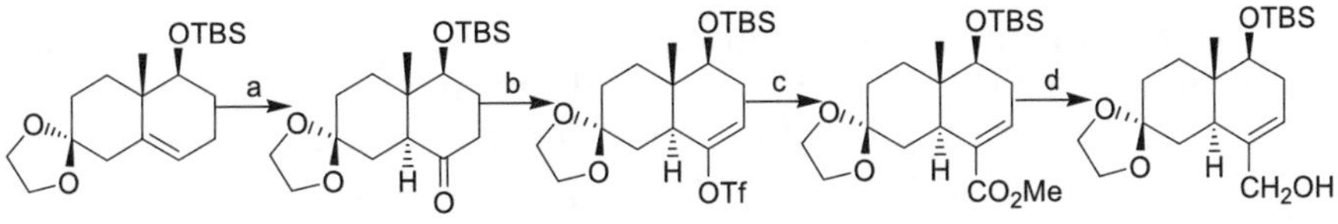

160 161 162 163 164

Scheme 7–50. Reagents and conditions: a: (1) $BH_3 \cdot THF$, THF, 0°C to room temperature (r.t.); (2) H_2O_2, NaOH, H_2O; (3) PDC, CH_2Cl_2, 0°C to r. t.; (4) NaOMe, MeOH (62%). b: (1) KHMDS, THF, −78°C; (2) $PhNTf_2$ (81%). c: $Pd(OAc)_2$, PPh_3, CO, Hünig's base, MeOH, DMF (73%). d: DIBAL, hexanes, −78°C (99%).

Hydroboration and oxidation of **160** yields an alcohol that is subsequently oxidized with PDC to give ketone compound **161**. Enolization and triflation converts this compound to enol triflate **162**, which can be further converted to α,β-unsaturated ester **163** upon palladium-mediated carbonylation methoxylation. The desired alcohol **164** can then be readily prepared from **163** via DIBAL reduction. Scheme 7–50 shows these conversions.

Danishefsky found that the method depicted in Scheme 7–50 was not suitable for large-scale synthesis of the intermediate, so another route involving

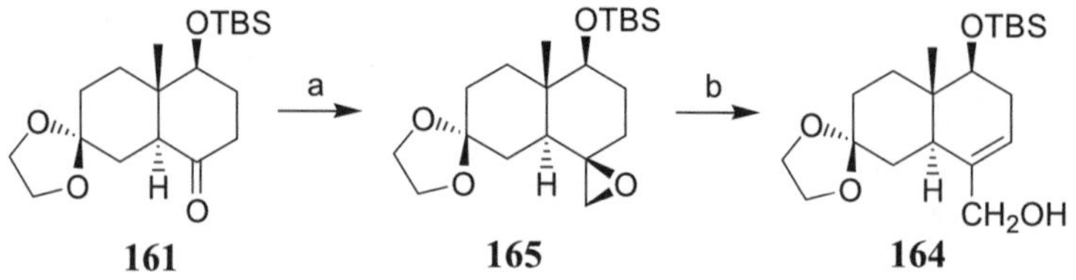

Scheme 7–51. Reagents and conditions: a: $Me_2S^+I^-$, KHMDS, THF, 0°C (99%). b: $Al(OPr^i)_3$, toluene, reflux (99%).

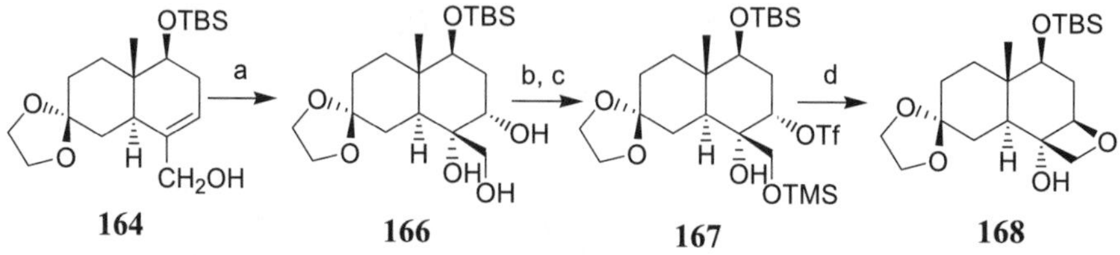

Scheme 7–52. Reagents and conditions: a: OsO_4, NMO, acetone, H_2O (66%). b: (1) TMSCl, pyridine, CH_2Cl_2, −78°C to room temperature (r. t.); (2) Tf_2O, −78°C to r. t. c: $(HOCH_2)_2$, 40°C (69%).

Corey's sulfonium ylide methodology[32] was then adopted. As shown in Scheme 7–51, conversion of ketone **161** to spiroepoxide **165** proceeded with high yield, and a Lewis acid–induced epoxide opening gave the desired allylic alcohol **164** with perfect yield.

Treating allylic alcohol **164** with osmium tetroxide generates triol **166** for construction of the D ring. The primary hydroxyl group of **166** is readily differentiated by silylation with TMS chloride, and the secondary hydroxyl group is then activated by triflation. Refluxing compound **167** in ethylene glycol produces the desired oxetane **168**, thus finishing the construction of the D ring of baccatin III (Scheme 7–52).

By protecting the C-4 secondary hydroxyl group and cleaving the ketal group, compound **168** can be converted to **169** (R = Bn or TBS) from which the functionalized C- and D-ring moiety can be obtained. Ketone enolization and α-hydroxylation furnishes compound **171**, which can be converted to the ring cleavage product **172** via oxidation and esterification. Subsequent functional group transformation furnishes compound **174**, a key intermediate for baccatin III (and taxol) synthesis (Scheme 7–53).

The A-ring moiety can be prepared from readily available compound **175** via routine chemisty, as shown in Scheme 7–54. Organolithium compound **176** is then coupled with **174** to produce the ABCD ring skeleton.

The coupling product **177** is subjected to epoxidation to give epoxide compound **178** on which the C-1 hydroxyl group will be generated via catalytic hydrogenation. After the dihydroxyl groups of the hydrogenation product **179** have been protected with cyclic carbonate, the C=C double bond between C-12

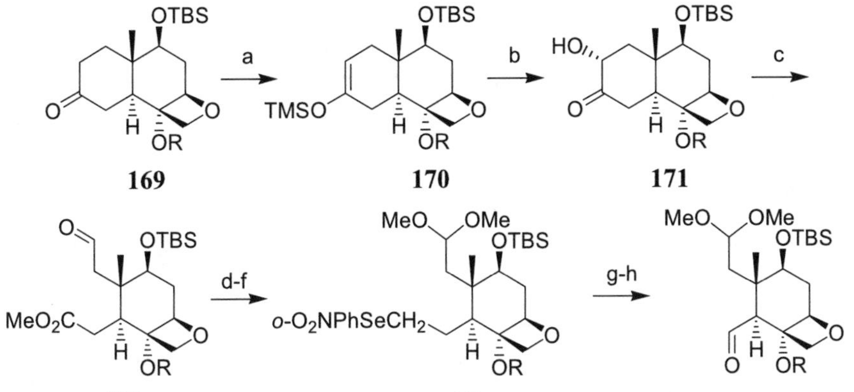

Scheme 7–53. Reagents and conditions: a: TMSOTf, Et₃N, CH₂Cl₂, −78°C. b: (1) 3,3-dimethyldioxirane, CH₂Cl₂, 0°C; (2) CSA, acetone, room temperature (r. t.) (89%). c: Pb(OAc)₄, MeOH, benzene, 0°C (97%). d: MeOH, CPTS, 70°C (97%). e: LiAlH₄, THF, 0°C (100%). f: o-NO₂C₆H₄SeCN, PBu₃, THF, r. t. (88%). g: 30% H₂O₂, THF, r. t. (90%). h: (1) O₃, CH₂Cl₂, −78°C; (2) PPh₃ (79% R = Bn or TBS).

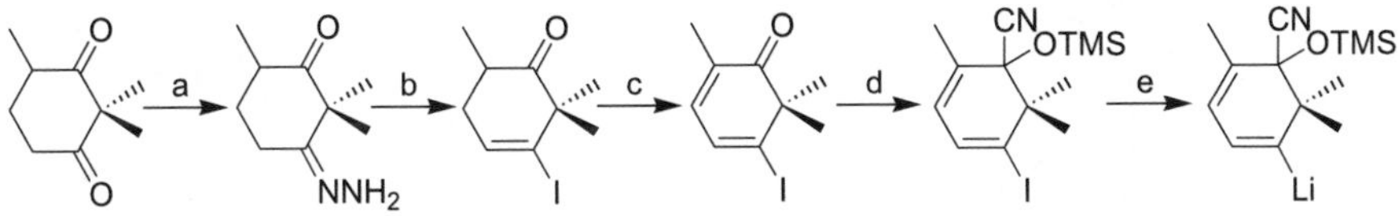

Scheme 7–54. Reagents and conditions: a: H₂NNH₂, Et₃N, EtOH (72%). b: I₂, DBN, THF (52%). c: I₂, DBN, THF (89%). d: TMSCN, cat. KCN, 18-crown-6, CH₂Cl₂ (89%). e: t-BuLi, THF −78°C.

and C-13 can be reduced, providing product **181** for the construction of the C ring (Scheme 7–55).

In Danishefsky's synthesis, a Heck reaction was chosen for the formation of the C-ring moiety. Thus, the C-11 carbonyl group in **181** is enolized, and the acetal group in the enolization product **182** is removed to give aldehyde **183**, from which the Heck reaction substrate **184** can be prepared through a Wittig reaction. The final intramolecular Heck reaction between the enol triflate and C=C, catalyzed by Pd(PPh₃)₄, provides **185** with the ABCD ring skeleton of baccatin III (Scheme 7–56).

The methylene group in compound **185** can now be converted to a ketone carbonyl group, providing compound **186** for installation of the C-9–ketone and C-10–acetate functional groups. Compound **187**, bearing all the necessary ABCD ring functional groups of baccatin III, can finally be synthesized using Holton's chemistry (Scheme 7–57). (see Section 7.5.1.1 for Holton's construction of the C-9–ketone and C-10–acetate functional groups.)

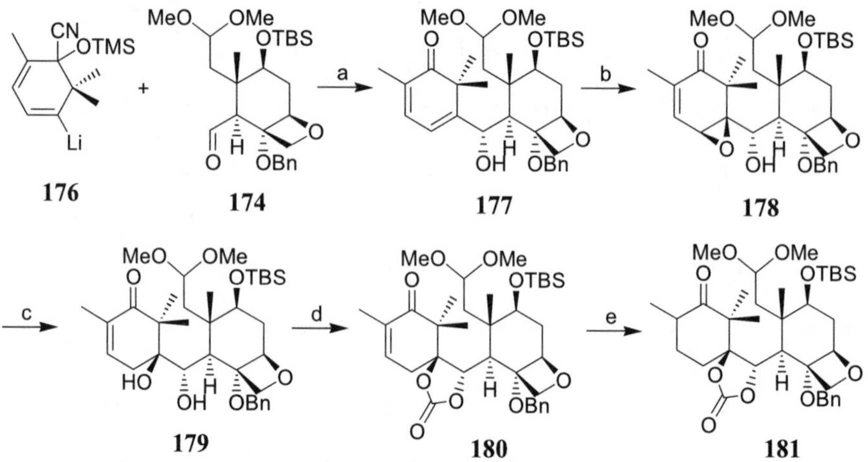

Scheme 7–55. Reagents and conditions: a: (1) THF, −78°C (93%); (2) TBAF, THF, −78°C (80%). b: *m*-CPBA, CH₂Cl₂, room temperature (80%). c: H₂, Pd/C, −5°C, EtOH (65%). d: CDI, NaH, DMF (81%). e: L-Selectride, THF, −78°C (93%).

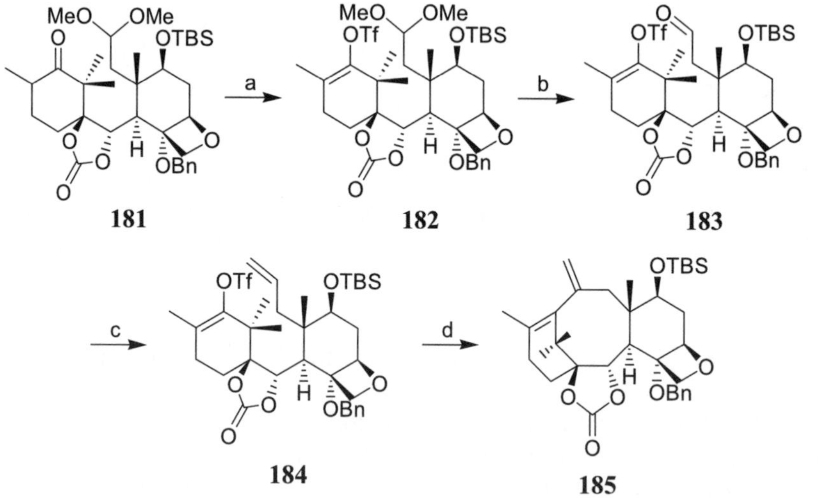

Scheme 7–56. Reagents and conditions: a: PhNTf₂, KHMDS, THF, −78°C (98%). b: PPTS, acetone, H₂O (96%). c: Ph₃P=CH₂, THF, −78° to 0°C (77%). d: Pd(PPh₃)₄, K₂CO₃, CH₃CN, 85°C (49%).

The final stage is to introduce the C-13 hydroxyl group via PCC oxidation of **187** and subsequent reduction of the carbonyl group, and this completes the total synthesis of baccatin III (Scheme 7–58). The synthesized compound is identical to a natural sample of baccatin III in all respects, including optical rotation.

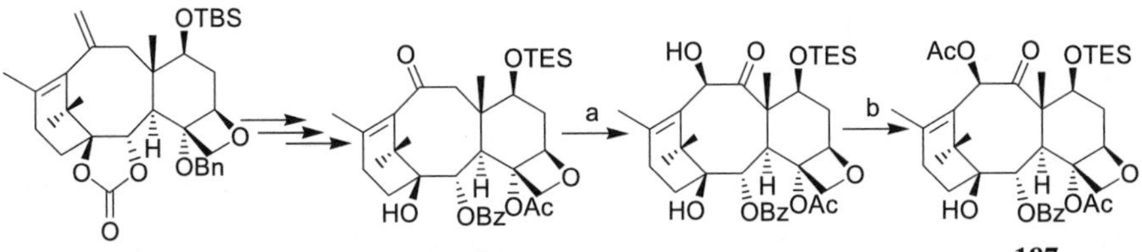

185 **186** **187**

Scheme 7–57. Reagents and conditions: a: (1) *t*-BuOK, $(PhSeO)_2O$, THF, −78°C; (2) *t*-BuOK, THF, −78°C (81%). b: Ac_2O, DMAP, pyridine (76%).

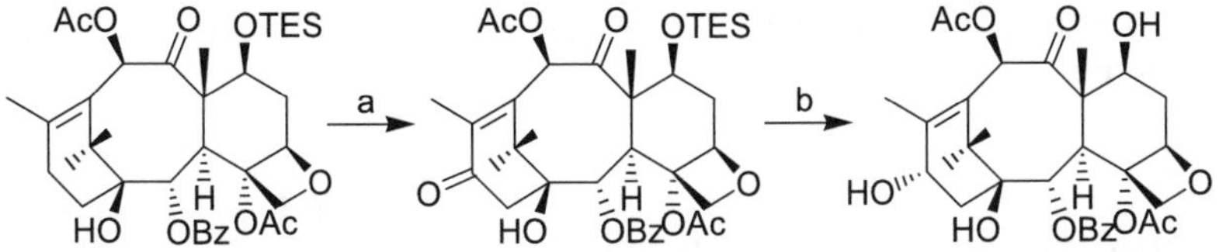

187 baccatin III

Scheme 7–58. Reagents and conditions: a: PCC, NaOAc, benzene, reflux (64%). b: (1) $NaBH_4$, MeOH (79%); (2) HF · pyridine, THF (85%).

7.5.1.5 Nicolaou's Synthesis. Nicolaou's synthesis also involves a B-ring closure to connect the A and C rings.[33] The retro synthetic approach is depicted in Scheme 7–59.

Nicolaou constructed the A ring moiety **188** and the C-ring moiety **192** through well-established chemistry, as shown in Schemes 7–60 and 7–61.

C-ring moiety **192** can be further converted to aldehyde **193** for the formation of the ABC ring skeleton (Scheme 7–62).

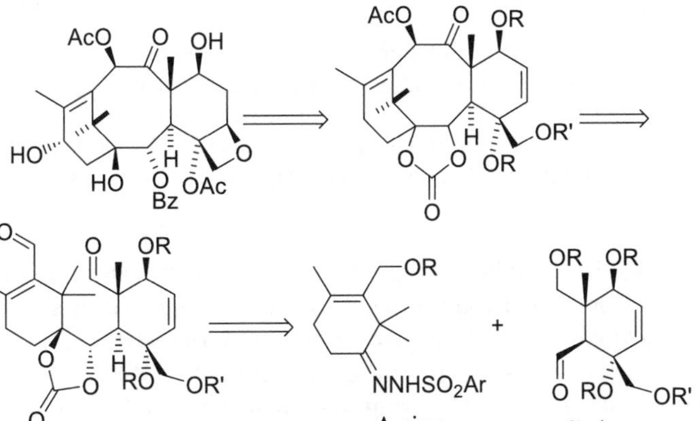

A-ring C-ring

Scheme 7–59

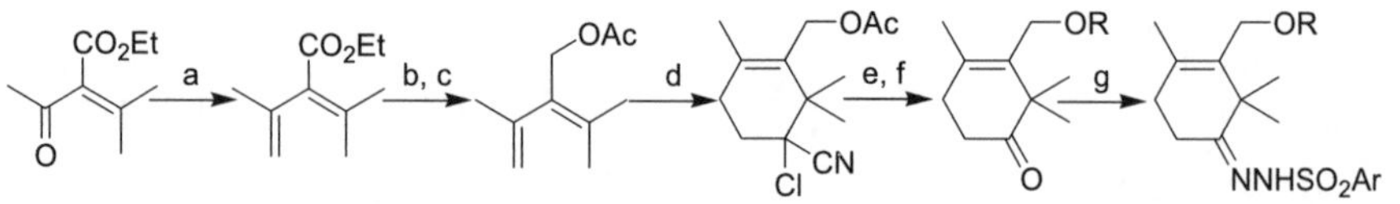

188 R = TBS, MEM

Scheme 7–60. Reagents and conditions: a: 1.2 eq. of MeMgBr, Et$_2$O, 0° to 25°C, 8 hours, then 0.2 eq. of p-TsOH, benzene, 65°C, 3 hours (70%). b: 2.2 eq. of DIBAL–H, CH$_2$Cl$_2$, −78° to 25°C, 12 hours (92%). c: 1.1 eq. of Ac$_2$O, 1.2 eq. of Et$_3$N, 0.2 eq. of DMAP, CH$_2$Cl$_2$, 0° to 25°C, 1 hour (96%). d: 1.5 eq. of 2-chloroacrylonitrile, 130°C, 72 hours (80%). e: 6.0 eq. of KOH, t-BuOH, 70°C. f: For R = TBS, 1.1 eq. of TBSCl, 1.2 eq. of imidazole, CH$_2$Cl$_2$, 25°C, 2 hours (85%); for R = MEM, 1.2 eq. of MEMCl, 1.3 eq. of (i-Pr)$_2$NEt, CH$_2$Cl$_2$, 25°C, 3 hours (95%). g: 2,4,6-Triisopropylbenzenesulfonyl)hydrazine.

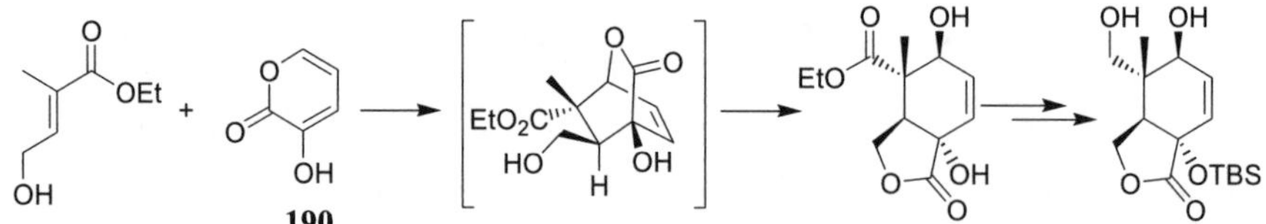

189   **190**   **191**   **192**

Scheme 7–61. Reagents and conditions for conversion from **189** and **190** to **191**: 1.4 eq. of **190**, 1.4 eq. of PhB(OH)$_2$, C$_6$H$_6$, reflux, 48 hours. Then 1.4 eq. of 2,2-dimethyl-1,3-propanediol, 25°C, 1 hour (79% based on 77% conversion of **189**).

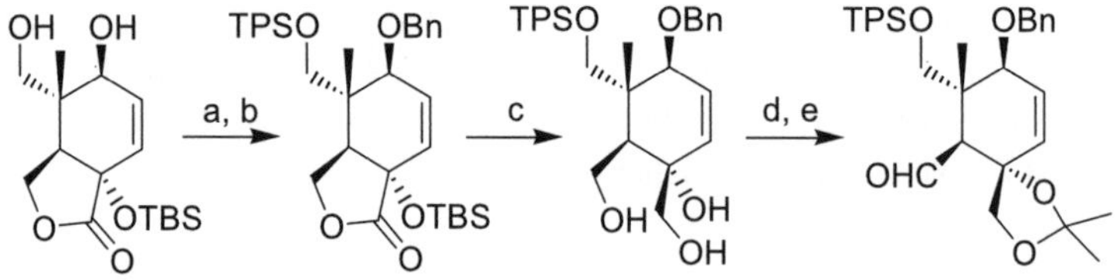

192   **193**

Scheme 7–62. Reagents and conditions: a: 1.3 eq. of TPSCl, 1.35 eq. of imidazole, DMF, 25°C, 12 hours (92%). b: 1.2 eq. of KH, 1.2 eq. of PhCH$_2$Br, 0.04 eq. of (n-Bu)$_4$NI, Et$_2$O, 25°C, 1 hour (88%). c: 3.0 eq. of LiAlH$_4$, Et$_2$O, 25°C, 12 hours (80%). d: 5.0 eq. of 2,2-dimethoxypropane, 0.05 eq. of CSA, CH$_2$Cl$_2$:Et$_2$O (98:2), 25°C, 7 hours (82%). e: 0.05 eq. of TPAP, 1.5 eq. of NMO, CH$_3$CN, 25°C, 2 hours (97%).

Next, a Shapiro coupling brings the A-ring moiety **188** and C-ring moiety **193** together. Compound **188** is treated with BuLi, followed by slow addition of **193**. The nucleophilic addition gives compound **194** as the Shapiro coupling product. Clearly, this strategy for constructing the ABC ring system and Danishefsky's strategy of ABC ring system formation are quite similar. The epoxidation of the C-1 and C-14 double bonds introduces an epoxide functional group, which can be converted to a C-1 hydroxyl group with reductive ring-

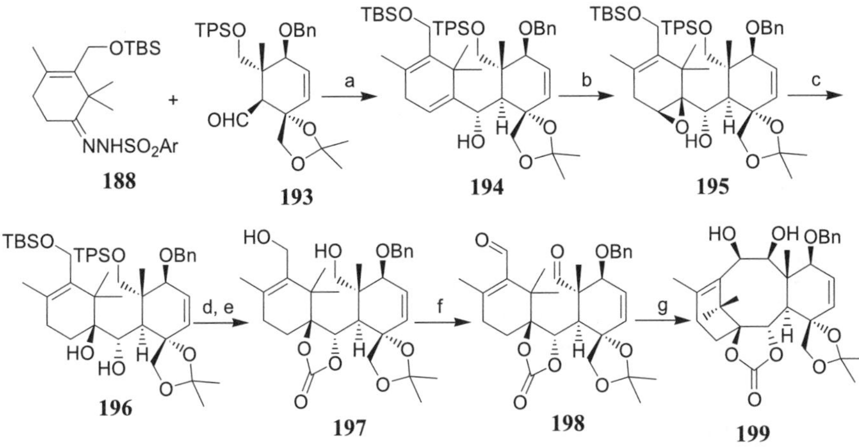

Scheme 7–63. Reagents and conditions: a: 1.1 eq. of **188**, 2.3 eq. of *n*-BuLi, THF, −78° to 0°C, 1.0 eq. of **193**, THF, −78°C, 0.5 hour (82%). b: 0.03 eq. of VO(acac)$_2$, 3.0 eq. of TBHP, C$_6$H$_6$, 4 Å MS, 25°C, 14 hours (87%). c: 5.0 eq. of LiAlH$_4$, 25°C, Et$_2$O, 7 hours (76%). d: 3.0 eq. of KH, Et$_2$O:HMPA = 3:1, 1.6 eq. of phosgene (20% in toluene), 25°C, 0.5 hour (86% based on 58% conversion). e: 3.8 eq. of TBAF, THF, 25°C, 14 hours (80%). f: 0.05 eq. of TPAP, 3.0 eq. of NMO, CH$_3$CN:CH$_2$Cl$_2$ = 2:1, 25°C, 2 hours (92%). g: 11 eq. of TiCl$_3$ · (DME)$_{1.5}$, 26 eq. of Zn–Cu, DME, reflux, 3.5 hours, then 70°C, then added **198** over 1 hour, 70°C for 0.5 hour.

opening reaction. The thus formed vicinal diol is protected by formation of a cyclic carbonate. The protected C-9 to C-10 primary hydroxyl groups are then deprotected and oxidized to give dialdehyde **198**. McMurry cyclization is then carried out in DME at 10°C in the presence of 11 eq. of TiCl$_3$ · (DME)$_{1.5}$ and 26 eq. of Zn–Cu couple. The coupling product, diol **199**, is generated with 25% yield. This yield, according to Nicolaou, was the optimal result (Scheme 7–63).

As this synthesis started from an achiral starting material, compound **199** must be resolved to secure enantiomerically pure intermediates for the synthesis of taxol. Treatment of (±)-diol **199** with excess (1*S*)-(−)-camphanic chloride in methylene chloride in the presence of Et$_3$N forms two diastereomeric mono-esters for chromatographic separation. Enantiomerically pure diol **199** can be regenerated from the ester in 90% yield with a specific rotation of +187 (*c* = 0.5, CHCl$_3$).

Next comes the adjustment of the C-9 and C-10 functional groups to their final form and the construction of the D ring. The higher reactivity of the allylic C-10 hydroxyl group in **199** makes it possible to selectively carry out the ace-tylation at the C-10 position. The resulting monoacetate can be oxidized, pro-viding the desired 9-keto–10-acetate **200**. Hydroboration at C-5 introduces a hydroxyl group, giving compound **201**, which can then be converted to ABCD ring compound **202** via 5-mesylate or 5-tosylate (Scheme 7–64).

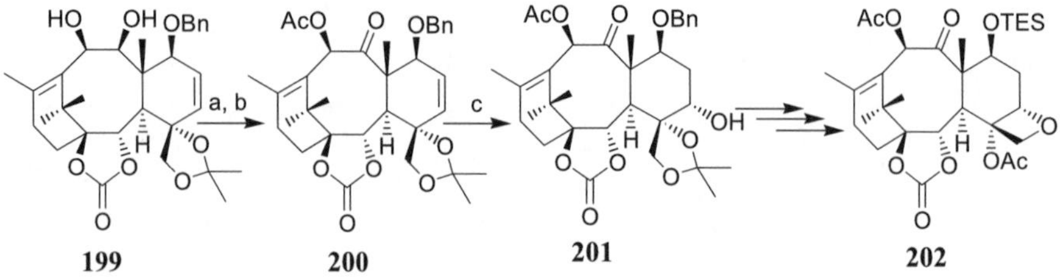

Scheme 7–64. Reagents and conditions: a: 1.5 eq. of Ac$_2$O, 1.5 eq. of DMAP, CH$_2$Cl$_2$, 25°C, 2 hours (95%). b: 0.1 eq. of TPAP, 3.0 eq. of NMO, CH$_3$CN, 25°C, 2 hours (93%). c: 10.0 eq. of BH$_3$·THF, THF, 0°C, 3 hours, then excess H$_2$O$_2$, saturated NaHCO$_3$, 25°C, 1 hour (42% plus 22% of C$_6$ hydroxy isomer).

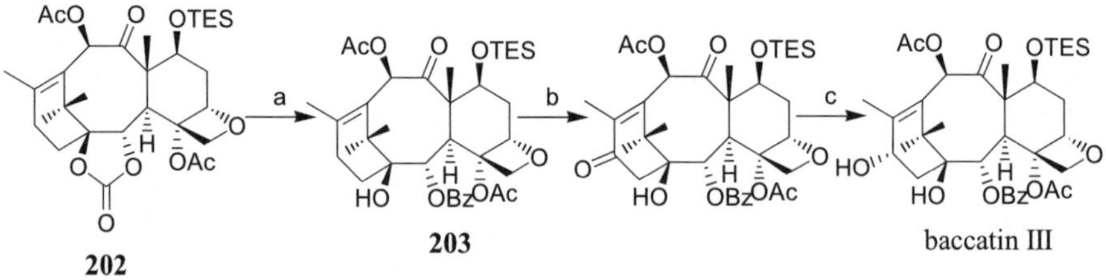

Scheme 7–65. Reagents and conditions: a: 5.0 eq. of PhLi, THF, −78°C, 10 minutes, then 10 eq. of Ac$_2$O, 5.0 eq. of DMAP, CH$_2$Cl$_2$, 2.5 hours (80%). b: 30 eq. of PCC, 30 eq. of NaOAC, celite, benzene reflux, 1 hour (75%). c: Excess NaBH$_4$, MeOH, 25°C, 3 hours.

Phenyl lithium attack on the cyclic carbonate convertes **202** to C-2 benzoate compound **203**. PCC oxidation and subsequent NaBH$_4$ reduction then furnishes the final baccatin III for the total synthesis of taxol (Scheme 7–65).

7.5.1.6 *Mukaiyama's Synthesis.*

Mukaiyama's synthesis starts with construction of the B-ring moiety, and aldol reactions are used extensively throughout the entire synthesis process.[34] The retro synthesis is outlined in Scheme 7–66.

The procedure begins with the oxidation of commercially available methyl 3-hydroxy-2,2-dimethyl propionate with Swern reagent to give the corresponding aldehyde. This is then converted to dimethyl acetal. Reduction of the ester functional group of this compound with LiAlH$_4$ followed by Swern oxidation gives the desired aldehyde **204**. Next, the asymmetric aldol reaction between **204** and ketene silyl acetal **205** is carried out in the presence of chiral diamine and the Lewis acid Sn(OTf)$_2$. The corresponding optically active ester **206** is obtained with good yield and selectivity (*anti/syn* = 80/20, 87–93% ee for the *anti*-isomer). See Scheme 7–67.

The secondary hydroxyl group of **206** is now protected with a PMB group. Reduction of the resulting diastereomeric mixture **207** gives the corresponding alcohol **208**, which is then separated to give a single stereoisomer. A silyl ether

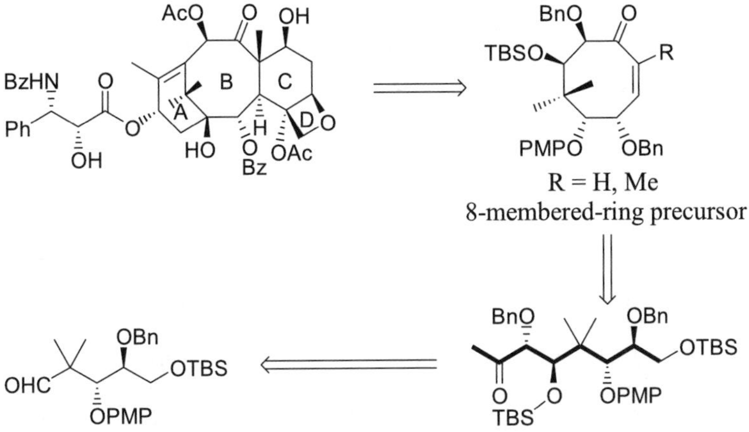

Scheme 7–66

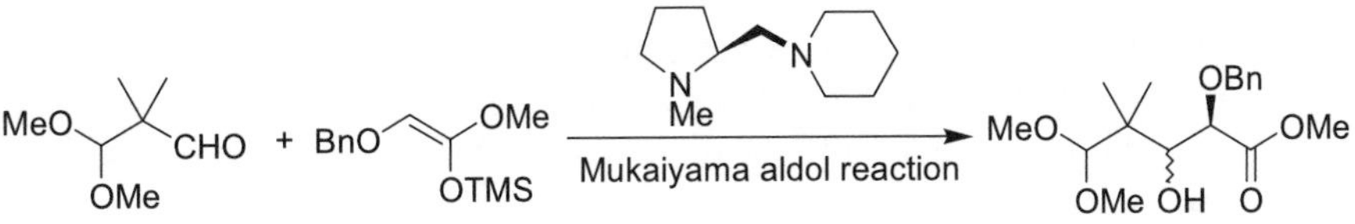

204 **205** **206**

Scheme 7–67. Reagents and conditions: $Sn(OTf)_2$, chiral diamine, $n\text{-}Bu_2Sn(OAc)_2$, CH_2Cl_2, −23°C (68%, *anti*/*syn* = 80/20).

209 is obtained from **208** upon treatment with *t*-butyldimethylsilyl chloride and imidazole. Finally, the acetal is deprotected by acetic acid to give the desired optically active aldehyde **210** (Scheme 7–68).

Alternatively, compound **210** can also be prepared from L-serine via routine conversions, as shown in Scheme 7–69.

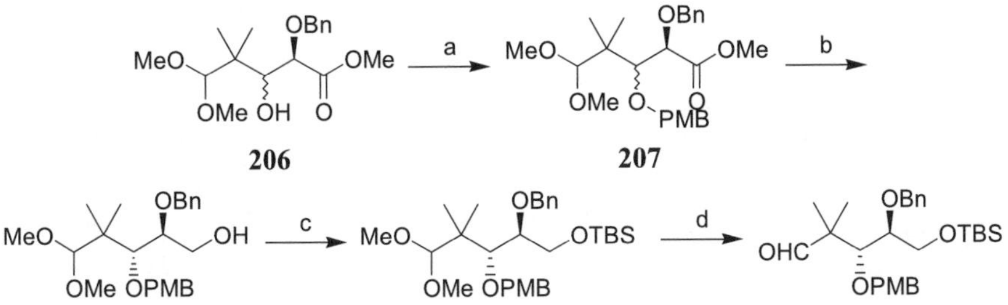

208 **209** **210**

Scheme 7–68. Reagents and conditions: a: $PMBOC(CCl_3)$=NH, TfOH, Et_2O, 0°C (95%, *anti*/*syn* = 80/20). b: $LiAlH_4$, THF, 0°C (86% from **207**-*anti*). c: TBSCl, imidazole, CH_2Cl_2, room temperature (93%). d: AcOH, H_2O, THF, room temperature (87%).

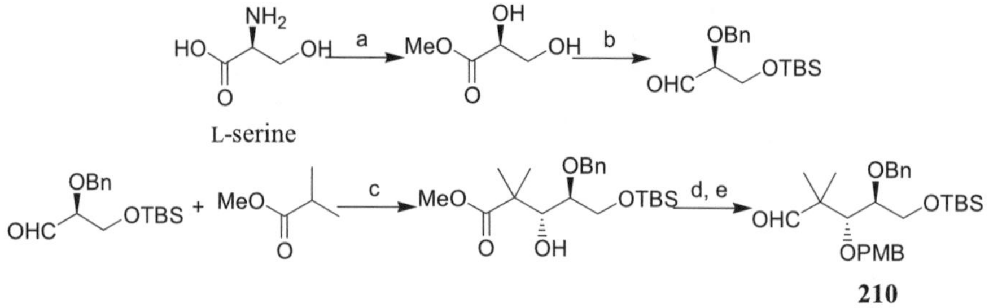

Scheme 7–69. Reagents and conditions: a: (1) NaNO₃, H₂SO₄, H₂O, room temperature (r. t.); (2) HC(OMe)₃, H₂SO₄, MeOH, 60°C (88%). b: (1) TBSCl, imidazole, DMF, 0°C (82%); (2) BnOC(CCl₃)=NH, TfOH, Et₂O, r. t. (100%); (3) DIBAL, hexane, −78°C, (95%). c: LDA, Et₂O, −78°C. d: PMBOC(CCl₃)₃=NH, TfOH, CH₂Cl₂, 0°C (99% based on 76% conversion). e: (1) DIBAL, hexane, −78°C (92%); (2) (COCl)₂, DMSO, Et₃N, CH₂Cl₂, −78°C to r. t. (97%).

Next is the chain extension to get compound **214** for construction of the C-ring moiety. Again, aldol reactions are the key steps. A mild Lewis acid MgCl₂ · Et₂O is the catalyst, and the desired *anti, anti, anti*-**211** can be obtained with good selectivity.

Treatment of the alcohol **211** with *t*-butyldimethylsilyl triflate and 2,6-lutidine affords disiloxyester **212** with high yield. Reduction of the ester function of **212** with DIBAL followed by Swern oxidation gives the corresponding aldehyde **213**, and subsequent alkylation with MeMgBr and Swern oxidation produce methyl ketone **214** (Scheme 7–70).

α-Bromoketo compound **215** is generated with high yield by bromination of the α-position of ketone **214**. The α-brominated intermediate **215** is methylated with LHMDS and MeI in THF. The TBS group is then removed, and Swern

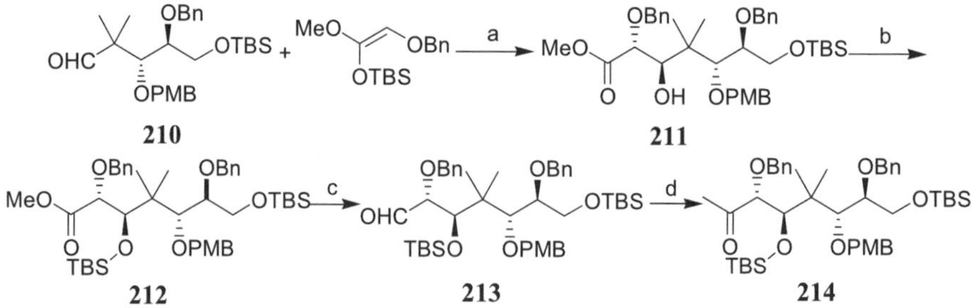

Scheme 7–70. Reagents and conditions: a: MgBr₂ · Et₂O, toluene, −15°C (87% based on 88% conversion, 71% of **211**-*anti, anti, anti*, 16% of **211**-*syn, anti, anti*). b: TBSOTf, 2,6-lutidine, CH₂Cl₂, 0°C (100%). c: (1) DIBAL, toluene, −78°C; (2) (COCl)₂, DMSO, Et₃N, CH₂Cl₂, −78°C to room temperature (94%). d: (1) MeMgBr, Et₂O, −78°C (99%); (2) (COCl)₂, DMSO, Et₃N, CH₂Cl₂, −78°C to room temperature (97%).

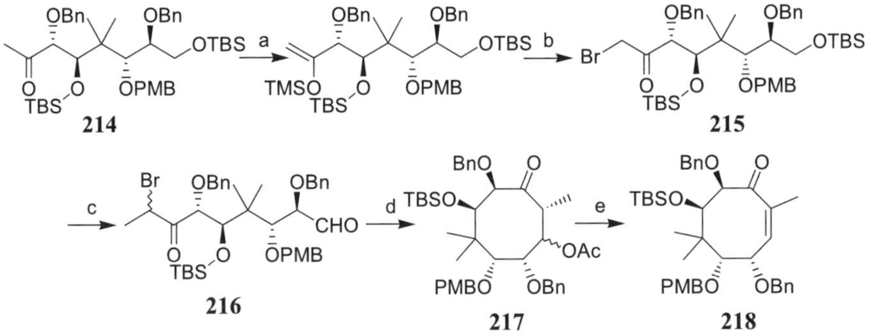

Scheme 7–71. Reagents and conditions: a: LHMDS, TMSCl, THF, −78°C. b: NBS, THF, 0°C (100%). c: (1) LHMDS, MeI, HMPA, THF, −78°C (100%); (2) 1N HCl, THF, room temperature (r. t.) (83%); (3) (COCl)$_2$, DMSO, Et$_3$N, CH$_2$Cl$_2$, −78°C to r. t. (95%). d: (1) SmI$_2$, THF, −78°C (70%); (2) Ac$_2$O, DMAP, pyridine, r. t. (87%). e: DBU, benzene, 60°C (91%).

oxidation produces aldehyde **216**. The ring-closure reaction of the optically active polyoxy unit **216** proceeds smoothly in the presence of excess SmI$_2$, giving the desired aldol product with high yield and good stereoselectivity. Subsequently, treating the acetylated product **217** with DBU generates the desired eight-membered ring enone **218** with high yield (Scheme 7–71). Compound **218** already possesses the B ring of baccatin III and can be used for the introduction of the C-ring moiety.

Michael addition of cuprate reagent, generated in situ from 7 eq. of 2-bromo-5-(triethylsiloxy)pentene, 14 eq. of *t*-BuLi, and 3.7 eq. of copper cyanide, to the enone **218** gives the eight-membered ring ketone **219** with high yield and perfect diastereoselectivity. Ketoaldehyde **220**, a precursor of the BC ring system of taxol, can be obtained in good yield by deprotection of the Michael adduct **219** with 0.5 N HCl and subsequent oxidation with a combination of TPAP and NMO. On treatment with a base, precursor **220** would be expected to generate an enolate anion, which in turn would form the desired bicyclic compound **221**. Indeed, when intramolecular aldol reaction of **220** is carried out in the presence of NaOMe at 0°C, the reaction proceeds smoothly, affording a mixture of bicyclic compounds **221** at nearly quantitative yield with good diastereoselectivity ($221\beta/221\alpha = 92/8$; see Scheme 7–72).

Diastereoselective reduction of the aldol **221**β can be achieved using AlH$_3$ in toluene at −78°C. The corresponding *cis*-diol is preferentially formed. The diol can be protected with isopropylidene acetal to provide tricyclic compound **222**. This can be converted to conformationally rigid C-1 ketone **223** by deprotection of the PMB group and successive oxidation with PDC (Scheme 7–73).

Alkylation of the C-1 carbonyl group of **223** with homoallyllithium reagent in benzene produces the desired bishomoallylic β-alcohol at high yield with perfect diastereoselectivity. Deprotection of the TBS group results in the for-

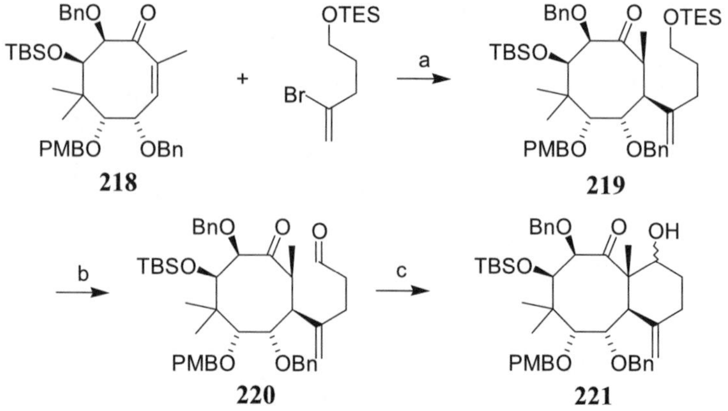

Scheme 7–72. Reagents and conditions: a: *t*-BuLi, CuCN, Et$_2$O, $-23°$C (99% based on 93% conversion). b: (1) 0.5 N HCl, THF, 0°C (97%); (2) TPAP, NMO, 4 Å MS, CH$_2$Cl$_2$, 0°C (92%). c: NaOMe, MeOH, THF, 0°C (98%, **221β/221α** = 92/8).

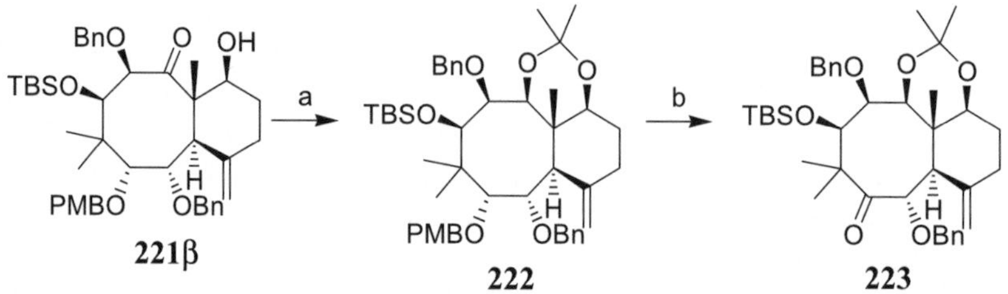

Scheme 7–73. Reagents and conditions: a: (1) AlH$_3$, toluene, $-78°$C (94%); (2) Me$_2$C(OMe)$_2$, CSA, CH$_2$Cl$_2$, room temperature (r. t.) (100%). b: (1) DDQ, H$_2$O, CH$_2$Cl$_2$, r. t. (97%); (2) PDC, CH$_2$Cl$_2$, r.t. (94% yield based on 96% conversion).

mation of C-1 to C-11 *cis*-diol **224**, and successive treatment with dialkylsilyl dichlorides yields the silylene compounds **225** with almost quantitative yields. Alkylation of these silylene compounds with methyllithium furnishes compounds **226** with the desired C-1 protection. Oxidation of the C-11 secondary alcohol gives the C-11 ketones **227** with good yields. The diketones **228** are obtained via oxygenation of the C-12 positions of **227**, and the A-ring moiety is then constructed via intramolecular pinacol coupling in the presence of a low-valence titanium reagent. Compound **230**, which will be used for further functionalization, can be obtained through debenzylation and desilylation of compound **229** (Scheme 7–74).

Regioselective protection of the pentaol **230** with bis-trichloromethyl carbonate and subsequent acetic anhydride treatment affords the C-10 acetoxy C-1, C-2 carbonate compound **231** in good yield. Deprotection of the acetonide function and regioselective silylation of the C-7 hydroxyl group, followed by

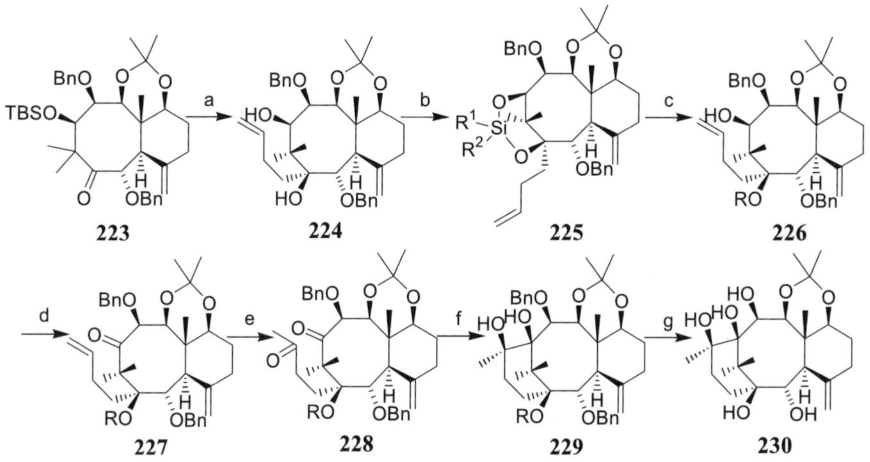

Scheme 7–74. Reagents and conditions: a: (1) homoallyl iodide, *s*-BuLi, *c*-hexane, benzene, $-23°C$ to $0°C$ (96%); (2) TBAF, THF, $50°C$ (100%). b: (1) *c*-HexMeSiCl$_2$, imidazole, DMF, room temperature (r. t.) (**225a**: 99%); (2) *c*-Hex$_2$Si(OTf)$_2$, pyridine, $0°C$ (**225b**: 100%); (3) *t*-BuMeSi(OTf)$_2$, pyridine, $0°C$ (**225c**: 100%). c: MeLi, HMPA, THF, $-78°C$ (**226a**: 96%); $-45°C$ (**226b**: 96%); $-45°C$ (**226c**: 96%). d: TPAP, NMO, 4 Å MS, CH$_2$Cl$_2$, CH$_3$CN, r. t. (80–91%). e: PdCl$_2$, H$_2$O, DMF, r. t. (91–98%). f: TiCl$_2$, LiAlH$_4$, THF, $40°C$ (**229a**: 43–71%); $45°C$ (**229b**: 51–63%); $35°C$ (**229c**: 42–52%). g: (1) Na, NH$_3$ (1), $-78°$ to $-45°C$; (2) TBAF, THF, r. t.

oxidation of the C-9 hydroxyl group with a combination of TPAP and NMO yields compound **232**. A novel taxoid **233** can finally be formed from **232** by desulfurization of the intermediate thionocarbonate with trimethylphosphite, thus furnishing the C-11 to C-12 double bond of baccatin III (Scheme 7–75).

Regioselective oxygenation at the C-13 position of **233** with PCC and NaOAc yields enone **234**, which can be stereoselectively reduced to the desired

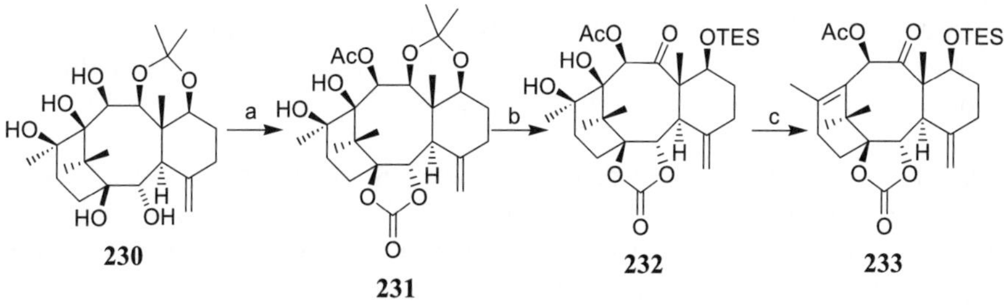

Scheme 7–75. Reagents and conditions: a: (1) (CCl$_3$O)$_2$CO, pyridine, CH$_2$Cl$_2$, $-45°C$ (100%); (2) Ac$_2$O, DMAP, benzene, $35°C$ (84%). b: (1) 3 N HCl, THF, $60°C$; (2) TESCl, pyridine, room temperature (r. t.) (83%); (3) TPAP, NMO, 4 Å MS, CH$_2$Cl$_2$, r. t. (76%). c: (1) TCDI, DMAP, toluene, $100°C$; (2) P(OMe)$_3$, $110°C$ (53%).

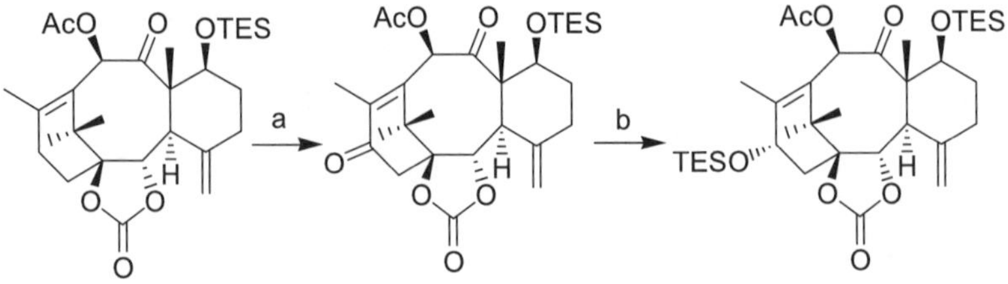

Scheme 7–76. Reagents and conditions: a: PCC, NaOAc, celite, benzene, 95°C (78%). b: (1) K-Selectride, THF, −23°C (87%); (2) TESOTf, pyridine, −23°C (98%).

α-alcohol upon K-Selectride treatment. Protection of the resulting α-hydroxyl group affords compound **235** possessing the ABC rings of baccatin III (Scheme 7–76).

Allylic bromination of **235** with excess CuBr and PhCO$_3$-t-Bu (1:1 molar ratio) gives the separable allylic bromides **236** and **237** with 62% and 15% yield, respectively. Further treating the allylic bromide **236** with CuBr in CH$_3$CN at 50°C leads to a mixture of **236** and **237**, from which 25% of **236** and 69% of **237** can be separated.

α-Face selective dihydroxylation of **237** with OsO$_4$ proceeds smoothly to give a dihydroxy bromide **238** with high yield and as a single stereoisomer. The desired oxetanol can then be obtained with good yield when this dihydroxy bromide **238** is treated with DBU at 50°C in toluene. The corresponding acetate **239** is then prepared by acetylation of the tertiary alcohol using acetic anhydride and DMAP in pyridine. Finally, benzoylation at the C-2 position of the C-1, C-2 carbonate **239** followed by desilylation of the benzoate with HF·pyridine generates baccatin III with high yield (Scheme 7–77).

7.5.2 Asymmetric Synthesis of the Taxol Side Chain

Several methods have been developed for the synthesis of the taxol side chain. We present here the asymmetric construction of this molecule via asymmetric epoxidation and asymmetric ring-opening reactions, asymmetric dihydroxylation and asymmetric aminohydroxylation reaction, asymmetric aldol reactions, as well as asymmetric Mannich reactions.

7.5.2.1 *Via Asymmetric Epoxidation and Related Reactions.* Denis et al.[35] synthesized the taxol side chain derivative via Sharpless epoxidation. Starting from *cis*-cinnamyl alcohol, the corresponding epoxide compound was prepared with 76–80% ee. Subsequent azide ring opening gives a product that possesses the side chain skeleton (Scheme 7–78).

Wang et al.[36] have used the chiral catalyst (DHQ)$_2$–PHAL (see Chapter 4 for the structure) for the asymmetric synthesis of the taxol side chain. Optically enriched diol was obtained at 99% ee via asymmetric dihydroxylation. Sub-

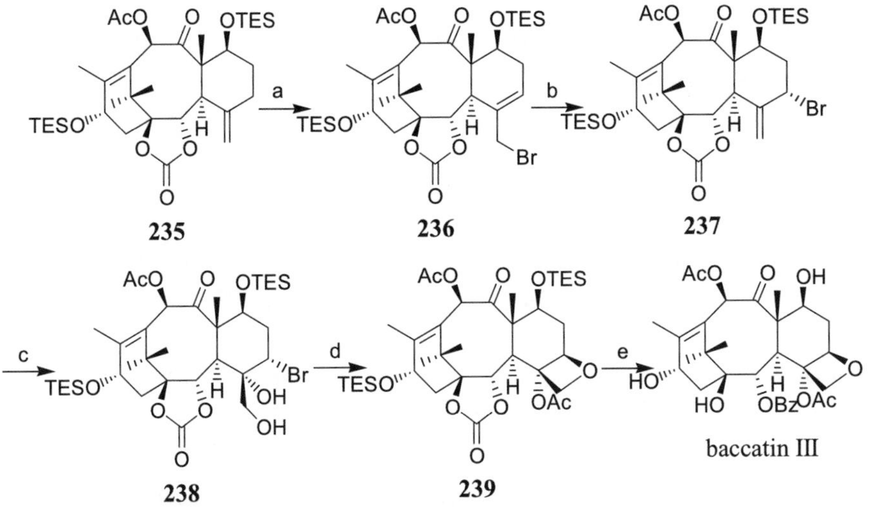

Scheme 7–77. Reagents and conditions: a: CuBr, PhCO₃-*t*-Bu, CH₃CN, −23°C (62% of **236**, 15% of **237**). b: CuBr, CH₃CN, 50°C (25% of **236**, 69% of **237**). c: OsO₄, pyridine, THF, room temperature (r. t.) (96% based on 96% conversion). d: (1) DBU, pyridine, toluene, 50°C (81% based on 52% conversion); (2) Ac₂O, DMAP, pyridine, r. t. (91%). e: (1) PhLi, THF, −78°C (94%); (2) HF · pyridine, THF, r. t. (96%).

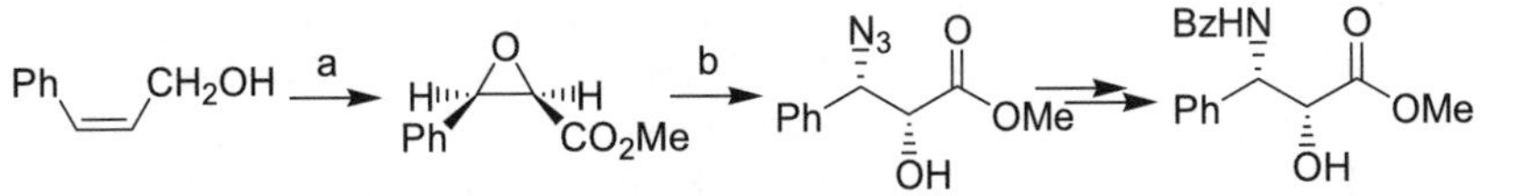

Scheme 7–78. Reagents and conditions: a: (1) TBHP, Ti(OPrⁱ)₄, L-(+)-DET, −30°C (61–65% yield, 76–80% ee); (2) RuCl₃, NaIO₄, NaHCO₃; (3) CH₂N₂ 84% yield. b: (1) Me₃SiN₃, ZnCl₂; (2) H₃O⁺ (97% yield).

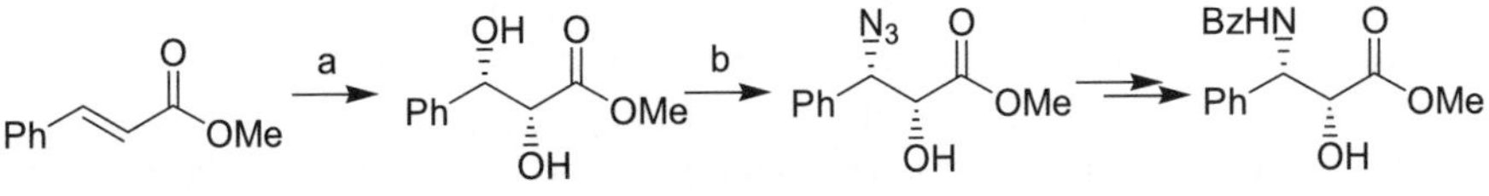

Scheme 7–79. Reagents and conditions: a: (DHQ)₂–PHAL, K₂[OsO₂(OH)₄], K₃[Fe(CN)₆], K₂CO₃, MeSO₂NH₂, 72% yield, 99% ee. b: (1) CH₃C(OMe)₃, TsOH; (2) CH₃COBr, −15°C; (3) NaN₃, DMF, 30°–40°C (71% yield).

sequent reaction converts the 3-hydroxy group to an azide group to furnish the asymmetric synthesis process (Scheme 7–79).

Sharpless' asymmetric aminohydroxylation can also be used for taxol side chain synthesis. For example, using DHQ as a chiral ligand, asymmetric aminohydroxylation of methyl *trans*-cinnamate provides compound **240** in high enantiomeric excess (Scheme 7–80).[37]

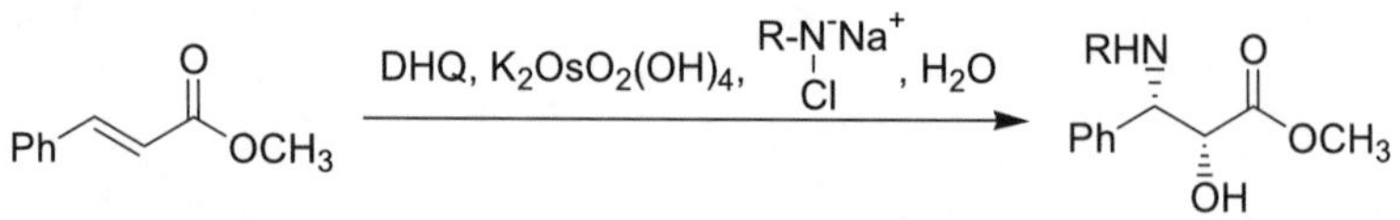

Scheme 7–80

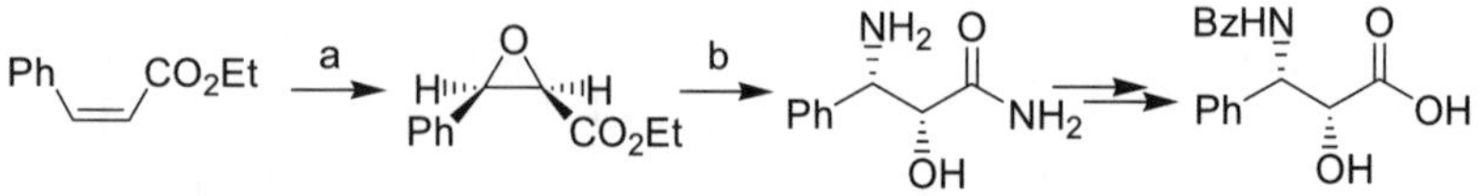

Scheme 7–81. Reagents and conditions: a: Mn–salen, 4-PPNO, NaOCl (56% yield, 95–97% ee). b: NH₃, EtOH, 100°C (65% yield).

Deng and Jacobsen[38] used Mn–salen complex for the asymmetric epoxidation of ethyl cinnamate. Over 95% ee was obtained for the epoxide compound (Scheme 7–81).

7.5.2.2 Via Aldol Reaction. Mukaiyama developed a different route to the taxol side chain, applying the asymmetric aldol reaction. Benzaldehyde is subjected to aldol reaction with an enol silyl ether **241** derived from *S*-ethyl benzyloxyethanethioate. In the presence of a chiral promoter consisting of chiral diamine, Sn(OTf)₂, and *n*-Bu₂Sn(OAc)₂, the aldol product **242** is obtained at high yield with almost perfect stereoselectivity (*anti/syn* = 99/1, 96% ee for the *anti*-aldol). A Mitsunobu reaction and subsequent reduction converts the C-3-*anti* hydroxyl group to a C-3-*syn* amino group, leading to compound **243** with the desired stereochemistry for the taxol side chain. Subsequent conversion steps furnish the full side chain for the final synthesis of taxol (Scheme 7–82).

Mukai et al.[39] used a chiral aryl chromium complex to synthesize the taxol side chain via substrate-controlled aldol reaction (Scheme 7–83).

Commerçon et al.[40] developed a method based on an Evans-type auxiliary-controlled aldol reaction. Subsequent treatment of the aldol product with base produced the standard epoxide compound for the asymmetric synthesis of the taxol side chain (Scheme 7–84).

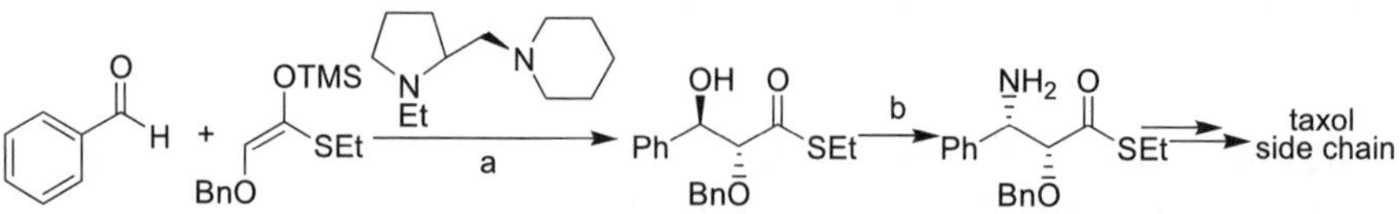

Scheme 7–82. Reagents and conditions: a: Sn(OTf)₂, chiral diamine, *n*-Bu₂Sn(OAc)₂, CH₂Cl₂, −78°C (95%). b: (1) HN₃, Ph₃P, DEAD, benzene, room temperature (82%); (2) Ph₃P, H₂O, THF, 55°C (90% yield based on 82% conversion).

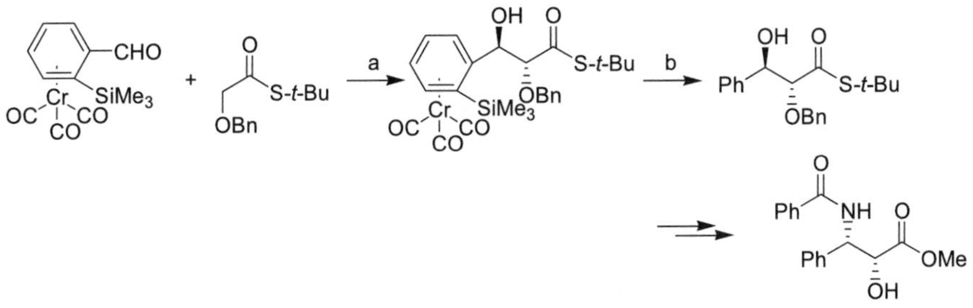

Scheme 7–83. Reagents and conditions: a: TiCl$_4$, Et$_3$N, CH$_2$Cl$_2$, −78°C (93%, ee, *anti/ syn* = 95:5). b: (1) *n*-Bu$_4$NF · HF, MeCN/THF; (2) hv.

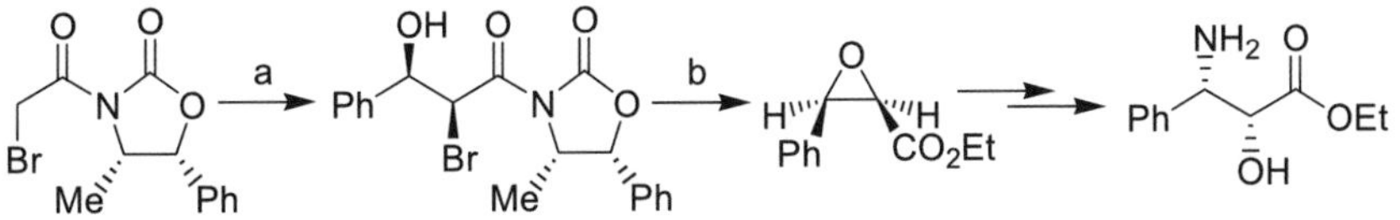

Scheme 7–84. Reagents and conditions: a: (1) *n*-Bu$_2$BOTf, Et$_3$N; (2) PhCHO (58%). b: EtOLi, THF (81%).

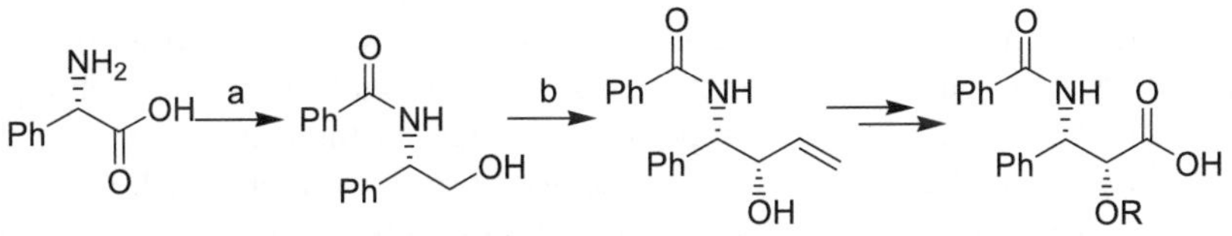

Scheme 7–85. Reagents and conditions: a: (1) LiAlH$_4$; (2) PhCOCl (79%). b: (1) Swern oxidation; (2) CH$_2$=CHMgBr (62%).

Denis et al.[41] also introduced a method for the synthesis of the side chain starting from the amino acid (*S*)-phenylglycine (Scheme 7–85).

7.5.3.2 *Via Mannich-Type Reaction.* Hattori et al.[42] used a Mannich-type reaction for constructing the taxol side chain. In the presence of a BINOL-containing boron compound, the asymmetric Mannich reaction proceeded smoothly, providing the product with good yield (Scheme 7–86).

Further details of the asymmetric construction of the taxol side chain or related compounds can be found elsewhere.[43]

7.6 SUMMARY

This chapter has introduced the asymmetric synthesis of several types of natural products: erythronolide A, 6-deoxyerythronolide, rifamycin S, prostaglandins and baccatin III, the polycyclic part of taxol, as well as the taxol side chain. The

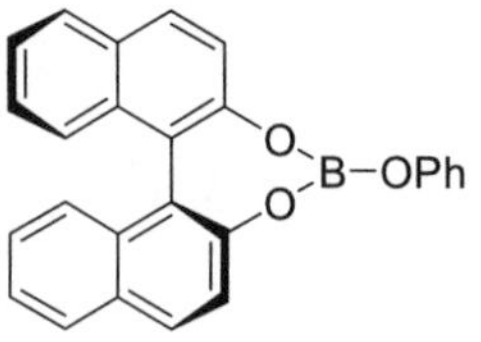

244

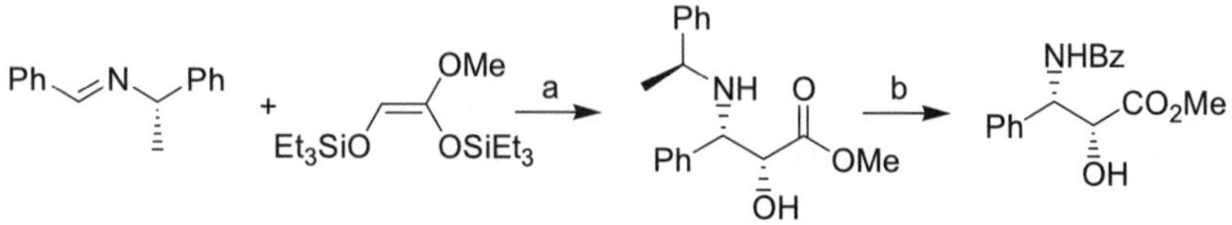

Scheme 7–86. Reagents and conditions: a: (1) **244**; (2) HCl (90–95%). b: (1) H_2, Pd/C; (2) BzCl (68%).

asymmetric syntheses of erythronolide A, 6-deoxyerythronolide, rifamycin S, and the prostaglandin compounds illustrate how the synthesis process can be simplified. The synthesis of baccatin III shows the opportunity and challenge in this field, as most of the synthetic work toward baccatin III did not fully exploit asymmetric synthesis as an efficient and economic tool for constructing optically active compounds. Some methods used traditional resolution techniques to isolate the optically active compounds very late in the procedure. As mentioned in the first chapter, this can be a tedious process, and as much as half of the synthesized product might be wasted. In the interest of speed in accomplishing the total synthesis of such an important and challenging compound, the strategy of using whatever available methodologies is correct and the results are admirable. However, once the target of total synthesis is achieved, future development of an economical process for taxol or its analogs will most likely depend on the development of new asymmetric synthetic methods. In future work, researchers are bound to encounter an increasing array of synthetic problems and opportunities in asymmetric synthesis.

7.7 REFERENCES

1. Masamune, S.; Choy, W.; Petersen, J. S.; Sita, L. R. *Angew. Chem. Int. Ed. Engl.* **1985**, *24*, 1.

2. Noyori, R. *Asymmetric Catalysis in Organic Synthesis*, John Wiley & Sons, New York, **1994**, p 298.

3. References on the synthesis of erythromycin A: (a) Woodward, R. B.; Logusch, E.; Nambiar, K. P.; Sakan, K.; Ward, D. E.; Au-Yeung, B. W.; Balaram, P.; Browne, L. J.; Card, P. J.; Chen, C. H.; Chênevert, R. B.; Fliri, A.; Frobel, K.; Gais, H.; Garratt, D. G.; Hayakawa, K.; Heggie, W.; Hesson, D. P.; Hoppe, D.; Hoppe, I.; Hyatt, J. A.; Ikeda, D.; Jacobi, P. A.; Kim, K. S.; Kobuke, Y.; Kojima, K.;

Krowicki, K.; Lee, V. J.; Leutert, T.; Malchenko, S.; Martens, J.; Matthews, R. S.; Ong, B. S.; Press,; J. B.; RajanBabu, T. V.; Rousseau, G.; Sauter, H. M.; Suzuki, M.; Tatsuta, K.; Tolbert, L. M.; Truesdale, E. A.; Uchida, I.; Ueda, Y.; Uyehara, T.; Vasella, A. T.; Vladuchick, W. C.; Wade, P. A.; Williams, R. M.; Wong, H. N. C. *J. Am. Chem. Soc.* **1981**, *103*, 3210. (b) Mulzer, J. *Angew. Chem. Int. Ed. Engl.* **1991**, *30*, 1452. (c) Stürmer, R.; Ritter, K.; Hoffmann, R. W. *Angew. Chem. Int. Ed. Engl.* **1993**, *32*, 101.

4. Woodward, R. B. in Todd, A. R. ed. *Perspectives in Organic Chemistry*, John Wiley & Sons, New York, **1956**, p 155.

5. Masamune, S.; Hirama, M.; Mori, S.; Ali, S. A.; Garvey, D. S. *J. Am. Chem. Soc.* **1981**, *103*, 1568.

6. (a) Nagaoka, H.; Rutsch, W.; Schmid, G.; Iio, H.; Johnson, M. R.; Kishi, Y. *J. Am. Chem. Soc.* **1980**, *102*, 7962. (b) Iio, H.; Nagaoka, H. Kishi, Y. *J. Am. Chem. Soc.* **1980**, *102*, 7965.

7. Nagaoka, H.; Kishi, Y. *Tetrahedron* **1981**, *37*, 3873.

8. Masamune, S.; Imperiali, B.; Garvey, D. S. *J. Am. Chem. Soc.* **1982**, *104*, 5528.

9. (a) Harada, T.; Kagamihara, Y.; Tanaka, S.; Sakamoto, K.; Oku, A. *J. Org. Chem.* **1992**, *57*, 1637. (b) Fraser-Reid, B.; Magdzinski, L.; Molino, B. F.; Mootoo, D. R. *J. Org. Chem.* **1987**, *52*, 4495. (c) Roush, W. R.; Palkowitz, A. D. *J. Am. Chem. Soc.* **1987**, *109*, 953. (d) Ziegler, F. E.; Clain, W. T.; Kneisley, A.; Stirchak, P.; Wester, R. T. *J. Am. Chem. Soc.* **1988**, *110*, 5442. (e) Roush, W. R.; Palkowitz, A. D.; Ando, K. *J. Am. Chem. Soc.* **1990**, *112*, 6348. (f) Nakata, M.; Akiyama, N.; Kamata, J.; Kojima, K.; Masuda, H.; Kinoshita, M.; Tatsuta, K. *Tetrahedron* **1990**, *46*, 4629. (g) Miyashita, M.; Yoshihara, K.; Kawamine, K.; Hoshino, M.; Irie, H. *Tetrahedron Lett.* **1993**, *34*, 6285. (h) Paterson, I.; McClure, C. K.; Schumann, R. C. *Tetrahedron Lett.* **1989**, *30*, 1293. (i) Born, M.; Tamm, C. *Helv. Chim. Acta* **1990**, *73*, 2242. (j) Harada, T.; Oku, A. *Synlett* **1994**, 95.

10. (a) Samuelssion, B. *Angew. Chem. Int. Ed. Engl.* **1983**, *22*, 805 (b) Bergström, S. *Angew. Chem. Int. Ed. Engl.* **1983**, *22*, 858.

11. See Chapdelaine, M. J.; Huke, M. "Tandem Vicinal Difunctionalization: β-Addition to α, β-Unsaturated Carbonyl Substrates Followed by α-Functionalization" in Paquette, L. A. ed. *Org. React.* John Willey & Sons, New York, **1990**, p. 38.

12. (a) Morita, Y.; Suzuki, M.; Noyori, R. *J. Org. Chem.* **1989**, *54*, 1785. (b) Suzuki, M.; Morita, Y.; Koyano, H.; Koga, M.; Noyori, R. *Tetrahedron* **1990**, *46*, 4809.

13. (a) Suzuki, M.; Yanagisawa, A.; Noyori, R. *J. Am. Chem. Soc.* **1985**, *107*, 3348. (b) Suzuki, M.; Yanagisawa, A.; Noyori, R. *J. Am. Chem. Soc.* **1988**, *110*, 4718. (c) Okamoto, S.; Shimazaki, T.; Kobayashi, Y.; Sato, F. *Tetrahedron Lett.* **1987**, *28*, 2033. (d) Sato, F.; Kobayashi, Y. *Synlett* **1992**, 849.

14. (a) Okamoto, S.; Shimazaki, T.; Kobayashi, Y.; Sato, F. *Tetrahedron Lett.* **1987**, *28*, 2033. (b) Sato, F.; Kobayashi, Y. *Synlett* **1992**, 849.

15. (a) Noyori, R.; Kitamura, M. *Angew. Chem. Int. Ed. Engl.* **1991**, *30*, 49. (b) Noyori, R.; Suga, S.; Kawai, K.; Okada, S.; Kitamura, M.; Oguni, N.; Hayashi, M.; Kaneko, T.; Matsuda, Y. *J. Organomet. Chem.* **1990**, *382*, 19.

16. Mukaiyama, T.; Suzuki, K. *Chem. Lett.* **1980**, 255.

17. Midland, M. M.; McDowell, D. C.; Hatch, R. L.; Tramontano, A. *J. Am. Chem. Soc.* **1980**, *102*, 867.

18. (a) Noyori, R.; Tomino, I.; Tanimoto, Y.; Nishizawa, M. *J. Am. Chem. Soc.* **1984**, *106*, 6709. (b) Noyori, R.; Tomino, I.; Yamada, M.; Nishizawa, M. *J. Am. Chem. Soc.* **1984**, *106*, 6717.

19. Kinato, Y.; Matsumoto, T.; Okamoto, S.; Shimazaki, T.; Kobayashi, Y. Sato, F. *Chem. Lett.* **1987**, 1523.

20. Ohta, T.; Takaya, H.; Kitamura, M.; Nagai, K.; Noyori, R. *J. Org. Chem.* **1987**, *52*, 3174.

21. For reviews, see (a) Kingston, D. G. I.; Molinero, A. A.; Rimoldi, J. M. *Prog. Chem. Org. Nat. Prod.* **1993**, *61*, 1. (b) Nicolaou, K. C.; Dai, W.-M.; Guy, R. K. *Angew. Chem. Int. Ed. Engl.* **1994**, *33*, 15.

22. Nicolaou, K. C.; Yang, Z.; Liu, J.-J.; Ueno, H.; Nantermet, P, G.; Guy, R. K.; Claiborne, C. F.; Renaud, J.; Couladouros, E. A.; Paulvannan, K.; Sorensen E. J. *Nature* **1994**, *367*, 630.

23. (a) Holton, R. A.; Somoza, C.; Kim, H.; Liang, F.; Biediger, R. J.; Boatman, P. D.; Shindo, M.; Smith, C. C.; Kim, S.; Nadizadeh, H.; Suzuki, Y.; Tao, C.; Vu, P.; Tang, S.; Zhang, P.; Murthi, K. K.; Gentile, L. N.; Liu, J. H. *J. Am. Chem. Soc.* **1994**, *116*, 1597. (b) Holton, R. A.; Kim, H.; Somoza, C.; Liang, F.; Biediger, R. J.; Boatman, P. D.; Shindo, M.; Smith, C. C.; Kim, S.; Nadizadeh, H.; Suzuki, Y.; Tao, C.; Vu, P.; Tang, S.; Zhang, P.; Murthi, K. K.; Gentile, L. N.; Liu, J. H. *J. Am. Chem. Soc.* **1994**, *116*, 1599.

24. (a) Holton, R. A.; Kennedy, R. M. *Tetrahedron Lett.* **1984**, *25*, 4455. (b) Holton, R. A. *J. Am. Chem. Soc.* **1984**, *106*, 5731. (c) Holton, R. A.; Juo, R. R.; Kim, H. B.; Williams, A. D.; Harusawa, S.; Lowenthal, R. E.; Yogai, S. *J. Am. Chem. Soc.* **1988**, *110*, 6558.

25. Lee, S. D.; Chan, T. H. Kwon, K. S. *Tetrahedron Lett.* **1984**, *25*, 3399.

26. Burgess, E. M.; Penton, H. R.; Taylor, E. A. *J. Org. Chem.* **1973**, *38*, 26.

27. (a) Wender, P. A.; Badham, N. F.; Conway, S. P.; Floreancig, P. E.; Glass, T. E.; Gränicher, C.; Houze, J. B.; Jänichen, J.; Lee, D.; Marquess, D. G.; McGrane, P. L.; Meng, W.; Mucciaro, T. P.; Mühlebach, M.; Natchus, M. G.; Paulsen, H.; Rawlins, D. B.; Satkofsky, J.; Shuker, A. J.; Sutton, J. C.; Taylor, R. E.; Tomooka, K. *J. Am. Chem. Soc.* **1997**, *119*, 2755. (b) Wender, P. A.; Badham, N. F.; Conway, S. P.; Floreancig, P. E.; Glass, T. E.; Houze, J. B.; Krauss, N. E.; Lee, D.; Marquess, D. G.; McGrane, P. L.; Meng, W.; Natchus, M. G.; Shuker, A. J.; Sutton, J. C.; Taylor, R. E. *J. Am. Chem. Soc.* **1997**, *119*, 2757.

28. Davis, F. A.; Vishwakarma, L. C.; Billmers, J. M.; Finn, J. *J. Org. Chem.* **1984**, *49*, 3241.

29. Crabtree, R. H.; Davis, M. W. *J. Org. Chem.* **1986**, *51*, 2655.

30. Hara, R.; Furukawa, T.; Horiguchi, Y.; Kuwajima, I. *J. Am. Chem. Soc.* **1996**, *118*, 9186.

31. (a) Masters, J. J.; Link, J. T.; Snyder, L. B.; Young, W. B.; Danishefsky, S. J. *Angew. Chem. Int. Ed. Engl.* **1995**, *34*, 1723. (b) Danishefsky, S. J.; Masters, J. J.; Young, W. B.; Link, J. T.; Snyder, L. B.; Magee, T. V.; Jung, D. K.; Isaacs, R. C. A.; Bornmann, W. G.; Alaimo, C. A.; Coburn, C. A.; Di Grandi, M. J. *J. Am. Chem. Soc.* **1996**, *118*, 2843.

32. Corey, E. J.; Chaykovsky, M. *J. Am. Chem. Soc.* **1965**, *87*, 1353.

33. (a) Nicolaou, K. C.; Nantermet, P. G.; Ueno, H.; Guy, R. K.; Couladouros, E. A.;

Sorensen, E. J. *J. Am. Chem. Soc.* **1995**, *117*, 624. (b) Nicolaou, K. C.; Liu, J.-J.; Yang, Z.; Ueno, H.; Sorensen, E. J.; Claiborne, C. F.; Guy, R. K.; Hwang, C.-K.; Nakada, M.; Nantermet, P. G. *J. Am. Chem. Soc.* **1995**, *117*, 634. (c) Nicolaou, K. C.; Yang, Z.; Liu, J.-J.; Nantermet, P. G.; Claiborne, C. F.; Renaud, J.; Guy, R. K.; Shibayama, K. *J. Am. Chem. Soc.* **1995**, *117*, 645. (d) Nicolaou, K. C.; Ueno, H.; Liu, J.-J.; Nantermet, P. G.; Yang, Z.; Renaud, J.; Paulvannan, K.; Chadha, R. *J. Am. Chem. Soc.* **1995**, *117*, 653.

34. Mukaiyama, T.; Shiina, I.; Iwadare, H.; Saitoh, M.; Nishimura, T.; Ohkawa, N.; Sakoh, H.; Nishimura, K.; Tani, Y.; Hasegawa, M.; Yamada, K.; Saitoh, K. *Chem. Eur. J.* **1999**, *5*, 121.

35. Denis, J.-N.; Greene, A. E.; Serra, A. A.; Luche, M. *J. Org. Chem.* **1986**, *51*, 46.

36. Wang, Z.-M.; Kolb, H. C.; Sharpless, K. B. *J. Org. Chem.* **1994**, *59*, 5104.

37. Goossen, L. J.; Liu, H.; Dress, K. R.; Sharpless, K. B. *Angew. Chem. Int. Ed. Engl.* **1999**, *38*, 1080.

38. Deng, L.; Jacobsen, E. N. *J. Org. Chem.* **1992**, *57*, 4320.

39. Mukai, C.; Kim, I. J.; Hanaoka, M. *Tetrahedron Asymmetry* **1992**, *3*, 1007.

40. Commerçon, A.; Bezard, D.; Bernard, F.; Bourzat, J. D. *Tetrahedron Lett.* **1992**, *33*, 5185.

41. a) Denis, J.-N.; Correa, A.; Greene, A. E. *J. Org. Chem.* **1991**, *56*, 6939. (b) Kanazawa, A. M.; Correa, A.; Denis, J.- N.; Luche, M.-J.; Greene, A. E. *J. Org. Chem.* **1993**, *58*, 255.

42. Hattori, K.; Miyata, M.; Yamamoto, H. *J. Am. Chem. Soc.* **1993**, *115*, 1151.

43. (a) Li, G.; Chang, H. T.; Sharpless, K. B. *Angew Chem. Int. Ed. Engl.* **1996**, *35*, 451. (b) Rudolph, J.; Sennhenn, P. C.; Vlaar, C. P.; Sharpless, K. B. *Angew. Chem. Int. Ed. Engl.* **1996**, *35*, 2810. (c) Reiser, O. *Angew. Chem. Int. Ed. Engl.* **1996**, *35*, 1308. (d) Kobayashi, S.; Ishitani, H.; Ueno, M. *J. Am. Chem. Soc.* **1998**, *120*, 431.

CHAPTER 8

Enzymatic Reactions and Miscellaneous Asymmetric Syntheses

Chapters 2 through 6 introduced many asymmetric organic reactions catalyzed by small molecules, such as C–C bond formation, reduction, and oxidation reactions. Chapter 7 provided further examples of how asymmetric reactions are used in organic synthesis. This chapter starts with a general introduction to enzyme-catalyzed asymmetric organic reactions.

Despite the diverse range of documented enzyme-catalyzed reactions, there are only certain types of transformations that have thus far emerged as synthetically useful. These reactions are the hydrolysis of esters, reduction/oxidation reactions, and the formation of carbon–carbon bonds. The first part of this chapter gives a brief overview by describing some examples of various biotransformations that can easily be handled and accessed by synthetic organic chemists. These processes are now attracting more and more attention from nonspecialists of enzymes.

In addition, there are an increasing number of methods in organic synthesis that involve C–C bond formation, oxidation, and reduction. Although these methods have been discussed in the previous chapters according to their categories, it would perhaps be desirable to detail them in the second part of this chapter. With the steady increase in the number of publications, some of the most advanced methodologies need to be classified, and this last chapter is the logical place for their inclusion.

The last part of this chapter presents some new concepts that are of great significance in the development of asymmetric synthesis.

8.1 ENZYMATIC AND RELATED PROCESSES

Enzymes are remarkable molecular devices that determine the pattern of chemical transformations in biological systems. The most striking characteristics of enzymes are their catalytic power and specificity. As a class of macromolecules, they are highly effective in catalyzing diverse chemical reactions because of their ability to specifically bind to a substrate and their ability to accelerate reactions by several orders of magnitude. Applying enzymes or organisms in organic synthesis has become one of the fastest growing areas for the development of new methodology, and the work in this area has been comprehensively reviewed.[1]

Most asymmetric reductions that can be enzymatically effected have been the reactions of ketones. These reactions can be conducted with whole cells as well as with isolated enzymes. In the latter case, of course, at least one equivalent of a cofactor such as NADH or NADPH (nicotinamide adenine dinucleotide) is required to serve as the actual reductant in the reaction system.

Enzyme-catalyzed reactions can provide a rich source of chiral starting materials for organic synthesis.[2] Enzymes are capable of differentiating the enantiotopic groups of prochiral and *meso*-compounds. The theoretical conversion for enzymatic desymmetrization of *meso*-compounds is 100%; therefore enzymatic desymmetrization of *meso*-compounds has gained much attention and constitutes an effective entry to the synthesis of enantiomerically pure compounds.

8.1.1 Lipase/Esterase-Catalyzed Reactions

Biocatalytic hydrolysis or transesterification of esters is one of the most widely used enzyme-catalyzed reactions. In addition to the kinetic resolution of common esters or amides, attention is also directed toward the reactions of other functional groups such as nitriles, epoxides, and glycosides. It is easy to run these reactions without the need for cofactors, and the commercial availability of many enzymes makes this area quite popular in the laboratory.

For the hydrolysis of esters, the general rule governing the selection of either a lipase or an esterase is shown in Figure 8–1. Esterase acts on type I esters, and lipase works on type II reactions. That is, hydrolysis or transesterification of an ester containing a large chiral carboxylic acid and a small alcohol molecule can best be effected by an esterase, and esters consisting of a small carboxylic acid and a large chiral alcohol may best be effected using a lipase. Thus esterase is used for the kinetic resolution of chiral acids, while lipase can be used for the

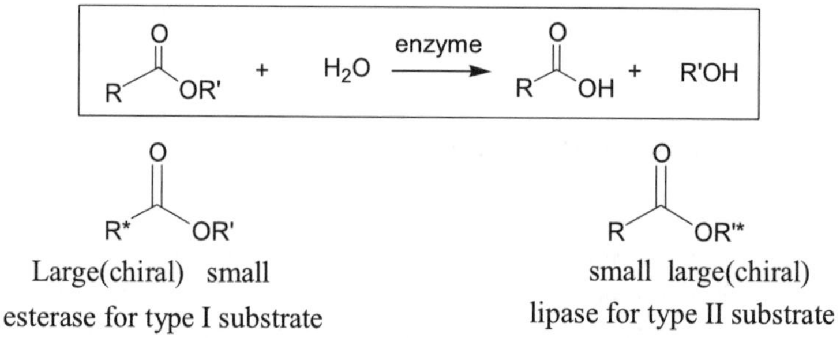

Large(chiral) small
esterase for type I substrate

small large(chiral)
lipase for type II substrate

Figure 8–1. Lipase/esterase-catalyzed ester hydrolysis.

kinetic resolution of chiral alcohols. Enzymes such as pig liver esterase (PLE), porcine pancreatic lipase (PPL), *Pseudomonas* sp. lipase, *Candida cylindracea* lipase, and *Mucor mienei* lipase are the more popular choices for these reactions.

In a lipase-catalyzed reaction, the acyl group of the ester is transferred to the hydroxyl group of the serine residue to form the acylated enzyme. The acyl group is then transferred to an external nucleophile with the return of the enzyme to its preacylated state to restart the catalytic cycle. A variety of nucleophiles can participate in this process. For example, reaction in the presence of water results in hydrolysis, reaction in alcohol results in esterification or transesterification, and reaction in amine results in amination. Kirchner et al.[3] reported that it was possible to use hydrolytic enzymes under conditions of limited moisture to catalyze the formation of esters, and this is now becoming very popular for the resolution of alcohols.[4]

Kinetic resolution reactions on C_2-symmetric substrates have important applications. Desymmetrization is just one example of such a kinetic resolution reaction. Enzymatic desymmetrization is outlined in Scheme 8–1.[5,6]

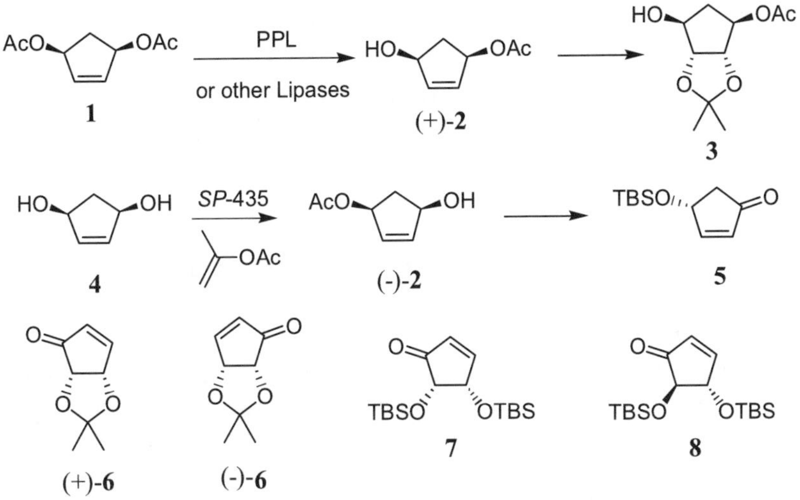

Scheme 8–1

A variety of enzymes (such as acetylcholine esterase, *Porcine pancreatic* lipase, *Pseudomonas cepacia* lipase, and *Candida antarcita* lipase) have been found useful in the preparation of enantiomerically pure cyclopentenol (+)-**2** from **1**. The enantiomeric (−)-**2** has been prepared from diol **4** by enzymatic acetylation catalyzed by *SP*-345 with isopropenyl acetate in an organic medium. The key intermediate cyclopentanones (+)-**6**, (−)-**6**, **7**, and **8**, which are useful in the preparation of many bioactive molecules, can be obtained from **3** and **5** via routine chemical transformations.[7]

8.1.2   Reductions

The asymmetric reduction of prochiral functional groups is an extremely useful transformation in organic synthesis. There is an important difference between isolated enzyme-catalyzed reduction reactions and whole cell–catalyzed transformations in terms of the recycling of the essential nicotinamide adenine dinucleotide (phosphate) [NAD(P)H] cofactor. For isolated enzyme-catalyzed reductions, a cofactor recycling system must be introduced to allow the addition of only a catalytic amount (5% mol) of NAD(P)H. For whole cell–catalyzed reductions, cofactor recycling is automatically achieved by the cell, and the addition of a cofactor to the reaction system is normally not required.

β-Hydroxy esters have been obtained successfully with baker's yeast (*Saccharomyces cerevisidae*), and this has shown a wide scope of application. The facial selectivity in the reduction of both isolated ketones and β-keto esters can be reliably determined by using Prelog's rule,[8] which predicts that the hydrogen addition by the yeast will occur from the front face (Scheme 8–2). *Anti*-Prelog microbial reduction of α-ketones with *Geotrichum* sp. 38 (G38) has been introduced by Gu et al.[9]

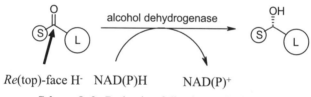

Re(top)-face H⁻   NAD(P)H   NAD(P)⁺

Scheme 8–2. Reduction following Prelog's rule.

The reduction of carbon–carbon double bonds can be achieved when one or more electron-withdrawing groups is attached to increase the reactivity of the double bond. For example, treating cyclohexendione **9** with baker's yeast gives compound **10**, reducing the double bond with high regioselectivity and stereoselectivity (>97%) (Scheme 8–3).[10] Recently, a novel C=C reductase has been isolated from the cells of baker's yeast. It shows good selectivity in reducing α,β-unsaturated ketones to (*S*)-ketones.[10b]

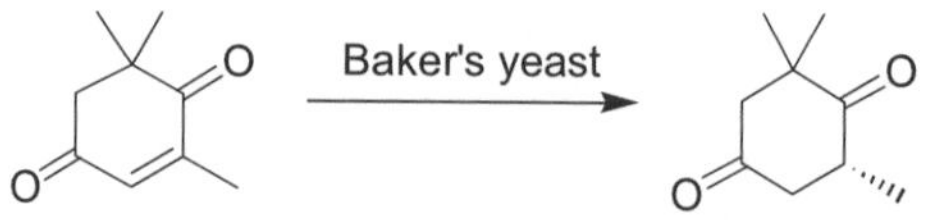

9   Baker's yeast   **10**

Scheme 8–3

8.1.3 Enantioselective Microbial Oxidation

It has been known since the 1950s that benzene and its derivatives can be oxidized to the corresponding cyclohexadienols in the presence of *Pseudomonas putida* (see Scheme 8–4 for an example).

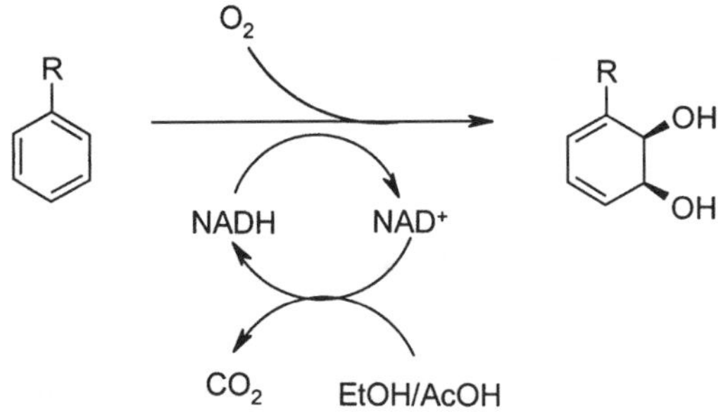

Scheme 8–4. Oxidation of substituted benzene with *Pseudomonas putida*.

The *Pseudomonas putidae* organisms were initially selected for their ability to use benzene as the sole carbon source and were thereafter mutated to prevent further metabolism of the cyclohexadienols produced. This provides an efficient approach to all four stereoisomers of sphingosine (Fig. 8–2)[11] (see Chapter 3 for the synthesis of sphingosine compounds).

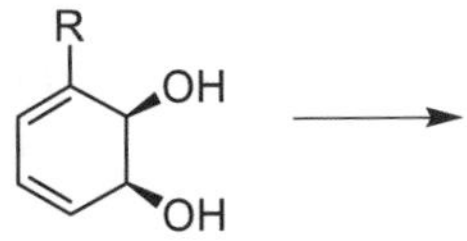

D-*erythro*-sphingosine
D-*threo*-sphingosine
L-*erythro*-sphingosine
L-*threo*-sphingosine

Figure 8–2. Preparation of sphingosine compounds via substituted benzene oxidation.

Bio-oxidation of bromobenzene **11** catalyzed by *Pseudomonas putidae* leads to diol **12**. Protection of diol **12**, followed by the addition of an acyl nitroso dienophile and subsequent reduction gives compound **14**. This compound can be used as the key intermediate in the preparation of (+)-1-deoxy-galacto-nojirimycin (**16**) and related indolizidine compounds (**15**) (Scheme 8–5).[12]

Another extensively studied bio-oxidation is the monooxygenase-catalyzed Baeyer-Villiger oxidation of cyclic ketones.[13] As the first example, cyclohexanone monooxygenase from *Acinetobacter* sp. NCIB 9871 was expressed in baker's yeast to create a general purpose reagent for asymmetric Baeyer-Villiger oxidation. This engineered yeast combines the advantages of purified enzymes with the benefits of whole-cell reactions. It has been used to oxidize a series of 2-, 3-, and 4-substituted cyclohexanones, providing the corresponding

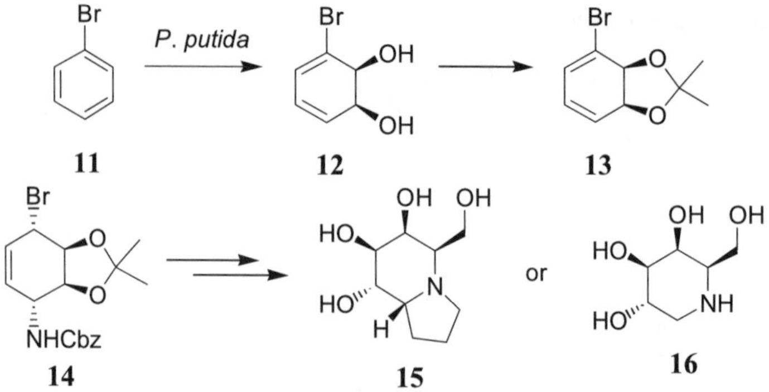

Scheme 8–5

TABLE 8–1. Yeast-Mediated Kinetic Resolution of 2-Substituted Ketones via Baeyer-Villiger Oxidation

Entry	R	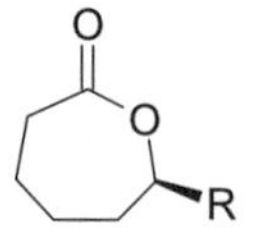 Yield (%)	ee (%)	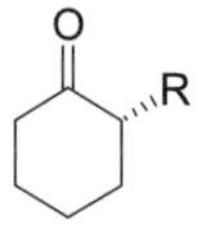 Yield (%)	ee (%)
1	Me	50	49	—	—
2	Et	79	95	69	98
3	n-Pr	54	97	66	92
4	i-Pr	41	98	46	96
5	Allyl	59	98	58	98
6	n-Bu	59	98	64	98

ee = Enantiomeric excess.

Reprinted with permission by Am. Chem. Soc., Ref. 13c.

lactones with good yields and high ee. Table 8–1 shows typical results for asymmetric Baeyer-Villiger oxidation of 2-substituted cyclohexanones.

It is only recently that isolated enzymes have been used in the presence of appropriate cofactor recycling systems.[14] Not long ago, application of the whole-cell system was the only way to get high yields and high ee in enzyme-catalyzed organic synthesis.

8.1.4 Formation of C–C Bond

Cyanohydrins are starting materials of widespread interest for preparing important compounds such as α-hydroxy acids/esters, α-amino acids, β-amino alcohols, α-hydroxy aldehydes, vicinal diols, and α-hydroxy ketones. Cyanohydrin compounds can be synthesized using various chiral catalysts such as cyclic dipeptides[15] and chiral complexes bearing Ti,[16] Al,[17] or B[18] as the central metal. The enantioselective synthesis of cyanohydrins has also been achieved by using enzymes, hydronitrilases (oxynitrilases), isolated from different plant sources.[19]

Oxynitrilases, isolated from either almond [(R)-specific][20] or a microorganism [(S)-specific],[21] catalyze the enantioselective addition of cyanide ion to a range of aromatic or aliphatic aldehydes, providing cyanohydrins with ee values of up to 99%.

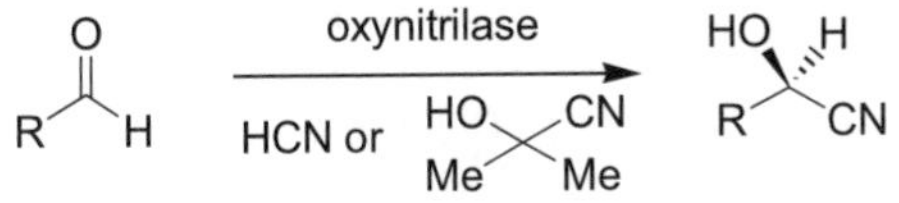

Scheme 8–6

To suppress the noncatalyzed reaction (which decreases the enantioselectivity) between acetone cyanohydrin and the substrate, ethyl acetate is required as a co-solvent, and a low reaction temperature is also essential. Han et al.[22] found that in organic solution with a trace amount of water the above reaction proceeds with the same high enantioselectivity as in the presence of an aqueous buffer. The reaction can be carried out at a wide range of temperatures from 0° to 30°C. To avoid using highly toxic potassium or sodium cyanide, acetone cyanohydrin is used as a cyano donor.

The following enzymes are useful for asymmetric organic synthesis:

PLE, pig liver esterase

PPL, pig pancreatic lipase

CCL, *Candida cylindracae* lipase

α-chym, α-chymotrypsin from bovine

p-lipase, from *Pseudomonas* species

PS lipase

A lipase

AF–Z

MY

DF 360

AnL, *Aspergilus niger*

PrL, *Penicillinm reguerferti*

CIL, *Candida lypolytica*

CSL, *Candida* sp 382

For discussions of some representative enzyme-catalyzed reactions, see elsewhere.[23]

8.1.5   Biocatalysts from Cultured Plant Cells

In recent years, extensive attention has been focused on finding cultured plant cells that can be used as catalysts for organic functional group transformations. A number of transformations employing freely suspended or immobilized plant cell cultures have been reported.[24] For example, Akakabe et al.[25] report that immobilized cells of *Daucus carota* from carrot can be used to reduce prochiral carbonyl substrates such as keto esters, aromatic ketones, and heterocyclic ketones to the corresponding secondary alcohols in (*S*)-configuration with enantiomeric excess of 52–99% and chemical yields of 30–63%.

$$R' = CH_2COOEt, CH_3$$
$$R = CH_3, C_2H_5, Ph, p\text{-}O_2NPh,$$
$$p\text{-}BrPh, \underset{N}{\bigcirc}, p\text{-}MePh, p\text{-}MeOPh$$

Other kinds of plant cell cultures such as immobilized tobacco cells have also been studied for the analogous transformation. The results show that plant cell cultures provide an accessible way of converting several prochiral ketones into the corresponding chiral secondary alcohols with reasonable chemical yield and high enantioselectivity.

8.2   MISCELLANEOUS METHODS

Recently, an increasing number of methods have become known in organic synthesis that are not easy to compile into an appropriate place in previous sections. They are hence grouped in this section.

8.2.1   Asymmetric Synthesis Catalyzed by Chiral Ferrocenylphosphine Complex

Chiral ferrocenylphosphine ligand[26] such as **17** bears a pendant side chain and a hydroxy group at the terminal position. When compound **17** is used in the palladium-catalyzed asymmetric allylic amination of allylic substrates containing a 1,3-disubstituted propenyl structure, product with good to excellent enantiomeric excess can be obtained (Scheme 8–7). Catalyst prepared in situ from Pd(dba)$_2$ and (*R*)- or (*S*)-**17** catalyzes the amination of substrate **18** with high yield and up to 97% ee. The proposed transition state of the active catalyst is shown in Figure 8–3. The hydroxyl group in the ligand can interact with the nucleophile. This facilitates its attack on the substrate and promotes the reaction.

Both enantiomers of racemic 2-propenyl acetate can be formed from *meso*-type π-alkyl palladium intermediates by oxidative addition. π-Allylpalladium complexes with two alkyl substituents at the 1- and 3-positons are known to

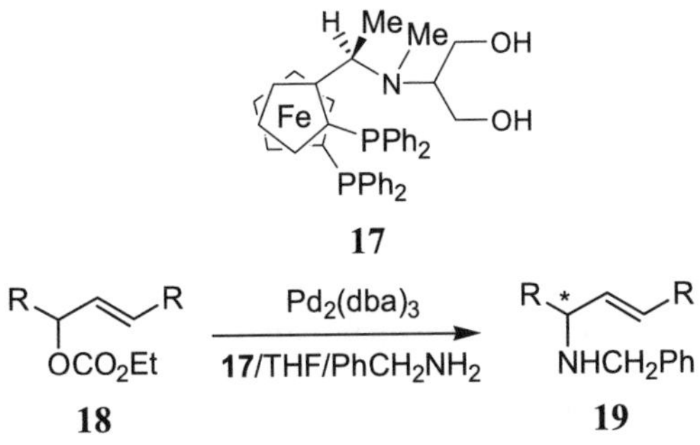

17

R = Ph   97% ee (*R*)
R = CH$_3$ 73% ee (*S*)
R = *n*-Pr 82% ee (*S*)
R = *i*-Pr 97% ee (*S*)

Scheme 8–7

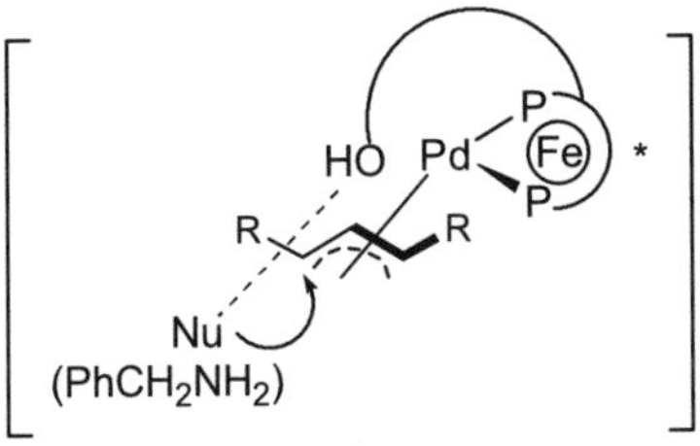

Figure 8–3. Transition state of **17** catalyzed reaction.

adopt the *syn*-conformation of both substituents. The attack by the nucleophile takes place from the preferred side on either of the two diastereotopic π-alkyl carbon atoms in the π-allyl palladium intermediate. The (*E*)-geometry amination product is finally formed with high selectivity.

8.2.2 Asymmetric Hydrosilylation of Olefins

Catalytic asymmetric hydrosilylation of prochiral olefins has become an interesting area in synthetic organic chemistry since the first successful conversion of alkyl-substituted terminal olefins to optically active secondary alcohols (>94% ee) by palladium-catalyzed asymmetric hydrosilylation in the presence of chiral monodentate phosphine ligand (MOP, **20**). The introduced silyl group can be converted to alcohol via oxidative cleavage of the carbon–silicon bond (Scheme 8–8).[27]

When norbornene is treated with trichlorosilane in this manner, quantitative yield of *exo*-2-trichlorosilylnorbornane is obtained, and (1*S*,2*S*,4*R*)-*exo*-2-norbornanol can be obtained in 96% ee upon hydrogen peroxide oxidation.[28] This reaction can be extended to other olefinic substrates such as 2,5-

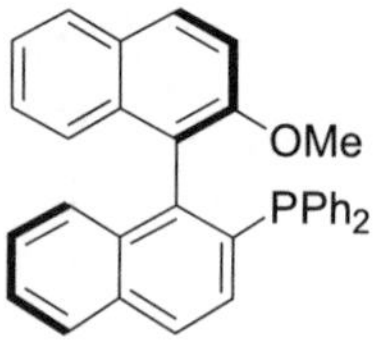

20 (*R*)-MOP
(*R*)-2-methoxy-2'-diphenylphosphino-1,1'-binaphthalene

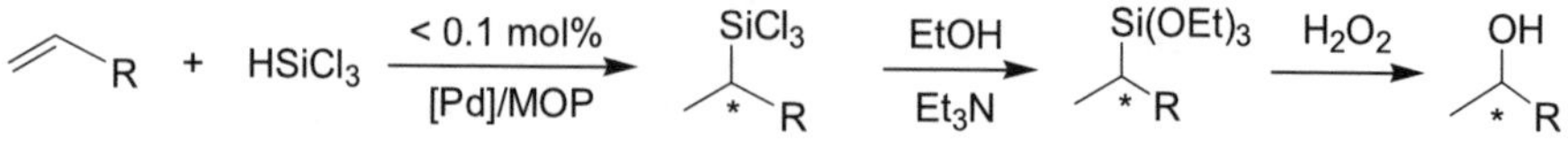

Scheme 8–8

dihydrofuran. As shown in Scheme 8–9, the catalytic asymmetric hydrosilylation of 2,5-dihydrofuran with trichlorosilane in the presence 0.1 mol% of
(*R*)-MOP (**20**)–coordinated palladium complex yields the corresponding hydrosilylation product **22** in up to 95% ee (Scheme 8–9).[29]

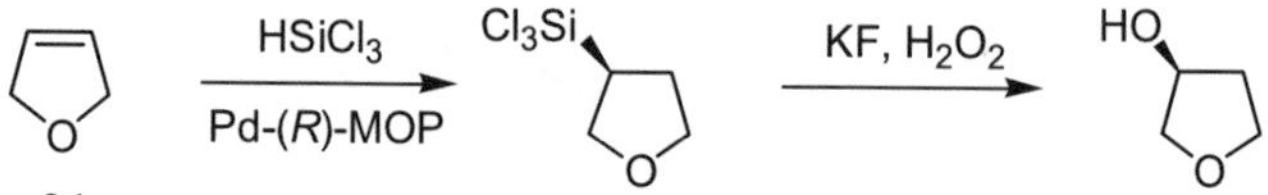

22
95% ee, 65% yield

Scheme 8–9

Other metal complexes such as titanium or ruthenium complexes can also be
used to catalyze the olefin hydrosilylation reactions. Further information is
provided elsewhere.[30]

8.2.3   Synthesis of Chiral Biaryls

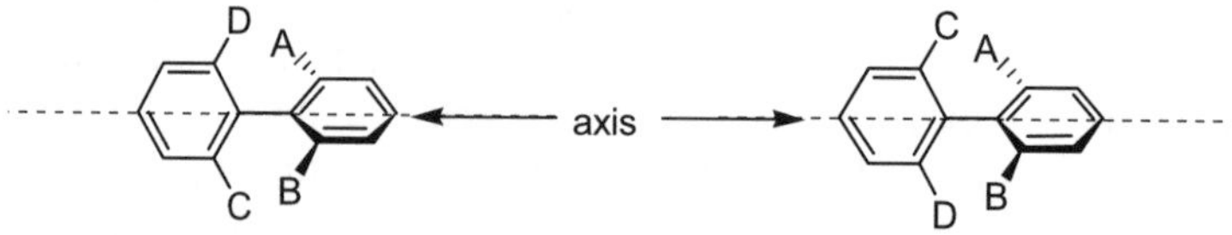

The hindered biaryls are examples of a different type of chiral compound due to
their rotational restriction around a C–C single bond. As long as the *ortho*-
substituents in a compound such as **23** are bulky enough, the compound can
exist in two forms, **23** and its enantiomer **23'**, which are not interconvertible.

Compounds with chirality caused by restriction of the C–C axial covalent bond include a range of naturally occurring as well as synthesized compounds. Not only do axially chiral biaryls constitute a structural feature of many products, but using biaryls possessing C_2 symmetry as chiral ligands in asymmetric synthesis has also been an area of considerable activity. Indeed, the synthesis of these axial chiral biaryls has gradually become one of the most interesting subjects in asymmetric organic synthesis. Here are some examples of these chiral biaryl compounds[31]:

(-)-ancistrocladine[32]   (-)-gossypol[33]   WS43708 A[34]

(-)-steganone[35]   Cram's crown ether[36]   (R)-BINAL-H[37]

In most cases, construction of axial nonracemic biaryls can be realized by metal-mediated intramolecular biaryl coupling in which the chirality is induced by *ortho*-substituted auxiliaries in an aromatic system.

Preparation of the chiral biphenyls and binaphthyls with high enantioselectivity can be achieved via substitution of an aromatic methoxyl group with an aryl Grignard reagent using oxazoline as the chiral auxiliary.[38] Schemes 8–10 and 8–11 outline the asymmetric synthesis of such chiral biaryl compounds.

As shown in Scheme 8–11, nucleophilic entry from the α-face (**24a**) may be hindered by the sterically bulky substituent R^2 on the oxazoline moiety; therefore entry from the β-face **24β** predominates. Free rotation of the magnesium methoxy bromide may be responsible for the sense of the axial chirality formed in the biaryl product. If the azaenolate intermediate **25** is re-aromatized with a 2′-methoxy substituent complexed to Mg, (*S*)-biphenyl product is obtained. Upon re-aromatization of azaenolate **25B**, (*R*)-product is obtained.

Another method for synthesizing chiral biaryls is the substrate-controlled

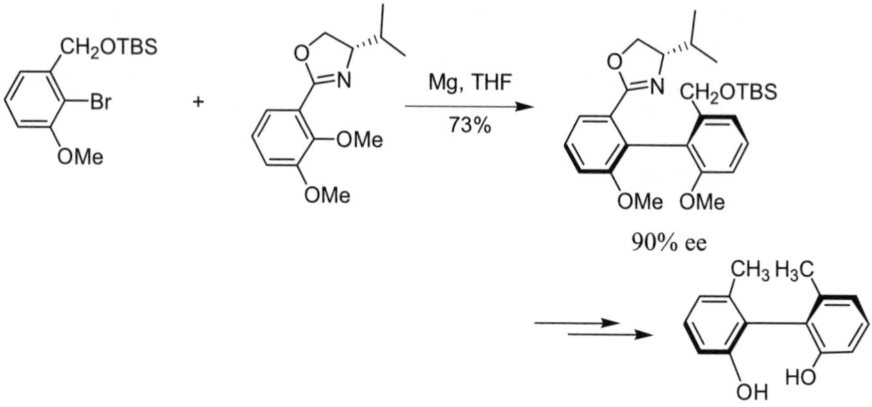

Scheme 8–10

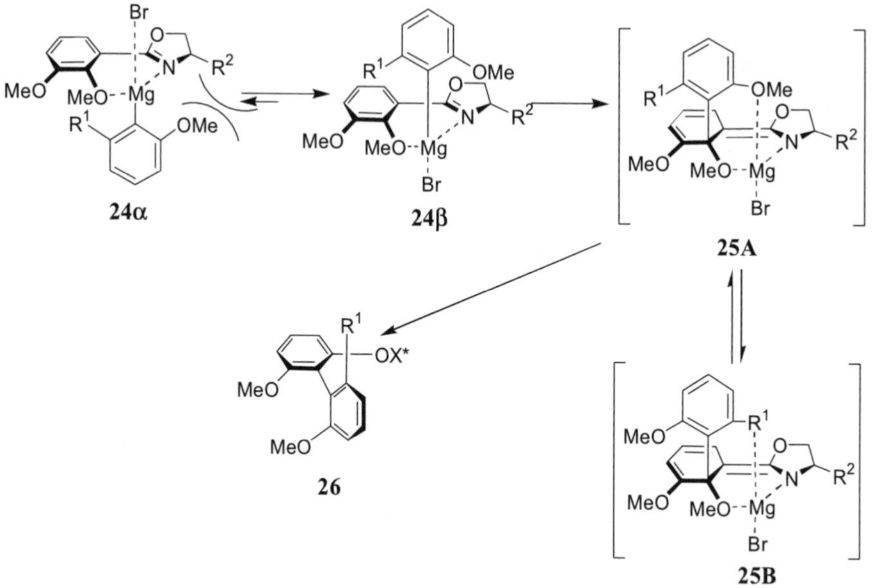

Scheme 8–11

asymmetric Ullmann reaction. Scheme 8–12 depicts the asymmetric synthesis of enantiomerically pure C_2-symmetric binaphthyls.[39]

When subjected to Ullmann biaryl reactions, 8-bromo-1-oxazolinylnaphthalene (S)-**27** can be converted to the 8,8′-substituted binaphthyl **28** in high diastereomeric excess (another isomer, the (aR,S)-diastereomer, is less than 3%).[40]

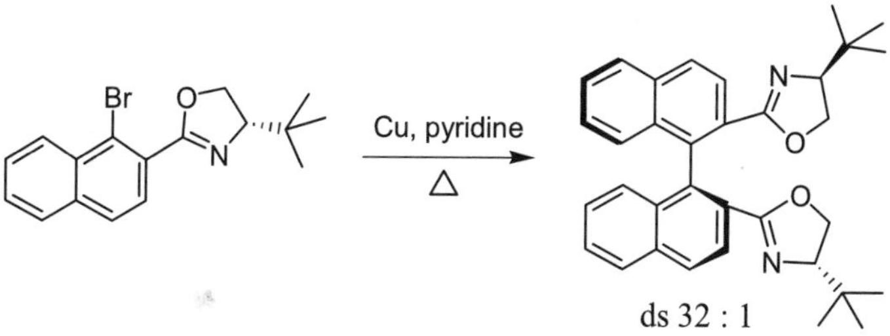

Scheme 8–12

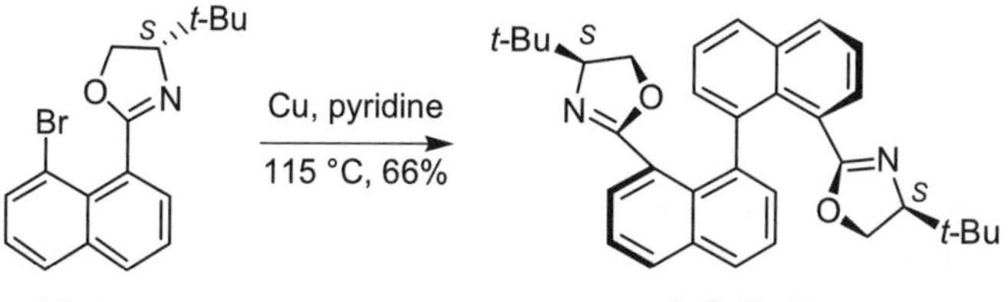

(S)-27   **(aS, S)-28a**

Based on this method,[39] several chiral axial aryls such as **29**, **30**, **31**, and **32** have been prepared.[31a,41] The diastereomeric alkaloids (−)-ancistrocladine **29** and (+)-hamatine **30** are also obtainable from the plants.

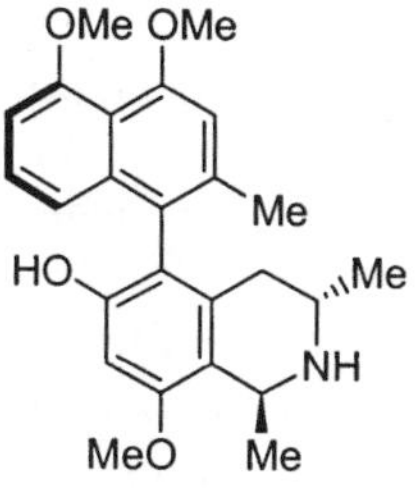

In fact, compounds **31** and **32** from *Ancistrocladus hamatus* have the same configuration in the axial stereogenic unit (atriousiners) but opposite configuration for the two stereogenic centers of the tetrahydroisoquinoline ring.

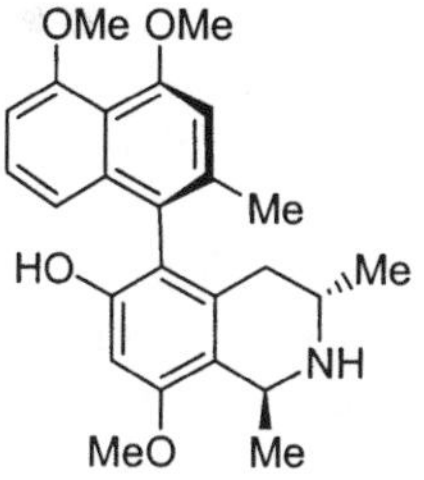

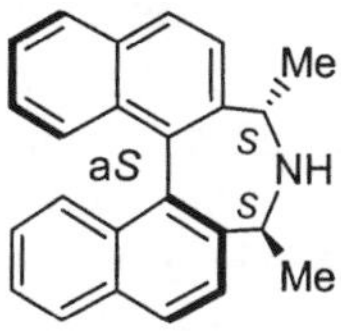

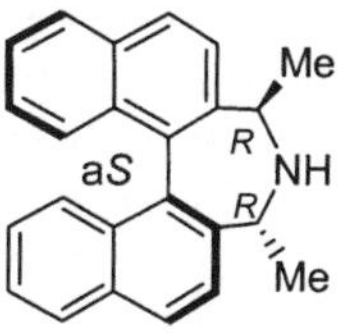

Lipshuz et al.[42] have developed a new approach to the chiral biaryls mediated by cyanocuprate. The diastereoselective coupling depends on the proper choice of the tether (Scheme 8–13).

The bromo-aryl groups are first linked by (S,S)-stilbene diol to form the dibromide **33**. Compound **33** is then dilithiated with *t*-BuLi at −78°C, followed by addition of CuCN. Intermediate **34** is presumably formed during the reaction. Reductive elimination promoted by molecular oxygen provides compound **35** at 77% yield with 93:7 diasteroselectivity. The final biaryl compound ellagi-

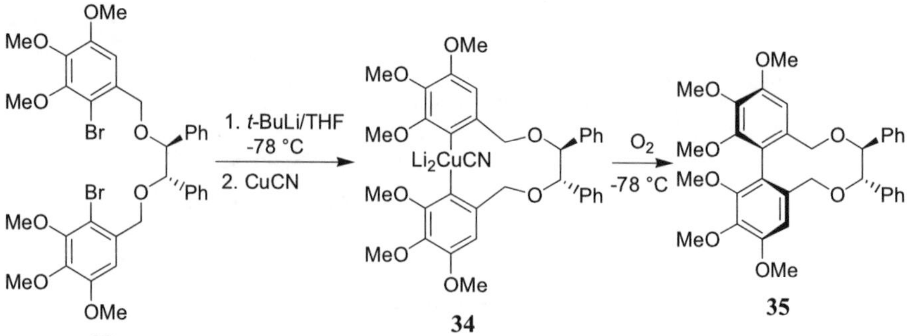

Scheme 8–13

tannin (tellimagrandrin II) can be obtained upon removal of the tether in **35** (the chiral stilbene diol) via catalytic hydrogenation over Pd/C in MeOH.

8.2.4 The Asymmetric Kharasch Reaction

In 1959, Kharasch et al.[43] reported an allylic oxyacylation of olefins. In the presence of *t*-butyl perbenzoate and a catalytic amount of copper salt in re-fluxing benzene, olefin was oxidized to allyl benzoate, which could then be converted to an allyl alcohol upon hydrolysis. It is desirable to introduce asymmetric induction into this allylic oxyacylation because allylic oxyacylation holds great potential for nonracemic allyl alcohol synthesis. Furthermore, this reaction can be regarded as a good supplement to other asymmetric olefinic reactions such as epoxidation and dihydroxylation.

The mechanism of the reaction involves initiating the reaction by forming a *t*-butoxy radical via copper(I)-mediated reductive homolysis of the perester O–O bond. The *t*-butoxy radical then abstracts an allylic hydrogen atom to give *t*-butanol and an allylic radical, followed by a rapid addition of copper(II) to the allylic radical to generate copper(III) benzoate with an allyl fragment. Rearrangement of the copper(III) intermediate gives the allylester product with the regeneration of Cu(I) catalyst. It has been found that copper(III) coordi-nated with a suitable ligand can induce the asymmetric formation of the allyl benzoate product (Scheme 8–14).

Andrus et al.[44] employed a C_2-symmetric bis(oxazoline) copper catalyst in the Kharasch reaction. When cyclohexene was used as the reaction substrate, yields ranging from 34% to 62% and ee from 30% to 81% were observed (Scheme 8–15).

DattaGupta and Singh[45] report the results of bis(oxazolinyl)pyridine in-duced asymmetric allylic oxidation. The reaction proceeds with 59% yield and 56% ee (Scheme 8–16).

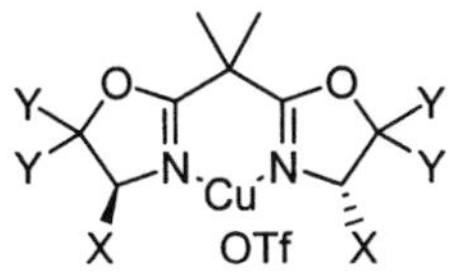

36a Y = H, X = *t*-Bu
36b Y = Me, X = *t*-Bu
36c Y = Me, X = Ph

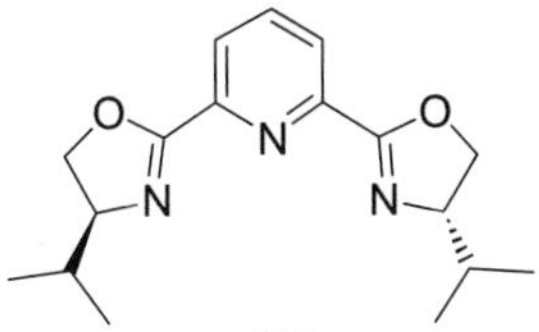

37

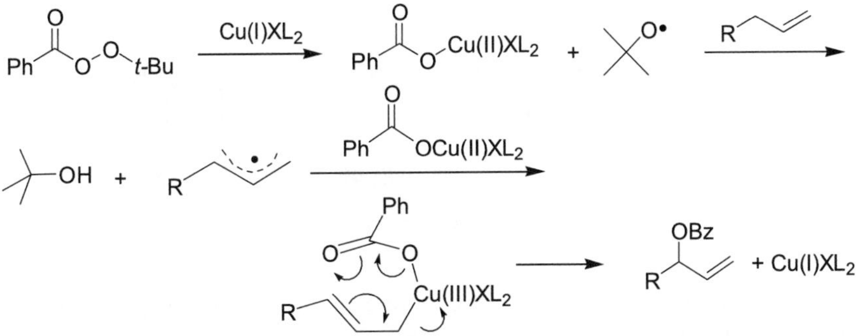

Scheme 8–14. Mechanism of allylic olefin oxidation with Cu(I) and perester.

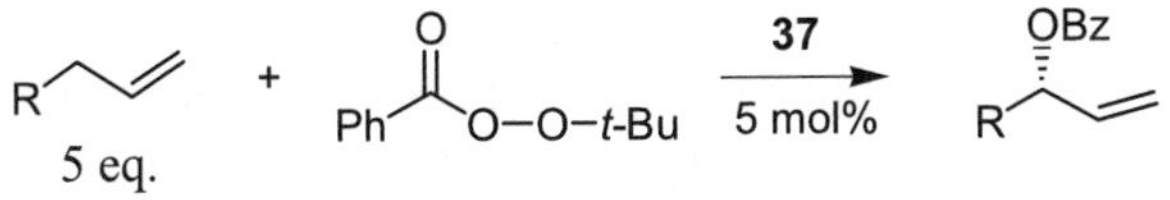

Scheme 8–15. Enantioselective allylic oxidation of olefins.

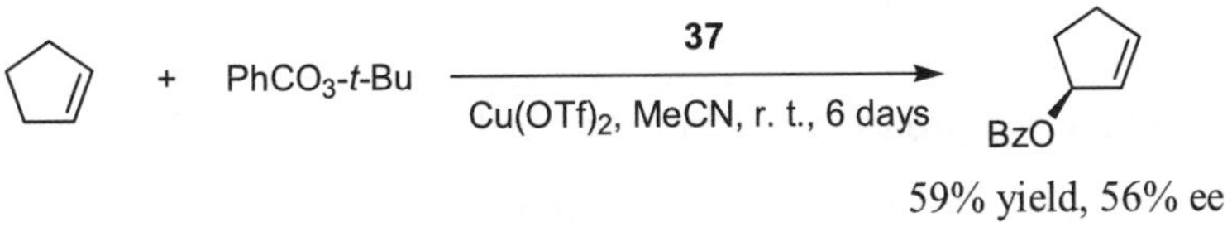

59% yield, 56% ee

Scheme 8–16

8.2.5 Optically Active Lactones from Metal-Catalyzed Baeyer-Villiger–Type Oxidations Using Molecular Oxygen as the Oxidant

The Baeyer-Villiger reaction refers to the oxidation of a ketone to an ester or lactone by peracids or other oxidants. Highly oxidized reagents such as peracids or other peroxy compounds are generally used as the typical oxygen source, but applying molecular oxygen as a simple oxidant in the asymmetric oxidation reaction has recently drawn much attention. Bolm et al.[46] were the first to develop the metal complex–catalyzed Baeyer-Villiger oxidation using molecular

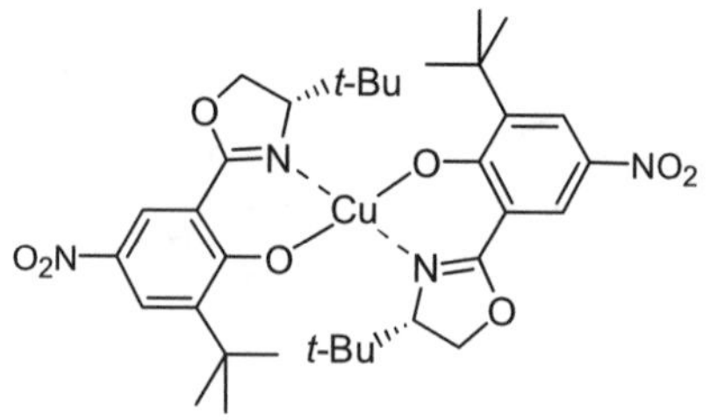

38

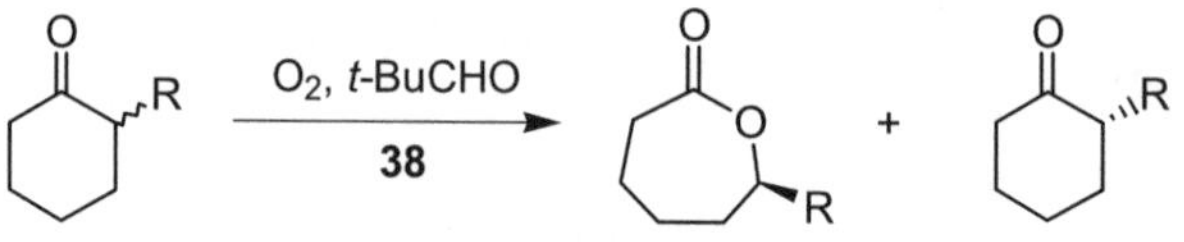

rac-**39**   (R)-**40**   (S)-**39**

Scheme 8–17

oxygen as oxidant to transform cyclic ketones to lactones in the presence of an aldehyde (Scheme 8–17).

Best results were obtained when the reaction was carried out in benzene-saturated water using p-nitro–substituted copper complex **38** as the catalyst and pivaldehyde as the oxygen acceptor (co-reducing reagent). Using 1 mol% of (S,S)-**38**, the reaction proceeds at 6°C under an oxygen atmosphere, providing the product lactone (R)-**40** at 41% yield with 69% ee. Because the unreacted ketone is thus enriched in (S)-configuration, this reaction can also be regarded as a kinetic resolution process. Dependence of the reactivity on ring size is found to be similar to that observed for peroxides. A six-membered ring substrate was oxidized faster, a five-membered ring reacted more slowly, and a seven-membered ring substrate did not react.

8.2.6 Recent Progress in Asymmetric Wittig-Type Reactions

The essence of asymmetric synthesis is producing a new stereogenic center in such a manner that the product consists of stereoisomers in unequal amount. In most cases, this can be achieved by the formation of a new sp^3 stereocenter. There is also another type of asymmetric reaction in which the employed substrates contain either a stereogenic unit or a pro-stereogenic unit apart from the functional group, and asymmetric synthesis occurs even though the nature of the reaction is not directly related to the newly formed sp^3 stereocenter. The Wittig reaction is invoked for the asymmetric synthesis of such molecules.[47]

Different types of the reagents (see Fig. 8–4) have been applied in asymmetric Wittig-type reactions. Because no new sp^3 stereocenter is formed in a Witting-type reaction, a substrate containing a stereogenic or pro-stereogenic unit apart from the carbonyl group is usually required to induce an asymmetric process.

A challenging goal is the development of catalytic asymmetric induction processes. Denmark et al.[48] have reported an asymmetric Wittig reaction using

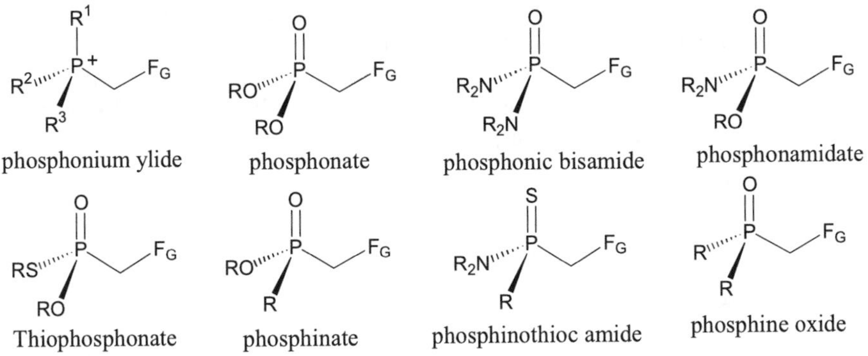

Figure 8–4. Some Wittig reagents. F_G = H, alkyl, aryl, or other functional groups.

phosphonamidates **41** and **42** as the chiral reagent, by which different dissymmetric alkenes are synthesized in good ee (Scheme 8–18).

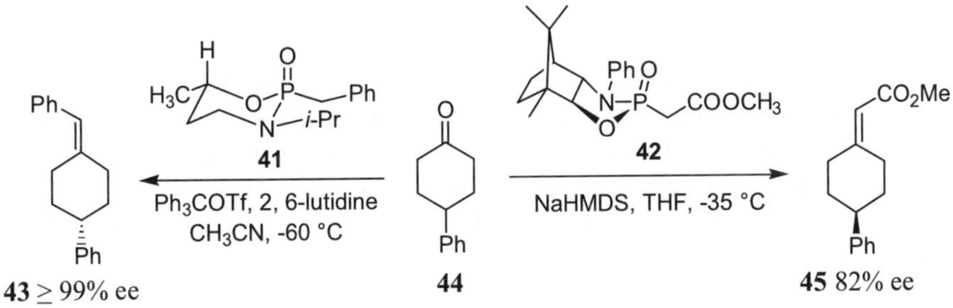

Scheme 8–18

Toda and Akai[49] reported that compound **48** reacted with the stable solid state inclusion compound of chiral host **46** and *meso*-ketone **47**, providing alkene **49** in 57% ee.

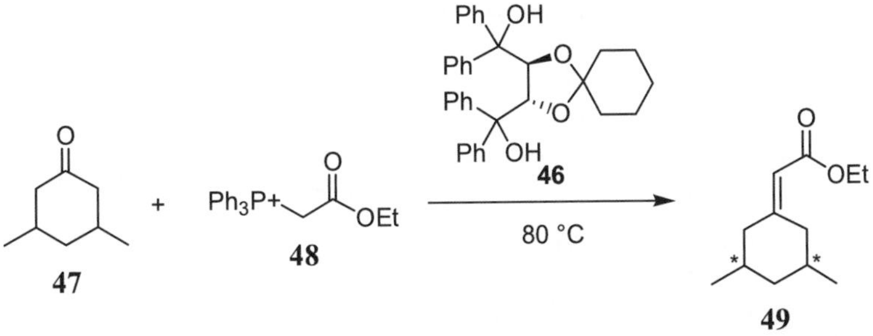

Tanaka et al.[50] also reported that chiral Horner-Wadsworth-Emmons reagent (*S*)-**51** reacted with an alternative carbonyl group of the *meso*-α-diketones in bicyclo-[2.2.1]-system **50**, giving nonracemic (Z)-**52** and (E)-**53**. The results are shown in Scheme 8–19. This is an example of desymmetrization of *meso*-compounds involving C–C bond formation. Reagent (S)-**51** differentiates the enantiotopic carbonyl groups of the α-diketones in the bicyclo[2.2.1]-system, leading to (Z)-**52** and (E)-**53**, respectively.

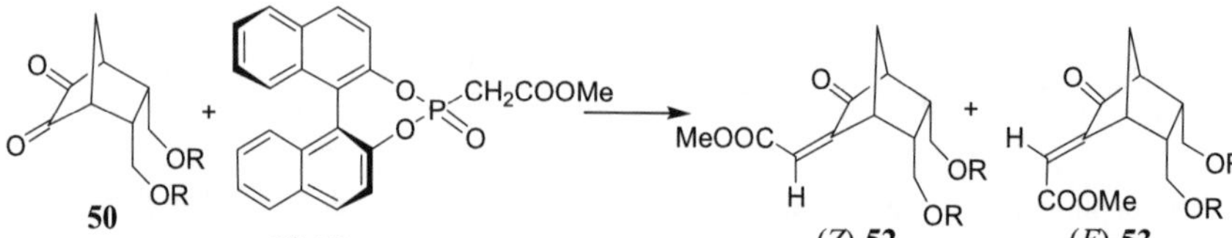

R in substrate **50**	Z/E	(Z)-**52**		(E)-**53**	
		yield	ee	yield	ee
R = Ac	72 : 28	58%	90%	23%	23%
R = Bn	75 : 25	66%	89%	23%	8%

Scheme 8–19

Arai et al.[51] reported that by using a catalytic amount of chiral quaternary ammonium salt as a phase transfer catalyst, a catalytic cycle was established in asymmetric HWE reactions in the presence of an inorganic base. Although catalytic turnover and enantiomeric excess for this reaction are not high, this is one of the first cases of an asymmetric HWE reaction proceeding in a catalytic manner (Scheme 8–20).

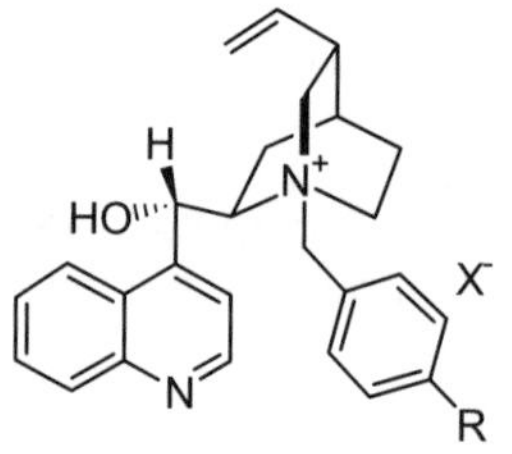

54 R = H, CF₃, X = Cl, Br

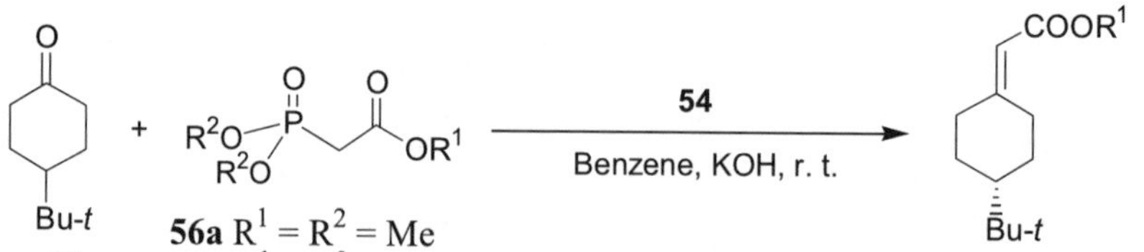

55

56a $R^1 = R^2 = $ Me
56b $R^1 = R^2 = $ Et
56c $R^1 = n\text{-Pr}, R^2 = $ Et
56d $R^1 = i\text{-Pr}, R^2 = $ Et
56e $R^1 = i\text{-Bu}, R^2 = $ Et

57

Scheme 8–20

8.2.7 Asymmetric Reformatsky Reactions

As a kind of nucleophilic addition reaction similar to the Grignard reaction, the Reformatsky reaction can afford useful β-hydroxy esters from alkyl haloacetate, zinc, and aldehydes or ketones. Indeed, this reaction may complement the aldol reaction for asymmetric synthesis of β-hydroxy esters.

The earliest enantioselective Reformatsky reaction was reported in 1973.[52] Compound (−)-spartein was used as the chiral ligand, but the reaction gave rather low yield. Almost 20 years later, in 1991, Soai and Kawase[53] reported an enantioselective Reformatsky reaction in which chiral amino alcohol **58** or **59** was used as the ligand. Aliphatic and aromatic β-hydroxy esters were obtained with moderate to good yields.

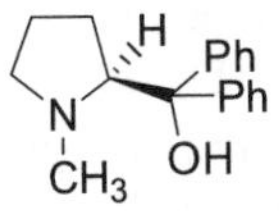

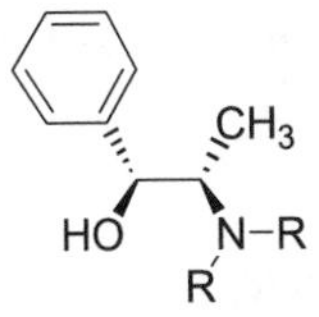

59a R = allyl
59b R = n-Bu
59c R = Me
59d R = n-octyl

58

(1S, 2R)-/(1R, 2S)-**59**

Chiral β-hydroxylester **61** cannot be satisfactorily obtained through the reaction between a prochiral ketone and the enolate. It can, however, be synthesized via the chiral ligand-induced asymmetric Reformatsky reaction of ketones (Scheme 8–21).

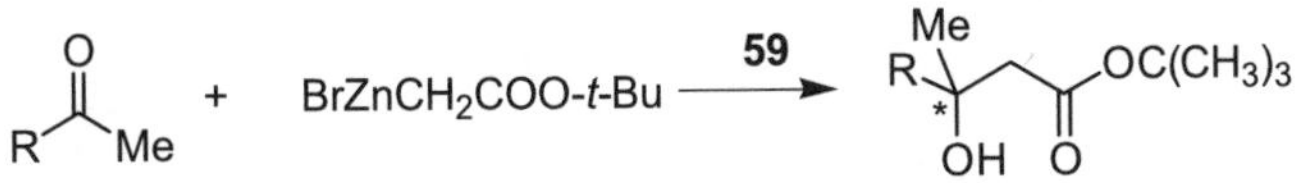

60a R = Ph
60b R = 2-naphthyl
60c R = 4-F-Ph

61

Scheme 8–21

Moreover, Soai et al.[53c] found that the enantioselective addition of Reformatsky reagents to prochiral ketones proceeds well when N,N-dialkylnorephedine **59** is used as the chiral ligand. When (1S,2R)-**59a** is used, the β-hydroxyl ester is obtained in 74% ee and 65% yield with (S)-configuration predominant. When (1R,2S)-**59a** is used, the product is obtained in 74% ee and at 47% yield with (R)-configuration prevailing.

Similarly, compound **58** [in both (R)- and (S)-forms] has been found to be a good ligand for the enantioselective addition of cyanomethyl zinc bromide to various aldehydes. Products with up to 93% ee are obtained.[53b] Scheme 8–22 shows one example. When R = Ph, product is obtained in 78% ee.

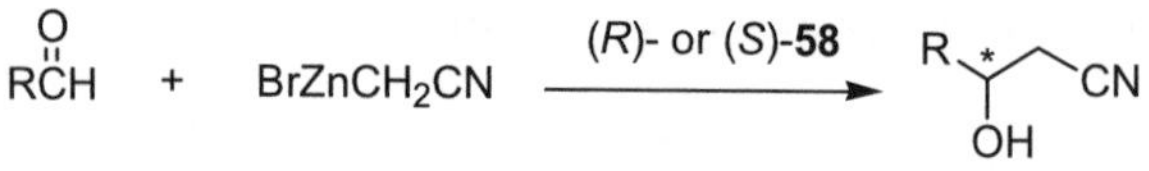

$$\underset{\text{RCH}}{\overset{\overset{O}{\|}}{}} \quad + \quad BrZnCH_2CN \quad \xrightarrow{\text{(R)- or (S)-58}} \quad \underset{OH}{\overset{R}{\underset{*}{}}}\diagdown CN$$

Scheme 8–22

Pini et al.[54] and Mi et al.[55] reported the promotion of the enantioselective Reformatsky reaction by using stoichiometric and/or catalytic amounts of amino alcohols. Moderate yields were obtained.

Samarium(II) iodide often exerts good chelation control with an oxygen or nitrogen moiety in organic molecules, and this results in highly stereoselective intra- or intermolecular 1,2- and 1,3-asymmetric inductions.[56] Fukuzawa et al.[57] demonstrated a highly diastereoselective intermolecular samarium iodide–mediated Reformatsky reaction using α-bromoacetyl-2-oxazolidinone **62** as the chiral auxiliary. This example is shown in Scheme 8–23. The yields for product **63** are good to excellent, and de values are high for the straight chain and branched aliphatic aldehydes as well as aromatic aldehydes. When X = i-Pr in **62**, the product obtained (RCHO = t-butyl aldehyde) has 99% de.

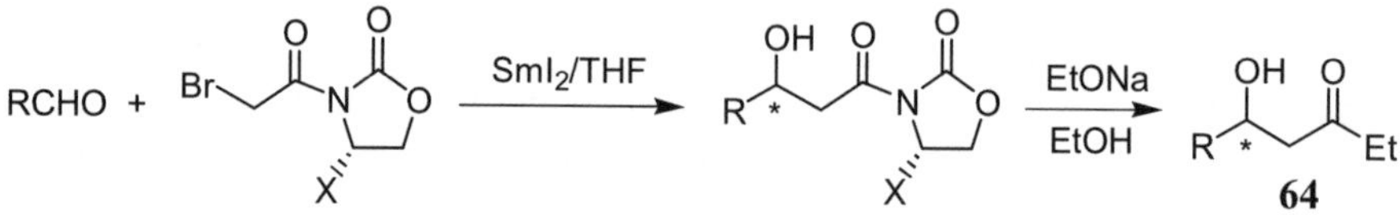

X = i-Pr, Ph, PhCH₂

Scheme 8–23

8.2.8 Catalytic Asymmetric Wacker Cyclization

Wacker cyclization has proved to be one of the most versatile methods for functionalization of olefins.[58,59] However, asymmetric oxidative reaction with palladium(II) species has received only scant attention. Using chiral ligand 1,1'-binaphthyl-2,2'-bis(oxazoline)–coordinated Pd(II) as the catalyst, high enantioselectivity (up to 97% ee) has been attained in the Wacker-type cyclization of o-alkylphenols (**66a–f**) (Scheme 8–24).

It is worth noting, however, that chiral phosphine–palladium complexes generated from palladium salts and BINAP or MOP cannot be used for this oxidation because phosphines will be readily oxidized to phosphine oxides under the reaction conditions, leading to the deactivation of the catalyst. As reaction without the chiral catalyst will give a racemic product, this deactivation of the catalyst will cause a drop in the enantioselectivity of the whole process.

Besides the function of the 1,1'-binaphthyl backbone, which is very important for high enantioselectivity, configuration of the central chirality on the

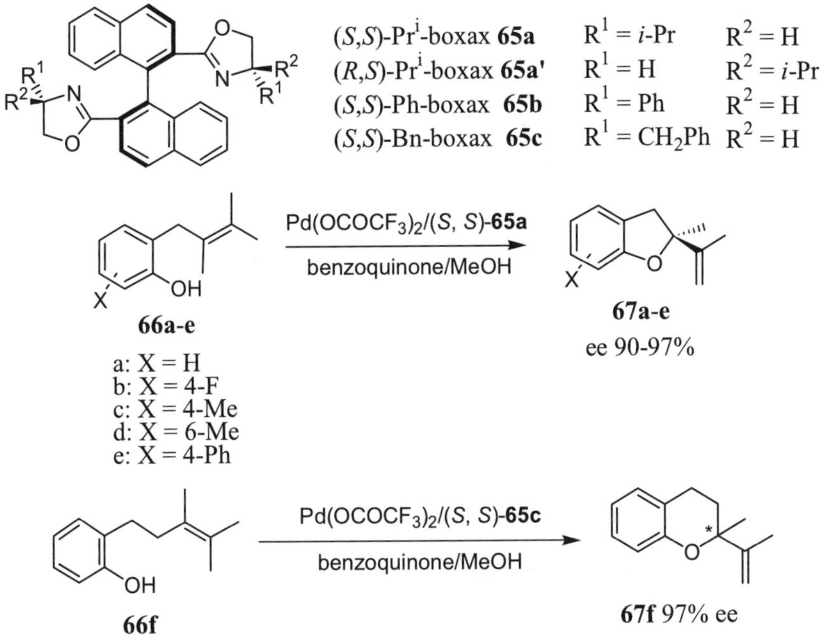

(S,S)-Pri-boxax 65a R1 = i-Pr R2 = H
(R,S)-Pri-boxax 65a' R1 = H R2 = i-Pr
(S,S)-Ph-boxax 65b R1 = Ph R2 = H
(S,S)-Bn-boxax 65c R1 = CH2Ph R2 = H
Pd(OCOCF3)2/(S, S)-65a
benzoquinone/MeOH
66a-e
a: X = H
b: X = 4-F
c: X = 4-Me
d: X = 6-Me
e: X = 4-Ph
67a-e
ee 90-97%
Pd(OCOCF3)2/(S, S)-65c
benzoquinone/MeOH
66f
67f 97% ee

Scheme 8–24

oxazoline attached to binaphthyl also has a great influence on enantioselectivity and catalytic activity. In Scheme 8–24, when (*S*,*S*)-*i*-Pr-boxax **65a** is used as the catalyst, the resulting products **67a–c** are obtained with (*S*)-configuration in 90–97% ee. The diastereoisomer (*R*,*S*)-*i*-Pr-boxax **65a′** is much less active and less enantioselective (3% yield, 18% ee (*R*)-configuration). Compounds **65b** and **65c** show almost the same high stereoselectivity as **65a**.

8.2.9 Palladium-Catalyzed Asymmetric Alkenylation of Cyclic Olefins

Palladium-catalyzed arylation of olefins and the analogous alkenylation (Heck reaction) are the useful synthetic methods for carbon–carbon bond formation.[60] Although these reactions have been known for over 20 years, it was only in 1989 that the asymmetric Heck reaction was pioneered in independent work by Sato et al.[60d] and Carpenter et al.[61] These scientists demonstrated that intramolecular cyclization of an alkenyl iodide or triflate yielded chiral cyclic compounds with approximately 45% ee. The first example of the intermolecular asymmetric Heck reaction was reported by Ozawa et al.[60c] Under appropriate conditions, the major product was obtained in over 96% ee for a variety of aryl triflates.[62]

Treating 2,3-dihydrofuran **68** with aryl triflate **69** in the presence of a base and a palladium catalyst generated in situ from $Pd(OAc)_2$ and (*R*)-BINAP leads to the formation of (R)-2-aryl-2,3-dihydrofuran **70** and its regioisomer (S)-**71** (Scheme 8–25).[60b] The best results are obtained in the presence of the proton sponge 1,8-bis(dimethyl amino)naphthalene (**72**) (Scheme 8–26).[60a,63]

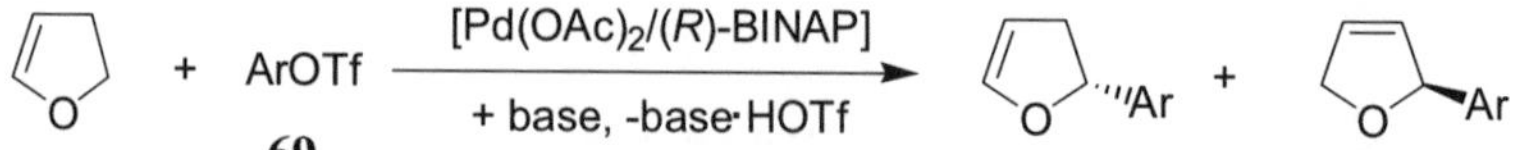

Scheme 8–25

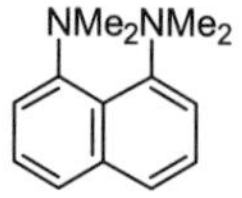

72

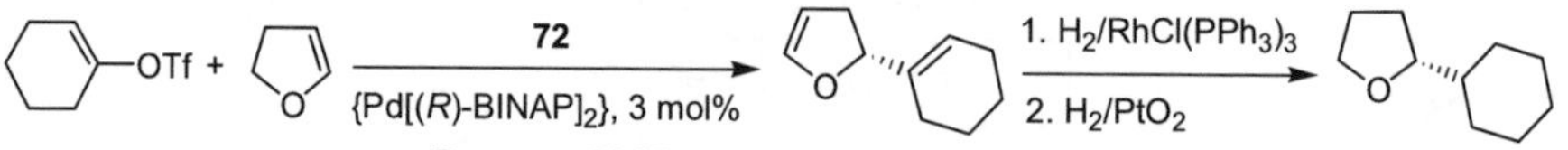

Triflate **73**	(R)-**74a**	(R)-**75**
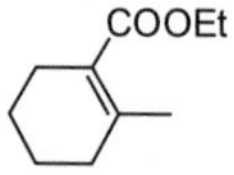 ⬡ COOEt	> 96% ee	62% yield
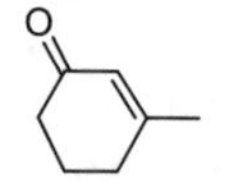 ⬡ O	> 96% ee	50% yield
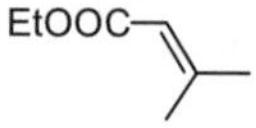 EtOOC	> 96% ee	50% yield
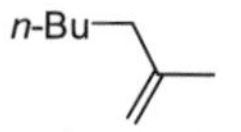 n-Bu	no eaction	

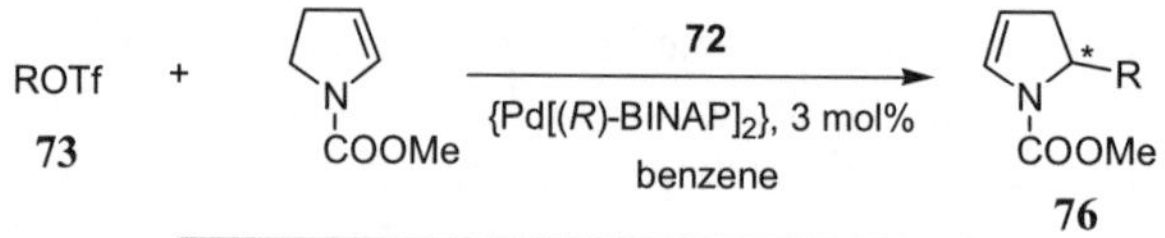

Triflate **73**	Yield (%)	ee (%)
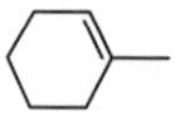 ⬡	45	96
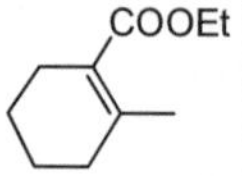 ⬡ COOEt	95	> 99

Scheme 8–26

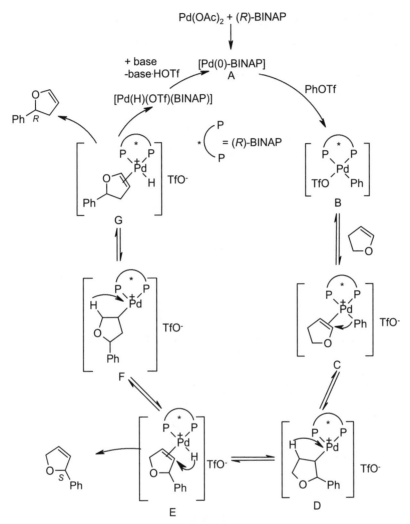

Figure 8–5. Proposed mechanism for the catalytic arylation of 2,3-dihydrofuran with phenyl triflate in the presence of Pd(OAc)$_2$-(R)-BINAP catalyst.

The proposed mechanism is illustrated in Figure 8–5.[60a] Oxidative addition of the phenyl triflate to the palladium(0)–BINAP species A gives phenylpalladium triflate B. Cleavage of the triflate and coordination of 2,3-dihydrofuran on B yields cationic phenyl palladium olefin species C. This species C bears a 16-electron square–planar structure that is ready for the subsequent enantioselective olefin insertion to complete the catalytic cycle (via D, E, F, and G). The base and catalyst precursor have profound effects on the regioselectivity and enantioselectivity.

8.2.10 Intramolecular Enyne Cyclization

Intramolecular enyne coupling of allylic enolates can be catalyzed by PdX_2, and this reaction can be used in the synthesis of various substituted γ-butyrolactones.[64] The strategy is outlined in Scheme 8–27.

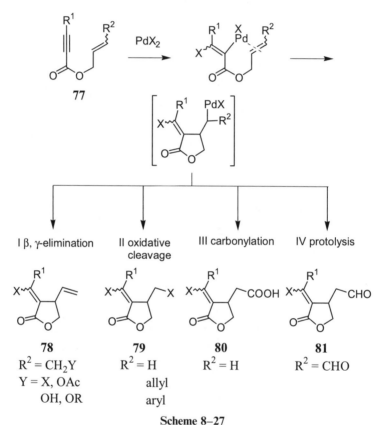

Scheme 8–27

The allylic 2-alkynoates **77** are a group of special enynes with an ester linkage between their double bond and triple bond. When halopalladation of the triple bond is followed by C–C double bond insertion and the cleavage of the carbon–palladium bond, a series of γ-lactones (**78–81**) can be obtained.

There are many advantages of this divalent palladium-catalyzed cyclization. First, oxygen-free conditions are no longer required. Second, the organic ligand-free catalyst can be more easily recovered under certain conditions. This reaction normally proceeds with high stereoselectivity, providing greater potential for the synthesis of biologically active compounds. Scheme 8–28 illustrates the synthesis of (−)-methylenolactocin.[65] The key step is the intramolecular cyclization of **82** [LiBr, Pd(OAc)$_2$, HOAc, room temperature], giving lactone **83** in

65% yield and extremely high diastereoselectivity (the other diastereomer cannot be detected) (Scheme 8–28).

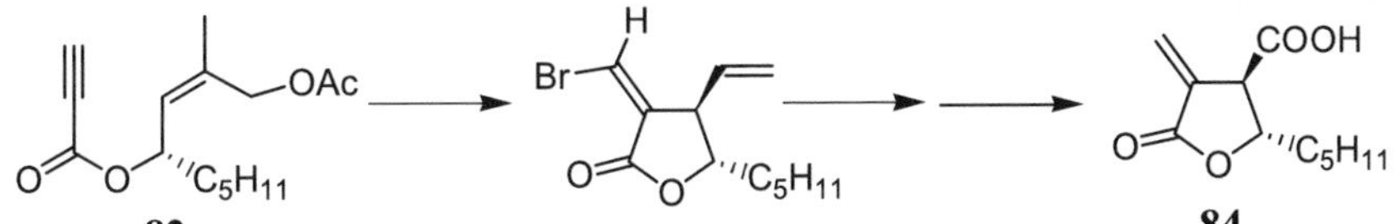

Scheme 8–28

8.2.11 Asymmetric Darzens Reaction

Compared with many other reactions for enantioselective formation of C–C bonds, the asymmetric Darzens condensation[66] has received less attention. Therefore, there is ample opportunity for chemists to improve the enantioselectivity of the reaction, as well as to develop the reaction itself.

Darzens reaction, the reaction between a carbonyl compound and an α-halo ester in the presence of a base, consists of an initial aldol-type addition and a subsequent intramolecular S_N2 reaction, forming an epoxide as its final product. Its high stereoselectivity thus relies on the stereoselectivity of the nucleophilic addition of an α-halo ester onto the carbonyl substrate, which can be either an aldehyde or a ketone.

Early work on the asymmetric Darzens reaction involved the condensation of aromatic aldehydes with phenacyl halides in the presence of a catalytic amount of bovine serum albumin. The reaction gave the corresponding epoxyketone with up to 62% ee.[67] Ohkata et al.[68] reported the asymmetric Darzens reaction of symmetric and dissymmetric ketones with (−)-8-phenylmenthyl α-chloroacetate as examples of a reagent-controlled asymmetric reaction (Scheme 8–29). When this (−)-8-phenyl menthol derivative was employed as a chiral auxiliary, Darzens reactions of acetone, pentan-3-one, cyclopentanone, cyclohexanone, or benzophenone with **86** in the presence of t-BuOK provided diastereomers of (2R,3R)-glycidic ester **87** with diastereoselectivity ranging from 77% to 96%.

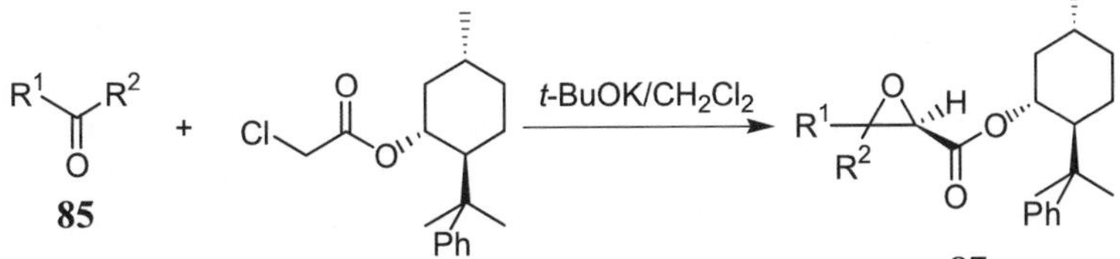

Scheme 8–29. (−)-Menthol derivative–mediated asymmetric Darzens reaction.

The Darzens reaction can also proceed in the presence of a chiral catalyst. When chloroacetophenone and benzaldehyde are subjected to asymmetric Darzens reaction, product **89** with 64% ee is obtained if chiral crown ether **88** is used as a phase transfer catalyst (Scheme 8–30).[69]

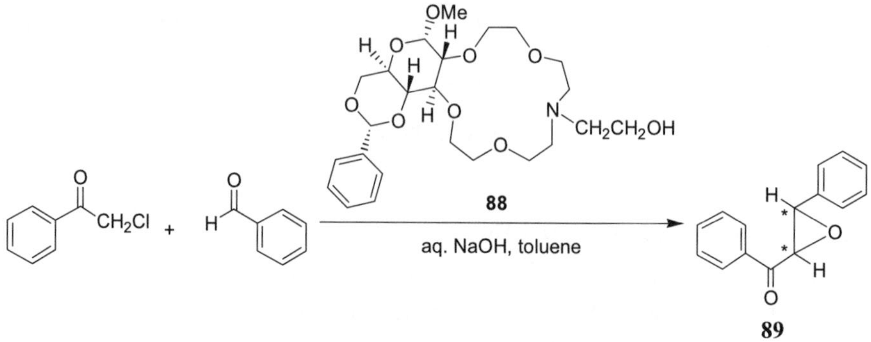

Scheme 8–30

8.2.12 Asymmetric Conjugate Addition

The conjugate addition of a nucleophile to α,β-unsaturated organic substrates is an important method for assembling structurally complex molecules. β-Substituted carbonyl compounds produced through 1,4-conjugate addition with organometallic reagents can also be versatile synthons for further organic transformations.

Enantioselective conjugate addition of an organometallic reagent to a pro-chiral substrate may be an attractive method for creating chiral centers in organic molecules. This can be achieved either by addition of a chiral organometallic reagent to α,β-unsaturated compounds or by addition of an achiral reagent in the presence of chiral catalysts. Organocopper compounds have played an important role in asymmetric conjugate additions.

In chiral ligand-modified organocopper compounds of the type RCu(L*)Li, the chiral ligand L* governs the stereoselectivity of the reaction. Good results can be obtained using these chiral cuprates in a stoichiometric manner, and naturally occurring alcohols and amines such as ephedrine and proline derivatives can be used as chiral ligands. However, these chiral cuprate-mediated reactions do entail two problems.

1. In solution, organocopper compounds may exist as an equilibrium of several species, and a loss of enantioselectivity may be inevitable if this equilibrium process produces some achiral but more reactive cuprate species. The way to overcome this problem is to develop a highly reactive chiral reagent to suppress the undesired, nonchiral species-mediated reactions.

2. Another problem is that most of the chiral organocopper reagents exhibit high substrate specificity. Good results with a specific chiral copper reagent may be observed for only one or a few substrates.

Both problems can be overcome via ligand-accelerated catalysis.[70] In this process, the presence of a suitable chiral ligand can lead to the formation of a highly reactive catalyst, and thus a stereoselective reaction may be favored over a nonselective one. In the presence of chiral copper, nickel, or rhodium complex, additions of organolithium, Grignard, or an organozinc reagent have all shown good to excellent enantioselectivity in asymmetric conjugate additions.

This work was initiated in 1988 when Villacorta et al.[71a] reported the asymmetric conjugate addition of a Grignard reagent to 2-cyclohexenone. This study showed that 1,4-adducts with 4–14% ee were obtained in the presence of aminotroponeimine copper complex.[71a] Enhanced results (74% ee) were obtained by adding HMPA or silyl halides.[71b] Several other copper complexes were also used for inducing asymmetric conjugate addition reactions. Moderate results were obtained in most cases when THF was used as the solvent and HMPA as the additive.

Based on the fact that hexamethylphosphoric triamide can greatly enhance the stereoselectivity of the reaction, chiral phosphorous amidites of type **90** have been synthesized and tested for inducing asymmetric conjugate additions, and indeed good results have been obtained. For example, Scheme 8–31 shows that product was obtained with 87% ee.[72]

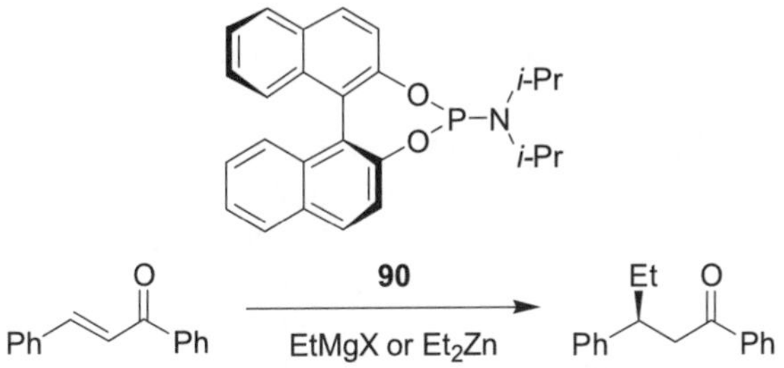

Scheme 8–31

In the conjugate addition of diethylzinc to enones catalyzed by copper reagent CuOTf or $Cu(OTf)_2$ in the presence of **90**, an ee of over 60% has been obtained. Study also shows that the actual catalyst in the reaction may be a Cu(I) species formed via in situ reduction of Cu(II) complexes.

The chiral phosphorous amidite was tested for asymmetric conjugate addition with other acyclic substrates, and again good results were obtained. The examples show that binaphthol-containing phosphorous amidites are good ligands for asymmetric conjugate additions. Further modifications have been carried out to enhance the stereoselectivity of the reaction. The bridging of bi-naphthol and a chiral amine through a phosphorous center may be a general feature of catalysts potentially suitable for highly enantioselective conjugate additions.

Chiral ligand **91**, which bears C_2-symmetric chiral ligand binaphthol and another C_2-symmetric chiral ligand bis(1-phenylethyl)amine, was synthesized.[73] This chiral ligand can be used in copper-catalyzed conjugate addition of di-alkylzinc reagents to numerous cyclic enones. Products with good yield and excellent enantioselectivity can be obtained through this process. For the reaction shown in Scheme 8–32, an ee of over 90% can generally be observed.

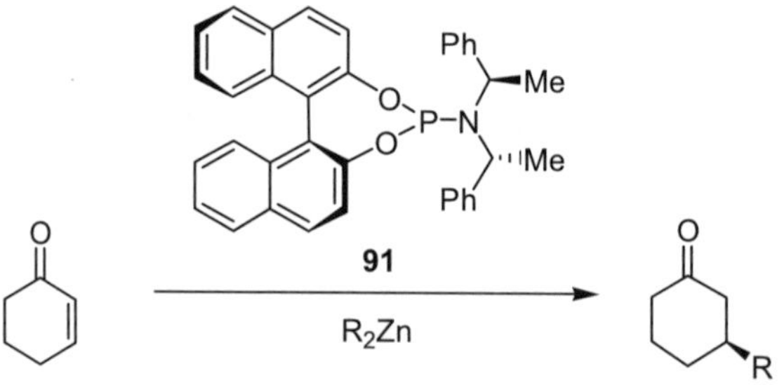

Scheme 8–32

Phosphite compounds, which have been discussed in the context of their application in asymmetric hydrogenation reactions (see Section 6.1.2.6), can also be used to effect the copper salt–mediated asymmetric conjugate addition of diethylzinc to enones.[74] As shown in Scheme 8–33, in the presence of di-phosphite **92** and copper salt $[Cu(OTf)_2]$, the asymmetric conjugate addition proceeds smoothly, giving the corresponding addition product with high conversion and ee. In contrast, the monophosphite **93** gave substantially lower ee.

Silylketene acetals and enolsilanes can also undergo conjugate addition to α,β-unsaturated carbonyl derivatives. This reaction is referred to as the Mukaiyama-Michael addition and can also be used as a mild and versatile method for C–C bond formation. As shown in Scheme 8–34, in the presence of C_2-symmetric Cu(II) Lewis acid **94**, asymmetric conjugate addition proceeds readily, giving product with high yield and enantioselectivity.[75]

Shibasaki's lanthanide–alkaline metal–BINOL system, discussed in Chapters 2 and 3, can also effect the asymmetric conjugate addition reaction. As shown in Scheme 8–35, enantioselective conjugate addition of thiols to α,β-unsaturated carbonyl compounds proceeds smoothly, leading to the corresponding products with high yield and high ee.[76]

It is worth noting that conjugate addition can also be effected by the addition of aryl- and alkenylboronic acid reagents to enones in the presence of rhodium catalysts. This new catalyst system has the following advantages:

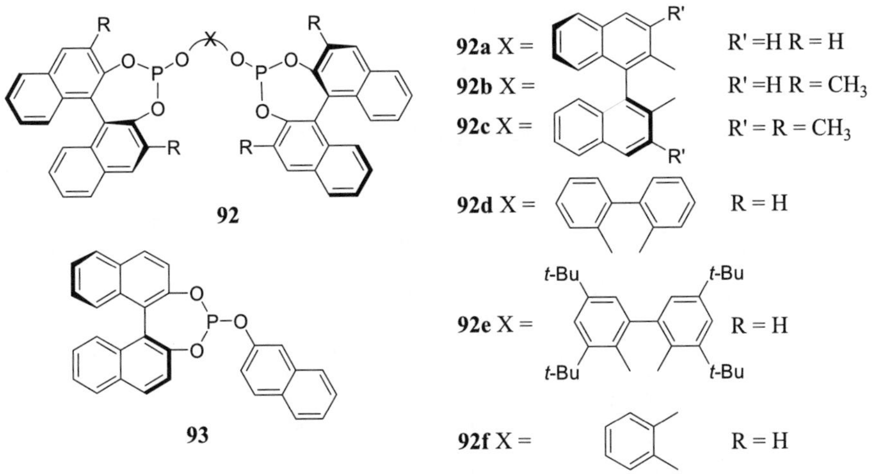

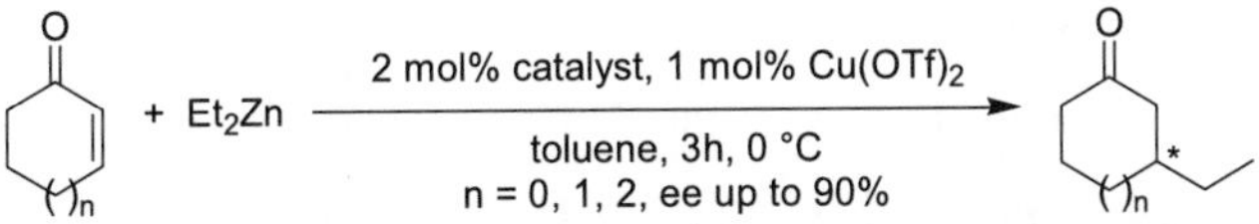

92

93

92a X =   R' =H R = H

92b X =   R' =H R = CH$_3$

92c X =   R' = R = CH$_3$

92d X =   R = H

92e X =   R = H

92f X =   R = H

$$\text{enone} + Et_2Zn \xrightarrow[\substack{\text{toluene, 3h, 0 °C} \\ n = 0, 1, 2, \text{ ee up to 90\%}}]{2 \text{ mol\% catalyst, 1 mol\% Cu(OTf)}_2} \text{product}$$

Entry	Ligand	Substrate (n)	Conver. (%)	Ee (%)
1	92a	1	100	90.2
2	92a	0	100	76.6
3	92b	1	100	42.1
4	92c	1	100	69.2
5	92e	1	100	89.3
6	92e	0	100	84.0
7	92f	1	100	88.3
8	92f	0	100	83.0
9	93	1	100	29.1

Scheme 8–33. Reprinted with permission by Pergamon Press Ltd., Ref. 74(b).

1. Organoboronic acids are stable in the presence of oxygen and moisture, permitting a protic or even aqueous reaction medium.

2. In the absence of rhodium catalyst, organoboronic acids are much less reactive toward enone substrates than the corresponding organometallic reagents used previously (such as organolithium reagents or Grignard reagents), and no 1,2-addition to enones takes place in the absence of the catalyst.

Thus, conjugate addition of PhB(OH)$_2$ to cyclohexenone proceeds smoothly in the presence of 3 mol% of Rh(acac)(C$_2$H$_4$)$_2$ and (S)-BINAP (1:1), providing the product in 97% ee with (S)-configuration predominant (Scheme 8–36).[77]

In addition to the asymmetric conjugate addition involving an enone substrate and a relatively inactive nucleophile, there exists another kind of reaction in which a deactivated substrate and a normal nucleophile are involved. For example, under proper conditions, ordinary organometallic compounds such as

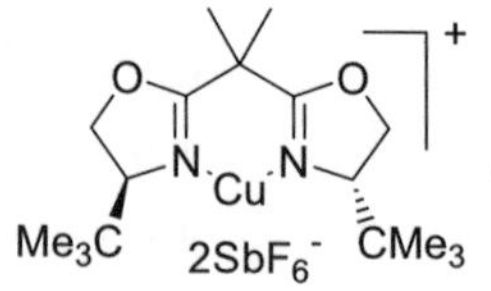

94

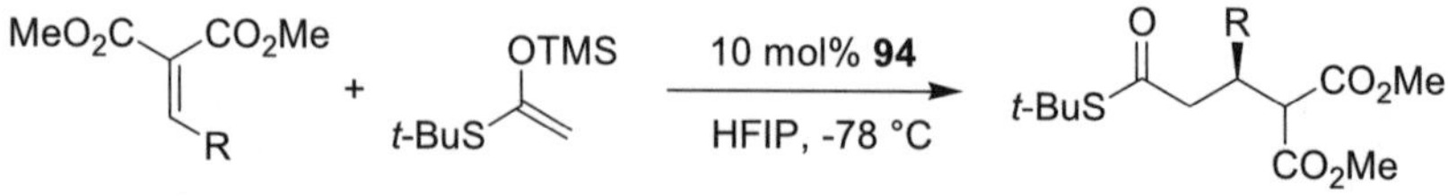

Entry	R	Time (h)	Isolated yield (%)	ee (%)
1	Phenyl	3	91	93
2	2-Furyl	5	88	94
3	2-Naphthyl	10	90	93
4	3-Ts-indolyl	48	99	86
5	2-MeOPh	12	92	99
6	c-Hexyl	5	95	95
7	c-Hexyl	12	96	93
8	c-Hexyl	20	99	95
9	i-Pr	6	93	93
10	t-Bu	8	89	90

Scheme 8–34. Reprinted with permission by Am. Chem. Soc., Ref. 75.

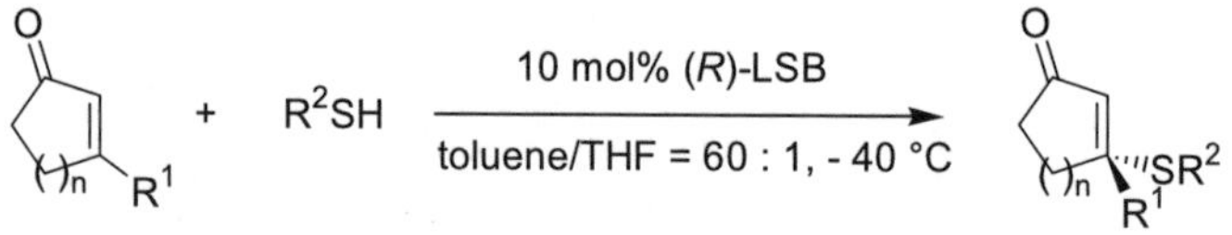

Entry	Substrate	R^2	Time	Yield (%)	Ee(%)
1	$R^1 = H, n = 2$	4-t-BuPhSH	20 min	93	84
2	$R^1 = H, n = 2$	PhSH	20 min	87	68
3	$R^1 = H, n = 2$	PhCH$_2$SH	14 h	86	90
4	$R^1 = H, n = 1$	PhCH$_2$SH	4 h	94	56
5	$R^1 = H, n = 3$	PhCH$_2$SH	41 h	87	83
6	$R^1 = CH_3, n = 2$	PhCH$_2$SH	43 h	56	85

Scheme 8–35. LSB-promoted catalytic asymmetric conjugate addition of thiols to enones. Reprinted with permission by Am. Chem. Soc., Ref. 76.

organolithium or Grignard reagent can also be used to achieve a similar result. As shown in Scheme 8–37, when activated by chiral Ni catalyst, ketal substrate can undergo a similar Michael-type addition, providing the substituted ketal with moderate to excellent ee. The resulting compound can be converted to the corresponding ketone or enol ether upon acid or base treatment. Chiral phosphine ligand can generally be utilized for this purpose.[78]

More information on asymmetric conjugate additions is provided elsewhere.[79,80]

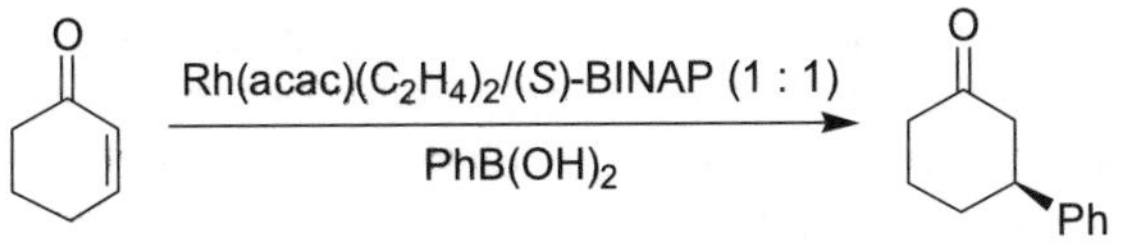

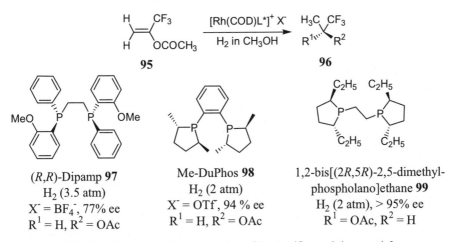

Scheme 8–36

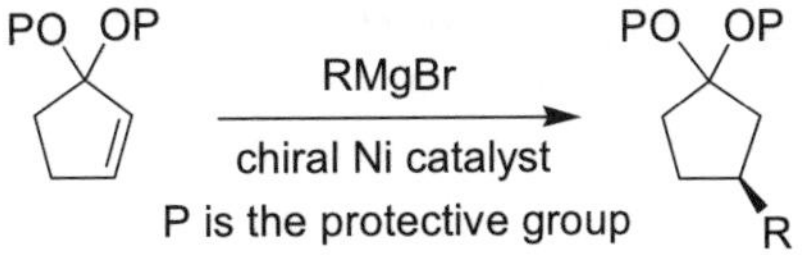

Scheme 8–37

8.2.13 Asymmetric Synthesis of Fluorinated Compounds

This section focuses on the preparation of fluorinated compounds through asymmetric hydrogenation/reduction reactions and nucleophilic additions by listing some examples. The first successful example of catalytic asymmetric hydrogenation of a fluoro-compound was reported by König et al.[81]

Scheme 8–38 presents some examples of asymmetric hydrogenations of fluorine-containing substrates catalyzed by Rh(I) complex. DIPAMP–Rh(I) complex [**97**-Rh(I)] catalyzes the enantioselective hydrogenation of fluorine-containing substrates with up to 77% ee.[82] C_2-symmetric chiral bidentate ligands such as bis(phospholanes), which were developed by Burk et al.,[83] have also proved suitable for the hydrogenation of fluorine-containing substrates. Thus, when **98** or **99** is applied in the hydrogenation of **95**, the corresponding hydrogenation product **96** is obtained with 94% ee and >95% ee, respectively.

(*R*,*R*)-Dipamp **97**
H_2 (3.5 atm)
$X^- = BF_4^-$, 77% ee
$R^1 = H$, $R^2 = OAc$

Me-DuPhos **98**
H_2 (2 atm)
$X^- = OTf^-$, 94 % ee
$R^1 = H$, $R^2 = OAc$

1,2-bis[(2*R*,5*R*)-2,5-dimethyl-phospholano]ethane **99**
H_2 (2 atm), > 95% ee
$R^1 = OAc$, $R^2 = H$

Scheme 8–38. Enantioselective hydrogenation of 1,1,1-trifluoro-2-(acetoxy)-2-propene.

Saburi et al.[84] found that ruthenium complexes containing chiral ligand can be used to catalyze the asymmetric hydrogenation of 2-fluoro-2-alkenoic acid (Z)- or (E)-**100**, providing the corresponding product **101** with good enantio-selectivity (Scheme 8–39 and Table 8–2).

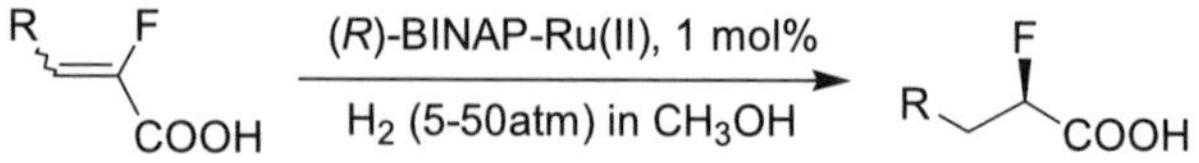

(Z)-**100**   **101**

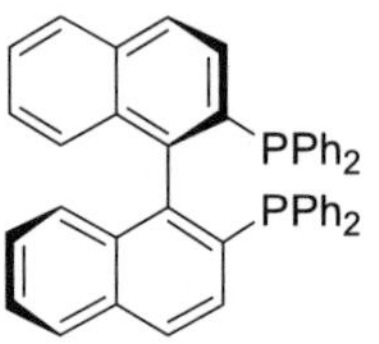

(R)-BINAP-Ru(II) catalyst
catalyst A = $Ru_2Cl_4[(R)\text{-BINAP}]_2N(C_2H_5)_3$
catalyst B = $Ru[OCOC(CH_3)_3]_2[(R)\text{-BINAP}]_2$

102

Scheme 8–39. Enantioselective hydrogenation of 2-fluoro-2-alkenoic acid.

TABLE 8–2. Catalytic Hydrogenation of Fluorine-Containing α,β-Unsaturated Carboxylic Acids

Entry	Catalyst	Substrate	R	Reaction Condition	ee (%)
1	A	(Z)-**100a**	C_3H_7	35°–80°C, 24 h	91
2	A	(Z)-**100b**	C_5H_{11}	35°–80°C, 24 h	89
3	B	(Z)-**100a**	C_3H_7	35°–80°C, 24 h	89
4	B	(Z)-**100b**	C_5H_{11}	35°–80°C, 24 h	89
5	A	(E)-**100a**	C_3H_7	50°C, 24 h	83

ee = Enantiomeric excess.
Reprinted with permission by Pergamon Press Ltd., Ref. 84.

Besides the above-mentioned catalytic asymmetric hydrogenation method for preparing fluorine-containing compounds, other reactions such as asymmetric reduction of achiral fluorine-containing ketones are also feasible methods for preparing chiral fluorinated compounds. For example, the oxazaborolidine system, which has been discussed in Chapter 6, can also be employed in the catalytic reduction of trifluoromethyl ketones. Scheme 8–40 depicts some examples.[85]

In Scheme 8–40, the reaction of 9-anthryl trifluoromethyl ketone **103** and mesityl trifluoromethyl ketone **104** with catecholborane **106** in the presence of 10 mol% of chiral catalyst **107** (CBS) provides (R)-carbinol **108** and **109** with 94% and 100% ee, respectively. When methyl ketone instead of trifluoromethyl ketone is used in the reaction, product **110** is obtained with (S)-configuration in 99.7% ee with over 95% yield.

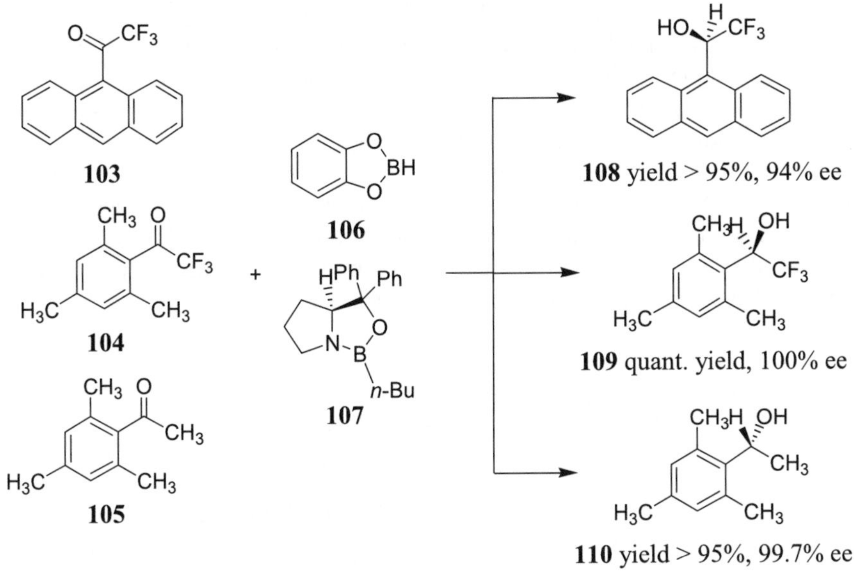

Scheme 8–40. Enantioselective hydroboration of trifluoromethyl ketones.

Fluorine-containing compounds can also be synthesized via enantioselective Reformatsky reaction using bromo-difluoroacetate as the nucleophile and chiral amino alcohol as the chiral-inducing agent.[86] As shown in Scheme 8–41, 1 equivalent of benzaldehyde is treated with 3 equivalents of **111** in the presence of 2 equivalents of **113**, providing α,α-difluoro-β-hydroxy ester **112** at 61% yield with 84% ee. Poor results are observed for aliphatic aldehyde substrates. For example, product **116** is obtained in only 46% ee.

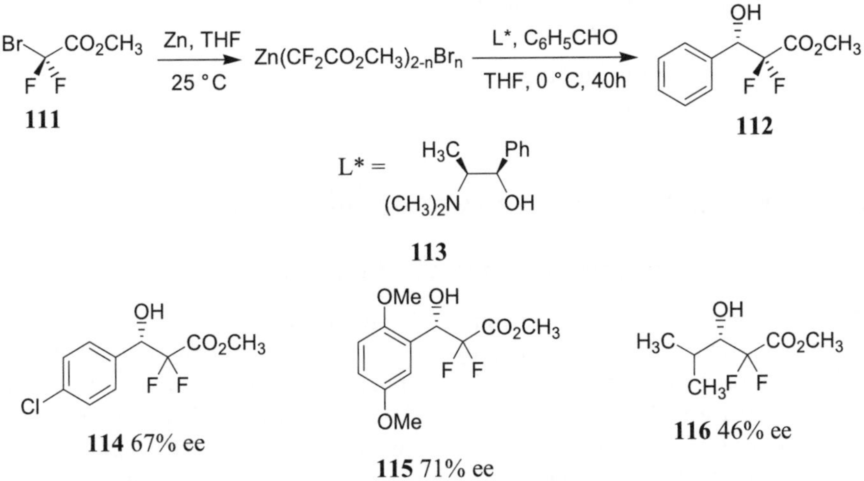

Scheme 8–41

The enantioselective addition of a nucleophile to a carbonyl group is one of the most versatile methods for C–C bond formation, and this reaction is discussed in Chapter 2. Trifluoromethylation of aldehyde or achiral ketone via addition of fluorinated reagents is another means of access to fluorinated compounds. Trifluoromethyl trimethylsilane [$(CH_3)_3SiCF_3$] has been used by Prakash et al.[87] as an efficient reagent for the trifluoromethylation of carbonyl compounds. Reaction of aldehydes or ketones with trifluoromethyltrimethylsilane can be facilitated by tetrabutyl ammonium fluoride (TBAF). In 1994, Iseki et al.[88] found that chiral quaternary ammonium fluoride **117a** or **117b** facilitated the above reaction in an asymmetric manner (Scheme 8–42).

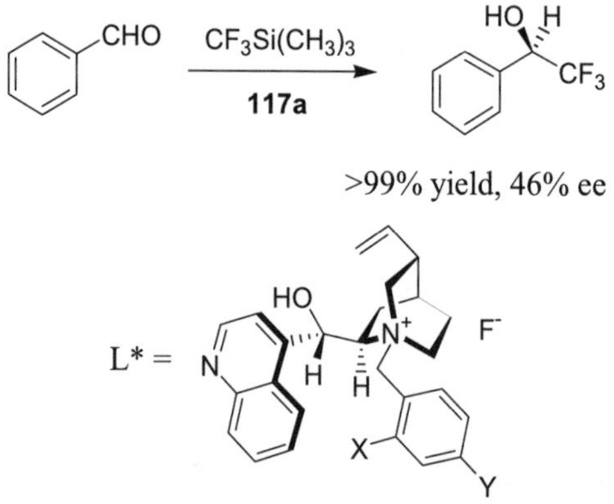

>99% yield, 46% ee

117a X = Y = CF_3
117b X = H, Y = CF_3

Entry	Substrate	Yield (%)	ee (%)
1	9-Anthrcarbaldehyde	98	45
2	Acetophenone	91	45
3	Phenyl *i*-propyl ketone	87	51
4	Benzaldehyde	99	35

Scheme 8–42. Asymmetric trifluoromethylation of carbonyl compounds by chiral quaternary ammonium fluorides.

8.3   NEW CONCEPTS IN ASYMMETRIC REACTION

8.3.1   Ti Catalysts from Self-Assembly Components

As mentioned in Chapter 1, ligand-accelerated catalysis occurs when a more effective chiral catalyst is obtained by replacing an achiral ligand with a chiral one. Mikami et al.[89] reported a different phenomenon in which a more active catalyst was formed by combining an achiral pre-catalyst with several chiral ligands. They found that the most active and enantioselective chiral catalyst was formed in preference to other possible ligand combinations (Scheme 8–43).

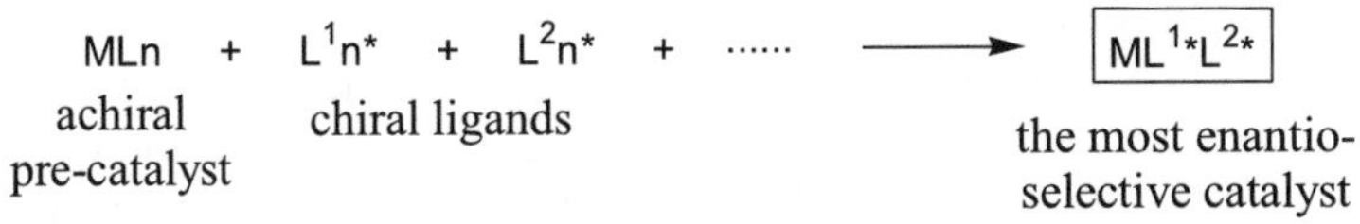

$$\text{MLn} \quad + \quad \text{L}^1\text{n*} \quad + \quad \text{L}^2\text{n*} \quad + \quad \cdots\cdots \quad \longrightarrow \quad \boxed{\text{ML}^{1*}\text{L}^{2*}}$$

achiral   chiral ligands

pre-catalyst

the most enantio-
selective catalyst

Scheme 8–43

The self-assembly of a chiral Ti catalyst can be achieved by using the achiral precursor $Ti(OPr^i)_4$ and two different chiral diol components, (R)-BINOL and (R,R)-TADDOL, in a molar ratio of 1:1:1. The components of "less basic" (R)-BINOL and the relatively more basic (R,R)-TADDOL assemble with $Ti(OPr^i)_4$ in a molar ratio of 1:1:1, yielding chiral titanium catalyst **118** in the reaction system. In the asymmetric catalysis of the carbonyl-ene reaction, **118** is not only the most enantioselective catalyst but also the most stable and the exclusively formed species in the reaction system.

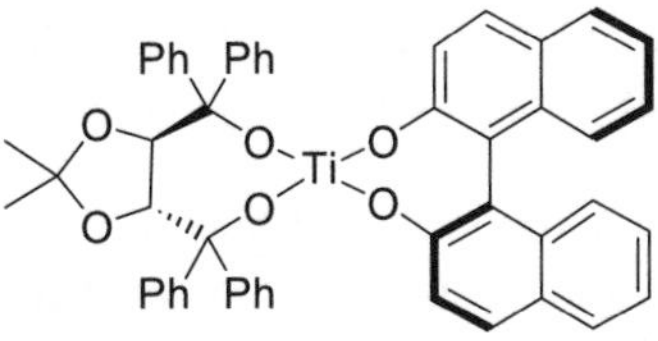

118

As shown in Scheme 8–44, (R)-TADDOL alone does not yield any carbonyl-ene product. In contrast, the reaction in the presence of **118** provides (R)-

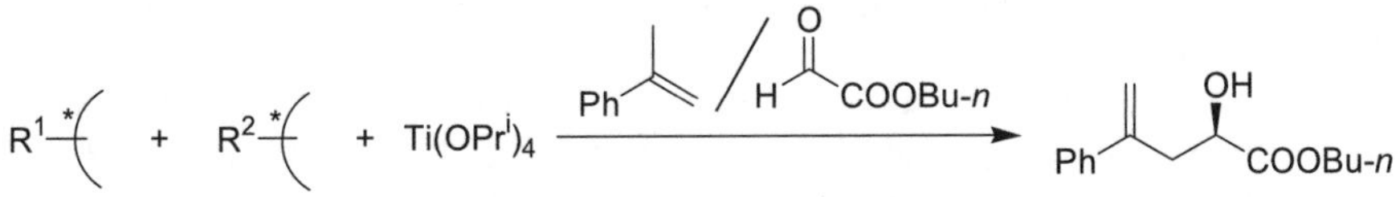

$$R^1 \ast \!\!\!\!\!\!- + \quad R^2 \ast \!\!\!\!\!\!- + \quad Ti(OPr^i)_4 \quad \xrightarrow{\quad Ph \diagup \quad \overset{O}{\underset{H}{\parallel}} \diagdown COOBu\text{-}n \quad} \quad Ph \diagdown \overset{OH}{\underset{COOBu\text{-}n}{|}}$$

Entry	$R^1\ast(OH)_2$	$R^2\ast(OH)_2$	Yield (%)	ee (%)
1	(R)-TADDOL	(R)-BINOL	50	91
2	(R)-TADDOL	-	0	-
3	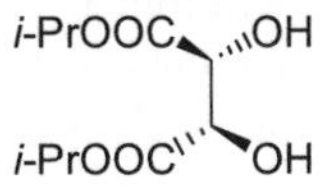 i-PrOOC,,,,OH i-PrOOC''''OH	(R)-BINOL	15	40
4	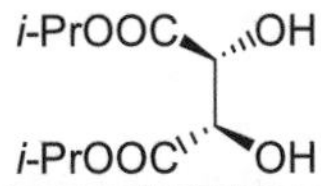 i-PrOOC,,,,OH i-PrOOC''''OH	-	4	0

Scheme 8–44. General scheme for Ti-catalyst asymmetric glyoxylate-ene reactions. Reprinted with permission by Wiley-VCH Verlag GmbH, Ref. 89.

product with 50% yield and 91% ee. In a similar reaction, when a catalyst derived from Ti(OPr^i)$_4$, (R)-BINOL, and DIPT is used, the product is obtained in 15% yield with 40% ee. In the reaction catalyzed by a combination of Ti(OPr^i)$_4$ with only DIPT, very poor yield is observed without enantioselectivity. A combination of (R)-5,5′-dichloro-2,2′-dihydorxy-1,1′-diphenyl and (R)-BINOL with Ti(OPr^i)$_4$ has been shown to be the best catalyst system for promoting highly selective and active carbonyl-ene reactions. The corresponding product can be obtained with 97% ee and 66% yield.

8.3.2 Desymmetrization

Desymmetrization, which refers to a process of efficiently desymmetrizing *meso*-molecules or achiral molecules to produce chiral ones, is a versatile method for preparing chiral nonracemic molecules.[90] Desymmetrization of *meso*-compounds generally leads to the formation of a C–C or a C–X (X is a hetero atom) bond. The reaction normally uses a functional group residing on the symmetric element (in most cases the C$_2$ axis or a plane) to differentiate two (or more) symmetrically equivalent functionalities elsewhere within the substrate molecule. This work was first reported by Hoye et al.[91] and Mislow and Siegel[92] in 1984.

Hoye demonstrated that the carbonyl group that lies on the C$_2$-symmetric axis in keto diacid **119** can be reduced to alcohol **120**. When followed by acid-catalyzed lactonization of the hydroxyl group with either a C-1 or C-9 carboxyl group, monolactone **121** or *ent*-**121** can be produced, thus realizing the desymmetrization of substrate **119** (Scheme 8–45).

In investigating catalytic desymmetrization, Trost et al.[93] demonstrated that palladium-catalyzed desymmetrization of *meso*-1,4-diol diesters (**122a–d**) gave monosubstituted products in high ee. In the course of desymmetrization, the cleavage of the leaving group is involved in the enantiodiscriminating step (Scheme 8–46).

Desymmetrization of compound **125** can be realized in the presence of a Pd complex containing **124**. Initial work showed[94] that oxazolidine-2-one (R = NTs) was obtained from bis-carbamates with a relatively low level of enantiomeric excess. In the presence of 1 equivalent of triethylamine, both ee and yield were enhanced dramatically (Scheme 8–47).

Miyafuji and Katsuki[95] reported the desymmetrization of *meso*-tetrahydrofuran derivatives via highly enantioselective C–H oxidation using Mn–salen catalysts. The optically active product lactols (up to 90% ee) are useful chiral building blocks for organic synthesis (Scheme 8–48).

8.3.3 Cooperative Asymmetric Catalysis

Cooperative catalysis between multiple metal centers is considered to be common in enzymatic systems,[96] and using this idea for designing catalytic systems

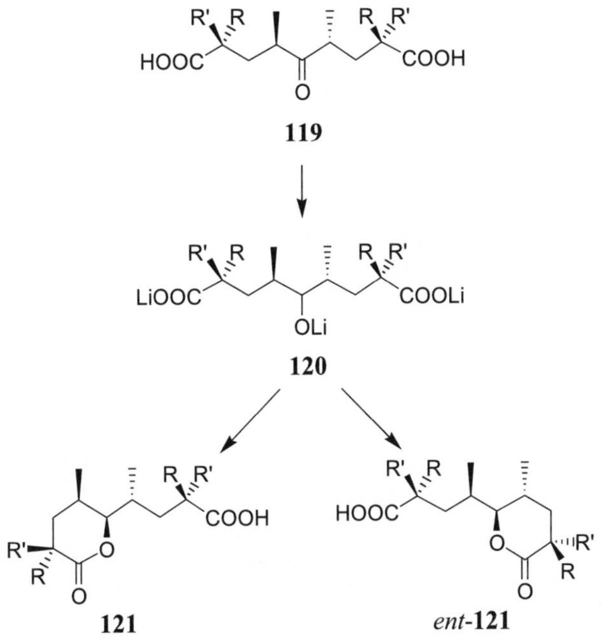

a R = R' = H
b R = Me, R' = H
c R = H, R' = Me

Scheme 8–45

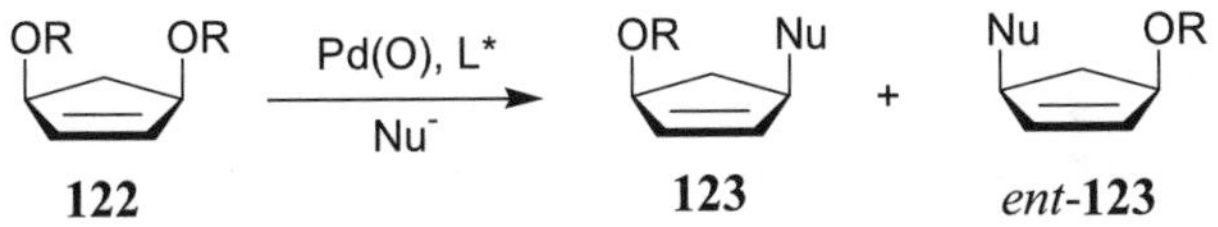

122a R = H
122b R = $COCH_3$
122c R = COPh
122d R = CONHTs

Scheme 8–46

has become an interesting development in asymmetric synthesis. In most of our previous discussions, the reaction systems contain one catalysis center by which substrate and reagent are oriented and activated and the asymmetric induction is finally realized. Recently, there have been some interesting catalysis systems that contain two kinds of catalytic center within one catalyst. These two catalytic centers are normally two metal centers functioning harmoniously, with one

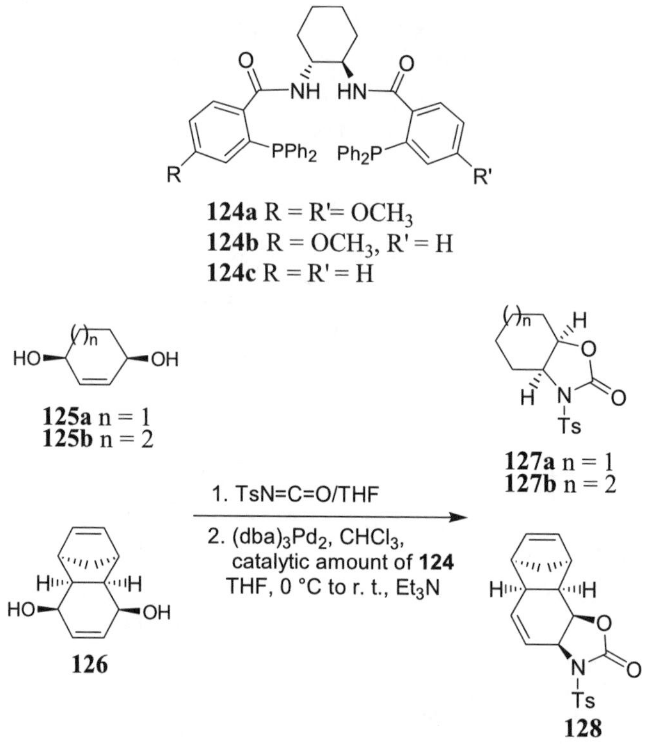

124a R = R'= OCH$_3$
124b R = OCH$_3$, R' = H
124c R = R' = H

125a n = 1
125b n = 2

127a n = 1
127b n = 2

126

1. TsN=C=O/THF

2. (dba)$_3$Pd$_2$, CHCl$_3$,
catalytic amount of **124**
THF, 0 °C to r. t., Et$_3$N

128

Diol	Oxazolidine-2-one	ee without Et$_3$N (%)	ee with Et$_3$N (%)
125a	127a	85	99
125b	127b	90	99
126	128	81	94

Scheme 8–47. Reprinted with permission by Am. Chem. Soc., Ref. 94.

center activating the substrate or the reagent and the other one directing the attack. These bimetallic catalytic systems may provide high chemoselectivity and/or stereoselectivity.

Reactions involving bimetallic catalysts, either homo-dinuclear or hetero-bimetallic complexes, and chemzymes were highlighted by Steinhagen and Helmchen[96c] in 1996. Some examples are discussed in Chapter 2. Among these examples, Shibasaki's reports have been of particular significance.[97] Shibasaki's catalyst is illustrated as **130**, which consists of one central metal M^1 (La^{+3}, Ba^{+2}, or Al^{+3}), three other metal ions (M^2)$^+$ [(M^2)$^+$ can be Li$^+$, Na$^+$, or K$^+$], and three bidentated ligands, such as (R)- or (S)-BINOL. The catalyst exhibits both Lewis acidic properties because of the existence of central metal and the Lewis basic properties because of the presence of the outer metal ions.

The multifunctional catalysts constitute a new class of widely applicable and

(R, R)-**129** R^1 = Ph

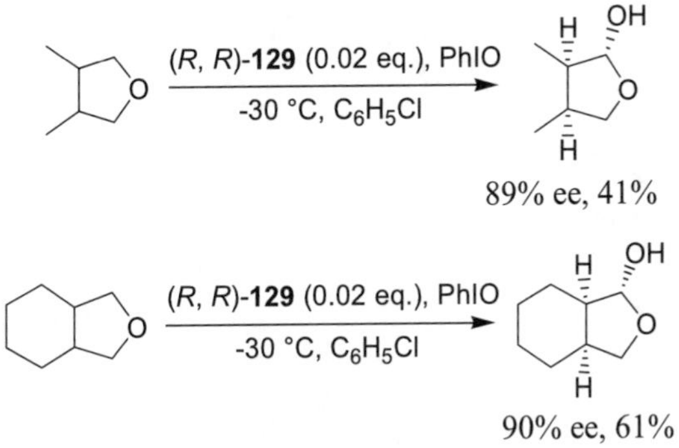

89% ee, 41%

90% ee, 61%

Scheme 8–48

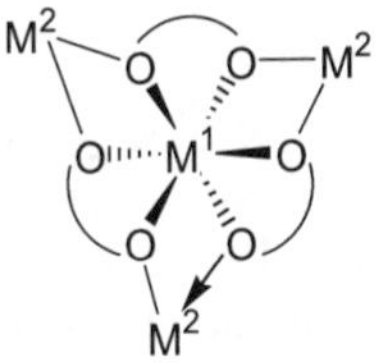

LLB : M^1 = La, M^2 = Li
LSB : M^1 = La, M^2 = Na
LPB : M^1 = La, M^2 = K
ALB : M^1 = Al, M^2 = Li
BaBM: M = Ba,
  one OH was replaced by OMe

130 M^1-M^2 BINOL complex

versatile chiral catalysts and can be used for many asymmetric reactions. It has been reported that asymmetric nitroaldol condensation (the Henry reaction, Chapter 3), aldol reaction (Chapter 3), hydrophosphonylation of imides (Chapter 2), Sharpless epoxidation (Chapter 4), desymmetrization of *meso*-epoxide (Chapter 4), BINAL–H reduction (Chapter 6), conjugate addition,[97f,g] tandem Michael-aldol addition,[97j] and 1,4-addition of Grignard reagent to enones[98] can all be catalyzed by this type of catalyst.

As shown in Scheme 8–49, this multifunctional catalyst can be applied in direct aldol reactions between an aldehyde R^1CHO and a ketone R^2COCH_3.[99]

LA represents Lewis acid in the catalyst, and M represents Brønsted base. In Scheme 8–49, Brønsted base functionality in the hetero-bimetalic chiral catalyst I can deprotonate a ketone to produce the corresponding enolate II, while at the same time the Lewis acid functionality activates an aldehyde to give intermediate III. Intramolecular aldol reaction then proceeds in a chelation-controlled manner to give β-keto metal alkoxide IV. Proton exchange between the metal alkoxide moiety and an aromatic hydroxy proton or an α-proton of a ketone leads to the production of an optically active aldol product and the regeneration of the catalyst I, thus finishing the catalytic cycle.

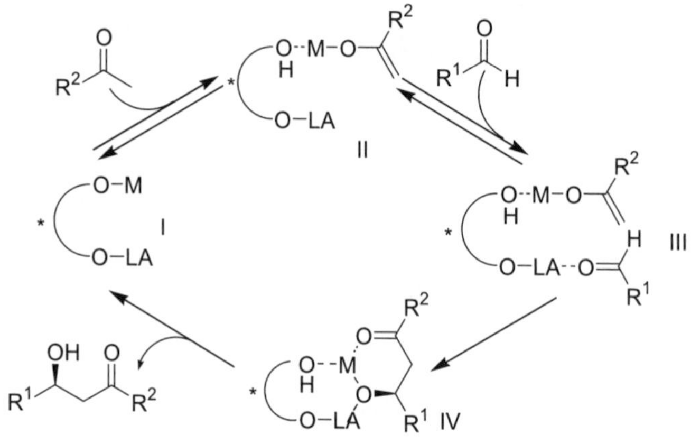

Scheme 8–49

Another example is the asymmetric cyanosilylation of aldehydes catalyzed by bifunctional catalyst **131**.[100] Compound **131** contains aluminum, the central metal, acting as a Lewis acid, and group X, acting as a Lewis base. The asymmetric cyanosilylation, as shown in Scheme 8–50, proceeds under the outlined

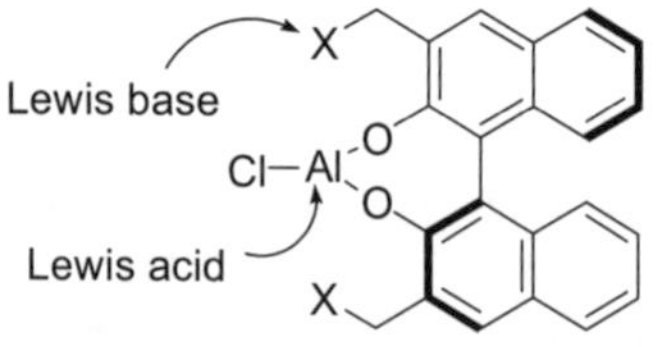

131

131a X = P(O)Ph$_2$
131b X = CH$_2$P(O)Ph$_2$
131c X = CHPh$_2$
131d X = P(O)[p-(CH$_3$)$_2$NPh]$_2$

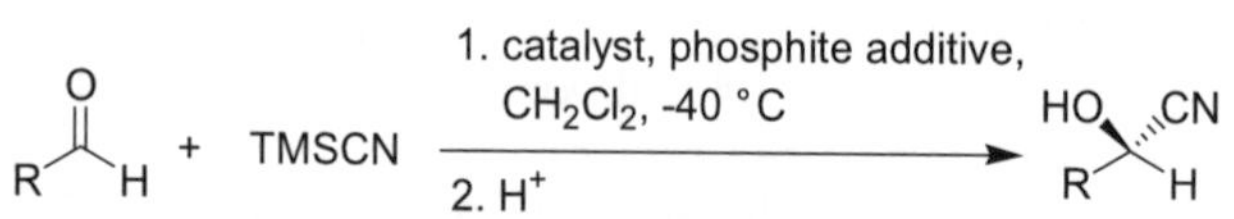

Scheme 8–50

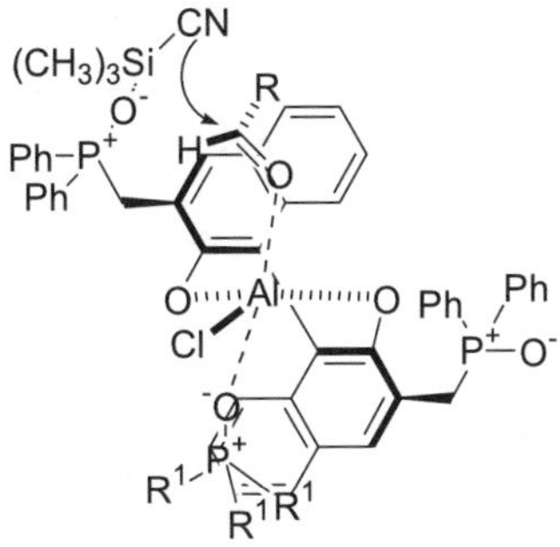

Figure 8–6. Transition state for **131** catalyzed asymmetric cyanosilylation.

conditions, giving the corresponding product with high yield and enantio-selectivity. The reaction transition state is shown in Figure 8–6. In the course of the reaction, the phosphine oxide group activates cyanosilylation reagent TMSCN, and the central Al atom activates the aldehyde, thus facilitating the reaction. As proposed by Shibasaki, there might be two competing reaction pathways in the case of more reactive aldehydes. The desired pathway, which gives enantioselective product, may involve the dual interaction between the aldehyde and the Lewis acid and also between the phosphine oxide and TMSCN, whereas the undesired pathway involves the interaction of the Lewis acid and aldehyde only. Shibasaki further proposed that the rate of the two pathways could be made to differ significantly from each other if the Lewis acidity of the catalyst were decreased. The purpose of the phosphite additive is to reduce the Lewis acidity of the central Al atom via phosphite coordination, thus increasing the enantioselectivity of the reaction. In Scheme 8–50, the cya-nosilylation of a series of aldehydes occurs at $-40°C$, and in most cases both the yield and the ee are over 90%.

Another example of this cooperative catalysis has been presented by Konsler et al.[101] in the course of their asymmetric ring-opening (ARO) study. They found that the ARO of *meso*-epoxides with TMS-N$_3$, catalyzed by Cr–salen compound **132**, showed a second-order kinetic dependence on the catalyst.[102] They then proposed that there might be cooperative, intramolecular bimetallic catalysis taking place, with one metal activating the substrate *meso*-epoxide and

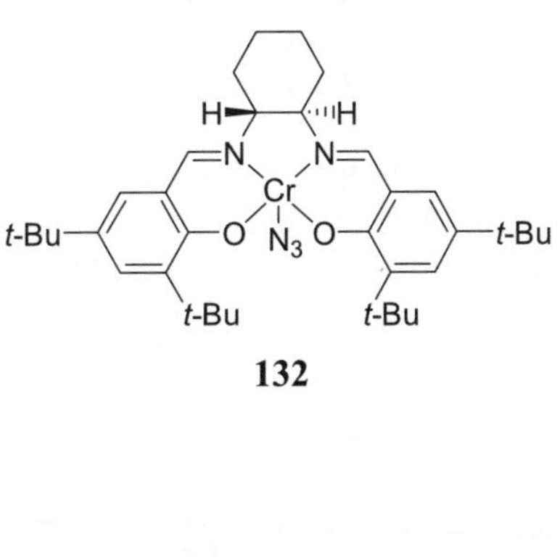

132

another metal activating the nucleophile TMS-N₃. Based on this assumption, new catalyst **133**, linking two Cr–salen moieties together, was synthesized and tested for its ability to catalyze ARO reactions.

133

133a n = 2, **133b** n = 4, **133c** n = 5, **133d** n = 6, **133e** n = 7, **133f** n = 8, **133g** n = 9

Their study showed that when compounds **133** catalyze asymmetric ring opening with cyclopentene oxide the reaction rate was significantly enhanced, especially when the reaction was catalyzed by **133c** (n = 5). Even if catalyst **133c** was used at a very low concentration, the reaction still proceeded very rapidly, providing the ring-opening product with comparable ee to that catalyzed by the monomeric analogs at much higher concentration. In designing the catalyst, it is essential to use a flexible tether of proper length so as to get the optimal transition state entropy and enthalpy.

8.3.4 Stereochemical Nonlinear Effects in Asymmetric Reaction

Nonlinear effects in asymmetric reaction (NLEs) refer to the nonlinear relationships between the ee value of the chiral auxiliary or ligand and the ee value in the product.[103]

The application of a chiral auxiliary or catalyst, in either stoichiometric or catalytic fashion, has been a common practice in asymmetric synthesis, and most of such auxiliaries are available in homochiral form. Some processes of enantiodifferentiation arise from diastereomeric interactions in racemic mixtures and thus cause enhanced enantioselectivity in the reaction. In other words, there can be a nonlinear relationship between the optical purity of the chiral auxiliary and the enantiomeric excess of the product. One may expect that a chiral ligand, not necessarily in enantiomerically pure form, can lead to high levels of asymmetric induction via enantiodiscrimination. In such cases, a nonlinear relationship (NLE) between the ee of the product and the ee of the chiral ligand may be observed.

In 1986, Puchot et al.[104] studied the nonlinear correlation between the enantiomeric excess of a chiral auxiliary and the optical yield in an asymmetric synthesis, either stoichiometric or catalytic. Negative NLEs [(−)-NLEs] were observed in the asymmetric oxidation of sulfide and in [S]-proline–mediated asymmetric Robinson annulation reactions, while a positive NLE [(+)-NLEs]

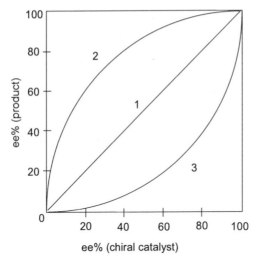

Figure 8–7. Relationship between the enantiomeric excess of the chiral catalyst and that of the product.

was observed for Sharpless epoxidation reactions. Many other reactions have since been found that show the nonlinear effect.[105]

Figure 8–1 depicts the relationship between the optical purity of the chiral catalyst and the ee of the product. In a simplified case, when two enantiomeric chiral ligands (L_R or L_S) are attached to a metal center (M), complexes ML_2 may be formed as the reactive species. Three complexes are possible: $ML_R L_S$, $ML_R L_R$, and $ML_S L_S$. Supposing that L_R is in excess and the stability constant for the *meso*-complex $ML_R L_S$ is greater than that of the chiral complexes, if *meso*-$ML_R L_S$ is the more active catalyst, a lower than expected ee will be obtained [(−)-NLEs, curve 3 in Fig. 8–7]. The ee will be higher than expected if the *meso*-catalyst is less reactive than $ML_R L_R$ or $ML_S L_S$ [(+)-NLEs, curve 2 in Fig. 8–7].

Diethylzinc addition to aldehyde is one of the earliest reactions showing nonlinear effects. The first paper on this transformation was published in 1986.[106] Diethylzinc is usually inert to aldehydes in hydrocarbon solvents. However, nucleophilic addition reactions proceed smoothly in the presence of β-dialkylamino alcohol, affording the corresponding secondary alcohol with high yield. Particularly noteworthy is that no alkylation occurs when 100 mol% of amino alcohol is added to organozinc, while the reaction affords the product with high yield in the presence of a catalytic amount (2–8 mol%) of amino alcohol. DAIB [(2S)-3-*exo*-(dimethylamino) isoborneol] is one of the best catalysts found thus far for catalyzing this reaction. Kagan and Oguni recognized this process as a chiral multiplication process, and a detailed study of its mechanism was provided by Kitamura et al.[107]

Reaction of diethylzinc and benzaldehyde in toluene containing a small amount of DAIB with 15% ee can afford a product with 95% ee, which is close to the 98% ee achieved with enantiomerically pure (2S)-DAIB. The reaction rate depends largely on the enantiomeric purity of the DAIB. Enantiomerically pure DAIB leads to a much faster reaction than racemic DAIB. Kitamura et al. contend that this is the result of auto association of the chiral intermediates generated by the reaction between the catalyst (−)-DAIB and the organozinc compound. When a mixture of (−)- and (+)-DAIB is used, two types of dimeric species are formed: homochiral [(−)-(−)-**134** and (+)-(+)-**134**] and heterochiral [(−)-(+)-**134**]. The signs refer to the enantiomer of the chiral ligand DAIB included in the complex. The enantiomeric monomers DAIB–ZnEt **135** are the active catalysts in this reaction, and each one produces predominantly one enantiomer of the product alcohol (Scheme 8–51 and Fig. 8–8).

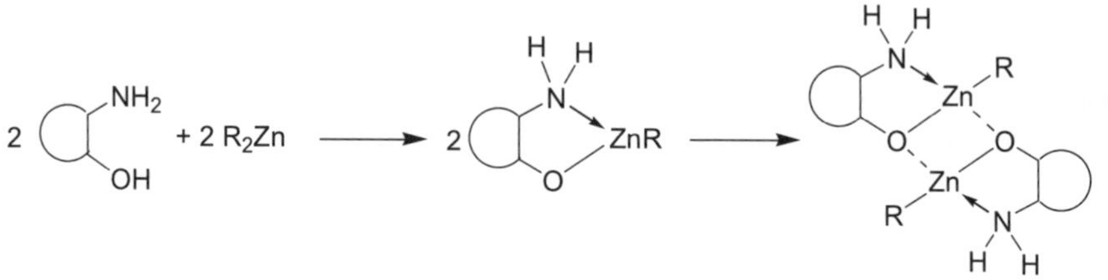

Scheme 8–51

There is an equilibrium between the dimer and monomer, and molecular orbital study suggests that the heterochiral dimer is more stable than the homochiral isomer. The existence and behavior of the dimeric species were well confirmed by experiments such as cryoscopic molecular weight and NMR measurement. In the NMR study of a DAIB-catalyzed dialkylzinc addition reaction, noticeable changes were observed in the spectrum of the homochiral dimer on the addition of benzaldehyde, while the spectrum of the heterochiral complex remained the same. This may imply that the heterochiral complex is very stable and does not react, and the homochiral dimer leads to the reaction product.

For a review of nonlinear effects in asymmetric synthesis, see Girard and Kagan.[108]

8.3.5 Chiral Poisoning

Another interesting issue is the possibility of creating optically active compounds with racemic catalysts. The term *chiral poisoning* has been coined for the situation where a chiral substance deactivates one enantiomer of a racemic catalyst. Enantiomerically pure (R,R)-chiraphos rhodium complex affords the (S)-methylsuccinate in more than 98% ee when applied in the asymmetric hydrogenation of a substrate itaconate.[109] An economical and convenient method

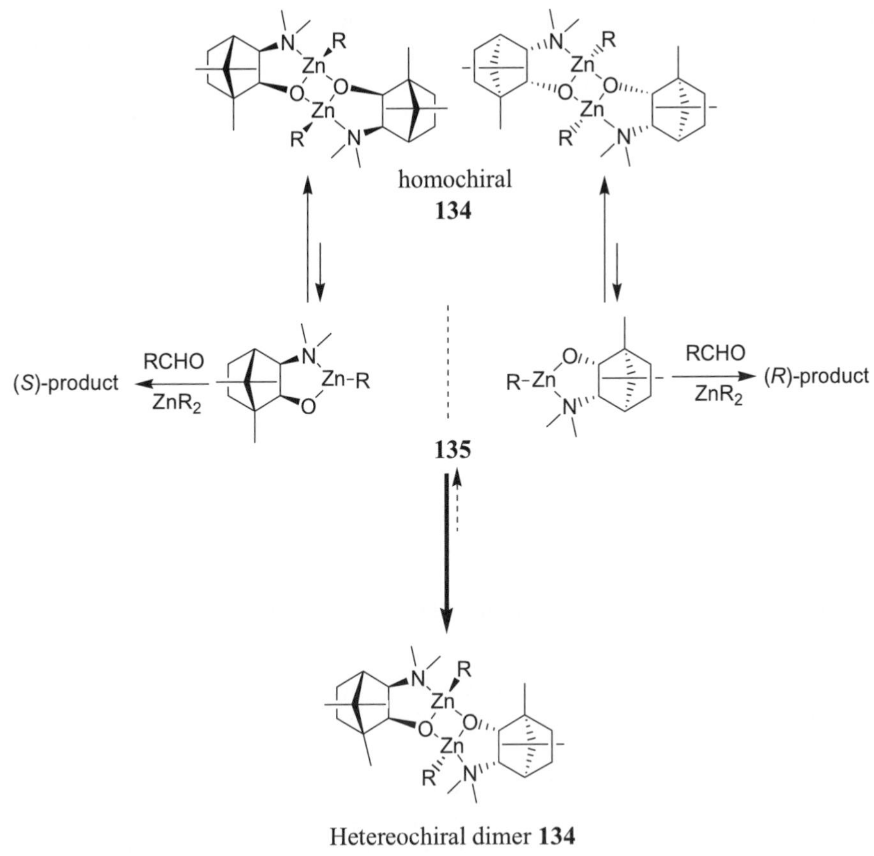

homochiral
134

(S)-product ←[RCHO / ZnR$_2$]— [structure] —[RCHO / ZnR$_2$]→ (R)-product

135

Hetereochiral dimer **134**

Figure 8–8. Association of chiral intermediates.

has then been proposed using racemic $[(chiraphos)Rh]_2^{2+}$ in the presence of a chiral poison. This chiral poison is added to preferentially block the functioning of one enantiomer of this catalyst and let only the other one work. Using racemic $[(chiraphos)Rh]_2^{2+}$ plus (S)-methophos as the catalytic system, product dimethyl succinate can be obtained in 49% ee (Scheme 8–52).[109] (S)-Methophos itself affords the hydrogenation product only in less than 2% ee.

Similarly,[110] (1R,2S)-ephedrine is an effective poison in the kinetic resolution of allylic alcohols using racemic BINAP instead of the expensive (R)-BINAP. (R)-2-cyclohexenol can thus be obtained in >95% ee using a racemic

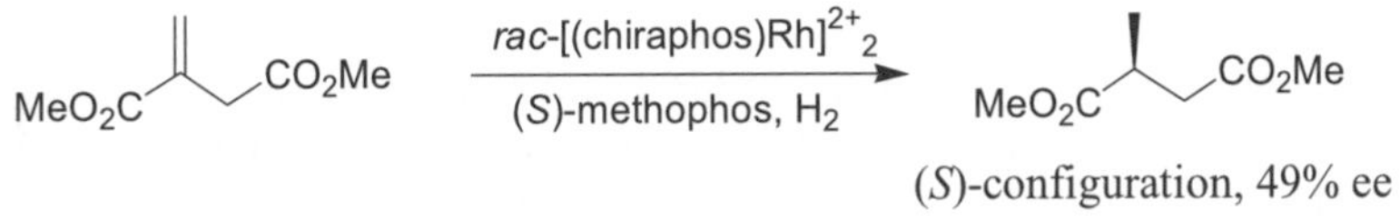

MeO$_2$C $\diagup\!\!\diagdown$ CO$_2$Me —[rac-$[(chiraphos)Rh]^{2+}_2$ / (S)-methophos, H$_2$]→ MeO$_2$C $\diagup\!\!\diagdown$ CO$_2$Me

(S)-configuration, 49% ee

Scheme 8–52

BINAP–Ru catalyst and $(1R,2S)$-ephedrine (Scheme 8–53). This result is similar to that obtained when catalyzed by pure (R)-BINAP. In pure (R)-BINAP complex–catalyzed hydrogenation, (S)-2-cyclohexenol can also be obtained with over 95% ee. This means that in the presence of (R)-BINAP–Ru catalyst, (R)-cyclohexenol is hydrogenated much faster than its (S)-enantiomer. When ephedrine is present, (R)-BINAP–Ru will be selectively deactivated, and the action of (S)-BINAP–Ru leads to the selective hydrogenation of (S)-2-cyclohexenol, leaving the intact (R)-2-cyclohexenol in high ee.

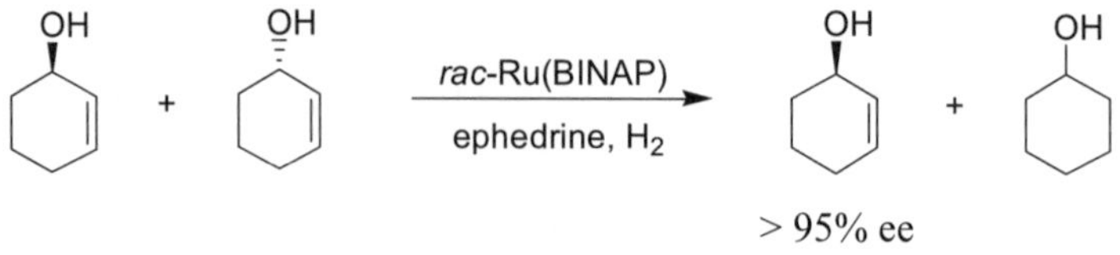

> 95% ee

Scheme 8–53

8.3.6 Enantioselective Activation and Induced Chirality

In contrast to chiral poisoning, the concept of chiral activation has also emerged recently. An additional activator selectively or preferentially activates one enantiomer of the racemic catalyst, resulting in much faster reaction, and gives products with high ee.

The idea of enantioselective activation was first reported by Mikami and Matsukawa[111] for carbonyl-ene reactions. Using an additional catalytic amount of (R)-BINOL or (R)-5,5′-dichloro-4,4′,6,6′-tetramethylbiphenyl as the chiral activator, (R)-ene products were obtained in high ee when a catalyst system consisting of rac-BINOL and $Ti(OPr^i)_4$ was employed for the enantioselective carbonyl ene reaction of glyoxylate (Scheme 8–54). Amazingly, racemic BINOL can also be used in this system as an activator for the (R)-BINOL–Ti catalyst, affording an enhanced level of enantioselectivity (96% ee).

$RuCl_2(TolBINAP)(DMF)_n$ in either racemic or enantiomerically pure form is a feeble catalyst for the hydrogenation of simple ketones. However, when (S,S)-1,2-diphenyl ethylenediamine $[(S,S)$-DPEN] is present in the reaction mixture, the asymmetric hydrogenation of the carbonyl group in 2,4,4-trimethyl-2-cyclohexenone takes place, providing the corresponding (S)-alcohol in 95% ee and with 100% yield. As shown in Scheme 8–55, in the presence of racemic $RuCl_2[(\pm)$-Tolbinap)(DMF)_n]$, (S,S)-DPEN, and KOH in a 7:1 mixture of isopropanol and toluene, the allylic alcohol is produced with high ee and high yield.

The complex TolBINAP–Ru dichloride existing in aggregate form is incapable of catalyzing the hydrogenation reaction. In the presence of (S,S)-DPEN, however, a monomeric diphosphine/diamine complex is readily formed to facilitate catalytic hydrogenation. In other words, under the hydrogenation conditions, one of the precatalysts, $RuCl_2[(R)$-Tolbinap)(DMF)_n]$, is selectively

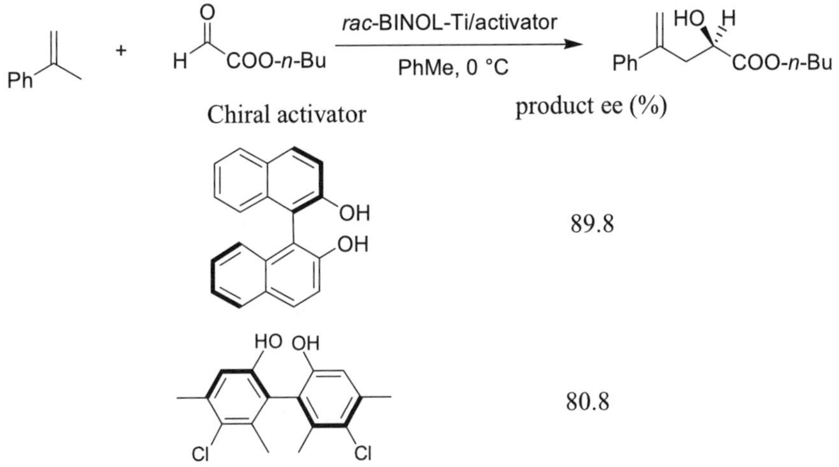

Scheme 8–54

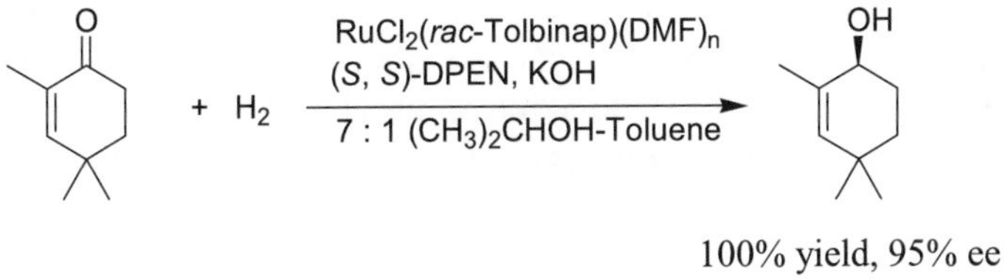

100% yield, 95% ee

Scheme 8–55

activated in the course of the reaction, providing the product with enantio-selectivity very close to the 96% that is obtained with a combination of enantio-merically pure (R)-Tolbinap and (S,S)-DPEN.

Another new strategy for catalytic asymmetric hydrogenation of prochiral ketones using achiral diphosphine ligand has also been demostrated.[112] In this design, a conformationally flexible bis(phosphanyl)biphenyl ligand (BIPHEP) was used for preparing the Ru catalyst. Following the activation by a chiral diamine, high enantioselectivity is obtained in the asymmetric hydrogenation of aromatic ketones.

The ability to rotate freely about the C–C' bond linking the two benzene rings makes possible a low energy barrier interconversion of the (R)- and (S)-configurations of biphenyl bidentate ligands **136** and **137**. This free rotation makes it possible for the complex to adopt a more favored configuration. Studies show that **136** and **137** are interconvertible, and equilibrium is estab-lished in the complex solution (CDCL₃/(CD₃)₂CDOD=1/2) after it stands for 3 hours at room temperature or 30 minutes at 80°C. An NMR study of a dilute solution of DM-BIPHEP/RuCl₂/(S,S)-DPEN shows that a 3:1 mixture of **136** and **137** is formed. The equilibrium is established through the cleavage of an Ru–P bond, rotation along the C–C' bond to invert the diphosphine configuration, and re-coordination of the P and Ru atom. When catalytic asymmetric hydrogenation of ketone is performed in the presence of this catalyst and a base, very good results are obtained.

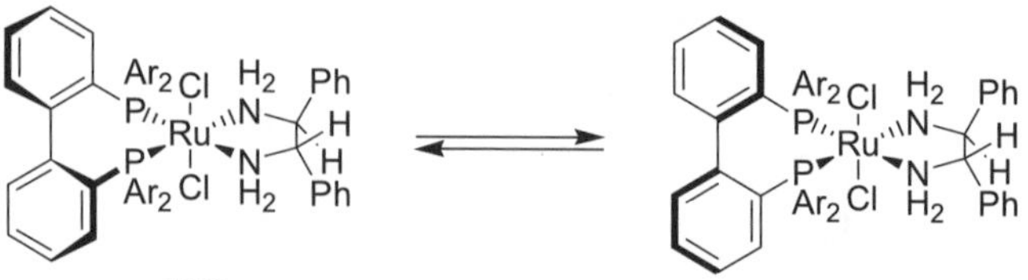

136   **137**

A similar reaction was reported by Reetz and Neugebauer[113] for the asymmetric hydrogenation of dimethyl itaconate (see Section 6.1.2.6). They used C_2-symmetric 1,4:3,6-dianhydro-D-mannite as the backbone (R^1 in **138**), together with 2,2'-dihydroxy-3,3'-dimethyl-1,1'-biphenyl (R^2 in **138**) to construct the diphosphite ligand **138**. Asymmetric hydrogenation using the **138**-coordinated Rh complex **139** leads to excellent enantioselectivity.[113] Studies by Reetz's group have shown that under the hydrogenation conditions, chirality of the biphenyl ligand is formed by the induction of chiral backbone HO-R^1-OH, and it is this dynamically formed chiral ligand that gives the high stereo induction in the catalytic hydrogenation. The results presented in Scheme 8–56 show that

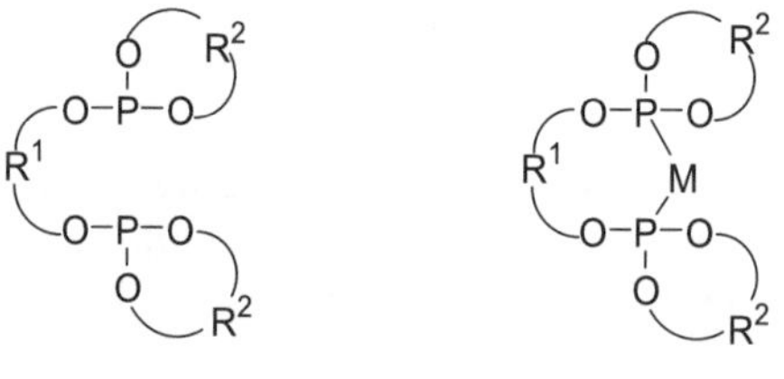

138   **139**

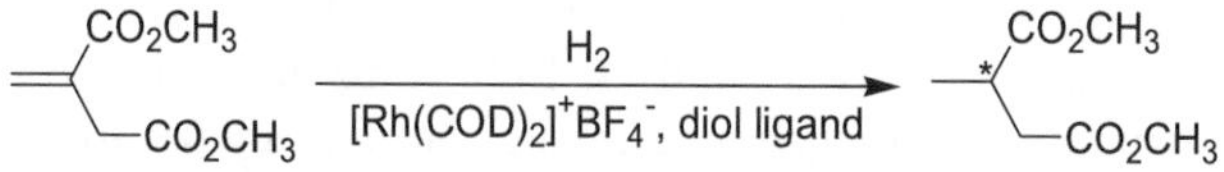

Entry	Diol	Temp. (°C)	Conver. (%)	ee (%)	Config.
1	2-Naphthol	20	65	21.0	(S)
2	(S)-BINOL	20	> 99	87.8	(S)
3	(R)-BINOL	20	> 99	94.5	(R)
4	(R)-BINOL	-10	> 99	96.2	(R)
5	2,2'-Dihydroxy-biphenyl	20	74	38.9	(S)
6	2,2'-Dihydroxy-3,3'-dimethyl biphenyl	20	> 99	96.8	(R)
7	2,2'-Dihydroxy-3,3'-dimethyl biphenyl	-10	> 99	98.2	(R)

Scheme 8–56. Reprinted with permission by Wiley-VCH Verlag GmbH, Ref. 113.

catalytic hydrogenation using achiral biphenyl ligand (Entry 6) can result in a product that shows comparable enantioselectivity to that obtained when using chiral BINAP ligand.

8.4 CHIRAL AMPLIFICATION, CHIRAL AUTOCATALYSIS, AND THE ORIGIN OF NATURAL CHIRALITY

Another achievement in recent asymmetric reaction study is the so-called chiral autocatalysis—where the product itself catalyzes its own asymmetric synthesis. In this process, the chiral catalyst and the products are the same in an asymmetric autocatalytic reaction. The separation of chiral catalyst from the product is not required, because the product itself is the catalyst. Starting from an optically active product with very low ee, this process allows the formation of a product with high ee values.[106,114]

Soai et al.[115] found that (S)-pyrimidyl alcohol **141** (20% mol, 94.8% ee) catalyzed its own synthesis in a reaction between the corresponding aldehyde **140** and diisopropyl zinc. The product eventually reached 48% yield and 95.7% ee (Scheme 8–57). In a similar manner, when the reaction was carried out starting from 20% of the (S)-**141** with only 2% ee, the first cycle gave the alcohol in 10% ee. Subsequent reaction cycles increased the ee up to 88%.

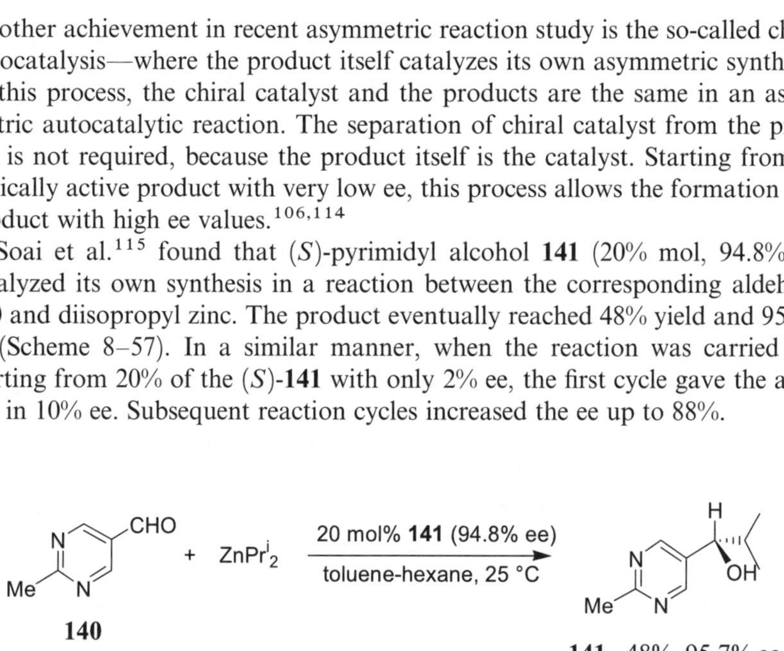

Scheme 8–57

A one pot asymmetric autocatalytic effect for the above reaction has also been shown, with remarkable amplification of enantiomeric excess (Scheme 8–58).[115c] Thus, a trace (about 3 mg) of 2-methylpyrimidyl alcohol **141** with only a slight enantiomeric excess (0.2–0.3% ee) can be automultiplied with

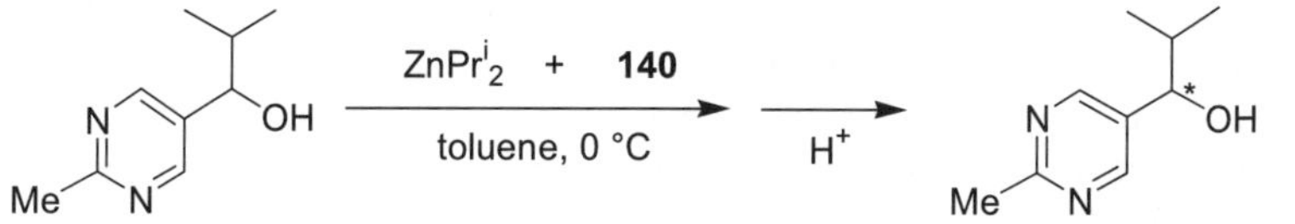

Scheme 8–58

dramatic amplification of enantiomeric excess (up to about 90% ee) in a one pot asymmetric autocatalytic reaction using diisopropylzinc and 2-methylpyrimidine-5-aldehyde.

Another practically perfect asymmetric catalysis has been observed in reactions using (2-alkynyl-5-pyrimidyl)alkanols as the catalyst. The asymmetric autocatalysis shown in Scheme 8–59 gives the corresponding product in high yield with over 99% ee.[116]

Scheme 8–59

The origin of chirality is an interesting issue that has attracted considerable attention. What is the origin of the chiral homogeneity in natural compounds such as L-α-amino acids? Several physical factors have been suggested as leading to the creation of this chirality. Moradpour et al.,[117] Bernstein et al.,[118] and Flores and Bonner[119] suggested that chirality could be induced in organic molecules by photosynthesis or photolysis using left or right circularly polarized light (CPL). However, the degree of enantiomeric imbalance caused by these physical factors is too small to be associated with the large enantiomeric imbalance in molecules found in nature. Shibata et al.[120] introduced a reaction system showing the possibility of amplification of enantiomeric imbalance starting from a trace amount of chiral initiator with low ee. This system suggests that slight symmetry breaking induced by the presence of a chiral initiator of very low ee can be dramatically amplified by asymmetric autocatalysis.

In Shibata's study, an amino acid such as leucine or valine with a very low ee is chosen as the initiator. The reason for using leucine or valine is that they are biologically important amino acids. Furthermore, some naturally occurring physical factor such as CPL can cause a slight imbalance of the enantiomers. This is important because a probiotic system might contain such amino acids, and CPL radiation over hundreds of thousands of years might then cause the enrichment of one isomer of the amino acid. Shibata's study shows that the first cycle of addition of diisopropylzinc to 2-methylpyrimidine-5-aldehyde **140** in the presence of an amino acid with slight enantiomeric imbalance can produce additional product **141** with small enantiomeric excess. Subsequent asymmetric autocatalysis then provides product **141** showing high enantiomeric excess.

Thus, it is possible that amino acids were first produced in a probiotic system. A slight enantiomeric imbalance in these amino acids might have been created by the action of some naturally occurring physical factors such as CPL. Alternatively, the imbalance might have been created in the presence of some physical factors at the time when these amino acids were formed. This imbalance might then have been amplified in other asymmetric reactions catalyzed by the amino acids, generating products with much higher enantiomeric excess and thus creating the natural chirality.

8.5 SUMMARY

This final chapter summarizes the enzyme-catalyzed asymmetric reactions and introduces some new developments in the area of asymmetric synthesis. Among the new developments, cooperative asymmetric catalysis is an important theme because it is commonly observed in enzymatic reactions. Understanding cooperative asymmetric catalysis not only makes it possible to design more enantioselective asymmetric synthesis reactions but also helps us to understand how mother nature contributes to the world.

Another question that has challenged the minds of scientists is the origin of chirality. Some scientists have argued that the breaking of the enantiomer balance was caused by some physical factors such as circularly polarized light, magnetic fields, or electric fields after the organic compounds were formed. Others thought that the imbalance was created during the formation of chiral organic compounds in the presence of the above-mentioned physical factors. Shibata's asymmetric autocatalysis and chiral amplification provided some interesting information about the possible origin of chirality in nature.

8.6 REFERENCES

1. (a) Whitesides, G. M.; Wong, C. H. *Angew. Chem. Int. Ed. Engl.* **1985**, *24*, 617. (b) Jones, J. B. *Tetrahedron* **1986**, *42*, 3351. (c) Chen, C. S.; Sih, C. J. *Angew. Chem. Int. Ed. Engl.* **1989**, *28*, 695. (d) Boland, W.; Frossl, C.; Lorenz, M. *Synthesis* **1991**, 1049. (e) Santaniello, E.; Ferraboschi, P.; Grisenti, P.; Manzocchi, A. *Chem. Rev.* **1992**, *92*, 1071. (f) Look, G. C.; Fotsch, C. H.; Wong, C. H. *Acc. Chem. Res.* **1993**, *26*, 182. (g) Mori, K. *Synlett* **1995**, 1097. (h) Schoffers, E.; Golebiowski, A.; Johnson, C. R. *Tetrahedron* **1996**, *52*, 3769. (i) Johnson, C. R. *Acc. Chem. Res.* **1998**, *31*, 333.

2. Hudlicky, T.; Olivo, H, F.; McKibben, B. *J. Am. Chem. Soc.* **1994**, *116*, 5108.

3. Kirchner, G.; Scollar, M. P.; Klibanov, A. M. *J. Am. Chem. Soc.* **1985**, *107*, 7072.

4. Weidner, J.; Theil, F.; Schick, H. *Tetrahedron Asymmetry* **1994**, *5*, 751.

5. Johnson, C. R.; Bis, S. J. *Tetrahedron Lett.* **1992**, *33*, 7287.

6. Johnson, C. R.; Braun, M. P. *J. Am. Chem. Soc.* **1993**, *115*, 11014.

7. (a) Johnson, C. R.; Penning, T. D. *J. Am. Chem. Soc.* **1988**, *110*, 4726. (b) Parry, R. J.; Haridas, K.; De Jong, R.; Johnson, C. R. *Tetrahedron Lett.* **1990**, *31*, 7549. (c) Parry, R. J.; Haridas, K.; De Jong, R.; Johnson, C. R. *J. Chem. Soc. Chem. Commun.* **1991**, 740. (d) Johnson, C. R.; Nerurker, B. M.; Golebiowski, A.; Sundram, H.; Esker, J. L. *J. Chem. Soc. Chem. Commun.* **1995**, 1139.

8. Prelog, V. *Pure Appl. Chem.* **1964**, *9*, 119.

9. (a) Gu, J.; Li, Z.; Lin, G. *Tetrahedron* **1993**, *49*, 5805. (b) Wei, Z.; Li, Z.; Lin, G. *Tetrahedron* **1998**, *54*, 13059.

10. (a) Leuenberger, H. G. W.; Boguch, W.; Barner, R.; Schmidt, M.; Zell, R. *Helv. Chim. Acta* **1976**, *59*, 1832. (b) Kawai, Y.; Hayashi, M.; Inaba, Y.; Saitou, K.; Ohno, A. *Tetrahedron Lett.* **1998**, *39*, 5225.

11. Nugent, T. C.; Hudlicky, T. *J. Org. Chem.* **1998**, *63*, 510.

12. Johnson, C. R.; Golebiowski, A.; Sundram, H.; Miller, M. W.; Dwaihy, R. L. *Tetrahedron Lett.* **1995**, *36*, 653.

13. (a) Stewart, J. D.; Reed, K. W.; Kayser, M. M. *J. Chem. Soc. Perkin Trans. 1* **1996**, 755. (b) Stewart, J. D.; Reed, K. W.; Zhu, J.; Chen, G.; Kayser, M. M. *J. Org. Chem.* **1996**, *61*, 7652. (c) Stewart, J. D.; Reed, K. W.; Martinez, C. A.; Zhu, J.; Chen, G.; Kayser, M. M. *J. Am. Chem. Soc.* **1998**, *120*, 3541.

14. (a) Robert, S. M.; Willetts, A. J. *Chirality* **1993**, *5*, 334. (b) Levitt, M. S.; Newton, R. F.; Robert, S. M.; Willetts, A. J. *J. Chem. Soc. Chem. Commun.* **1990**, 619.

15. (a) Oku, J.; Inoue, S. *J. Chem. Soc. Chem. Commun.* **1981**, 229. (b) Mattews, B. R.; Jackson, W. R.; Jayatilake, G. S.; Wilshire, C.; Jacobs, H. A. *Aust. J. Chem.* **1988**, *41*, 1697.

16. (a) Minamikawa, H.; Hayakawa, S.; Yamada, T.; Iwasawa, N.; Nazasaka, K. *Bull. Chem. Soc. Jpn.* **1988**, *61*, 4379. (b) Narasaka, K.; Yamada, T.; Minamikawa H. *Chem. Lett.* **1987**, 2073.

17. Mori, A.; Ohno, H.; Nitta, H.; Tanaka, K.; Inoue, S. *SynLett* **1991**, 563.

18. Reetz, M. T.; Kunisch, F.; Heitmann, P. *Tetrahedron Lett.* **1986**, *27*, 4721.

19. (a) Effenberger, F.; Ziegler, T.; Förster, S. *Angew. Chem. Int. Ed. Engl.* **1987**, *26*, 458. (b) Effenberger, F.; Hörsch, B.; Förster, S.; Ziegler, T. *Tetrahedron Lett.* **1990**, *31*, 1249. (c) Effenberger, F.; Hörsch, B.; Weingart, F.; Ziegler, T.; Kühner, S. *Tetrahedron Lett.* **1991**, *32*, 2605. (d) Brussee, J.; Loos, W. T.; Kruse, C. G.; Van der Gen, A. *Tetrahedron* **1990**, *46*, 979. (e) Niedermeyer, U.; Kula, M. *Angew. Chem. Int. Ed. Engl.* **1990**, *29*, 386. (f) Klempier, N.; Griengl, H.; Hayn, M. *Tetrahedron Lett.* **1993**, *34*, 4769. (g) Klempier, N.; Pichler, U.; Griengl, H. *Tetrahedron Asymmetry* **1995**, *6*, 845. (h) Schmidt, M.; Hervé, S.; Klempier, N.; Griengl, H. *Tetrahedron* **1996**, *52*, 7833.

20. Effenberger, F.; Ziegler, T.; Förster, S. *Angew. Chem. Int. Ed. Engl.* **1987**, *26*, 458.

21. Klempier, N.; Griengl, H.; Hayn, M. *Tetrahedron Lett.* **1993**, *34*, 4769.

22. (a) Han, S.; Lin, G.; Li, Z. *Tetradedron Asymmetry* **1998**, *9*, 1835. (b) Lin, G.; Han, S.; Li, Z. *Tetrahedron* **1999**, *55*, 3531.

23. (a) Nakamura, K.; Miyai, T.; Kawai, Y.; Nakajima, N.; Ohno, A. *Tetrahedron Lett.* **1990**, *31*, 1159. (b) Ozegowski, R.; Kunath, A.; Schick, H. *Tetrahedron Asymmetry* **1993**, *4*, 695. (c) Ferraboschi, P.; Casati, S.; Grisenti, P.; Santaniello, E. *Tetrahedron Asymmetry* **1993**, *4*, 9. (d) Howell, J. A. S.; Palin, M. G.; El Hafa, H.; Top, S.; Jaouen, G. *Tetrahedron Asymmetry* **1992**, *3*, 1355. (e) Kamal, A.; Damayanthi, Y.; Rao, M. V. *Tetrahedron Asymmetry* **1992**, *3*, 1361. (f) Kawai, Y.; Takanobe, K.; Tsujimoto, M.; Ohno, A. *Tetrahedron Lett.* **1994**, *15*, 147. (g) Tsuji, K.; Terao, Y.; Achiwa, K. *Tetrahedron Lett.* **1989**, *30*, 6189. (h) Bevinakatti, H. S.; Newadkar, R. V. *Tetrahedron Asymmetry* **1993**, *4*, 773. (i) Chenevert, R.; Courchesne, G. *Tetrahedron Asymmetry* **1995**, *6*, 2093. (j) Barnier, J.; Blanco, L.; Rousseau, G.; Grubé-Jampel, E.; Fresse, I. *J. Org. Chem.* **1993**, *58*, 1570. (k) Förster, S.; Roos, J.; Effenberger, F.; Wajant, H.; Sprauer, A. *Angew. Chem. Int. Ed. Engl.* **1996**, *35*, 437. (l) Roberts, S. M.; Willetts, A. J. *Chirality* **1993**, *5*, 334. (m) Levitt,

M. S.; Newton, R. F.; Roberts, S. M.; Willetts, A. J. *J. Chem. Soc. Chem. Commun.* **1990**, 619. (n) Hudlicky, T.; Price, J. D. *Synlett.* **1990**, 159.

24. (a) Suga, T.; Hamada, H.; Hirata, T. *Chem. Lett.* **1987**, 471. (b) Hamada, H. *Bull. Chem. Soc. Jpn.* **1988**, *61*, 869.

25. Akakabe, Y.; Takahashi, M.; Kamezawa, M.; Kikuchi, K.; Tachibana, H.; Ohtani, T.; Naoshima, Y. *J. Chem. Soc. Perkin Trans. 1* **1995**, 1295.

26. Hayashi, T.; Yamamoto, A.; Ito, Y.; Nishioka, E.; Miura, H.; Yanagi, K. *J. Am. Chem. Soc.* **1989**, *111*, 6301.

27. Uozumi, T.; Hayashi, T. *J. Am. Chem. Soc.* **1991**, *113*, 9887.

28. Uozumi, Y.; Lee, S.; Hayashi, T. *Tetradedron Lett.* **1992**, *33*, 7185.

29. Uozumi, Y.; Hayashi, T. *Tetrahedron Lett.* **1993**, *34*, 2335.

30. (a) Bode, B. M.; Day, P. N.; Gordon, M. S. *J. Am. Chem. Soc.* **1998**, *120*, 1552. (b) Maruyama, Y.; Yamamura, K.; Nakayama, I.; Yoshiuchi, K.; Ozawa, F. *J. Am. Chem. Soc.* **1998**, *120*, 1421.

31. For reviews: (a) Bringmann, G.; Walter, R.; Weirich, R. *Angew. Chem. Int. Ed. Engl.* **1990**, *29*, 977. (b) Kagan, H. B.; Riant, O. *Chem. Rev.* **1992**, *92*, 1007. (c) Rosini, C.; Franzini, L.; Raffaelli, A.; Salvadori, P. *Synthesis* **1992**, 503.

32. Govindachari, T. R.; Nagarajan, K.; Parthasarathy, P. C.; Rajagopalan, T. G.; Desai, H. K.; Kartha, G.; Chen, S. L.; Nakanishi, K. *J. Chem. Soc. Perkin Trans. 1* **1974**, 1413.

33. (a) Huang, L.; Si, Y.; Snatzeke, G.; Zheng, D.; Zhou, J. *Collect. Czech. Chem. Commun.* **1988**, *53*, 2664. (b) Recent synthesis: Meyers, A. I.; Willemsen, J. J. *Tetrahedron* **1998**, *54*, 10493.

34. (a) Uchida, I.; Ezaki, M.; Shigematsu, N.; Hashimoto, M. *J. Org. Chem.* **1985**, *50*, 1341. (b) Kannan, R.; Williams, D. H. *J. Org. Chem.* **1987**, *52*, 5435.

35. Tomioka, K.; Ishiguro, T.; Koga, K. *Tetrahedron Lett.* **1980**, *23*, 2973.

36. Cram, D. J.; Cram, J. M. *Science* (Washington D. C.) **1974**, *183*, 803.

37. Noyori, R.; Tomino, I.; Tanimoto, Y.; Nishizawa, M. *J. Am. Chem. Soc.* **1984**, *106*, 6709.

38. Moorlag, H.; Meyers, A. I. *Tetrahedron Lett.* **1993**, *34*, 6989.

39. Nelson, T. D.; Meyers, A. I. *J. Org. Chem.* **1994**, *59*, 2655.

40. Meyers, A. I.; McKennon, M. J. *Tetrahedron Lett.* **1995**, *36*, 5869.

41. Meyers, A. I.; Nguyen, T. H. *Tetrahedron Lett.* **1995**, *36*, 5873.

42. Lipshuz, B. H.; Liu, Z.; Kayser, F. *Tetrahedron Lett.* **1994**, *35*, 5567.

43. Kharasch, M. S.; Sosnovsky, G.; Yang, N. C. *J. Am. Chem. Soc.* **1958**, *81*, 5819.

44. Andrus, M. B.; Argade, A. B.; Chen, X.; Pammet, M. G. *Tetrahedron Lett.* **1995**, *36*, 2945.

45. DattaGupta, A.; Singh, V. K. *Tetrahedron Lett.* **1996**, *37*, 2633.

46. Bolm, C.; Schlingloff, G.; Weickhardt, K. *Angew. Chem. Int. Ed. Engl.* **1994**, *33*, 1848.

47. Rein, T.; Reiser, O. *Acta Chem. Scand.* **1996**, *50*, 369.

48. (a) Denmark, S. E.; Chen, C. *J. Am. Chem. Soc.* **1992**, *114*, 10674. (b) Denmark, S. E.; Rivera, I. *J. Org. Chem.* **1994**, *59*, 6887.

49. Toda, F.; Akai, H. *J. Org. Chem.* **1990**, *55*, 3446.

50. Tanaka, K.; Watanabe, T.; Ohta, Y.; Fuji, K. *Tetrahedron Lett.* **1997**, *38*, 8943.

51. Arai, S.; Hamaguchi, S.; Shioiri, T. *Tetrahedron Lett.* **1998**, *39*, 2997.

52. Guetté, M.; Capillon, J.; Guetté, J. *Tetrahedron* **1973**, *29*, 3659.

53. (a) Soai, K.; Kawase, Y. *Tetrahedron Asymmetry* **1991**, *2*, 781. (b) Soai, K.; Hirose, Y.; Sakata, S. *Tetrahedron Asymmetry* **1992**, *3*, 677. (c) Soai, K.; Oshio, A.; Saito, T. *J. Chem. Soc. Chem. Commun.* **1993**, 811.

54. Pini, D.; Mastantuono, A.; Salvadori, P. *Tetradedron Asymmetry* **1994**, *5*, 1875.

55. Mi, A.; Wang, Z.; Chen, Z.; Jiang, Y.; Chan, A. S. C.; Yang, T. K. *Tetrahedron Asymmetry* **1995**, *6*, 2641.

56. For a review, see Molander, G. A.; Harris, C. R. *Chem. Rev.* **1996**, *96*, 307.

57. Fukuzawa, S.; Tatsuzawa, M.; Hirano, K. *Tetrahedron Lett.* **1998**, *39*, 6899.

58. Uozumi, Y.; Kato, K.; Hayashi, T. *J. Am. Chem. Soc.* **1997**, *119*, 5063.

59. For reviews, see Tsuji, J. *Palladium Reagents and Catalysts*, John Willey & Sons, Chichester, **1995**, pp 19–124.

60. (a) For explanation of the mechanism, see Ozawa, F.; Kubo, A.; Matsumoto, Y.; Hayashi, T. *Organometallics* **1993**, *12*, 4188. For catalytic asymmetric arylation of cyclic olefins, see (b) Hayashi, T.; Kubo, A.; Ozawa, F. *Pure Appl. Chem.* **1992**, *64*, 421. (c) Ozawa, F.; Kubo, A.; Hayashi, T. *J. Am. Chem. Soc.* **1991**, *113*, 1417. For other examples of asymmetric Heck reactions, see (d) Sato, Y.; Sodeoka, M.; Shibasaki, M. *J. Org. Chem.* **1989**, *54*, 4738. (e) Kagechika, K.; Shibasaki, M. *J. Org. Chem.* **1991**, *56*, 4093. (f) Sakamoto, T.; Kondo, Y.; Yamanaka, H. *Tetrahedron Lett.* **1992**, *33*, 6845. (g) Ashimori, A.; Overman, L. E. *J. Org. Chem.* **1992**, *57*, 4571.

61. Carpenter, N. E.; Kucera, D. J.; Overman, L. E. *J. Org. Chem.* **1989**, *54*, 5846.

62. Ozawa, F.; Kubo, A.; Hayashi, T. *Tetrahedron Lett.* **1992**, *33*, 1485.

63. Ozawa, F.; Kobatake, Y.; Hayashi, T. *Tetrahedron Lett.* **1993**, *34*, 2505.

64. Lu, X.; Zhu, G.; Wang, Z. *Synlett* **1998**, 115.

65. (a) Zhu, G.; Lu, X. *J. Org. Chem.* **1995**, *60*, 1087. (b) Zhu, G.; Lu, X. *Tetradedron Asymmetry* **1995**, *6*, 885.

66. (a) Newman, M. S.; Magerlein, B. *J. Org. React.* **1945**, *5*, 413. (b) Ballester, M. *Chem. Rev.* **1955**, *55*, 283.

67. Annunziata, R.; Banfi, S.; Colonna, S. *Tetrahedron Lett.* **1985**, *26*, 2471.

68. Ohkata, K.; Kimura, J.; Shinohara, Y.; Takagi, R.; Hiraga, Y. *J. Chem. Soc. Chem. Commun.* **1996**, 2411.

69. Bakó, P.; Szöllosy, Á.; Bombicz, P.; Töke, L. *Synlett* **1997**, 291.

70. Berrisford, D. J.; Bolm, C.; Sharpless, K. B. *Angew. Chem. Int. Ed. Engl.* **1995**, *34*, 1059.

71. (a) Villacorta, G. M.; Rao, C. P.; Lippard, S. J. *J. Am. Chem. Soc.* **1988**, *110*, 3175. (b) Ahn, K. H.; Klassen, R. B.; Lippard, S. J. *Organometallics* **1990**, *9*, 3178.

72. De Vries, A. H. M.; Meetsma, A.; Feringa, B. L. *Angew. Chem. Int. Ed. Engl.* **1996**, *35*, 2374.

73. (a) Feringa, B. L.; Pineschi, M.; Arnold, L. A.; Imbos, R.; De Vries, A. H. M. *Angew. Chem. Int. Ed. Engl.* **1997**, *36*, 2620. (b) Naasz, R.; Arnold, L. A.; Pineschi, M.; Keller, E.; Feringa, B. L. *J. Am. Chem. Soc.* **1999**, *121*, 1104.

74. (a) Yan, M.; Yang, L.; Wong, K.; Chan, A. S. C. *J. Chem. Soc. Chem. Commun.* **1999**, 11. (b) Yan, M.; Chan, A. S. C. *Tetrahedron Lett.* **1999**, *40*, 6645.

75. Evans, D. A.; Rovis, T.; Kozlowski, M. C.; Tedrow, J. S. *J. Am. Chem. Soc.* **1999**, *121*, 1994.

76. Emori, E.; Arai, T.; Sasai, H.; Shibasaki, M. *J. Am. Chem. Soc.* **1998**, *120*, 4043.

77. Takaya, Y.; Ogasawara, M.; Hayashi, T. *J. Am. Chem. Soc.* **1998**, *120*, 5579.

78. Gomez-Bengoa, E.; Heron, N. M.; Didiuk, M. T.; Luchaco, C. A.; Hoveyda, A. H. *J. Am. Chem. Soc.* **1998**, *120*, 7649.

79. (a) Sibi, M. P.; Shay, J. J.; Liu, M.; Jasperse, C. P. *J. Am. Chem. Soc.* **1998**, *120*, 6615. (b) Niu, D.; Zhao, K. *J. Am. Chem. Soc.* **1999**, *121*, 2456.

80. (a) Krause, N.; Gerold, A. *Angew. Chem. Int. Ed. Engl.* **1997**, *36*, 187. (b) Krause, N. *Angew. Chem. Int. Ed. Engl.* **1998**, *37*, 283.

81. König, K. E.; Bachman, G. L.; Vineyard, B. D. *J. Org. Chem.* **1980**, *45*, 2362.

82. (a) Roucoux, A.; Devocelle, M.; Carpentier, J.; Agbossou, F.; Mortreux, A. *Synlett* **1995**, 358. (b) Zhang, X.; Taketomi, T.; Yoshizumi, T.; Kumobayashi, H.; Akutagawa, S.; Mashima, K.; Takaya, H. *J. Am. Chem. Soc.* **1993**, *115*, 3318. (c) Mashima, K.; Akutagawa, T.; Zhang, X.; Takaya, H.; Taketomi, T.; Kumobayashi, H.; Akutagawa, S. *J. Organomet. Chem.* **1992**, *428*, 213.

83. (a) Burk, M. J.; Feaster, J. E.; Harlow, R. L. *Organometallics* **1990**, *9*, 2653. (b) Burk, M. J.; Harlow, R. L. *Angew. Chem. Int. Ed. Eng.* **1990**, *29*, 1462. (c) Burk, M. J.; Feaster, J. E.; Harlow, R. L. *Tetradedron Asymmetry* **1991**, *2*, 569.

84. Saburi, M.; Shao, L.; Sakurai, T.; Uchida, Y. *Tetrahedron Lett.* **1992**, *33*, 7877.

85. (a) Corey, E. J.; Bakshi, R. K. *Tetrahedron Lett.* **1990**, *31*, 611. (b) Corey, E. J.; Cheng, X.; Cimprich, K. A.; Sarshar, S. *Tetrahedron Lett.* **1991**, *32*, 6835.

86. Braun, M.; von Der Hagen, A.; Waldmüller, D. *Liebigs Ann. Chem.* **1995**, 1447.

87. (a) Prakash, G. K. S.; Krishnamurti, R.; Olah, G. A. *J. Am. Chem. Soc.* **1989**, *111*, 393. (b) Krishnamurti, R.; Bellew, D. R.; Prakash, G. K. S. *J. Org. Chem.* **1991**, *56*, 984. (c) Prakash, G. K. S.; Yudin, A. K. *Chem. Rev.* **1997**, *97*, 757.

88. Iseki, K.; Nagai, T.; Kobayashi, Y. *Tetrahedron Lett.* **1994**, *35*, 3137.

89. Mikami, K.; Matsukawa, S.; Volk, T.; Terada, M. *Angew. Chem. Int. Ed. Engl.* **1997**, *36*, 2768.

90. For the terminology of desymmetrization, see Mikami, K.; Narisawa, S.; Shimizu, M.; Terada, M. *J. Am. Chem. Soc.* **1992**, *114*, 6566.

91. Hoye, T. R.; Peck, D. R.; Swanson, T. A. *J. Am. Chem. Soc.* **1984**, *106*, 2738.

92. Mislow K.; Siegel, J. *J. Am. Chem. Soc.* **1984**, *106*, 3319.

93. (a) Trost, B. M.; van Vranken, D, L.; Bingel, C. *J. Am. Chem. Soc.* **1992**, *114*, 9327. (b) Trost, B. M.; Li, L.; Guile, S. D. *J. Am. Chem. Soc.* **1992**, *114*, 8745. (c) Trost, B. M.; Pulley, S. R. *J. Am. Chem. Soc.* **1995**, *117*, 10143. (d) Trost, B. M.; Shi, Z. *J. Am. Chem. Soc.* **1996**, *118*, 3037. (e) Trost, B. M.; Madsen, R.; Guile, S. D. *Tetrahedron Lett.* **1997**, *38*, 1707.

94. Trost, B. M.; Patterson, D. E. *J. Org. Chem.* **1998**, *63*, 1339.

95. Miyafuji, A.; Katsuki, T. *Tetrahedron* **1998**, *54*, 10339.

96. (a) Sträter, N.; Lipscomb, W. N.; Klabunde, T.; Krebs, B. *Angew. Chem. Int. Ed. Engl.* **1996**, *35*, 2024. (b) Wilcox, D. E. *Chem. Rev.* **1996**, *96*, 2435. (c) Steinhagen, H.; Helmchen, G. *Angew. Chem. Int. Ed. Engl.* **1996**, *35*, 2339.

97. (a) Sasai, H.; Tokunaga, T.; Watanabe, S.; Suzuki, T.; Itoh, N.; Shibasaki, M. *J. Org. Chem.* **1995**, *60*, 7388. (b) Sasai, H.; Yamada, Y. M. A.; Suzuki, T.; Shibasaki, M. *Tetrahedron* **1994**, *50*, 12313. (c) Sasai, H.; Kim, W.; Suzaki, T.; Shibasaki, M.; Mitusda, M.; Hasegawa, J.; Ohashi, T. *Tetrahedron Lett.* **1994**, *35*, 6123. (d) Sasai, H.; Suzuki, T.; Arai, S.; Arai, T.; Shibasaki, M. *J. Am. Chem. Soc.* **1992**, *114*, 4418. (e) Sasai, H.; Suzuki, T.; Itoh, N.; Tanaka, K.; Date, T.; Okamura, K.; Shibasaki, M. *J. Am. Chem. Soc.* **1993**, *115*, 10372. (f) Sasai, H.; Arai, T.; Shibasaki, M. *J. Am. Chem. Soc.* **1994**, *116*, 1571. (g) Sasai, H.; Arai, T.; Satow, Y.; Houk, K. N.; Shibasaki, M. *J. Am. Chem. Soc.* **1995**, *117*, 6194. (h) Sasai, H.; Arai, S. Tahara, Y.; Shibasaki, M. *J. Org. Chem.* **1995**, *60*, 6656. (i) Arai, T.; Bougauchi, M.; Sasai, H.; Shibasaki, M. *J. Org. Chem.* **1996**, *61*, 2926. (j) Arai, T.; Sasai, H.; Aoe, K.; Okamura, K.; Date, T.; Shibasaki, M. *Angew. Chem. Int. Ed. Engl.* **1996**, *35*, 104. (k) Yamada, Y. M. A.; Shibasaki, M. *Tetrahedron Lett.* **1998**, *39*, 5561.

98. Van Koten, G. *Pure Appl. Chem.* **1994**, *66*, 1455.

99. Yoshikawa, N.; Yamada, Y. M. A.; Das, J.; Sasai, H.; Shibasaki, M. *J. Am. Chem. Soc.* **1999**, *121*, 4168.

100. Hamashima, Y.; Sawada, D.; Kanai, M.; Shibasaki, M. *J. Am. Chem. Soc.* **1999**, *121*, 2641.

101. Konsler, R. G.; Karl, J.; Jacobsen, E. N. *J. Am. Chem. Soc.* **1998**, *120*, 10780.

102. Hansen, K. B.; Leighton, J. L.; Jacobsen, E. N. *J. Am. Chem. Soc.* **1996**, *118*, 10924.

103. For a review, see Avalos, M.; Babiano, R.; Cintas, P.; Jimemez, J. L.; Palacios, J. C. *Tetradedron Asymmetry* **1997**, *8*, 2997.

104. Puchot, C.; Samuel, O.; Duñach, E.; Zhao, S.; Agami, C.; Kagan, H. B. *J. Am. Chem. Soc.* **1986**, *108*, 2353.

105. These reactions include (1) addition of organozincs to aldehydes: (a) Noyori, R. *Chem. Soc. Rev.* **1989**, *18*, 187. (2) 1,4-Addition of organozinc compounds: (b) Bolm, C. *Tetradedron Asymmetry* **1991**, *2*, 701. (c) Bolm, C.; Ewald, M.; Felder, M. *Chem. Ber.* **1992**, *125*, 1205. (d) Bolm, C.; Felder, M.; Müller, J. *Synlett* **1992**, 439. (3) Asymmetric glyoxylate-ene reaction: (e) Terada, M.; Mikami, K.; Nakai, T. *J. Chem. Soc. Chem. Commum.* **1990**, 1623. (4) Trimethylsilylcyanation of carbonyl compounds: (f) Hayashi, M.; Matsuda, T.; Oguni, N. *J. Chem. Soc. Chem. Commun.* **1990**, 1364. (5) Allylation and aldol condensation on aldehydes: (g) Bedeschi, P.; Casolari, S.; Costa, A. L.; Tagliavini, E.; Umani-Ronchi, A. *Tetrahedron Lett.* **1995**, *36*, 7897. (h) Keck, G. E.; Krishnamurthy, D. *J. Am. Chem. Soc.* **1995**, *117*, 2363. (6) Asymmetric epoxidation: (i) Guillaneux, D.; Zhao, S. H.; Samuel, O.; Rainford, D.; Kagan, H. B. *J. Am. Chem. Soc.* **1994**, *116*, 9430. (7) Asymmetric ring-opening of *meso*-epoxides: (j) Hansen, K. B.; Leighton, J. L.; Jacobsen, E. N. *J. Am. Chem. Soc.* **1996**, *118*, 10924. (8) Diels-Alder reaction, hetero-Diels-Alder reaction and 1,3-dipolar cycloaddition reaction: (k) Seebach, D.; Dahinden, R.; Marti, R. E.; Beck, A. K.; Plattner, D. A.; Kühnle F. N. M. *J. Org. Chem.* **1995**, *60*, 1788.

106. Kitamura, M.; Suga, S.; Kawai, K.; Noyori, R. *J. Am. Chem. Soc.* **1986**, *108*, 6071.

107. (a) Kitamura, M.; Okada, S.; Suga, S.; Noyori, R. *J. Am. Chem. Soc.* **1989**, *111*, 4028. (b) Kitamura, M.; Suga, S.; Oka, H.; Noyori, R. *J. Am. Chem. Soc.* **1998**, *120*, 9800.

108. Girard, C.; Kagan, H. B. *Angew. Chem. Int. Ed. Engl.* **1998**, *37*, 2922.

109. Faller, J. W.; Parr, J. *J. Am. Chem. Soc.* **1993**, *115*, 804.

110. Faller, J. W.; Tokunaga, M. *Tetrahedron Lett.* **1993**, *34*, 7359.

111. Mikami, K.; Matsukawa, S. *Nature* **1997**, *385*, 613.

112. Mikami, K.; Korenaga, T.; Terada, M.; Ohkuma, T.; Pham, T.; Noyori, R. *Angew. Chem. Int. Ed. Engl.* **1999**, *38*, 495.

113. Reetz, M. T.; Neugebauer, T. *Angew. Chem. Int. Ed. Engl.* **1999**, *38*, 179.

114. (a) Matsuda, Y.; Kaneko, T.; Oguni, N. *J. Am. Chem. Soc.* **1988**, *110*, 7877. (b) Kitamura, M.; Okada, S.; Suga, S.; Noyori, R. *J. Am. Chem. Soc.* **1989**, *111*, 4028.

115. (a) Soai, K.; Shibata, T.; Morioka, H.; Choji, K. *Nature* **1995**, *378*, 767. (b) Shibata, T.; Morioka, H.; Hayase, T.; Choji, K.; Soai, K. *J. Am. Chem. Soc.* **1996**, *118*, 471. (c) Shibata, T.; Hayase, T.; Yamamoto, J.; Soai, K., *Tetradedron Asymmetry* **1997**, *8*, 1717.

116. Shibata, T.; Yonekubo, S.; Soai, K. *Angew. Chem. Int. Ed. Engl.* **1999**, *38*, 659.

117. Moradpour, A.; Nicoud, J. F.; Balavoine, G.; Kagan, H. B.; Tsoucaris, G. *J. Am. Chem. Soc.* **1971**, *93*, 2353.

118. (a) Bernstein, W. J.; Calvin, M.; Buchardt, O. *J. Am. Chem. Soc.* **1972**, *94*, 494. (b) Bernstein, W. J.; Calvin, M.; Buchardt, O. *J. Am. Chem. Soc.* **1973**, *95*, 527.

119. Flores, J. J.; Bonner, W. A.; Massey, G. A. *J. Am. Chem. Soc.* **1977**, *99*, 3622.

120. Shibata, T.; Yamamoto, J.; Matsumoto, N.; Yonekubo, S.; Osanai, S.; Soai, K. *J. Am. Chem. Soc.* **1998**, *120*, 12157.